NOTATION AND SYMBOLS USED IN THIS BOOK

Notation		Symbols	
$n!$	Factorial of n	f, ω	Analog frequency variables
$\log(x)$	Logarithm to the base 10	F, Ω	Digital frequency variables
$\ln(x)$	Logarithm to the base e	ν	Normalized frequency variable
$\mathrm{abs}(x)$	Absolute value of x	P	Power
$\lvert x \rvert$	Absolute value of x	E	Energy
$\mathbb{A}[x(t)]$	Area of $x(t)$ for all time	∇	First difference
$\mathbb{A}[x(t)](a,b)$	Area of $x(t)$ over limits (a,b)	T	Time period
		t_s	Sampling interval
$\int_T$	Integral over duration T, e.g., from α to $\alpha + T$	dB	Decibel
		$x(t), y(t)$	CT signals
$\angle x$	Phase angle of x	$x[n], y[n]$	DT signals
$\mathrm{Re}[x]$	Real part of x	$\{x[n]\}$	Numeric values of DT signal $x[n]$, e.g., $\{8\ 2\ 2\ 1\}$
$\mathrm{Im}[x]$	Imaginary part of x		
$\mathrm{int}(x)$	Integer part of x	$\uparrow$	Marker for the index $n = 0$ in $\{x[n]\}$, e.g., $\{8\ 3\ 5\ 2\ 1\ 4\}$ $\uparrow$
$\mathrm{sinc}(x)$	$\sin(\pi x)/(\pi x)$		
$\mathrm{rect}(x)$	Rectangular pulse between $-\frac{1}{2}$ and $\frac{1}{2}$ with unit height and area		
		$h(t), h[n]$	Impulse response
		$H(.)$	Transfer function
$\mathrm{tri}(x)$	Triangular pulse between -1 and 1 with unit height and area	$\star$	Symbol for the convolution operation
		$\star\star$	Symbol for the correlation operation
$x°$	x degrees	$\bullet$	Symbol for periodic convolution
x^*	Complex conjugate of x	$\bullet\bullet$	Symbol for periodic correlation
		$*$	MATLAB symbol for multiplication

Special Functions

(We follow the notation in *Handbook of Mathematical Functions* by Abramowitz and Stegun, Dover Publications, 1964)

$\Gamma(\nu)$	Gamma function
$\text{si}(x)$	Sine integral
$T_n(x)$	Chebyshev polynomial of the first kind
$U_n(x)$	Chebyshev polynomial of the second kind
$C(x)$	Fresnel cosine integral
$S(x)$	Fresnel sine integral
$J_n(x)$	Bessel function of integer order n
$j_\nu(x)$	Spherical Bessel function of fractional order ν

Analog and Digital Signal Processing

Ashok Ambardar
Michigan Technological University

PWS Publishing Company

 An International Thomson Publishing Company

Boston • Albany • Bonn • Cincinnati • Detroit • London • Madrid • Melbourne • Mexico City • New York • Paris • San Francisco • Singapore • Tokyo • Toronto • Washington

PWS PUBLISHING COMPANY
20 Park Plaza, Boston, MA 02116-4324

International Thomson Publishing
The trademark ITP is used under license

For more information contact:

PWS Publishing Co.
20 Park Plaza
Boston, MA 02116

International Thomson Editores
Campos Eliseos 385, Piso 7
Col. Polanco
11560 Mexico C.F., Mexico

International Thomson Publishing Europe
Berkshire House 168–173
High Holborn
London WC1 V7AA
England

International Thomson Publishing GmbH
Königswinterer Strasse 418
53227 Bonn, Germany

Thomas Nelson Australia
102 Dodds Street
South Melbourne, 3205
Victoria, Australia

International Thomson Publishing Asia
221 Henderson Road
#05-10 Henderson Building
Singapore 0315

Nelson Canada
1120 Birchmount Road
Scarborough, Ontario
Canada M1K 5G4

International Thomson Publishing Japan
Hirakawacho Kyyowa Building, 31
2-2-1 Hirakawacho
Chiyoda-ku, Tokyo 102
Japan

Sponsoring Editor: Tom Robbins
Assistant Editor: Ken Morton
Production and Interior Design:
  Pamela Rockwell
Marketing Manager: Nathan Wilbur
Manufacturing Coordinator: Lisa Flanagan
Compositor: Eigentype Compositors
Cover Printer: Henry Sawyer Company
Text Printer and Binder: RR Donnelley,
  Crawfordsville

Library of Congress Cataloging-in-Publication Data

Ambardar, Ashok
  Analog and digital signal processing / Ashok Ambardar.
    p.   cm.
  Includes bibliographical references (p.   ) and index.
  ISBN 0-534-94086-2
  1. Signal processing.   2. MATLAB.   I. Title.
TK5102.9.A45   1995
621.382'2--dc20   94-45288
                                                                      CIP

Printed and bound in the United States of America.
95 96 97 98 99 – 10 9 8 7 6 5 4 3 2

To Nancy and Shyamaji, for keeping the memories alive!

CONTENTS

Software routines for sections marked (*M) appear in the MATLAB appendices.

1 Overview

2 Signals

3 Systems

LIST OF TABLES

PREFACE

This book on analog and digital signal processing is intended to serve both as a text for students and as a source of basic reference for professionals across various disciplines. As a text, it is geared to junior/senior electrical engineering students and details the material covered in a typical undergraduate curriculum. As a reference, it attempts to provide a broader perspective by introducing additional special topics towards the later stages of each chapter. Complementing this text, but deliberately not integrated into it, is a set of powerful software routines (running under MATLAB) that can be used not only for reinforcing and visualizing concepts but also for problem solving and advanced design.

The text stresses the fundamental principles and applications of signals, systems, transforms and filters. It deals with concepts that are crucial to a full understanding of time-domain and frequency-domain relationships. Our ultimate objective is that the student be able to think clearly in both domains and switch from one to the other with relative ease. It is based on the premise that what might often appear obvious to the expert may not seem so obvious to the budding expert. Basic concepts are, therefore, explained and illustrated by worked examples to bring out their importance and relevance.

Scope

The text assumes familiarity with elementary calculus, complex numbers, basic circuit analysis and (in a few odd places) the elements of matrix algebra. It covers the core topics in analog and digital signal processing taught at the undergraduate level. The links between analog and digital aspects are explored and emphasized throughout. The topics covered in this text may be grouped into the following broad areas:

1. An introduction to **signals** (Chapter 2) and **systems** (Chapter 3), their representation and their classification.
2. **Convolution**, a method of time-domain analysis, which also serves to link the time domain and the frequency domain (Chapter 4).
3. **Fourier series** (Chapter 5) and **transforms** (Chapter 6), which provide a spectral description of analog signals and their applications (Chapter 7).
4. The **Laplace transform** (Chapter 8), which forms a useful tool for system analysis and its applications (Chapter 9).

5. Applications of Fourier and Laplace techniques to **analog filter design** (Chapter 10).
6. **Sampling** and the discrete-time Fourier transform (**DTFT**) (Chapter 11) of sampled signals, and the **DFT** and the **FFT** (Chapter 12), all of which reinforce the central concept that sampling in one domain leads to a periodic extension in the other.
7. The **z-transform** (Chapter 13), which extends the DTFT to the analysis of discrete-time systems.
8. Applications of digital signal processing to the design of **digital filters** (Chapter 14).

We have tried to preserve a rational approach and include all the necessary mathematical details, but we have also emphasized heuristic explanations whenever possible. Each chapter is more or less structured as follows:

1. A short opening section outlines the objectives and topical coverage and points to the required background.
2. Central concepts are introduced in early sections and illustrated by worked examples. Special topics are developed only in later sections.
3. Within each section, the material is broken up into bite-sized pieces. Results are tabulated and summarized for easy reference and access.
4. Whenever appropriate, concepts are followed by remarks, which highlight essential features or limitations.
5. The relevant software routines and their use are outlined in MATLAB appendices to each chapter. Sections that can be related to the software are specially marked in the table of contents.
6. End-of-chapter problems include a variety of drills and exercises. MATLAB code to generate answers to many of these appears on the supplied disk.

A solutions manual for instructors is available from the publisher.

How to Use This Text

This text contains sufficient material for a two-semester or two- to three-quarter sequence. It is set up to allow sequential or parallel coverage of analog and digital aspects. In either case, the ties that bind and separate the two are always stressed throughout.

Naturally, the manner in which the material is covered will depend on the syllabus requirements and the individual preference of the instructor. At the risk of sounding presumptuous, we offer two possible approaches. The first is to cover analog aspects of Chapters 1–4 and all of Chapters 5–9 in the first course, followed by digital aspects of Chapters 1–4 and Chapters 10–14 in the sequel. The second is to cover all of Chapters 1–7 in the first course, followed by Chapters 8–14 in the sequel.

Software

A unique feature of this text is the *analog and digital signal processing* (ADSP) software toolbox for signal processing and analytical and numerical computation designed to run under all versions of MATLAB. The routines are self-demonstrating and can be used to reinforce essential concepts, validate the results of analytical paper and pencil solutions, and solve complex problems that might, otherwise, be beyond the skills of analytical computation demanded of the student.

The toolbox includes programs for generating and plotting signals, regular and periodic convolution, symbolic and numerical solution of differential and difference equations, Fourier analysis, frequency response, asymptotic Bode plots, symbolic results for system response, inverse Laplace and inverse z-transforms, design of analog, IIR and FIR filters by various methods, and more.

Since our primary intent is to present the principles of signal processing, not software, we have made no attempt to integrate MATLAB into the text. Software related aspects appear only in the appendices to each chapter. This approach also maintains the continuity and logical flow of the textual material, especially for users with no inclination (or means) to use the software. In any case, the self-demonstrating nature of the routines should help you to get started even if you are new to MATLAB. As an aside, all the graphs for this text were generated using the supplied ADSP toolbox.

We hasten to provide two disclaimers. First, our use of MATLAB is not to be construed as an endorsement of this product. We just happen to like it. Second, our routines are supplied in good faith; we fully expect them to work on your machine, but provide no guarantees!

Acknowledgements

This book has gained immensely from the incisive, sometimes provoking, but always constructive, criticism of Dr. J.C.Mandojana. Many other individuals have also contributed in various ways to this effort. Special thanks are due, in particular, to

Drs. R.W. Bickmore and R.T. Sokolov, who critiqued early drafts of several chapters and provided valuable suggestions for improvement.

Dr. A.R. Hambley, who willingly taught from portions of the final draft in his classes.

Drs. D.B. Brumm, P.H. Lewis and J.C. Rogers, for helping set the tone and direction in which the book finally evolved.

Mr. Scott Ackerman, for his invaluable computer expertise in (the many) times of need.

At PWS Publishing, the editor Mr. Tom Robbins, for his constant encouragement, and Ms. Pam Rockwell for her meticulous attention to detail during all phases of editing and production, and Ken Morton, Lai Wong, and Lisa Flanagan for their behind-the-scenes help.

The students, who tracked down inconsistencies and errors in the various drafts, and provided extremely useful feedback.

The Mathworks, for permission to include modified versions of a few of their *m-files* with our software.

We would also like to thank Dr. Mark Thompson, Dr. Hadi Saadat and the following reviewers for their useful comments and suggestions:

Doran Baker, *Utah State University*
Ken Sauer, *University of Notre Dame*
George Sockman, *SUNY—Binghamton*
James Svoboda, *Clarkson University*
Kunio Takaya, *University of Saskatchewan*

And now for something completely different!

We have adopted the monarchic practice of using words like *we*, *us*, and *our* when addressing you, the reader. This seems to have become quite fashionable, since it is often said that no book can be the work of a single author. In the present instance, the reason is even more compelling because my family, which has played an important part in this venture, would never forgive me otherwise! In the interests of fairness, however, when the time comes to accept the blame for any errors, I have been graciously awarded sole responsibility!

We close with a limerick that, we believe, actually bemoans the fate of all textbook writers. If memory serves, it goes something like this

A limerick packs laughs anatomical,
In space that's quite economical,
But the good ones I've seen,
So seldom are clean,
And the clean ones so seldom are comical.

Campus lore has it that students complain about texts prescribed by their instructors as being too highbrow or tough and not adequately reflecting student concerns, while instructors complain about texts as being low-level and, somehow, less demanding. We have consciously tried to write a book that both the student and the instructor can tolerate. Whether we have succeeded remains to be seen and can best be measured by your response. And, if you have read this far, and are still reading, we would certainly like to hear from you.

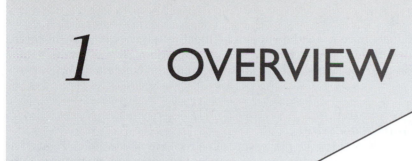

1 OVERVIEW

1.1 Introduction

"I listen and I forget,
I see and I remember, I do and I learn."
A Chinese Proverb

This book comes with no sound effects (which will, no doubt, be provided by your instructor!), but there is much to see and remember. And if you do what needs doing, there is also a lot to learn.

This book is about signals and their processing by systems. This first chapter provides an overview of the topics and concepts covered and the connections between them. We hope you return periodically to fill in the missing details and get a feel for how all the pieces fit together.

1.2 Signals

Our world is full of signals, both natural and man-made. Examples are the variation in air pressure when we speak, the daily highs and lows in temperature and the periodic electrical signals (EKG) generated by the heart. Signals represent information. Often, signals may not convey the required information directly and may not be free from disturbances. It is in this context that signal processing forms the basis for enhancing, extracting, storing or transmitting useful information. Electrical signals perhaps offer the widest scope for such manipulations. In fact, it is commonplace to convert signals to electrical form for processing.

A signal that varies continuously in time and amplitude is called an **analog signal**. If time assumes discrete values and amplitudes are restricted to a finite number of levels, we get a **digital signal**.

Analog signals have been the subject of much study in the past. In recent decades, digital signals have received widespread attention. Being numbers, they can be processed by the same logic circuits used in digital computers.

1.2.1   *Signal Processing*

Two conceptual schemes for the processing of signals are illustrated in Figure 1.1. The digital processing of analog signals requires an analog-to-digital converter (ADC) for sampling the analog signal and a digital-to-analog converter (DAC) to convert the processed digital signal back to analog form.

Figure 1.1   Analog and digital signal processing.

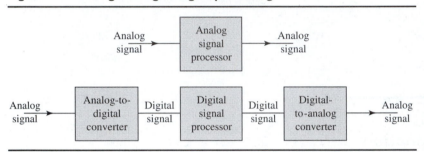

Few other technologies have revolutionized the world as profoundly as those based on digital signal processing. For example, the technology of recorded music was, until recently, completely analog from end to end, and the most important commercial source of recorded music used to be the LP (long-playing) record. The advent of the digital compact disc (CD) has changed all that in the span of just a few short years, and made the LP practically obsolete. Signal processing, both analog and digital, forms the core of this application and many others.

1.3
Systems

Systems may process analog or digital signals. All systems obey energy conservation. Loosely speaking, the **state** of a system refers to variables, such as capacitor voltages and inductor currents, which yield a measure of the system energy. The initial state is described by the initial value of these variables or initial conditions. A system is **relaxed** if initial conditions are zero. In this book, we study only **linear systems** whose input-output relation is linear. If a complicated input can be split into simpler forms, linearity allows us to find the response as the sum of the response to each of the simpler forms. This is **superposition**.

Many systems are actually nonlinear. The study of nonlinear systems often involves making simplifying assumptions, such as linearity.

1.4
Time and Frequency

The concept of frequency is intimately tied to sinusoids. The familiar form of a sinusoid is the oscillating, periodic time waveform (Figure 1.2). An alternative is to visualize the same sinusoid in the **frequency domain** in terms of its **magnitude** and **phase** at the given frequency (Figure 1.2). Each sinusoid has a unique representation in the frequency domain.

Figure 1.2   A sinusoid and its frequency-domain representations.

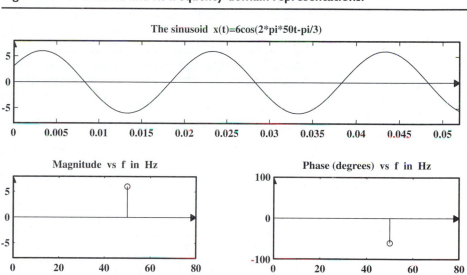

1.4.1   *Transformed Domain and Time-Domain Analysis*

One approach to system analysis relies on **transformations,** which map signals and systems into a transformed domain such as the frequency domain. This results in simpler mathematical operations to evaluate system behavior, but there is a price to pay. Since the response is evaluated in the transformed domain, we must have the means to remap this response to the time domain through an **inverse transformation.** Examples of this method include phasor analysis, Laplace transforms, z-transforms and Fourier transforms.

If the analysis is tractable in the time domain itself, no remapping of the output is necessary. Examples of time-domain methods include **state variable analysis, differential equations** and **convolution.**

Different methods of system analysis allow different perspectives on both the system and the analysis results. Some are more suited to particular systems. Others are more amenable to numerical computation.

1.4.2   *An Example*

As an example, consider the *RC* circuit of Figure 1.3. The capacitor voltage is governed by the differential equation

$$\frac{dv_0}{dt} + \frac{1}{\tau}v_0(t) = \frac{1}{\tau}v_i(t)$$

Assuming zero initial conditions, the response of this circuit to various inputs is summarized in Table 1.1 and sketched in Figure 1.3.

Note how the dc and cosine input yield only the **steady-state** response. The response to the step and switched sine also includes a decaying exponential representing the **transient** component.

Figure 1.3   An *RC* circuit and its response to various inputs.

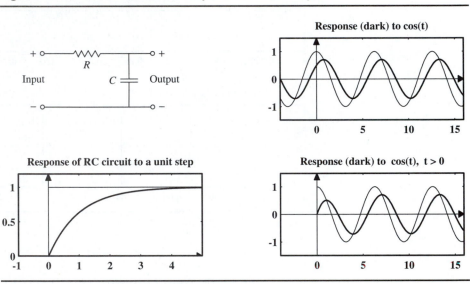

Table 1.1   General Response of an *RC* Lowpass Filter.

Input $v_i(t)$	General Response $v_0(t)$	$v_0(t)$ for $\tau = 1$, $\omega_0 = 1$, $V_i = 1$
$V_i, t \geq 0$	$V_i[1 - \exp(-t/\tau)], t > 0$	$1 - \exp(-t)$
$V_i \cos(\omega_0 t)$	$\dfrac{V_i \cos(\omega_0 t - \theta)}{[1 + \omega_0^2\tau^2]^{1/2}} \quad \theta = \tan^{-1}(\omega_0\tau)$	$\dfrac{1}{\sqrt{2}} \cos(t - \tfrac{1}{4}\pi)$
$V_i \cos(\omega_0 t), t > 0$	$\dfrac{V_i \cos(\omega_0 t - \theta)}{[1 + \omega_0^2\tau^2]^{1/2}} + \dfrac{V_i \omega_0 \tau}{1 + \omega_0^2\tau^2} \exp(-t/\tau)$	$\dfrac{1}{\sqrt{2}} \cos(t - \tfrac{1}{4}\pi) - \dfrac{1}{2}\exp(-t)$

The **time constant** $\tau = RC$ is an important measure of system characteristics. Figure 1.4 reveals that for small τ, the output resembles the input and follows the rapid variations in the input more closely. For large τ, the system smooths the sharp details in the input. This smoothing effect is typical of a **lowpass filter**. The time constant thus provides one index of performance in the time domain.

Figure 1.4   Step response and rise time of the *RC* circuit for various τ.

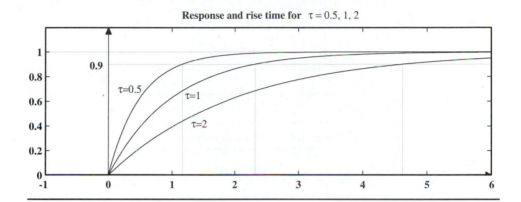

Another commonly used index is the **rise time**, which indicates the rapidity with which the response reaches its final value. As evident from Figure 1.4, a smaller time constant τ also implies a smaller rise time.

1.5.1   *Measures in the Frequency Domain*

The response to the cosine depends on $\omega_0 \tau$ (Figure 1.5). For $\omega_0 \tau \ll 1$, both the magnitude and phase of the output approach the applied input. For a given system (or τ), the output magnitude decreases for higher frequencies. Such a system describes a **lowpass filter**. Rapid time variations or sharp features in a signal thus correspond to higher frequencies.

1.5.2   *Spectral Characteristics*

A plot of the magnitude and phase **spectrum** versus frequency (Figure 1.6, p. 6) gives us the **frequency response** of the system. The magnitude spectrum clearly shows the effects of attenuation at high frequencies.

At the frequency $\omega_0 = \omega_B = 1/\tau$, the magnitude and phase of the transfer function are $1/\sqrt{2}$ and $-45°$. This frequency defines the **bandwidth**. More generally, the bandwidth is a measure of the frequency duration or the range of frequencies past which the output faces rapid attenuation.

There are measures analogous to bandwidth that describe the time **duration** of a signal over which much of the signal is concentrated.

Figure 1.5   Response of the *RC* circuit to a cosine for various $\omega_0\tau$.

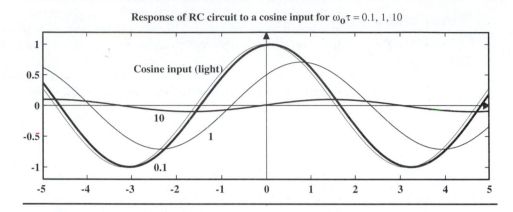

Response of RC circuit to a cosine input for $\omega_0\tau = 0.1, 1, 10$

Figure 1.6   Frequency response of the *RC* circuit.

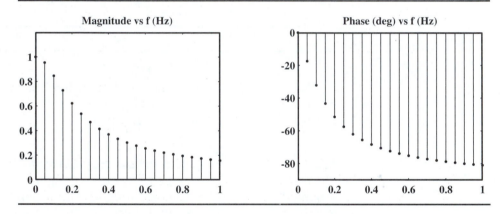

The relation $\omega_B\tau = 1$ clearly brings out the reciprocity in time and frequency. The smaller the "duration" τ or the more localized the time signal, the larger its bandwidth or frequency spread. The quantity $\omega_B\tau$ is a measure of the **time-bandwidth product**, a relation analogous to the uncertainty principle of quantum physics. We cannot simultaneously make both duration and bandwidth arbitrarily small.

1.5.3   *The Transfer Function and Impulse Response*

To find the frequency response, we must apply pure cosines with unit magnitude at all possible frequencies. Conceptually, we could apply all of them together and use superposition to find the response in one shot. What does such a combination represent? Since all cosines equal unity at $t = 0$, the combination approaches infinity at the origin. As Figure 1.7 suggests, the combination also approaches zero elsewhere due to the cancellation of the infinitely many equal positive and negative values at all other instants. We give this signal (subject to unit area) the name unit **impulse**. We can find the response of a relaxed system over the entire frequency range by applying a single impulse as the input.

Figure 1.7   Genesis of the impulse function as a sum of sinusoids.

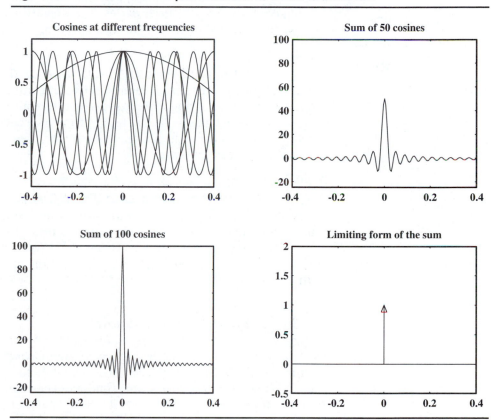

A system is completely characterized in the frequency domain by its frequency response or system **transfer function**. The time-domain response to an impulse is called the **impulse response**. A system is completely characterized in the time domain by its impulse response. Naturally, the transfer function and impulse response are two equivalent ways of looking at the same system.

1.6
Convolution

The idea of decomposing a complicated signal into simpler forms is very attractive for both signal and system analysis. One approach to time domain analysis describes the input as a sum of weighted impulses and finds the response as a sum of weighted impulse responses. This describes the process of **convolution**. Since the response is, in theory, a cumulative sum of infinitely many impulse responses, the convolution operation is actually an integral. One of the most useful results is that convolution is replaced by the much simpler operation of multiplication when we move to a transformed domain.

1.7 Frequency Description of Periodic Signals

A periodic signal with period T may be described by a weighted combination of sinusoids at its fundamental frequency $f_0 = 1/T$ and integer multiples kf_0. This combination is called a **Fourier series**. The weights are the magnitude and phase of each sinusoid and describe a discrete spectrum in the frequency domain.

1.7.1   *Frequency Domain Description of Arbitrary Signals*

For large T, the frequencies kf_0 in the discrete spectrum of a periodic signal become more closely spaced. A nonperiodic signal may be conceived, in the limit, as a single cycle of a periodic signal as $T \to \infty$. Its spectrum then becomes continuous and leads to a frequency domain representation of arbitrary signals (periodic or otherwise) in terms of **Fourier transforms**.

1.7.2   *System Analysis Using Transforms*

A Fourier series only allows us to find the response of relaxed systems to periodic signals. It yields just the steady-state response. The Fourier transform, on the other hand, allows us to analyze relaxed systems with arbitrary inputs. The **Laplace transform** uses a complex frequency to extend the analysis both to a larger class of inputs and to systems with nonzero initial conditions.

1.7.3   *Numerical Approximations*

In engineering practice, we are often forced to make compromises, no matter how elegant the results based on theory. System complexity often demands a numerical approach to analysis. This imposes its own limitations. We compute the response at discrete intervals rather than a time continuum and can thus only approximate the true response. A smaller time interval is better but more computation-intensive. Its choice is a compromise based on computation time, how much error we can tolerate, how the error propagates and the stability of the numerical algorithm used.

1.8 Discrete-Time Representations

The **sampling** of analog signals is often a matter of practical necessity. It is also the first step in **digital signal processing** (DSP). The good news is that a signal can be sampled without loss of information if it is **bandlimited** to a highest-frequency f_B and sampled at intervals less than $\frac{1}{2f_B}$. This is the celebrated **sampling theorem.** The bad news is that most signals are not bandlimited and even a small sampling interval may not be small enough. If the sampling interval exceeds the critical value $\frac{1}{2f_B}$, a phenomenon known as **aliasing** manifests itself. Components of the analog signal at high frequencies appear at (alias to) lower frequencies in the sampled signal. This results in a sampled signal with a smaller highest frequency. Aliasing effects are impossible to undo once the samples are acquired. It is thus commonplace to bandlimit the signal before sampling.

Numerical processing using digital computers requires finite data with finite precision. We must limit signal amplitudes to a finite number of levels. This process, called **quantization,** produces nonlinear effects that can be described only in statistical terms. Quantization also leads to an irreversible loss of information and is, therefore, considered only in the final stage in any design. As a result, the terms digital signal processing (DSP) and discrete-time signal processing are often used synonymously.

1.8.1   *Discrete Transforms*

It turns out that sampling in one domain leads to periodicity in the other. Just as a periodic time signal has a discrete spectrum, a discrete-time signal has a periodic spectrum. This duality forms the basis for the **discrete-time Fourier transform** (DTFT). As an extension, if the time signal is both discrete and periodic, its spectrum is also discrete and periodic. This idea leads to the **discrete Fourier transform** (DFT), whose evaluation is speeded up by the so-called **fast Fourier transform** (FFT) algorithms.

Yet another transform for discrete signals is the **z-transform**, which may be regarded as a discrete version of the Laplace transform.

1.9
From Concept
to Application

A system may be viewed as an entity that processes, or **filters**, the input signal. The term *filter* is often used to denote systems that tailor the response in a specified way, suppressing certain frequencies, for example. In practice, no measurement process is perfect or free from disturbances or **noise**. In a very broad sense, a signal may be viewed as something that is desirable, and noise may be regarded as an undesired characteristic that tends to degrade the signal. In this context, filtering describes a signal-processing operation that allows signal enhancement, noise reduction or increased **signal-to-noise ratio**.

Analog filters are used both in analog signal processing and to bandlimit the input to digital signal processors. **Digital filters** are used for tasks such as interpolation, extrapolation, smoothing and prediction.

Digital signal processing continues to play an increasingly important role in diverse applications that range literally from A (astronomy) to Z (zeumatography).

1.10
A Synopsis of
the Overview

Signals may be analog or digital and form inputs and outputs of systems. Undesired signals or disturbances are labeled noise (Chapter 2).

A system is an interconnection of components that processes signals. Linear systems allow the use of superposition. The impulse response yields system information in the time domain for relaxed systems (Chapter 3).

The time response of a relaxed system is the convolution of the input and the system impulse response (Chapter 4).

A periodic signal can be described by a Fourier series, a combination of sinusoids at harmonically related frequencies. Periodic signals have a discrete spectrum (Chapter 5).

Aperiodic signals have a continuous spectrum and are described by their Fourier transform (Chapter 6).

There is inherent reciprocity and duality between time and frequency. The effective duration and the effective bandwidth of a signal cannot both be made arbitrarily small simultaneously (Chapter 7).

The Laplace transform allows us to find the response of systems with nonzero initial conditions (Chapters 8 and 9).

The transfer function describes a system in the transformed domain. In this domain, the response equals the product of the transformed input and the transfer function.

An analog signal can be sampled without loss of information only if it is band-limited (Chapter 11).

Sampling in one domain leads to periodicity in the other. The spectrum of sampled signals is periodic. The spectrum of periodic signals is discrete. The spectrum of discrete and periodic signals is also periodic and discrete (Chapter 12).

The z-transform allows us to find the response of discrete-time systems with nonzero initial conditions (Chapter 13).

Analog and digital filters are important examples of signal processing that find numerous applications (Chapters 10 and 14).

Appendix 1A MATLAB Demonstrations and Routines

1A.1   *How to Use the Supplied Software*

The software supplied with this book runs under MATLAB. To use it, you must copy it to a directory in the MATLAB path. For details on running MATLAB, consult the MATLAB user's guide or the documentation on the disk.

The routines are in *m*-files such as *afn.m* with extension *.m*. Once MATLAB is running, help on a routine, say *afn.m*, is available by entering one of the following (without the extension *.m*) at the MATLAB prompt >>:

```
>>help afn
```

This brings up the syntax and information on how to run *afn.m*:

```
>>afn
```

For most supplied routines, this also runs a demonstration of *afn.m*. If you are new to MATLAB, we suggest you run *tutor1.m* and *tutor2.m* for a quick overview of its main features. To run *tutor1.m* simply type

```
>>tutor1
```

1A.2   *Getting Started on the Supplied MATLAB Routines*

Two supplied routines for plotting signals are *funplot.m* and *lines.m*.

Commands for Plotting Signals

```
>>t=-3:0.02:3;plot(t,sinc(t))    %Plot is a built-in MATLAB command.
```

This plots the function sinc(t) over $[-3\ 3]$ at intervals of 0.02.

```
>>funplot('sinc(t)',[-5 5])
```

This plots the function 'sinc(t)' over time limits $[-5\ 5]$.

```
>>t=-3:0.2:3;lines(t,sinc(t),'*')
```

This plots sinc(t) over $[-3\ 3]$ at intervals of 0.2 as lines capped by *.

```
>>t=-1:3;plot(t,2*t),axesn
```

Plots $2t$ over $[-1\ 3]$ and draws x and y axes with arrows.

Demonstrations and Utility Functions

```
>>tutor1    %Explains simple MATLAB commands and syntax.
>>tutor2    %Explains function and script files.
>>axesn     %Draws x and y axes (with arrows) on an existing plot.
>>tour      %A guided tour of all the supplied routines.
```

2 SIGNALS

2.0
Scope and Objectives

Signals convey information. They include physical quantities such as voltage, current and intensity. They may be functions of (one or more) independent variables such as position, time or frequency. This chapter is largely a compendium of facts, definitions and concepts. Its objectives are twofold:

1. To classify and catalog the signals to be encountered in this book.
2. To collect, under one roof, concepts that are central to understanding many of the techniques of signals and system analysis. These concepts will be encountered in many different contexts in subsequent chapters.

Here and elsewhere in this book, we confine ourselves only to **deterministic** signals whose behavior, unlike that of **random** signals, is always predictable.

The chapter is divided into four major parts:

Part 1 An introduction to the classification of analog signals.
Part 2 A catalog of commonly encountered analog signals.
Part 3 An introduction to sampled and discrete-time (*DT*) signals.
Part 4 A catalog of commonly encountered *DT* signals.

Useful Background   The concepts described here require some familiarity with the fundamentals of calculus (integration and differentiation).

2.1
Signals

The study of signals allows us to assess how they might be processed to extract useful information, which is what signal processing is all about. This book deals with one-dimensional signals that are functions of time or frequency. They may be described by mathematical expressions or graphically by curves or as tabulated values. At any instant, the value of a signal corresponds to its (instantaneous) amplitude. Time may assume a continuum of values, t, or discrete values, nt_s, with n an integer. The amplitude may assume a continuum of values or be quantized to a finite number of discrete levels between its extremes. This results in four possible kinds of signals, as shown in Figure 2.1 and listed in Table 2.1.

Figure 2.1   Analog, sampled, quantized and digital signals.

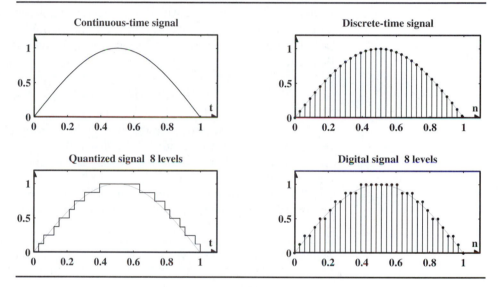

Table 2.1   Analog and Digital Signals.

Signal	Notation	Characteristics
Analog	$x(t)$	Continuous-time (CT) and continuous amplitude
Sampled	$x_s[n]$	Discrete-time (DT) and continuous amplitude
Quantized	$x_Q(t)$	Continuous-time (CT) and discrete amplitude
Digital	$x_Q[n]$	Discrete-time (DT) and discrete amplitude

Quantization is typically considered only as the final step in any design since it results in errors that are difficult to analyze quantitatively. The terms *discrete-time*, *sampled* and *digital* are, therefore, often used synonymously.

Real signals, alas, are not easy to describe quantitatively. They must often be approximated by idealized forms or models amenable to mathematical manipulation. It is these models that we concentrate on in this chapter.

2.2
Signal Classification Based on Duration

Figure 2.2 illustrates various signals and how they are classified. Signals can be of finite or infinite duration. Finite duration signals are called **timelimited**. Signals of semi-infinite extent may be

1. **Rightsided** if they are zero for $t < t_0$, where t_0 is finite.
2. **Leftsided** if they are zero for $t > t_0$, where t_0 is finite.

Signals of infinite extent $(-\infty < t < \infty)$ are called **twosided**.

Signals that are zero for $t < 0$ are often called **causal**. Signals that are zero for $t > 0$ are called **anticausal**. All others are termed **noncausal**. The term *causal* is used in analogy with causal systems discussed in Chapter 3. A causal signal $x(t)$, $t \geq 0$ may also be expressed as $x(t)u(t)$, where $u(t)$ is the **unit step function** (Figure 2.2) defined by

$$u(t) = \begin{cases} 0 & t < 0 \\ 1 & t > 0 \end{cases}$$

Piecewise continuous signals possess different expressions over different intervals. Such signals may also possess discontinuities or *jumps*. Examples of piecewise signals include $x(t) = u(t)$, which is discontinuous at $t = 0$, and the twosided exponential $x(t) = \exp(-|t|)$, which is continuous at $t = 0$.

The derivative at a discontinuity is not defined in the usual sense. If a signal or any of its derivatives is discontinuous, it describes a **singularity signal**. Examples

Figure 2.2   Examples of signals classified by duration and area.

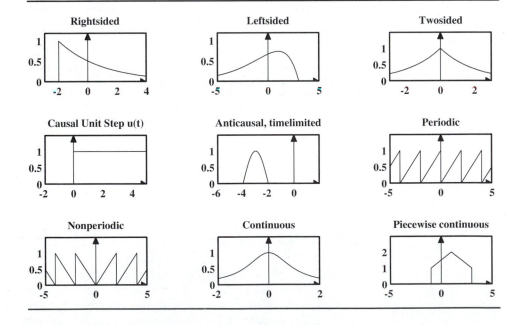

of singularity signals include the onesided ramp $x(t) = tu(t)$, and the twosided exponential $x(t) = \exp(-|t|)$.

Continuous signals are defined by a single expression over all time. Thus $x(t) = \sin(t)$ and $x(t) = t$ are both continuous, but $x(t) = |t|$ is *piecewise continuous* (it equals t for $t > 0$ and $-t$ for $t < 0$).

Periodic signals repeat the same pattern endlessly. The *smallest* repetition duration is called the **time period** T and leads to the formal definition

$$x_p(t) = x_p(t \pm nT) \qquad \text{for integer } n$$

where n is an integer. Signals lacking periodicity are termed **nonperiodic** or **aperiodic**. Onesided or timelimited signals can never be periodic.

2.2.1   *Signal Classification Based on Areas and Integrals*

The area of a signal $x(t)$ or $x^2(t)$ provides useful measures of its *size*. Mathematically, the area of $x(t)$ is found from the integral

$$\mathbb{A}[x(t)] = \int_{-\infty}^{\infty} x(t)dt$$

Notation: We use the notation $\mathbb{A}[x(t)]$ for the integral of $x(t)$ to coax you to visualize it in terms of its physical meaning (the area of $x(t)$). The area of $x(t)$ over the interval (a, b) is denoted by $\mathbb{A}[x(t)](a, b)$.

Absolutely integrable or **finite area signals** possess finite absolute area:

$$\mathbb{A}[|x(t)|] = \int_{-\infty}^{\infty} |x(t)| \, dt < \infty$$

Notation: We use the term *finite area* to mean *finite absolute area*.

All *timelimited* functions of finite amplitude are finite area signals, but a finite area signal does not necessarily imply a timelimited signal.

Example 2.1
A Finite Area Signal

The decaying exponential $x(t) = -\exp(-t/\tau)u(t)$ is of infinite extent. Its area is $\mathbb{A}[x(t)] = -\tau$. Its absolute area is $\mathbb{A}[|x(t)|] = \tau$.

2.3
Operations on
Signals

In addition to pointwise sums, differences and products, the commonly used operations on signals include shifting and scaling. These are illustrated in Figure 2.3 and summarized with examples in Table 2.2.

Figure 2.3   Examples of operations on signals.

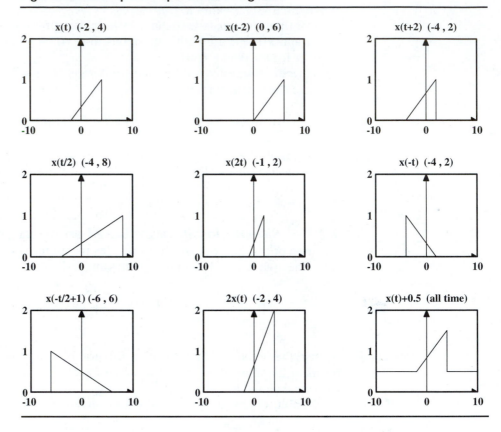

Table 2.2   Operations on Continuous-Time Signals.

Operation	Example	Explanation
Time shift	$x(t-2)$	Shift $x(t)$ right by 2 (delay).
	$x(t+2)$	Shift $x(t)$ left by 2 (advance).
Time scale	$x(2t)$	Compress $x(t)$ by factor of 2 (speed up).
	$x(\frac{1}{2}t)$	Stretch $x(t)$ by factor of 2 (slow down).
Folding	$x(-t)$	Fold $x(t)$ about origin.
Combinations	$x(-\frac{1}{2}t+1)$	Shift $x(t)$ left by 1, fold and stretch by 2.
or	$x[-\frac{1}{2}(t-2)]$	Fold & stretch $x(t)$ by 2 and shift right by 2.
Amplitude scale	$2x(t)$	Multiply ordinate by factor of 2.
Amplitude shift	$x(t)+\frac{1}{2}$	Add dc offset of $\frac{1}{2}$ to $x(t)$ *everywhere*.

A **time shift** displaces a function $x(t)$ in time without changing its shape. We represent the operation symbolically as $t \to t - \beta$. Shifting $x(t)$ to the *right* by 3 units *delays* $x(t)$ to produce $x(t - 3)$. Shifting $x(t)$ to the *left* by 3 units *advances* $x(t)$ to yield $x(t + 3)$.

Time scaling by α shrinks or stretches $x(t)$ to $x(\alpha t)$. It speeds up or slows down time. In symbolic notation, we have $t \to \alpha t$. The signal $x(3t), (\alpha > 1)$ is a *compressed* version of $x(t)$ since t is speeded up to $3t$, whereas $x(\frac{1}{3}t), (\alpha < 1)$ represents a *stretched* version of $x(t)$. **Reflection** or **folding** is just a scaling operation with $\alpha = -1$. It creates a mirror image of $x(t)$ about the vertical axis.

An **amplitude shift** adds a constant K to $x(t)$ *everywhere*, even where $x(t)$ is zero, such that $x(t) \to x(t) + K$. This transforms timelimited signals into infinite-duration signals. The converse does not hold in general. **Amplitude scaling** by C amplifies all function values by C, without affecting the duration of $x(t)$, such that $x(t) \to Cx(t)$.

2.3.1   *Operations in Combination*

The signal $y(t) = x(\alpha t - t_0)$ may be related to $x(t)$ in one of two ways:

1. $y(t) = x(\alpha t - t_0)$ using shifting by t_0, followed by scaling

$$x(t) \frac{\text{Shift}}{t \to t - t_0} \to \quad x(t - t_0) \quad \frac{\text{Scale}}{t \to \alpha t} \to x(\alpha t - t_0)$$

2. $y(t) = x[\alpha(t - t_0/\alpha)]$ using scaling by α, followed by shifting by t_0/α

$$x(t) \frac{\text{Scale}}{t \to \alpha t} \to \quad x(\alpha t) \quad \frac{\text{Shift}}{t \to t - t_0/\alpha} \to x(\alpha t - t_0)$$

In each sequence, the shift and scaling is on the unscaled variable t.
    In the form $y(t) = x[\alpha(t - \tau)]$ we also observe that

1. A change in sign of t results in a reflection about $t = 0$ (the origin).
2. A change in sign of α results in a reflection about $t = \tau$.

2.4 Signal Classification Based on Symmetry

The concept of folding or reflection is the basis for signal symmetry. We shall examine *even, odd* and *halfwave* symmetry, the last being defined only for periodic signals. Examples of symmetric signals are shown in Figure 2.4 and their characteristics are listed in Table 2.3.

Figure 2.4   Illustrating various types of symmetry.

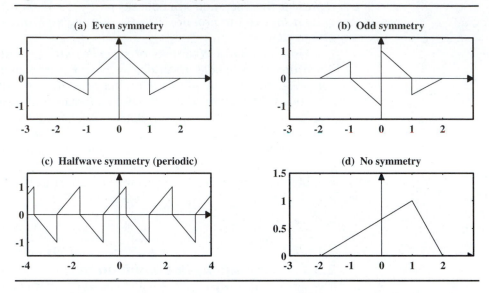

(a) Even symmetry

(b) Odd symmetry

(c) Halfwave symmetry (periodic)

(d) No symmetry

Table 2.3   Implications of Symmetry.

Type	Definition	Characteristics
Even	$x_e(t) = x_e(-t)$	$\mathbb{A}[x_e(t)](-\alpha, \alpha) = 2\mathbb{A}[x_e(t)](0, \alpha)$
		$x_p(t)$ also equals $x_p(T - t)$ if periodic
Odd	$x_o(t) = -x_o(-t)$	$\mathbb{A}[x_o(t)](-\alpha, \alpha) = 0$ and $x_o(0) = 0$
		$x_p(t)$ also equals $-x_p(T - t)$ if periodic
No symmetry	$x(t) = x_e(t) + x_o(t)$	$x_e(t) = \frac{1}{2}[x(t) + x(-t)], x_o(t) = \frac{1}{2}[x(t) - x(-t)]$
Halfwave	$x_p(t) = -x_p(t \pm \frac{1}{2}T)$	$x_{av} = 0$. Defined for periodic signals only.

2.4.1   Even Symmetry

If $x(t)$ is identical to its folded version (Figure 2.4), we have *mirror symmetry* about the vertical axis and $x(t)$ is called **even symmetric** or simply **even**. For even symmetry, we require $x_e(t) = x_e(-t)$. Even symmetry is invariant to scaling or an amplitude shift. The area under $x_e(t)$ is twice the area on either side of the origin. *Periodic*, even symmetric signals also show even symmetry about $\frac{1}{2}T$. As a result, $x_p(t) = x_p(T - t)$.

2.4.2   Odd Symmetry

If $x(t)$ and its folded version $x(-t)$ differ only in sign, we have *symmetry about the origin* (Figure 2.4) and $x(t)$ is called **antisymmetric**, **odd symmetric** or simply **odd**. For odd symmetry, we require $x_o(t) = -x_o(-t)$. For an odd signal, $x_o(0) = 0$ and the area under $x_o(t)$ over limits that are symmetric about the origin is always zero. We thus have

$$\int_{-\alpha}^{\alpha} x_o(t)\, dt = 0$$

Odd symmetry remains invariant to amplitude or time scaling but not to an amplitude shift. *Periodic*, odd symmetric signals also show odd symmetry about $\frac{1}{2}T$. As a result, $x_p(t) = -x_p(T - t)$.

Remark: The product of an even and odd signal is always odd. The product of two even signals or two odd signals is always even.

2.4.3   *Halfwave Symmetry*

Halfwave symmetry is defined only for a *periodic signal* $x_p(t)$. If $x_p(t)$ and its value $x_p(t \pm \frac{1}{2}T)$ half a period away are equal in magnitude but differ in sign, $x_p(t)$ is said to possess halfwave symmetry (Figure 2.4). Thus

$$x_p(t) = -x_p(t \pm \tfrac{1}{2}T) = -x_p(t \pm nT \pm \tfrac{1}{2}T) \qquad \text{for integer } n$$

Halfwave symmetric signals always show *two half-cycles over one period with each half-cycle an inverted replica of the other*. Their average value (average area per period) always equals zero.

2.4.4   *Even and Odd Parts of Signals*

An unsymmetric signal $x(t)$ may always be expressed as the sum of an even part $x_e(t)$ and an odd part $x_o(t)$. This is based on the observations that

1. The sum $y_e(t) = x(t) + x(-t)$ is always even.
2. The difference $y_o(t) = x(t) - x(-t)$ is always odd.

If we add $\frac{1}{2}y_e(t)$ and $\frac{1}{2}y_o(t)$, we recover $x(t)$ (Figure 2.5, p. 20) and thus

$$x_e(t) = \tfrac{1}{2}[x(t) + x(-t)]$$
$$x_o(t) = \tfrac{1}{2}[x(t) - x(-t)]$$

We can implement these relations graphically (Figure 2.5) or analytically.

Example 2.2
Even and Odd Parts of
Signals

(a) If $x(t) = \exp(-t)u(t)$, the even and odd parts of $x(t)$ are
$x_e(t) = \frac{1}{2}[\exp(-t)u(t) + \exp(t)u(-t)]$, $x_o(t) = \frac{1}{2}[\exp(-t)u(t) - \exp(t)u(-t)]$.
(b) If $x(t) = [\sin(t) + 1]^2$, expand it to give $x(t) = \sin^2(t) + 2\sin(t) + 1$. We then recognize the even and odd parts as $x_e(t) = \sin^2(t) + 1$ and $x_o(t) = 2\sin(t)$.

Figure 2.5   **Even and odd parts of an unsymmetric signal.**

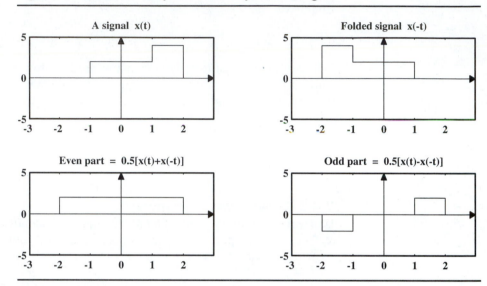

The **instantaneous power** delivered to a resistor R having a voltage $v(t)$ across it due to a current $i(t)$ may be expressed as

$$p(t) = v(t)i(t) = i^2(t)R = \frac{v^2(t)}{R}$$

The total energy delivered to the resistor equals

$$E = \int_{-\infty}^{\infty} p(t)\,dt = R \int_{-\infty}^{\infty} i^2(t)\,dt = \frac{1}{R} \int_{-\infty}^{\infty} v^2(t)\,dt$$

2.5.1 *Normalized Signal Energy and Signal Power*

To make meaningful comparisons, we normalize this relation to a 1-Ω resistor and denote the voltage or current by $x(t)$ to give

$$E_x = \int_{-\infty}^{\infty} |x(t)|^2\,dt = \mathbb{A}[|x(t)|^2]$$

We call E_x the **normalized signal energy** or just the **energy** in $x(t)$. We use $|x(t)|$ to admit complex valued signals and obtain a real result.

The **normalized signal power** is the *time average* of the signal energy *over all time*. If $x(t)$ is periodic with period T, the signal power is simply the *average energy per period* and we have

$$P_x = \frac{1}{T} \int_T |x(t)|^2\,dt \quad \text{for periodic signals}$$

Notation: $\int_T$ means integration over any convenient one-period duration.

2.5
Signal Classification Based on Energy and Power

More generally, we can compute the power in any signal $x(t)$ by averaging its energy over a longer and longer finite stretch T_0. As $T_0 \to \infty$, we obtain

$$P_x = \lim_{T_0 \to \infty} \frac{1}{T_0} \int_{T_0} |x(t)|^2 \, dt \quad \text{for nonperiodic signals}$$

This more general form is useful only for nonperiodic signals.

The area and energy for some common signals (Figure 2.6) are summarized in Table 2.4 and are quite useful in computing energy and power for many other signals that can be expressed as their combinations.

Figure 2.6   Area and energy for various pulse shapes listed in Table 2.4.

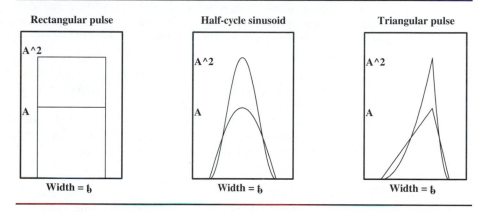

Table 2.4   Area and Energy of Common Signals.

Shape	$\mathbb{A}[x(t)]$	$E_x = \mathbb{A}[x^2(t)]$
Rectangular pulse [height $= A$, width $= t_0$]	At_0	$A^2 t_0$
Half-cycle sinusoid [height $= A$, width $= t_0$]	$2At_0/\pi$	$\frac{1}{2}A^2 t_0$
Triangular pulse [height $= A$, width $= t_0$]	$\frac{1}{2}At_0$	$\frac{1}{3}A^2 t_0$
Decaying exponential [$A\exp(-\alpha t)u(t)$]	A/α	$A^2/2\alpha$

NOTE: For a combination $y(t) = x_1(t) + x_2(t)$, $\mathbb{A}[y] = \mathbb{A}[x_1] + \mathbb{A}[x_2]$. However, $\mathbb{A}[y^2] = \mathbb{A}[x_1^2] + \mathbb{A}[x_2^2]$ only if x_1 and x_2 show no common overlap. If overlap exists, $\mathbb{A}[y^2] = \mathbb{A}[x_1^2] + \mathbb{A}[x_2^2] + 2\mathbb{A}[x_1 x_2]$, where $x_1 x_2$ is the product of x_1 and x_2.

Example 2.3
Effect of Operations

For a signal $x(t)$ (from Table 2.4, for example), convince yourself that

1. Shifting or folding $x(t)$ will not change its area or energy.

2. Compressing $x(t)$ to $x(\alpha t)$ will reduce its area and energy by $|\alpha|$.
3. The energy $\mathbb{A}[x^2(t)]$ does not equal $\{\mathbb{A}[x(t)]\}^2$.
4. The energy $\mathbb{A}[x^2(t)]$ is always positive.

2.5.2   *Energy Signals and Power Signals*

The definitions for power and energy serve to classify signals as either energy signals or power signals, as summarized in Table 2.5.

Table 2.5   Energy Signals and Power Signals.

Signal $x(t)$	Criterion	Implication	Examples
Energy signal	$E_x < \infty$	$P_x = 0$	Timelimited pulses
Power signal	$P_x < \infty$	$E_x = \infty$	Periodic signals ($P_x = x_{\text{rms}}^2$)

Energy Signals   A signal with finite energy is called an **energy signal** or a **square integrable**. Energy signals have zero average power since we are averaging finite energy over all (infinite) time.

Examples of energy signals include onesided or twosided decaying and damped exponentials, damped sinusoids, sinc, Lorentzian and Gaussian signals (see Sec. 2.7) and all timelimited signals of finite amplitude.

Example 2.4
Energy Signals

(a) The signal of Figure 2.4b shows four nonoverlapping triangular pulses of unit width and heights $1, -1, \frac{1}{2}$ and $-\frac{1}{2}$. Using Table 2.4, we find its signal energy as $E = \frac{1}{3}(1)^2(1) + \frac{1}{3}(-1)^2(1) + \frac{1}{3}(\frac{1}{2})^2(1) + \frac{1}{3}(-\frac{1}{2})^2(1) = \frac{5}{6}$.

(b) The signal $x(t) = [\exp(-t) - \exp(-\frac{1}{2}t)]u(t)$ is an energy signal. Since $E_x = \mathbb{A}[x^2(t)]$ and $x^2(t) = [\exp(-2t) + \exp(-t) - 2\exp(\frac{3}{2}t)]u(t)$, we get

$$E_x = \mathbb{A}[\exp(-2t)](0, \infty) + \mathbb{A}[\exp(-t)](0, \infty) - 2\mathbb{A}[\exp(-3t/2)](0, \infty)$$

$$= \frac{1}{2} + 1 - \frac{4}{3} = \frac{1}{6}$$

Comment: As a consistency check, ensure that the energy is always positive!

Power Signals   Signals with finite power are called **power signals**. Such signals possess finite (nonzero) average power and infinite energy.

Examples of power signals are periodic signals and their onesided forms.

Remark: Power signals and energy signals are mutually exclusive because energy signals have zero power and power signals have infinite energy. A signal may, of course, be neither and have infinite energy and infinite power (e.g., t^n), or infinite energy and zero power (e.g., $1/\sqrt{t}, t \geq 1$).

2.6
Periodic Signals

Periodic signals are power signals that are characterized by several measures. The **duty ratio** of a periodic signal $x_p(t)$ equals the ratio of its pulse width and period. The **average value** x_{av} equals its average area per period. The signal power P_x is its average energy per period. The **rms value** x_{rms} equals $\sqrt{P_x}$ and corresponds to a dc signal with the same power as $x_p(t)$. None of these change if $x_p(t)$ is time scaled or time shifted. The average value x_{av}, signal power P_x and rms value x_{rms} are formally defined by

$$x_{av} = \frac{1}{T}\int_T x(t)\,dt \qquad P_x = \frac{1}{T}\int_T |x(t)|^2\,dt \qquad x_{rms} = \sqrt{P_x}$$

The average value can never exceed the rms value and thus $x_{av} \le x_{rms}$.

Example 2.5
Power Signals

(a) The signal of Figure 2.4c has a period $T = 2$ and two half-cycles per period. Table 2.4 gives $\mathbb{A}[x^2(t)] = \frac{1}{3}(1)^2(1) + \frac{1}{3}(-1)^2(1) = \frac{2}{3}$. Thus $P_x = \mathbb{A}[x^2(t)]/T = \frac{1}{3}$.

(b) For the constant signal $x(t) = A$, $x_{av} = A$, $P_x = A^2$ and $x_{rms} = A = x_{av}$.

(c) Let $x(t) = A\exp(j\omega t)$. Since $x(t)$ is complex valued, we work with $|x(t)|$. And since $|x(t)| = A$, the power in $x(t)$ equals $P_x = A^2$ as in part (b).

2.6.1 Combinations of Periodic Signals

The **common period** or **time period** T of a combination of periodic signals is the *smallest duration* over which each component completes an integer number of cycles. It is given by the LCM (least common multiple) of the individual periods. The **fundamental frequency** f_0 is the reciprocal of T and equals the GCD (greatest common divisor) of the individual frequencies. We can find a common value only for a **commensurate** combination in which the periods (frequencies) or their ratios are **rational fractions** (ratio of integers with common factors canceled out). For example, the periods $\frac{3}{7}$, 4 and $\frac{5}{8}$ are commensurate. So are the periods $\frac{3}{7}\pi$, 4π and $\frac{5}{8}\pi$ (because any of their ratios is a rational fraction).

Example 2.6
Periodic Combinations

Consider the signal $x(t) = 2\sin(\frac{2}{3}t) + 3\cos(\frac{1}{2}t) + 4\cos(\frac{1}{3}t - \frac{1}{5}\pi)$.

1. The periods of the individual components are 3π, 4π and 6π. The period T of $x(t)$ is their LCM and equals 12π. Thus, $\omega_0 = 2\pi/T = \frac{1}{6}$.

2. The frequencies of the individual components are $\frac{2}{3}$, $\frac{1}{2}$ and $\frac{1}{3}$ rad/s. The fundamental frequency ω_0 is their GCD and equals $\frac{1}{6}$. Thus, $T = 2\pi/\omega_0 = 12\pi$.

2.6.2 Almost Periodic Signals

For noncommensurate combinations such as $x(t) = 2\cos(\pi t) + 4\sin(3t)$, where the ratios of the periods (or frequencies) are not rational, we simply cannot find a common period (or frequency) and there is no repetition! Such combinations are called **almost periodic** or **quasi-periodic**.

Example 2.7
Almost Periodic Signals

(a) The signal $x(t) = \sin(t) + \sin(\pi t)$ is almost periodic. The frequencies 1 and π are not commensurate.

(b) The signal $y(t) = \sin(t)\sin(\pi t)$ is also almost periodic. We rewrite $y(t)$ as $y(t) = \sin(t)\sin(\pi t) = \frac{1}{2}\cos[(1 - \pi)t] - \frac{1}{2}\cos[(1 + \pi)t]$. Once again, the frequencies $(1 - \pi)$ and $(1 + \pi)$ are not commensurate.

2.6.3   *Power in Combinations of Periodic Signals*

For a combination of periodic signals, say $y(t) = x_1(t) + x_2(t)$, whose periods may or may not be commensurate, it is easily shown that the average value y_{av} is given by the sum of the individual average values. Thus,

$$y_{av} = x_{1av} + x_{2av}$$

If the individual components comprise *sinusoids at different frequencies* or pulse trains with no overlap over their pulse widths, the power P_y, likewise, equals the sum of the individual powers and the rms value equals $\sqrt{P_y}$. Thus,

$$P_y = P_{x1} + P_{x2} \qquad y_{rms} = \sqrt{P_y}$$

Remark: If these conditions are not met, the power and rms value must be found by other (usually tedious) means such as the limiting form.

Switched Periodic Signals   If $x(t)$ is a periodic signal $x_p(t)$ switched on or off at t_0, its power equals only half the power in $x_p(t)$.

$$P[x(t)] = P[x_p(t)u(t - t_0)] = P[x_p(t)u(t_0 - t)] = \tfrac{1}{2}P[x_p(t)]$$

If we use the limiting relation for power and average it over equal durations about t_0, say $(t_0 - \frac{1}{2}T_0, t_0 + \frac{1}{2}T_0)$, the area of $x^2(t)$ equals only half the area of $x_p^2(t)$ since $x(t)$ is zero for $t \le t_0$. And since the averaging duration is T_0 for both $x(t)$ and $x_p(t)$, the power in $x(t)$ equals half the power in $x_p(t)$!

Example 2.8
Power in Periodic Signals

(a) For the full rectified sine wave $x(t) = |\sin(t)|$, $T = \pi$. Using Table 2.4, we find $P_x = \mathbb{A}[x^2(t)]/T = \frac{1}{2}$. This is identical to the power in $\sin(t)$ itself!

(b) The switched periodic signal $y(t) = |\sin(t)|u(t - 5)$ starts at $t = 5$. Its power equals half the power in $x(t) = |\sin(t)|$ and thus $P_y = \frac{1}{2}P_x = \frac{1}{4}$.

(c) The signal $x(t) = \sin(t)\sin(\pi t)$ equals $\frac{1}{2}\cos[(1 - \pi)t] - \frac{1}{2}\cos[(1 + \pi)t]$. It is almost periodic and contains different frequencies. The average power is the sum of the individual powers and thus $P_x = \frac{1}{2}(\frac{1}{2})^2 + \frac{1}{2}(\frac{1}{2})^2 = \frac{1}{4}$.

(d) Let $x(t) = 2 + 3\sin(t) - 4\cos(t) + \cos(10t - \frac{1}{3}\pi)$. We *must* combine terms at like frequencies before finding P_x or x_{rms}. The easiest way is phasors. We write $3\sin(t) - 4\cos(t) \rightarrow 3\angle-90° - 4\angle 0 = -j3 - 4 = 5\angle\theta \rightarrow 5\cos(t + \theta)$, where θ need not be evaluated. Then $P_x = (2)^2 + \frac{1}{2}(5)^2 + \frac{1}{2}(1)^2 = 17$. The rms value equals $x_{rms} = \sqrt{17} = 4.123$.

(e) Consider $x(t) = 2\sin(2t)u(t) + 2\sin(t)$. One way to find P_x is to note that the frequencies are different and obtain $P_x = \frac{1}{2}(\frac{1}{2})(2)^2 + \frac{1}{2}(2)^2 = 3$. Another way is to express $x(t)$ as a sum of nonoverlapping functions $x_1(t)$ and $x_2(t)$, where $x_1(t) = 2\sin(2t)$, $t < 0$ and $x_2(t) = 2\sin(2t) + 2\sin(t)$, $t > 0$ and get

$$P_x = \tfrac{1}{2}(\tfrac{1}{2})(2)^2 + \tfrac{1}{2}[\tfrac{1}{2}(2)^2 + \tfrac{1}{2}(2)^2] = 3$$

as before.

(f) Let $x(t) = 2\sin(t)u(t) + 2\sin(t)$. Since the frequencies are identical, we first split $x(t)$ into the two nonoverlapping functions $x_1(t) = 2\sin(t)$, $t < 0$ and $x_2(t) = 4\sin(t)$, $t > 0$. We then obtain $P_x = \frac{1}{2}(\frac{1}{2})(2)^2 + \frac{1}{2}(\frac{1}{2})(4)^2 = 1 + 4 = 5$.

2.6.4   *Harmonic Signals and Sinusoids*

Sinusoids and **harmonic signals** and are among the most useful periodic signals. They are described by the general forms (Figure 2.7)

$$x_p(t) = A\cos(2\pi f_0 t + \theta) \qquad x(t) = A\exp[j(2\pi f_0 t + \theta)]$$

Figure 2.7 A sinusoid and its onesided and twosided spectra.

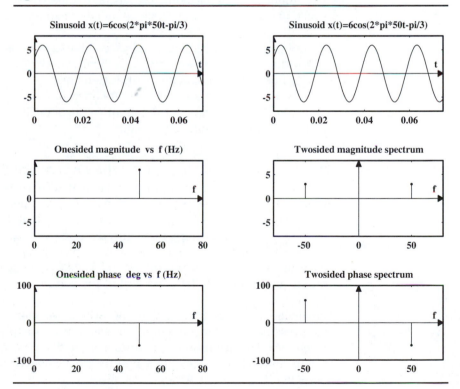

The two forms are, of course, related by Euler's identity such that

$$x_p(t) = \text{Re}\{A \exp[j(2\pi f_0 t + \theta)]\} = \tfrac{1}{2}A \exp[j(2\pi f_0 t + \theta)] + \tfrac{1}{2}A \exp[-j(2\pi f_0 t + \theta)]$$

The various time and frequency measures are also related by

$$f_0 = 1/T \qquad \omega_0 = 2\pi/T = 2\pi f_0 \qquad \theta = \omega_0 t_0 = 2\pi f_0 t_0 = 2\pi t_0/T$$

The quantity A is called the **magnitude** or **peak value** of the harmonic. The time t_0 or the **phase** θ measures the shift with respect to the pure cosine $\cos(2\pi f_0 t)$ or the harmonic $\exp(j2\pi f_0 t)$ (the accepted standard in engineering work). The form $x_p(t) = \text{Re}\{A \exp[j(2\pi f_0 t + \theta)]\}$ describes a rotating vector or **phasor** of constant length A whose angle $(2\pi f_0 t + \theta)$ varies with time.

Remarks: 1. For a sinusoid $x_p(t) = A \cos(\omega t + \theta)$, $x_{av} = 0$, $P_x = \tfrac{1}{2}A^2$ and $x_{rms} = A/\sqrt{2}$.
2. For a harmonic signal $x(t) = A \exp(j\omega t + \theta)$, $x_{av} = 0$ and $P_x = A^2$.
3. An analog sinusoid or harmonic signal is *always periodic and unique* for any choice of period or frequency (quite in contrast to digital sinusoids, which we study in Section 2.16). Combinations of sinusoids are periodic only if their frequencies are commensurate.

2.6.5   *Importance of Harmonic Signals*

The importance of harmonic signals and sinusoids stems from the following aspects, which are discussed in detail in later chapters.

1. Any signal can be represented by a combination of harmonics—periodic ones by harmonics at discrete frequencies (Fourier series), aperiodic ones by harmonics at all possible frequencies (Fourier transform).
2. The response of a linear system (defined in Chapter 3) to a harmonic input is also a harmonic signal at the input frequency. This forms the basis for system analysis in the frequency domain.

2.6.6   *Harmonic Signals in the Frequency Domain*

The very concept of frequency is intimately tied to harmonic signals, which link the time and frequency description of signals. This link permeates all aspects of signal processing and, even though discussed extensively in later chapters, is important enough to warrant an immediate introduction.

The familiar way of visualizing $x_p(t) = A \cos(2\pi f_0 t + \theta)$ graphically is as an oscillating periodic waveform with period T. Its frequency is indicated only indirectly on the graph through the reciprocal of the period T.

Magnitude and Phase Spectra   Another way of visualizing the same sinusoid is by suppressing time and sketching the magnitude A and phase θ separately versus frequency. The **magnitude spectrum** is plotted as a point with height A at $f = f_0$ in the magnitude–frequency plane and the **phase spectrum** as a point with height θ

at $f = f_0$ in the phase-frequency plane (Figure 2.7). Vertical lines are often drawn but only for better visualization; they have no other significance. The plots are also called **line spectra** or **discrete spectra** and represent the sinusoid in the **frequency domain**. The time period is indicated only indirectly through the reciprocal of f_0. Each sinusoid has a unique representation in this frequency domain.

We can describe the spectrum of the harmonic signal $x(t) = A \exp[j(2\pi f_0 t + \theta)]$ in quite the same way, with magnitude A and phase θ at the frequency f_0.

Twosided Spectra   Frequency is a positive number but we can also interpret negative frequencies! Using the identity $\cos(-\alpha) = \cos(\alpha)$, a sinusoid at the frequency $-f_0$ may be expressed as

$$A \cos(-2\pi f_0 t + \theta) = A \cos(2\pi f_0 t - \theta)$$

In other words, negative frequencies simply amount to a phase reversal!

Using this result, we can now write $x_p(t) = A \cos(2\pi f_0 t + \theta)$ as

$$x_p(t) = A \cos(2\pi f_0 t + \theta) = \tfrac{1}{2} A \cos(2\pi f_0 t + \theta) + \tfrac{1}{2} A \cos(-2\pi f_0 t - \theta)$$

Using Euler's relation, we also have the alternate form

$$x_p(t) = A \cos(2\pi f_0 t + \theta) = \tfrac{1}{2} A \exp[j(2\pi f_0 + \theta)] + A \exp[-j(2\pi f_0 + \theta)]$$

Either form describes a spectrum with equal magnitudes $\tfrac{1}{2} A$ (having even symmetry) and a phase of $+\theta$ and $-\theta$ (with odd symmetry) at f_0 and $-f_0$, respectively (Figure 2.7) and describes a **twosided spectrum**. For real sinusoids and their combinations, the *twosided spectra show even symmetry for magnitude and odd symmetry for phase*.

A combination of sinusoids at f_0 and its integer multiples $k f_0$ (or its harmonics) is periodic with period $T = 1/f_0$. Its discrete spectrum shows magnitude and phase components at $k f_0$.

Bandlimited Signals   If the highest frequency present in a signal is f_B, it is said to be **bandlimited** to f_B. Audio signals, for example, contain frequencies only up to (about) 20 kHz and are thus bandlimited to 20 kHz.

Example 2.9
Onesided and Twosided Spectra

Consider $x(t) = 2 \sin(2\pi 20t) - 4 \cos(2\pi 30t + 50°) + 6 \cos(2\pi 50t - \tfrac{1}{3}\pi)$. First, we rewrite $x(t)$ using a cosine reference, angles in degrees and positive magnitudes (absorbing negative signs by adding $\pm 180°$ to the phase), to obtain

$$x(t) = 2 \cos(2\pi 20t - 90°) + 4 \cos(2\pi 30t - 130°) + 6 \cos(2\pi 50t - 60°)$$

This leads to the onesided and twosided spectra of Figure 2.8.

Comment: The signal $x(t)$ is bandlimited to 50 Hz (its highest frequency).

Figure 2.8 **Spectra of $x(t) = 2\sin(2\pi 20t) - 4\cos(2\pi\,30t + 50°) + 6\cos(2\pi\,50t - \frac{1}{3}\pi)$.**

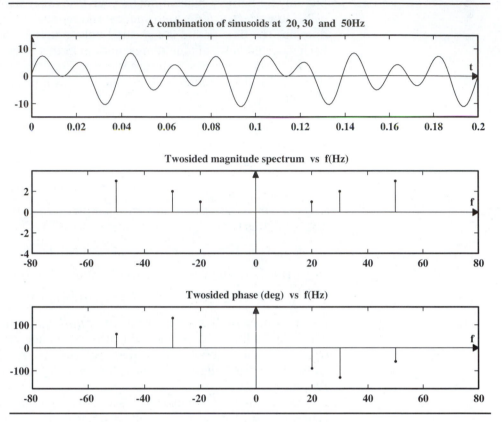

2.6.7 *Signal Processing—A Sneak Preview*

Signal processing typically involves concepts related to both the time domain and frequency domain. By way of an example, we briefly describe a communication system based on **amplitude modulation** with reference to Figure 2.9.

High-frequency signals are easier to transmit over long distances. The key to efficient transmission lies in converting low-frequency signals to high-frequency signals. One way is to multiply the low-frequency **message signal** $x_S(t)$ by a high-frequency sinusoid, or **carrier**, $x_C(t)$. If $x_S(t) = \cos(2\pi f_B t)$ and $x_C(t) = \cos(2\pi f_C t)$, we obtain the **amplitude modulated signal** $x_M(t)$ as

$$x_M(t) = x_S(t)x_C(t) = \cos(2\pi f_B t)\cos(2\pi f_C t)$$
$$= \tfrac{1}{2}\cos[2\pi(f_C + f_B)t] + \tfrac{1}{2}\cos[2\pi(f_C - f_B)t]$$

Note how the shape of $x_S(t)$ is preserved in the **envelope** of $x_M(t)$. The spectrum of $x_M(t)$ shows only high-frequency components at $f_C \pm f_B$. Antennas convert the modulated signal to electromagnetic energy, which is transmitted through space and converted back to an electrical signal at the receiver.

Figure 2.9 Illustrating the concepts involved in amplitude modulation.

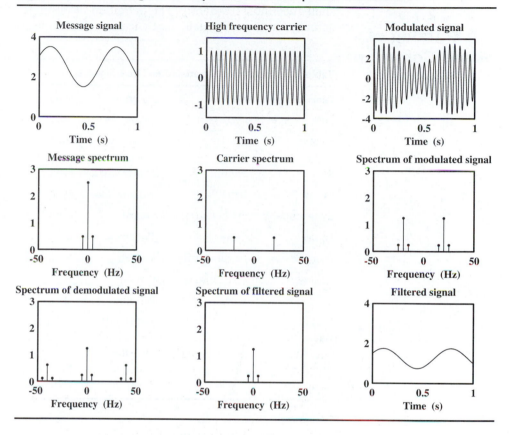

One way to recover the message at the receiver is to multiply, or **demodulate**, $x_M(t)$ by the same carrier $x_C(t)$ to obtain the **demodulated signal** $x_D(t)$. We obtain

$$x_D(t) = x_M(t)x_C(t) = \cos(2\pi f_C t)\{\tfrac{1}{2}\cos[2\pi(f_C + f_B)t] + \tfrac{1}{2}\cos[2\pi(f_C - f_B)t]\}$$

Using trigonometric identities, this simplifies to

$$x_D(t) = \tfrac{1}{4}\{\cos(2\pi f_C t) + \cos[2\pi(2f_C - f_B)t]\}$$
$$+ \tfrac{1}{4}\{\cos(2\pi f_C t) + \cos[2\pi(2f_C + f_B)t]\}$$

The spectrum of $x_D(t)$ clearly suggests that the message $x_S(t) = \cos(2\pi f_B t)$ can be recovered by feeding $x_D(t)$ into a **lowpass filter**, which passes $x_S(t)$ but blocks the two high-frequency components at $2f_C \pm f_B$.

2.7
A Catalog of Some Common Signals

Table 2.6 and Figure 2.10 summarize the major characteristics of signals that we shall encounter frequently. The definition of all finite area signals in this table is actually based on normalizing their area to unity.

Table 2.6   Common Continuous-Time Signals.

Function	Notation	Duration	Symmetry	$\mathbb{A}[\lvert x(t)\rvert]$	Energy/Power
Step	$u(t)$	Causal	None	∞	$P = \frac{1}{2}$
Signum	$\text{sgn}(t)$	Twosided	Odd	∞	$P = 1$
Ramp	$r(t)$	Causal	None	∞	Neither
Rect	$\text{rect}(t)$	Timelimited	Even	1	$E = 1$
Tri	$\text{tri}(t)$	Timelimited	Even	1	$E = \frac{2}{3}$
Exponential	$\exp(-t)u(t)$	Causal	None	1	$E = \frac{1}{2}$
Laplace	$\frac{1}{2}\exp(-\lvert t\rvert)$	Twosided	Even	1	$E = \frac{1}{4}$
Damped exp	$t\exp(-t)u(t)$	Causal	None	1	$E = \frac{1}{4}$
Cosine	$\cos(2\pi f_0 t)$	Periodic	Even	∞	$P = \frac{1}{2}$
Sine	$\sin(2\pi f_0 t)$	Periodic	Odd	∞	$P = \frac{1}{2}$
Sinc	$\sin(\pi t)/\pi t$	Twosided	Even	∞	$E = 1$
Gaussian	$\exp(-\pi t^2)$	Twosided	Even	1	$E = 1/\sqrt{2}$
Lorentzian	$1/[\pi(1 + t^2)]$	Twosided	Even	1	$E = 1/(2\pi)$
Impulse	$\delta(t)$	Zero	Even	1	Neither
Doublet	$\delta'(t)$	Zero	Odd	∞	Neither

The **unit step function** $u(t)$ is discontinuous at $t = 0$, where its value is undefined. With $u(0) = \frac{1}{2}$, $u(t)$ is called the **Heaviside unit step**. In its general form, $u[f(t)]$ equals 1 if $f(t) > 0$ and 0 if $f(t) < 0$. The folded step $u(\alpha - t)$ is leftsided and terminates at $t = \alpha$.

The **signum function** $\text{sgn}(t)$ is defined by

$$\text{sgn}(t) = \begin{cases} -1 & t < 0 \\ 1 & t > 0 \end{cases} = u(t) - u(-t) = 2u(t) - 1$$

It is characterized by a *sign change* at $t = 0$. Its value at $t = 0$ is also undefined and chosen as zero.

The **unit ramp** $tu(t)$ or $r(t)$ is also the running integral of the unit step

$$tu(t) = r(t) = \int_0^t u(\tau)\, d\tau = \int_{-\infty}^t u(\tau)\, d\tau \qquad u(t) = r'(t)$$

The step and signum allow a signal $x(t)$ to be turned on or off or have its polarity switched after some instant. Figure 2.11 (p. 32) illustrates that

1. $x(t)u(t - \beta)$ represents a signal $x(t)$ turned on at $t = \beta$.
2. $x(t)u(\beta - t)$ represents a signal $x(t)$ turned off at $t = \beta$.
3. $x(t)\text{sgn}(\beta - t)$ reverses the polarity of $x(t)$ at $t = \beta$.

Figure 2.10 Illustrating the analog signals of Table 2.6.

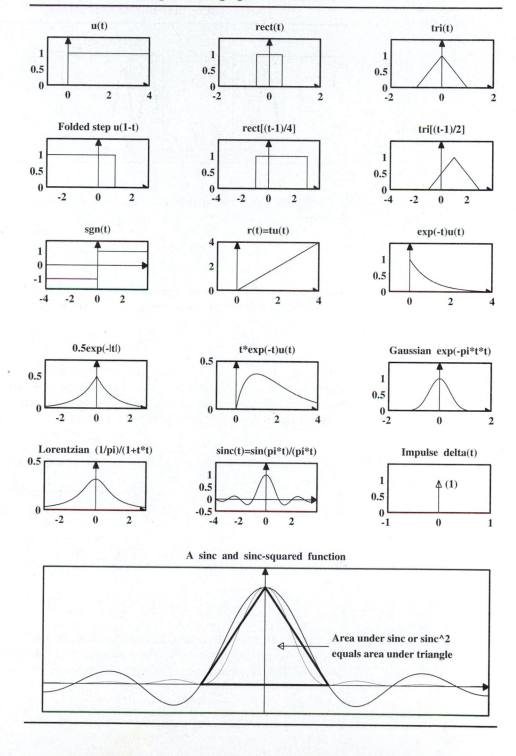

Figure 2.11   A signal, say 2 sin (4t), can be turned on (or off) by a step while a signum function can be used to switch its polarity.

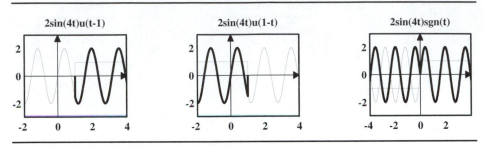

Pulse Signals   The signal **rect**(t) is an even rectangular pulse of unit width and height:

$$\text{rect}(t) = \begin{cases} 1 & |t| < \frac{1}{2} \\ 0 & |t| > \frac{1}{2} \end{cases} = u(t + \tfrac{1}{2}) - u(t - \tfrac{1}{2})$$

The form rect$[(t - \beta)/\alpha]$ is a rectangular pulse of width α, centered at $t = \beta$.

The even triangular pulse **tri**(t) extends from -1 to 1 and has unit height

$$\text{tri}(t) = \begin{cases} 1 - |t| & |t| \le 1 \\ 0 & |t| \ge 1 \end{cases} = r(t + 1) - 2r(t) + r(t - 1)$$

The form tri$[(t - \beta)/\alpha]$ is a triangular pulse of width 2α centered at $t = \beta$. These **pulse signals** serve as **windows** to limit and shape arbitrary signals. Thus, $x(t)\text{rect}(t)$ equals $x(t)$ *abruptly truncated* past $|t| = \frac{1}{2}$, whereas $x(t)\text{tri}(t)$ equals $x(t)$ *linearly tapered* about $t = 0$ and zero past $t = \pm 1$.

Example 2.10
Signal Representation

The function $y(t)$ of Figure 2.12 may be represented using

1. Window functions: $y(t) = 4\,\text{rect}[\frac{1}{2}(t - 4)] + 2\,\text{tri}(t - 4)$
2. Steps and ramps: $y(t) = 4u(t - 3) + 2r(t - 3) - 4r(t - 4) + 2r(t - 5) - 4u(t - 5)$

3. An interval description: $y(t) = \begin{cases} -2 + 2t & 3 \le t \le 4 \\ 14 - 2t & 4 \le t \le 5 \\ 0 & \text{elsewhere} \end{cases}$

Continuous Signals   The **sinc, Gaussian** and **Lorentzian** functions are smooth energy signals. All possess unit area. The Gaussian and Lorentzian are bell-shaped and positive. The sinc function is encountered in many contexts. It is defined as

$$\text{sinc}(t) = \frac{\sin(\pi t)}{\pi t}$$

Figure 2.12   Example 2.10 shows various ways of describing the signal y(t).

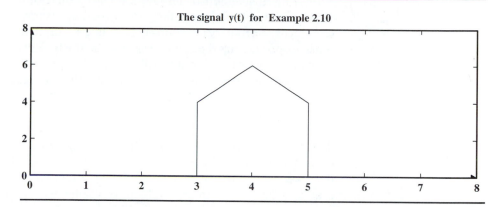

Since the sine term oscillates while the factor $1/t$ decreases with time, $\text{sinc}(t)$ shows decaying oscillations. At $t = 0$, the sinc function produces the indeterminate form $0/0$. Using the approximation $\sin(\alpha) \approx \alpha$, or even l'Hôpital's rule, we establish that $\text{sinc}(t) = 1$ in the limit as $t \to 0$

$$\lim_{t \to 0} \frac{\sin(\pi t)}{\pi t} \approx \frac{\pi t}{\pi t} = 1 \quad \text{or} \quad \lim_{t \to 0} \frac{\sin(\pi t)}{\pi t} = \lim_{t \to 0} \frac{\pi \cos(\pi t)}{\pi} = 1$$

A sketch of $\text{sinc}(t)$ in Figure 2.10 shows an even function with unit height at the origin, a central **mainlobe** between the first zero on either side of the origin and progressively decaying negative and positive **sidelobes**. The factor π in the definition ensures both unit area and unit distance between zero crossings, which occur at $t = \pm 1, \pm 2, \pm 3, \dots$. The signal $\text{sinc}^2(t)$ also has unit area with zeros at unit intervals but is entirely positive.

A useful rule: The area under $\text{sinc}(\alpha t)$ or $\text{sinc}^2(\alpha t)$ equals the area under the triangle inscribed within its central lobe (Figure 2.10).

Remark: The sinc is not a finite area signal. The area under $|\,\text{sinc}(t)\,|$ is actually infinite because its sidelobes do not decay fast enough.

2.8
The Impulse Function

Loosely speaking, an **impulse** is a tall, narrow spike with finite area. An informal definition of the **unit impulse function**, denoted by $\delta(t)$ and also called a **delta function** or **Dirac function**, is

$$\delta(t) = 0, \quad t \neq 0 \qquad \mathbb{A}[\delta(t)] = \int_{-\infty}^{\infty} \delta(\tau)\, d\tau = 1$$

It says that $\delta(t)$ is of *zero duration* but possesses *finite area*. To put the best face on this, we introduce a third, equally bizarre, criterion which says that $\delta(t)$ is infinite (or is unbounded) at $t = 0$!

Mathematicians will, no doubt, wince at this definition.

2.8.1   *Impulses as Limiting Representations*

A rigorous discussion of impulses involves distribution theory and the concept of **generalized functions**, which are beyond our scope. What emerges is that many sequences of ordinary functions behave like an impulse in their limiting form, in the sense of possessing the same *properties* as the impulse. Signals such as the pulses $\frac{1}{\tau}\mathrm{rect}(t/\tau)$ and $\frac{1}{\tau}\mathrm{tri}(t/\tau)$, the exponentials $\frac{1}{\tau}\exp(-t/\tau)u(t)$ and $\frac{2}{\tau}\exp(-|t|/\tau)$, the sinc functions $\frac{1}{\tau}\mathrm{sinc}(t/\tau)$ and $\frac{1}{\tau}\mathrm{sinc}^2(t/\tau)$, the Gaussian $\frac{1}{\tau}\exp[-\pi(t/\tau)^2]$ and the Lorentzian $\tau/[\pi(\tau^2 + t^2)]$ are all equivalent to the impulse $\delta(t)$ as $\tau \to 0$.

2.8.2   *Limiting Form of the Rectangular Pulse*

Consider the rectangular pulse $(1/\tau)\mathrm{rect}(t/\tau)$ of width τ and height $1/\tau$. As we decrease τ, its width shrinks and the height increases proportionately to maintain unit area. As $\tau \to 0$, we get a tall, narrow spike with unit area that satisfies all the criteria associated with an impulse (Figure 2.13).

Figure 2.13   The impulse as a limiting form of the rectangular pulse and a sketch of the signal $x(t) = 2\delta(t - 2) - 4\delta(t + 3) + 6\delta(t - 4)$.

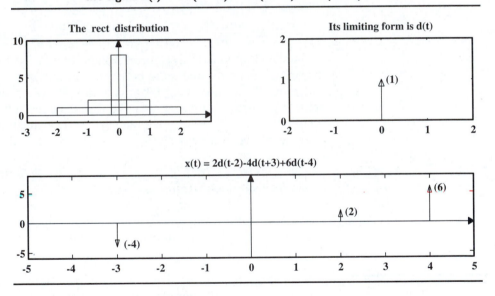

Remarks:
1. The area under the impulse $A\delta(t)$ equals A and is called its **strength**.
2. An impulse is absolutely integrable but not square integrable. If you try to obtain the form for $\delta^2(t)$ by limiting operations on a squared pulse with height $1/\tau^2$, it will actually approach infinity as $\tau \to 0$ and possess infinite area. The function $\delta^2(t)$ is, therefore, not defined.
3. The function $A\delta(t)$ is shown as an arrow with its area A labeled next to the tip. For visual appeal, we make its height proportional to A. Remember, however,

that its "height" at $t = 0$ is infinite or undefined. An impulse with negative area is shown as an arrow directed downward.

4. The area under an impulse $\delta(t - t_0)$ may be evaluated using any lower and upper limits enclosing its time of occurrence, t_0. With τ_1 and τ_2 denoting instants just prior to and just past t_0, we may write

$$\mathbb{A}[\delta(t - t_0)] = \int_{\tau_1}^{\tau_2} \delta(\tau - t_0)\, d\tau = \begin{cases} 1 & \tau_1 < t_0 < \tau_2 \\ 0 & \text{otherwise} \end{cases}$$

Example 2.11
Sketching Impulses

Let $x(t) = 2\delta(t - 2) - 4\delta(t + 3) + 6\delta(t - 4)$. This signal is shown in Figure 2.13. The total area under $x(t)$ equals the algebraic sum of the areas under each impulse to give $\mathbb{A}[x(t)] = 2 - 4 + 6 = 4$.

2.8.3   Relating the Impulse and Step Function

Consider the *finite ramp* $x(t)$ of Figure 2.14 whose derivative $x'(t)$ is the rectangular pulse $\frac{1}{\tau}\text{rect}(\frac{t}{\tau})$ of unit area (Figure 2.14). We may write

$$x'(t) = \frac{1}{\tau}\text{rect}\left(\frac{t}{\tau}\right)$$

As $\tau \to 0$, $x(t)$ approaches the unit step $u(t)$ (Figure 2.14). Its derivative $x'(t)$ thus

Figure 2.14   The limiting form of the finite ramp is a step. The limiting form of the derivative of the finite ramp is an impulse.

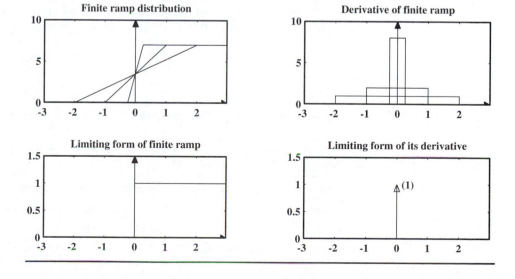

approaches $u'(t)$. Now, since $x'(t)$ is a rectangular pulse with unit area that approaches $\delta(t)$ (Figure 2.14), $u'(t)$ also equals $\delta(t)$!

$$\lim_{\tau \to 0} = x'(t) = u'(t) = \delta(t)$$

Using the defining integral for $\delta(t)$ with an arbitrary upper limit, we get

$$\int_{-\infty}^{t} \delta(\tau)\, d\tau = \begin{cases} 1 & t > 0 \\ 0 & t < 0 \end{cases} = u(t) \quad \text{and} \quad \frac{du(t)}{dt} = \delta(t)$$

The unit impulse may thus be regarded as the derivative of $u(t)$, and the unit step as the running integral of $\delta(t)$.

2.8.4   *Derivative of Signals with Jumps (Discontinuities)*

The derivative of $x(t) = Au(t) + Bu(t - \alpha)$ is given by $x'(t) = A\delta(t) + B\delta(t - \alpha)$. This describes two impulses at $t = 0$ and $t = \alpha$. The strengths A and B of these impulses equal the heights of the two steps.

We can generalize this result and state that *the derivative at a jump or discontinuity results in an impulse whose strength equals the jump.* This leads to a simple rule for finding the *generalized derivative* of an arbitrary signal $x(t)$:

1. For piecewise continuous portions, find the ordinary derivative $x'(t)$.
2. At discontinuities, include impulses whose strength equals the jump.

Example 2.12
Derivative of Signals with Discontinuities

(a) The derivative of $y(t) = \text{rect}(t/2) = u(t + 1) - u(t - 1)$ is simply (Figure 2.15)

$$y'(t) = \delta(t + 1) - \delta(t - 1)$$

(b) The derivative of $x(t) = 4\exp(-\frac{1}{2}t)[u(t) - u(t - 2)]$ shown in Figure 2.15 comprises the derivative of $4\exp(-\frac{1}{2}t)$ (which equals $-2\exp(-\frac{1}{2}t)$) and two impulses at the discontinuities ($t = 0$ and $t = 2$) whose strengths equal $4\exp(0) = 4$ at $t = 0$ and $-4\exp(-1) = -4/e$ at $t = 2$.

2.8.5   *Three Properties of the Impulse*

Three useful properties of the impulse function are summarized in Table 2.7.

Table 2.7   Properties of the Impulse Function.

Property	Expression	Result	Meaning
Scaling	$\delta[\alpha(t - \beta)]$	$\left\|\frac{1}{\alpha}\right\| \delta(t - \beta)$	An impulse of strength $1/\|\alpha\|$
Product	$x(t)\delta(t - \alpha)$	$x(\alpha)\delta(t - \alpha)$	An impulse of strength $x(\alpha)$
Sifting	$\int_{-\infty}^{\infty} x(t)\delta(t - \alpha)\, dt$	$x(\alpha)$	Area under the product

Figure 2.15   A rectangular and exponential pulse and their derivatives.

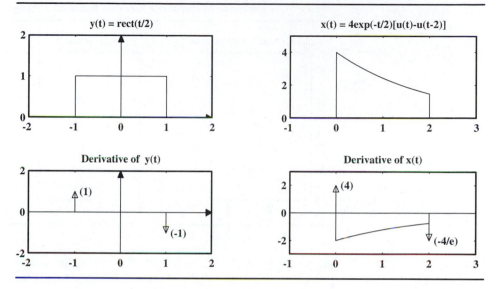

The Scaling Property   Since $\delta(t)$ has unit area, the "compressed" impulse $\delta(\alpha t)$ should have an area of $1/|\alpha|$. Since $\delta(\alpha t)$ still occurs at $t = 0$, it may be regarded as an unscaled impulse $|1/\alpha|\,\delta(t)$ whose area equals $1/|\alpha|$. A formal proof is based on a change of variable in the defining integral.

Since a time shift does not affect areas, we have the more general result

$$\delta[\alpha(t - \beta)] = \left|\frac{1}{\alpha}\right|\delta(t - \beta)$$

The Product Property   The product of a signal $x(t)$ with a tall, narrow pulse (in the limit, an impulse) centered at $t = \alpha$ is also a tall, narrow pulse whose height is scaled by $x(\alpha)$ (Figure 2.16). In the limit, this describes an impulse with strength $x(\alpha)$ and leads to the product property

$$x(t)\delta(t - \alpha) = x(\alpha)\delta(t - \alpha) \qquad x(t)\delta(t) = x(0)\delta(t)$$

The Sifting Property   The product property immediately suggests that the area under the product $x(t)\delta(t - \alpha) = x(\alpha)\delta(t - \alpha)$ equals $x(\alpha)$. In other words, $\delta(t - \alpha)$ sifts out the value of $x(t)$ at the impulse location $t = \alpha$ and thus

$$\int_{-\infty}^{\infty} x(t)\delta(t - \alpha)\,dt = x(\alpha) \qquad \int_{-\infty}^{\infty} x(t)\delta(t)\,dt = x(0)$$

This extremely important result is called the **sifting property**. It is the sifting action of an impulse (what it does) that purists actually regard as a formal definition of the impulse.

Figure 2.16   **The product of (the limiting form of) a unit impulse at $t = \alpha$ and $x(t)$ is an impulse at the same location but with strength $x(\alpha)$.**

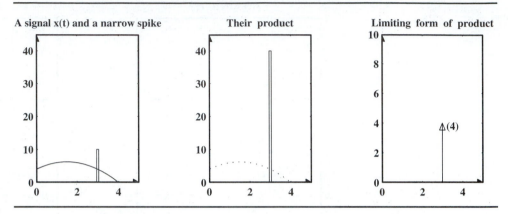

A signal x(t) and a narrow spike

Their product

Limiting form of product

Example 2.13

Properties of the Impulse Function

(a) Consider $z(t) = 4t^2\delta(2t - 4)$. The scaling property gives $z(t) = (4t^2)\frac{1}{2}\delta(t - 2) = 4(2)^2(\frac{1}{2})\delta(t - 2) = 8\delta(t - 2)$. Thus, $z(t)$ is an impulse of strength 8 located at $t = 2$, a function we can sketch.

(b) $I_1 = \int_0^\infty 4t^2\delta(t + 1)\, dt = 0$ since $\delta(t + 1)$ lies outside the limits of integration.

(c) We evaluate $I_2 = \int_{-4}^2 \cos(2\pi t)\delta(2t + 1)\, dt$, using scaling and sifting to give

$$I_2 = \int_{-4}^2 \cos(2\pi t)\left[\tfrac{1}{2}\delta(t + \tfrac{1}{2})\right] dt = \tfrac{1}{2}\cos(2\pi t)\,\Big|_{t=-\frac{1}{2}} = \tfrac{1}{2}\cos(-\pi) = -\tfrac{1}{2}$$

2.8.6   *The Comb Signal and Ideally Sampled Signal*

A periodic unit impulse train with unit period is called a **comb signal** and is denoted by comb(t). The scaled signal comb(t/t_s) describes a periodic impulse train with period t_s and impulse strengths t_s. Mathematically,

$$\text{comb}(t) = \sum_{k=-\infty}^{\infty} \delta(t - k) \qquad \text{comb}(t/t_s) = \sum_{k=-\infty}^{\infty} t_s\delta(t - kt_s)$$

A signal $x(t)$ multiplied by a periodic unit impulse train with period t_s is called an **ideally sampled signal**, $x_S(t)$, and is described by (Figure 2.17)

$$x_S(t) = x(t) \sum_{k=-\infty}^{\infty} \delta(t - kt_s) = \sum_{k=-\infty}^{\infty} x(kt_s)\delta(t - kt_s)$$

Note that $x_S(t)$ is also an impulse train *but it is not periodic*. The strength of each impulse equals the signal value $x(kt_s)$. This form actually provides a link between analog and digital signals. The ideally sampled signal $x_S(t)$ may also be written in terms of the comb signal as

$$x_S(t) = (1/t_s)x(t)\text{comb}(t/t_s)$$

Figure 2.17 **The signal comb(t/t_s) whose product with x(t) yields the ideally sampled signal $x_S(t)$. A staircase approximation of x(t) yields the impulse train $\hat{x}(t)$. This approximation to x(t) improves as we decrease t_s.**

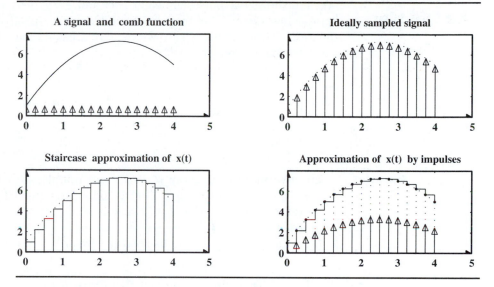

2.8.7   *Signal Approximation by Impulses*

In practice, any tall, narrow function may be regarded as an impulse whose strength equals the area under the function. This idea lies at the heart of approximating a smooth signal $x(t)$ by a summation of weighted, shifted impulses. We section $x(t)$ into narrow rectangular strips of width t_s and replace each strip at the location kt_s by an impulse $t_s x(kt_s)\delta(t - kt_s)$ whose strength equals the area $t_s x(kt_s)$ under the strip (Figure 2.17). This yields the impulse approximation $\hat{x}(t)$ for $x(t)$ as

$$\hat{x}(t) = \sum_{k=-\infty}^{\infty} t_s x(kt_s)\delta(t - kt_s) = x(t)\,\mathrm{comb}(t/t_s)$$

Naturally, this approximation improves as t_s decreases. As $t_s \to d\lambda \to 0$, kt_s takes on a continuum of values λ and the summation approaches an integral from $-\infty$ to $+\infty$ to yield the actual signal $x(t)$. Formally,

$$x(t) = \lim_{t_s \to 0} \sum_{k=-\infty}^{\infty} t_s x(kt_s)\delta(t - kt_s) = \int_{-\infty}^{\infty} x(\lambda)\delta(t - \lambda)\,d\lambda$$

The integral equals $x(t)$ by the sifting property. We may thus represent a signal $x(t)$ as a weighted, infinite sum of shifted impulses.

2.9
Special Topics:
The Doublet

The derivative of an impulse is called a **doublet** and denoted by $\delta'(t)$. To see what it represents, consider the triangular pulse $x(t) = \frac{1}{\tau}\text{tri}\left(\frac{t}{\tau}\right)$. As $\tau \to 0$, $x(t)$ approaches $\delta(t)$, and its derivative $x'(t)$ (Figure 2.18) should thus approach $\delta'(t)$. Now, $x'(t)$ is odd and shows two pulses of height $1/\tau^2$ and $-1/\tau^2$ with zero area. As $\tau \to 0$, $x'(t)$ approaches $+\infty$ and $-\infty$ from below and above, respectively. Thus $\delta'(t)$ is an odd function characterized by zero width, zero area and magnitudes of $+\infty$ and $-\infty$ at $t = 0$. Formally, we write

$$\delta'(t) = \begin{cases} 0 & t \neq 0 \\ \text{undefined} & t = 0 \end{cases} \qquad \int_{-\infty}^{\infty} \delta'(t)\, dt = 0 \qquad \delta'(-t) = -\delta'(t)$$

Remarks: 1. The two infinite spikes in $\delta'(t)$ are not impulses (their area is not constant), nor do they cancel. In fact, $\delta'(t)$ is indeterminate at $t = 0$. The signal $\delta'(t)$ is, therefore, sketched as a set of two spikes (Figure 2.18), which leads to the name *doublet*.
2. Even though $\mathbb{A}[\delta'(t)] = 0$, its absolute area $\mathbb{A}[|\delta'(t)|]$ is infinite.

2.9.1   *Three Properties of the Doublet*
Three useful properties of doublets are summarized in Table 2.8.

Figure 2.18   **The limiting forms of a tri signal and its derivative yield an impulse and the derivative of the impulse (doublet).**

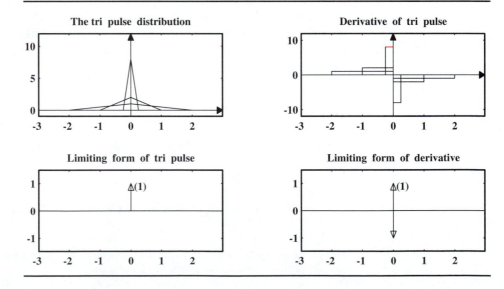

Table 2.8   Properties of the Doublet.

Property	Expression	Result
Scaling	$\delta'[\alpha(t-\beta)]$	$\frac{1}{\alpha\|\alpha\|}\delta'(t-\beta)$ and $\delta'(-t)=-\delta'(t)$
Product	$x(t)\delta'(t-\alpha)$	$x(\alpha)\delta'(t-\alpha)-x'(\alpha)\delta(t-\alpha)$
Sifting	$\int_{-\infty}^{\infty}x(t)\delta'(t-\alpha)\,dt$	$-x'(\alpha)$

The Scaling Property   The scaling property comes about if we write the form for the scaled impulse and take derivatives

$$\delta(\alpha t-\beta)=\frac{1}{|\alpha|}\delta(t-\beta)\quad \alpha\delta'(\alpha t-\beta)=\frac{1}{|\alpha|}\delta'(t-\beta)\quad \delta'(\alpha t-\beta)=\frac{1}{\alpha|\alpha|}\delta'(t-\beta)$$

With $\alpha=-1$, we get $\delta'(-t)=-\delta'(t)$. This implies that $\delta'(t)$ is an odd function.

The Product Property   The derivative of $x(t)\delta(t-\alpha)$ may be described in one of two ways. First, using the rule for derivatives of products, we have

$$\frac{d}{dt}[x(t)\delta(t-\alpha)]=x'(t)\delta(t-\alpha)+x(t)\delta'(t-\alpha)=x'(\alpha)\delta(t-\alpha)+x(t)\delta'(t-\alpha)$$

Second, using the product property of impulses, we also have

$$\frac{d}{dt}[x(t)\delta(t-\alpha)]=\frac{d}{dt}[x(\alpha)\delta(t-\alpha)]=x(\alpha)\delta'(t-\alpha)$$

Equating the two and rearranging, we get the rather unexpected result

$$x(t)\delta'(t-\alpha)=x(\alpha)\delta'(t-\alpha)-x'(\alpha)\delta(t-\alpha)$$

Unlike impulses, $x(t)\delta'(t-\alpha)$ does not just equal $x(\alpha)\delta'(t-\alpha)$!

The Sifting Property   Integrating this result yields the sifting property

$$\int_{-\infty}^{\infty}x(t)\delta'(t-\alpha)=\int_{-\infty}^{\infty}x(\alpha)\delta'(t-\alpha)-\int_{-\infty}^{\infty}x'(\alpha)\delta(t-\alpha)=-x'(\alpha)$$

A doublet, $\delta'(t-\alpha)$, sifts out the *negative* derivative of $x(t)$ at $t=\alpha$.

Remarks: 1. Higher derivatives of $\delta(t)$ obey $\delta^{(n)}(t)=(-1)^{n}\delta^{(n)}(t)$, are alternately odd and even, and possess zero area. All are limiting forms of the same sequences that generate impulses provided their ordinary derivatives up to the required order exist. None are absolutely integrable.

2. *The impulse is unique in being the only absolutely integrable function from among all its derivatives and integrals* (the step, ramp, etc.).

Example 2.14
Properties of the Doublet

(a) Figure 2.19 shows how to sketch the signal $x(t) = 2\delta'(t-1) - 3\delta'(t-2)$.

Figure 2.19   The representation of $x(t) = 2\delta'(t-1) - 3\delta'(t-2)$.

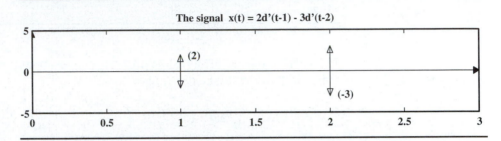

The signal $x(t) = 2d'(t-1) - 3d'(t-2)$

(b) Consider $x(t) = \exp(-t)\delta'(t)$. Using the product property yields

$$x(t) = \delta'(t) - (-1)\delta(t) = \delta'(t) + \delta(t)$$

(c) Let $I = \int_{-2}^{2}[(t-3)\delta(2t+2) - t\delta(t+4) + (8\cos\pi t)\delta'(t - \frac{1}{2})]\,dt$. The second impulse occurs at $t = -4$ and does not contribute to the integral. With $\delta(2t+2) = \frac{1}{2}\delta(t+1)$, the sifting property of impulses and doublets gives

$$I = \frac{1}{2}(t-3)\Big|_{t=-1} - 8\frac{d}{dt}(\cos\pi t)\Big|_{t=\frac{1}{2}} = \frac{1}{2}(-1-3) + 8\pi\sin\frac{1}{2}\pi = -2 + 8\pi$$

2.10
Special Topics: Moments

Moments are general measures based on area. The nth **moment** is defined as $m_n = \mathbb{A}[t^n x(t)]$ and $m_0 = \mathbb{A}[x(t)]$ is just the area under $x(t)$. The normalized first moment $m_x = m_1/m_0 = \mathbb{A}[tx(t)]/\mathbb{A}[x(t)]$ equals the **mean**. **Central moments** (about the mean) are defined by $\mu_n = \mathbb{A}[(t-m_x)^n x(t)]$.

The two moments m_n and μ_n are given by

$$m_n = \int_{-\infty}^{\infty} t^n x(t)\,dt \qquad \mu_n = \int_{-\infty}^{\infty} (t-m_x)^n x(t)\,dt$$

The second central moment μ_2 is called the **variance**. It is often denoted by σ^2 and equals

$$\mu_2 = \sigma^2 = \frac{m_2}{m_0} - m_x^2$$

To account for complex signals or sign changes, it is often more useful to define moments in terms of the absolute quantities $|x(t)|$ or $|x(t)|^2$.

The first few moments are widely used in both physics and engineering, as summarized in Table 2.9.

Table 2.9   Moments and Their Utility.

Note: In physics, $x(t)$ may represent the mass density.
In probability, $x(t)$ is the density function of a random variable.
For random or power signals, $x(t)$ may also be the density function.
For systems, $x(t)$ may be the response $h(t)$ to an impulse input.

Moment	Physics	Probability	Power Signals	Energy Signals
m_0	Area	Sum	Area	Area
$m_x = \frac{m_1}{m_0}$	Centroid	Mean	Average	Delay
m_2/m_0		Mean square	Total power	
$\sigma^2 = \frac{m_2}{m_0} - m_x^2$	Moment of inertia	Variance	ac power	
σ	Radius of gyration	Standard deviation		Half-width (duration)

Moments and Power Signals   For power signals, the normalized second moment, m_2/m_0, equals the total power. The variance, $\sigma^2 = m_2/m_0 - m_x^2$, is the difference between the total power and the dc power. The variance may thus be regarded as the power in a signal with its dc offset removed.

Moments and Energy Signals   For an energy signal, m_x is a measure of the effective signal **delay** while σ measures its effective width, or **duration**.

Example 2.15
Moments of Energy Signals

(a) Consider the signal $x(t) = \frac{1}{3}, 0 \le t \le 3$. Its various moments are

$$m_0 = \mathbb{A}[x(t)] = 1 \qquad m_1 = \int_{-\infty}^{\infty} tx(t)\,dt = \int_0^3 \tfrac{1}{3}t\,dt = \tfrac{3}{2}$$

$$m_2 = \int_{-\infty}^{\infty} t^2 x(t)\,dt = \int_0^3 \tfrac{1}{3}t^2\,dt = 3$$

This gives $m_x = m_1/m_0 = \frac{3}{2}$ and $\sigma^2 = (m_2/m_0) - m_x^2 = 3 - \frac{9}{4} = \frac{3}{4}$. We then obtain the effective delay as $m_x = \frac{3}{2}$ and the effective duration as $\sigma = \sqrt{\frac{3}{4}}$.

(b) For the signal $x(t) = \exp(-t)u(t)$, the various moments equal

$$m_0 = \int_0^{\infty} e^{-t}\,dt = 1 \quad m_1 = \int_0^{\infty} te^{-t}\,dt = 1 \quad m_2 = \int_0^{\infty} t^2 e^{-t}\,dt = 2$$

The effective delay is $m_x = m_1/m_0 = 1$. Further, σ^2 equals

$$\sigma^2 = \int_0^{\infty} (t - m_x)^2 e^{-t}\,dt = \frac{m_2}{m_0} - m_x^2 = 2 - 1 = 1$$

Thus, the effective duration is given by $\sigma = 1$.

2.11
Discrete-Time
Signals

Continuous-time (CT) signals are defined for every possible time instant. In contrast to this is the situation where signals can, in fact, be defined only at selected instants. Such signals are called **discrete-time** (DT) **signals** and may arise in one of two ways:

1. Some situations (such as annual rainfall records or daily stock market prices) lead naturally to discrete-time signals.
2. In other situations, we sample a CT signal $x(t)$ at intervals t_s and disregard $x(t)$ at all other times to get $x(nt_s)$. If we assign values of $x(nt_s)$ to a discrete sequence $x_S[n]$, we get a **sampled signal**.

   In either case, we shall assume a uniform interval t_s between samples. This is both practically sound and mathematically convenient.

Remarks:

1. The term *discrete* suggests that $x_S[n]$ describes a **time series,** an *ordered sequence* of numbers which embodies the time history of the signal.
2. A sampled or discrete signal is just a sequence of values. It contains no direct information about t_s unless separately specified.
3. When comparing CT and sampled signals, we shall assume that the origin $t = 0$ also corresponds to the sampling instant $n = 0$.
4. In analogy with CT signals, DT signals may be leftsided, rightsided, twosided or timelimited, and causal, anticausal or noncausal.

2.11.1   *Notation*

1. We denote a *sampled signal* by $x_S[n]$, where the subscript suggests an arbitrary sampling interval t_s . The quantity $S_F = 1/t_s$ then defines the **sampling rate** or **sampling frequency** in **samples per second** (or hertz).
2. We denote a DT signal or sequence by $x[n]$ if $t_s = 1$ or $S_F = 1$. This implies a normalization of the time interval or sampling frequency.
3. We use the notation $\{x[n]\}$ to describe a DT signal $x[n]$ as a numeric sequence. We use a marker (↑) to show the origin (*but only if the sequence does not start at n = 0*). We use an ellipsis (...) to show infinite extent on either side.

Here are some examples:

$\{x[n]\} = \{1, 2, 4, 8,...\}$ A causal sequence for $x[n] = 2^n, n \geq 0$

$\{x[n]\} = \{0, 0, 0, 6, 1, 3, 4, 4\}$ A sequence with $x[0] = x[1] = x[2] = 0$

$\{x_S[n]\} = \{...1, 2, \underset{\uparrow}{6}, 4\}, t_s = \frac{1}{2}$ A leftsided sampled sequence with $x_S[0] = 6$

$\{x[n]\} = \{\underset{\uparrow}{2}, 4, 6, 8...\}$ A rightsided sequence with $x[0] = 6$

$\{x[n]\} = \{...2, 4, \underset{\uparrow}{6}, 8...\}$ A twosided sequence with $x[0] = 6$

A comparison of signal measures based on duration and area is provided in Table 2.10 for CT, sampled and DT signals. Note how the summations for sampled signals involve the sampling interval t_s .

Table 2.10   Duration- and Area-Based Signal Measures: A Comparison.

Measure	CT	Sampled	DT
Notation	$x(t)$	$x_S[n]$	$x[n]$
Duration			
Rightsided	$x(t) = 0$ for $t < t_0$	$x_S[n] = 0$ for $n < k$	$x[n] = 0$ for $n < k$
Leftsided	$x(t) = 0$ for $t > t_0$	$x_S[n] = 0$ for $n \geq k$	$x[n] = 0$ for $n \geq k$
Twosided	both sides of $t = 0$	both sides of $n = 0$	both sides of $n = 0$
Timelimited	finite duration	finite duration	finite duration
Casual	$x(t) = 0$ for $t < 0$	$x_S[n] = 0$ for $n < 0$	$x[n] = 0$ for $n < 0$
Area			
Areas	$\int x(t)$	$t_s \sum x_S[n]$	$\sum x[n]$
Finite area	$\mathbb{A}[|x(t)|] < \infty$	$t_s \sum |x_S[n]| < \infty$	$\sum |x[n]| < \infty$
Finite energy	$\mathbb{A}[|x(t)|^2] < \infty$	$t_s \sum |x_S[n]|^2 < \infty$	$\sum |x[n]|^2 < \infty$

2.12.1   *Areas and the Discrete Sum*

Summation is the discrete-time equivalent of integration. The integral $y(t)$ or area under a signal $x(t)$ over the duration $(0, t)$ is defined as

$$y(t) = \int_0^t x(\lambda)\, d\lambda$$

A simple way to evaluate $y(t)$ is to section $x(t)$ into rectangular strips of uniform width t_s and sum their areas. This is equivalent to a staircase approximation or step interpolation for $x(t)$ with step heights $x(nt_s)$, $n = 0, 1, 2, \ldots$, corresponding to the sampled signal $x_S[n]$ (Figure 2.20). We thus obtain the sampled equivalent of integration or area under $x_S[n]$ as

$$\mathbb{A}\{x_S[n]\} = \sum_{k=0}^{n} x_S[k]t_s = t_s \sum_{k=0}^{n} x_S[k]$$

If we normalize the sampling interval t_s to unity, we obtain the DT equivalent of integration, called the **discrete sum**, given by

$$\mathbb{A}\{x[n]\} = \sum_{k=0}^{n} x[k]$$

Signals for which $\mathbb{A}\{|x_S[n]|\}$ is finite are called **absolutely summable**. Signals for which $\mathbb{A}\{|x_S[n]|^2\}$ is finite are called **square summable**.

Figure 2.20   A signal x(t) and its straircase approximation.

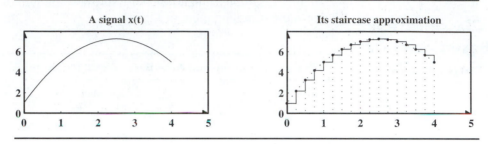

**2.13
Operations on
DT Signals**

The operations on *DT* and sampled signals are summarized in Table 2.11 and Figure 2.21. Most operations are analogous to their analog versions. We concentrate on the time-scaling operation, which stretches or compresses a signal. The problems associated with the "time scaling" of *DT* signals arise not in what happens but how it happens. We explain with reference to Figure 2.21.

Table 2.11   Operations on *DT* Signals.

Operation	Example	Explanation
Sum	$x[n] + y[n]$	Pointwise sum of $x[n]$ and $y[n]$.
Product	$x[n]y[n]$	Pointwise product of $x[n]$ and $y[n]$.
Time shift	$x[n-3]$	Shift $x[n]$ right by 3 (delay).
	$x[n+3]$	Shift $x[n]$ left by 3 (advance).
Folding	$x[-n]$	Fold $x[n]$ about the origin.
Amplitude scale	$3x[n]$	Multiply the ordinate by factor of 3.
Amplitude shift	$x[n] + 3$	Add dc offset of 3 to $x[n]$ everywhere.
Decimation	$x[2n]$	Compress $x[n]$ by factor of 2 (speed up). (Remove alternative samples.)
Interpolation	$x[n/3]$	Stretch $x[n]$ by a factor of 3 (slow down). (Two interpolated values between samples.)
Combinations	$x[-3n+6]$	Shift $x[n]$ left by 6, fold and decimate by 3.
	$x[\frac{2}{3}n - 1]$	Delay $x[n]$ by 1, insert $(3-1) = 2$ interpolated values between samples and decimate by 2.

2.13.1   *Decimation*

Suppose $x_S[n]$ corresponds to the *CT* signal $x(t)$ sampled at intervals t_s. The signal $y_S[n] = x_S[2n]$ then corresponds to the compressed signal $x(2t)$ sampled at t_s and has only half the length (alternate samples) of $x_S[n]$.

Figure 2.21 Illustrating operations on discrete-time signals.

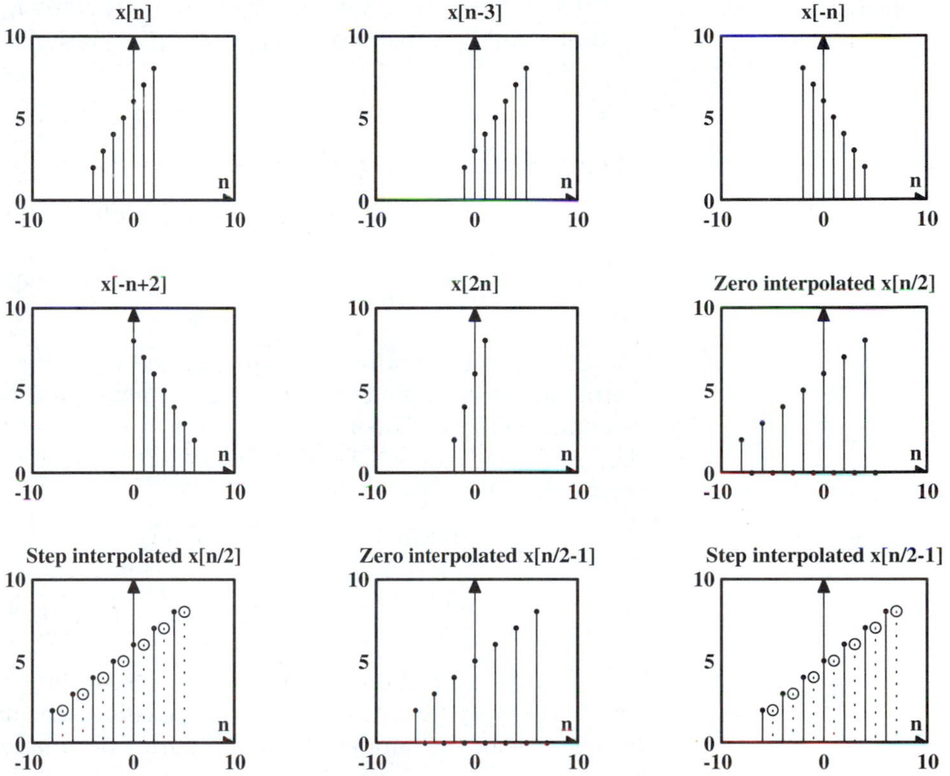

We can also get $y_S[n]$ from $x(t)$ if we sample it at intervals $2t_s$ (or at a sampling rate $1/2t_s$). Decimation by a factor of N is equivalent to sampling $x(t)$ at intervals Nt_s and implies an *N-fold reduction in the sampling rate*.

The *DT* signal $x[Nn]$ is generated from $x[n]$ by retaining every Nth sample corresponding to the indices $k = Nn$ and discarding all others (Figure 2.21).

2.13.2   *Interpolation*

By analogy, if $x_S[n]$ corresponds to $x(t)$ sampled at intervals t_s , then $x_S[\frac{1}{2}n]$ corresponds to $x(t)$ sampled at $\frac{1}{2}t_s$ with twice the length of $x_S[n]$. Interpolation by a factor of N is equivalent to sampling $x(t)$ at intervals t_s/N and implies an *N-fold increase in the sampling rate*.

The *DT* signal $x[\frac{1}{2}n]$ shows one new sample between adjacent samples of $x[n]$ and is twice the length of $x[n]$. How do we get the extra samples from just $x[n]$? By interpolation, of course (Figure 2.21). We may choose each new sample as zero (*zero interpolation*), a constant equal to the previous sample (*step interpolation*) or the average of adjacent samples (*linear interpolation*).

Example 2.16
Decimation and
Interpolation

Let $\{x[n]\} = \{1, 2, 6, 4, 8\}$. To generate $x[2n]$, we remove every other sample and obtain $x[2n] = \{1, 6, 8\}$.

To obtain $x[\frac{1}{2}n]$ from $x[n]$ by step interpolation, we use each previous value between adjacent samples to give $\{x[\frac{1}{2}n]\} = \{1, 1, 2, 2, 6, 6, 4, 4, 8, 8\}$. ___

Some Caveats   Consider the two sets of operations shown below.

$$x[n] \xrightarrow[n \to 2n]{\text{Decimate}} x[2n] \xrightarrow[n \to n/2]{\text{Interpolate}} x[n]$$

$$x[n] \xrightarrow[n \to n/2]{\text{Interpolate}} x[n/2] \xrightarrow[n \to 2n]{\text{Decimate}} x[n]$$

On the face of it, both start with $x[n]$ and recover $x[n]$. Does this mean that interpolation and decimation are inverse operations? Not really. In truth, only the second sequence of operations recovers $x[n]$ exactly. We illustrate by an example. Let $x[n] = \{1, 2, 6, 4, 8\}$. Using step interpolation, the two sequences of operations result in

$$\{1, 2, 6, 4, 8\} \xrightarrow[n \to 2n]{\text{Decimate}} \{1, 6, 8\} \xrightarrow[n \to n/2]{\text{Interpolate}} \{1, 1, 6, 6, 8, 8\}$$

$$\{1, 2, 6, 4, 8\} \xrightarrow[n \to n/2]{\text{Interpolate}} \{1, 1, 2, 2, 6, 6, 4, 4, 8, 8\} \xrightarrow[n \to 2n]{\text{Decimate}} \{1, 2, 6, 4, 8\}$$

Clearly, decimation is indeed the inverse of interpolation, but the converse is not necessarily true. After all, no interpolation scheme can recover or predict the exact value of the samples discarded during decimation.

2.14
Classification Based on Periodicity and Symmetry

The classification of DT signals follows a scheme analogous to that for CT signals leading to results of Table 2.12.

2.14.1   *Periodicity*

A periodic DT signal repeats after every N samples and is described by

$$x[n] = x[n \pm kN] \ldots k = 0, 1, 2, 3, \ldots$$

Note that N is an integer. The fundamental frequency F_0 of $x[n]$ equals

$$F_0 = \frac{1}{N} \text{ Hz} \qquad \Omega_0 = 2\pi F_0 \text{ rad/s}$$

Combinations   A combination of DT periodic signals is periodic only if every frequency is a rational quantity (ratio of integers). If so, the common period N of the combination is given by the LCM of the individual periods.

Table 2.12   Signal Classification: A Comparison.

Measure	CT	Sampled	DT
Symmetry			
Even	$x(t) = x(-t)$	$x_S[n] = x_S[-n]$	$x[n] = x[-n]$
Odd	$x(t) = -x(-t)$	$x_S[n] = -x_S[-n]$	$x[n] = -x[-n]$
Halfwave	$x(t) = -x(t \pm \frac{1}{2}T)$	$x_S[n] = -x_S[n \pm \frac{1}{2}N]$	$x[n] = -x[n \pm \frac{1}{2}N]$
Periodicity	$x(t) = x(t + T)$	$x_S[n] = x_S[n + N]$	$x[n] = x[n + N]$
Period	T	Nt_S	N
Average	$\frac{1}{T}\int x(t)\,dt$	$\frac{1}{N}\sum x_S[n]$	$\frac{1}{N}\sum x[n]$
Energy	$\int |x(t)|^2\,dt$	$t_S \sum |x_S[n]|^2$	$\sum |x[n]|^2$
Power	$\frac{1}{T}\int |x(t)|^2\,dt$	$\frac{1}{N}\sum |x_S[n]|^2$	$\frac{1}{N}\sum |x[n]|^2$

Combinations of Periodic Signals
CT: Commensurate frequencies $T = LCM$ or periods
DT: Rational frequencies $N = LCM$ of periods

Quasiperiodic Signals
CT: Noncommensurate frequencies
DT: Noncommensurate or irrational frequencies

2.14.2   Symmetry

Even and odd symmetry of DT signals is described by

$$x_e[n] = x_e[-n] \qquad x_o[n] = -x_o[-n]$$

For an odd signal, we must have $x_o[0] = 0$. The discrete sum for an odd time function over symmetric limits is zero (Figure 2.22, p. 50).

$$\sum_{k=-M}^{M} x_o[k] = 0$$

Symmetric periodic signals also show symmetry about $n = \frac{1}{2}N$. We then have

$$x_p[n] = x_p[N - n] \text{ (even)} \qquad x_p[n] = -x_p[N - n] \text{ (odd)}$$

Halfwave symmetry is defined only for periodic DT signals and requires

$$x_{\text{hw}}[n] = -x_{\text{hw}}[n \pm \tfrac{1}{2}N]$$

2.14.3   Odd and Even Parts of DT Signals

The even part $x_e[n]$ and odd part $x_o[n]$ of a DT signal $x[n]$ are given by

$$x_e[n] = \tfrac{1}{2}\{x[n] + x[-n]\}$$
$$x_o[n] = \tfrac{1}{2}\{x[n] - x[-n]\}$$

Example 2.17

Even and Odd Parts

Figure 2.22 shows the steps involved in finding the even and odd parts of a *DT* signal $x[n]$.

Figure 2.22   Symmetry and the odd and even parts of unsymmetric signals.

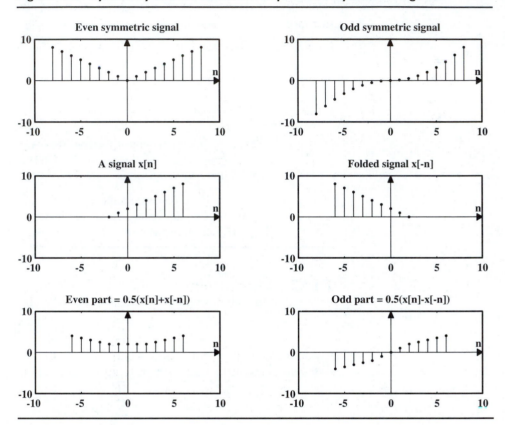

Remark: For a signal of the form $x[n]u[n]$ where $x[0]$ is not zero, $x_e[0]$ equals $x[0]$. This is unlike *CT* signals of the form $x(t)u(t)$, whose even part at $t = 0$ equals $\frac{1}{2}x(0)$. This ambiguity results from the way in which we define the *CT* and *DT* step function at the origin.

2.14.4   *Energy and Energy Signals*

The signal energy of a *DT* signal is defined by

$$E_x = \sum_{k=-\infty}^{\infty} |x[k]|^2$$

The absolute value allows us to use this relation even for complex-valued signals. For sampled signals, this relation must be multiplied by t_s. Signals with finite energy are called energy (or square summable) signals.

2.14.5   *Average Value, Power and Power Signals*

For periodic signals, the average value and power may be found by averaging $x[n]$ or $|x[n]|^2$ over one period. Thus

$$x_{av} = \frac{1}{N}\sum_{k=0}^{N-1} x[k] \qquad P_x = \frac{1}{N}\sum_{k=0}^{N-1} |x[k]|^2$$

We use only N samples in the summation since the last sample of one period serves as the first sample of the next. The limits are thus $(0, N-1)$.

For nonperiodic power signals, we use the limiting forms

$$x_{av} = \lim_{M\to\infty} \frac{1}{2M+1}\sum_{k=-M}^{M} x[k] \qquad P_x = \lim_{M\to\infty} \frac{1}{2M+1}\sum_{k=-M}^{M} |x[k]|^2$$

Signals with finite power are called power signals. All periodic signals are power signals.

Remarks:

1. For sampled signals, the factor t_s is not required in the summations for power and average value. The reason is that averaging over one period $T = Nt_s$ cancels out the effect of multiplication by t_s!
2. The limiting form is useful primarily for power signals that are not periodic (such as switched periodic signals and random signals).

Example 2.18
Energy and Power

(a) Consider the signal $x[n] = (\frac{1}{2})^n u[n]$. This describes a onesided decaying exponential with $\{x[n]\} = \{1, \frac{1}{2}, \frac{1}{4}, \frac{1}{8}, \ldots\}$. The signal energy is given by

$$E_x = \sum x^2[n] = \sum(\tfrac{1}{2})^{2n}u[n] = \sum(\tfrac{1}{4})^n u[n] = 1 + \tfrac{1}{4} + \tfrac{1}{16} + \cdots$$

This geometric series sums to $E_x = 1/(1 - \frac{1}{4}) = \frac{4}{3}$.

(b) Consider the signal $x[n] = 4\cos(2\pi n/4)$. This is periodic with $N = 4$. Now $\{x[n]\} = \{\ldots 4, 0, -4, 0, 4, 0, -4, 0, \ldots\}$ and we have $x_{av} = \frac{1}{4}\sum x[n](0,3) = 0$.

The signal power equals $P_x = \frac{1}{4}\sum\{x^2[n]\}(0,3) = \frac{1}{4}[16 + 16] = 8$.

(c) Let $x[n] = \exp(j2\pi n/4)$. This signal is also periodic with period $N = 4$. Since $|x[n]| = 1$, the signal power equals $P_x = \frac{1}{4}\sum|x[n]|^2, (0,3) = \frac{1}{4}[4] = 1$. ___

2.15
A Catalog of
Discrete-Time
Signals

Sampled and *DT* signals are much like their *CT* versions in many cases. Figure 2.23 illustrates many of these signals, and Table 2.13 presents a comparison. There are some subtle differences and a few major surprises.

Figure 2.23   Illustrating some discrete-time signals of Table 2.13.

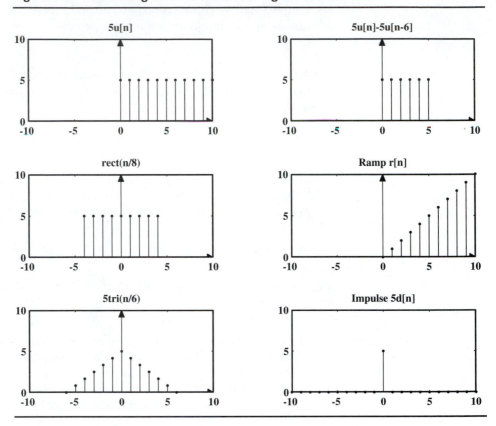

Table 2.13   *CT*, Sampled and *DT* Signals: A Comparison.

Function	CT	Sampled at t_s	DT
Step	$u(t)$	$u_S[n]$	$u[n]$
Value at origin	$u(0)$ undefined Choose $u(0) = \frac{1}{2}$	$u_S[0] = 1$	$u[0] = 1$
Ramp	$tu(t) = r(t)$	$nt_s u_S[n] = r_s[n]$	$nu[n] = r[n]$
Impulse	$\delta(t)$	$\delta_S[n]$	$\delta[n]$
Area	$\mathbb{A}[\delta(t)] = 1$	$\mathbb{A}\{\delta_S[n]\} = t_s$	$\mathbb{A}\{\delta[n]\} = 1$
Value at origin	$\delta(0) \to \infty$ or undefined	$\delta_S[0] = 1$	$\delta[0] = 1$
Scaling	$\delta(\alpha t) = \left\lvert \frac{1}{\alpha} \right\rvert \delta(t)$	$\delta_S[\alpha n] = \delta_S[n]$	$\delta[\alpha n] = \delta[n]$
Product property	$x(t)\delta(t - t_0)$ $= x(t_0)\delta(t - t_0)$	$x_S[n]\delta_S[n - k]$ $= x_S[k]\delta_S[n - k]$	$x[n]\delta[n - k]$ $= x[k]\delta[n - k]$

Table 2.13   (continued).

Function	CT	Sampled at t_s	DT
Sifting property	$\int x(t)\delta(t-t_0)$ $= x(t_0)$	$t_s \sum x_S[n]\delta_S[n-k]$ $= t_s x_S[k]$	$\sum x[n]\delta[n-k]$ $= x[k]$
x as sum of impulses	$x(t) =$ $\int x(\lambda)\delta(t-\lambda)d\lambda$	$x_S[n] =$ $t_s \sum x_S[k]\delta_S[n-k]$	$x[n] =$ $\sum x[k]\delta[n-k]$
Exponential	$e^{-t}u(t)$ Exponential decay for $t \geq 0$	$a^n u_S[n], a > 0$ Decay if $a < 1$ Growth if $a > 1$ Constant if $a = 1$	$a^n u[n], a > 0$ Decay if $a < 1$ Growth if $a > 1$ Constant of $a = 1$

The Step Function   The *DT* unit step function $u[n]$ is defined as

$$u[n] = \begin{cases} 0 & n < 0 \\ 1 & n \geq 0 \end{cases}$$

Remarks: **1.** Unlike $u(t)$ whose value is subject to ambiguity (or not defined) at $t = 0$, $u[n]$ has a well-defined, unique value of $u[0] = 1$.

2. Because of this, the pulse $x[n] = u[n] - u[n - N]$ contains only N and not $N + 1$ unit samples, which lie at $n = 0, 1, \ldots, N - 1$.

The Rect Function   We denote the *DT* version of the rectangular pulse by $\text{rect}(n/2N)$. It has $2N + 1$ unit samples over $-N \leq n \leq N$ and is defined by

$$\text{rect}(n/2N) = 1, \quad |n| \leq N$$

We use the factor $2N$ because it gets around the problem of having to deal with half-integer values of n when N is an odd number.

The Ramp Function   The unit ramp $r[n]$ is defined as:

$$r[n] = nu[n] = \begin{cases} 0 & n < 0 \\ n & n \geq 0 \end{cases}$$

The signal $x[n] = Anu[n] = Ar[n]$ describes a *DT* ramp whose slope A is given by $x[k] - x[k - 1]$, the difference between adjacent sample values.

The Triangular Pulse   The *DT* version of the triangular pulse is $\text{tri}(n/N)$:

$$\text{tri}(n/N) = 1 - |n|/N, \quad n = 0, 1, 2, \ldots, N$$

It has $2N + 1$ samples over $-N \leq n \leq N$, with both $x[N]$ and $x[-N]$ being zero.

2.15.1   The DT Impulse (Sample) Function

The *unit* impulse (*unit* sample) function $\delta[n]$ is the *DT* counterpart of the *CT* unit impulse function $\delta(t)$ and is defined as

$$\delta[n] = \begin{cases} 0 & n \neq 0 \\ 1 & n = 0 \end{cases}$$

Remark: Note that $\delta[0] = 1$. Thus $\delta[n]$ is always well defined and completely free of the kind of ambiguities associated with the function $\delta(t)$ at $t = 0$.

A *DT* impulse delayed by k units is represented as $\delta[n - k]$. Its nonzero element occurs at $n = k$.

2.15.2   Properties of the DT Impulse

The Product Property   The product of a signal $x[n]$ with the *DT* impulse $\delta[n - k]$ results in a *DT* impulse with strength $x[k]$. Thus,

$$x[n]\delta[n - k] = x[k]\delta[n - k]$$

The Sifting Property   The product property leads directly to

$$\sum_{n=-\infty}^{\infty} x[n]\delta[n - k] = x[k]$$

The impulse extracts the value $x[k]$ from $x[n]$ at the impulse location $n = k$. The product and sifting properties are analogous to their *CT* counterparts.

Scaling   The scaled *DT* impulse $\delta[\alpha n]$ equals $\delta[n]$! This is quite unlike the scaling property for a scaled *CT* impulse $\delta(\alpha t)$, which equals $|\frac{1}{\alpha}|\delta(t)$.

Example 2.19
Properties of the *DT* Impulse

Consider the signals $x[n] = (2)^n$ and $y[n] = \delta[n - 3]$.

(a) The product $z[n] = x[n]y[n] = (2)^3\delta[n - 3] = 8\delta[n - 3]$ is an impulse.
(b) The sum $\sum z[n]$ is a value given by $\sum (2)^n\delta[n - 3] = (2)^3 = 8$.
(c) The scaled impulse $\delta[3n]$ simply equals $\delta[n]$.

2.15.3   DT Signal Representation by DT Impulses

A *DT* signal $x[n]$ may be expressed as a sum of shifted *DT* impulses $\delta[n - k]$ whose strengths $x[k]$ correspond to the signal values at $n = k$. Thus

$$x[n] = \cdots + x[-1]\delta[n + 1] + x[0]\delta[n] + x[1]\delta[n - 1] + x[2]\delta[n - 2] + \cdots$$
$$= \sum_{k=-\infty}^{\infty} x[k]\delta[n - k]$$

The *DT* step $u[n]$ may also be expressed as the running sum of $\delta[n]$:

$$u[n] = \sum_{k=-\infty}^{n} \delta[k] \quad \text{or} \quad u[n] = \sum_{k=-\infty}^{\infty} u[k]\delta[n - k] = \sum_{k=0}^{\infty} \delta[n - k]$$

Example 2.20
Describing Sequences and Signals

(a) The sequence $\{x[n]\} = \{4, 2, 1, -3, 6\}$ may also be written as:

$$x[n] = 4\delta[n+1] + 2\delta[n] + \delta[n-1] - 3\delta[n-2] + 6\delta[n-3]$$

(b) The signal $x[n]$ sketched in Figure 2.24 may be represented variously as

1. A numeric sequence $\{x[n]\} = \{0, 0, 0, 2, 4, 6, 8, 8, 8, 8\}$
2. A sum of impulses described by

$$x[n] = 2\delta[n-3] + 4\delta[n-4] + 6\delta[n-5]$$
$$+8\delta[n-6] + 8\delta[n-7] + 8\delta[n-8] + 8\delta[n-9]$$

3. A sum of steps and ramps described by

$$x[n] = 2r[n-2] - 2r[n-6] - 8u[n-10]$$

Comment: Note that the argument of the step function is $[n-10]$ not $[n-9]$.

Figure 2.24   Example 2.20 shows various ways of describing the signal $x[n]$.

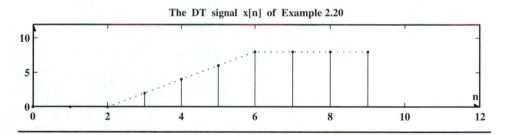

The DT signal x[n] of Example 2.20

(c) The signal $x[n] = \text{tri}(n/4)$ may also be described in various ways as

1. A numeric sequence $\{x[n]\} = \{0, \frac{1}{4}, \frac{1}{2}, \frac{3}{4}, 1, \frac{3}{4}, \frac{1}{2}, \frac{1}{4}, 0\}$
2. A sum of impulses described by

$$x[n] = \tfrac{1}{4}\delta[n+3] + \tfrac{1}{2}\delta[n+2] + \tfrac{3}{4}\delta[n+1] + \delta[n]$$
$$+\tfrac{3}{4}\delta[n-1] + \tfrac{1}{2}\delta[n-2] + \tfrac{1}{4}\delta[n-3]$$

3. A sum of steps and ramps described by

$$x[n] = \tfrac{1}{4}r[n+4] - \tfrac{1}{2}r[n] + \tfrac{1}{4}r[n-4]$$

2.15.4   *The DT Sinc Function*

In analogy with the *CT* sinc function, the *DT* sinc function is defined by

$$\text{sinc}(n/N) = \frac{\sin(n\pi/N)}{n\pi/N} \qquad \text{sinc}(0) = 1$$

We make the following observations:

1. The signal $\text{sinc}(n/N)$ equals zero at $n = kN, k = \pm1, \pm2, \dots$
2. At $n = 0$, $\text{sinc}(0) = 0/0$ and cannot be evaluated in the limit since n can take on only integer values. *We therefore define* $\text{sinc}(0) = 1$.
3. The definition of $\text{sinc}(n/N)$ implies that $\text{sinc}(n) = \delta[n]$.

2.15.5   *DT Exponentials*

A onesided *DT* exponential is often described using the base a as

$$x[n] = a^n u[n]$$

Comparison with the continuous-time form $e^{-\alpha t}u(t)$ suggests that $e^{-\alpha} \to a$.

If a is real and positive, we observe the following (Figure 2.25):

1. $a > 1$: Exponential growth. (Example: $\{2^n u[n]\} = \{1\ 2\ 4\ 8\dots\}$).
2. $a < 1$: Exponential decay. (Example: $\{(\frac{1}{2})^n u[n]\} = \{1\ \frac{1}{2}\ \frac{1}{4}\ \frac{1}{8}\dots\}$).
3. $a = 1$: A unit constant. Clearly, $x[n] = u[n]$ in this case.

If a is real and negative, we observe the following (Figure 2.25):

1. $a < -1$: Exponential growth but with values alternating in sign.
2. $a > -1$: Exponential decay but with values alternating in sign.
3. $a = -1$: A constant alternating between 1 (even n) and -1 (odd n).

The twosided *complex* exponential $x[n] = a^n$ can be described using the various formulations of a complex number as

$$x[n] = a^n = [r\exp(j\theta)]^n = r^n\exp(jn\theta) = r^n[\cos(n\theta) + j\sin(n\theta)]$$

This function requires separate plots for the real and imaginary parts or for magnitude and phase. As before, with increasing n, three cases arise (Figure 2.25):

1. For $0 < r < 1$ we have the familiar damped sinusoid with real and imaginary parts as exponentially decaying cosines and sines, respectively.
2. With $r = 1$, there is no damping and the real and imaginary parts are pure cosines and sines with a peak value of unity.
3. For $r > 1$, we obtain exponentially growing sinusoids.

Figure 2.25   Illustrating the discrete-time exponentials of Section 2.15.

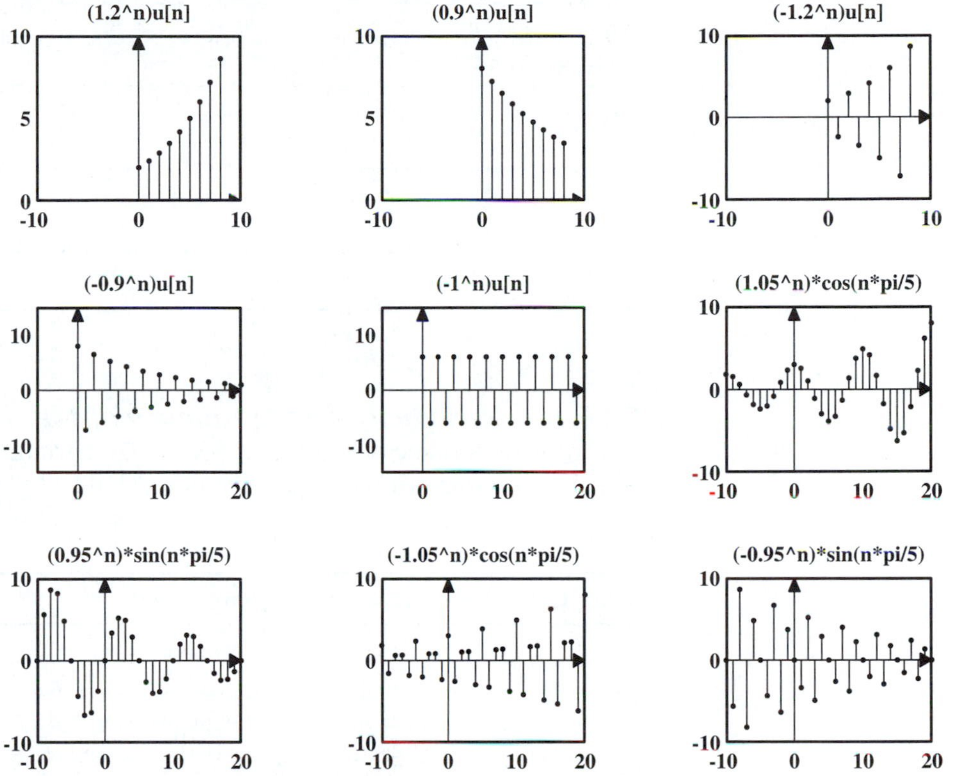

Sampled and *DT* harmonic signals and sinusoids have the form

$$x_S[n] = \exp(j2\pi fnt_s) = \exp(j\omega nt_s) \qquad x[n] = \exp(j2\pi Fn) = \exp(j\Omega n)$$
$$x_S[n] \;= \cos(2\pi fnt_s) = \cos(\omega nt_s) \qquad x[n] = \cos(2\pi Fn) = \cos(\Omega n)$$

The quantities ω and f describe analog frequencies.

2.16
DT Harmonics
and Sinusoids

2.16.1   *Digital Frequency*

The quantity $S_F = 1/t_s$ defines the *sampling rate* in samples/second or the *sampling frequency* in hertz. The normalized frequency, $F = f/S_F$, defines the **digital frequency** denoted by F. It is measured in units of *cycles/sample* or *cycles/cycle*. The digital frequency $\Omega = 2\pi F = 2\pi f/S_F$ is then measured in *radians/sample* or in *radians/cycle*.

The various analog and digital frequencies are shown in Figure 2.26 for comparison. Note that the analog frequency $f = S_F$ corresponds to the digital frequency $F = 1$ or $\Omega = 2\pi$.

Figure 2.26   A comparison of analog and digital frequencies.

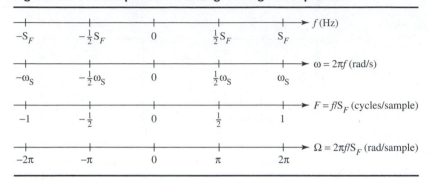

2.16.2   *Periodicity and Frequency of DT Harmonic Signals*

The ideas of periodicity and frequency for *DT* harmonics are quite different from the *CT* case. These differences are summarized in Table 2.14.

Table 2.14   Harmonic Signals: A Comparison.

	CT	Sampled	DT
Sinusoid	$\cos(2\pi f_0 t + \theta)$	$\cos(2\pi n f_0/S_F + \theta)$	$\cos(2\pi F_0 n + \theta)$
Harmonic	$\exp(2\pi f_0 t + \theta)$	$\exp(2\pi n f_0/S_F + \theta)$	$\exp(2\pi F_0 n + \theta)$
Periodicity	$x(t) = x(t \pm kT)$	$x_S[n] = x_S[n \pm kN]$	$x[n] = x[n \pm kN]$
Period	T	Nt_s	N
Periodic for	any f_0	rational f_0/S_F	rational F_0
Unique for	$-\infty < f_0 < \infty$	$-\frac{1}{2}S_F < f < f\frac{1}{2}S_F$	$-\frac{1}{2} < F < \frac{1}{2}$

Periodicity   *Not all DT harmonics are periodic.* For a *CT* periodic signal with period T, we require $x(t) = x(t + T)$. For a *DT* periodic signal $x[n]$ with period N, we require $x[n] = x[n + N]$ if repetition occurs every N samples. For a *DT* harmonic with a digital frequency F, we may write

$$x[n] = \exp(j2\pi F n) \qquad x[n + N] = \exp[j2\pi F(n + N)] = \exp(j2\pi F n)\exp(j2\pi F N)$$
$$x[n] = \cos(2\pi F n) \qquad x[n + N] = \cos[2\pi F(n + N)] = \cos(2\pi F n + 2\pi F N)$$

For periodicity, FN must be an integer because if $FN = k$, then $\exp(j2\pi k) = 1$, $\cos(\alpha) = \cos(\alpha + 2k\pi)$ and we thus satisfy $x[n] = x[n + N]$.

The requirement $FN = k$ means that a sampled or *DT* harmonic is periodic only if its digital frequency $F = f/S_F$ is a *rational fraction of the form* k/N. If it is, the period equals its denominator N (Figure 2.27).

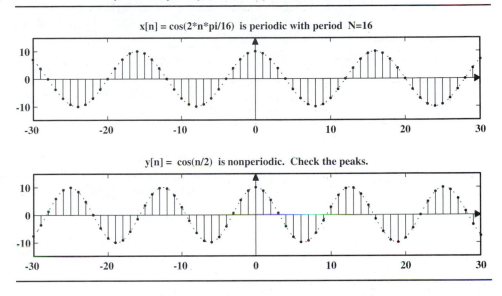

Figure 2.27 A periodic sinusoid shows repetition. An almost periodic sinusoid does not (note the peaks), but has a periodic envelope.

Remarks: **1.** A rational fraction is one that can be expressed as the ratio of two reduced integers. The word *reduced* suggests that common factors in the numerator and denominator, if any, have been canceled out.

2. If F is not a rational fraction, there is no periodicity. The sinusoid is classified as nonperiodic or almost periodic (Figure 2.27).

3. Even though a *DT* sinusoid may not always be periodic, it will always have a periodic envelope.

Frequency *Not all DT harmonics are distinct.* If the digital frequency $F = k/N$ is augmented by any integer m, it results in the same sinusoid. This should be obvious since

$$\exp[2\pi(m + k/N)n] = \exp(2\pi mn)\exp(2\pi nk/N) = \exp(2\pi nk/N)$$
$$\cos[2\pi(m + k/N)n] = \cos(2\pi mn + 2\pi nk/N) = \cos(2\pi nk/N)$$

There are thus only N distinct versions of a sinusoid with period N. These correspond to the digital frequencies $F = k/N$, $k = 1, 2, \ldots, N$, all of which lie in the range $(0, 1)$. This also corresponds to the range $(-\frac{1}{2}, \frac{1}{2})$ since any sinusoid at a frequency $|F| > \frac{1}{2}$ may always be described as a sinusoid at a frequency $|F| < \frac{1}{2}$ by subtracting out integer values. The frequency range $(-\frac{1}{2}, \frac{1}{2})$ is called the **principal range**.

Remarks: **1.** A signal $\cos(2\pi nk/N)$ may be treated as the kth harmonic of $\cos(2\pi n/N)$ with fundamental frequency $F_0 = 1/N$. The common period of both is N. Unlike

CT harmonics, *the common period N is k times the true period of* $\cos(2\pi nk/N)$ *only if N/k is an integer.*

2. The harmonic signals $x[n] = \exp(j2\pi nk/N)$, $k = 1, \ldots, N$ may be shown on an Argand diagram as unit vectors at angles $2\pi k/N$ whose tips lie on a unit circle.

Example 2.21

DT Harmonics

Let $x[n] = \cos[\frac{1}{5}n\pi]$. Its digital frequency is $F = \frac{1}{10}$. Its period is $N_x = 10$. Its 2nd harmonic $x_2[n] = \cos[\frac{2}{5}n\pi]$, with $F = \frac{2}{10}$, has a period of $N_2 = 5$ (not 10). Its 3rd harmonic $x_3[n] = \cos[\frac{3}{5}n\pi]$, with $F = \frac{3}{10}$, has a period $N_3 = 10$. The common period N of the combination $x[n] + x_2[n] + x_3[n]$ equals 10.

2.16.3   *The Principal Range*

Consider a *DT* sinusoid $x[n] = \cos(2\pi Fn + \theta)$ whose digital frequency F has an integer part M and a fractional part F_r such that $|F_r| < \frac{1}{2}$. Then

$$x[n] = \cos[2\pi(M + F_r)n + \theta] = \cos(2\pi F_r n + \theta)$$

Its digital frequency thus equals F_r. If F_r is negative, $x[n]$ may be written as $x[n] = \cos(-2\pi F_r n + \theta) = \cos(2\pi F_r n - \theta)$. We then have a phase reversal.

Example 2.22

The Principal Range

(a) If $x[n] = \cos[2\pi\frac{10}{3}n + \theta]$, then $F = \frac{10}{3}$ and $x[n]$ can be folded back to $x_A[n] = \cos[2\pi(\frac{10}{3} - 3)n + \theta] = \cos(2\pi\frac{1}{3}n + \theta)$ with $F = \frac{1}{3}$. The period of both is $N = 3$.

(b) If $x[n] = \cos[2\pi\frac{8}{3}n + \theta]$, then $F = \frac{8}{3}$ and $x[n]$ can also be folded back to $x_A[n] = \cos[2\pi(\frac{8}{3} - 3)n + \theta] = \cos(-2\pi\frac{1}{3}n + \theta)$ with $F = -\frac{1}{3}$. This also equals $x_A[n] = \cos(2\pi\frac{1}{3}n - \theta)$ with $F = \frac{1}{3}$ and reversed phase.

2.16.4   *Aliasing*

If a *CT* signal $x(t) = \cos(2\pi f_0 t + \theta)$ is sampled at the rate S_F, the sampled signal is given by $x_S[n] = \cos(2\pi n f_0/S_F + \theta) = \cos(2\pi n F_0 + \theta)$. Only a frequency $F_0 < \frac{1}{2}$ ensures a unique sampled representation of $x(t)$. This implies that $f_0 < \frac{1}{2}S_F$ or $S_F > 2f_0$, and places a lower bound on the sampling frequency.

The frequency $F = \frac{1}{2}$ or $f = \frac{1}{2}S_F$ is called the **folding frequency**. If $F_0 > \frac{1}{2}$ or $f_0 > \frac{1}{2}S_F$, it can be folded back to a lower frequency $|f_a| < \frac{1}{2}S_F$ by subtracting out integers or multiples of S_F. The signal $x[n]$ then describes a sampled version of a *CT* signal at the lower frequency f_a, and we say that **aliasing** has occurred. The lower frequency f_a, which impersonates or mimics f_0, is called the **aliased frequency**, and the *CT* signal at this aliased frequency f_a describes an **aliased signal**. To prevent aliasing, we require a sampling frequency S_F exceeding $2f_0$, twice the frequency of the sinusoid.

The Sampling Theorem   The **sampling theorem** generalizes the above result to arbitrary signals bandlimited to some highest frequency f_B. It asserts that to avoid aliasing and retain their unique identity, such signals must be sampled at a rate

$S_F > 2f_B$. The critical sampling rate $S_N = 2f_B$ is called the **Nyquist rate**. We study the sampling theorem and its implications in detail in later chapters.

2.17 Continuous-Time Signals: A Synopsis

Measure	Explanation
Amplitude	Analog signals show a continuum of amplitudes.
	Quantized signals are restricted to a finite number of levels.
Duration	Rightsided signals are zero for $t < t_0$.
	Leftsided signals are zero for $t > t_0$.
	Twosided signals continue forever on either side of $t = 0$.
	Timelimited signals have finite duration.
	Causal signals are zero for $t < 0$.
Areas	For absolutely integrable (finite area) signals, $\mathbb{A}[|x(t)|] < \infty$.
	For square integrable or finite energy signals, $\mathbb{A}[|x(t)|^2] < \infty$.
	The areas $\mathbb{A}[x(t)]$ and $\mathbb{A}[x^2(t)]$ when $x(t) = x_1 + x_2 + \cdots$ comprises linear segments or sinusoidal pulses may be found using Table 2.4 and $\mathbb{A}[x(t)] = \mathbb{A}[x_1] + \mathbb{A}[x_2] + \cdots$ and $\mathbb{A}[x^2(t)] = \mathbb{A}[x_1^2] + \mathbb{A}[x_2^2] + 2\mathbb{A}[x_1 x_2] + \cdots$
Periodicity	For periodic signals, $x(t) = x(t \pm kT)$ for integer k.
	The period T of sinusoidal combinations = LCM of all periods.
	Quasiperiodic signals show noncommensurate periods.
	Nonperiodic signals show no periodicity.
Symmetry	For even symmetry, $x(t) = x(-t)$ and $\mathbb{A}[x(t)] = 2\mathbb{A}[x(t)](0, \infty)$.
	For odd symmetry, $x(t) = -x(-t)$. $\mathbb{A}[x(t)](-\alpha, \alpha) = 0$ and $x(0) = 0$.
	For halfwave (periodic) signals, $x(t) = -x(t \pm \frac{1}{2}T)$ and $x_{av} = 0$.
Energy	For nonperiodic signals with finite energy, $E_x = \mathbb{A}[|x(t)|^2]$.
	Examples: Most finite area signals, exponentially damped signals.
	and timelimited signals of finite amplitude.
Power	A periodic signal $x(t)$ is a power signal with $P_x = x_{rms}^2$.
	For a switched periodic signal $y(t) = x_p(t)u(t - t_0)$, $P_y = \frac{1}{2}P_x$.
	For a nonperiodic power signal $x(t)$, $P_x = \lim_{T_0 \to \infty} \frac{1}{T_0} \int_{T_0} |x(t)|^2 \, dt$.

2.18 Discrete-Time Signals: A Synopsis

Measure	Explanation
Amplitude	DT signals show a continuum of amplitude values.
	Digital signals are quantized to a finite number of levels.
Duration	Rightsided signals are zero for $n < n_0$.
	Leftsided signals are zero for $n > n_0$.
	Twosided signals continue forever on either side of $n = 0$.
	Timelimited signals have finite duration.
	Causal signals are zero for $n < 0$.

Areas	For absolutely *summable* (finite area) signals: $\sum \lvert x[n] \rvert < \infty$. For square *summable* or finite energy signals: $\sum \lvert x[n] \rvert^2 < \infty$.
Periodicity	For periodic signals, $x[n] = x[n \pm kN]$ for integer k and period N. The period N of sinusoidal combinations = LCM of all periods. Quasiperiodic signals show noncommensurate periods. Nonperiodic signals show no periodicity.
Symmetry	For even symmetry, $x[n] = x[-n]$. For odd symmetry, $x[n] = -x[-n]$, $\sum x[n](-k, k) = 0$ and $x[0] = 0$. For halfwave (periodic) signals, $x[n] = -x[n \pm \frac{1}{2}N]$ and $x_{av} = 0$.
Energy	For nonperiodic signals with finite energy, $E_x = \sum \lvert x[n] \rvert^2$. Examples: Most finite area signals and timelimited signals.
Power	Periodic signals are power signals with $P_x = \frac{1}{N} \sum_{n=0}^{N-1} \lvert x[n] \rvert^2$. For a nonperiodic power signal $x[n]$, $P_x = \lim_{N \to \infty} \frac{1}{2N+1} \sum_{n=-N}^{N} \lvert x[n] \rvert^2$.

Appendix 2A
MATLAB
Demonstrations
and Routines

Getting Started on the Supplied MATLAB Routines

Examples of supplied MATLAB routines related to concepts described in this chapter are listed below. See Appendix 1A for how to get more information.

NOTE: The % sign and comments following it need not be typed.

CT Signals

```
>>t=0:0.02:5;y=sinc(2*t)          %Generates sinc(2t) over [0 5]
>>t=0:0.02:5;y=sinc2(t)           %Generates sinc^2(t) over [0 5]
>>t=-3:0.02:3;y=tri(2*t+1)        %Generates tri(2t+1) over [-3 3]
>>t=-3:0.02:3;y=uramp(t/2)        %Generates r(t/2) over [-3 3]
>>t=-3:0.02:3;y=urect(t/2,.5)     %rect(t/2) [-3 3] endpoints = 1/2
>>t=0:0.02:5;y=sinc2(t)           %Generates sinc^2(t) over [-3 3]
>>t=0:0.01:5;y=periodic('tri(t-1)',t,2) %Periodic triangular wave T=2
```

Use *plot*(t, y) to plot any of the above *CT* signals.

DT Signals

```
>>n=-10:10;y=ustep(n)     %Generates values of u[n] over [-10 10]
>>n=-10:10;y=urect(n/4)   %Generates values of rect[n/4] over [-10 10]
>>n=-10:10;y=udelta(n)    %Generates values of delta[n] over [-10 10]
```

Use *lines*(n, y) to plot any of the above *DT* signals.

Power/Energy

```
>>enerpwr('sin(pi*t)',[0 1])     %Computes energy in sin(pi*t) over [0 1]
>>enerpwr('sin(pi*t)',[0 1],3)   %Power in sin(pi*t) duration = [0 1], T=3
```

Operations

```
>>[t1,x1]=operate(t,x,a,b)  %Creates time and function arrays for the
                            %signal x(at+b) from values of t and x(t)
>>[eu,o]=evenodd(x)         %Generates even and odd part of signal x
```

Use $plot(t, x, t1, x1)$ to plot $x(t)$ and $x(at + b)$ on the same set of axes.

Utility Functions

```
>>y=simpson('t.*t+1',[0 2])  %Computes the area under 1+t^2 over [0 2]
```

The routine *simpson* uses string functions of t in single quotes as shown.

```
>>y=lcm1([8 12 18])     %LCM of INTEGERS 8,12,18 (or rational fractions)
>>x=gcd1([8 3],[18 4])  %GCD of rational fractions 8/18, 3/4 (or integers)
>>y=alog(5.6)           %antilog of 5.6 (base e is default 2nd argument)
>>y=alog(5.6,6)         %antilog of x (base 6)
>>y=alog(5.6,0)         %gain corresponding to xdB where x=20log10(y)
```

CT Signals

C2.1 **(Signal Duration)** For each signal of Figure C2.1 (p. 64):
(a) Classify as timelimited, twosided, leftsided or rightsided.
(b) Classify as causal, anticausal or noncausal.

C2.2 **(Operations on Signals)** For each signal $x(t)$ in Figure C2.2, sketch
(a) $x(t + 2)$ **(b)** $x(-t - 2)$ **(c)** $x(2t - 2)$ **(d)** $x(-2t + 2)$
(e) $x(-2t - 2)$ **(f)** The even part of $x(t)$ **(g)** The odd part of $x(t)$

C2.3 **(Symmetry)** Find the even and odd parts of
(a) $x(t) = \exp(-t)u(t)$ **(b)** $x(t) = (1 + t)^2$ **(c)** $x(t) = [\sin(t) + \cos(t)]^2$

C2.4 **(Symmetry)** Evaluate the following using concepts of symmetry.
(a) $\int_{-3}^{3}(4 - t^2)\sin(5t)\,dt$ **(b)** $\int_{-2}^{2}[4 - t^3\cos(\frac{1}{2}\pi t)]\,dt$

C2.5 **(Classification)** For each signal of Figure C2.1:
(a) Classify in terms of symmetry, periodicity and power/energy.
(b) For each symmetric signal, evaluate $\mathbb{A}[x(t)]$.
(c) For each periodic signal, evaluate x_{rms}.
(d) For each power signal, evaluate the power.
(e) For each energy signal, evaluate the energy.

Figure C2.1

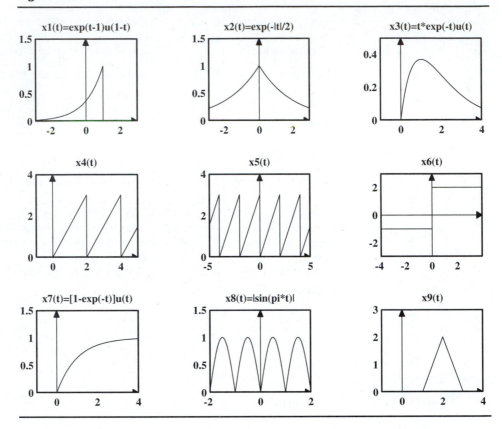

Figure C2.2

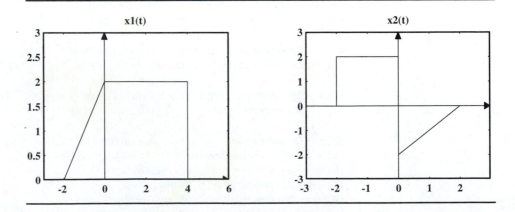

C2.6 (Signal Classification) Sketch each signal, classify it as a power signal, energy signal or neither and find its power or energy as and where applicable.

(a) $t\exp(-t)u(t)$ **(b)** $\exp(t)[u(t) - u(t-1)]$ **(c)** $t\exp(-|t|)$

(d) $[1 - \exp(-t)]u(t)$ **(e)** $\exp(-t)$ **(f)** $10\exp(-t)\sin(t)u(t)$

C2.7 (Periodic Signals)
(a) Justify that the area under the product of two periodic signals is infinite or zero. Justify the following for a periodic signal $x_p(t)$ with period T:
(b) If $x_p(t)$ is both even and halfwave symmetric, it must show even symmetry about $t = \frac{1}{2}T$ and odd symmetry about $t = \frac{1}{4}T$.
(c) If $x_p(t)$ is both odd and halfwave symmetric, it must show odd symmetry about $t = \frac{1}{2}T$ and even symmetry about $t = \frac{1}{4}T$.
(d) If $x_p(t)$ is shifted to $x_p(t - \alpha)$ or scaled to $x_p(\alpha t)$, the average value, signal power and rms value do not change.

C2.8 (Periodic Signals) Classify each signal as periodic, nonperiodic or almost periodic and find its power. For each periodic signal, find the fundamental frequency and fundamental period and sketch its magnitude and phase spectra.

(a) $x_1(t) = 4 - 3\sin(12\pi t) + \sin(30\pi t)$ **(b)** $x_2(t) = \cos(10\pi t)\cos(20\pi t)$

(c) $x_3(t) = \cos(10\pi t) - \cos(20t)$ **(d)** $x_4(t) = \cos(10\pi t)\cos(10t)$

(e) $x_5(t) = 2\cos(8\pi t) + \cos^2(6\pi t)$ **(f)** $x_6(t) = \cos(2t) - \sqrt{2}\cos(2t - \frac{1}{4}\pi)$

C2.9 (RMS Value) Find the signal power and rms value for a periodic pulse train with peak value A and duty ratio D if the pulse shape is

(a) Rectangular   **(b)** A half-sinusoid   **(c)** A sawtooth   **(d)** Triangular

C2.10 (Signal Description) Represent each signal in Figure C2.10 using
(a) An interval by interval description.
(b) Steps and/or ramps.
(c) Rect and/or tri functions.

C2.11 (Sketching Signals) Sketch each of the following signals.

(a) $x_1(t) = 3\sin(4\pi t - 30°)$ **(b)** $x_2(t) = 2\,\text{tri}(t - 1) + 2\,\text{tri}[\frac{1}{2}(t - 2)]$

(c) $x_3(t) = 9t\exp(-3|t|)$ **(d)** $x_4(t) = t\,\text{rect}(\frac{1}{2}t)$

(e) $x_5(t) = \text{rect}(\frac{1}{2}t)\text{sgn}(t)$ **(f)** $x_6(t) = \exp(-2|t - 1|)$

(g) $x_7(t) = \text{sinc}[\frac{1}{3}(t - 3)]$ **(h)** $x_8(t) = u(t - 1)u(3 - t)$

(i) $x_9(t) = u(1 - |t|)$ **(j)** $x_{10}(t) = u(1 - |t|)\text{sgn}(t)$

(k) $x_{11}(t) = \text{rect}(t/6)\text{rect}[\frac{1}{4}(t - 2)]$ **(l)** $x_{12}(t) = \text{rect}(t/6)\text{tri}[\frac{1}{2}(t - 2)]$

C2.12 (The Impulse) Sketch the following signals.

(a) $3\delta(t - 2)$ **(b)** $3\delta(2t - 2)$ **(c)** $3t\delta(t - 2)$

(d) $3t\delta(2t - 2)$ **(e)** $\text{comb}(t/2)$ **(f)** $\text{comb}(t/2)\text{rect}[(t - 5)/5]$

C2.13 (The Sifting Property) Evaluate the following integrals.

(a) $\int_{-\infty}^{\infty}(4 - t^2)\delta(t + 3)\,dt$ **(b)** $\int_{-3}^{6}(4 - t^2)\delta(t + 4)\,dt$

(c) $\int_{-3}^{6}(6 - t^2)[\delta(t + 4) + 2\delta(2t + 4)]\,dt$ **(d)** $\int_{-\infty}^{\infty}\delta(t - 2)\delta(x - t)\,dt$

(e) $\int_{-3}^{6}(4 - t^2)\delta'(t - 4)\,dt$ **(f)** $\int_{-\infty}^{t}[\delta(t + 2) - \delta(t - 2)]\,dt$

Figure C2.10

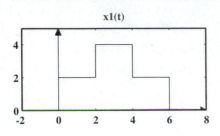

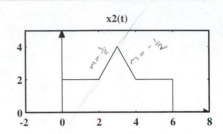

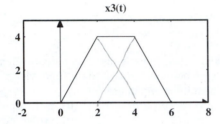

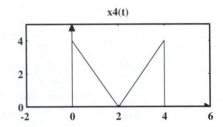

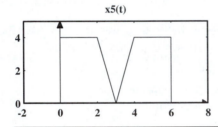

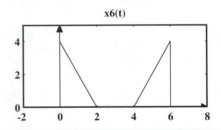

C2.14 (Generalized Derivatives) Sketch $x(t)$, $x'(t)$ and $x''(t)$ for
(a) $x(t) = 4\,\mathrm{tri}[\frac{1}{2}(t - 2)]$
(b) $x(t) = \exp(-t)u(t)$
(c) $x(t) = 2\,\mathrm{rect}(\frac{1}{2}t) + \mathrm{tri}(t)$
(d) $x(t) = \exp(-\,|\,t\,|)$
(e) $x(t) = [1 - \exp(-t)]u(t)$
(f) $x(t) = \exp(-2t)\mathrm{rect}[(t - 2)/2]$

C2.15 (Ideally Sampled Signals) Sketch the ideally sampled signal and the impulse approximation for each signal assuming $t_s = 0.5$.
(a) $x_1(t) = \mathrm{rect}(t/4)$
(b) $x_2(t) = r(t + 2) - 2r(t) + r(t - 2)$
(c) $x_3(t) = \sin(\pi t)\mathrm{rect}[(t - 2)/2]$
(d) $x_4(t) = t\,\mathrm{rect}(\frac{1}{2}t)$

C2.16 (Finite Area) Use $\delta(t)$ and $\mathrm{sinc}(t)$ as examples to confirm that
(a) If $\mathbb{A}[|\,x(t)\,|]$ is finite, $\mathbb{A}[|\,x(t)\,|^2]$ need not be finite.
(b) If $\mathbb{A}[|\,x(t)\,|^2]$ is finite, $\mathbb{A}[|\,x(t)\,|]$ need not be finite.
(c) If $\mathbb{A}[|\,x(t)\,|^2]$ is finite, $\mathbb{A}[x(t)]$ is also finite.

C2.17 (Energy) Consider an energy signal $x(t)$, $(-3 \le t \le 3)$ with energy $E = 12$ J. Find the limits and the signal energy for
(a) $x(3t)$ **(b)** $2x(t)$ **(c)** $x(t - 4)$ **(d)** $x(-t)$ **(e)** $x(-\frac{1}{3}t - 2)$

C2.18 (Power) Consider a periodic signal $x_p(t)$ with time period $T = 6$ and power $P = 4$ W. Find the period and signal power for

(a) $x_p(3t)$ (b) $2x_p(t)$ (c) $x_p(t-4)$ (d) $x_p(-t)$ (e) $x_p(-\frac{1}{3}t - 2)$

C2.19 (Power/Energy) If $y(t) = x_1(t) + x_2(t)$, the area under $y^2(t)$ equals

$$\mathbb{A}[y^2(t)] = \mathbb{A}[x_1^2(t)] + \mathbb{A}[x_2^2(t)] + 2\mathbb{A}[x_1(t)x_2(t)]$$

(a) Use this result to show that for any energy (or power) signal, the signal energy (or power) equals the sum of the energy (or power) in its odd and even parts.
(b) Over one period, a periodic signal shows a linear increase from A to B in T_1 s and a linear decrease from B to A in T_2 s and stays constant for the rest of the period. What is the signal power if the period equals $2(T_1 + T_2)$s?

C2.20 (Power/Energy) Use $u(t)$, $u(t-1)$, $\exp(-t)u(t)$ and others as examples to argue for or against the following statements.
(a) The sum of energy signals is an energy signal.
(b) The sum of a power and an energy signal is a power signal.
(c) The algebraic sum of two power signals can be an energy signal or a power signal or zero!
(d) The product of two energy signals is zero or an energy signal.
(e) The product of a power and energy signal is an energy signal or zero.
(f) The product of two power signals is a power signal or zero.

C2.21 (Periodicity) The sum of two periodic signals is periodic if their periods T_1 and T_2 are commensurate. Under what conditions will their product be periodic? Use sinusoids as examples to prove your point.

C2.22 (Periodicity) Let $x(t) = \sin(2\pi t)$.
(a) Sketch $x(t)$ and find its period and power.
(b) Is $y(t) = \exp[x(t)]$ periodic? If so, find its period and sketch.
(c) Is $z(t) = \exp[jx(t)]$ periodic? If so, find its period and power.

C2.23 (Halfwave Symmetry) Argue that if a halfwave symmetric signal $x(t)$ with period T is made up of several sinusoidal components, each component is also halfwave symmetric over one period T. Which of the following signals show halfwave symmetry?
(a) $x_1(t) = \cos(2\pi t) + \cos(6\pi t) + \cos(10\pi t)$
(b) $x_2(t) = 2 + \cos(2\pi t) + \sin(6\pi t) + \sin(10\pi t)$
(c) $x_3(t) = \cos(2\pi t) + \cos(4\pi t) + \sin(6\pi t)$

C2.24 (Impulses as Limiting Forms)
(a) Argue that in the limit as $\alpha \to 0$, the Gaussian $x(t) = (1/\alpha)\exp(-\pi t^2/\alpha^2)$ describes the impulse $\delta(t)$.
(b) Argue that the limiting form of the Lorentzian $g(x) = \alpha/(\alpha^2 + x^2)$ as $\alpha \to 0$ is the impulse $\pi\delta(x)$.

C2.25 (Impulses) It is possible to show that the signal $\delta[f(t)]$ is a string of impulses at the roots t_k of $f(t) = 0$ whose strengths equal $1/\,|\,f'(t_k)\,|$. Use this result to sketch the following signals.
(a) $x_1(t) = \delta(t^2 - 3t + 2)$ (b) $x_2(t) = \delta[\sin(\pi t)]$

DT Signals

D2.1 (DT Signal Duration) For each signal of Figure D2.1:
(a) Classify it as timelimited, twosided, leftsided or rightsided.
(b) Which of the signals are causal?
(c) Which of the signals are absolutely summable?
(d) Which of the signals are square summable?

Figure D2.1

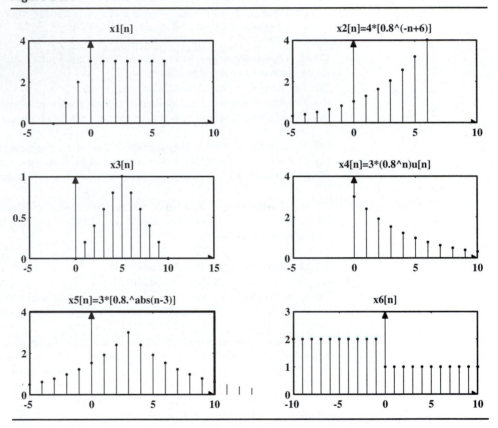

D2.2 (DT Signal Duration) Use examples to show that the product of a rightsided and a leftsided signal is always timelimited or zero.

D2.3 (DT Sequences) Sketch the following sequences.
(a) $x_1[n] = \{0, 2, 4, 6\}, t_s = 2$
(b) $x_2[n] = \{6, 4, 2, 0\}$
(c) $x_3[n] = \{\ldots, -3, -2, -1, 0, 1, \ldots\}$
(d) $x_4[n] = \{3, 2, 1, 1, 2\}, t_s = \frac{1}{2}$
(e) $x_5[n] = \{3, 2, 0, 1, 0, \overset{\uparrow}{1}, 0, 2\}$
(f) $x_6[n] = \{4, 0, \overset{\uparrow}{2}, 0, 0, 0, 2, \ldots\}$

D2.4 (DT Operations) Let $x[n] = \{6, 4, 2, 0\}$. Find and sketch
(a) $x[n-2]$ **(b)** $x[n+2]$ **(c)** $x[-n+2]$ **(d)** $x[-n-2]$

D2.5 (Interpolation) Let $x[n] = \{4, 0, 2, 0, 0, 0, 2, \ldots\}$
                                        $\uparrow$
(a) Find and sketch $x[2n]$.
(b) Find and sketch $x[\frac{1}{2}n]$ using zero interpolation.
(c) Find and sketch $x[\frac{1}{2}n]$ using step interpolation.
(d) Find and sketch $x[\frac{1}{2}n]$ using linear interpolation.

D2.6 (Interpolation and Decimation) Let $x[n] = 4\,\mathrm{tri}(n/4)$. Carefully sketch the following signals:
(a) $x[\frac{2}{3}n]$ using zero interpolation followed by decimation.
(b) $x[\frac{2}{3}n]$ using step interpolation followed by decimation.
(c) $x[\frac{2}{3}n]$ using decimation followed by zero interpolation.
(d) $x[\frac{2}{3}n]$ using decimation followed by step interpolation.

D2.7 (Symmetry) Sketch each signal and its even and odd parts.
(a) $x_1[n] = \exp(-n)u[n]$ **(b)** $x_2[n] = u[n]$ **(c)** $x_3[n] = 1 + u[n]$
(d) $x_4[n] = u[n] - u[n-5]$ **(e)** $x_6[n] = \mathrm{tri}[(n-3)/3]$

D2.8 (Sketching DT Signals) Sketch each of the following signals:
(a) $x_1[n] = r[n+2] - r[n-2] + r[n-4]$ for $-3 \le n \le 8$
(b) $x_2[n] = \mathrm{rect}[\frac{1}{6}(n-3)]$ **(c)** $x_3[n] = \mathrm{tri}[n/5]$
(d) $x_4[n] = \mathrm{tri}[(n-2)/4]$ **(e)** $x_5[n] = \mathrm{rect}[\frac{1}{8}n]\mathrm{sinc}[\frac{1}{4}n]$

D2.9 (Sketching DT Signals) Sketch the signals below. How are they related?
(a) $x_1[n] = \delta[n]$ **(b)** $x_2[n] = \mathrm{rect}[n]$ **(c)** $x_3[n] = \mathrm{tri}[n]$ **(d)** $x_4[n] = \mathrm{sinc}[n]$

D2.10 (DT Signal Description) For each signal $x[n]$ shown in Figure D2.10
(a) Write the sequence $\{x[n]\}$. Mark the index $n = 0$ by an arrow.
(b) Represent each using impulses.
(c) Represent each using steps and/or ramps.
(d) Find the signal energy.
(e) Find the signal power assuming the sequence repeats itself.

Figure D2.10

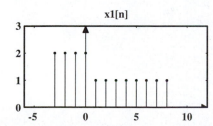
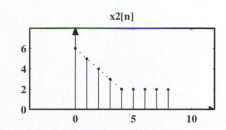

D2.11 (*DT* **Harmonics and Sinusoids**) Classify each of the following as periodic, nonperiodic or almost periodic. For each periodic signal, compute the period N.

(a) $x_1[n] = \cos[\frac{1}{2}\pi n]$ **(b)** $x_2[n] = \cos[\frac{1}{2}n]$

(c) $x_3[n] = \sin[\frac{1}{4}\pi n] - 2\cos[\frac{1}{6}\pi n]$ **(d)** $x_4[n] = 2\cos[\frac{1}{4}\pi n] + \cos^2[\frac{1}{4}\pi n]$

(e) $x_5[n] = 4u[n] - 3\sin[7n\pi/4]u[n]$ **(f)** $x_6[n] = \cos[5n\pi/12] + \cos[4n\pi/9]$

(g) $x_7[n] = \cos[8n\pi/3] + \cos[8n/3]$ **(h)** $x_8[n] = \cos[8n\pi/3]\cos[n\pi/2]$

D2.12 (**Folding and Aliasing**) Consider the following *DT* signals:

(1) $x[n] = \cos[4n\pi/3]$ (2) $x[n] = \sin[4n\pi/3] + 3\sin[8n\pi/3]$.

(a) Find a version with a digital frequency $|F| \le \frac{1}{2}$ for each.

(b) Find two versions with a digital frequency $|F| > 1.5$ for each.

D2.13 (**Aliasing**) Each of the following sinusoids is sampled at $S_F = 100$ Hz.

(1) $x_1(t) = \cos(320\pi t + \frac{1}{4}\pi)$ (2) $x_2(t) = \sin(140\pi t - \frac{1}{4}\pi)$ (3) $x_3(t) = \sin(60\pi t)$

(a) Does the sampled signal show aliasing?

(b) Find its sampled version with digital frequency $|F| \le \frac{1}{2}$.

D2.14 (**Norms**) Norms provide a measure of the *size* of a signal. The **p-norm** or **Holder norm** $\|x\|_p$ for *DT* signals is defined by $\{\sum |x|^p\}^{1/p}$, where $0 < p < \infty$ is a positive integer. For $p = \infty$, we also define $\|x\|_\infty$ as the peak absolute value $|x|_{max}$.

(a) Let $x[n] = [3, -j4, 3 + j4]$. Find $\|x\|_1$, $\|x\|_2$ and $\|x\|_\infty$.

(b) What is the significance of each of these norms?

3 SYSTEMS

Systems process signals. In fact, the description of systems relies heavily on how they respond to arbitrary or specific signals. In the time domain, many systems can be described by their response to arbitrary signals in terms of differential or difference equations. The class of *linear, time-invariant* systems (discussed in this chapter) can also be described in terms of their impulse response, the response to an impulse input.

This chapter deals with systems, their classification and their time-domain representation. It also introduces the all-important concept of the impulse response, which forms a key ingredient in system analysis. The chapter is divided into four major parts:

Part 1 Classification of continuous-time systems.
Part 2 Analysis of continuous-time systems.
Part 3 Classification of discrete-time systems.
Part 4 Analysis of discrete-time systems.

The objectives of Parts 2 and 4 are to familiarize you with system analysis based on the solution of differential and difference equations.

Useful Background   This chapter requires familiarity with basic calculus (integration and differentiation) and some of the signals described in Chapter 2.

3.1 Introduction

A system processes signals. In its broadest sense, a physical system is an interconnection of devices and elements subject to physical laws. A system that processes analog signals is referred to as an **analog system** or a *CT* **system**. A system that processes sampled or *DT* signals is called a **sampled data system,** or **discrete-**

time *(DT)* **system**. The signal to be processed forms the **excitation** or **input** to the system. The processed signal is termed the **response** or **output**.

For our purposes, a system may be represented by a *black box* (Figure 3.1), whose response $y(t)$ is governed by the input $x(t)$. When we analyze a system, we presume that the output has been specified as one of the many possible system variables. In many cases a system may be excited by more than one input, and this leads to the more general idea of *multiple-input* systems. We address only *single-input, single-output* systems in this text.

Figure 3.1 Black box representation of a system.

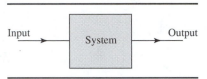

The study of systems involves the input, output and system specifications. Conceptually, we can determine any one of these in terms of the others.

Analysis   **System analysis** implies a study of the response subject to known inputs and system formulations.

Synthesis   Known input-output specifications, on the other hand, usually allow us to identify, or synthesize, the system. **System identification** or **synthesis** is much more difficult because many system formulations are possible for the same input-output relationship.

Models   Most real-world systems are quite complex and almost impossible to analyze quantitatively. Of necessity, we are forced to use **models** or abstractions that retain the essential features of the system and simplify the analysis, while still providing meaningful results. The analysis of systems refers to the analysis of the models that, in fact, describe such systems, and it is customary to treat the system and its associated models synonymously. In the context of signal processing, a system that processes the input signal in some fashion is also called a **filter**.

As with signals, systems may be classified in several ways. This allows us to make informed decisions on the choice of a given method of analysis over others, depending on the context in which the system is considered.

3.1.1   *Terminology of Systems*

A system requires two separate descriptions for a complete specification, one in terms of its components or external structure and the other in terms of its energy level or internal **state**.

State Variables   The *state* of a system is described by a set of **state variables** that allow us to establish the energy level of the system at any instant. Such variables may represent physical quantities or may have no physical significance whatever. Their choice is governed primarily by what the analysis requires. For example, capacitor voltages and inductor currents are often used as state variables since they provide an instant measure of the system energy (through their own energy $\frac{1}{2}Cv_C^2$ and $\frac{1}{2}Li_L^2$).

The Relaxed State   Any inputs applied to the system result in a change in the energy or state of the system. All physical systems are, by convention, referenced to a zero-energy state (variously called the **ground state**, the **rest state**, the **relaxed state** or the **zero state**) at $t = -\infty$.

The Initial State   The behavior of a system is governed not only by the input but also by the state of the system at the instant at which the input is applied. The initial values of the state variables define the **initial conditions** or **initial state**. This initial state, which must be known before we can establish the complete system response, embodies the past history of the system. It allows us to predict the future response due to any input *regardless of how the initial state was arrived at.*

3.1.2   *System Representation in the Time Domain*

At the quantitative level, CT systems are usually modeled by differential equations. These relate the output $y(t)$ and input $x(t)$ through absolute constants (such as 2 or ϵ_0), parameters (arbitrary constants such as R and C), and the independent coordinate variable t. The general form of a differential equation is

$$a_0 \frac{d^n y}{dt^n} + a_1 \frac{d^{n-1} y}{dt^{n-1}} + \cdots + a_{n-1} \frac{dy}{dt} + a_n y = b_0 \frac{d^m x}{dt^m} + b_1 \frac{d^{m-1} x}{dt^{m-1}} + \cdots + b_{m-1} \frac{dx}{dt} + b_m x.$$

The coefficients a_k and b_k may be functions of x and/or y and/or t. It is customary to normalize the coefficient of the highest derivative of y to 1.

Order   The **order** n of the differential equation refers to the order of the *highest derivative of the output $y(t)$.*

3.1.3   *Operators*

Any equation is based on a set of operations. An **operator** is a rule or a set of directions—a recipe if you will—that shows us how to transform one function to another. For example, the derivative operator $s \equiv d/dt$ transforms a function $x(t)$ to $y(t) = s[x(t)]$ or $dx(t)/dt$. If an operator or a rule of operation is represented by the symbol $\mathcal{O}$, the equation

$$\mathcal{O}[x(t)] = y(t)$$

implies that if the function $x(t)$ is treated exactly as the operator $\mathcal{O}$ requires, we obtain the function $y(t)$.

Example 3.1
Illustrating Operations

The operation $\mathcal{O}[\ ] = 4s[\ ] + 6$ says that to get $y(t)$, we must take the derivative of $x(t)$, multiply by 4 and then add 6 to the result:

$$4s[x(t)] + 6 = 4\frac{dx}{dt} + 6 = y(t)$$

3.1.4   *Linear Operators*

An operator $\mathcal{L}$ is termed a **linear operator** if and only if

$$\mathcal{L}[ax(t) + bx_2(t)] = a\mathcal{L}[x_1(t)] + b\mathcal{L}[x_2(t)]$$

Here $x_1(t)$ and $x_2(t)$ represent functions (real or complex) that can legitimately be operated upon by $\mathcal{L}$, and a and b are any two scalar constants (real or complex). If an operation performed on a linear combination of x_1 and x_2 produces the same results as a linear combination of operations on x_1 and x_2 separately, the operation is linear. If not, it is nonlinear. Examples of linear operators include derivatives and integrals.

Example 3.2
Testing for Linear
Operations

(a) Consider the operator $\mathcal{L}[\ ] = \log[\ ]$. Applying this operation to two quantities x_1 and x_2 gives the left-hand side of the linearity relation as

$$\mathcal{L}[ax_1 + bx_2] = \log[ax_1 + bx_2]$$

The right-hand side, using the properties of logarithms, yields

$$a\mathcal{L}[x_1] + b\mathcal{L}[x_2] = a\log[x_1] + b\log[x_2] = \log[x_1^a x_2^b]$$

Clearly, the results differ and the log operator is nonlinear.

(b) Consider the operator $\mathcal{O}[x] = C[x] + D$. Applying this operation to the left-hand side of the linearity relation we obtain

$$\mathcal{O}[ax_1 + bx_2] = C[ax_1 + bx_2] + D$$

With the same operation applied to the right-hand side, we get

$$a\mathcal{O}[x_1] + b\mathcal{O}[x_2] = a[Cx_1 + D] + b[Cx_2 + D] = C[ax_1 + bx_2] + D[a + b]$$

The two clearly differ and make the operation $C[x] + D$ nonlinear.

Comment: The operation $C[x] + D$ *looks* linear because we are mentally trying to equate the concept of linearity in the algebraic sense to that of linearity in an operational sense.

3.1.5   *Superposition*

Linearity actually implies two important constraints:

1. *Additivity:* A linear operation on the sum of two functions is equivalent to the sum of linear operations applied to each separately. This is the **additive property.**

2. *Homogeneity:* A linear operation on $Kx(t)$ is equivalent to K times the linear operation on $x(t)$. This is the **homogeneity property.**

Together, the two describe the **principle of superposition.** An important concept that forms the basis for the study of linear systems is that *the superposition of linear operators is also linear.*

3.2 System Classification

The classification of systems is based on notions such as linearity, time-invariance, causality and memory, as summarized in Table 3.1.

An nth-order system may be described by the differential equation

$$a_0 \frac{d^n y}{dt^n} + a_1 \frac{d^{n-1}y}{dt^{n-1}} + \cdots + a_{n-1}\frac{dy}{dt} + a_n y = b_0 \frac{d^m x}{dt^m} + b_1 \frac{d^{m-1}x}{dt^{m-1}} + \cdots + b_{m-1}\frac{dx}{dt} + b_m x$$

Table 3.1   *CT System Classification.*

Classification	Characteristics of System Equation
Linear	No constant terms (with respect to x or y). Coefficients may depend on t but not on x or y. Examples: $y''(t) + 3t^2 y(t) = x'(t)$, $y''(t) + 3y(t) = 2x(t)$
Nonlinear	Constant terms in addition to terms in x and y. Coefficients that depend on x or y. Examples: $y'(t) + 3t = x(t)$, $y''(t) + x'(t)y(t) = x(t)$
Time invariant	Coefficients may depend on x or y but not on t. Examples: $y'(t) + 3^{y(t)}y(t) = x(t)$, $y(t) = 2x^2(t-2)$
Time-varying	Coefficients are explicit functions of t. x and y have arguments of the form $x(2t)$, $y(e^t)$. Examples: $y'(t) + 3^t y(t) = x(t)$, $y''(t) + y(t) = x'(2t) + x(t)$
Causal	Present response does not depend on future values. Examples: $y''(t) = x(t)$, $y(t+2) = x^2(t+1)$, $y(t) = x(\frac{1}{2}t), t > 0$
Noncausal	Present response depends on future inputs/outputs. Examples: $y'(t) = x'(t+1)$, $y(t+2) = x(t+3), y(t) = x(2t)$
Dynamic (Memory)	System equation is a differential equation or variables x and y are of the form $x(2t-3), y(t+1)$. Examples: $y''(t) = x^2(t)$, $y(t) = 4x(2t)$, $y(t) = x(t+1)$
Instantaneous	System equation is algebraic. Variables occur as $x(t)$, $y(t)$ with argument t. Examples: $y(t) = x^2(t) - 2x(t), y(t) + 3 = 4x(t)$

Using the *derivative operator* $s^k \equiv d^k/dt^k$ with $s^0 \equiv 1$, we may recast this equation in **operator notation** as

$$[a_0 s^n + a_1 s^{n-1} + \cdots + a_{n-1} s + a_n]y = [b_0 s^m + b_1 s^{m-1} + \cdots + b_{m-1} s + b_m]x$$

The important class of **linear time-invariant** (LTI) systems are described by differential equations with *constant coefficients* (constant both with respect to the system variables x and y and with respect to t). *For nonlinear systems, the coefficients are functions of x or y while for time-varying systems, they are explicit functions of t.*

The differential equation for a relaxed LTI system may be expressed as

$$[s^n + a_1 s^{n-1} + \cdots + a_{n-1} s + a_n]y = [b_0 s^m + b_1 s^{m-1} + \cdots + b_{m-1} s + b_m]x$$

This, in turn, leads to its **operational transfer function** as

$$H(s) = \frac{P_m(s)}{Q_n(s)} = \frac{b_0 s^m + b_1 s^{m-1} + \cdots + b_{m-1} s + b_m}{s^n + a_1 s^{n-1} + \cdots + a_{n-1} s + a_n}$$

If $m < n, H(s)$ is called **strictly proper.** If $m \le n, H(s)$ is called **proper.** If $m > n$, $H(s)$ is called **improper.**

3.2.1   *Linear Systems*

A linear system is one for which superposition applies (Figure 3.2). Only if a system is both additive and homogeneous can it be termed linear.

A (single-input) linear system is actually subject to three constraints:

1. The system equation must involve only linear operators.
2. The system must contain no internal independent sources.
3. The system must be relaxed (with zero initial conditions).

Homogeneity implies that scaling the input leads to an identical scaling of the output. In particular, this means zero output for zero input and a linear input-output relation passing through the origin. This is possible only if each system element obeys a similar relationship at its own terminals. Independent sources have terminal characteristics that are constant and do not pass through the origin. A system that includes such sources is, therefore, nonlinear. Formally, a linear system must also be relaxed (with zero initial conditions) if superposition is to hold.

3.2.2   *Incrementally Linear Systems*

A system is called **incrementally linear** if *changes* in the input are related linearly to *changes* in the response. This allows us to use superposition even for a system with internal sources and nonzero initial conditions that is otherwise linear. We treat it as a multiple-input system by including the sources or initial conditions as additional inputs. The output then equals the superposition of the outputs due

Figure 3.2 Linearity and time invariance.

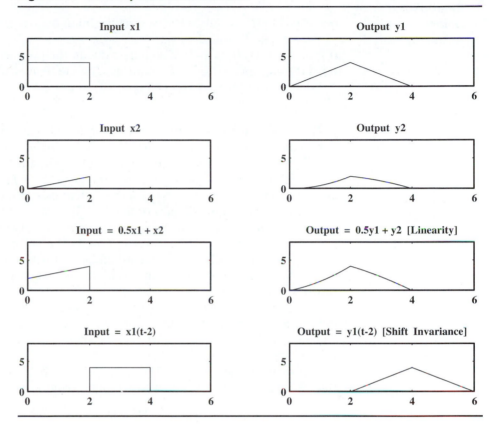

to each input acting alone. On this basis, we define a system to be (incrementally) linear if:

1. Its response can be written as a sum of the *zero-input response* (due to the initial conditions alone) and the *zero-state response* (due to the input alone). This is called the **principle of decomposition**.
2. The zero-input response obeys superposition.
3. The zero-state response obeys superposition.

3.2.3 Testing for Linearity

To test for linearity, we can use the linearity operation or examine the system equation. A nonlinear system has coefficients that are functions of x and y. Terms such as t^2, $y^2(t)$, or $a^{y(t)}$ make a system nonlinear, but terms such as $tx(t)$ or $\sin(t)y(t)$ do not. Constant terms also make a system nonlinear.

Example 3.3

Recognizing Linear and Nonlinear Systems

Here is how the following systems stack up:

(a) $y(t) = |x(t)|$ is nonlinear. Scaling the input by a negative factor will not result in a similarly scaled output.

(b) $y(t) = x(t)y(t) + x'(t)$ is nonlinear due to the product term $x(t)y(t)$.

(c) $y(t) = 2\sin(t)x(t) + x'(t)$ is linear. The coefficient of $x(t)$ is $2\sin(t)$, which does not depend on $x(t)$ or $y(t)$.

3.2.4 Time-Invariant (Shift-Invariant) Systems

In essence, time invariance implies that the shape of the response $y(t)$ depends only on the shape of the input $x(t)$ and not on the time when it is applied. If the input is shifted to $x(t - \alpha)$, the response equals $y(t - \alpha)$ and is shifted by the same amount (Figure 3.2). In other words, the system does not change with time. Such a system is also called **stationary** or **fixed.**

Every element of a time-invariant system must also be time invariant with a value that is constant *with respect to time.* If the value depends on the input or output, this only makes the system nonlinear. Coefficients in a system equation that depend on the element values cannot show explicit dependence on time for time-invariant systems.

In a time-varying system, the value of at least one system element is a function of time. In physical systems this could happen due to aging. As a result, the system equation contains time-dependent coefficients. An example is a system containing a time-varying resistor.

Formally, if the operator $\mathcal{O}$ transforms the input $x(t)$ to the output $y(t)$ such that $\mathcal{O}[x(t)] = y(t)$, a time-invariant system requires that

$$\mathcal{O}[x(t - t_0)] = y(t - t_0)$$

Remarks: **1.** A cascade of two time-invariant systems is also time invariant.

            2. Time-varying systems are not amenable to closed-form solutions except in the simplest of cases and often require a numerical solution.

3.2.5 Testing for Time Invariance

We can test for time invariance by applying the time-invariance relation. We can also examine the system equation directly. The coefficients of a time-varying system depend *explicitly* on t. Terms such as $e^{-t}x(t)$ make a system time-varying but $e^{y(t)}x(t)$ does not. Terms with *scaled* system variables such as $x(2t)$, $x(2t - 3)$ or $x(t^2)$, also make a system time-varying.

Example 3.4

Testing for Time Invariance

We examine the following systems for time invariance:

(a) $y(t) = tx(t)$. The operator that transforms $x(t)$ to $y(t)$ is $\mathcal{O}[\ ] = t[\ ]$. Applying this to $x(t - t_0)$ gives $\mathcal{O}[x(t - t_0)] = tx(t - t_0)$, but the system equation predicts $y(t - t_0) = (t - t_0)x(t - t_0)$. Thus, the system is time-varying.

(b) $y'(t) + y(t) = t \exp(-t)$: Time invariant. The input here is $x(t) = t \exp(-t)$. Only if the output is regarded as $\exp(-t)$ or t will it be time varying.

(c) $y(t) = x(2t + 1) + x'(t)$: Time-varying. The argument of $x(2t + 1)$ is scaled.

(d) $y'(t) = 2x'(t) + \sin[y(t)]x(t)$: Time invariant. The coefficient $\sin[y(t)]$ of $x(t)$ depends only on $y(t)$ and is not an explicit function of t. ___

3.2.6 *Implications of Linearity and Time Invariance*

In most practical cases, if an input $x(t)$ to a relaxed LTI system undergoes a linear operation, the output $y(t)$ undergoes the same linear operation. Thus

1. The input $dx(t)/dt$ results in the response $dy(t)/dt$.
2. The input $\int x\, dt$ results in the response $\int y\, dt$.

Relating the Step Response and the Impulse Response As a consequence of the above results, if a step input $u(t)$ leads to the step response $w(t)$, the impulse $\delta(t) = u'(t)$ leads to the impulse response $h(t) = w'(t)$. This result is the basis for the assertion that the impulse response $h(t)$ equals the derivative of the step response $w(t)$, a concept we use quite often in practice.

Superposition and System Analysis The superposition property is what makes the analysis of linear systems so much more tractable than nonlinear systems. Often, an arbitrary function may be decomposed into its simpler constituents, the response due to each analyzed separately and more effectively, and the total response found using superposition. This approach forms the basis for several methods of system analysis:

1. Representing a signal $x(t)$ as a weighted sum of impulses is the basis for the convolution method (Chapter 4).
2. Representing a periodic signal $x(t)$ as a linear combination of harmonic signals is the basis for the Fourier series (Chapter 5).
3. Representing a signal $x(t)$ as a weighted sum of complex exponentials is the basis for Fourier and Laplace transforms (Chapters 6 and 8).

3.2.7 *Causal and Noncausal Systems*

A **causal** or **nonanticipating** system is one whose present response does not depend on future values of the input. The system cannot anticipate future inputs in order to *generate or alter* the present response. Systems whose present response is affected by future inputs are termed **noncausal** or **anticipating.** A formal definition of causality requires identical inputs up to a given duration to produce identical responses over the same duration. Causality depends not only on the system details but also the specification of the input and response. A system may be causal for one input-output set and noncausal for another.

1. If a causal system is also linear (with no internal sources) and relaxed, there is no response until an input is applied.

2. Ideal systems (such as ideal filters) often turn out to be noncausal but also form the yardstick by which the performance of many practical systems, implemented to perform the same task, is usually assessed.

Example 3.5

Recognizing Causal and Noncausal Systems

We examine the following systems for causality:

(a) $y(t) = x(t) + 3$: A zero input leads to a nonzero response but the system is causal because the present response is not affected by future values.

(b) $y(t + 1) = x(t + 5)$: Noncausal. Output at $t = 0$ requires input at $t = 4$.

(c) $y(t + 7) = x(t + 5)$: Causal. Output at $t = 0$ requires input at $t = -2$.

(d) $y(t) = x^2(t)$: Causal.

(e) $y(t) = [x(t)]^{\frac{1}{2}}$: Noncausal. If $x_1(t) = x(t) = t^2$ for $t < t_0$, $y(t)$ could be either t or $-t$ for $t < t_0$. The response differs for two identical inputs.

3.2.8   *Instantaneous and Dynamic Systems*

A **dynamic system**, or a system with **memory**, is characterized by *differential* equations. Its present response depends on both present and past inputs. The memory or past history is due to energy storage elements that lead to the differential form.

In contrast, the response of resistive circuits or circuits operating in the steady state depends only on the instantaneous value of the input, not on past or future values. Such systems are also termed **instantaneous, memoryless** or **static.** All instantaneous systems are also causal.

The system equation of an instantaneous system is *algebraic* and x and y are of the form $x(t)$ and $y(t)$ with no scaled or shifted arguments.

Example 3.6

Recognizing Instantaneous and Dynamic Systems

We examine the following systems for memory:

(a) $y(t) = x'(t)$: Dynamic. A differential equation. Note that we cannot find the derivative $x'(t_0)$ from a knowledge of $x(t_0)$ alone.

(b) $y(t) = x(at)$: Instantaneous only if $a = 1$.

(c) $y(t) = x(t + 2)$: Dynamic, since x is of the form $x(t + 2)$, not $x(t)$.

(d) $y(t) = x(t) + 3$: Instantaneous and also causal, but nonlinear.

3.2.9   *Sundry Classifications*

Lumped and Distributed Systems   Electrical signals are just electromagnetic waves. Their propagation over a finite distance (between the two ends of an element, say) requires a finite time. The signal value is thus a function of spatial variables, in addition to time. The propagation time t_p to travel a distance d at a velocity v is simply $t_p = vd$. The signal velocity v equals c/λ, the ratio of its speed c through the propagation medium, and its wavelength λ. The propagation time thus equals cd/λ and is inversely proportional to the wavelength λ. If λ is much larger than the dimensions of the element through which the signal propagates, we may assume instantaneous propagation, and ignore the effects of the spatial distribution of the signal within the element. This assumption is the basis for the lumped parameter description of an element. A **lumped parameter** system includes only such lumped elements. It is described by ordinary differential equations with lumped (spatially independent) coefficients. If spatial dependence cannot be ignored, the system elements must be regarded as distributed in space and lead to a **distributed parameter** system. Partial differential equations are required for the description and analysis of distributed systems such as transmission lines.

Deterministic and Stochastic Systems   If we get the same response each time the same input is applied and such a response can be predicted by specific physical laws, the system is termed **deterministic.** The response of a **random** or **stochastic** system to the same input cannot be precisely predicted and must be described, on the average, in probabilistic terms, using statistical rather than physical laws.

3.3
Analysis of LTI Systems

LTI systems can be analyzed in the time domain using a system description based on any one of the three models listed in Table 3.2.

Table 3.2   System Analysis in the Time Domain.

System Model	Typical Restrictions	Comments and Applications
Differential equation	Very few	Useful only for low order.
State variables	Linear system for matrix formulation	Useful for complex systems and multiple-input/multiple-input systems and numerical solutions.
Impulse response	Relaxed LTI system	Amenable to numerical solution; provides link between time and frequency domains.

The Differential Equation Representation   This representation is quite general and applies even to nonlinear and time-varying systems. For LTI systems it also allows the computation of the response using superposition even in the presence of initial conditions. Two major disadvantages of this approach are that as the system order and complexity increases, both the formulation of the differential equations and the evaluation of the initial conditions become quite difficult.

The State Variable Representation   This representation describes an nth-order system by n simultaneous first-order equations, called **state equations,** in terms of n state variables. It is quite useful for complex or nonlinear systems and those with multiple inputs and outputs. For linear systems, the state equations can be solved using matrix methods.

The Impulse Response Representation   This representation normally applies to *relaxed* LTI systems (even though it can be extended to include the effect of initial conditions). It describes such a system in terms of its impulse response $h(t)$. Unlike differential or state representations, the system response $y(t)$ appears explicitly in the governing relation called the *convolution integral*. It also provides the formal ties between the time domain and the transformed domain that allow us to relate time-domain and transformed-domain methods of system analysis. We discuss this method in detail in Chapter 4.

3.4
LTI Systems Described by Differential Equations

Consider an LTI system described by the differential equation

$$[a_0 s^n + a_1 s^{n-1} + \cdots + a_{n-1}s + a_n]y(t) = [b_0 s^m + b_1 s^{m-1} + \cdots + b_{m-1}s + b_m]x(t)$$

The response $y(t)$ of such a system is a combination of:

1. The **zero-state response** $y_{zs}(t)$, which arises if the input is applied to a relaxed system (with zero initial conditions).
2. The **zero-input response** $y_{zi}(t)$, which results if the system is allowed to relax from its given state to the zero state in the absence of any applied input.

Initial Conditions   Differential equations for LTI systems are typically solved subject to specified initial conditions (IC). An nth-order equation requires n initial conditions. For a prescribed set of initial conditions, there is a *unique* solution for the output. This is the **uniqueness theorem**, which guarantees that if we are, by some means, able to find a solution to such an equation, then that is the only solution possible.

   If the initial conditions are specified at $t = t_0$, the resulting solution is then valid for $t \geq t_0$. The standard choice is $t = 0$ and the initial conditions are usually specified as the values of the response $y(t)$ and its $n - 1$ successive derivatives $y'(t), y''(t), \ldots$ at $t = 0$.

3.4.1   *The Method of Undetermined Coefficients*
Linear differential equations with constant coefficients are best solved by the method of undetermined coefficients. We evaluate the total response as the sum of the **natural response** $y_N(t)$ and the **forced response** $y_F(t)$. Note that $y_N(t)$ and $y_F(t)$ do not, in general, correspond to the zero-input and zero-state response, even though each pair adds up to the total response.

Remarks: 1. For stable systems, the natural response is also called the **transient response**, since it decays to zero with time.
2. For systems with harmonic or switched harmonic inputs, the forced response is a harmonic at the input frequency, and is termed the **steady-state response**.

3.4.2 The Natural Response

The form of the natural response depends only on the system details and is independent of the nature of the input. It comprises exponentials whose exponents are the roots (real or complex) of the so-called *characteristic equation* or *characteristic polynomial*. The **homogeneous differential equation** (whose right-hand side is zero) is given by

$$[a_0 s^n + a_1 s^{n-1} + a_2 s^{n-2} + \cdots + a_{n-2} s^2 + a_{n-1} s + a_n]y = 0$$

The **characteristic equation** is defined by

$$a_0 s^n + a_1 s^{n-1} + a_2 s^{n-2} + \cdots + a_{n-2} s^2 + a_{n-1} s + a_n = 0$$

Its n roots, s_k, $k = 1, 2, \ldots, n$, define the form of the natural response as

$$y_N(t) = K_1 \exp(s_1 t) + K_2 \exp(s_2 t) + \cdots + K_n \exp(s_n t)$$

Here, $K_1, K_2, \ldots, K_n$ are, as yet, undetermined constants. This form must be modified for multiple roots. Table 3.3 summarizes the various forms. For combinations, we use superposition. Since complex roots occur in conjugate pairs, their associated constants also form conjugate pairs to ensure that $y_N(t)$ is real. Algebric details lead to the preferred form with two real constants, as listed in Table 3.3.

Table 3.3 Form of the Natural Response for CT LTI Systems.

Root of Characteristic Equation	Form of Natural Response
Real and distinct root s_k	$C_k \exp(s_k t)$
Complex conjugate $\beta \pm j\omega$	$[C_1 \cos(\omega t) + C_2 \sin(\omega t)] \exp(\beta t)$
Real repeated root $(s_k)^p$	$(K_0 + K_1 t + K_2 t^2 + \cdots + K_p t^p) \exp(s_k t)$
Complex, repeated $(\beta \pm j\omega)^p$	$(C_0 + C_1 t + C_2 t^2 + \cdots + C_p t^p) \cos(\omega t) \exp(\beta t) +$
	$(D_0 + D_1 t + D_2 t^2 + \cdots + D_p t^p) \sin(\omega t) \exp(\beta t)$

Example 3.7
The Natural Response

(a) Consider the homogeneous differential equation $[s^2 + 3s + 2]y = 0$. The characteristic equation is $s^2 + 3s + 2 = 0$ with roots: $s_1 = -1, s_2 = -2$. The natural response is $y_N(t) = K_1 \exp(s_1 t) + K_2 \exp(s_2 t) = K_1 \exp(-t) + K_2 \exp(-2t)$.

(b) Consider the homogeneous equation $[(s + 4)(s^2 + 4s + 5)]y = 0$.

Its characteristic equation has the roots: $s_1 = -4, s_2 = -2 + j, s_3 = -2 - j$. The natural response may therefore be expressed as:

$$y_N(t) = K_1 \exp(-4t) + \exp(-2t)[A_1 \cos(t) + B_1 \sin(t)], \qquad t \geq 0$$

3.4.3 The Forced Response

The **forced response** arises due to the interaction of the system with the input and thus depends on both the input and the system details. It can be found uniquely from the actual system differential equation, independent of the natural response or initial conditions. The forced response has the same general form as the **forcing function,** the right-hand side of the differential equation, which is, of course, a function of the input.

Table 3.4 summarizes the form of the forced response for various forcing functions. For combinations, we simply use superposition. The constants in the forced response are found by satisfying the actual, not homogeneous, differential equation.

Table 3.4   Form of the Forced Response for *CT* LTI Systems.
Note: If the RHS involves the form $\exp(-\alpha t)$ where α is also a root of the characteristic equation repeated r times, the forced response form must be multiplied by t^r.

Forcing Function (RHS)	Form of Forced Response
C (constant)	C_1 (constant)
$\exp(-\alpha t), \quad \alpha \neq$ root of characteristic equation	$K\exp(-\alpha t)$
$\cos(\omega t + \beta)$	$K_1\cos(\omega t) + K_2\sin(\omega t)$ or $K\cos(\omega t + \theta)$
$\exp(-\alpha t)\cos(\omega t + \beta)$	$[K_1\cos(\omega t) + K_2\sin(\omega t)]\exp(-\alpha t)$
t	$K_0 + K_1 t$ (general first-order polynomial)
t^p	$K_0 + K_1 t + K_2 t^2 + \cdots + K_p t^p$
$t\exp(-\alpha t)$	$(K_0 + K_1 t)\exp(-\alpha t)$
$t^p\exp(-\alpha t)$	$(K_0 + K_1 t + K_2 t^2 + \cdots + K_p t^p)\exp(-\alpha t)$
$t\cos(\omega t + \beta)$	$(K_1 + K_2 t)\cos(\omega t) + (K_3 + K_4 t)\sin(\omega t)$

The Total Response   The **total response** is found by *first* adding the forced and natural response and then evaluating the undetermined constants (in the natural component) using the prescribed initial conditions.

Example 3.8
Natural and Forced
Responses

(a) Consider a system governed by $[s^2 + 3s + 2]y = 4\exp(-3t)$ subject to the initial conditions $y(0) = 3$, $y'(0) = 4$. Its natural response equals

$$y_N(t) = K_1\exp(s_1 t) + K_2\exp(s_2 t) = K_1\exp(-t) + K_2\exp(-2t)$$

From Table 3.4, for the forced response we select $y_F(t) = K\exp(-3t)$.
To find K, we satisfy $[s^2 + 3s + 2]y_F = 4\exp(-3t)$.
Since $[s]y_F = -3K\exp(-3t)$, $[s^2]y_F = 9K\exp(-3t)$, we obtain
$[9K - 9K + 2K]\exp(-3t) = 4\exp(-3t)$ and $K = 2$. Thus $y_F(t) = 2\exp(-3t)$.
Then $y(t) = y_N(t) + y_F(t) = K_1\exp(-t) + K_2\exp(-2t) + 2\exp(-3t)$.
To find K_1 and K_2, we invoke initial conditions on $y(t)$ to obtain
$y(0) = K_1 + K_2 + 2 = 3$, $y'(0) = -K_1 - 2K_2 - 6 = 4$.
This gives $K_2 = -11$, $K_1 = 12$ and leads to

$$y(t) = 12\exp(-t) - 11\exp(-2t) + 2\exp(-3t), \qquad t \geq 0$$

Comment: As a check, we confirm that $y(0) = 3$ and $y'(0) = -12 + 22 - 6 = 4$.

Comment: The constants K_1 and K_2 must be found from the *total response.*

(b) Consider a system governed by $[s^2 + 3s + 2]y = 4\exp(-2t)$ with initial conditions $y(0) = 3, y'(0) = 4$. The natural response is $y_N(t) = K_1\exp(-t) + K_2\exp(-2t)$ (see Example 3.7a). Since the forcing function $4\exp(-2t)$ has the same exponent as a root of the characteristic equation, we choose $y_F(t) = Kt\exp(-2t)$.

To find K, we satisfy $[s^2 + 3s + 2]y_F = 4\exp(-2t)$. Now
$[s]y_F = -2Kt\exp(-2t) + K\exp(-2t),$
$[s^2]y_F = -2K(1 - 2t)\exp(-2t) - 2K\exp(-2t).$
Thus $[-2K + 4Kt - 2K - 6Kt + 3K + 2Kt]\exp(-2t) = 4\exp(-2t)$ and $K = -4$.
We then obtain $y_F(t) = -4\exp(-2t)$ and $y(t) = y_N(t) + y_F(t) = K_1\exp(-t) + K_2\exp(-2t) - 4t\exp(-2t)$. Using initial conditions, we get $y(0) = K_1 + K_2 = 3$, $y'(0) = -K_1 - 2K_2 - 4 = 4$. Thus $K_2 = -11, K_1 = 14$ and
$y(t) = 14\exp(-t) - 11\exp(-2t) - 4t\exp(-2t), \qquad t \geq 0.$

Comment: As a check, we confirm that $y(0) = 3$ and $y'(0) = -14 + 22 - 4 = 4$.

3.4.4 *The Zero-State and Zero-Input Response*
The zero-input response requires only the homogeneous equation since the input is assumed to be zero. It equals the natural response $y_N(t)$ subject to the *prescribed initial conditions.* Similarly, the zero-state response $y_{zs}(t)$ is the complete response, *assuming zero initial conditions.*

Example 3.9
Zero-Input and
Zero-State Response

Let $[s^2 + 3s + 2]y = 4\exp(-3t)$ with $y(0) = 3, y'(0) = 4$. Its natural response is

$$y_N(t) = K_1\exp(s_1t) + K_2\exp(s_2t) = K_1\exp(-t) + K_2\exp(-2t)$$

The zero-input response is found from $y_N(t)$ and the prescribed IC:

$$y_{zi}(t) = K_1\exp(-t) + K_2\exp(-2t), y_{zi}(0) = K_1 + K_2 = 3, y'_{zi}(0) = -K_1 - 2K_2 = 4$$

This yields $K_2 = -7, K_1 = 10$ and $y_{zi}(t) = 10\exp(-t) - 7\exp(-2t)$.

Similarly, $y_{zs}(t)$ is found from the general form of $y(t)$ but with zero IC. We start with $y_{zs}(t) = K_1\exp(-t) + K_2\exp(-2t) + 2\exp(-3t)$ and obtain

$$y_{zs}(0) = K_1 + K_2 + 2 = 0, \qquad y'_{zs}(0) = -K_1 - 2K_2 - 6 = 0$$

This yields $K_2 = -4, K_1 = 2$ and $y_{zs}(t) = 2\exp(-t) - 4\exp(-2t) + 2\exp(-3t)$. The total response, as expected, is the sum

$$y(t) = y_{zi}(t) + y_{zs}(t) = 12\exp(-t) - 11\exp(-2t) + 2\exp(-3t), \qquad t \geq 0$$

3.4.5   *Using Linearity and Superposition in System Analysis*

Here is how we find the response of the general system described by

$$[a_0 s^n + a_1 s^{n-1} + \cdots + a_{n-1} s + a_n] y(t) = [b_0 s^m + b_1 s^{m-1} + \cdots + b_{m-1} s + b_m] x(t)$$

1. First, compute the zero-state response $y_0(t)$ to

$$[a_0 s^n + a_1 s^{n-1} + \cdots + a_{n-1} s + a_n] y_0(t) = x(t)$$

2. Next, use linearity and superposition to find $y_{zs}(t)$ as

$$y_{zs}(t) = [b_0 s^m + b_1 s^{m-1} + \cdots + b_{m-2} s^2 + b_{m-1} s + b_m] y_0(t)$$

3. Find the zero-input response $y_{zi}(t)$.
4. Evaluate the total response as $y(t) = y_{zs}(t) + y_{zi}(t)$.

Note that the zero-state response requires superposition, but the *zero-input response needs to be computed and included just once.*

Example 3.10
Using Concepts Based on Linearity

Consider the system $[s^2 + 3s + 2]y = [2s + 1]x$, with $x(t) = 4\exp(-3t)$, and initial conditions $y(0) = 0, y'(0) = 1$.

1. From Example 3.8, the relaxed system $[s^2 + 3s + 2]y = x$ has the response

$$y_0(t) = K_1 \exp(-t) + K_2 \exp(-2t) + 2\exp(-3t)$$

Using zero IC, $K_1 + K_2 + 2 = 0$ and $-K_1 - 2K_2 - 6 = 0$ and $K_1 = 2, K_2 = -4$. Thus $y_0(t) = 2\exp(-t) - 4\exp(-2t) + 2\exp(-3t)$.
2. Since $[s]y_0 = -2\exp(-t) + 8\exp(-2t) - 6\exp(-3t)$, we get

$$y_{zs}(t) = [2s + 1]y_0 = -2\exp(-t) + 12\exp(-2t) - 10\exp(-3t), \qquad t \geq 0$$

3. $y_{zi}(t) = C_1 \exp(-t) + C_2 \exp(-2t)$ with initial conditions $y(0) = 0, y'(0) = 1$. This yields $C_1 + C_2 = 0$, and $-C_1 - 2C_2 = 1$. Solving these, we find $C_1 = 1, C = -1$ and thus $y_{zi}(t) = \exp(-t) - \exp(-2t)$.
4. Finally, $y(t) = y_{zs}(t) + y_{zi}(t) = -\exp(-t) + 11\exp(-2t) - 10\exp(-3t), \quad t \geq 0$.

3.5
LTI Systems Described by the Impulse Response

The **impulse response,** denoted by $h(t)$, is simply the response of a relaxed, linear system to a unit impulse $\delta(t)$. It provides us with a basis for evaluating the system response to any arbitrary input (using the convolution method of Chapter 4). The impulse response is also related to the *transfer function*, which is used extensively for system analysis in the transformed domain (Chapters 5–8). It is often easier to find the impulse response of a system by resorting to the transformed domain. Here we consider approaches that rely only on time-domain methods and on the concepts of linearity.

Direct methods for finding the impulse response in the time domain are often tedious. Here is why. Since $\delta(t) = 0, t > 0$, we must apparently solve a homogeneous differential equation (zero input) subject to zero initial conditions (relaxed system), meaning zero response. (We shall see later that the impulse input actually forces the initial conditions to change.)

Before examining the direct approach, we present two indirect methods.

3.5.1   *Impulse Response from the Step Response*

The impulse response $h(t)$ is the derivative of the step response $w(t)$. This suggests a simple way of finding $h(t)$ for a relaxed system. We evaluate the step response $w(t)$ and then let $h(t) = w'(t)$.

Example 3.11
Impulse Response from
the Step Response

Consider a relaxed lowpass RC filter (Figure 3.3) excited by a unit step input. The differential equation for the step response $w(t)$ is found by writing a node equation at node A, to give $w'(t) + w(t)/\tau = 1/\tau$, where $\tau = RC$.

Figure 3.3   A lowpass RC filter and its response to a rectangular pulse and exponential input.

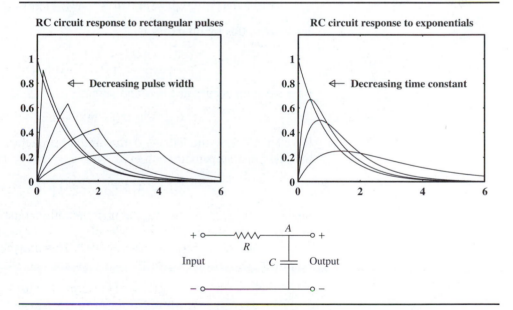

The characteristic equation $(s + 1/\tau) = 0$ yields the natural response as

$$w_n(t) = K \exp(-t/\tau), \qquad t \geq 0$$

Since the input is constant for $t \geq 0$, we assume the forced response to be $w_F(t) = B$. The governing differential equation yields $0 + B/\tau = 1/\tau$ or $B = 1$.

The total response equals $w(t) = w_f(t) + w_n(t) = 1 + K \exp(-t/\tau), t \geq 0$. Since $w(0) = 0$, we get $0 = 1 + K$ or $K = -1$. The step response is thus

$$w(t) = 1 - \exp(-\frac{t}{\tau}), \quad t > 0 \quad \text{or} \quad w(t) = [1 - \exp(-\frac{t}{\tau})]u(t)$$

The impulse response $h(t)$ equals the derivative of the step response. Thus

$$h(t) = w'(t) = (\frac{1}{\tau}) \exp(\frac{-t}{\tau}), \quad t \geq 0 \quad \text{or} \quad h(t) = (\frac{1}{\tau}) \exp(-\frac{t}{\tau})u(t)$$

3.5.2   *Impulse Response from Limiting Forms of the Impulse*

Conceptually, if an input $x(t)$ approaches a unit impulse $\delta(t)$, the response $y(t)$ approaches the impulse response $h(t)$. How we use this concept is illustrated in the following example.

Example 3.12

The Concept of the Impulse Response

(a) From Example 3.11, the step response $w(t)$ of the RC filter of Figure 3.3 equals $w(t) = [1 - \exp(-t/\tau)]u(t)$. Its response to $v_i(t) = Au(t) - Au(t - t_0)$, a pulse of width t_0 and height A, is found using linearity and superposition to give

$$v_0(t) = A\{1 - \exp(-t/\tau)\}u(t) - A\{1 - \exp[-(t - t_0)/\tau]\}u(t - t_0)$$

We rewrite this by intervals as

$$v_0(t) = \begin{cases} A[1 - \exp(-t/\tau)] & t \leq t_0 \\ -A\exp(-t/\tau)\} + A\exp[-(t - t_0)/\tau] & t \geq t_0 \end{cases}$$

Its maximum occurs at $t = t_0$ and equals

$$v_{max} = v_0(t_0) = A[1 - \exp(-t_0/\tau)]$$

If we let $A = 1/t_0$ and let $t_0 \to 0$, the input approaches $\delta(t)$. If $t_0/\tau \ll 1$, we may use the linear approximation $\exp(\pm x) = 1 \pm x, (x \ll 1)$ to obtain

$$v_{max} = A\left[1 - \left(1 - \frac{t_0}{\tau}\right)\right] = \frac{At_0}{\tau} = \frac{1}{\tau}$$

Since $t_0 \approx 0$, $v_0(t)$ reaches v_{max} almost instantaneously. The response for $t \geq t_0$ equals $v_0(t) = -A\exp(-t/\tau) + A\exp[-(t - t_0)/\tau] = (1/\tau)\exp(-t/\tau)$ and effectively describes the response for $t \geq 0$. This may be regarded as the impulse response $h(t)$ and we have

$$h(t) = (1/\tau)\exp(-t/\tau)u(t)$$

(b) If the input is $(A/t_0)\exp(-t/t_0)u(t)$ instead, the response equals

$$v_0(t) = \frac{A}{t_0 - t}[\exp(-t/t_0) - \exp(-t/\tau)]u(t)$$

With $A = 1$ and $t_0 \ll \tau$, as $t_0 \to 0$, the input approaches $\delta(t)$ and $v_0(t)$ also reduces to $v_0(t) = h(t) = (1/\tau)\exp(-t/\tau)u(t)$ (Figure 3.3).

Comment: Any tall, narrow spike may thus be used to estimate $h(t)$.

3.5.3 *Impulses and Sudden Changes in State*

Example 3.12 reveals a key concept. The impulse response $h(t)$ of the system

$$v_c'(t) + \left(\frac{1}{\tau}\right) v_c(t) = \frac{v_1(t)}{\tau}$$

where $v_1(t) = \delta(t)$, is entirely equivalent to the output of a source-free (zero-input) system, but subject to the nonzero initial condition $v_0(0) = 1/\tau$. In other words, *the impulse results in a sudden change in the initial conditions (state)* and then disappears.

To formalize this result, consider the relaxed first-order system

$$y'(t) + a_1 y(t) = x(t)$$

The response to $x(t) = \delta(t)$ is then equivalent to the response of a **source-free** (zero-input) system governed by

$$y'(t) + a_1 y(t) = 0 \quad \text{with} \quad y(0) = 1$$

3.5.4 *An Operational Method for Finding the Impulse Response*

The results for the first-order system can be extended to the nth-order system

$$[s^n + a_1 s^{n-1} + \cdots + a_{n-1} s + a_n] y = x$$

Its impulse response $h(t)$ equals the response of the source-free system

$$[s^n + a_1 s^{n-1} + \cdots + a_{n-1} s + a_n] y = 0$$

with the *highest-order* initial condition $y^{(n-1)}(0) = 1$ and all others zero. Now, consider the general system described by

$$[s^n + a_1 s^{n-1} + \cdots + a_{n-1} s + a_n] y = [b_0 s^m + b_1 s^{m-1} + \cdots + b_{m-1} s + b_m] x$$

Recall that its operational transfer function may be written as

$$H(s) = \frac{P_m(s)}{Q_n(s)} = \frac{b_0 s^m + b_1 s^{m-1} + \cdots + b_{m-1} s + b_m}{s^n + a_1 s^{n-1} + \cdots + a_{n-1} s + a_n}$$

If $H(s)$ is *strictly proper* $(m < n)$, we find $h(t)$ by a two-step approach:

1. Compute $h_1(t)$ for the system $[s^n + a_1 s^{n-1} + \cdots + a_{n-1} s + a_n] y = x$.
2. Invoke linearity to obtain $h(t) = [b_0 s^m + b_1 s^{m-1} + \cdots + b_{m-1} s + b_m] h_1$.

If $m \geq n, h(t)$ will contain impulses and its derivatives. In this case, we use long division to deflate $P_m(s)$ to a polynomial $P_r(s)$ whose degree r is less than n. With $m - n = p$, we get

$$H(s) = \frac{P_m(s)}{Q_n(s)} = c_p s^p + \cdots + c_1 s + c_0 + \frac{P_r(s)}{Q_n(s)} = c_p s^p + \cdots + c_1 s + c_0 + H_r(s)$$

Since $H_r(s)$ is now strictly proper, the impulse response $h(t)$ equals the sum of the impulse response $h_r(t)$ and the impulsive terms and has the form

$$h(t) = h_r(t) + c_0\delta(t) + c_1\delta^{(1)}(t) + \cdots + c_p\delta^{(p)}(t)$$

Example 3.13
Operational Method for Finding $h(t)$

(a) Consider a system described by $[s^2 + 3s + 2]y = x$. The characteristic equation $s^2 + 3s + 2 = 0$ has the roots $s_1 = -1$, $s_2 = -2$. This leads to the solution $h_1(t) = K_1\exp(-t)h_1'(0) + K_2\exp(-2t)$. Using $h_1(0) = 0$ and $h_1'(0) = 1$, we have $h_1(0) = K_1 + K_2 = 0$, $h_1'(0) = -K_1 - 2K_2 = 1$. This set of equations leads to $K_1 = 1$, $K_2 = -1$ and $h(t) = [\exp(-t) - \exp(-2t)]u(t)$.

(b) Let $[s^2 + 3s + 2]y = [s^2]x$. Since this is not strictly proper, we set up

$$H(s) = \frac{P(s)}{Q(s)} = \frac{s^2}{s^2 + 3s + 2} = 1 + \frac{-3s - 2}{s^2 + 3s + 2} = 1 + H_r(s)$$

From part (a), with $[s^2 + 3s + 2]y = x$, $\quad h_1(t) = \exp(-t) - \exp(-2t)$.
The response $h_r(t)$ of $[s^2 + 3s + 2]y = [-3s - 2]x$ is

$$h_r(t) = [-3s - 2]h_1 = 3[\exp(-t) - 2\exp(-2t)] - 2[\exp(-t) - \exp(-2t)]$$

This simplifies to $h_r(t) = \exp(-t) - 4\exp(-2t)$ and we finally obtain

$$h(t) = (1)\delta(t) + h_r(t) = \delta(t) + [\exp(-t) - 4\exp(-2t)]u(t) \quad \rule[0.5ex]{1.2em}{0.15ex}$$

3.5.5   *Impulse Response by Impulse Matching*

The operational method suggests an alternative formulation based on the so-called concept of **impulse matching.** For a linear system governed by

$$[a_0s^n + a_1s^{n-1} + \cdots + a_{n-1}s + a_n]h = [b_0s^m + b_1s^{m-1} + \cdots + b_{m-1}s + b_m]\delta$$

$h(t)$ may be expressed in terms of its natural response as

$$h(t) = h_n(t)u(t) + \sum_{i=0}^{p}\alpha_i\delta^{(i)}(t)$$

If $m < n$, $h(t)$ equals just the natural response $h_n(t)u(t)$ and contains no impulses. For $m \geq n$, the difference $m - n = p$ establishes the number $p + 1$ of impulsive terms in $h(t)$. To evaluate $h(t)$ in this case, we use a two step-process:

1. Find the initial conditions $h_n(0)$, $h_n'(0)$, ... and the constants α_i using *impulse matching.* We match the coefficients of terms containing impulses and their derivatives on either side of the equation.
2. Use the initial conditions to find the undetermined constants in the *natural response* $h_n(t)$ (and not $h(t)$, which includes impulses).

The process is best illustrated by an example.

(a) Consider the differential equation $i''(t) + 3i'(t) + 2i(t) = \frac{1}{2}\delta'(t)$. Since $m < n$, $i(t) = i_n(t)u(t)$. Using properties of the impulse, we get

$$i' = i'_n(t)u(t) + i_n(t)\delta(t) = i'_n(t)u(t) + i_n(0)\delta(t)$$
$$i'' = i''_n(t)u(t) + i'_n(t)\delta(t) + i_n(0)\delta'(t) = i''_n(t)u(t) + i'_n(0)\delta(t) + i_n(0)\delta'(t)$$

Substituting for $i(t)$ and collecting coefficients of the impulsive terms,

$$[i''_n + 3i'_n + 2i_n]u(t) + [i'_n(0) + 3i_n(0)]\delta(t) + i_n(0)\delta'(t) = \frac{1}{2}\delta'(t)$$

On comparing coefficients of δ and δ' on either side, we obtain

$$i_n(0) = \frac{1}{2} \qquad i'_n(0) + 3i_n(0) = 0 \qquad \text{or} \qquad i'_n(0) = -\frac{3}{2}$$

The natural response is found by solving the homogeneous equation

$$i''(t) + 3i'(t) + 2i(t) = 0$$

This yields $i_n(t) = K_1 \exp(-t) + K_2 \exp(-2t)$, $t \geq 0$. Since $i_n(0) = K_1 + K_2 = \frac{1}{2}$ and $i'_n(0) = -K_1 - 2K_2 = -\frac{3}{2}$, we get $K_1 = -\frac{1}{2}$ and $K_2 = 1$. Thus $i_n(t) = i(t) = h(t) = -(\frac{1}{2})\exp(-t)u(t) + \exp(-2t)u(t)$. This establishes the impulse response if $i(t)$ is regarded as the output.

(b) Consider the differential equation $v''(t) + 3v'(t) + 2v(t) = \delta''(t)$. Since $m = n = 2$, the impulse response must now be described by

$$v(t) = v_n(t)u(t) + \alpha_0\delta^{(0)}(t) = v_n(t)u(t) + \alpha_0\delta(t)$$

Using properties of the impulse, the various derivatives of $v(t)$ yield

$$v' = v'_n(t)u(t) + v_n(0)\delta(t) + \alpha_0\delta'(t)$$
$$v'' = v''_n(t)u(t) + v'_n(0)\delta(t) + v_n(0)\delta'(t) + \alpha_0\delta''(t)$$

Substituting for $v(t)$ and collecting coefficients of the impulsive terms gives

$$[v''_n + 3v'_n + 2v_n]u(t) + [v'_n(0) + 3v_n(0) + 2\alpha_0]\delta + [v_n(0) + 3\alpha_0]\delta' + \alpha_0\delta'' = \delta''$$

Comparing coefficients of δ, δ', and δ'' on either side, we obtain

$$\alpha_0 = 1 \qquad v_n(0) + 3\alpha_0 = 0 \qquad v'_n(0) + v_n(0)3 + 2\alpha_0 = 0$$

These yield the three unknowns as $\alpha_0 = 1$, $v_n(0) = -3$, $v'_n(0) = 7$. The natural response is $v_n(t)u(t) = K_1 \exp(-t)u(t) + K_2 \exp(-2t)u(t)$. With $v_n(0) = -3 = K_1 + K_2$ and $v_n(0) = 7 = -K_1 - 2K_2$, we get $K_1 = 1$, $K_2 = -4$. Finally, $h(t) = v(t) = v_n(t)u(t) + \alpha_0\delta(t) = [\exp(-t) - 4\exp(-2t)]u(t) + \delta(t)$.

3.6 System Stability

Stability is important in practical design. It is defined in many contexts. In the time domain, stability involves constraints on the nature of the system response. **Bounded-input, bounded-output** (BIBO) **stability** implies that *every* bounded input results in a bounded output. For LTI systems described by differential equations in which the highest derivative of the input never exceeds that of the output (meaning a *proper* operational transfer function), Tables 3.3 and 3.4 lead to the following necessary and sufficient condition for BIBO stability:

> *For an LTI system to be stable, every root of its characteristic equation must have a negative real part.*

Roots with negative real parts ensure that the natural and zero-input response always decay with time (see Table 3.3), and the forced and zero-state response always remain bounded for *every* bounded input.

What if the real parts of the roots are zero? Simple (nonrepeated) roots with zero real parts produce a constant or sinusoidal (bounded) natural response. If the input is also constant or a sinusoid at the same frequency, the forced response is a ramp or growing sinusoid (unbounded). Repeated roots with zero real parts result in a natural response that is itself a growing sinusoid or polynomial (unbounded).

Remarks:
1. If the highest derivative of the input exceeds that of the output (i.e., the operational transfer function is not proper), the system is unstable. For example, if $y(t) = dx(t)/dt$, a step input (bounded) produces an impulse (unbounded at $t = 0$).
2. The stability condition is equivalent to having an impulse response $h(t)$ which is absolutely integrable. We develop this concept this in Chapter 4.
3. The stability of nonlinear or time-varying systems must usually be checked by other means.

Example 3.15
Concepts Based on Stability

(a) The system $[s^3 + 3s + 2]y = x$ is stable since the roots of its characteristic equation are $s = -1, -2$.
(b) The system $[s^2 + 3s]y = x$ with roots $s = 0, -3$ is unstable since one of the roots does not have a negative real part. Its natural response has the form $y_N(t) = Au(t) + B\exp(-3t)u(t)$ and is bounded. But for an input $x(t) = u(t)$, the forced response has the form $Ctu(t)$, which becomes unbounded.
(c) The system $[s^3 + 3s^2]y = x$ is unstable. The two roots at $s = 0$ produce a natural response $y_N(t) = Au(t) + Btu(t) + C\exp(-3t)u(t)$, which is unbounded.

3.7
Digital Systems

CT systems are usually characterized by differential equations where the system variables represent continuous functions of time. Systems for which both the input and output represent discrete-time or digital signals are referred to as **digital systems** or **discrete-time systems** and are typically described by *difference* equations. Digital computers are prime examples of such systems. In digital systems, changes in the variables can occur only at selected time instants. We also have *hybrid* systems such as digital to analog converters (DAC) that convert digital inputs to analog outputs.

3.7.1   DT Operators

An operator allows us to transform one function to another. In general, if an operator is represented by the symbol $\mathcal{O}$, the equation

$$\mathcal{O}\{x[n]\} = y[n]$$

implies that if the function $x[n]$ is treated exactly as the operator $\mathcal{O}$ tells us, we obtain the function $y[n]$. The **backward shift operator,** denoted by z^{-1}, transforms $x[n]$ to $x[n-1]$ and we then express this transformation in operator notation as $z^{-1}\{x[n]\} = x[n-1] = y[n]$.

Example 3.16

The Concept of an Operator

The operation $\mathcal{O}[\ ] = 4z^3[\ ] + 6$ says that to get $y[n]$, we must shift $x[n]$ three units, multiply by 4 and then add 6 to the result:

$$4z^3\{x[n]\} + 6 = 4x[n+3] + 6 = y[n]$$

3.7.2   Sums, Differences and Difference Equations

In Chapter 2 we defined the discrete-time equivalent of the integral $y(t)$ over $(0, t)$ of the function $x(t)$ as the summation

$$y[nt_s] \approx t_s \sum_{k=0}^{n} x[kt_s] \qquad y[n] = \sum_{k=0}^{n} x[k]$$

This representation is based on a *step interpolation* algorithm that finds $y[k]$ from $y[k-1]$ by adding the area $t_s x(kt_s)$ of the rectangular strip to give the result $y[k] = y[k-1] + t_s x[k]$.

The derivative of a continuous function $x(t)$ at some instant may be regarded as the slope of $x(t)$ at that instant. The **forward Euler algorithm** numerically approximates the slope in terms of the present and next value (Figure 3.4). The more commonly used **backward Euler algorithm** approximates the slope in terms of the present and previous value (Figure 3.4). It is an *implicit* method, requiring no future values, and defines the slope as

$$y_1(t) = x'(t) \approx [x(t) - x(t - \Delta t)]/\Delta t$$

For a sampled signal, Δt equals the sampling interval t_s, and we have

$$y_1[nt_s] = x'[nt_s] \approx \{x[nt_s] - x[(n-1)t_s]\}/t_s$$

Similarly, the second derivative yields

$$x''(t) \approx [y_1(t) - y_1(t - t_s)]/t_s = \{x[nt_s] - 2x[(n-1)t_s] + x[(n-2)t_s]\}/t_s^2$$

For DT signals, since $t_s = 1$, the first and second derivatives result in the **first difference** denoted by $\nabla(y)$ and the **second difference** denoted by $\nabla^2(y)$. Using the shift operator z^{-1}, these *backward* differences may be expressed as

$$\nabla(y) = y_1[n] = x[n] - x[n-1] = (1 - z^{-1})\{x[n]\}$$
$$\nabla^2(y) = \nabla[n] - \nabla[n-1] = x[n] - 2x[n-1] + x[n-2] = (1 - z^{-1})^2\{x[n]\}$$

Figure 3.4   Approximating the derivative by the backward and forward difference.

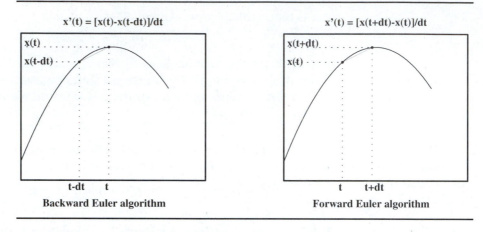

By induction, the mth **backward difference** may be written in operator form as

$$\nabla^m(y) = (1 - z^{-1})^m \{x[n]\}$$

Remark: For the *forward difference algorithm,* we have $y_1[n] = x[n + 1] - x[n]$. In operator form, $y_m[n] = (z - 1)\{x[n]\}$. Successive application of this rule leads to the mth **forward difference** $y_m[n] = (z - 1)^m \{x[n]\}$.

Conversion of an Nth-order differential equation using backward differences leads to the general form of an Nth-order *difference equation*:

$$A_0y[n] + A_1y[n - 1] + A_2y[n - 2] + \cdots + A_Ny[n - N]$$
$$= B_0x[n] + B_1x[n - 1] + \cdots + B_Mx[n - M]$$

It is customary to normalize A_0 to unity. The **order** N describes the highest difference $y[n - N]$ in the output. The normalized operator form then becomes

$$[1 + A_1z^{-1} + A_2z^{-2} + \cdots + A_Nz^{-N}]y = [B_0 + B_1z^{-1} + B_2z^{-2} + \cdots + B_Mz^{-M}]x$$

3.7.3   *DT Linear Operators and Superposition*
A *DT* operator $\mathcal{L}$ is termed a *DT* **linear operator** if and only if

$$\mathcal{L}\{ax_1[n] + bx_2[n]\} = a\mathcal{L}\{x_1[n]\} + b\mathcal{L}\{x_2[n]\}$$

Here, a and b are scalar constants. If an operation performed on a linear combination of x_1 and x_2 produces the same results as a linear combination of operations on x_1 and x_2 separately, the operation is linear. If not, it is nonlinear. As with *CT* systems, linearity implies *superposition,* and the superposition of linear operators is also linear.

Example 3.17
Testing for Linear
Operators

(a) Consider the operator $\mathcal{O}[x] = C[x] + D$. Applying the operation to the left-hand side (LHS) of the linearity relation, we obtain

$$\mathcal{O}[ax_1 + bx_2] = C[ax_1 + bx_2] + D$$

With the same operation applied to the right-hand side (RHS), we get

$$a\mathcal{O}[x_1] + b\mathcal{O}[x_2] = a[Cx_1 + D] + b[Cx_2 + D] = C[ax_1 + bx_2] + D[a + b]$$

The two clearly differ, so the operation $C[x] + D$ is nonlinear.

(b) Consider the operator $\mathcal{O}[\ ] = n[\ ]$, which transforms $x[n]$ to $nx[n]$. Using the operation on the left-hand side of the linearity relation yields

$$\mathcal{O}[ax_1 + bx_2] = n[ax_1 + bx_2] = anx_1[n] + bnx_2[n]$$

With the same operation applied to the right-hand side, we get

$$a\mathcal{O}[x_1] + b\mathcal{O}[x_2] = a[nx_1] + b[nx_2] = anx_1[n] + bnx_2[n]$$

The two are identical and make the operation $n[\ ]$ a linear operation.

3.8
Classification of
DT Systems

The classification of *DT* systems exactly parallels that of *CT* systems and is summarized in Table 3.5 (p. 96).

The important class of *linear time-invariant* (LTI) systems are described by difference equations with *constant coefficients* (constant with respect to x and y and n). *For nonlinear systems, the coefficients are functions of x or y while for time-varying systems, they are explicit functions of n.*

3.8.1   *Linear Systems*

A linear system satisfies all of the following constraints:

(a) Its response can be written as a sum of the *zero-input response* (due to the initial conditions alone) and the *zero-state response* (due to the input alone). This is called the **principle of decomposition**.

(b) The zero-input response obeys superposition.

(c) The zero-state response obeys superposition.

We can test a system for linearity by using the linearity relation or examining the system equation. For a linear system, the coefficients may be explicit functions of n, but not of x and y. Terms such as $y^2[n]$, $x[n]y[n-1]$ and $a^{y[n]}x[n]$ make a system nonlinear, but terms such as $n^2x[n]$ and $\sin[n]y[n]$ do not.

Example 3.18
Linear and Nonlinear
Systems

We examine the following systems for linearity:

(a) $y[n] = x^2[n]$ is clearly nonlinear. The coefficient of $x[n]$ is $x[n]$.

(b) $y[n] = 3^n + 2x[n]$ is nonlinear since 3^n is constant (with respect to x and y).

Table 3.5   Classification of *DT* Systems.

Classification	Characteristics of System Equation
Linear	No constant terms (with respect to x or y).
	Coefficients may depend on t but not on x or y.
	Examples: $y[n] + 3^n y[n-1] = x[n]$, $y[n] = x[n] - x[n-2]$
Nonlinear	Constant terms in addition to terms in x and y.
	Coefficients that depend on x or y.
	Examples: $y[n] + 3^n = x[n-1]$, $y[n] = x^2[n]$, $3^{y[n]} y[n] = x[n]$
Time invariant	Coefficients may depend on x or y but not on n.
	Examples: $y^2[n] + 3x[n]y[n-1] = x[n-1]$, $3^{y[n]} y[n] = x[n]$
Time-varying	Coefficients are explicit functions of n.
	x and y have scaled arguments, e.g., $x[2n]$.
	Examples: $y[n] + 3^n y[n-1] = x[n]$, $y[n] + y[n-1] = x[2n]$
Causal	Present response does not depend on future values.
	Examples: $y[n] = x[n-1] - x[n-2]$, $y[n] = (n+1)^2 x[n]$
Noncausal	Present response depends on future inputs/outputs.
	Examples: $y[n] = x[n+1] - x[n-1]$, $y[n] = x[2n]$
Dynamic	System equation is a difference equation.
	variables occur with scaled arguments, e.g., $x[2n]$.
	Examples: $y[n] = x^2[n] - x[n-1]$, $y[n] = 4x[2n]$, $y[n] = x[n+1]$
Instantaneous	System equation is algebraic.
	Variables occur as $x[n], y[n]$ with argument n.
	Examples: $y[n] = x^2[n] - 2x[n]$, $y[n] + 3 = 4x[n]$

(c) $y[n] = 2\sin[n]x[n]$ is linear since $2\sin[n]$ is a function of n alone.

(d) $y[n] = 2^{x[n]}x[n]$ is nonlinear since $2^{x[n]}$ depends on $x[n]$.

3.8.2   *Time-Invariant (Shift-Invariant) Systems*

For a time-invariant system, a time displacement of the input results in an identical displacement of the response. Such a system is also called **stationary** or **fixed**. If the operator $\mathcal{O}$ transforms the input $x[n]$ to the output $y[n]$ such that $\mathcal{O}[x[n]] = y[n]$, time invariance requires that

$$\mathcal{O}\{x[n - n_0]\} = y[n - n_0]$$

We can test for time invariance by applying the time-invariance relation. We can also examine the system equation directly. The coefficients of a time-varying system depend explicitly on n. Terms such as $x[2n]$, with scaled system variables, also make a system time-varying.

Example 3.19
Testing for Time
Invariance

We test the following systems for time invariance:

(a) $y[n] = nx[n]$: The operator that transforms $x[n]$ to $y[n]$ is $\mathcal{O}[\ ] = n[\ ]$. The shifted input $x[n - n_0]$ gives $\mathcal{O}\{x[n - n_0]\} = nx[n - n_0]$, but the system equation predicts $y[n - n_0] = [n - n_0]x[n - n_0]$. The system is thus time-varying.

(b) $y[n] = 2^{x[n]}x[n]$ is time invariant ($2^{x[n]}$ does not depend explicitly on n).

(c) $y[n] = 2^n x[n] + x[n - 2]$ is time-varying (2^n depends explicitly on n).

(d) $y[n] = 2x[n] + x[n]y[n - 1]$ is time invariant (but nonlinear).

(e) $y[n] = 2x[2n] + y[n - 1]$ is time-varying (due to $x[2n]$).

3.8.3 Implications of Linearity and Time-Invariance

The superposition property forms the key to the analysis of linear systems. An arbitrary function may be decomposed into its simpler constituents, the response due to each analyzed separately and the total response found using superposition. This leads to several approaches for system analysis:

1. Representing an arbitrary signal $x[n]$ as a weighted sum of shifted *DT* impulses is the basis for the discrete convolution method (Chapter 4).

2. Representing a *DT* periodic signal $x[n]$ as a linear combination of *DT* harmonics is the basis for the Discrete Fourier Series (Chapter 12).

3. Representing a signal $x[n]$ as a weighted sum of *DT* exponentials is the basis for the Discrete-Time Fourier Transform and the z-transform (Chapters 11 and 13).

3.8.4 Causal and Noncausal Systems

A **causal** or **nonanticipating** system cannot anticipate future inputs in order to *generate or alter* the present response. The present response does not depend on future values of the input. Systems whose present response is affected by future inputs are termed **noncausal** or **anticipating**.

We can check for causality from the system difference equation. For example, the system $y[n] + A_1 y[n - 1] + A_2 y[n - 2] + \cdots + A_N y[n - N] = x[n + K]$ is causal for $K \leq 0$, and $y[n + N] + A_1 y[n + N - 1] + A_2 y[n + N - 2] + \cdots + A_N y[n] = x[n + K]$ is causal for $K \leq N$.

We can also use the **operational transfer function** $H(z)$ derived from the difference equation in operator form. Its general form may be expressed as

$$H(z) = \frac{B_0 z^M + B_1 z^{M-1} + \cdots + B_{M-1} z + B_M}{A_0 z^N + A_1 z^{N-1} + \cdots + A_{N-1} z + A_N} = \frac{P_M(z)}{Q_N(z)}$$

If $H(z)$ is not proper ($M > N$), the system is noncausal.

Example 3.20
Causal and Noncausal
Systems

(a) $y[n + 2] + y[n + 1] = x[n + 5]$ is noncausal. To find $y[0]$, we need $x[3]$.

(b) $y[n + 7] + y[n + 5] = x[n + 4]$ is causal. It is, in fact, identical to the system $y[n] + y[n - 2] = x[n - 3]$.

(c) $y[n] = 2x[\alpha n]$ is causal only for $\alpha = 1$. For $\alpha > 1$, outputs for $n > 0$ depend on future values of the input while for $\alpha < 1$, outputs for $n < 0$ require knowledge of future values of the input. ⎯⎯

3.8.5   Instantaneous and Dynamic Systems

Systems whose present response depends not only on the present value of the input but its past (and future) values are termed **dynamic** or systems with memory. The representation of such systems requires difference equations. If the response at n_0 depends only on the input at n_0 and not at any other times, the system is called **instantaneous, static** or **memoryless.** The system equation of an instantaneous system is *algebraic* and x and y are of the form $x[n]$ and $y[n]$ with no scaled or shifted arguments. *All instantaneous systems are also causal.*

Example 3.21
Instantaneous and
Dynamic Systems

We test the following systems for memory:

(a) $y[n] = 2^n x[n]$ is instantaneous (but time-varying).
(b) $y[n] = x[n + 2]$ is dynamic since $y[n_0]$ depends on $x[n_0 + 2]$, not on $x[n_0]$.
(c) $y[n] = 2x[n] + x[n - 1]$ is dynamic (it is a difference equation). ⎯⎯

3.9
Analysis of *DT*
LTI Systems

DT systems can be analyzed in the time domain using a system description based on any one of the three models as listed in Table 3.6.

Table 3.6   System Analysis in the Time Domain.

System Model	Typical Restrictions	Comments and Applications
Difference equation	Very few	Useful only for low order.
State variables	Linear system for matrix formulation	Useful for complex systems and multiple-input/multiple-output systems and numerical solution.
Impulse response	Relaxed LTI system	Useful for numerical solution; provides link between time and frequency domains.

The Difference Equation Representation   This representation applies even to nonlinear and time-varying systems. For LTI systems, it allows computation of the response using superposition even if initial conditions are present.

The State Variable Representation   This representation describes an nth-order system by n simultaneous first-order difference equations called **state equations** in terms of n state variables. It is useful for complex or nonlinear systems and those with multiple inputs and outputs. For linear systems, state equations can be solved using matrix methods. The state variable form is readily amenable to numerical solution.

The Impulse Response Representation This representation normally applies to *relaxed* LTI systems. It describes such a system in terms of its impulse response $h[n]$. Unlike difference or state representations, the system response $y[n]$ appears explicitly in the governing relation called the *convolution sum.* It also allows us to relate time-domain and transformed-domain methods of system analysis. We discuss the details of this method in Chapter 4.

3.10 Digital Filters and Their Response

In the parlance of digital signal processing, discrete-time systems are also called **digital filters.** From now on, we shall use this terminology extensively. Consider the general difference equation

$$y[n] + A_1 y[n-1] + A_2 y[n-2] + \cdots + A_N y[n-N]$$
$$= B_0 x[n] + B_1 x[n-1] + \cdots + B_M x[n-M]$$

This describes an Nth-order **recursive filter** whose present output depends on its own past values $y[n-k]$ and on the past and present values of the input. Since its impulse response $h[n]$ is usually of infinite duration, it is also called an **infinite impulse response** (IIR) **filter.**

If all the A_K equal zero, we obtain a **nonrecursive filter** described by

$$y[n] = B_0 x[n] + B_1 x[n-1] + \cdots + B_M x[n-M]$$

The present response depends only on the present and past values of the input and shows no dependence (recursion) on past values of the response. The response may be thought of as a weighted sum (moving average) of the present input and its delayed versions. The system impulse response is of finite duration and describes a **finite impulse response** (FIR) **filter.**

It is common to refer to the largest input delay M as the order of an FIR filter (even though technically the order is zero) and $M + 1$ as the **filter length.**

3.10.1 *Toward the Solution of Difference Equations*

For nonrecursive systems, whose output depends only on the input, the solution is the system equation itself. For recursive systems, whose difference equation incorporates dependence on past values of the input, the solution may be implemented in one of two ways:

1. We can use recursion and successively compute the values of the output as far as desired. This approach is simple but it is not always easy to discern a closed-form solution for the output.
2. We can use an analytical approach that exactly parallels the solution of differential equations.

Initial Conditions   As with differential equations, we need N initial conditions for a complete solution to an Nth-order difference equation. Typically, we specify $y[-1]$ for a first-order equation, $y[-1]$ and $y[-2]$ for a second-order system, and $y[-1], y[-2], \ldots, y[-N]$ for an Nth-order system.

3.10.2   Solution by Recursion

The solution of a difference equation by recursion is straightforward. Given an Nth-order difference equation subject to the initial conditions $y[-1], y[-2], \ldots, y[-N]$, we successively generate values of $y[0], y[1], \ldots$, as far as desired. This approach is best illustrated by some examples.

Example 3.22
System Response Using Recursion

(a) Consider a system described by $y[n] = a_1 y[n-1] + b_0 u[n]$. Let the initial condition be $y[-1] = 0$. We then successively compute:

$$y[0] = a_1 y[-1] + b_0 u[0] = b_0$$
$$y[1] = a_1 y[0] + b_0 u[1] = a_1 b_0 + b_0 = b_0[1 + a_1]$$
$$y[2] = a_1 y[1] + b_0 u[2] = a_1[a_1 b_0 + b_0] + b_0 = b_0[1 + a_1 + a_1^2]$$

The form of $y[n]$ may be discerned as

$$y[n] = b_0[1 + a_1 + a_1^2 + \cdots + a_1^{n-1} + a_1^n]$$

Using the closed form for the geometric sequence (Appendix A) results in

$$y[n] = \frac{b_0(1 - a_1^{n+1})}{(1 - a_1)}$$

(b) Consider a system described by $y[n] = a_1 y[n-1] + b_0 n u[n]$. Let the initial condition be $y[-1] = 0$. We then successively compute:

$$y[0] = a_1 y[-1] = 0$$
$$y[1] = a_1 y[0] + b_0 u[1] = b_0$$
$$y[2] = a_1 y[1] + 2b_0 u[2] = a_1 b_0 + 2b_0$$
$$y[3] = a_1 y[2] + 3b_0 u[3] = a_1[a_1 b_0 + 2b_0] + 3b_0 = a_1^2 + 2a_1 b_0 + 3b_0$$

The general form is thus $y[n] = a_1^{n-1} + 2b_0 a_1^{n-2} + 3b_0 a_1^{n-3} + (n-1)b_0 a_1 + nb_0$. We can find a more compact form for this, but not without some effort. By adding and subtracting $b_0 a_1^{n-1}$ and factoring out a_1^n, we obtain

$$y[n] = a_1^n - b_0 a_1^{n-1} + b_0 a_1^n[a_1^{-1} + 2a_1^{-2} + 3a_1^{-3} + \cdots + na_1^{-n}]$$

Using the closed form for $\Sigma k x^k$, $(1, N)$ where $x = a^{-1}$ (Appendix A), we get

$$y[n] = a_1^n - b_0 a_1^{n-1} + b_0 a_1^n \frac{a^{-1}[1 - (n+1)a^{-n} + na^{-(n+1)}]}{(1 - a^{-1})^2}$$

Comment: What a chore! More elegant ways of solving difference equations are described later in this chapter.

(c) Consider the recursive system $y[n] = y[n-1] + x[n] - x[n-3]$. If $x[n]$ equals $\delta[n]$ and $y[-1] = 0$, we successively obtain

$$y[0] = y[-1] + \delta[0] - \delta[-3] = 1 \qquad y[1] = y[0] + \delta[1] - \delta[-2] = 1$$
$$y[2] = y[1] + \delta[2] - \delta[-1] = 1 \qquad y[3] = y[2] + \delta[3] - \delta[0] = 1 - 1 = 0$$
$$y[4] = y[3] + \delta[4] - \delta[1] = 0 \qquad y[5] = y[4] + \delta[5] - \delta[2] = 0$$

Comment: The impulse response of this "recursive" filter is zero after the first three values and has a finite length. It is actually a nonrecursive filter in disguise! ▬▬

3.10.3 *Formal Solution of Difference Equations*

In analogy with differential equations, the formal solution of difference equations may be expressed as a superposition of the **natural response** and the **forced response**. Consider the Nth-order difference equation

$$y[n] + A_1 y[n-1] + A_2 y[n-2] + \cdots + A_N y[n-N] = x[n]$$

subject to the initial conditions $y[-1], y[-2], y[-3], \ldots, y[-N]$.

Its solution requires:

1. The form of the natural response $y_N[n]$ from the homogeneous equation.
2. The forced response $y_F[n]$ from the entire difference equation.
3. The total response as the superposition $y[n] = y_N[n] + y_F[n]$.
4. Evaluation of the constants in $y[n]$ using initial conditions.

Remarks: **1.** For stable systems, the natural response is also called the **transient response** since it decays to zero with time.
2. For a system with harmonic or switched harmonic inputs, the forced response is also a harmonic at the input frequency and is called the **steady-state response**.
3. If the input is just $x[n-K]$, the response $y[n]$ up to $n = K-1$ (i.e., $y[0], y[1], \ldots, y[K-1]$) depends on the initial conditions $y[-1], \ldots, y[-N]$ alone and must be found by recursion. To find the complete response, we must make use of the initial conditions $y[K-1], \ldots, y[0], \ldots, y[K-N-1]$.

3.10.4 *The Natural Response*

The natural response uses only the *homogeneous* equation. It contains only exponentials whose exponents correspond to the roots (real or complex) of the *characteristic equation*. The **homogeneous difference equation** reads

$$y[n] + A_1 y[n-1] + A_2 y[n-2] + \cdots + A_N y[n-N] = 0$$

The **characteristic equation** is defined by

$$1 + A_1 z^{-1} + A_2 z^{-2} + \cdots + A_N z^{-N} = z^N + A_1 z^{N-1} + \cdots + A_N = 0$$

This equation has N roots, $z_k, k = 1, 2, \ldots, N$. The form of the natural response contains N discrete-time exponentials of the form $(z_k)^n$. This form must be modified for multiple roots. Table 3.7 summarizes the various forms. For combinations, we use superposition.

Table 3.7   Form of the Natural Response for _DT_ LTI Systems.

Root of Characteristic Equation	Form of Natural Response
Real and distinct root r	$C(r)^n$
Complex conjugate $r \exp(j\theta)$	$[C_1 \cos(n\theta) + C_2 \sin(n\theta)](r)^n$
Real repeated root $(r)^p$	$(K_0 + K_1 n + K_2 n^2 + \cdots + K_p n^p)(r)^n$
Complex, repeated $[r \exp(j\theta)]^p$	$(C_0 + C_1 n + C_2 n^2 + \cdots + C_p n^p) \cos(n\theta)(r)^n$ $+ (D_0 + D_1 n + D_2 n^2 + \cdots + D_p n^p) \sin(n\theta)(r)^n$

Example 3.23
Finding the Natural Response

(a) Consider the homogeneous difference equation

$$y[n] - \tfrac{1}{6} y[n-1] - \tfrac{1}{6} y[n-2] = 0$$

The characteristic equation is $1 - \tfrac{1}{6} z^{-1} - \tfrac{1}{6} z^{-2} = 0$ or $z^2 - \tfrac{1}{6} z - \tfrac{1}{6} = 0$ with roots $z_1 = \tfrac{1}{2}$ and $z_2 = -\tfrac{1}{3}$. Thus $y_N[n] = K_1(z_1)^n + K_2(z_2)^n = K_1(\tfrac{1}{2})^n + K_2(-\tfrac{1}{3})^n$.

(b) Consider the homogeneous equation

$$y[n+2] - 2y[n+1] + 2y[n] = 0$$

Its characteristic equation is $z^2 - 2z + 2 = 0$ with roots $z_1 = 1 + j$ and $z_2 = 1 - j$. Since $z_1, z_2 = \sqrt{2} \exp(\pm j \tfrac{1}{4} \pi)$, Table 3.7 suggests the natural response

$$y_N[n] = (\sqrt{2})^n [A_1 \cos(\tfrac{1}{4} n\pi) + B_1 \sin(\tfrac{1}{4} n\pi)]$$

Comment: The response is actually a *growing* sinusoid due to the factor $(\sqrt{2})^n$.

3.10.5   *The Forced Response*

The forced response can be found uniquely and independently of the natural response or initial conditions. The forced response has the same form as the forcing function (the RHS) of the difference equation. Table 3.8 summarizes these forms for various forcing functions. For combinations, we simply use superposition.

Example 3.24
Forced and Natural Response

(a) Consider the difference equation $y[n] - \tfrac{1}{6} y[n-1] - \tfrac{1}{6} y[n-2] = 4u[n]$ with the given initial conditions $y[-1] = 0, y[-2] = 12$. From Example 3.23, the natural response equals $y_N[n] = K_1(z_1)^n + K_2(z_2)^n = K_1(\tfrac{1}{2})^n + K_2(-\tfrac{1}{3})^n$. Since the

---

Table 3.8   Form of the Forced Response for DT LTI Systems.
Note: If the RHS involves the form $(\alpha)^n$ where α is also a
root of the characteristic equation repeated r times, the forced
response form must be multiplied by n^r.

Forcing function (RHS)	Form of forced response
C (constant)	C_1 (constant)
$(\alpha)^n, \alpha \neq$ root of characteristic equation	$K(\alpha)^n$
$\cos(n\theta + \beta)$	$K_1 \cos(n\theta) + K_2 \sin(n\theta)$ or $K \cos(n\theta + \phi)$
$(\alpha)^n \cos(n\theta + \beta)$	$[K_1 \cos(n\theta) + K_2 \sin(n\theta)](\alpha)^n$
n	$K_0 + K_1 n$ (general first-order polynomial)
n^p	$K_0 + K_1 n + K_2 n^2 + \cdots + K_p n^p$
$n(\alpha)^n$	$(K_0 + K_1 n)(\alpha)^n$
$n^p(\alpha)^n$	$(K_0 + K_1 n + K_2 n^2 + \cdots + K_p n^p)(\alpha)^n$
$n \cos(n\theta + \beta)$	$(K_1 + K_2 n) \cos(n\theta) + (K_3 + K_4 n) \sin(n\theta)$

---

forcing function is $4u[n]$, the forced response $y_F[n]$ is constant and we choose $y_F[n] = C$. Then $y_F[n-1] = C, y_F[n-2] = C$ and

$$y_F[n] - \tfrac{1}{6}y_F[n-1] - \tfrac{1}{6}y_F[n-2] = C - \tfrac{1}{6}C - \tfrac{1}{6}C = 4 \qquad \text{or} \qquad C = 6$$

Thus $y_F[n] = 6$. The total response $y[n]$ now equals

$$y[n] = y_N[n] + y_F[n] = K_1(\tfrac{1}{2})^n + K_2(-\tfrac{1}{3})^n + 6$$

Using the IC, we find $y[-1] = 0 = 2K_1 - 3K_2 + 6$ and $y[-2] = 12 = 4K_1 + 9K_2 + 6$. Solving for the constants, we obtain $K_1 = -1.2$ and $K_2 = 1.2$. Thus

$$y[n] = -1.2(\tfrac{1}{2})^n + 1.2(-\tfrac{1}{3})^n + 6, \qquad n \geq 0$$

As a check, $y[-1] = -2.4 - 3.6 + 6 = 0$ and $y[-2] = -4.8 + 10.8 + 6 = 12$.

Comment: Since $y[n]$ satisfies $y[-1], y[-2]$, it is actually valid for $n \geq -2$.

(b) Let $y[n] - y[n-1] + \tfrac{1}{4}y[n-2] = x[n-2]$ with $x[n] = 3(\tfrac{1}{2})^n u[n]$, and initial conditions $y[-1] = 0$, $y[-2] = -4$.

The characteristic equation $z^2 - z + \tfrac{1}{4} = 0$ has two equal roots $z_1 = z_2 = \tfrac{1}{2}$. Thus $y_N[n] = (A + Bn)(\tfrac{1}{2})^n$. Here is how we proceed now:

1. The RHS has the form $(\tfrac{1}{2})^n$. Since $\tfrac{1}{2}$ is a root of the characteristic equation repeated twice, we use $y_F[n] = Kn^2(\tfrac{1}{2})^n$ (see note in Table 3.8). Then, $y_F[n-1] = K(n-1)^2(\tfrac{1}{2})^{n-1}$ and $y_F[n-2] = K(n-2)^2(\tfrac{1}{2})^{n-2}$. Using these in the original difference equation,

$$Kn^2(\tfrac{1}{2})^n - K(n-1)^2(\tfrac{1}{2})^{n-1} + \tfrac{1}{4}K(n-2)^2(\tfrac{1}{2})^{n-2} = 3(\tfrac{1}{2})^{n-2}$$

The constant K is found by evaluating this for any n. With $n = 2$, we get $4K(\frac{1}{2})^2 - K(\frac{1}{2}) + 0 = 3$ or $K = 6$. Thus $y[n] = y_N[n] + y_F[n] = (A + Bn + 6n^2)(\frac{1}{2})^n$.

2. The input form $x[n - 2]$ requires the IC $y[0], y[1]$. By recursion, we find $y[0] = y[1] - \frac{1}{4}y[-2] + 3u[-2] = 1$ and $y[1] = y[0] - \frac{1}{4}y[-1] + 3u[-1] = 1$. This gives $y[0] = 1 = A$ and $y[1] = 1 = (A + B + 6)(\frac{1}{2})$ or $A = 1$, $B = -5$. Finally, $y[n] = (1 - 5n + 6n^2)(\frac{1}{2})^n$. This is valid for $n \geq 0$ (not just $n \geq 2$).

Comment: The derived IC were needed only because the input started at $n = 2$. An alternative is to use the approach described later in Section 3.10.7.

3.10.6   *The Zero-Input and Zero-State Response*

The zero-input response is found from the natural response $y_N[n]$ and the *prescribed initial conditions*. Similarly, the zero-state response $y_{zs}[n]$ is found from the complete solution $y[n]$ *assuming zero initial conditions*.

Note that the natural and forced response do not, in general, correspond to the zero-input and zero-state response even though each pair adds up to the total response.

Example 3.25
Zero-Input and
Zero-State Response

Consider the difference equation $y[n] - \frac{1}{6}y[n - 1] - \frac{1}{6}y[n - 2] = 4$, $n \geq 0$, with the initial conditions $y[-1] = 0$, $y[-2] = 12$. From Example 3.22, the natural response equals $y_N[n] = K_1(z_1)^n + K_2(z_2)^n = K_1(\frac{1}{2})^n + K_2(-\frac{1}{3})^n$.

We find the zero-input response $y_{zi}[n]$ by evaluating K_1 and K_2 from $y_N[n]$ using the given initial conditions. We obtain

$$0 = K_1(\tfrac{1}{2})^{-1} + K_2(-\tfrac{1}{3})^{-1} = 2K_1 - 3K \qquad 12 = K_1(\tfrac{1}{2})^{-2} + K_2(-\tfrac{1}{3})^{-2} = 4K_1 + 9K_2$$

Thus $K_1 = 1.2$, $K_2 = 0.8$ and $y_{zi}[n] = 1.2(\frac{1}{2})^n + 0.8(-\frac{1}{3})^n$. To find the zero-state response, we assume zero initial conditions. The forced response found in Example 3.24 equals $y_F[n] = 6$. The response $y_{zs}[n]$ equals

$$y_{zs}[n] = y_N[n] + y_F[n] = K_1(\tfrac{1}{2})^n + K_2(-\tfrac{1}{3})^n + 6$$

With $y[-1] = 0$ and $y[-2] = 0$, we find

$$y[-1] = 0 = 2K_1 - 3K_2 + 6 \qquad y[-2] = 0 = 4K_1 + 9K_2 + 6$$

We obtain $K_1 = -2.4$, $K = 0.4$ and $y_{zs}[n] = -2.4(\frac{1}{2})^n + 0.4(-\frac{1}{3})^n + 6$. As a check, $y[n] = y_{zi}[n] + y_{zs}[n] = -1.2(\frac{1}{2})^n + 1.2(-\frac{1}{3})^n + 6$, as in Example 3.24.

3.10.7   *Solution of the General Difference Equation*

To find the response of the general linear system described by

$$y[n] + A_1y[n - 1] + A_2y[n - 2] + \cdots + A_Ny[n - N]$$
$$= B_0x[n] + B_1x[n - 1] + \cdots + B_Mx[n - M]$$

we use a four-step approach based on linearity and superposition:

1. Compute the zero-state response $y_0[n]$ to

$$y_0[n] + A_1 y_0[n-1] + A_2 y_0[n-2] + \cdots + A_N y_0[n-N] = x[n]$$

2. Use linearity and superposition to find $y_{zs}[n]$ as

$$y_{zs}[n] = B_0 y_0[n] + B_1 y_0[n-1] + \cdots + B_M y_0[n-M]$$

3. Find the zero-input response $y_{zi}[n]$ using initial conditions.
4. Find the total response as $y[n] = y_{zs}[n] + y_{zi}[n]$.

Note that the zero-input response is computed and included just once.

Example 3.26
Response of a General System

Let $y[n] - \frac{1}{6}y[n-1] - \frac{1}{6}y[n-2] = 10u[n] - 20u[n-1]$ with the initial conditions $y[-1] = 0, y[-2] = 30$.

1. We start with the *relaxed* system $y_0[n] - \frac{1}{6}y_0[n-1] - \frac{1}{6}y_0[n-2] = u[n]$. We find $y_0[n] = K_1(\frac{1}{2})^n + K_2(-\frac{1}{3})^n + 1.5$ (following Example 3.24, say). With $y[-1] = y[-2] = 0$, we get $2K_1 - 3K_2 + 1.5 = 0$, $4K_1 + 9K_2 + 1.5 = 0$. This yields $K_1 = -0.6$, $K_2 = 0.1$, and $y_0[n] = [-0.6(\frac{1}{2})^n + 0.1(-\frac{1}{3})^n + 1.5]u[n]$.
2. The zero-state response then equals $y_{zs}[n] = 10y_0[n] - 20y_0[n-1]$ and thus

$$y_{zs}[n] = [-6(\tfrac{1}{2})^n + (-\tfrac{1}{3})^n + 15]u[n] - [-12(\tfrac{1}{2})^{n-1} + 2(-\tfrac{1}{3})^{n-1} + 30]u[n-1]$$

3. Now, $y_{zi}[n] = C_1(\frac{1}{2})^n + C_2(-\frac{1}{3})^n$ with $y[-1] = 0$, $y[-2] = 30$. This results in $2K_1 - 3K_2 = 0, 4K_1 + 9K_2 = 30$, $K_1 = 3$, $K_2 = 2$ and $y_{zi}[n] = 3(\frac{1}{2})^n + 2(-\frac{1}{3})^n$. As a check, $y_{zi}[-1] = 3(2) + 2(-3) = 0$ and $y_{zi}[-2] = 3(4) + 2(9) = 30$. This shows that $y_{zi}[n]$ actually holds for $n \geq -2$ and we write $y_{zi}[n] = [3(\frac{1}{2})^n + 2(-\frac{1}{3})^n]u[n+2]$.
4. Finally, the total response $y[n] = y_{zi}[n] + y_{zs}[n]$ equals

$$y[n] = [3(\tfrac{1}{2})^n + 2(-\tfrac{1}{3})^n]u[n+2] + [-6(\tfrac{1}{2})^n + (-\tfrac{1}{3})^n + 15]u[n]$$
$$-[-12(\tfrac{1}{2})^{n-1} + 2(-\tfrac{1}{3})^{n-1} + 30]u[n-1]$$

Comment: For $n \geq 1$, this simplifies to $y[n] = 21(\frac{1}{2})^n + 9(-\frac{1}{3})^n - 15$.

3.11
Impulse Response of Digital Filters

In analogy with relaxed *CT* systems, the *DT* impulse response is simply the response to a unit *DT* impulse $\delta[n]$, but is much easier to find.

3.11.1 *Impulse Response of Nonrecursive Filters*

A nonrecursive filter of length $M + 1$ is described by

$$y[n] = B_0 x[n] + B_1 x[n-1] + \cdots + B_M x[n-M]$$

With $x[n] = \delta[n]$, its impulse response $h[n]$ is an $M + 1$ term sequence which may be expressed as

$$h[n] = B_0\delta[n] + B_1\delta[n-1] + \cdots + B_M\delta[n-M] \quad \text{or} \quad \{h[n]\} = \{B_0, B_1, \ldots, B_M\}$$

3.11.2   *Impulse Response of Causal Recursive Filters*

Recursion provides a simple means of obtaining as many terms of the impulse response $h[n]$ of a recursive filter as we please, though we may not always be able to discern a closed form from the results. Here is an example.

Example 3.27
Impulse Response by Recursion

For the system described by $y[n] = ay[n-1] + bx[n]$, we find the impulse response $h[n]$ as the solution to $h[n] = ah[n-1] + b\delta[n]$ subject to the initial conditions $y[-1] = 0$. We recursively obtain:

$$h[0] = ah[-1] + b\delta[0] = b \qquad h[1] = ah[0] = ab$$
$$h[2] = ah[1] = a^2b \qquad h[3] = ah[2] = a^3b$$

In this case, the general form of $h[n]$ is easy to discern as $h[n] = a^n bu[n]$.

A Formal Approach   To find the impulse response of the general system

$$y[n] + A_1y[n-1] + A_2y[n-2] + \cdots + A_Ny[n-N]$$
$$= B_0x[n] + B_1x[n-1] + \cdots + B_Mx[n-M]$$

we use a two-step approach based on linearity and superposition.

1. With $h[-1] = h[-2] = \cdots = h[-N] = 0$, find the impulse response $h_0[n]$ to

$$h_0[n] + A_1h_0[n-1] + A_2h_0[n-2] + \cdots + A_Nh_0[n-N] = \delta[n]$$

2. Use linearity and superposition to find $h[n]$ as

$$h[n] = B_0h_0[n] + B_1h_0[n-1] + \cdots + B_Mh_0[n-M]$$

Remark: The IC given above always yield a trivial solution. We must use recursion to get *at least one* new nonzero initial condition, say $y[K]$, and compute $h_0[n]$ using $y[K], y[K-1], \ldots, y[K-N+1]$.

Example 3.28
Impulse Response Computation

Let $y[n] - \frac{1}{6}y[n-1] - \frac{1}{6}y[n-2] = 4x[n] - x[n-1]$.
To find $h[n]$, start with $h_0[n] - \frac{1}{6}h_0[n-1] - \frac{1}{6}h_0[n-2] = \delta[n]$.
Its characteristic equation $z^2 - \frac{1}{6}z - \frac{1}{6}$ has the roots $z_1 = \frac{1}{2}, z_2 = -\frac{1}{3}$.
Thus $h_0[n] = K_1(\frac{1}{2})^n + K_2(-\frac{1}{3})^n$. With $h_0[-1] = 0$ and $h_0[-2] = 0$, we get the trivial result $K_1 = 0, K_2 = 0$. We must therefore recursively compute $h_0[0], h_0[1], \ldots,$ until we get a nonzero value.

We find $h_0[0] = \frac{1}{6}h_0[-1] + \frac{1}{6}h_0[-2] + \delta[0] = 1$.
We now use $h_0[0]$ and $h_0[-1]$ to obtain

$$h_0[0] = 1 = K_1 + K_2 \qquad h_0[-1] = 0 = 2K_1 - 3K_2$$

This yields $K_1 = 0.6, K_2 = 0.4$ and $h_0[n] = [0.6(\frac{1}{2})^n + 0.4(-\frac{1}{3})^n]u[n]$.
Finally, since $h[n] = 4h_0[n] - h_0[n-1]$, we get

$$h[n] = [2.4(\tfrac{1}{2})^n + 1.6(-\tfrac{1}{3})^n]u[n] - [0.6(\tfrac{1}{2})^{n-1} + 0.4(-\tfrac{1}{3})^{n-1}]u[n-1]$$

3.11.3 *Impulse Response of Noncausal Recursive Filters*

For a noncausal system, the impulse response $h[n]$ does not simply equal the natural response $h_N[n]$, $n \geq 0$ but involves impulses for $n < 0$. For a formal solution, we proceed in analogy with *CT* systems. We set up the operational transfer function $H(z) = P_M(z)/Q_N(z)$ from the difference equation. Here $P_M(z)$ and $Q_N(z)$ are polynomials in z of degree M and N. For $M \leq N$, $H(z)$ is a proper transfer function and describes a causal system. If $M > N$, we use long division to deflate the degree of $P_M(z)$ to N. With $M - N = p$, we get

$$H(z) = \frac{P(z)}{Q(z)} = c_1 z + \cdots + c_p z^p + \frac{R_N(z)}{Q_N(z)} = c_1 z + \cdots + c_p z^p + H_r(z)$$

Here, $H_r(z)$ is a reduced *proper* transfer function corresponding to a causal system. The impulse response $h[n]$ then equals the sum of the causal impulse response $h_r[n]$ of the reduced system and impulsive terms. It has the form

$$h[n] = h_r[n] + \sum_{k=1}^{P} c_k \delta[n+k] = h_r[n] + c_1\delta[n+1] + \cdots + c_p\delta[n+P]$$

The difference $M - N = P$ establishes the number of impulsive terms in $h[n]$.

Example 3.29

Impulse Response of a Noncausal Filter

Consider the noncausal filter $y[n+2] - \frac{1}{4}y[n] = 8x[n+3]$. Its operational transfer function is $H(z) = 8z/(1 - \frac{1}{4}z^{-2}) = 8z^3/(z^2 - \frac{1}{4})$.

Using long division, we rewrite this as

$$H(z) = \frac{8z + 2z}{z^2 - \frac{1}{4}} = \frac{8z + 2z^{-1}}{1 - \frac{1}{4}z^{-2}} = 8z + H_r(z)$$

The reduced system has the difference equation $y_r[n] - \frac{1}{4}y_r[n-2] = 2x[n-1]$. Its characteristic equation $z^2 - \frac{1}{4}$ has the roots $z_1 = \frac{1}{2}, z_2 = -\frac{1}{2}$. Thus $h_r[n] = K_1(\frac{1}{2})^n + K_2(-\frac{1}{2})^n$, $n \geq 0$. We now use $h_r[n] - \frac{1}{4}h_r[n-2] = 2\delta[n-1]$ and the IC $h_r[-2] = h_r[-1] = 0$ to recursively compute

$$h_r[0] = \tfrac{1}{4}h_r[-2] + 2\delta[-1] = 0, \qquad h_r[1] = \tfrac{1}{4}h_r[-1] + 2\delta[0] = 2$$

Thus $h_r[0] = 0$ and $h_r[1] = 2$. Using these initial conditions, we find

$$h_r[0] = 0 = K_1 + K_2, \qquad h_r[1] = 2 = \tfrac{1}{2}K_1 - \tfrac{1}{2}K_2$$

Thus $K_1 = 2$, $K_2 = -2$ and $h_r[n] = 2(\frac{1}{2})^n u[n] - 2(-\frac{1}{2})^n u[n]$. Finally, $h[n] = h_r[n] + 8\delta[n+1] = 2(\frac{1}{2})^n u[n] - 2(-\frac{1}{2})^n u[n] + 8\delta[n+1]$.   ▬

3.11.4   *Recursive Form for Nonrecursive Digital Filters*

The terms FIR and nonrecursive are synonymous. A nonrecursive filter always has a finite impulse response. The terms IIR and recursive are often, but not always, synonymous. Not all recursive filters have an infinite impulse response. In fact, nonrecursive filters can often be implemented in recursive form.

Remark: A recursive filter may also be described by a nonrecursive filter of the form $y[n] = B_0 x[n] + B_1 x[n-1] + \cdots + B_M x[n-M]$ if we know all the past inputs. In general, this implies $M \to \infty$.

Example 3.30
Recursive Form for
Nonrecursive Filters

Consider the nonrecursive filter $y[n] = x[n] + x[n-1] + x[n-2]$. Its impulse response is simply $h[n] = \delta[n] + \delta[n-1] + \delta[n-2]$. Clearly, $h[n]$ describes the finite sequence $\{h[n]\} = \{1, 1, 1\}$. To cast this filter in recursive form, we first compute $y[n-1]$ as $y[n-1] = x[n-1] + x[n-2] + x[n-3]$. Upon subtraction from the original equation, we obtain the recursive form

$$y[n] - y[n-1] = x[n] - x[n-3]$$

Its response can now be found by recursion.   ▬

3.12
**Stability of *DT*
LTI Systems**

As with *CT* systems, the conditions for bounded-input, bounded-output (BIBO) stability of *DT* LTI systems involve the roots of the characteristic equation. The results of Tables 3.7 and 3.8 lead to the following necessary and sufficient condition:

> *For a DT LTI system to be stable, every root of its characteristic equation must have a magnitude less than unity.*

Root magnitudes *less than* unity ensure that the natural and zero-input response contain terms that always decay with time (see Table 3.7) and the forced and zero-state response remain bounded for *every* bounded input.

What about roots with unit magnitude? If such roots are simple, the natural response contains sinusoids or constants (bounded), but if the input has the same frequency, the response is a growing sinusoid or ramp and thus unbounded. If such roots are repeated, the natural response itself is unbounded because it contains growing sinusoids or polynomials.

Remarks: **1.** Unlike *CT* systems, the operational transfer function can be improper. This only makes the system noncausal and not necessarily unstable.
2. The stability condition is equivalent to having an absolutely summable impulse response $h[n]$. We develop this concept in Chapter 4.

3. The stability of nonlinear or time-varying systems usually must be checked by other means.

Example 3.31
Concepts Based on Stability

(a) The system $y[n] - \frac{1}{6}y[n-1] - \frac{1}{6}y[n-2] = x[n]$ is stable since the roots of its characteristic equation are $z = \frac{1}{2}, -\frac{1}{3}$ and their magnitudes are less than 1.

(b) The system $y[n] - y[n-1] = x[n]$ is unstable. With the root $z = 1$, the natural response $y_N = Ku[n]$ is actually bounded. But for an input $x[n] = u[n]$ with the same form, the forced response has the form $Cnu[n]$, which becomes unbounded.

(c) The system $y[n] - 2y[n-1] + y[n-2] = x[n]$ is unstable. The two roots at $z = 1$ produce an unbounded natural response $y_N[n] = Au[n] + Bnu[n]$.

(d) The system $y[n] - \frac{1}{2}y[n-1] = nx[n]$ is linear but time varying. It is also unstable: The bounded unit step input $x[n] = u[n]$ results in a response that includes the ramp $nu[n]$, which becomes unbounded.

3.13 System Classification and Analysis: A Synopsis

Systems	Many systems are described by their response to arbitrary signals in terms of differential/difference equations or by their impulse response. Most are quite complex and are modeled by simpler forms.
State Variables	Variables such as inductor currents allow us to establish the energy level or state of a system. Applied inputs change this energy or state. Systems are defined to be in the zero-energy state or relaxed state at $t = -\infty$. The behavior of a system depends on the input and initial state (conditions).
Operators	Operators are rules that show us how to transform one function to another. An operator is termed linear if it is both additive and homogeneous.
Linear Systems	These systems involve linear operators, contain no internal sources, and imply superposition. Based on linearity, the impulse response equals the derivative of the step response.
Time-Invariant Systems	For a time-invariant, fixed or stationary system, if the input is shifted, the response is shifted by the same amount.
Causal Systems	A causal or nonanticipating system is one whose present response is not affected by future values of the input.
Dynamic Systems	These systems have a response that depends on present and past inputs and its system equation is differential rather than algebraic.
LTI (linear time-invariant) Systems	These systems can be analyzed using (a) differential/difference equations (subject to initial conditions), (b) state variable representations using simultaneous first-order equations, or (c) the impulse response representation for relaxed systems.

Analysis	The response of an LTI system is the sum of (a) the zero-state and zero-input response or (b) the natural and forced response. The form of the natural response depends only on the homogeneous equation and comprises exponentials and damped exponentials. The forced response has the same general form as the forcing function/input. It can be found uniquely from the actual system differential equation.
Stability	For BIBO stability, every bounded input must result in a bounded output. For a stable CT LTI system, the highest derivative of the input must never exceed that of the input, and the real part of every root of its characteristic equation must be negative. For a stable DT LTI system, the magnitude of every root of its characteristic equation must be less than 1. Stability of nonlinear/time-varying systems must be checked by other means.
Impulse Response	The impulse response is often easier to find in the transformed domain. In the time domain, we find it from the step response or by using operational methods or recursion.
Digital Filters	Systems characterized by difference equations are also called digital filters. For a recursive (IIR) filter, the present output depends on its own past values and on past and present values of the input. Its impulse response $h[n]$ is usually of infinite duration. For a nonrecursive (FIR) filter, the present response depends only on the present and past values of the input. Its impulse response is of finite duration. Nonrecursive filters can often be implemented in recursive form.

Appendix 3A
MATLAB
Demonstrations
and Routines

Getting Started on the MATLAB Demonstrations
We present the MATLAB routines related to concepts in this chapter, along with some examples of their use. See Appendix 1A for introductory remarks.

NOTE: All polynomial coefficients must be entered in descending order.

Analytical Response of CT and DT Systems in Symbolic Form

```
[yt,yzs,yzi]=sysresp1(ty,b,a,x,ic)   %Computes system response
```

    For CT systems: $ty = 's'$, and b, a = coefficients arrays of differential equation

        $x = [K, a, p, w, r]$ for CT inputs of the form $K \exp(-at)t^p \cos(wt + r)$

For DT systems: $ty = 'z'$, and b, a = coefficients arrays of difference equation

$x = [K, a, p, w, r, n_0]$ for DT inputs of the form $K a^{n-n_0}(n - n_0)^p \cos[w(n - n_0) + r]$

Let $[s^2 + 4s + 3]y = [s + 2]x$ with $x = 2t \exp(-2t)u(t)$, $y(0) = 3$, $y'(0) = -1$. To find the total zero-state and zero-input response use:

```
>>[yt,yzs,yzi]=sysresp1('s',[1 2],[1 4 3],[2 2 1],[3 -1])
```

MATLAB will respond with

```
yt=-2*exp(-2*t)+5*exp(-t)              %Total response
yzs=1*exp(-3*t)-2*exp(-2*t)+1*exp(-t)  %Zero-state response
yzi=-1*exp(-3*t)+4*exp(-t)             %Zero-input response
```

These are string functions that can be evaluated using *eval*.
    For example, to plot the response, use

```
t=0:0.01:4;y=eval(yt);plot(t,y)
```

Numerical Response of CT Systems

```
y=ctsim(b,a,x,t,ic,ty)     %Numerical solution of differential equations
```

Here, x is a string input, e.g., 'exp$(-2*t)$', t is a time array such as $t = 0{:}0.01{:}2$, and ty is the integration algorithm ('rk1', 'rk2', 'rk4' or 'simp') for various Runge-Kutta orders and Simpson's algorithm.

Numerical Response of DT Systems

```
y=dtsim(b,a,x,ic)     %Recursive solution of difference equations
```

Here, x is an array of input values, e.g., $x = 0{:}0.01{:}2$ or $x = \exp(-t)$.

Utility Functions

```
[a,b,ic]=ode2de(a,b,ts,ic,ty) %Converts differential to difference equation
```

Here, ts = time step, ty = 'f' or 'b' for forward or backward Euler algorithm.

PROBLEMS

CT Systems

C3.1 (Operators) Which of the following describe linear operators?
(a) $\mathcal{O}[\ ] = 4[\ ]$ **(b)** $\mathcal{O}[\ ] = 4[\ ] + 3$ **(c)** $\mathcal{O}[\ ] = \int_{-\infty}^{t}[\ ]\,dt$
(d) $\mathcal{O}[\ ] = \sin[\ ]$ **(e)** $\mathcal{O}[\ ] = 4d[\ ]/dt + 3[\ ]$

C3.2 (System Classification) In each of the systems below, $x(t)$ is the input and $y(t)$ is the output. Classify each in terms of linearity, shift invariance, memory and causality.
(a) $y''(t) + 3y'(t) = 2x'(t) + x(t)$ **(b)** $y''(t) + 3y(t)y'(t) = 2x'(t) + x(t)$
(c) $y''(t) + 3tx(t)y'(t) = 2x'(t)$ **(d)** $y''(t) + 3y'(t) = 2x^2(t) + x(t+2)$
(e) $y(t) + 3 = x^2(t) + 2x(t)$ **(f)** $y(t) = 2x(t+1) + 5$
(g) $y''(t) + e^{-t}y'(t) = |\,x'(t-1)\,|$ **(h)** $y(t) = x^2(t) + 2x(t+1)$

(i) $y''(t) + y'[\cos(t)] = x'(t+1)$
(k) $y'(t) + \int_0^t y(t)\, dt = |\, x'(t)\,| - x(t)$

(j) $y(t) + t \int_{-\infty}^t y(t)\, dt = 2x(t)$
(l) $y''(t) + t \int_0^{t+1} y(t)\, dt = x'(t) + 2$

C3.3 (Classification) Explain why the system $y''(t) + y'(t) = t\sin(t)$ is time invariant while the system $y''(t) + y'(t) = tx(t)$, where $x(t) = \sin(t)$, is time-varying.

C3.4 (Classification) Consider the two systems described below:
(1) $y(t) = x(\alpha t)$ (2) $y'(t) = x(t+\alpha)$

(a) For what values of α is each system linear?
(b) For what values of α is each system causal?
(c) For what values of α is each system time invariant?
(d) For what values of α is each system instantaneous?

C3.5 (Classification) Classify the following systems:
(a) The modulation system $y(t) = x(t)\cos(2\pi f_0 t)$
(b) The modulation system $y(t) = [A + x(t)]\cos(2\pi f_0 t)$
(c) The sampling system $y(t) = (1/t_s)\,\text{comb}\,(t/t_s)x(t)$

C3.6 (System Response) For each of the following, evaluate the natural, forced, zero-state, zero-input and total response. Assume $y'(0) = 1$ and all other initial conditions zero.
(a) $[s^2 + 5s + 6]y = 6u(t)$
(b) $[s^2 + 5s + 6]y = 2\exp(-t)u(t)$
(c) $[s^2 + 4s + 3]y = 36tu(t)$
(d) $[s^2 + 4s + 4]y = 2\exp(-2t)u(t)$
(e) $[s^2 + 4s + 4]y = 8\cos(2t)u(t)$
(f) $[(s+1)^2(s+2)]y = \exp(-2t)u(t)$

C3.7 (Impulse Response) Find the step response and impulse response of each circuit shown in Figure C3.7.

C3.8 (Impulse Response) Find the impulse response of the systems described by the following:
(a) $[s+3]y = x$
(b) $[s^2 + 4s + 3]y = x$
(c) $[s^2 + 4s + 4]y = x$
(d) $[s^2 + 4s + 3]y = [2s - 1]x$
(e) $[s+4]y = [s^2 + 1]x$

C3.9 (Impulse Response) The input-output relation for a hypothetical LTI system is shown in Figure C3.9. What is its impulse response?

C3.10 (Stability) Which of the systems of Problem C3.8 is stable and why?

C3.11 (Impulse Response) The voltage input to a series RC circuit with a time constant τ is $(1/\alpha)\exp(-t/\alpha)$.
(a) Find the analytical form for the capacitor voltage.
(b) Show that as $\alpha \to 0$, we obtain the impulse response $h(t)$ of the RC circuit.

C3.12 (System Response) The step response of a linear system is given by
$w(t) = [1 - \exp(-t)]u(t)$.
(a) Establish its impulse response $h(t)$.
(b) Evaluate and sketch the response to $x(t) = \text{rect}(t - \frac{1}{2})$.

Figure C3.7

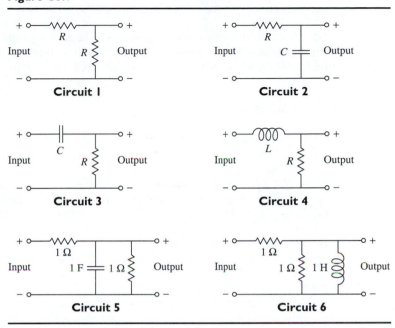

Circuit 1

Circuit 2

Circuit 3

Circuit 4

Circuit 5

Circuit 6

Figure C3.9

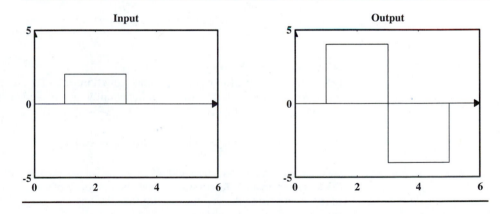

Input

Output

C3.13 (System Response) Consider two series RC circuits with $\tau_1 = 0.5$ and $\tau_2 = 5$. The input to both is a 5-V rectangular pulse of 1 s starting at $t = 0$.

(a) Find and sketch the capacitor voltage of each circuit.

(b) At what time after the input pulse is switched off will the output of both systems attain the same value?

C3.14 (Classification and Stability) Argue for or against the following statements assuming relaxed systems and constant element values. You may validate your arguments using simple circuits.

(a) A system with only resistors is always instantaneous and stable.

(b) A system with only inductors and/or capacitors is always stable.

(c) An RLC system with at least one resistor is always linear, causal and stable.

DT Systems

D3.1 (DT Operators) Which of the following describe linear operators?

(a) $\mathcal{O}[\ ] = 4[\ ]$ **(b)** $\mathcal{O}[\ ] = 4[\ ] + 3$ **(c)** $\mathcal{O}[\ ] = \exp[\ ]$

D3.2 (DT System Classification) In each of the systems below, $x[n]$ is the input and $y[n]$ is the output. Check each for linearity, shift invariance, memory and causality.

(a) $y[n] - y[n-1] = x[n]$ **(b)** $y[n] + y[n+1] = nx[n]$

(c) $y[n] - y[n+1] = x[n+2]$ **(d)** $y[n+2] - y[n+1] = x[n]$

(e) $y[n+1] - x[n]y[n] = nx[n+2]$ **(f)** $y[n] + y[n-3] = x^2[n] + x[n+6]$

(g) $y[n] - 2^n y[n] = x[n]$ **(h)** $y[n] = x[n] + x[n-1] + x[n-2]$

D3.3 (Difference Equations)

(a) Explain how the following difference equations for LTI systems are related, how they differ and what initial conditions are needed for their solution.

(1) $y[n] - 2y[n-1] + 3y[n-2] = x[n]$

(2) $y[n+2] - 2y[n+1] + 3y[n] = x[n]$

(b) Repeat part (a) for the following difference equations:

(1) $y[n] - 2y[n-1] + 3y[n-2] = x[n]$

(2) $y[n+2] - 2y[n+1] + 3y[n] = x[n+2]$

D3.4 (Operator Forms)

(a) Explain how the following difference equations for LTI systems in operator form differ, how they are related and what initial conditions are needed for their solution.

(1) $[1 + 2z^{-1} + 3z^{-2}]y = 2^n u[n]$ (2) $[z^2 + 2z + 3]y = 2^n u[n]$

(b) Repeat part (a) for the following difference equations:

(1) $[1 + 2z^{-1} + 3z^{-2}]y = 2^n u[n]$ (2) $[z^2 + 2z + 3]y = 2^{n+2} u[n+2]$

D3.5 (Response by Recursion) Find the response of the following systems by recursion to $n = 4$ and try to discern the general form for $y[n]$.

(a) $y[n] - ay[n-1] = \delta[n]$ with $y[-1] = 0$

(b) $y[n] - ay[n-1] = u[n]$ with $y[-1] = 1$

(c) $y[n] - ay[n-1] = nu[n]$ with $y[-1] = 0$

(d) $y[n] + 4y[n-1] + 3y[n-2] = u[n-2]$ with $y[-1] = 0, y[-2] = 1$

D3.6 (DT System Response) Let $y[n] - \frac{1}{2}y[n-1] = x[n]$ with $y[-1] = -1$. Find the response of this system for the following choices of $x[n]$.

(a) 2

(b) $(\frac{1}{4})^n$

(c) $n(\frac{1}{4})^n$

(d) $(\frac{1}{2})^n$

(e) $n(\frac{1}{2})^n$

(f) $(\frac{1}{2})^n \cos(\frac{1}{2}n\pi)$

D3.7 (*DT System Response*) For each system, set up a difference equation and evaluate the natural, forced and total response. Assume $y[-1] = 0, y[-2] = 1$. Check your answer for the total response by comparing its first four values with those obtained by recursion.

(a) $[1 + 4z^{-1} + 3z^{-2}]y = u[n]$

(b) $[1 + 4z^{-1} + 4z^{-2}]y = 2^n u[n]$

(c) $[1 + 4z^{-1} + 8z^{-2}]y = \cos(n\pi)u[n]$

(d) $[(1 + 2z^{-1})^2]y = 2^n n u[n]$

(e) $[1 + \frac{3}{4}z^{-1} + \frac{1}{8}z^{-2}]y = (\frac{1}{3})^n u[n]$

(f) $[1 + \frac{1}{2}z^{-1} + \frac{1}{4}z^{-2}]y = \cos(\frac{1}{2}n\pi)u[n]$

(g) $[z^2 + 4z + 4]y = 2^n u[n]$

(h) $[1 - \frac{1}{2}z^{-1}]y = (\frac{1}{2})^n \cos(\frac{1}{2}n\pi)u[n]$

D3.8 (*DT System Response*) For each system, set up a difference equation and compute the zero-state, zero-input and total response assuming $x[n] = u[n]$ and $y[-1] = y[-2] = 1$.

(a) $[1 - z^{-1} - 2z^{-2}]y = x$

(b) $[z^2 - z - 2]y = x$

(c) $[1 - \frac{3}{4}z^{-1} + \frac{1}{8}z^{-2}]y = [z^{-1}]x$

(d) $[1 - \frac{3}{4}z^{-1} + \frac{1}{8}z^{-2}]y = [1 + z^{-1}]x$

(e) $[1 - \frac{1}{4}z^{-2}]y = x$

(f) $[z^2 - \frac{1}{4}]y = [2z^2 + 1]x$

D3.9 (**Impulse Response by Recursion**) Find the impulse response $h[n]$ by recursion up to $n = 4$ for each of the following systems.

(a) $y[n] - y[n-1] = 2x[n]$

(b) $y[n] - 3y[n-1] + 6y[n-2] = x[n-1]$

(c) $y[n] - 2y[n-3] = x[n-1]$

(d) $y[n] - y[n-1] + 6y[n-2] = nx[n-1] + 2x[n-3]$

D3.10 (**Analytical Form for Impulse Response**) For each digital filter:

(1) Classify as IIR (recursive) or FIR (nonrecursive),

(2) Classify as causal or noncausal,

(3) Find the general form for the impulse response and

(4) Check your results by comparing the first few sample values using recursion for each causal IIR filter.

(a) $y[n] = x[n] + x[n-1] + x[n-2]$

(b) $y[n] = x[n+1] + x[n] + x[n-1]$

(c) $y[n] + 2y[n-1] = x[n]$

(d) $y[n] + 2y[n-1] = x[n-1]$

(e) $y[n] + 2y[n-1] = 2x[n] + 6[n-1]$

(f) $y[n] + 2y[n-1] = x[n+1] +$
$4x[n] + 6x[n-1]$

(g) $[1 + 4z^{-1} + 3z^{-2}]y = [z^{-2}]x$

(h) $[z^2 + 4z + 4]y = [z + 3]x$

(i) $[z^2 + 4z + 8]y = x$

(j) $y[n] + 4y[n-1] + 4y[n-2] =$
$x[n] - x[n+2]$

D3.11 (**Stability**) Which of the systems of Problem D3.7 is stable? Explain.

D3.12 (**Nonlinear Systems**) One way to solve nonlinear difference equations is by recursion. Consider $y[n]y[n-1] - \frac{1}{2}y^2[n-1] = \frac{1}{2}Ku[n]$.

(a) What makes this system nonlinear?

(b) Using $y[-1] = 2$, recursively obtain $y[0]$, $y[1]$ and $y[2]$.

(c) Use $K = 2$, $K = 4$ and $K = 9$ in the results of part (b) to confirm that this system finds the square root of the input K.

(d) Repeat parts (b) and (c) with $y[-1] = 1$ to check whether the choice of the initial condition affects system operation.

D3.13 (Difference Equations from Differential Equations) Consider the differential equation $y''(t) + 3y'(t) + 2y(t) = 2u(t)$.

(a) Confirm that this describes a stable analog system.

(b) Convert this to a difference equation using the backward Euler algorithm and check the stability of the resulting digital filter.

(c) Convert this to a difference equation using the forward Euler algorithm and check the stability of the resulting digital filter.

(d) Which algorithm is better in terms of preserving stability? Can the result of part (c) be generalized to any arbitrary analog system?

D3.14 (LTI Concepts and Stability)

(a) Argue that neither of the following describes an LTI system.

  (1) $y[n] + 2y[n-1] = x[n] + x^2[n]$ (2) $y[n] - 0.5y[n-1] = nx[n] + x^2[n]$

(b) How can you check for the stability of each system?

D3.15 (Response of Causal and Noncausal Systems) A difference equation may describe a causal or noncausal system depending on how the initial conditions are prescribed. Consider a first-order system governed by $y[n] + \alpha y[n-1] = x[n]$.

(a) With $y[n] = 0, n < 0$, this describes a causal system. Assume $y[-1] = 0$ and find the first few terms $y[0], y[1], \ldots$ of the impulse response and step response using recursion and establish the general form for $y[n]$.

(b) With $y[n] = 0, n > 0$, we have a noncausal system. Assume $y[0] = 0$ and rewrite the difference equation as $y[n-1] = \{-y[n] + x[n]\}/\alpha$ to find the first few terms $y[0], y[-1], y[-2], \ldots$ of the impulse response and step response using recursion and establish the general form for $y[n]$.

4 CONVOLUTION

4.0 Scope and Objectives

Convolution is a central concept in relating the time and frequency domains. In the time domain, convolution is just a method of system analysis. Frequency domain viewpoints will be presented later, both in this chapter and others. This chapter is divided into four parts:

Part 1 The *convolution integral,* its evaluation and its properties.
Part 2 Special topics in continuous-time convolution and correlation.
Part 3 Numerical, sampled and discrete-time convolution.
Part 4 Special topics in discrete convolution, including deconvolution.

We also establish connections between time-domain and transformed-domain methods for signal and system analysis based on the convolution operation.

Useful Background   The following topics provide a useful background:

1. The sifting property of impulse functions (Chapter 2).
2. The concept of describing arbitrary signals as a superposition of shifted impulses (Chapter 2).
3. The concepts of linearity, time invariance and causality (Chapter 3).
4. The concept of the impulse response (Chapter 3).

4.1 Introduction

The convolution method for finding the system response $y(t)$ to an arbitrary input $x(t)$ relies on two constraints (necessary conditions):

1. The system must be relaxed and linear (superposition applies).
2. The system must be shift invariant.
We also assume that the system is described by its impulse response $h(t)$.

An informal way to establish a mathematical form for $y(t)$ is as follows. We start with a relaxed, linear, shift-invariant system whose impulse response is $h(t)$. Its response to a unit impulse $\delta(t)$ is then $h(t)$. Its response to a weighted, shifted impulse $K\delta(t - t_s)$, using shift invariance, is $Kh(t - t_s)$.

As described in Chapter 2, we can approximate $x(t)$ by an impulse train. We divide $x(t)$ into narrow rectangular strips of width t_s at kt_s, $k = 0, \pm1, \pm2, \ldots$ and replace each strip by an impulse whose strength $t_s x(kt_s)$ equals the area under each strip. We may thus approximate

$$x(t) \approx \sum_{k=-\infty}^{\infty} t_s x(kt_s)\delta(t - kt_s) \qquad \text{sum of shifted impulses}$$

Since $x(t)$ is a sum of weighted shifted impulses, the response $y(t)$, by superposition, is a sum of the weighted shifted impulse responses

$$y(t) = \sum_{k=-\infty}^{\infty} t_s x(kt_s)h(t - kt_s) \qquad \text{sum of shifted } h(t)$$

This process is illustrated in Figure 4.1.

Figure 4.1   The process of convolution. We approximate $x(t)$ by an impulse train and sum the responses to each impulse to obtain $y(t)$.

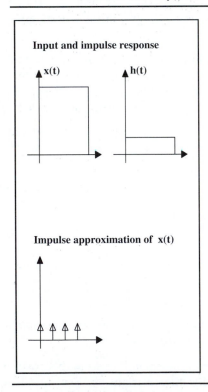
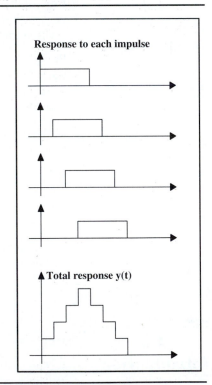

In the limit as $t_s \to d\lambda \to 0, kt_s$ describes a continuous variable λ and both $x(t)$ and $y(t)$ may be represented by integral forms to give

$$x(t) = \int_{-\infty}^{\infty} x(\lambda)\delta(t - \lambda)\,d\lambda \qquad y(t) = \int_{-\infty}^{\infty} x(\lambda)h(t - \lambda)\,d\lambda$$

The integral for $x(t)$ is simply based on the sifting property of impulses. The result

$$y(t) = \int_{-\infty}^{\infty} x(\lambda)h(t - \lambda)\,d\lambda$$

describes the **convolution integral** for finding system response. It forms the defining relation for all convolution operations.

4.1.1   Notation
The convolution operation is represented symbolically as follows:

$$y(t) = x(t) \star h(t)$$

Convolution involves the input and the impulse response as functions of the variable λ (Figure 4.2). Due to the folded, shifted function $h(t - \lambda)$, the convolution integral is also called the **folding integral.** The integrand $x(\lambda)h(t - \lambda)$ is called the **convolution kernel.**

4.1.2   The Convolution Process
In the convolution integral, t determines the relative location of $h(t - \lambda)$ with respect to $x(\lambda)$. The convolution will yield a nonzero result only for those values of t over which $h(t - \lambda)$ and $x(\lambda)$ overlap. The response $y(t)$ for all time requires the convolution for every value of t. We must evaluate the area under the product $x(\lambda)h(t - \lambda)$ as t varies. Varying t amounts to sliding the folded function $h(-\lambda)$ past $x(\lambda)$ by the chosen values of t.

4.1.3   What the Convolution Relation Tells Us
For most physical systems, $h(t), t \geq 0$ is usually infinite and decays to zero for large t. At any instant t_0, the response $y(t_0)$ may be regarded as the *cumulative interaction* of the input $x(\lambda)$ with the folded impulse response $h(t_0 - \lambda)$ as sketched in Figure 4.2. Due to the folding, later values of the input are multiplied or weighted by correspondingly earlier (and larger) values of the impulse response and summed to yield $y(t_0)$. We note that

1. The response at t_0 depends on both the present and past values of the input and impulse response. Recent inputs are weighted more heavily by the (folded) impulse response, which "remembers" recent inputs better than less recent ones as its memory fades (decays) into the past.
2. Future values of the input or impulse response have absolutely no influence on the present or previous response up to t_0.
3. For a system with no memory of the past, the response will depend only on the present value. Its impulse response is just an impulse.

Figure 4.2   **The functions $x(\lambda)$ and $h(t_0 - \lambda)$. Later values of the input are multiplied by earlier values of the impulse response to yield $y(t_0)$ as the area under the product $x(\lambda)\,h(t_0 - \lambda)$.**

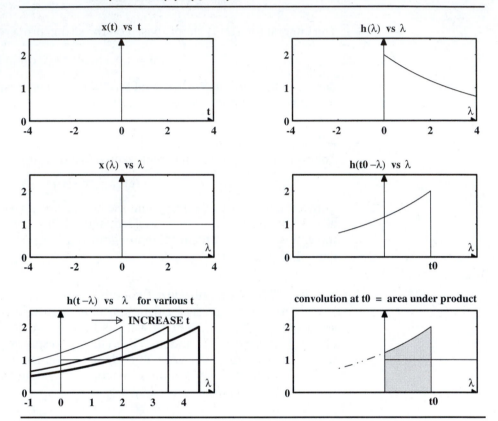

4.1.4   *Convolution as a Mathematical Operation*

Apart from its physical significance in representing the system response, the convolution integral is just another mathematical operation. A useful result is that it is symmetric with respect to the arguments of its kernel. It takes only a change of variable $\xi = t - \lambda$ to show that

$$x(t) \star h(t) = \int_{-\infty}^{\infty} x(\lambda)h(t - \lambda)\,d\lambda = -\int_{\infty}^{-\infty} x(t - \xi)h(\xi)\,d\xi$$

$$= \int_{-\infty}^{\infty} x(t - \xi)h(\xi)\,d\xi = h(t) \star x(t)$$

This is the *commutative property,* one where the order is unimportant. It says that, at least mathematically, we can switch the roles of the input and the impulse response for any system.

The convolution integral for two causal signals $x(t)u(t)$ and $h(t)u(t)$ involves the product $x(\lambda)u(\lambda)h(t - \lambda)u(t - \lambda)$. Since $u(\lambda)$ is zero for $\lambda < 0$ and $u(t - \lambda)$ is a leftsided step, which is zero for $\lambda > t$, the integrand is nonzero only over the range $0 \le \lambda \le t$. And since both $u(\lambda)$ and $u(t - \lambda)$ are unity in this range, the limits on the convolution integral simplify to

$$y(t) = \int_0^t x(\lambda) h(t - \lambda) \, d\lambda \qquad x(t) \text{ and } h(t) \text{ zero for } t < 0$$

This and other results are summarized in Table 4.1.

Table 4.1   The Nature of Convolution.

Nature of signals	Nature of convolution
Both twosided	Twosided
Both leftsided	Leftsided
Both rightsided	Rightsided
Both causal	Causal

4.2
Convolution of Some Common Signals

Convolution is an integral operation that may be evaluated analytically, graphically or numerically. The result depends on the nature of the signals being convolved. For example, we should expect to see

1. Linear forms if both signals are piecewise constant.
2. Quadratic forms if one is linear and the other is piecewise constant.
3. Cubic forms if both are piecewise linear.

Table 4.2 and Figure 4.3 illustrate some useful convolution results.

Table 4.2   Convolution of Common Signals.

Convolution	Result
$\delta(t) \star \delta(t)$	$\delta(t)$
$\delta(t) \star x(t)$	$x(t)$
$u(t) \star u(t)$	$tu(t) = r(t)$
$u(t) \star tu(t)$	$\frac{1}{2} t^2 u(t)$
$\text{rect}(t) \star \text{rect}(t)$	$\text{tri}(t)$
$u(t) \star e^{-t} u(t)$	$[1 - e^{-t}] u(t)$
$r(t) \star e^{-t} u(t)$	$r(t) - [1 - e^{-t}] u(t)$
$e^{-at} u(t) \star e^{-at} u(t)$	$t e^{-at} u(t)$
$e^{-at} u(t) \star e^{-bt} u(t)$	$\frac{1}{b-a} [e^{-at} - e^{-bt}] u(t), \qquad a \ne b$
$e^{-at} u(t) \star t e^{-at} u(t)$	$\frac{1}{2} t^2 e^{-at} u(t)$

Figure 4.3   Illustrating some of the convolutions listed in Table 4.2.

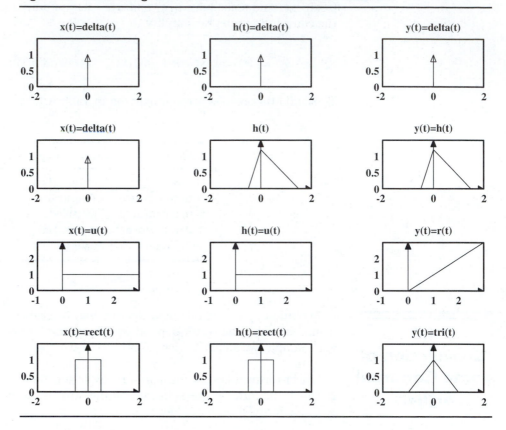

4.2.1   *Convolution with Impulses*

Since an impulse input $\delta(t)$ results in $h(t)$, we have the obvious result:

$$\delta(t) \star h(t) = h(t)$$

This says that the convolution of any signal $h(t)$ with an impulse reproduces the signal $h(t)$. We can use the sifting property to prove the same result.

With $h(t) = \delta(t)$, we have the less obvious result $\delta(t) \star \delta(t) = \delta(t)$.

4.2.2   *Analytical Evaluation of Convolution*

In analytical convolution, we describe $x(t)$ and $h(t)$ by single expressions using step functions. The result is also obtained as a single expression. In problem solving, it is important that you keep the following in mind:

I. In the convolution integral, t is a constant with respect to λ.

Figure 4.3   (continued)

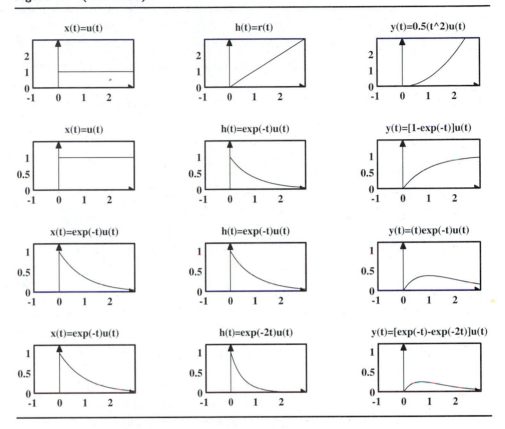

2. Both $x(\lambda)$ and $h(t - \lambda)$ are functions of λ (and not t).
3. Piecewise functions should be described in terms of step functions. The step functions with each term will determine the integration limits and the range of the result. Only after each result has been written out, also using steps, can they be combined into a single expression.

Example 4.1

Analytical Convolution of Some Simple Signals

(a) Consider the convolution of $x(t) = u(t)$ and $h(t) = tu(t)$. With $h(\lambda) = \lambda u(\lambda)$ and $x(t - \lambda) = u(t - \lambda)$, we have

$$y(t) = x(t) \star h(t) = \int_{-\infty}^{\infty} x(t - \lambda) h(\lambda) \, d\lambda = \int_{-\infty}^{\infty} \lambda u(t - \lambda) u(\lambda) \, d\lambda$$

If $t \geq 0$, the product $u(\lambda)u(t - \lambda)$ equals 1 from 0 to t. If $t < 0$, $u(\lambda)u(t - \lambda)$ is zero. We may then express the convolution as

$$y(t) = \int_0^t t \, d\lambda = \tfrac{1}{2}t^2 \quad t \geq 0 \quad \text{and} \quad y(t) = 0, \quad t < 0$$

This may be recast into a single expression as $y(t) = \tfrac{1}{2}t^2 u(t)$.

Comment: Note that $y(t)$ is zero for $t < 0$ because the product of $u(\lambda)$ and $u(t - \lambda)$ in the kernel is zero. Restated, the nonzero result holds as long as the upper limit exceeds the lower limit in the integration. The argument for the step $u(.)$ in the final result is the difference

$$u(\text{arg}) = u(\text{upper limit} - \text{lower limit}) = u(t - 0) = u(t)$$

Comment: We treat t as a constant with respect to the variable λ during integration.

(b) Consider an RC lowpass filter with $h(t) = (1/\tau)\exp(-t/\tau)u(t)$. The response $y(t)$ due to a unit step input $u(t)$ equals the convolution

$$y(t) = \int_{-\infty}^{\infty} x(t - \lambda)h(\lambda)\, d\lambda$$

$$= \frac{1}{\tau}\int_{-\infty}^{\infty} u(t - \lambda)\exp(-\lambda/\tau)u(\lambda)\, d\lambda = \frac{1}{\tau}\int_{0}^{t}\exp(-\lambda/\tau)\, d\lambda \qquad t \geq 0$$

Upon integration, we obtain the result

$$y(t) = -\exp(-\lambda/\tau)\Big|_{0}^{t} = 1 - \exp(-t/\tau) \qquad (t \geq 0) = [1 - \exp(-t/\tau)]u(t)$$

(c) Consider the convolution of $x(t) = u(t + 1) - u(t - 1)$ with itself. Changing the arguments to $x(\lambda)$ and $x(t - \lambda)$ results in the convolution

$$y(t) = \int_{-\infty}^{\infty} [u(\lambda + 1) - u(\lambda - 1)][u(t - \lambda + 1) - u(t - \lambda - 1)]\, d\lambda$$

This involves four products: $u(\lambda + 1)u(t - \lambda + 1)$, $-u(\lambda + 1)u(t - \lambda - 1)$, $-u(\lambda - 1)u(t - \lambda + 1)$ and $u(\lambda - 1)u(t - \lambda - 1)$. These must be integrated separately. The limits on each integral may be simplified by noting that

1. $u(t - \lambda + 1)$ is a leftsided step that terminates at $\lambda = t + 1$.
2. $u(t - \lambda - 1)$ is a leftsided step that terminates at $\lambda = t - 1$.

Upon using these simplifications, we get

$$y(t) = \int_{-1}^{t+1} d\lambda - \int_{-1}^{t-1} d\lambda - \int_{1}^{t+1} d\lambda + \int_{1}^{t-1} d\lambda$$

The four integrals equal $t + 2$, $-t$, $-t$ and $t - 2$, respectively. The ranges over which these hold are $t + 1 > -1$, $t - 1 > -1$, $t + 1 > 1$ and $t - 1 > 1$, respectively. Using these ranges and then combining the results gives the convolution $y(t)$ as

$$y(t) = (t + 2)u(t + 2) - tu(t) - tu(t) + (t - 2)u(t - 2) = r(t + 2) - 2r(t) + r(t - 2)$$

Comment: If we were to simply add the results without regard to their range of validity, we would obtain $y(t) = (t + 2) - t - t + (t - 2) = 0$, which holds only for $t \geq 2$.

Example 4.1 reveals some important results that can be generalized to the convolution $y(t)$ of any arbitrary signals $x(t)$ and $h(t)$ as follows:

1. If $x(t)$ is zero for $t < \alpha$ and $h(t)$ is zero for $t < \beta$, $y(t)$ starts (being nonzero) at $t = (\alpha + \beta)$.
2. The duration of $y(t)$ equals the sum of the durations of $x(t)$ and $h(t)$.
3. The area of $y(t)$ equals the product of the areas under $x(t)$ and $h(t)$.

Example 4.2
The Start of Convolution

(a) If both $x(t)$ and $h(t)$ start at $t = 0$, so does their convolution.
(b) If $x(t)$ starts at $t = 3$ and $h(t)$ at $t = -5$, the convolution starts at $t = -2$.
(c) The convolution $\delta(t - 3) \star \delta(t + 5)$ equals $\delta(t + 2)$.

4.3
Convolution by Ranges (Graphical Convolution)

Analytical convolution is tedious to express and sketch range by range. It is often easier to find such range-by-range results directly. The convolution over each range has a unique form. In some cases, the results may even be obtained by graphical evaluation of areas without solving integrals, which leads to the name **graphical convolution**.

In this approach, we start with $h(\lambda)$ and $x(t - \lambda)$ described by ranges. As we increase t and slide $x(t - \lambda)$ to the right from a position of no overlap, the overlap region changes, as do the expressions for their product and the resulting convolution (area). We have a new range for each new overlap region or product expression. The choice of which function to fold and slide is arbitrary. But choosing the one with the simpler representation leads to simpler integrals. Here is an illustration.

Example 4.3
The Concept of Convolution by Ranges

Consider the convolution of $x(t)$ and $h(t)$ described by

$$x(t) = \begin{cases} A & 0 \le t \le t_0 \\ 0 & \text{otherwise} \end{cases} \qquad h(t) = \begin{cases} \dfrac{1}{\tau} \exp(-t/\tau) & 0 \le t \le t_1 \\ 0 & \text{elsewhere} \end{cases}$$

Assuming $t_1 > t_0$, these functions, sketched in Figure 4.4, lead to $h(\lambda)$ and $x(t - \lambda)$, also sketched in Figure 4.4 and described by

$$h(\lambda) = \begin{cases} \dfrac{1}{\tau} \exp(-\lambda/\tau) & 0 \le \lambda \le t_1 \\ 0 & \text{elsewhere} \end{cases} \qquad x(t - \lambda) = \begin{cases} A & 0 \le t - \lambda \le t_0 \\ 0 & \text{otherwise} \end{cases}$$

The overlap between $x(t - \lambda)$ and $h(\lambda)$ starts only at $t = 0$. Thus, $y(t) = 0, t < 0$. As we shift $x(t - \lambda)$ past $h(\lambda)$, we observe (Figure 4.4)

1. Increasing, partial overlap over $0 \le t \le t_0$ with overlap limits $(0, t)$.
2. Total overlap over $t_0 \le t \le t_1$ with overlap limits $(t - t_0, t)$.
3. Partial, decreasing overlap over $t_1 \le t \le t_1 + t_0$ with limits $(t - t_0, t_1)$.
4. No overlap for $t \ge t_1 + t_0$.

During overlap, $x(t - \lambda)h(\lambda)$ always equals $(A/\tau)\exp(-\lambda/\tau)$ and the convolution $y(t)$ is described by using appropriate overlap limits for each range as

$$0 \le t \le t_0: y(t) = \int_0^t \frac{A}{\tau}\exp(-\lambda/\tau)\,d\lambda = A - A\exp(-t/\tau)$$

$$t_0 \le t \le t_1: y(t) = \int_{t-t_0}^t \frac{A}{\tau}\exp(-\lambda/\tau)\,d\lambda = A\exp[-(t-t_0)/\tau] - A\exp(-t/\tau)$$

$$t_1 \le t \le t_1 + t_0: y(t) = \int_{t-t_0}^{t_1} \frac{A}{\tau}\exp(-\lambda/\tau)\,d\lambda = A\exp[-(t-t_0)/\tau] - A\exp(-t_1/\tau)$$

The convolution $y(t)$ is thus nonzero for the following ranges (Figure 4.4)

$$y(t) = \begin{cases} A - A\exp(-t/\tau) & 0 \le t \le t_0 \\ A\exp[-(t-t_0)/\tau] - A\exp(-t/\tau) & t_0 \le t \le t_1 \\ A\exp[-(t-t_0)/\tau] - A\exp(-t_1/\tau) & t_1 \le t \le t_1 + t_0 \end{cases}$$

Comments: **1.** The convolution $y(t)$ matches at the endpoints of each range.

Figure 4.4   Illustrating convolution by ranges for Example 4.3.

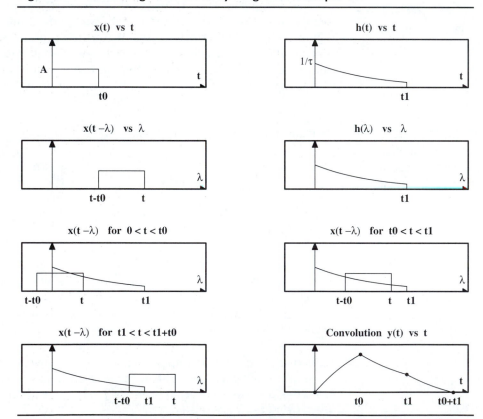

---

2. Full overlap lasts for $t_1 - t_0$, the *difference* of the pulse widths.
3. Both partial overlap regions last for t_0, the *smaller* pulse width.

4.3.1 The Recipe for Convolution by Ranges

The mechanics of convolution by ranges may be viewed as a bookkeeping operation. As we fold and slide, we must keep track of changes in the function shape, ranges, integration limits, and so on. The key is to be able to establish the correct ranges. Each range represents the largest duration over which the convolution is described by the same expression.

4.3.2 Convolution Ranges from the Pairwise Sum

A new range begins each time a range endpoint of one function slides past a range endpoint of the other. This is the basis for the following foolproof **pairwise sum rule** for obtaining the convolution ranges.

Pairwise Sum Rule

1. Set up two sequences containing the range endpoints of $x(t)$ and $h(t)$.
2. Form their pairwise sum. This is a sequence obtained by summing each value from one sequence with every value from the other.
3. Arrange the pairwise sum in increasing order and discard duplications.

The resulting sequence yields the endpoints for the convolution ranges.

Example 4.4
Illustrating The Pairwise Sum Rule

The pairwise sum of the sequences $\{0, 1, 3\}$ and $\{-2, 0, 2\}$ gives $\{-2, 0, 2, -1, 1, 3, 1, 3, 5\}$. The ordered sequence becomes $\{-2, -1, 0, 1, 1, 2, 3, 3, 5\}$.

Discarding duplications, we get $\{-2, -1, 0, 1, 2, 3, 5\}$. The ranges for nonzero convolution are then $-2 \le t \le -1$, $-1 \le t \le 0$, $0 \le t \le 1$, $1 \le t \le 2$, $2 \le t \le 3$ and $3 \le t \le 5$.

4.3.3 Graphical Convolution of Piecewise Constant Signals

The convolution of *piecewise constant* signals is linear over every range. All we then need is the convolution at the endpoints of each range, and we then connect the dots, so to speak. Since the product function is also piecewise constant, these convolution values can easily be calculated graphically.

Example 4.5
Convolution of Piecewise Constant Signals

Let $h(t) = u(t) + u(t - 1) + u(t - 2) - 3u(t - 3)$ and $x(t) = h(-t)$ (Figure 4.5). To find $y(t) = x(t) \star h(t)$, we identify the start of the convolution at $t = -3$ and convolution ranges at unit intervals up to $t = 3$. The product $h(\lambda)x(t - \lambda)$ with t at each endpoint yields the following convolution results (Figure 4.5):

$$y(-3) = 0 \quad y(-2) = 3 \quad y(-1) = 8 \quad y(0) = 14 \quad y(1) = 8 \quad y(2) = 3 \quad y(3) = 0$$

Note: The convolution $x(t) \star x(-t)$ is called the *autocorrelation* of $x(t)$ and is always even symmetric, with a maximum at the origin. We discuss autocorrelation in Section 4.9.

Figure 4.5   Illustrating convolution by ranges for Example 4.5.

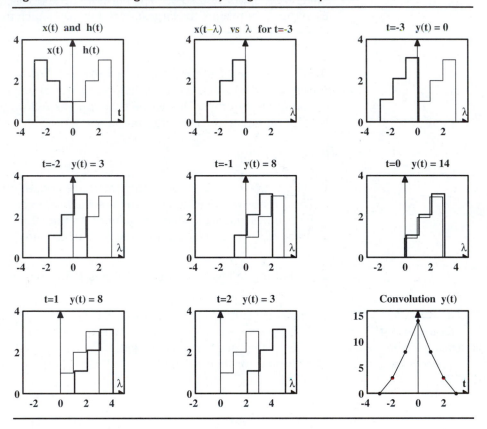

4.3.4   *Convolution by Ranges for Arbitrary Signals*

Here now is the recipe for convolution by ranges for arbitrary signals:

1. Establish the convolution ranges using the pairwise sum rule.
2. Sketch $h(\lambda)$ versus λ and determine its expressions.
3. Sketch $x(t - \lambda)$ versus λ and determine its expressions and range endpoints.
4. Select a convenient value for t in each range (t_L, t_H) (the midpoint, say) to locate $x(t - \lambda)$ relative to $h(\lambda)$. Since t could actually vary between t_L and t_H, *never substitute this numerical value for t in the sketches, range endpoints or expressions for $x(t - \lambda)$.*

5. Establish the overlap region and find the convolution as the integral (or area) of the product $x(t - \lambda)h(\lambda)$ to obtain $y(t)$ over this range.
6. *Consistency check*: The convolution value at endpoints of each range must match (unless convolving impulses and/or their derivatives).

Example 4.6
Convolution by Ranges

Consider the signals $x(t)$ and $h(t)$ (Figure 4.6 p. 130) given by

$$x(t) = \begin{cases} 2 & -2 \leq t \leq 1 \\ 0 & \text{elsewhere} \end{cases} \qquad h(t) = \begin{cases} t & 0 \leq t \leq 1 \\ 1 & 1 \leq t \leq 3 \\ 0 & \text{elsewhere} \end{cases}$$

The endpoints of $x(t)$ and $h(t)$ are $\{-2, 1\}$ and $\{0, 1, 3\}$. Their pairwise sum gives $\{-2, -1, 1, 1, 2, 4\}$. Discarding duplications, we get the range endpoints $\{-2, -1, 1, 2, 4\}$. The convolution ranges are $(-2, -1), (-1, 1), (1, 2)$ and $(2, 4)$. The interval-by-interval expressions for $x(t - \lambda)$ and $h(\lambda)$ are

$$x(t - \lambda) = \begin{cases} 2 & t - 1 \leq \lambda \leq t + 2 \\ 0 & \text{elsewhere} \end{cases} \qquad h(\lambda) = \begin{cases} \lambda & 0 \leq \lambda \leq 1 \\ 1 & 1 \leq \lambda \leq 3 \\ 0 & \text{elsewhere} \end{cases}$$

We fold $x(\lambda)$ and add t to each ordinate to get the range endpoints of $x(t - \lambda)$ as $\{t - 1, t + 2\}$. The range endpoints for $h(\lambda)$ are simply $\{0, 1, 3\}$.

We sketch $h(\lambda)$ and $x(t - \lambda)$ and label the range endpoints (Figure 4.6). For each range, we pick a value of t within that range (say, the midpoint) and position and sketch $x(t - \lambda)$ relative to $h(\lambda)$ (Figure 4.6). Note that we *must* label the range endpoints of $x(t - 2)$ as $\{t - 1, t + 2\}$ in all the sketches.

The convolution is found by setting up integrals with appropriate limits based on the expressions and overlap region of $h(\lambda)$ and $x(t - \lambda)$ for each range (Figure 4.6). We obtain the following results:

Range	Convolution
$-2 \leq t \leq -1$	$\int_0^{t+2} 2(\lambda)\, d\lambda = (\lambda^2)\Big\|_0^{t+2} = (t + 2)^2$
$-1 \leq t \leq 1$	$\int_0^1 2(\lambda)\, d\lambda + \int_1^{t+2} 2\, d\lambda = (\lambda^2)\Big\|_0^1 + (2\lambda)\Big\|_1^{t+2} = 2t + 3$
$1 \leq t \leq 2$	$\int_{t-1}^1 2(\lambda)\, d\lambda + \int_1^3 2\, d\lambda = (\lambda^2)\Big\|_{t-1}^1 + (2\lambda)\Big\|_1^3 = -t^2 + 2t + 4$
$2 \leq t \leq 4$	$\int_{t-1}^3 2\, d\lambda = (2\lambda)\Big\|_{t-1}^3 = -2t + 8$

Since $x(t)$ is constant while $h(t)$ is piecewise linear, their convolution must yield only linear or quadratic forms. Our results confirm this.

Consistency Checks: Since $h(t)$ is 3 units wide and starts at $t = 0$, and $x(t)$ is 3 units wide and starts at $t = -2$, the starting time of $y(t)$ is -2 and the duration of $y(t)$ is 6.

The convolution results do match at the endpoints of the various regions with $y(-2) = 0$, $y(-1) = 1$, $y(1) = 5$, $y(0) = -7$, $y(2) = 4$ and $y(4) = 0$.

Figure 4.6   Illustrating convolution by ranges for Example 4.6.

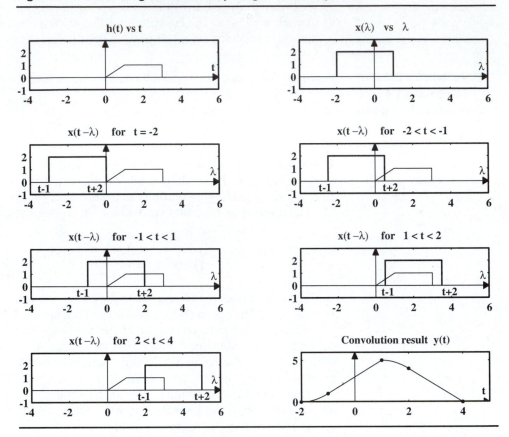

4.3.5   *An Application: Signal Averaging*

A **signal-averaging filter** is used for smoothing rapid fluctuations in a signal. It has an impulse response of the form $h(t) = (1/T)[u(t) - u(t - T)]$. To find the response $y(t)$ to a signal $x(t)$, we create $h(t - \lambda)$ and slide it past $x(\lambda)$. After the first T units of partial overlap, $y(t)$ may be written

$$y(t) = \frac{1}{T} \int_{t-T}^{t} x(\lambda)d\lambda \qquad t \geq T$$

The response equals the average of the input over the past T units and is, therefore, a smoother version of the input (Figure 4.7). A signal-averaging filter is also called a **smoothing filter** or **moving average filter**.

Figure 4.7 A signal-averaging filter.

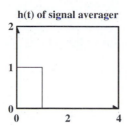
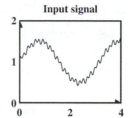
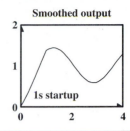

h(t) of signal averager · Input signal · Smoothed output

4.4 Some Properties of Convolution

The concept of linearity and shift invariance lie at the heart of many of the properties of convolution. Table 4.3 provides a useful list.

Table 4.3 A Summary of Convolution Properties.

Property	Result
Amplitude scaling	$Kx(t) \star h(t) = x(t) \star Kh(t) = Ky(t)$
Addition	$[x_1(t) + x_2(t)] \star h(t) = y_1(t) + y_2(t)$
Superposition	$[K_1 x_1(t) + K_2 x_2(t)] \star h(t) = K_1 y_1(t) + K_2 y_2(t)$
Delay	$x(t) \star h(t-\alpha) = x(t-\alpha) \star h(t) = y(t-\alpha)$
	$x(t-\alpha) \star h(t-\beta) = x(t-\beta) \star h(t-\alpha) = y(t-\alpha-\beta)$
Impulse response	$\delta(t) \star h(t) = h(t)$
Step response	$h(t) = y'_u(t)$
	$u(t) \star x(t) = \int_{-\infty}^{t} x(t)\,dt$
Derivatives	$x(t) \star h'(t) = x'(t) \star h(t) = y'(t)$
	$x'(t) \star h'(t) = y''(t)$
	$x^{(m)}(t) \star h(t) = x(t) \star h^{(m)}(t) = y^{(m)}(t)$
	$x^{(m)}(t) \star h^{(n)}(t) = x^{(n)}(t) \star h^{(m)}(t) = y^{(m+n)}(t)$
Time scaling	$x(\alpha t) \star h(\alpha t) = \left|\dfrac{1}{\alpha}\right| y(\alpha t)$
Areas	$\mathbb{A}[x(t)]\mathbb{A}[h(t)] = \mathbb{A}[y(t)]$
Duration	$T_x + T_h = T_y$
Symmetry	$x_e(t) \star h_e(t) = y_e(t)$
	$x_o(t) \star h_o(t) = y_e(t)$
	$x_o(t) \star h_e(t) = y_o(t)$
	$x_e(t) \star h_o(t) = y_o(t)$

4.4.1 *Properties Based on Linearity*

A relaxed linear system obeys superposition and suggests the following:

1. A scaling of the input results in an identical scaling of the output.

$$Kx(t) \star h(t) = Ky(t)$$

2. The sum of weighted inputs results in the sum of the weighted outputs

$$[K_1x_1(t) + K_2x_2(t)] \star h(t) = K_1y_1(t) + K_2y_2(t)$$

4.4.2 Properties Based on Shift Invariance

If the input is delayed by t_0, so too is the response. We then write

$$x(t - t_0) \star h(t) = y(t - t_0)$$

Mathematically, we can reverse the roles of $x(t)$ and $h(t)$ and argue that

$$x(t) \star h(t - t_0) = y(t - t_0) = x(t - t_0) \star h(t)$$

With $y(t) = x(t) \star h(t)$, a formal proof is as follows:

$$x(t) \star h(t - t_0) = \int_{-\infty}^{\infty} x(\lambda)h(t - t_0 - \lambda)\, d\lambda = y(t - t_0)$$

If both x and h are delayed, we use this property in succession to obtain

$$x(t - \alpha) \star h(t - \beta) = y(t - \alpha - \beta)$$

Example 4.7
Using the Shifting Property

Consider an RC lowpass filter with $h(t) = (1/\tau)\exp(-t/\tau)u(t)$. Its response to the step input $u(t)$ is, from Example 4.1b, $y_u(t) = [1 - \exp(-t/\tau)]u(t)$.

Using the shifting property and superposition, the response $y(t)$ of the filter to the input $x(t) = A[u(t) - u(t - t_0)]$ equals $y(t) = A[y_u(t) - y_u(t - t_0)]$ or

$$y(t) = A[1 - \exp(-t/\tau)]u(t) - A[1 - \exp\{-(t - t_0)/\tau\}]u(t - t_0)$$

Example 4.8
The Convolution of Two
Rectangular Pulses

Let $x_1(t) = 2[u(t + 1) - u(t - 1)]$ and $x_2(t) = [u(t + 1) - u(t - 1)]$. Their convolution $y(t) = x_1(t) \star x_2(t)$ may be described using superposition as

$$y(t) = 2[u(t + 1) \star u(t + 1) - u(t + 1) \star u(t - 1) - u(t - 1) \star u(t + 1)$$
$$+ u(t - 1) \star u(t - 1)]$$

Since $u(t) \star u(t) = r(t)$, we invoke time invariance for each term to give

$$y(t) = 2r(t + 2) - 4r(t) + 2r(t - 2) = 4\text{tri}(t/2)$$

4.4.3   *The Area Property*

The area under the convolution equals the product of the areas under each of the convolved functions. A formal proof involves interchanging the order of integration and noting that areas are shift invariant. We have

$$\mathbb{A}[y(t)] = \int_{-\infty}^{\infty} \int_{-\infty}^{\infty} x(\lambda)h(t-\lambda)\, d\lambda\, dt$$

$$= \int_{-\infty}^{\infty} \left[\int_{-\infty}^{\infty} h(t-\lambda)\, dt\right] x(\lambda)\, d\lambda = \mathbb{A}[h(t)]\mathbb{A}[x(t)]$$

Example 4.9

Illustrating the Area Property

Consider an *RC* lowpass filter with $h(t) = (1/\tau)\exp(-t/\tau)u(t)$. Its response to the pulse input $x(t) = A[u(t) - u(t-t_0)]$ was found in Example 4.7 as

$$y(t) = A[1 - \exp(-t/\tau)]u(t) - A[1 - \exp\{-(t-t_0)/\tau\}]u(t-t_0)$$

We can assert that the area of $y(t)$ equals $\mathbb{A}[h(t)]\mathbb{A}[x(t)] = (1)(At_0) = At_0$.

Comment: Try integrating $y(t)$ at your own risk to arrive at the same answer!

4.4.4   *The Time-Scaling Property*

If both $x(t)$ and $h(t)$ are scaled by α to $x(\alpha t)$ and $h(\alpha t)$, the duration property suggests that the convolution $y(t)$ is also scaled by α. We may write the scaled convolution as $Ky(\alpha t)$. To find K, we note that each of the areas, $\mathbb{A}[x(\alpha t)]$, $\mathbb{A}[h(\alpha t)]$ and $\mathbb{A}[y(\alpha t)]$, is scaled by $|\frac{1}{\alpha}|$. The product $\mathbb{A}[x(\alpha t)]\mathbb{A}[h(\alpha t)]$ is thus scaled by $1/\alpha^2$. Clearly, to satisfy the area property, we must choose $K = |\frac{1}{\alpha}|$. The scaling property then becomes $x(\alpha t) \star h(\alpha t) = |\frac{1}{\alpha}| y(\alpha t)$.

A formal proof uses the change of variable $\alpha\lambda = \xi$ as follows:

$$x(\alpha t)h(\alpha t) = \int_{-\infty}^{\infty} x(\alpha t - \alpha\lambda)h(\alpha\lambda)\, d\lambda = |\frac{1}{\alpha}| \int_{-\infty}^{\infty} x(\alpha t - \xi)h(\xi)\, d\xi = |\frac{1}{\alpha}| y(\alpha t)$$

Remarks: 1. The time-scaling property is valid only when *both functions are scaled by the same factor.*
2. If both functions are folded ($\alpha = -1$), so is their convolution.
3. The convolution of $x(t)$ with its folded version $x(-t)$ is always an *even* function with a maximum at $t = 0$. The convolution $x(t) \star x(-t)$ is called the **autocorrelation** of $x(t)$ and is discussed in Section 4.9.

Example 4.10

Using the Scaling Property

Consider the convolution $u(t) \star \exp(-t)u(t) = [1 - \exp(-t)]u(t)$

(a) Using the time-scaling property with scale factor $= \alpha$, we can write $u(\alpha t) \star \exp(-\alpha t)u(\alpha t) = (1/\alpha)[1 - \exp(-\alpha t)]u(\alpha t)$. Since $u(\alpha t) = u(t)$, we have the more general result $u(t) \star \exp(-\alpha t)u(t) = (1/\alpha)[1 - \exp(-\alpha t)]u(t)$.

(b) With $\alpha = -1$ in the scaling property (folding), we obtain

$$u(-t) \star \exp(t)u(-t) = [1 - \exp(t)]u(-t)$$

Comment: This confirms that the convolution of leftsided signals is also leftsided.

4.4.5   *Convolution of Symmetric Signals*

With $\alpha = -1$, the scaling property yields $x(-t) \star h(-t) = y(-t)$. As a result:

1. If $x(t) = x(-t)$ and $h(t) = h(-t)$ or $x(t) = -x(-t)$ and $h(t) = -h(-t)$, we have $y(-t) = y(t)$. Thus, the convolution of two even or two odd signals is even.
2. If $x(t) = x(-t)$ and $h(t) = -h(-t)$ or $x(t) = -x(-t)$ and $h(t) = h(-t)$, we have $y(-t) = -y(t)$. Thus, the convolution of an odd and an even signal is odd.

4.4.6   *Derivatives and Integrals*

Derivatives and integrals are linear operations. A linear operation on the input to a system results in a similar operation on the response. Thus, the input $x'(t)$ results in the response $y'(t)$ and we have $x'(t) \star h(t) = y'(t)$. The symmetry in the convolution kernel also allows us to write

$$x(t) \star h'(t) = x'(t) \star h(t) = y'(t)$$

Repeated differentiation of either $x(t)$ or $h(t)$ leads to the general results

$$x^{(m)}(t) \star h(t) = x(t) \star h^{(m)}(t) = y^{(m)}(t) \qquad x^{(m)}(t) \star h^{(n)}(t) = y^{(m+n)}(t)$$

4.4.7   *Relating the Step Response and Impulse Response*

The derivative property suggests that since the impulse is the derivative of the step, the impulse response is the derivative of the step response. If the step response is denoted $y_u(t)$, we have

$$y'_u(t) = u'(t) \star h(t) = \delta(t) \star h(t) = h(t)$$

We remark that finding the impulse response from the step response is not a very practical alternative because the derivative operation is very sensitive to noise.

Example 4.11
Impulse Response from
Step Response

The step response of a lowpass RC filter equals $y_s(t) = [1 - \exp(-t/\tau)]u(t)$. Its impulse response is then $h(t) = y'_u(t) = (1/\tau) \exp(-t/\tau)u(t)$.

Integration   Integration of $x(t)$ results in integration of the response. In particular, if $y'(t)$ represents the convolution $x'(t) \star h(t)$, then $y(t)$ may be obtained as the running integral of $y'(t)$:

$$y(t) = \int_{-\infty}^{t} y'(\xi)\, d\xi$$

If $y'(t)$ is defined over different ranges, $y(t)$ may also be expressed range by range. For the range $t \geq \alpha$, it takes the form

$$y(t) = \int_{-\infty}^{\alpha} y'(\xi)\, d\xi + \int_{\alpha}^{t} y'(\xi)\, d\xi = y(\alpha) + \int_{\alpha}^{t} y'(\xi)\, d\xi \qquad t \geq \alpha$$

Convolution with Steps   From this result, the convolution $x(t) \star u(t)$ equals

$$y(t) = x(t) \star u(t) = \int_{-\infty}^{\infty} x(\lambda) u(t-\lambda)\, d\lambda = \int_{-\infty}^{t} x(\lambda)\, d\lambda$$

The convolution of $x(t)$ with a step is simply the *running integral* of $x(t)$.

Example 4.12
Convolution with Steps and Superposition

(a) The convolution of $u(t)$ and the pulse $x(t) = \sin(t)[u(t) - u(t-4\pi)]$ equals

$$u(t) \star x(t) = \int_{-\infty}^{t} x(\xi)\, d\xi = \int_{0}^{t} \sin(\xi)\, d\xi = \begin{cases} 1 - \cos(t) & 0 \le t \le 4\pi \\ 0 & t \ge 4\pi \end{cases}$$

For $t \ge 4\pi$, $u(t) \star x(t)$ equals the area under $x(t)$, which is zero, because $x(t)$ describes two full cycles of $\sin(t)$.

(b) Using the shifting property, the convolution $u(t-\pi) \star x(t)$ is given by

$$u(t-\pi) \star x(t) = \begin{cases} 1 - \cos(t-\pi) & \pi \le t \le 5\pi \\ 0 & t \ge 5\pi \end{cases} = \begin{cases} 1 + \cos(t) & \pi \le t \le 5\pi \\ 0 & t \ge 5\pi \end{cases}$$

(c) If $h(t) = u(t) - u(t-\pi)$ (Figure 4.8), we can find $y(t) = h(t) \star x(t)$ using superposition. The result (Figure 4.8), expressed range by range, equals

$$h(t) \star x(t) = \begin{cases} 1 - \cos(t) & 0 \le t \le \pi \\ -2\cos(t) & \pi \le t \le 4\pi \\ -1 - \cos(t) & 4\pi \le t \le 5\pi \end{cases}$$

Figure 4.8   The convolution result for Example 4.12c.

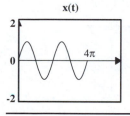
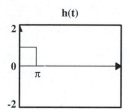
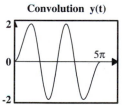

4.4.8   *Repeated Convolution*

Convolution is a commutative operation, and the convolution of $x(t)$ and $h(t)$ can be carried out in any order. Repeated convolution of *energy* signals is also associative, and order is unimportant. With three signals, for example,

$$x_1 \star (x_2 \star x_3) = (x_1 \star x_2) \star x_3 = (x_1 \star x_3) \star x_2 = (x_3 \star x_2) \star x_1 = \cdots$$

For energy signals, we also observe that

1. If all signals are onesided, the convolution is also onesided.
2. The convolution is smoother than any of the signals convolved.

Example 4.13
Repeated Convolution

Consider the convolution $y(t) = r(t) \star \exp(-t)u(t) = u(t) \star u(t) \star \exp(-t)u(t)$. Using the associative property and results from Table 4.2, we obtain

$$y(t) = u(t) \star [u(t) \star \exp(-t)u(t)] = u(t) \star [1 - \exp(-t)]u(t)$$

Finally, expanding this using superposition, we get

$$y(t) = u(t) \star u(t) - u(t) \star \exp(-t)u(t) = r(t) - [1 - \exp(-t)]u(t)$$

4.4.9   *Impulse Response of Systems in Cascade and Parallel*

For two systems in cascade, as shown in Figure 4.9, the response of the first system is simply $y_1(t) = x(t) \star h_1(t)$. The response $y(t)$ is then

$$y(t) = y_1(t) \star h_2(t) = [x(t) \star h_1(t)] \star h_2(t) = x(t) \star [h_1(t) \star h_2(t)]$$

Figure 4.9   Impulse response of systems in cascade and in parallel.

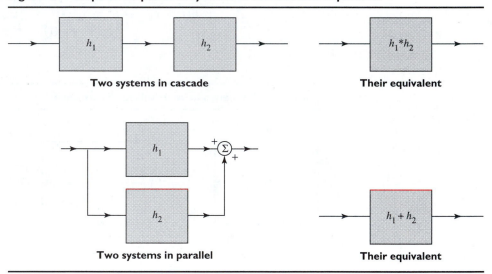

Two systems in cascade   Their equivalent

Two systems in parallel   Their equivalent

Clearly, if we wish to replace the cascaded system by an equivalent system with impulse response $h(t)$ such that $y(t) = x(t) \star h(t)$, it follows that

$$h(t) = h_1(t) \star h_2(t)$$

Generalizing this result, the impulse response $h(t)$ of N cascaded systems is simply the convolution of the N individual impulse responses

$$h(t) = h_1(t) \star h_2(t) \star \cdots \star h_N(t)$$

Remarks: 1. If the $h_k(t)$ are energy signals, the order of cascading is unimportant.
2. The overall impulse response of N systems in parallel (Figure 4.9) equals the sum of the N individual impulse responses.

4.5
The Response to Periodic Inputs

Consider a system with impulse response $h(t) = \exp(-t)u(t)$ excited by a harmonic input $x(t) = \exp(j\omega_0 t)$. The response $y(t)$ then equals

$$y(t) = \int_{-\infty}^{\infty} \exp[-(t-\lambda)]u(t-\lambda)\exp(j\omega_0\lambda)\,d\lambda$$

$$= \int_{-\infty}^{t} \exp(-t)\exp[\lambda(1+j\omega_0)]\,d\lambda = \frac{\exp(-t)}{1+j\omega_0}[\exp[\lambda(1+j\omega_0)]]\Big|_{-\infty}^{t}$$

$$= \frac{\exp(-t)}{1+j\omega_0}\exp[t(1+j\omega_0)] = \frac{1}{1+j\omega_0}\exp(j\omega_0 t) = Kx(t)$$

The response $y(t) = Kx(t)$, where $K = 1/(1+j\omega_0)$ is a constant, is also a harmonic at the input frequency ω_0. More generally, the response of LTI systems to any periodic input is also periodic with the same period as the input. In the parlance of convolution, *the convolution of two signals, one of which is periodic, is also periodic and has the same period.* Most periodic inputs are not as easy to handle as harmonic signals or sinusoids. We now present a formal approach to analyzing systems with periodic inputs.

4.5.1   *Periodic Extension and Wraparound*

If we know the response to one period of the input, we can find the entire response using superposition. For a system with $h(t) = (1/\tau)\exp(-t/\tau)u(t)$, the response to the pulse $x_1(t) = A[u(t) - u(t_0)]$ is (see Example 4.7)

$$y_1(t) = x_1(t) \star h(t) = \begin{cases} A[1 - \exp(-t/\tau)] & 0 \le t \le t_0 \\ A\exp(-t/\tau)[\exp(t_0/\tau) - 1] & t \ge t_0 \end{cases}$$

Now, if the input is a periodic rectangular pulse train with period T, the response $y_p(t)$ is the sum of $y_1(t)$ and its shifted versions:

$$y_p(t) = \cdots + y_1(t+2T) + y_1(t+T) + y_1(t) + y_1(t-T) + \cdots = \sum y_1(t+kT)$$

This is a periodic signal and called the **periodic extension** (PE) of $y_1(t)$.

More formally, the sum of a *finite area* signal or energy signal $x(t)$ and *all* its replicas shifted by multiples of T yields its periodic extension $x_{pe}(t)$, which is a periodic signal with period T. Thus

$$x_{pe} = \sum_{k=-\infty}^{\infty} x(t+kT)$$

Wraparound   Rather than add $x(t)$ and its shifted versions, we can chop $x(t)$ into one-period segments, line them up at $t = 0$, and add them (Figure 4.10). This **wraparound** gives us one period of the periodic extension. Table 4.4 lists the periodic extensions of several useful signals.

Figure 4.10   Illustrating periodic extension and wraparound.

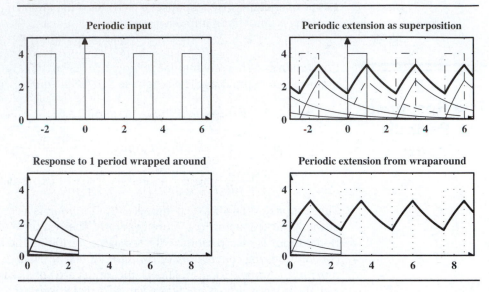

Table 4.4   Periodic Extensions of Various Signals.

Signal $x(t)$	Periodic extension $x_{\text{pe}}(t)$ for $(0,T)$
$\exp(-t/\tau)u(t)$	$\dfrac{\exp(-t/\tau)}{1-\exp(-T/\tau)}$
$t\exp(-t/\tau)u(t)$	$\dfrac{t\exp(-t/\tau)}{1-\exp(-T/\tau)} + \dfrac{T\exp[-(t+T)/\tau]}{[1-\exp(-T/\tau)]^2}$
$\exp(-\lvert t\rvert/\tau)$	$\dfrac{\exp(-\lvert t\rvert/\tau)+\exp(\lvert t\rvert/\tau)\exp(-T/\tau)}{1-\exp(-T/\tau)}\qquad (-\tfrac{1}{2}T, \tfrac{1}{2}T)$
$\operatorname{sinc}(Nt/T)$	$\operatorname{sinc}(Nt/T)/\operatorname{sinc}(t/T)$ (for odd integer N)

Remark: The area under one period of the periodic extension $x_{\text{pe}}(t)$ equals the total area under $x(t)$ and serves as a useful consistency check.

Example 4.14
Periodic Extension

(a) If $x(t)$ is a pulse of duration $\alpha \le T$, its periodic extension $x_{\text{pe}}(t)$ is just $x(t)$ with period T. If $\alpha > T$, $x_{\text{pe}}(t)$ involves wrapping around one-period segments of $x(t)$ past T units and adding to $x(t)$ (Figure 4.11).

(b) The periodic extension of $x(t) = \exp(-t/\tau)u(t)$ with period T using wraparound may be expressed as (Figure 4.11)

$$x_{\text{pe}}(t) = \exp(-t/\tau) + \exp[-(t+T)/\tau] + \exp[-(t+2T)/\tau] + \cdots$$

$$= \frac{\exp(-t/\tau)}{1 - \exp(-T/\tau)}$$

Figure 4.11   Periodic extension using wraparound for Example 4.14.

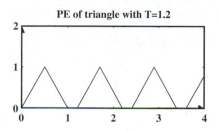

PE of triangle with T=1.2

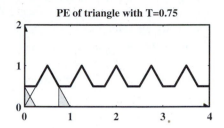

PE of triangle with T=0.75

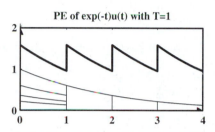

PE of exp(-t)u(t) with T=1

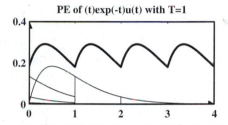

PE of (t)exp(-t)u(t) with T=1

(c)   Consider $x(t) = t \exp(-t/\tau)u(t)$. Wrapping (shifting left) one-period portions of $x(t)$ (Figure 4.11) results in $x_{\mathrm{pe}}(t)$ over $(0, T)$ as

$$x_{\mathrm{pe}}(t) = t \exp(-t/\tau) + (t + T) \exp[-(t + T)/\tau] + (t + 2T) \exp[-(t + 2T)/\tau] + \cdots$$
$$= t \exp(-t/\tau)[1 + \exp(-T) + \exp(-2T) + \cdots]$$
$$+ \exp(-t/\tau)[T \exp(-T/\tau) + 2T \exp(-2T/\tau) + \cdots]$$

Replacing the two infinite series by their closed-form sums (Appendix A)

$$x_{\mathrm{pe}}(t) = \frac{t \exp(-t/\tau)}{1 - \exp(-T/\tau)} + \exp(-t/\tau)\frac{T \exp(-T/\tau)}{[1 - \exp(-T/\tau)]^2}$$

4.5.2   *Response to Periodic Inputs Using Wraparound*

The response $y_p(t)$ to a periodic input $x_p(t)$ can also be found much more easily using periodic extension and wraparound as follows:

1. Create the periodic extension $h_{\mathrm{pe}}(t)$ of $h(t)$.
2. Find the convolution of one period of $h_{\mathrm{pe}}(t)$ and one period of $x_p(t)$.
3. Wrap this past one period to generate one period of $y_p(t)$.

Remark: The wraparound in the last step is necessary because the convolution of one period of $h_{\mathrm{pe}}(t)$ and one period of $x_p(t)$ extends for *two periods*.

Example 4.15

Convolution with
Periodic Signals Using
Wraparound

Consider a lowpass filter with $h(t) = (1/\tau)\exp(-t/\tau)u(t)$ excited by a rectangular pulse train with height A, pulse width t_0 and period T.

From Table 4.4, one period of the periodic extension of $h(t)$ equals

$$h_1(t) = \frac{(1/\tau)\exp(-t/\tau)}{1 - \exp(-T/\tau)}$$

Based on the results of Example 4.3, the convolution $x_1(t) \star h_1(t)$ yields

$$x_1(t) \star h_1(t) = \begin{cases} K[1 - \exp(-t/\tau)] & 0 \le t \le t_0 \\ K\{\exp[-(t - t_0)/\tau] - \exp(-t/\tau)\} & t_0 \le t \le T \\ K\{\exp[-(t - t_0)/\tau] - \exp(-T/\tau)\} & T \le t \le T + t_0 \end{cases}$$

where $K = A/[1 - \exp(-T/\tau)]$.

We wrap around this result past $t = T$ to get one period of $y_p(t)$ as

$$y_p(t) = \begin{cases} K[1 - \exp(-t/\tau)] + K\{\exp[-(t + T - t_0)/\tau] - \exp(-T/\tau)\} & 0 \le t \le t_0 \\ K\{\exp[-(t - t_0)/\tau] - \exp(-t/\tau)\} & t_0 \le t \le T \end{cases}$$

4.5.3   The Cyclic Approach

Yet another way to find the system response is to use only one period of the periodic extension $h_{pe}(\lambda)$ and the *entire* periodic signal $x_p(t - \lambda)$. Since $x_p(t - \lambda)$ is periodic, as portions of $x_p(t - \lambda)$ slide out to the right of the one period window, identical portions come into view (are wrapped around) from the left. This is the *cyclic* method with wraparound built in. After $x_p(t - \lambda)$ slides one full period, the convolution replicates.

Example 4.16

Convolution with
Periodic Signals Using the
Cyclic Method

Let $h(t) = (1/\tau)\exp(-t/\tau)u(t)$ and $x_1(t) = t/T, 0 \le t \le T$. Lining up $x_p(t - \lambda)$ with one period of $h_{pe}(\lambda)$ (Figure 4.12) shows portions of two pulses in partial view as $x_p(t - \lambda)$ is slid to the right for T units. The periodic output $y_p(t)$ over one period $(0, T)$ may then be written as

$$y_p(t) = \int_0^t \left[\frac{t - \lambda}{T}\right]\left[\frac{(1/\tau)\exp(-\lambda/\tau)}{1 - \exp(-T/\tau)}\right]d\lambda + \int_t^T \left[\frac{t - \lambda}{T} + 1\right]\left[\frac{(1/\tau)\exp(-\lambda/\tau)}{1 - \exp(-T/\tau)}\right]d\lambda$$

Simplification (some tedious algebra!) leads to the result (Figure 4.12)

$$y_p(t) = \frac{t - \tau}{T} + \frac{\exp(-t/\tau)}{1 - \exp(-T/\tau)}$$

Table 4.5 summarizes the methods for finding the response to periodic inputs. Note that all rely on periodic extension in one form or another.

Figure 4.12 Illustrating cyclic convolution for Example 4.16.

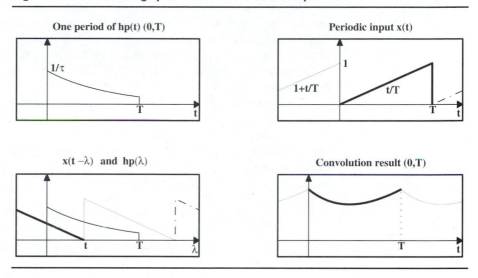

Table 4.5 System Response to Periodic Inputs.
Note: h_{pe} = periodic extension of h; h_1 = one period of h_{pe}

Method	Explanation
Superposition	Find $y_1(t) = x_1(t) \star h(t)$. Find its periodic extension $y_p(t)$.
Wraparound	Create $h_{pe}(t)$. Find $y_1(t) = x_1(t) \star h_1(t)$ and wrap around.
Cyclic	Create $h_{pe}(t)$. Shift $x_p(t - \lambda)$ past $h_1(\lambda)$ to give $y_p(t)$.

4.5.4 *Periodic Convolution of Two Periodic Signals*

Convolution involves finding the area under the product of one signal and a folded shifted version of another. Periodic signals are of infinite extent. For such signals, this product will be either *infinite or zero.* Clearly, *the convolution of two periodic signals does not exist in the usual sense.* Here is where the concept of **periodic convolution** comes in. The distinction between regular and periodic convolution is rather like the distinction between energy and power. For signals with infinite energy, a much more useful measure is the signal power, defined as the energy averaged over all time. Similarly, for signals whose convolution is infinite, the average convolution is a much more useful measure. This average is called periodic convolution. Formally, it is found by averaging the convolution of one periodic signal with a *finite stretch* T_0 of the other, as we let $T_0 \rightarrow \infty$.

Remark: Periodic convolution is also called **cyclic convolution** or **circular convolution.** If only one (or none) of the signals is periodic, we just perform their regular (also called **aperiodic** or **linear**) convolution.

4.5.5   *Periodic Convolution for Identical Periods*

For two periodic signals $x_p(t)$ and $h_p(t)$ with identical or common period T, the periodic convolution $y_p(t)$ for one period T may be found by convolving one period T of each signal and wrapping around the result past T. This is rather like finding the system response to periodic inputs, except that no periodic extension is required. We can even use the cyclic method (but not superposition) to compute the periodic convolution $y_p(t)$.

If we convolve one period of $x_p(t)$ with an N-period window of $h_p(t)$ and wrap around the result past one period T, the convolution is still periodic with period T but equals $Ny_p(t)$. Averaging or normalizing the convolution by N or NT is the way to eliminate the dependence of convolution on the window width. With a one-period window as the standard choice, we define periodic convolution for signals with identical (or common) period T as

$$y_p(t) = x_p(t) \bullet h_p(t) = \frac{1}{T} \int_0^T x_{1p}(\lambda) h_p(t - \lambda)\, d\lambda \qquad 0 \le t \le T$$

Notation: Periodic convolution is represented symbolically as $x_p(t) \bullet h_p(t)$.

Remark: The "periodic" convolution of other power signals with noncommensurate periods must be found from a limiting form (see Section 4.8.1).

Example 4.17
Periodic Convolution

Consider two periodic rectangular pulse trains with pulse widths of 3 and 2 units (Figure 4.13) and identical period T. Their regular convolution $y_L(t)$ is a trapezoid of width 5 units. Various choices for T yield the following results for their periodic convolution (Figure 4.13). For clarity, we have ignored the normalizing factor $1/T$ in the results.

T	Need wraparound?	Result
6	No	Identical to $y_L(t)(0 < t < 5)$ and zero for $5 < t < 6$
5	No	Identical to $y_L(t)$
4	Yes, past $t = 4$	Add $y_L(t)(0,1)$ to $y_L(t)(4,5)$
3	Yes, past $t = 3$	Add $y_L(t)(0,2)$ to $y_L(t)(3,5)$

4.6
Connections:
Convolution and
Transform
Methods

Not only is convolution an important method of system analysis in the time domain, it also leads to concepts that form the basis of every transform method described in this text. Its role in linking the time domain and the transformed domain is intimately tied to the concept of eigensignals and eigenvalues.

4.6.1   *Eigensignals and Eigenvalues*

The analysis of linear systems using convolution is based on expressing an input $x(t)$ as a weighted sum of shifted impulses, finding the response to each such impulse and using superposition to obtain the total response. The signal $x(t)$ can also

Figure 4.13 Periodic convolution results for Example 4.17.

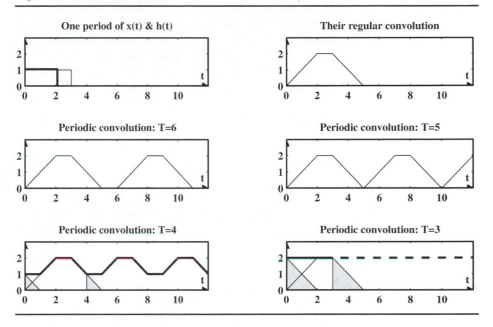

be described by a weighted combination of other useful signals besides impulses. Let $\phi_k(t)$ form a class of functions that result in the response $\psi_k(t)$. If $x(t)$ can be expressed as a weighted sum of these $\phi_k(t)$, the response $y(t)$ is a corresponding weighted sum of $\psi_k(t)$

$$x(t) = \sum_k \alpha_k \phi_k \qquad y(t) = \sum_k \beta_k \psi_k$$

What better choice for $\phi_k(t)$ than one that yields a response that is just a scaled version of itself such that $\psi_k(t) = A_k \phi_k(t)$? Then, $y(t)$ equals

$$y(t) = \sum_k \beta_k \psi_k = \sum_k \beta_k A_k \phi_k$$

Finding the output thus reduces to finding just the scale factors β_k, which may be real or complex. Signals ϕ_k that are preserved in form by a system except for a scale factor β_k are called **eigensignals, eigenfunctions** or **characteristic functions** because they are intrinsic (in German, *eigen*) to the system. The factor β_k by which the eigensignal is scaled is called the **eigenvalue** of the system or the **system function.** We may write

$$response = (eigensignal)(system\ function)$$

4.6.2 *Harmonics as Eigensignals of Linear Systems*

Sinusoids or harmonic signals form the most important class of eigensignals of linear systems. The system response to a harmonic signal $\exp(j\omega t)$ is another harmonic signal $K\exp(j\omega t)$ at the same frequency ω but scaled in magnitude and phase by the (possibly complex) constant K.

More generally, signals of the form $\exp(st)$, where $s = \sigma + j\omega = s + j2\pi f$ is a complex quantity, are also eigensignals of every linear system.

4.6.3 *Transformed Domain Relations*

Consider the eigensignal $x(t) = \exp(st)$ as the input to a linear system with impulse response $h(t)$. The response $y(t)$ equals the convolution

$$y(t) = x(t) \star h(t) = \int_{-\infty}^{\infty} \exp[s(t-\lambda)]h(\lambda)\,d\lambda = \exp(st)\int_{-\infty}^{\infty} h(\lambda)\exp(-s\lambda)\,d\lambda$$

The response equals the eigensignal $\exp(st)$ times the system function, which is a function only of s. If we denote this system function $H(s)$, we have

$$H(s) = \int_{-\infty}^{\infty} h(\lambda)\exp(-s\lambda)\,d\lambda$$

This is also called the **transfer function**. It is actually a description in terms of a weighted sum of complex exponentials and is, in general, also complex. Now, the signal $x(t)$ also yields a similar description

$$X(s) = \int_{-\infty}^{\infty} x(\lambda)\exp(-s\lambda)\,d\lambda$$

This defines the **twosided Laplace transform** $X(s)$ of $x(t)$.

The transform of the response $y(t) = x(t)H(s)$ may also be written as

$$Y(s) = \int_{-\infty}^{\infty} y(t)\exp(-st)\,dt = \int_{-\infty}^{\infty} H(s)x(t)\exp(-st)\,dt = H(s)X(s)$$

Since $y(t)$ also equals $x(t) \star h(t)$, *convolution in the time domain is equivalent to multiplication in the transformed domain.*

With only slight modifications, we can describe several other transformed-domain relations, as summarized in Table 4.6 and explained below:

1. For causal signals and systems, the lower limit becomes zero and we obtain the **onesided Laplace transform**, or simply the **Laplace transform**.
2. With $s = j2\pi f$, we use $\exp(j2\pi f)$ as the eigensignals and transform $h(t)$ to the frequency domain in terms of its **steady-state transfer function** $H(f)$ and the signal $x(t)$ to its **Fourier transform** $X(f)$.
3. For a single harmonic $x(t) = \exp(j2\pi f_0 t)$, the impulse response $h(t)$ transforms to a complex number $H(f_0) = K\exp(j\theta)$. This produces the response $K\exp[j(2\pi f_0 t + \theta)]$ and describes the method of **phasor analysis**.
4. For a periodic signal $x_p(t)$ described by a combination of harmonically related exponentials $\exp(jk2\pi f_0 t)$ at the discrete frequencies kf_0, we obtain a frequency domain description of $x_p(t)$ over one period in terms of its **Fourier series** coefficients $X_S[k]$.

Table 4.6   Eigensignals and System Functions.

Eigensignal	System function	Transform method
$\exp(st)$	$H(s) = \int_{-\infty}^{\infty} h(\lambda)\exp(-s\lambda)\,d\lambda$	Twosided Laplace transform
$\exp(st)u(t)$	$H(s) = \int_{0}^{\infty} h(\lambda)\exp(-s\lambda)\,d\lambda$	Laplace transform
$\exp(j2\pi ft)$	$H(f) = \int_{-\infty}^{\infty} h(\lambda)\exp(-j2\pi f\lambda)\,d\lambda$	Fourier transform
$\exp(j2\pi f_0 t)$	$H(f_0) = K\exp(j\theta)$	Phasor analysis

4.7
Stability and Causality

In Chapter 3, we examined the stability of LTI systems that were described by differential equations. Bounded-input, bounded-output (BIBO) stability of such systems required every root of the characteristic equation to have negative real parts. For systems described by their impulse response, this actually translates to an equivalent condition that involves constraints on the nature of the system impulse response $h(t)$.

If $x(t)$ is bounded such that $|x(t)| < M$, then its shifted version $x(t - \lambda)$ is also bounded. Using the *fundamental theorem of calculus* (the absolute value of any integral cannot exceed the integral of the absolute value of its integrand), the convolution integral yields the following inequality

$$|y(t)| < \int_{-\infty}^{\infty} |h(\tau)|\,|x(t-\tau)|\,d\tau < M\int_{-\infty}^{\infty} |h(\tau)|\,d\tau$$

It follows immediately that for $y(t)$ to be bounded, we require

$$\int_{-\infty}^{\infty} |h(\tau)|\,d\tau < \infty$$

For BIBO stability, therefore, $h(t)$ must be absolutely integrable. This is both a necessary and sufficient condition. If satisfied, we are guaranteed a stable system.

Remarks: **1.** We always have a stable system if $h(t)$ is an energy signal.
2. Since the impulse response is defined only for linear systems, the stability of nonlinear systems usually must be checked by other means.

4.7.1   *The Impulse Response of Causal Systems*

For a causal system, we require an impulse response of the form $h(t)u(t)$ which equals zero for $t < 0$. This ensures that an input $x(t)u(t - t_0)$ that starts at $t = t_0$ results in a response $y(t)$ that also starts at $t = t_0$. We see this by setting up the convolution integral as

$$y(t) = \int_{-\infty}^{\infty} x(\lambda)u(\lambda - t_0)h(t - \lambda)u(t - \lambda)\,d\lambda = \int_{t_0}^{t} x(\lambda)h(t - \lambda)\,d\lambda$$

Remarks: **1.** This innocuous result actually imposes a powerful constraint on $h(t)$. The even and odd parts of the $h(t)$ are not independent and $h(t)$ can, in fact, be found from its even (or odd) part alone. We pursue this further in Chapter 7.
2. Systems with $h(t) = 0, t < 0$ are also called **physically realizable**.

Example 4.18
Stability of a
Differentiating System

The system $y(t) = x'(t)$ has the impulse response $h(t) = \delta'(t)$. It is not BIBO stable since $\mathbb{A}[|h(t)|]$ is not finite. The (bounded) input $u(t)$, for example, results in the response $\delta(t)$, which is clearly unbounded (at $t = 0$).

Comment: Extending this concept, for a stable system we require the highest derivative of the input not to exceed the highest derivative of the output. This implies an operational transfer function that is not proper.

———

Example 4.19
Stability of an Ideal Filter

In Chapter 7, we shall learn that an ideal lowpass filter is described by an impulse response of the form $h(t) = \text{sinc}(f_B t)$. Since $h(t)$ is not zero for $t < 0$, this filter is physically unrealizable. Since $|h(t)| = |\text{sinc}(f_B t)|$ is not absolutely integrable, it is also unstable.

———

Signals for which the convolution does not exist are not hard to find. The convolution of the step $u(t)$ and the folded step $u(-t)$ is one such example. Unfortunately, the question of existence can be answered only in terms of sufficient conditions that do not preclude the convolution of functions for which such conditions are violated. These conditions are related primarily to signal energy, area, or onesidedness. The results are summarized in Table 4.7. Note that the convolution of energy signals and the convolution of two same-sided signals always exist.

Table 4.7   Existence of Convolution.

Nature of $x(t)$ and $h(t)$	Example
Both energy signals	$\exp(-t)u(t) \star \text{rect}(t)$
Both timelimited	$\text{rect}(t) \star \text{tri}(t)$
Both rightsided	$u(t) \star u(t-1)$
Both leftsided	$tu(-t) \star u(-t)$
$x(t)$ or $h(t)$ timelimited	$\text{rect}(t) \star \exp(-|t|)$
Both absolutely integrable over $(-\infty, 0)$	$tu(t+2) \star u(t+5)$
Both absolutely integrable over $(0, \infty)$	$tu(2-t) \star u(5-t)$
$x(t)$ or $h(t)$ twosided and absolutely integrable over $(-\infty, \infty)$	$\text{rect}(t) \star \text{sgn}(t)$
Both twosided and absolutely integrable over $(-\infty, \infty)$	$\exp(-|t|) \star \exp(-|t|)$

Signals for which convolution does not exist include
$1 \star 1, 1 \star u(t), \cos(t) \star u(t), \cos(t) \star \sin(t)u(t), \cos(t) \star \cos(t), \exp(t) \star \exp(t)$

4.8.1   *Existence of Convolution for Periodic Signals*

The regular convolution of two periodic functions is identically zero if the periods are noncommensurate and one of them has a zero average value. Otherwise the regular convolution does not exist (is infinite). This is why we turn to averages to obtain the periodic convolution. We define "periodic" convolution by using a folded version of one signal as it slides past the other over a longer and longer duration T_0. As $T_0 \rightarrow \infty$, we obtain

$$y(t) = x_p(t) \bullet h_p(t) = \lim_{T_0 \to \infty} \frac{1}{T_0} \int_{T_0} x_T(\lambda) h(t - \lambda) \, d\lambda$$

This form is typically used for almost periodic and random signals. If both signals are periodic with identical period T, we replace T_0, the averaging duration, by the period T to yield the much simpler form defined previously in Section 4.5.5.

4.8.2   *Convolution Properties Based on Moments*

The nth moment of a signal $x(t)$ is defined as $m_n = \mathbb{A}[t^n x(t)]$. Thus $m_0 = \mathbb{A}[x(t)]$ is just the area under $x(t)$. The area property suggests the following obvious result for the zeroth moments:

$$\mathbb{A}[y] = \mathbb{A}[x]\mathbb{A}[h] \qquad \text{or} \qquad m_0(y) = m_0(x) m_0(h)$$

The quantity $D_x = m_x = m_1/m_0 = \mathbb{A}[tx(t)]/\mathbb{A}[x(t)]$ corresponds to the **effective delay** or mean for an energy signal. The delays are also related by

$$D_y = D_x + D_h \qquad \text{or} \qquad \frac{m_1(y)}{m_0(y)} = \frac{m_1(h)}{m_0(h)} + \frac{m_1(x)}{m_0(x)}$$

Central moments (about the mean) are defined as $\mu_n = \mathbb{A}[(t - m_x)^n x(t)]$. The variance $\mu_2 = \mathbb{A}[(t - m_x)^2 x(t)] = [m_2(x)/m_0(x)] - m_x^2 = T_x^2$ is a measure of the **effective duration** for an energy signal. The durations are related by

$$T_y = \{T_x^2 + T_h^2\}^{\frac{1}{2}} \qquad \text{or} \qquad \mu_2(y) = \mu_2(x) + \mu_2(h)$$

Cascaded Systems   The impulse response of a cascade of linear, shift-invariant systems with impulse responses $h_1(t), h_2(t), \ldots$ is described by the repeated convolution

$$h(t) = h_1(t) \star h_2(t) \star h_3(t) \star \cdots$$

The moment properties suggest that the overall system delay D equals the sum of delays of the individual systems and the overall duration T equals the square root of the sum of squares of the individual durations.

Example 4.20
Moment Properties and
Cascaded Systems

Consider a lowpass filter with $h(t) = \exp(-t)u(t)$. The moments of $h(t)$ are

$$m_0(h) = \mathbb{A}[h(t)] = 1 \quad m_1(h) = \mathbb{A}[th(t)] = 1 \quad m_2(h) = \mathbb{A}[t^2 h(t)] = 2 \quad D_h = 1$$

(a) For an input $x(t) = h(t)$, $y(t) = t\exp(-t)u(t)$. The moments of $y(t)$ are

$$m_0(y) = \mathbb{A}[y(t)] = 1 \qquad m_1(y) = \mathbb{A}[ty(t)] = 2 \qquad D_y = 2$$

We see that $m_0(y) = m_0(x) m_0(h) = 1$ and $D_y = D_x + D_h = 2$.

(b) We compute the effective filter duration T_h as

$$T_h^2 = \mathbb{A}[(t - m_h)^2 h(t)] = [m_2(h)/m_0(h)] - m_h^2 = 2 - 1 = 1$$

For a cascade of N identical lowpass filters, the overall effective delay D equals $ND_h = N$ and the overall effective duration T equals $[NT_h^2]^{\frac{1}{2}} = \sqrt{N}$. ____

4.8.3   *Repeated Convolution and the Central Limit Theorem*

Repeated convolution of a function with itself or other functions is much smoother than any of the functions convolved. For many energy signals, repeated convolution begins to take on the bell-shaped Gaussian form as the number n of functions convolved becomes large and may be expressed as

$$y_N(t) = x_1(t) \star x_2(t) \star \cdots \star x_N(t) \approx \frac{K}{\sqrt{2\pi\sigma_N^2}} \exp\left[\frac{(t - m_N)^2}{2\sigma_N^2}\right]$$

Here, m_N is the sum of the individual means (delays), σ_N^2 is the sum of the individual variances and the constant K equals the product of the areas under each of the convolved functions.

$$K = \prod_{k=1}^{n} \mathbb{A}[x_k(t)]$$

This result is one manifestation of the **central limit theorem**. It allows us to assert that the response of a complex system comprising many subsystems is Gaussian, since its response is based on repeated convolution. The individual responses need not be Gaussian and need not even be known.

Remarks: **1.** The central limit theorem fails if any function has zero area, making $K = 0$. Sufficient conditions for it to hold require a finite average variance and finite absolute third moments. All timelimited functions and many others satisfy these rather weak conditions.

2. The system function $H(f)$ of a large number of cascaded systems is also a Gaussian because convolution in the time domain is equivalent to multiplication in the frequency domain.

3. In probability theory, the central limit theorem asserts that the sum of N statistically independent random variables approaches a Gaussian for large N, regardless of the nature of their distributions.

____

Example 4.21
Illustrating the Central
Limit Theorem

(a) Consider a cascade of identical RC lowpass filters with $h(t) = \exp(-t)u(t)$. The impulse response $y_n(t)$ of the cascaded system is simply the repeated convolution of $h(t)$ with itself. The first few convolutions yield

$$y_1(t) = t\exp(-t)u(t) \qquad y_2(t) = \tfrac{1}{2}t^2\exp(-t)u(t) \qquad y_3(t) = (\tfrac{1}{3})t^3\exp(-t)u(t)$$

If the convolution is repeated n times, the result generalizes to

$$y_n(t) = \exp(-t)u(t) \star \exp(-t)u(t) \star \cdots \star \exp(-t)u(t) = \frac{t^n}{n!}\exp(-t)u(t)$$

The peak shifts to the right by one unit for each successive convolution. The results (Figure 4.14) reveal the progression toward the bell-shaped Gaussian with increasing n. To find the Gaussian form as $n \to \infty$, we start with the mean m_h, variance σ^2 and area A for $\exp(-t)u(t)$. From Example 4.20,

$$A = m_0 = \mathbb{A}[h(t)] = 1 \qquad m_h = \frac{m_1}{m_0} = 1 \qquad \sigma^2 = 1$$

For n cascaded systems, we have $m_N = nm_h = n$, $\sigma_N^2 = n\sigma^2 = n$, and $K = A^n = 1$.

These values lead to the Gaussian approximation for $y_n(t)$ as

$$y_n(t) \approx \frac{1}{\sqrt{2\pi n}} \exp\left[-\frac{(t-n)^2}{2n}\right]$$

(b) An even more striking example is provided by the convolution of several even symmetric rectangular pulses (Figure 4.14), which begins to take on a Gaussian look after only a few repeated convolutions.

Figure 4.14   Illustrating the central limit theorem for repeated convolution of exponentials and rect pulses as it approaches a Gaussian form.

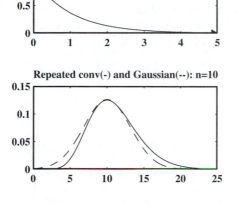
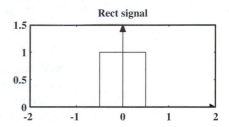
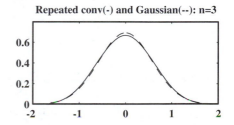
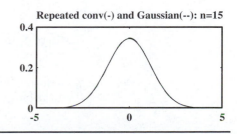

4.9
Special Topics: Correlation

Correlation is an operation similar to convolution. It involves sliding one function past the other and finding the area under the resulting product. Unlike convolution, however, no folding is performed. The correlation $R_{xx}(t)$ of two identical functions $x(t)$ is called **autocorrelation**. For two different functions $x(t)$ and $y(t)$, the correlation $R_{xy}(t)$ or $R_{yx}(t)$ is referred to as **cross correlation**.

Using the symbol $\star\star$ to denote correlation, we define the two operations as

$$R_{xx}(t) = x(t) \star\star x(t) = \int_{-\infty}^{\infty} x(\lambda)x(\lambda - t)\, d\lambda$$

$$R_{xy}(t) = x(t) \star\star y(t) = \int_{-\infty}^{\infty} x(\lambda)y(\lambda - t)\, d\lambda$$

$$R_{yx}(t) = y(t) \star\star x(t) = \int_{-\infty}^{\infty} y(\lambda)x(\lambda - t)\, d\lambda$$

A change of variable leads to the equivalent formulations

$$R_{xx}(t) = \int_{-\infty}^{\infty} x(\lambda + t)x(\lambda)\, d\lambda$$

$$R_{xy}(t) = \int_{-\infty}^{\infty} x(\lambda + t)y(\lambda)\, d\lambda$$

$$R_{yx}(t) = \int_{-\infty}^{\infty} y(\lambda + t)x(\lambda)\, d\lambda$$

Remarks: **1.** The variable t is often referred to as the **lag**.
2. Some authors prefer to switch the definitions of $R_{xy}(t)$ and $R_{yx}(t)$.

4.9.1   *Some Properties of Correlation*

Cross correlation is a measure of similarity between different functions. Since the order matters in cross correlation, we have two cross correlation functions given by $R_{xy}(t)$ and $R_{yx}(t)$ as

$$R_{xy}(t) = \int_{-\infty}^{\infty} x(\lambda)y(\lambda - t)\, d\lambda \qquad R_{yx}(t) = \int_{-\infty}^{\infty} y(\lambda)x(\lambda - t)\, d\lambda$$

At $t = 0$, we have

$$R_{xy}(0) = \int_{-\infty}^{\infty} x(\lambda)y(\lambda)\, d\lambda = R_{yx}(0)$$

Thus $R_{xy}(0) = R_{yx}(0)$. The cross correlation also satisfies the inequality

$$|R_{xy}(t)| \le [R_{xx}(0)R_{yy}(0)]^{\frac{1}{2}} = [E_x E_y]^{\frac{1}{2}} \text{ or } [P_x P_y]^{\frac{1}{2}}$$

where E and P represent the signal energy and signal power, respectively.

Area and Duration   Since folding does not affect the area or duration, the area and duration properties for convolution also apply to correlation.

Commutation   The absence of folding means that the correlation depends on which function is shifted and, in general, $x(t) \star\star y(t) \neq y(t) \star\star x(t)$. Since shifting one function to the right is actually equivalent to shifting the other function to the left by an equal amount, the correlation $R_{xy}(t)$ is related to $R_{yx}(t)$ by $R_{xy}(t) = R_{yx}(-t)$.

Correlation as Convolution   The absence of folding actually implies that the correlation of $x(t)$ and $y(t)$ is equivalent to the convolution of $x(t)$ with the folded version $y(-t)$, and we have $R_{xy}(t) = x(t) \star\star y(t) = x(t) \star y(-t)$.

Existence   Correlation of $x(t)$ and $h(-t)$ follows the existence conditions for convolution provided they are now applied to $x(t)$ and $h(-t)$ (or $x(-t)$ and $h(t)$).

Periodic Correlation   The correlation of two periodic functions or power signals is defined in the same sense as periodic convolution

$$R_{xy}(t) = \frac{1}{T}\int_T x(\lambda)y(\lambda - t)\, d\lambda \qquad R_{xy}(t) = \lim_{T_0 \to \infty} \frac{1}{T_0}\int_{T_0} x(\lambda)y(\lambda - t)\, d\lambda$$

The first form defines the correlation of periodic functions with identical periods T, and is also periodic with the same period T. The second form is typically reserved for nonperiodic power signals or random signals.

4.9.2   *Autocorrelation*

The autocorrelation operation involves identical functions. It can thus be performed in any order and represents a commutative operation.

What Autocorrelation Means   Autocorrelation may be viewed as a measure of similarity, or *coherence*, between a function $x(t)$ and its shifted version. Clearly, under no shift, the two functions "match" and result in a maximum for the autocorrelation (Figure 4.15). But with increasing shift, it would be natural to expect the similarity and hence the correlation between $x(t)$ and its shifted version to decrease. And as the shift approaches infinity, all traces of similarity vanish and the autocorrelation decays to zero.

Remark: If we shift a periodic signal past itself, the two line up after every period and make the autocorrelation periodic with the same period.

Symmetry   Since $R_{xy}(t) = R_{yx}(-t)$, we have $R_{xx}(t) = R_{xx}(-t)$. This means that *the autocorrelation of a real function is even.* The autocorrelation of an even function $x(t)$ also equals the convolution of $x(t)$ with itself, because the folding operation leaves an even function unchanged.

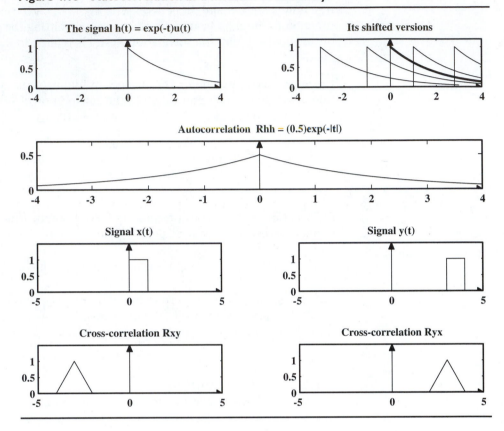

Figure 4.15   Autocorrelation as a measure of similarity.

Maximum Value   It turns out that the autocorrelation function is symmetric about the origin where it attains its maximum value. It thus satisfies

$$R_{xx}(t) \leq R_{xx}(0)$$

It follows that *the autocorrelation $R_{xx}(t)$ is finite and nonnegative for all t.*

Signal Energy and Power   The value of the autocorrelation function at the origin is related to the signal energy (or power for periodic signals) by

$$R_{xx}(0) = \int_{-\infty}^{\infty} x(\lambda) x(\lambda - 0)\, d\lambda = \int_{-\infty}^{\infty} x^2(\lambda)\, d\lambda \qquad R_{xx}(0) = \frac{1}{T} \int_{T} x_p^2(\lambda)\, d\lambda$$

Example 4.22
Autocorrelation and
Cross Correlation

(a) For $t > 0$, the autocorrelation of $h(t) = \exp(-t)u(t)$ (Figure 4.15) equals

$$R_{hh}(t) = \int_{t}^{\infty} e^{-\lambda} e^{-(\lambda-t)}\, d\lambda = e^{t} \int_{t}^{\infty} e^{-2\lambda}\, d\lambda = \tfrac{1}{2} e^{-t} \qquad t > 0$$

Since $R_{hh}(t) = R_{hh}(-t)$, we have, $R_{hh}(t) = \tfrac{1}{2}\exp(-|t|)$ (Figure 4.15). As ex-

Figure 4.16   The autocorrelation of x(t) for Example 4.22b.

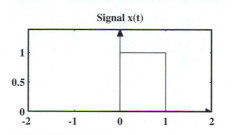
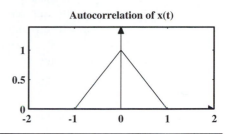

pected, $R_{hh}(t)$ shows a maximum at $t = 0$.

(b) Let $x(t) = \text{rect}(t - 0.5)$ and $y(t) = \text{rect}(t - 3.5)$ (Figure 4.15). If we shift $y(t)$, we obtain $R_{xy}(t) = \text{tri}(t + 3)$ (Figure 4.15). If we shift $x(t)$, we obtain $R_{yx}(t) = \text{tri}(t - 3)$ (Figure 4.15). The two are folded versions of each other. The autocorrelation R_{xx} is shown in Figure 4.16.

4.10
The Numerical Approach to Convolution

The idea of numerical convolution arises naturally from situations where the functions to be convolved cannot be represented by simple mathematical expressions. Numerical convolution requires numerical evaluation of the integral or area under the product function $x(\lambda)h(t - \lambda)$ from values of $x(t)$ and $h(t)$, usually at uniformly spaced intervals t_s (Figure 4.17).

Figure 4.17   Illustrating the numerical approach to convolution.

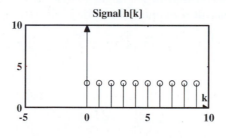
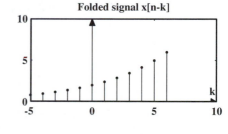
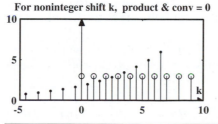
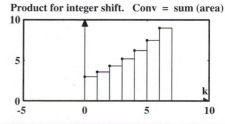

The Step Interpolation Algorithm   The simplest algorithm for finding the area under a function is the **rectangular rule**. We approximate the area as the sum of areas under rectangular strips of width t_s and heights that equal the function values (the product $x(\lambda)h(t - \lambda)$ in our case) at the instants nt_s, as illustrated in Figure 4.17. It results in a *staircase*, or stepwise approximation. Since we are, in effect, interpolating to a constant value between samples, the algorithm is often referred to as the **step interpolation** or **constant interpolation** algorithm.

4.10.1   *The Convolution Sum*

The area under the rectangular strips of the product function may now be written as $t_s x_s[k]h_s[n - k]$, and their sum (the total area) represents the **sampled convolution** $y_s[n]$ of two sampled sequences $x_S[n]$ and $h_S[n]$ at $t = nt_s$:

$$y_S[n] \approx \sum_{k=-\infty}^{\infty} x_S[k]h_S[n - k]t_s = t_s \sum_{k=-\infty}^{\infty} x_S[k]h_S[n - k]$$

This relation, also called the **convolution sum,** provides us with the value of the convolution $y_s[n]$ at any arbitrary time instant nt_s . As in the continuous case, we can interchange the arguments of x and h without affecting the result.

$$y_S[n] = t_s \sum_{k=-\infty}^{\infty} x_S[n - k]h_S[k]t_s = t_s \sum_{k=-\infty}^{\infty} x_S[k]h_S[n - k]$$

For causal $h_S[n]$ and $x_S[n]$, this relation simplifies to

$$y_S[n] = t_s \sum_{k=0}^{n} x_S[k]h_S[n - k] = t_s \sum_{k=0}^{n} x_S[n - k]h_S[k]$$

4.10.2   *Discrete Convolution*

The numerical convolution of *DT* signals with $t_s = 1$ using step interpolation is called **discrete convolution** and defined by the relation

$$y[n] = x[n] \star h[n] = \sum_{k=-\infty}^{\infty} x[k]h[n - k]$$

4.10.3   *Analytical Evaluation of DT Convolution*

Table 4.8 and Figure 4.18 provide a list of discrete convolution results. The procedure for analytical convolution mimics the continuous case and may be implemented quite readily if $x[n]$ and $h[n]$ are described by simple enough analytical expressions. Often, we must resort to a table of closed-form solutions for finite or infinite series.

In keeping with the results of *CT* convolution, the convolution of $x[n]$ with a *DT* impulse replicates $x[n]$, whereas the convolution of $x[n]$ with a *DT* step results in the running sum of $x[n]$:

$$x[n] \star \delta[n] = x[n] \qquad x[n] \star u[n] = \sum_{k=-\infty}^{n} x[k]$$

Figure 4.18   Some discrete convolution results of Table 4.8.

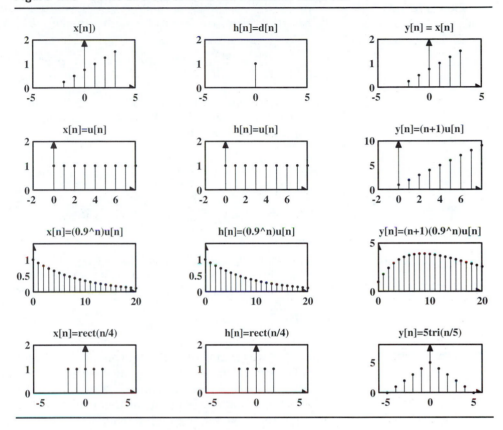

**Table 4.8   Discrete Convolution of Some
            Common Signals.**

Convolution	Result
$\delta[n] \star \delta[n]$	$\delta[n]$
$\delta[n] \star x[n]$	$x[n]$
$u[n] \star u[n]$	$(n+1)u[n]$
$r[n] \star u[n]$	$\frac{1}{2}n(n+1)u[n]$
$r[n] \star r[n]$	$\frac{1}{6}n(n^2-1)u[n]$
$a^n u[n] \star u[n]$	$\frac{1-a^{n+1}}{1-a}u[n]$
$a^n u[n] \star a^n u[n]$	$(n+1)a^n u[n]$
$a^n u[n] \star b^n u[n]$	$\frac{a^{n+1}-b^{n+1}}{a-b}u[n] \quad a \ne b$
$a^n u[n] \star r[n]$	$\frac{a(a^n-1)+n(1-a)}{(1-a)^2}u[n]$
$\text{rect}(n/2N) \star \text{rect}(n/2N)$	$(2N+1)\text{tri}[n/(2N+1)]$

Example 4.23
Analytical Evaluation of
DT Convolution

(a) Let $x[n] = h[n] = a^n u[n]$, $a < 1$. Then $x[k] = a^k u[k]$ and $h[n-k] = a^{n-k} u[n-k]$. The lower limit on the convolution sum simplifies to $k = 0$ (due to $u[k]$), the upper limit to $k = n$ (due to $u[n-k]$), and we get

$$y[n] = \sum_{k=-\infty}^{\infty} a^k a^{n-k} u[k] u[n-k] = \sum_{k=0}^{n} a^k a^{n-k} = a^n \sum_{k=0}^{n} 1 = (n+1) u[n]$$

(b) Let $x[n] = nu[n+1], h[n] = a^{-n} u[n], a < 1$. Since $h[n-k] = a^{-(n-k)} u[n-k]$ and $x[k] = ku[k+1]$, the lower and upper limits on the convolution sum become $k = -1$ and $k = n$. Then

$$y[n] = \sum_{k=-1}^{n} k a^{-(n-k)} = -a^{-n-1} + a^{-n} \sum_{k=0}^{n} k a^k$$

$$= -a^{-n-1} + \frac{a^{-n+1}}{(1-a)^2} [1 - (n+1) a^n + n a^{n+1}]$$

Comment: The results of analytical DT convolution often involve finite or infinite summations. In part (b) we used results from Appendix A to generate the closed-form solution. This may not always be feasible, however.

4.11 Convolution of Finite Sequences

In practice, we often work with sequences of finite length. In keeping with the process of convolution, we fold one sequence and sum the products as this folded sequence is successively shifted, one index at a time, past the other. The sampled convolution is found by multiplying the results by t_s. This process is so much simpler than the convolution of CT signals. We now describe the various methods of implementing discrete convolution.

4.11.1   *The Sliding Strip Method*

One method of computing $y[n]$ is to list the values of the folded function on a strip of paper and slide it along the stationary function, to better visualize the process. This technique has prompted the name **sliding strip method**. We simulate this method by showing the successive positions of the stationary and folded sequence along with the resulting products, the convolution sum and the actual convolution.

Example 4.24
Convolution by the
Sliding Strip Method

Let $h_S[n] = \{2, 5, 0, 4\}, x_S[n] = \{4, 1, 3\}, t_s = \frac{1}{2}$. If both sequences start at $n = 0$, the folded sequence $x_S[-k]$ equals $x[-k] = \{3, 1, \overset{\uparrow}{4}\}$.

We line up the folded sequence to begin overlap and shift it successively, summing the product sequence as we go, to obtain the discrete convolution. To get the numerical convolution, we multiply by t_s . Here are the results:

	$t = 0$	$t = t_s$	$t = 2t_s$
x	2 5 0 4	0 2 5 0 4	2 5 0 4
h	3 1 4	3 1 4	3 1 4
xh	0 0 8 0 0 0	0 2 20	6 5 0
	SUM 8	SUM 22	SUM 11
	$y_s[0]$ 4	$y_s[1]$ 11	$y_s[2]$ 5.5

	$t = 3t_s$	$t = 4t_s$	$t = 5t_s$
x	2 5 0 4	2 5 0 4	2 5 0 4
h	3 1 4	3 1 4	3 1 4
xh	15 0 16	0 4	12
	SUM 31	SUM 4	SUM 12
	$y_s[3]$ 15.5	$y_s[4]$ 2	$y_s[5]$ 6

The discrete convolution $y[n]$ is $\{8, 22, 11, 31, 4, 12\}$. The numerical convolution $y_S[n]$ is $\{4, 11, 5.5, 15.5, 2, 6\}$.

4.11.2   The Duration Property
The preceding example reveals two important concepts in discrete convolution.

Start   The starting index of the convolution equals the sum of the starting indices of the signals being convolved. If the two signals start at the indices $n = n_0$ and $n = n_1$, the convolution starts at the index $n = n_0 + n_1$.

Duration   The convolution extends from 0 to 2.5 units and satisfies the duration property. For sequences of length M and N, the durations equal $(M - 1)t_s$ and $(N - 1)t_s$ (since there are two endpoints). Their convolution extends for $(M + N - 2)t_s$ units, and thus corresponds to a sequence with $M + N - 1$ samples. Denoting the number of samples in the convolution sequence by S, we have the simple relation

$$S = M + N - 1$$

4.11.3   Multiplication and Discrete Convolution
Discrete convolution is tied intimately to the operations of both polynomial and numeric multiplication. We show this by considering the results of the previous example:

$$h[n] = \{2, 5, 0, 4\} \qquad x[n] = \{4, 1, 3\} \qquad y[n] = \{8, 22, 11, 31, 4, 12\}$$

Convolution and Polynomial Multiplication Let $h[n]$ and $x[n]$ represent the coefficients of two polynomials $h(w)$ and $x(w)$ in descending order such that

$$h(w) = 2w^3 + 5w^2 + 0w + 4 \qquad x(w) = 4w^2 + 1w + 3$$

It should not take a lot to convince you that the product $y(w) = x(w)h(w)$ is

$$y(w) = 8w^5 + 22w^4 + 11w^3 + 31w^2 + 4w + 12$$

whose coefficients correspond exactly to the discrete convolution sequence.

Convolution and Numeric Multiplication What may surprise you is that if we let w represent the number 10, we get

$$y(w) = 8(10)^5 + 22(10)^4 + 11(10)^3 + 31(10)^2 + 4(10) + 12 = 1{,}034{,}152$$

which equals just the product of the two numbers 2,504 and 413.

Here is an easier way to obtain this product. We rewrite the convolution sequence as numbers with *carries*, and then, starting from the right, add the carries to the numbers on the left as we go.

$$8 \quad 22 \quad 11 \quad 31 \quad 4 \quad 12 \quad \longrightarrow \quad 8 \quad 2^2 \quad 1^1 \quad 1^3 \quad 4 \quad 2^1$$

Then starting from the right, we keep adding the carries to their neighbors to generate the identical result:

$$8 \quad 22 \quad 11 \quad 31 \quad 4 \quad 12 \quad \longrightarrow \quad 8 \quad 2^2 \quad 1^1 \quad 1^3 \quad 4 \quad 2^1 \quad \longrightarrow \quad 1 \quad 0 \quad 3 \quad 4 \quad 1 \quad 5 \quad 2$$

4.11.4   *Algorithms for Discrete Convolution*

There are several methods for evaluating the discrete convolution based on algorithms for polynomial and numeric multiplication or the sliding strip method presented earlier. We describe these below.

The Sum-by-Column Method We set up a row of index values beginning with the starting index of the convolution and $h[n]$ and $x[n]$ below it. We regard $x[n]$ as a sequence of weighted shifted impulses. Each element (impulse) of $x[n]$ generates a shifted impulse response (product with $h[n]$) starting at its index (to indicate the shift). Summing the response (by columns) gives the discrete convolution. Note that none of the sequences is folded.

It is better (if only to save paper) to let $x[n]$ be the shorter sequence. For sampled convolution, we multiply the convolution sequence by t_s. Even the numerical convolution may be obtained by using numerical integration formulas to the columnwise values which represent the product sequences.

Example 4.25
The Sum-by-Column
Method of Discrete
Convolution

(a) Let $h[n] = \{2, 5, 0, 4\}$ and $x[n] = \{4, 1, 3\}$. Their convolution starts at $n = -3$. The following diagrams illustrate how the algorithm can generate both the convolution and the numeric product of the two sequences.

Convolution   Product

n	-3	-2	-1	0	1	2
h	2	5	0	4		
x	4	1	3			
	08	20	00	16		
		02	05	00	04	
			06	15	00	12
y	08	22	11	31	04	12

$$
\begin{array}{ccccccc}
& 2 & 5 & 0 & 4 & \times & 4\ 1\ 3 \\
& & 7 & 5 & 1 & 2 & \\
& 2 & 5 & 0 & 4 & & \\
\hline
1 & 0 & 0 & 1 & 6 & & \\
1 & 0 & 3 & 4 & 1 & 5 & 2 \\
\hline
\end{array}
$$

$$8 \quad 22 \quad 11 \quad 31 \quad 4 \quad 12 \quad \longrightarrow \quad 8 \quad 2^2 \quad 1^1 \quad 1^3 \quad 4 \quad 2^1 \quad \longrightarrow \quad 1\ 0\ 3\ 4\ 1\ 5\ 2$$

Convolution   Use carries & carry over   Multiplication

(b) If both $x[n]$ and $h[n]$ started at $n = 0$, $y[n]$ would also start at $n = 0$.

(c) If $t_s = 2$, the sampled convolution would equal $\{16, 44, 22, 63, 8, 24\}$.

(d) Using $t_s = 1$ and the trapezoidal algorithm $\mathbb{A} = t_s[\frac{1}{2}y_0 + y_1 + y_2 + \cdots + \frac{1}{2}y_N]$ to compute the area for each column, we obtain the numerical convolution as

$$\{\tfrac{1}{2}(8), \quad \tfrac{1}{2}(20+2), \quad \tfrac{1}{2}(0) + 5 + \tfrac{1}{2}(6), \quad \tfrac{1}{2}(16) + 0 + \tfrac{1}{2}(15), \quad \tfrac{1}{2}(4+0), \quad \tfrac{1}{2}(12)\}$$

This yields the numerical convolution sequence $y_S[n] = \{4, 11, 8, 15.5, 2, 6\}$.

The Lattice Method   Another algorithm relies on generating an $M \times N$ lattice with values for one sequence entered at the top and for the other on the side of this lattice. The squares are then filled in with the corresponding products. Values along the off-diagonals now represent the product function at each time instant, and the sum along these, when read from left to right, yields the discrete convolution sequence. Again, no folding is required.

Example 4.26
The Lattice Method for
Discrete Convolution

We illustrate the lattice algorithm for $h[n] = \{2, 5, 0, 4\}$ and $x[n] = \{4, 1, 3\}$.

Note: The lattice algorithm can also generate the product of two sequences directly if we split each cell and enter the products as shown. Summation along the off-diagonals proceeds from right to left and uses carries. The final result is read from left to right, as usual.

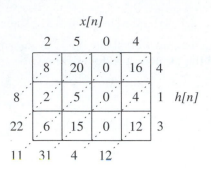

Convolution = [8 22 11 31 4 12]

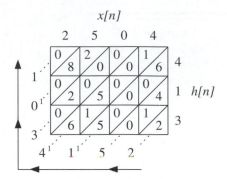

Product = 1,034,152

Example 4.27
The Response of an
Averaging Filter

Given the input $\{x[n]\} = \{2, 4, 6, 8, 10, 12, \ldots\}$, what system will result in the response $\{y[n]\} = \{1, 3, 5, 7, 9, 11, \ldots\}$? At each instant, the response is the average of the input and its previous value. This system describes an averaging or moving average filter. Its difference equation is simply $y[n] = \frac{1}{2}\{x[n] + x[n-1]\}$. Its impulse response is thus $h[n] = \frac{1}{2}[\delta[n] + \delta[n-1]]$, or $\{h[n]\} = \{\frac{1}{2}, \frac{1}{2}\}$. Using discrete convolution, we find the response as follows

x:	2	4	6	8	10	12	$\ldots$	
h:	$\frac{1}{2}$	$\frac{1}{2}$						
		1	2	3	4	5	6	$\ldots$
			1	2	3	4	5	6	$\ldots$
y:	1	3	5	7	9	11	$\ldots$	

This is indeed what we expected.

4.12
**Properties of
DT Convolution**

Table 4.9 compares the properties of continuous-time and discrete-time convolution. The definitions involve integration for analog signals and summation for discrete sequences.

Step Response   For continuous time functions, the impulse response is the derivative of the step response, whereas for discrete signals it is the *first difference* of the step response. The convolution of a signal with a step is the running integral (running sum) of the signal itself.

Derivatives   The results for discrete sequences involve differences and superposition. The first derivative (difference) of both functions results in the second derivative (difference) of the convolution

$$x'(t) \star h'(t) = y''(t)$$
$$\{x[n] - x[n-1]\} \star \{h[n] - h[n-1]\} = y[n] - 2y[n-1] + y[n-2]$$

Table 4.9 **Properties of Continuous and Discrete Convolution.**

Property	Continuous convolution	Discrete convolution
Basic result	$y(t) = \int_{-\infty}^{\infty} x(\lambda)h(t-\lambda)\,d\lambda$	$y[n] = \sum_{k=-\infty}^{\infty} x[k]h[n-k]$
Causal signals	$\int_0^t x(\lambda)h(t-\lambda)\,d\lambda$	$\sum_{k=0}^{n} x[k]h[n-k]$
Amplitude scaling	$Kx(t) \star h(t) = Ky(t)$	$Kx[n] \star h[n] = Ky[n]$
Addition	$[x_1 + x_2] \star h = y_1 + y_2$	$[x_1 + x_2] \star h = y_1 + y_2$
Superposition	$[Ax_1 + Bx_2] \star h = Ay_1 + By_2$	$[Ax_1 + Bx_2] \star h = Ay_1 + By_2$
Delay	$x(t) \star h(t-\alpha) = x(t-\alpha) \star h(t) = y(t-\alpha)$	$x[n] \star h[n-K] = x[n-K] \star h[n] = y[n-K]$
	$x(t-\alpha) \star h(t-\beta) = y(t-\alpha-\beta)$	$x[n-M] \star h[n-K] = y[n-M-K]$
Impulse response	$\delta(t) \star h(t) = h(t)$	$\delta[n] \star h[n] = h[n]$
Step response	$h(t) = y_u'(t)$	$h[n] = y_u[n] - y_u[n-1]$
Convolution with steps	$u(t) \star x(t) = \int_{-\infty}^{t} x(t)\,dt$	$u[n] \star x[n] = \sum_{k=-\infty}^{n} x[k]$
Areas (sums)	$\left[\int_{-\infty}^{\infty} x(t)\,dt\right]\left[\int_{-\infty}^{\infty} h(t)\,dt\right] = \int_{-\infty}^{\infty} y(t)\,dt$	$\left[\sum_{k=-\infty}^{\infty} x[k]\right]\left[\sum_{k=-\infty}^{\infty} h[k]\right] = \sum_{k=-\infty}^{\infty} y[k]$
Duration	$T_y = T_x + T_h$	$N_y = N_x + N_h - 1$
Derivatives	$x'(t) \star h(t) = y'(t)$	$\{x[n] - x[n-1]\} \star h[n] = y[n] - y[n-1]$
Symmetry	$x_e(t) \star h_e(t) = y_e(t)$	$x_e[n] \star h_e[n] = y_e[n]$
	$x_o(t) \star h_o(t) = y_e(t)$	$x_o[n] \star h_o[n] = y_e[n]$
	$x_o(t) \star h_e(t) = y_o(t)$	$x_o[n] \star h_e[n] = y_o[n]$
	$x_e(t) \star h_o(t) = y_o(t)$	$x_e[n] \star h_o[n] = y_o[n]$

Area In analogy with *CT* signals, the sum of the convolution samples equals the product of the sums of samples of the two functions convolved. This is a useful consistency check for finite-length sequences.

4.12.1 *Zero Interpolation*
Based on the results of polynomial multiplication, if we insert m zeros between each sample of the convolved sequences, the regular convolution sequence also shows m zeros between its samples.

Example 4.28
The Effect of Zero Insertion

From Example 4.24, we have $h[n] = \{2, 5, 0, 4\}, x[n] = \{4, 1, 3\}$, whose convolution equals $y[n] = \{8, 22, 11, 31, 4, 12\}$. With one zero between each sample, we obtain

$$h_z[n] = \{2, 0, 5, 0, 0, 0, 4\} \qquad x_z[n] = \{4, 0, 1, 0, 3\}$$

To find their convolution, we set up the polynomials

$$h_z(w) = 2w^6 + 5w^4 + 0w^2 + 4 \qquad x_z(w) = 4w^4 + 1w^2 + 3$$

Replacing w^2 by v, we get polynomials identical to the unpadded sequences

$$h_z(v) = 2v^3 + 5v^2 + 0v + 4 \qquad x_z(v) = 4v^2 + 1v + 3$$

The product equals $h_z(v) = 8v^5 + 22v^4 + 11v^3 + 31v^2 + 4v + 12$
Substituting w^2 for v, and inserting zeros for the missing coefficients,

$$h_z(w) = 8w^{10} + 0w^9 + 22w^8 + 0w^7 + 11w^6 + 0w^5 + 31w^4 + 0w^3 + 4w^2 + 0w + 12$$

The convolution is then $y_z[n] = \{8, 0, 22, 0, 11, 0, 31, 0, 4, 0, 12\}$. This is just $y[n]$ with one zero inserted between each sample. $\underline{\underline{\quad\quad}}$

4.13 Convolution of Discrete Periodic Signals

The convolution of sampled or discrete sequences, one or both of which are periodic, is also periodic and relies on periodic extension. In analogy with analog signals, we define the periodic extension of a discrete-time energy signal $x[n]$ over one period $(0, N)$ by

$$x_{\text{pe}}[n] = \sum_{k=-\infty}^{\infty} x[n + kN]$$

If $x[n]$ is shorter than N, we obtain one period of its periodic extension simply by padding $x[n]$ with zeros (to increase its length to N).

The methods for discrete convolution with periodic inputs are identical to the continuous case (see Table 4.5). We illustrate by an example.

Example 4.29 Convolution When One Sequence Is Periodic

We find the convolution $y_p[n]$ of a periodic signal $x_p[n] = \{2, 1, 3\}$ with $N = 3$ and the aperiodic signal $h[n] = \{2, 1, 1, 3, 1\}$ using two methods.

(a) We find the regular convolution $y[n] = x_p[n] \star h[n]$ to obtain

$$y[n] = \{2, 1, 3\} \star \{2, 1, 1, 3, 1\} = \{4, 4, 9, 10, 8, 10, 3\}$$

To find $y_p[n]$, values past $N = 3$ are wrapped around and summed to give

$$\{4, 4, 9, 10, 8, 10, 3\} \quad \text{Wrap around} \quad \begin{Bmatrix} 4, 4, & 9 \\ 10, 8, & 10 \\ 3 \end{Bmatrix} \quad \text{Sum} \quad \{17, 12, 19\}$$

(b) We create the periodic extension $h_p[n]$ with $N = 3$ using wraparound to get $h_p[n] = \{5, 2, 1\}$. We find the regular convolution of one period of each sequence to give $y[n] = \{2, 1, 3\} \star \{5, 2, 1\} = \{10, 9, 19, 7, 3\}$. Finally, wraparound of $y[n]$ past these samples yields $y_p[n] = \{17, 12, 19\}$, as before. $\underline{\underline{\quad\quad}}$

4.13.1 *Periodic Convolution of Periodic DT Signals*

In analogy with analog signals, if both sequences are periodic, so is their convolution. For two periodic sequences $x_p[n]$ and $h_p[n]$ with period N, the discrete periodic convolution over one period is given by

$$y_p[n] = x_p[n] \bullet h_p[n] = \sum_{k=0}^{N-1} x_p[k]h_p[n - k]$$

An optional normalizing factor $1/N$ ensures identical results, no matter how long we choose the duration. The normalized convolution of nonperiodic power signals may be found by using the limiting form of the above definition

$$y_p[n] = x_p[n] \bullet h_p[n] = \lim_{N \to \infty} \frac{1}{2N + 1} \sum_{k=-N}^{N} x_p[k]h_p[n - k]$$

4.13.2 *Periodic Convolution from Regular Convolution*

As with CT signals, the periodic convolution of DT signals from their linear convolution involves wraparound. For two sequences $x[n]$ and $h[n]$ of equal length N, we find the periodic convolution $y_p[n]$ in two steps.

1. We find the regular convolution $y_R[n] = x[n] \star h[n]$, which has $2N - 1$ samples.
2. We wrap around $y_R[n]$ past N samples, and form the sum to obtain $y_p[n]$.

Example 4.30
Periodic Convolution Using Wraparound

Consider the periodic signals $x_p[n] = \{1, 2, 3\}, h_p[n] = \{1, 0, 2\}$ with period $N = 3$. Their regular convolution is easily found to be $y_R[n] = \{1, 2, 5, 4, 6\}$. Using wraparound past three samples, the periodic convolution equals $y_p[n] = \{5, 8, 5\}$.

4.13.3 *Periodic Convolution Using the Cyclic Method*

In the cyclic method, we fold one sequence and restrict our attention to just one period of the stationary sequence as the *periodic extension* of the folded sequence moves past it. For example, if $y[n] = \{1, 2, 3, 4\}$ for the first period, starting at $n = 0$, its *folded periodic extension* for the first period corresponds to $\{1, 4, 3, 2\}$ also starting at $n = 0$. Convolution values are then found as before, but no wraparound is required. There are several ways to implement this process.

The Sliding Strip Method First, we line up the folded sequence with a one-period window (N samples) of the stationary sequence to start the convolution. Next, we shift the *periodic extension* of the folded signal N times, one sample at a time, to generate the periodic convolution.

Example 4.31
Periodic Convolution Using the Sliding Strip Method

Let $x[n] = \{1, 2, 3\}, h[n] = \{1, 0, 2\}$. We position $h[-n] = \{1, 2, 0\}$ and shift its periodic extension successively to obtain the periodic convolution as

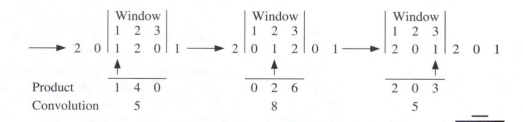

	Window	Stationary sequence
	1 2 3	Folded sequence ready to start the convolution
2 0 1		

The sliding strip method is equivalent to wrapping around each sample of the folded periodic extension as it slides out of the one-period window of the stationary sequence. For the previous example, we have

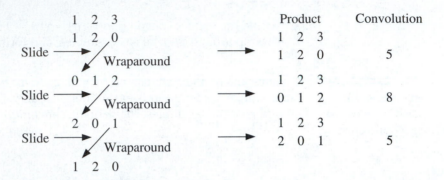

The Circular Method   We line up $x[k]$ clockwise around a circle and $h[k]$ counterclockwise (i.e., folded), on a concentric circle positioned to start the convolution (Figure 4.19). Shifting the folded sequence turns it clockwise. At each turn, the convolution equals the sum of the pairwise products (Figure 4.19). This approach clearly brings out the cyclic nature of periodic convolution.

4.13.4   *Periodic Convolution and the Circulant Matrix*

Periodic convolution may also be expressed as a matrix multiplication. We set up an $N \times N$ matrix whose columns equal $x[n]$ and its cyclically shifted versions (and whose rows equal successively shifted versions of the *first* period of the folded signal $x[-n]$). This is called the **circulant matrix** or **convolution matrix**. An $N \times N$ circulant matrix has the general form

$$
C_N = \begin{bmatrix}
x[0] & x[N-1] & \ldots x[2] & x[1] \\
x[1] & x[0]\ldots & & x[2] \\
x[2] & x[1] & \ldots & x[3] \\
\vdots & \vdots & & \vdots \\
x[N-2] & & \ldots x[0] & x[N-1] \\
x[N-1] & x[N-2] & \ldots x[1] & x[0]
\end{bmatrix}
$$

Note that each diagonal of the circulant matrix has equal values. Such a *constant diagonal matrix* is also called a **Toeplitz matrix**. Its product with an $N \times 1$ column matrix describing $h[n]$ yields the periodic convolution as an $N \times 1$ column matrix. This result may be normalized by N, if required.

Figure 4.19 The circular method of periodic convolution.

x[n](*) is cw   h[n](o) is ccw

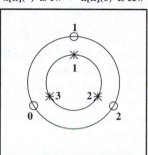

(1) Find y[0] and rotate h[n] cw

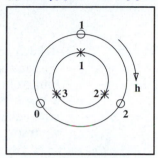

y[0]=1*1+2*2+0*3=5

(2) Find y[1] and rotate h[n] cw

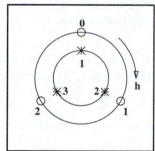

y[1]=0*1+1*2+2*3=8

(3) Find y[2]

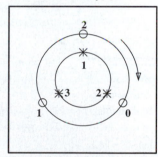

y[2]=2*1+0*2+1*3=5

Example 4.32

Periodic Convolution Using the Circulant Matrix

Consider $x[n] = \{1, 0, 2\}$ and $h[n] = \{1, 2, 3\}$, described over one period ($N = 3$).

(a) The circulant matrix C_X and convolution are given by

$$C_X = \begin{bmatrix} 1 & 2 & 0 \\ 0 & 1 & 2 \\ 2 & 0 & 1 \end{bmatrix} \qquad H = \begin{bmatrix} 1 \\ 2 \\ 3 \end{bmatrix} \qquad y_{p1}[n] = \begin{bmatrix} 1 & 2 & 0 \\ 0 & 1 & 2 \\ 2 & 0 & 1 \end{bmatrix}\begin{bmatrix} 1 \\ 2 \\ 3 \end{bmatrix} = \begin{bmatrix} 5 \\ 8 \\ 5 \end{bmatrix}$$

Normalizing $y_{p1}[n]$ by $N = 3$, we obtain $y[n] = y_{p1}[n]/3 = \{5/3, 8/3, 5/3\}$.

(b) The convolution of $x[n]$ and $h[n]$ over a two-period window yields

$$C_X = \begin{bmatrix} 1 & 2 & 0 & 1 & 2 & 0 \\ 0 & 1 & 2 & 0 & 1 & 2 \\ 2 & 0 & 1 & 2 & 0 & 1 \\ 1 & 2 & 0 & 1 & 2 & 0 \\ 0 & 1 & 2 & 0 & 1 & 2 \\ 2 & 0 & 1 & 2 & 0 & 1 \end{bmatrix} \qquad H = \begin{bmatrix} 1 \\ 2 \\ 3 \\ 1 \\ 2 \\ 3 \end{bmatrix} \qquad y_{p2}[n] = \begin{bmatrix} 10 \\ 16 \\ 10 \\ 10 \\ 16 \\ 10 \end{bmatrix}$$

Note that $y_{p2}[n]$ is twice the one-period result $y_{p1}[n]$, but still periodic with $N = 3$. Normalizing $y_{p2}[n]$ by the window width 6, we obtain $y_{p2}[n]/6 = \{5/3, 8/3, 5/3, 5/3, 8/3, 5/3\}$. For one period, this result is identical to $y_N[n]$.

4.13.5   Regular Convolution from Periodic Convolution

We can also find regular convolution from periodic convolution. The regular convolution of two sequences of length M and N yields a convolution sequence of length $S = M + N - 1$. But, periodic convolution requires sequences of equal length. To ensure equal lengths, and obtain the regular convolution from periodic convolution, we resort to **zero padding** as follows:

1. We generate *zero-padded sequences* of length $M + N - 1$ by padding each with enough (trailing or leading) zeros. This simply introduces additional (trailing or leading) zeros in their regular convolution result.
2. The *regular* convolution of the original, unpadded sequences equals the *periodic* convolution of the zero-padded sequences.

Example 4.33
Regular Convolution
Using the Circulant Matrix

Let $x[n] = \{2, 5, 0, 4\}$ and $h[n] = \{4, 1, 3\}$. Their regular convolution has $S = M + N - 1 = 6$ samples. Using trailing zeros, we create the padded sequences

$$x_z[n] = \{2, 5, 0, 4, 0, 0\} \qquad h_z[n] = \{4, 1, 3, 0, 0, 0\}$$

The periodic convolution $x_z[n] \bullet h_z[n]$, using the circulant matrix, equals

$$C_x = \begin{bmatrix} 2 & 0 & 0 & 4 & 0 & 5 \\ 5 & 2 & 0 & 0 & 4 & 0 \\ 0 & 5 & 2 & 0 & 0 & 4 \\ 4 & 0 & 5 & 2 & 0 & 0 \\ 0 & 4 & 0 & 5 & 2 & 0 \\ 0 & 0 & 4 & 0 & 5 & 2 \end{bmatrix} \qquad H = \begin{bmatrix} 4 \\ 1 \\ 3 \\ 0 \\ 0 \\ 0 \end{bmatrix} \qquad y_p[n] = \begin{bmatrix} 8 \\ 22 \\ 11 \\ 31 \\ 4 \\ 12 \end{bmatrix}$$

This equals the regular convolution $y[n] = x[n] \star h[n]$ obtained previously by several other methods (see Examples 4.24–4.26).

4.14
Connections: Discrete Convolution and Transform Methods

Just as *CT* convolution provides a connection between time-domain and frequency-domain methods of system analysis for *CT* signals, *DT* convolution provides a similar connection for *DT* signals. It forms the basis for every transform method described in this text. Its role in linking the time-domain and the transformed domain is intimately tied to the concept of discrete eigensignals and eigenvalues.

4.14.1   DT Harmonics, Eigensignals and Discrete Transforms

Just as the *CT* harmonic $\exp(j2\pi ft)$ is an eigensignal of analog systems, the *DT* harmonic $\exp(j2\pi nF)$ (periodic or otherwise) is an eigensignal of *DT* systems. The response of an LTI system to such a harmonic is also a harmonic at the same frequency but with changed magnitude and phase. This key concept leads to idea of discrete transforms analogous to Fourier and Laplace transforms, as summarized in Table 4.10.

Table 4.10   Discrete-Time Eigensignals and System Functions.

Eigensignal	System function	Transform method
$\exp(j2\pi nF)$	$H_p(F) = \sum_{k=-\infty}^{\infty} h[k]\exp(-j2\pi kF)$	Discrete-time Fourier transform
z^n	$H(z) = \sum_{k=-\infty}^{\infty} h[k]z^{-k}$	z-transform

4.14.2   The DTFT

For the harmonic input $x[n] = \exp(j2\pi nF)$, the response $y[n]$ equals

$$y[n] = \sum_{k=-\infty}^{\infty} \exp[j2\pi(n-k)F]h[k] = \exp(j2\pi nF) \sum_{k=-\infty}^{\infty} h[k]\exp(-j2\pi kF) = x[n]H_p(F)$$

This is just the input modified by the system function $H_p(F)$, where

$$H_p(F) = \sum_{k=-\infty}^{\infty} h[k]\exp(-j2\pi kF)$$

The quantity $H_p(F)$ describes the **discrete-time Fourier transform** (DTFT) or **discrete-time frequency response** of $h[n]$. Unlike the CT system function $H(f)$, the DT system function $H_p(F)$ is periodic with a period of unity, since $\exp(-j2\pi kF)$ equals $\exp[-j2\pi k(F+1)]$. This periodicity is also a direct consequence of the discrete nature of $h[n]$.

As with the impulse response $h[n]$, which is just a signal, any signal $x[n]$ may be described by its DTFT $X_p(F)$. The response $y[n] = x[n]H_p[F]$ may then be transformed to its DTFT $Y_p[n]$ to give

$$Y_p(F) = \sum_{k=-\infty}^{\infty} y[k]\exp(-j2\pi Fk) = \sum_{k=-\infty}^{\infty} x[k]H_p(F)\exp(-j2\pi Fk) = H_p(F)X_p(F)$$

Once again, convolution in time corresponds to multiplication in frequency.

4.14.3   The z-Transform

For an input $x[n] = r^n\exp(j2\pi nF) = [r\exp(j2\pi F)]^n = z^n$, where z is complex, with magnitude $|z| = r$, the response may be written as

$$y[n] = x[n] \star h[n] = \sum_{k=-\infty}^{\infty} z^{n-k}h[k] = z^n \sum_{k=-\infty}^{\infty} h[k]z^{-k} = x[n]H(z)$$

The response equals the input modified by the system function $H(z)$, where

$$H(z) = \sum_{k=-\infty}^{\infty} h[k]z^{-k}$$

The complex quantity $H(z)$ describes the **z-transform** of $h[n]$ and is not, in general, periodic in z.

Denoting the z-transform of $x[n]$ and $y[n]$ by $X(z)$ and $Y(z)$, we write

$$Y(z) = \sum_{k=-\infty}^{\infty} y[k]z^{-k} = \sum_{k=-\infty}^{\infty} x[k]H(z)z^{-k} = H(z)X(z)$$

Convolution in time corresponds to multiplication in the z-domain.

If $r = 1, x[n] = [\exp(j2\pi F)]^n = \exp(j2\pi nF)$ and the z-transform reduces to the DTFT. We can thus obtain the DTFT of $x[n]$ from its z-transform $X(z)$ by letting $z = \exp(j2\pi F)$ or $|z| = 1$ to give

$$H(F) = H(z)|_{z=\exp(j2\pi F)} = H(z)|_{|z|=1}$$

The DTFT is thus the z-transform evaluated on the unit circle $|z| = 1$.

Remark: For sampled signals, the summations for $H_p(F)$ and $H(z)$ should actually be multiplied by t_s. This is not common practice, however.

4.14.4   *What to Keep in Mind*

Convolution plays a central role in signal processing and system analysis. The main reason for the popularity of transformed domain methods of system analysis is that *convolution in one domain corresponds to multiplication in the other.* The type of convolution required depends on the type of signals under consideration, as summarized in Table 4.11. For example, we perform *discrete* convolution when dealing with *DT* signals. The frequency-domain representation of such signals gives rise to periodic spectra and requires *periodic* convolution in the frequency domain.

Table 4.11   Nature of Convolution for Transform Methods.

Transform	Time-domain convolution	Transformed-domain convolution
Laplace transform	Linear	Complex
Fourier transform	Linear	Linear
Fourier series	Periodic	Discrete
DTFT	Discrete	Periodic
z-Transform	Discrete	Complex
DFT and DFS	Discrete and periodic	Discrete and periodic

Another key concept is that *sampling in the time domain leads to a periodic extension in the frequency domain and vice versa.* This implies that the spectrum (DTFT) of a discrete-time signal is periodic, whereas a discrete spectrum (the Fourier series) corresponds to a periodic time signal. If we sample the periodic DTFT, we obtain the **discrete Fourier transform** (DFT), which corresponds to a *periodic, discrete-time* signal. If we sample a periodic time signal, we obtain the **discrete Fourier series** (DFS), which is also discrete and periodic.

4.15
Stability and
Causality of *DT*
Systems

From Chapter 3, the BIBO stability of LTI systems described by difference equations requires every root of the characteristic equation to have a magnitude less than unity. For systems described by their impulse response, this is entirely equivalent to requiring the impulse response $h[n]$ to be absolutely summable. Here is why. If $x[n]$ is bounded such that $|x[n]| < M$, so too is its shifted version $x[n-k]$. The convolution sum then yields the following inequality:

$$|y[n]| < \sum_{k=-\infty}^{\infty} |h[k]||x[n-k]| < M \sum_{k=-\infty}^{\infty} |h[k]|$$

and if $|y[n]| < \infty$, then

$$\sum_{k=-\infty}^{\infty} |h[k]| < \infty$$

This is both a necessary and sufficient condition and if met we are assured a stable system. We always have a stable system if $h[n]$ is an energy signal. Since the impulse response is defined only for linear systems, the stability of nonlinear systems must be investigated by other means.

Causality   In analogy with *CT* systems, causality of *DT* time systems implies a nonanticipatory system with an impulse response $h[n] = 0, n < 0$.

Example 4.34
The Concept of Stability

(a) The FIR filter described by $y[n] = x[n+1] - x[n]$ has the impulse response sequence $\{h[n]\} = \{1, -1\}$. It is a stable system, since $\sum |h[n]| = |1| + |-1| = 2$. It is also noncausal, since $h[n] = \delta[n+1] - \delta[n]$ is not zero for $n < 0$.

Comment: FIR filters are always stable because $\sum |h[n]|$ is the absolute sum of a finite sequence and is thus always finite.

(b) A filter described by $h[n] = (-\frac{1}{2})^n u[n]$ is causal. It describes a system with the difference equation $y[n] = x[n] + ay[n-1]$. It is also stable because $\sum |h[n]|$ is finite. We see that $\sum |h[n]| = 1 + |-\frac{1}{2}| + |-\frac{1}{2}|^2 + \cdots = 1/(1 - |-\frac{1}{2}|) = 2$.

(c) A filter described by the difference equation $y[n] = nx[n] - \frac{1}{2}y[n-1]$ is causal but time-varying. It is also unstable. If we apply a step input $u[n]$ (bounded input), then $y[n] = nu[n] + \frac{1}{2}y[n-1]$. The term $nu[n]$ grows without bound and makes this system unstable. We caution you that this approach is not a formal way of checking for the stability of time-varying systems.   ▬▬

4.16
Special Topics:
Deconvolution

Given the system impulse response $h(t)$, the response $y(t)$ of the system to an input $x(t)$ is simply the convolution of $x(t)$ and $h(t)$. Given $x(t)$ and $y(t)$ instead, how do we find $h(t)$? This situation arises very often in practice and is referred to as **deconvolution** or **system identification**.

For *DT* systems, we have a partial solution to this problem. Since discrete convolution may be thought of as polynomial multiplication, *discrete* deconvolution may be regarded as polynomial division. One approach to discrete deconvolution is to use the idea of *long division*, a familiar process, illustrated in the following example.

Example 4.35
Deconvolution Using
Polynomial Division

Consider $x[n] = \{2, 5, 0, 4\}$ and $y[n] = \{8, 22, 11, 31, 4, 12\}$. We regard these as being the coefficients, in descending order, of the polynomials

$$x(w) = 2w^3 + 5w^2 + 0w + 4 \qquad y(w) = 8w^5 + 22w^4 + 11w^3 + 31w^2 + 4w + 12$$

The polynomial $h(w)$ may be deconvolved out of $x(w)$ and $y(w)$ by performing the division $y(w)/x(w)$. This yields

$$
\begin{array}{r}
4w^2 + 1w + 3 \\
\hline
2w^3 + 5w^2 + 0w + 4 \overline{)\ 8w^5 + 22w^4 + 11w^3 + 31w^2 + 4w + 12} \\
8w^5 + 20w^4 + \ \ 0w^3 + 16w^2 \\
\hline
2w^4 + 11w^3 + 15w^2 + 4w + 12 \\
2w^4 + \ \ 5w^3 + \ \ 0w^2 + 4w \\
\hline
6w^3 + 15w^2 + 0w + 12 \\
6w^3 + 15w^2 + 0w + 12 \\
\hline
0
\end{array}
$$

The coefficients of the polynomial describe the sequence $h[n] = \{4, 1, 3\}$.

4.16.1   *Deconvolution by Recursion*

This method can also be recast as a recursive algorithm. The convolution

$$y[n] = x[n] \star h[n] = \sum_{k=0}^{n} h[k]x[n-k]$$

when evaluated at $n = 0$, provides the seed value of $h[0]$ as

$$y[0] = x[0]h[0] \qquad h[0] = y[0]/x[0]$$

We now separate the term containing $h[n]$ in the convolution relation

$$y[n] = \sum_{k=0}^{n} h[k]x[n-k] = h[n]x[0] + \sum_{k=0}^{n-1} h[k]x[n-k]$$

and evaluate $h[n]$ for successive values of $n > 0$ from

$$h[n] = \frac{1}{x[0]} \left[y[n] - \sum_{k=0}^{n-1} h[k]x[n-k] \right]$$

If all goes well, we need to evaluate $h[n]$ only at $M - N + 1$ points where M and N are the lengths of $y[n]$ and $x[n]$, respectively.

Naturally, problems arise if a remainder is involved. This may well happen in the presence of noise, which could modify the values in the output sequence even slightly. In other words, the approach is quite susceptible to noise or roundoff error and not very practical.

Example 4.36
Deconvolution Using Recursion

Let $x[n] = \{2, 5, 0, 4\}$ and $y[n] = \{8, 22, 11, 31, 4, 12\}$. We need only $6 - 4 + 1$ or 3 evaluations of $h[n]$. The starting value is $h[0] = y[0]/x[0] = 4$ and the two successive values for $h[n]$ result in

$$h[1] = \frac{1}{x[0]} \left[y[1] - \sum_{k=0}^{0} h[k]x[1-k] \right] = \frac{1}{x[0]} \{y[0] - h[0]x[1]\} = 1$$

$$h[2] = \frac{1}{x[0]} \left[y[2] - \sum_{k=0}^{1} h[k]x[2-k] \right] = \frac{1}{x[0]} \{y[0] - h[0]x[2] - h[1]x[1]\} = 3$$

4.17
Special Topics: Sampling Considerations

The digital processing of signals requires finite sequences. Signals like exponentials and sinusoids must be viewed for a finite time and lead to *truncation errors*, even though the convolution result is exact over some duration, as we have pointed out earlier.

Example 4.37
The Effects of Truncation

Consider the signals $x(t) = \exp(-t)u(t)$ and $h(t) = u(t)$. We examine their convolution $y(t) = x(t) \star h(t)$ subject to the following truncations:

1. Let $x(t)$ be truncated past T_x units and $h(t)$ past T_h units. If you picture the folding and sliding operation mentally, you should realize that the convolution of the truncated signals will be identical to $y(t)$ over the range $0 \le t \le T_{\min}$, where $T_{\min}$ equals the smaller of T_x and T_h (such that neither of the truncated ends figures in the product).
2. If $x(t)$ is truncated on both sides to lie between T_{x1} and T_{x2} and $h(t)$ is, likewise, truncated to lie between T_{h1} and T_{h2}, there is no range over which the convolution result is identical to $y(t)$.

Choice of Sampling Interval   If a signal is bandlimited to a frequency f_B, the sampling theorem predicts a sampling interval $t_s < \frac{1}{2f_B}$. Theory also predicts that bandlimited signals can never be timelimited, and vice versa. In theory, then, bandlimited signals must be sampled for infinite time and timelimited signals require $t_s = 0$, which brings us right back to the question of how best to choose the sampling interval. There are no easy answers. Naturally, we must assign reasonable criteria. One approach, for example, is to make the difference in the energy between the sampled and actual signal as small as desired.

Sampling Discontinuous Functions   If the sampling instant falls exactly on a discontinuity, how do we assign function values? There are several ways:

1. Choose a sampling interval that avoids points of discontinuity.

2. Shift the sampling instants to avoid points of discontinuity.
3. Choose the function value as the average value at the discontinuity. This is equivalent to using the trapezoidal rule for convolution.

Roundoff Errors Roundoff errors arise from the sampling process itself and include:

1. **jitter** caused by slight variations in the sampling interval and
2. **quantization error** of amplitudes to a finite number of levels.

These considerations are discussed further in Chapter 11.

4.18 Special Topics: Discrete Correlation

In analogy with CT correlation, the DT correlation of two energy signals $x[n]$ and $y[n]$ is defined by

$$R_{xy}[n] = \sum_{k=-\infty}^{\infty} x[k]y[k-n] = \sum_{k=-\infty}^{\infty} x[k+n]y[k] \qquad n = 0, \pm 1, \pm 2, \ldots$$

$$R_{yx}[n] = \sum_{k=-\infty}^{\infty} y[k]x[k-n] = \sum_{k=-\infty}^{\infty} y[k+n]x[k] \qquad n = 0, \pm 1, \pm 2, \ldots$$

Remark: Some authors prefer to switch the definitions of $R_{xy}[n]$ and $R_{yx}[n]$.

The two forms are folded versions of each other in that $R_{xy}[n] = R_{yx}[-n]$. The
index n is often called the **lag**.

The link between convolution and correlation is given by the relation

$$R_{xy}[n] = x[n] \star y[-n]$$

To obtain the DT autocorrelation $R_{xx}[n]$, we simply replace y by x in the above
relations. For real sequences, $R_{xx}[n]$ is even-symmetric:

$$R_{xx}[n] = R_{xx}[-n]$$

The autocorrelation attains a maximum at $n = 0$, and satisfies the inequality

$$|R_{xx}[n]| \le R_{xx}[0] = E_x$$

Example 4.38
DT Autocorrelation and Cross Correlation

(a) Let $x[n] = a^n u[n]$, $|a| < 1$. Since $x[k-n] = a^{k-n}u[k-n]$ starts at $k = n$, we ob-
tain the autocorrelation $R_{xx}[n]$ for $n \ge 0$ as

$$R_{xx}[n] = \sum_{k=-\infty}^{\infty} x[k]x[k-n] = \sum_{k=n}^{\infty} a^k a^{k-n} = \sum_{m=0}^{\infty} a^{m+n}a^m = a^n \sum_{m=0}^{\infty} a^{2m} = a^n \frac{1}{1-a^2}$$

Since the autocorrelation is an even function, $R_{xx}[n] = a^{|n|}/(1-a^2)$. Its maxi-
mum value is $R_{xy}[0] = 1/(1-a^2)$.

(b) Let $x[n] = a^n u[n]$, $|a| < 1$ and $y[n] = \text{rect}(n/2N)$. To find $R_{xy}[n]$, shift $y[k]$ and find the sum of the products over different ranges. Since $y[k - n]$ shifts the pulse to the right over the limits $-N + n, N + n$, $R_{xy}[n]$ equals zero until $n = -N$. We then obtain

$$-N \le n \le N - 1 \text{ (partial overlap):} \quad R_{xy}[n] = \sum_{k=-\infty}^{\infty} x[k]y[k - n] = \sum_{k=0}^{N+1} a^k = \frac{1 - a^{N+n+1}}{1 - a}$$

$$n \ge N \text{ (total overlap):} \quad R_{xy}[n] = \sum_{k=-N+1}^{N+1} a^k = \sum_{m=0}^{2N} a^{m-N+1} = a^{-N+1}\frac{1 - a^{2N+1}}{1 - a}$$

4.18.1 *Periodic DT Correlation*

For periodic sequences with identical period N, the periodic *DT* correlation is defined in analogy with periodic *CT* correlation as

$$R_{xy}[n] = \frac{1}{N}\sum_{k=0}^{N-1} x[k]y[k - n] \qquad R_{yx}[n] = \frac{1}{N}\sum_{k=0}^{N-1} y[k]x[k - n]$$

We can also find the periodic correlation $R_{xy}[n]$ using convolution and wraparound, provided we use one period of the *folded, periodic extension* of the sequence $y[n]$.

4.19 Continuous-Time Convolution: A Synopsis

Convolution is a means of finding the system response in the time domain.

The Concept Start with the response $h(t)$ to an impulse $\delta(t)$. Represent the input $x(t)$ as a summation of shifted impulses. The response $y(t)$ is just the superposition of the responses to each shifted impulse. The result is

$$y(t) = x(t) \star h(t) = \int_{-\infty}^{\infty} x(t - \lambda)h(\lambda)\, d\lambda$$

The Kernel The convolution kernel involves $x(t - \lambda)$ and $h(\lambda)$ which are functions of λ. We replace t by λ to create $x(\lambda)$ and $h(\lambda)$. We fold $x(\lambda)$ to get $x(-\lambda)$. Sliding $x(-\lambda)$ (shifting it right by t) generates $x(t - \lambda)$.

The Process Start from a position of no overlap with $x(t - \lambda)$ to the left of $h(\lambda)$. The convolution process involves shifting the folded function $x(t - \lambda)$ past $h(\lambda)$ and computing the areas under the product $x(t - \lambda)h(\lambda)$ as we go.

Analytical Convolution This convolution requires $x(t)$ and $h(t)$ as single expressions in terms of steps and yields a single expression for $y(t)$. The step functions serve to simplify the integration limits and establish the duration over which each solution is valid.

Convolution by Ranges	This requires interval-by-interval expressions for $x(t)$ and $h(t)$. The convolution $y(t)$ is also expressed range by range. Convolution ranges are found using the pairwise sum rule. For each range, $x(t - \lambda)$ is positioned with t lying within that range. This sets the overlap between $x(t - \lambda)$ and $h(\lambda)$. The convolution is the area under $x(t - \lambda)h(\lambda)$.
Properties	$x(t) \star h(t - \alpha) = y(t - \alpha)$ $\delta(t) \star x(t) = x(t)$ $u(t) \star x(t) = \int x(t)\,dt$, $x(t) \star h'(t) = y'(t)$. Duration: $T_y = T_x + T_h$. Area: $\mathbb{A}[y] = \mathbb{A}[x]\mathbb{A}[h]$.
Eigenvalues	Harmonic inputs are eigenvalues of linear systems. Only their magnitude and phase is changed by the system.
Smoothing	The response $y(t)$ is a function of t and is often smoother and broader than either of the functions $x(t)$ and $h(t)$.
Periodic Convolution	The convolution of periodic signals is not defined in the usual sense. The convolution of periodic signals with identical periods T involves wraparound. We find the convolution of their one-period segments and wrap the portion past T to the beginning and sum all the portions.
Stability	For stability, we require $\sum \lvert h(t) \rvert$ to be finite.
Causality	For a causal system, we require $h(t) = 0$, $t < 0$.
Central Limit Theorem	Repeated convolution leads to a Gaussian form for many energy signals.
Correlation	Correlation is just convolution without folding. Correlation provides a measure of "similarity" between two signals. The autocorrelation $R_{xx}(t)$ of $x(t)$ is the convolution of $x(t)$ with $x(-t)$. The cross correlations of $x(t)$ and $y(t)$ are related by $R_{xy}(t) = R_{yx}(-t)$.

4.20 Discrete-Time Convolution: A Synopsis

Convolution is a means of finding the system response in the time domain.

The Concept	Start with the response $h[n]$ to an impulse $\delta[n]$. Represent the input $x[n]$ as a summation of shifted impulses. The response $y[n]$ is just the superposition of the responses to each shifted impulse. The result is

$$y[n] = x[n] \star h[n] = \sum_{k=-\infty}^{\infty} x[n-k]h[k]$$

Discrete convolution is equivalent to polynomial multiplication.

The Process	Start from a position of no overlap with $x[n-k]$ to the left of $h[k]$. The convolution process involves shifting the folded function $x[n-k]$ past $h[n]$ and computing the sum of the product $x[n-k]h[k]$ as we go.
Numerical Convolution	To obtain the numerical convolution, the results of discrete convolution must be multiplied by the sampling interval t_s.

Analytical Convolution	This requires $x[n]$ and $h[n]$ as single expressions in terms of step functions and yields a single expression for $y[n]$. The step functions serve to simplify the summation limits and establish the duration over which each solution is valid.
Properties	$x[n] \star h[n - n_0] = y[n - n_0]$, $\delta[n] \star x[n] = x[n]$, $u[n] \star x[n] = \sum x[n]$
Duration	For sequences with M and N samples, the convolution sequence has $M + N - 1$ samples. Area: $\sum y[n] = \sum x[n] \sum h[n]$. Starting index of $y[n]$ equals the sum of starting indices of $x[n]$ and $h[n]$.
Periodic Convolution	The convolution of periodic signals is not defined in the usual sense. The convolution of periodic signals with identical periods N involves wraparound. We find the convolution of their one-period segments and wrap the portion past N to the beginning and sum all the portions.
Regular from Periodic Convolution	The regular convolution of two sequences (lengths L_1 and L_2) is equivalent to the periodic convolution of the two sequences, each zero padded to length $L_1 + L_2 - 1$.
Eigenvalues	DT harmonic inputs (periodic or otherwise) are eigenvalues of linear systems. Only their magnitude and phase is changed by the system.
Stability	For stability, we require $\sum |h[n]|$ to be finite.
Causality	For a causal system, we require $h[n] = 0$, $n < 0$.
Correlation	Correlation is just convolution without folding. Correlation provides a measure of "similarity" between two signals. The autocorrelation $R_{xx}[n]$ of $x[n]$ is the convolution of $x[n]$ with $x[-n]$. The cross correlations of $x[n]$ and $y[n]$ are related by $R_{xy}[n] = R_{yx}[-n]$.

Appendix 4A
MATLAB
Demonstrations
and Routines

Getting Started on the Supplied MATLAB Routines

This appendix presents MATLAB routines related to concepts in this chapter, and some examples of their use. See Appendix 1A for introductory remarks.

NOTE: The % sign and comments following it need not be typed!

Convolution

To find the convolution of two functions, try

```
>>t1=0:0.02:1;x=sin(pi*t1); t2=-1:0.02:1;y=tri(t2);
>>y=convnum(x,h,0.02)            %Convolution of x and y, ts = 0.02
```

To plot y, we note that the convolution starts at $t = -1$. We generate a time axis starting at $t = -1$ whose length is equal that of y and plot as follows:

```
>>start =-1;l=length(y);t3=start+(0:l-1)*0.02;plot(t3,y)
>>y=convnum(x,h,0.02,'t')  %Numerical convolution (trapezoidal algorithm).
```

```
>>ty=pairsum(xa,ha)
```

Finds convolution range endpoints using the pairwise sum rule if *xa* and *ha* are arrays of the range endpoints of $x(t)$ and $h(t)$.

Periodic Convolution

```
>>y=convmat([1 3 4],5)    %Circulant matrix for x={1,3,4},N=5
>>y=convp(x,h,ts,ty)
```

Periodic convolution using $ty = 'w'$ for wraparound, $'c'$ for circulant matrix or $'f'$ for FFT.

Correlation

```
>>y=corrxy(x,h,ts)    %Correlation of x and h with sampling interval ts
>>y=corrp(x,h,ts)     %Periodic correlation of x and h
```

Demonstrations and Utility Functions

```
>>democonv            %Demonstration of convolution concepts
>>democorr            %Demonstration of correlation concepts
>>convplot('sin (pi*t)','tri(t)',[0 1],[-1 1],0.01)
```

Graphical movie of evolving products and convolution of $\sin(\pi t), [0, 1]$ and $\mathrm{tri}(t), [-1, 1]$ at intervals of 0.01.

```
>>clt('exp',10)
```

Compares 10 repeated convolutions of $\exp(-t)$ with the expected Gaussian result (central limit theorem). The type may also be 'rect' or 'tri'.

```
>>y=perext(x,n)              %Periodic extension of array x with period n
```

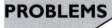

CT Convolution

C4.1 (The Convolution Concept) Consider the signals $x(t) = 2[u(t) - u(t - 5)]$ and $h(t) = u(t) - u(t - 5)$.

(a) Approximate $x(t)$ by 5 impulses located at $t = 0, 1, 2, 3, 4$. Find the convolution of $h(t)$ with each impulse, and use superposition to find the approximate form of $y(t) = x(t) \star h(t)$.

(b) How long does the convolution last, and what is its maximum?

(c) If $x(t)$ were approximated by 10 impulses, how long would the convolution last, and what would be its maximum value?

(d) What would you expect the exact form of $y(t)$ to approach?

C4.2 (Convolution Kernel) For each signal $x(t)$, sketch $x(\lambda)$ versus λ and $x(t - \lambda)$ versus λ, and label each expression.
(a) $x(t) = r(t)$ **(b)** $x(t) = u(t - 2)$
(c) $x(t) = 2\text{tri}[\frac{1}{2}(t - 1)]$ **(d)** $x(t) = \exp(-|t|)$

C4.3 (Convolution Concepts) Using the defining relation, evaluate the analytical convolution of the signals $x(t)$ and $h(t)$ at $t = 0$.
(a) $x(t) = u(t - 1)$, $h(t) = u(t + 2)$ **(b)** $x(t) = u(t)$, $h(t) = tu(t - 1)$
(c) $x(t) = tu(t + 1)$, $h(t) = (t + 1)u(t)$ **(d)** $x(t) = u(t)$, $h(t) = \cos(\frac{1}{2}\pi t)\text{rect}(\frac{1}{2}t)$

C4.4 (Analytical Convolution) Evaluate the convolution $y(t) = x(t) \star h(t)$ analytically, and sketch $y(t)$.
(a) $x(t) = \exp(-t)u(t)$, $h(t) = r(t)$
(b) $x(t) = t\exp(-\alpha t)u(t)$, $h(t) = u(t)$
(c) $x(t) = \exp(-t)u(t)$, $h(t) = \cos(t)u(t)$
(d) $x(t) = \exp(-t)u(t)$, $h(t) = \cos(t)$
(e) $x(t) = 2t[u(t + 2) - u(t - 2)]$, $h(t) = u(t) - u(t - 4)$
(f) $x(t) = 2tu(t)$, $h(t) = \text{rect}(t/2)$
(g) $x(t) = r(t)$, $h(t) = (1/t)u(t - 1)$

C4.5 (Convolution by Ranges) For the $x(t)$ and $h(t)$ of Figure C4.5 (p. 178):
(a) Establish the convolution ranges.
(b) Sketch $x(\lambda)h(t - \lambda)$ to evaluate the convolution at $t = 0$.
(c) Find and sketch the convolution over each range.

C4.6 (Convolution by Ranges) Consider a series RC circuit with $\tau = 1$. Find the capacitor voltage, its maximum value and time of maximum, if the input voltage equals:
(a) $\text{rect}(t - \frac{1}{2})$ **(b)** $t\,\text{rect}(t - \frac{1}{2})$ **(c)** $(1 - t)\text{rect}(t - \frac{1}{2})$

C4.7 (Convolution and Its Properties)
(a) Starting with $u(t) \star u(t) = r(t)$, find and sketch the convolution
  $\text{rect}(t/2) \star \text{rect}(t/4)$.
(b) Starting with $\text{rect}(t) \star \text{rect}(t) = \text{tri}(t)$, find and sketch the convolution
  $[\text{rect}(t + \frac{1}{2}) - \text{rect}(t - \frac{1}{2})] \star \text{rect}(t - \frac{1}{2})$.

C4.8 (Operations on the Impulse) Explain the difference between each of the following operations on the impulse $\delta(t - 1)$:
(a) $[e^{-t}u(t)]\delta(t - 1)$ **(b)** $\int_{-\infty}^{\infty} e^{-t}\delta(t - 1)\,dt$ **(c)** $[e^{-t}u(t)] \star \delta(t - 1)$

C4.9 (Properties) The step response of a system is $\exp(-t)u(t)$. Compute the response of this system to the following inputs:
(a) $x(t) = r(t)$ **(b)** $x(t) = \text{rect}(t/2)$
(c) $x(t) = \text{tri}[(t - 2)/2]$ **(d)** $x(t) = \delta(t + 1) - \delta(t - 1)$

C4.10 (Properties) Given the step response $w(t)$ of two systems, find the response to the input $x(t)$.
(a) $w(t) = r(t) - r(t - 1)$, $x(t) = \sin(2\pi t)u(t)$
(b) $w(t) = \exp(-t)u(t)$, $x(t) = \exp(-t)u(t)$

Figure C4.5

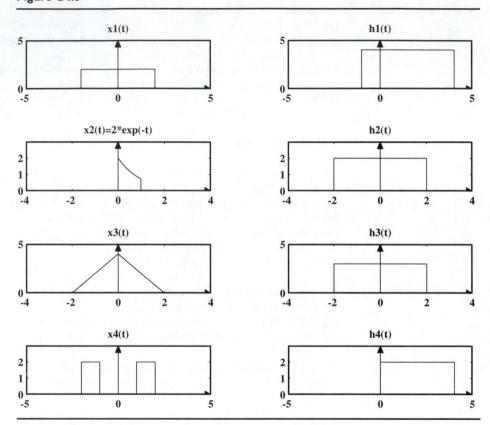

C4.11 (Eigenvalues) The input $x(t)$ and response $y(t)$ of two systems are given. Which of the systems are linear and why?

(a) $x(t) = \cos(t), \quad y(t) = \frac{1}{2}\sin(t - \frac{1}{4}\pi)$

(b) $x(t) = \cos(t), \quad y(t) = \cos(2t)$

C4.12 (Cascaded Systems)

(a) Let $h_1(t) = \exp(-t)u(t)$ and $h_2(t) = \exp(-t)u(t)$. What is the response of the cascaded system to $x(t) = u(t)$?

(b) Let $h_1(t) = \exp(-t)u(t)$ and $h_2(t) = \delta(t) - \exp(-t)u(t)$. What is the response of the cascaded system to the input $x(t) = \exp(-t)u(t)$?

C4.13 (Cascading) The impulse response of two cascaded systems equals the convolution of their impulse responses. Does the step response $w_C(t)$ of two cascaded systems equal $w_1(t) \star w_2(t)$, the convolution of their step responses? If not, how is $w_C(t)$ related to $w_1(t)$ and $w_2(t)$?

C4.14 (Cascading) Consider the three circuits of Figure C4.14.
(a) Find their impulse responses $h_1(t)$, $h_2(t)$ and $h_3(t)$.
(b) What is the impulse response $h_{12}(t)$ of the ideal cascade of the first two circuits? Does this equal the impulse response $h_3(t)$? Explain.
(c) What value of R will ensure an impulse response $h_3(t)$ that differs from $h_{12}(t)$ by no more than 1% at $t = 0$?

Figure C4.14

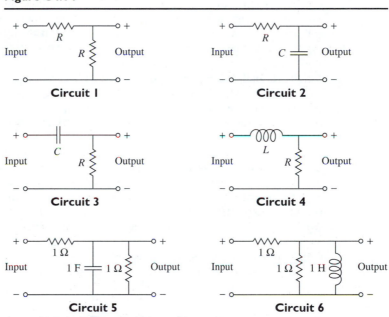

C4.15 (Stability) Which of the following systems are stable? Which are causal?
(a) $h(t) = \exp(t+1)u(t-1)$ **(b)** $h(t) = \exp(t)u(t+1)$
(c) $h(t) = \delta(t)$ **(d)** $h(t) = [1 - \exp(-t)]u(t)$
(e) $h(t) = \delta(t) - \exp(-t)u(t)$ **(f)** $h(t) = \text{sinc}(t-1)$

C4.16 (Stability)
(a) Is the system $dy(t)/dt = x(t)$ stable? Explain.
(b) Is the system $dy(t)/dt + \alpha y(t) = x(t)$ stable for $\alpha > 0$? Explain.
(c) Is the system $y^{(n)}(t) = x(t)$ stable for any $n \geq 1$? Explain.
(d) Is the system $y(t) = x^{(n)}(t)$ stable for any $n \geq 1$? Explain.

C4.17 (Causality)
(a) Argue that the impulse response $h(t)$ of a causal system must be zero for $t < 0$.
(b) If the input to a causal system starts at $t = t_0$, at what time does the response start?

C4.18 (Convolution and Smoothing) Convolution is usually a smoothing operation unless one signal is an impulse or its derivative, but exceptions occur even for smooth signals. Evaluate and comment on the duration and smoothing effects of the following convolutions:

(a) $\text{rect}(t) \star \text{tri}(t)$ **(b)** $\text{rect}(t) \star \delta(t)$ **(c)** $\text{rect}(t) \star \delta'(t)$

C4.19 (Convolution and System Classification) Consider the systems:
(1) $h(t) = 2\delta(t)$ (2) $h(t) = \delta(t) + \delta(t-3)$ (3) $h(t) = \exp(-t)u(t)$

(a) Find the response of each to the input $x(t) = u(t) - u(t-1)$.
(b) For system 1, the input is zero at $t = 2$ and so is the response. Does the statement "zero output if zero input" apply to dynamic or instantaneous systems or both? Explain.
(c) Argue that system 1 is instantaneous. What about the other two?
(d) What must be the form of $h(t)$ for an instantaneous system?

C4.20 (Signal-Averaging Filter) Consider a signal-averaging filter whose impulse response is $h(t) = (1/T)\text{rect}[(t - \frac{1}{2}T)/T]$.
(a) What is the response of this system to the input $u(t)$?
(b) What is the response of this system to a periodic sawtooth wave with height A, duty ratio D and period T?

C4.21 (Periodic Extension) The periodic extension $x(t)$ has the same form and area as $x(t)$. Use this concept to find the constants in the following assumed form for $x_{pe}(t)$ of each signal and confirm the entries of Table 4.4:
(a) $x(t) = \exp(-t/\tau)u(t)$ assuming $x_{pe}(t) = K\exp(-t/\tau)$, $(0, T)$
(b) $x(t) = t\exp(-t/\tau)u(t)$ assuming $x_{pe}(t) = (A + Bt)\exp(-t/\tau)$, $(0, T)$

C4.22 (Periodic Extension) Sketch the periodic extension of each of the signals in Figure C4.22 for one period T with $T = 6$ and $T = 4$.

Figure C4.22

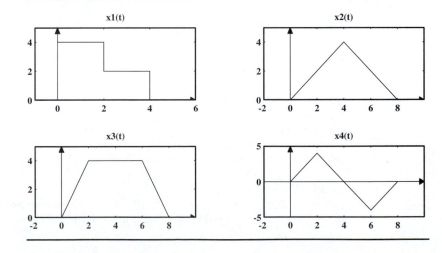

C4.23 (Convolution and Periodic Inputs) The voltage input to a series RC circuit with $\tau = 1$ is a rectangular pulse train starting at $t = 0$. The pulses are of unit width and height and occur every 2 s.
(a) What is the capacitor voltage at $t = 1$ s?
(b) What is the capacitor voltage at $t = 2$ s?
(c) Assuming that the input has been applied for a long enough time, what is the steady-state capacitor voltage?

C4.24 (Periodic Convolution) Carefully sketch the periodic convolution of the periodic signals $x(t)$ and $h(t)$ shown in Figure C4.24.

C4.25 (The Duration Property) The convolution duration usually equals the sum of the durations of the convolved signals. But consider:
(1) $u(t) \star \sin(\pi t)[u(t) - u(t - 2)]$ (2) $\text{rect}(t) \star \cos(2\pi t)u(t)$

(a) Evaluate each convolution and find its duration. Is the duration infinite? If not, what causes it to be finite?
(b) In the first convolution, replace the sine pulse by an arbitrary signal $x(t)$ of zero area and finite duration T_d and argue that the convolution is nonzero only for a duration T_d.
(c) In the second convolution, replace the cosine by an arbitrary periodic signal $x_p(t)$ with

---

Figure C4.24

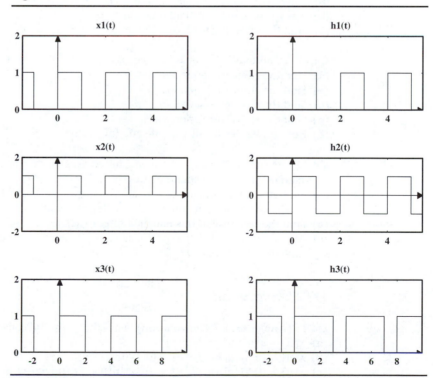

zero average value and a period $T = 1$. Argue that the convolution is nonzero for only for 1 unit.

C4.26 (Convolution and Moments) For each of the following signal pairs, find the moments m_0, m_1 and m_2 and verify each of the convolution properties based on moments.
(a) $x(t) = h(t) = \text{rect}(t)$
(b) $x(t) = \exp(-t)u(t)$, $h(t) = \exp(-2t)u(t)$

C4.27 (Central Limit Theorem) Recall that the repeated convolution of the signal $h(t) = \exp(-t)u(t)$ has the form $h_n(t) = (t^n/n!) \exp(-t)u(t)$.
(a) Show that $h_n(t)$ has a maximum at $t = n$.
(b) Assume a Gaussian approximation $h_n(t) \approx g_n(t) = K \exp[-\alpha(t - n)^2]$. Equating $h_n(t)$ and $g_n(t)$ at $t = n$, show that $K = (n^n/n!) \exp(-n)$.
(c) Using the **Stirling limit** defined as

$$\lim_{n \to \infty} \left[\sqrt{n} \frac{n^n}{n!} \exp(-n) \right] = \frac{1}{\sqrt{2\pi}}$$

show that $K = \sqrt{1/(2\pi n)}$.
(d) Equating the areas of $h_n(t)$ and $g_n(t)$, show that $\alpha = 1/(2n)$.

C4.28 (Inverse Systems) Given $h(t) = \exp(-t)u(t)$, we wish to find the inverse system such that $h(t) \star h_I(t) = \delta(t)$. The form that we require for the inverse system $h_I(t)$ is $h_I(t) = K_1\delta(t) + K_2\delta'(t)$.
(a) For what values of K_1 and K_2 will $h(t) \star h_I(t) = \delta(t)$?
(b) Is the inverse system stable? Causal?
(c) Find the inverse system corresponding to $h(t) = 2\exp(-3t)u(t)$.

C4.29 (Correlation) Let $x(t) = \text{rect}(t + \frac{1}{2})$ and $h(t) = \text{trect}(t - \frac{1}{2})$.
(a) Find the autocorrelation R_{xx}.
(b) Find the autocorrelation R_{hh}.
(c) Find the cross correlation R_{hx}.
(d) Find the cross correlation R_{xh}.
(e) How are the results of parts (c) and (d) related?

C4.30 (Matched Filters) A folded, shifted version of $x(t)$ defines the impulse response of a **matched filter** corresponding to $x(t)$.
(a) Find and sketch the impulse response of a causal matched filter for the signal $x(t) = \text{rect}(t - 3.5)$ assuming the smallest delay necessary.
(b) Find the response of this matched filter to $x(t)$.
(c) At what time t_m does the response attain its maximum value, and how is t_m related to time of occurrence of $x(t)$?

DT Convolution

D4.1 (Analytical DT Convolution) Find the analytical convolution for
(a) $x[n] = u[n]$, $h[n] = u[n]$
(b) $x[n] = (0.8)^n u[n]$, $h[n] = (0.4)^n u[n]$
(c) $x[n] = (0.5)^n u[n]$, $h[n] = (0.5)^n \{u[n + 3] - u[n - 4]\}$

(d) $x[n] = (a)^n u[n]$, $y[n] = (a)^n u[n]$
(e) $x[n] = (a)^n u[n]$, $y[n] = (b)^n u[n]$
(f) $x[n] = (a)^n u[n]$, $y[n] = \text{rect}[n/2N]$

D4.2 (Convolution Algorithms) For the following signals $x[n]$ and $h[n]$:
(1) Find the convolution using the sum-by-column method.
(2) Confirm your results using the lattice method.

(a) $x[n] = \{1, 2, 0, 1\}$, $h[n] = \{2, 2, 3\}$
(b) $x[n] = \{0, 2, 4, 6\}$, $h[n] = \{6, 4, 2, 0\}$
(c) $x[n] = \{-3, -2, -1, 0, 1\}$, $h[n] = \{4, 3, 2\}$
(d) $x[n] = \{3, 2, 1, 1, \overset{\uparrow}{2}\}$, $h[n] = \{4, 2, 3, 2\}$
(e) $x[n] = \{3, 0, \overset{\uparrow}{2}, 0, 1, 0, 1, 0, 2\}$, $h[n] = \{4, 0, 2, 0, \overset{\uparrow}{3}, 0, 2\}$
(f) $x[n] = \{0, 0, 0, 3, \overset{\uparrow}{1}, 2\}$, $h[n] = \{4, 2, 3, \underset{\uparrow}{2}\}$

D4.3 (Properties) Let $x[n] \star h[n] = \{1, 3, 4, \overset{\uparrow}{2}, 7, 6\}$.

(a) Find $x[n/3] \star h[n/3]$.   **(b)** Find $x[n-1] \star h[n+4]$.

D4.4 (Numerical Convolution) Find the numerical convolution of each pair of signals using (1) step interpolation (rectangular rule) and, (2) linear interpolation (trapezoidal rule).
(a) $x[n] = \text{rect}[n/4]$, $h[n] = \text{rect}[n/6]$, $t_s = 2$
(b) $x[n] = 2\text{tri}[n/2]$, $h[n] = 4\text{tri}[n/4]$, $t_s = 1$

D4.5 (DT Convolution) Let $x[n] = \text{rect}[n/2]$ and $h[n] = \text{rect}[n/4]$.
(a) Find $x[n] \star x[n]$ and $h[n] \star h[n]$ and express each result in terms of $\text{tri}[n]$.
(b) Generalize this to show that $y[n] = \text{rect}[n/2N] \star \text{rect}[n/2N]$ may be expressed as $y[n] = M\text{tri}[n/M]$, where $M = 2N + 1$.
(c) Do these results satisfy the area property of convolution?

D4.6 (Application) Consider a two-point averaging filter whose present output equals the average of the present and previous input.
(a) Set up a difference equation for this system.
(b) What is the impulse response of this system?
(c) What is the response of this system to the sequence $\{1, 2, 3, 4, 5\}$?
(d) Use convolution to show that the system performs the required averaging operation.

D4.7 (Causality)
(a) Argue that the impulse response $h[n]$ of a causal system must be zero for $n < 0$.
(b) If the input to a causal system starts at $n = n_0$, when does the response start?

D4.8 (Stability) Are the following systems stable? Causal?
(a) $h[n] = (2)^n u[n-1]$ **(b)** $y[n] = 2x[n+1] + 3x[n] - x[n-1]$
(c) $h[n] = (-\frac{1}{2})^n u[n]$ **(d)** $\{h[n]\} = \{3, 2, 1, 1, \underset{\uparrow}{2}\}$

D4.9 (Periodic Convolution) Consider the following $x[n]$ and $h[n]$:
(1) $x[n] = \{1, 2, 0, 1\}$, $h[n] = \{2, 2, 3\}$

(2) $x[n] = \{0, 2, 4, 6\}$, $h[n] = \{6, 4, 2, 0\}$
(3) $x[n] = \{-3, -2, -1, 0, 1\}$, $h[n] = \{4, 3, 2\}$
(4) $x[n] = \{3, 2, 1, 1, \overset{\uparrow}{2}\}$, $h[n] = \{4, 2, 3, 2, 0\}$

(a) Find the periodic convolution $y_p[n]$ from $y[n]$ using wraparound.
(b) Find the periodic convolution from the circulant matrix for $x[n]$.
(c) What is the *minimum* number of zeros we must append to each in order to find their regular convolution from periodic convolution?

D4.10 (Convolution and Interpolation)   Convolution may be viewed as interpolation. Consider a system with impulse response $h[n] = [\frac{1}{2}, \frac{1}{2}]$. Let the input to the system be the sequence defined by $x[n] = \{0, 2, 4, 6, 8, 10, 12, 14\}$.
(a) Find the output $y[n] = x[n] \star h[n]$.
(b) Show that, except for end effects, the output describes a linear interpolation between the samples of $x[n]$.
(c) What output would result if $h[n]$ were $\{\frac{1}{4}, \frac{1}{4}, \frac{1}{4}, \frac{1}{4}\}$?
(d) What should $h[n]$ be if the present output equals the average of the past N input values? (This describes an N-point moving average filter.)

D4.11 (Impulse Response of Difference Algorithms)   Two systems to compute the forward and backward difference are described by:
(1) $y_F[n] = x[n+1] - x[n]$ (2) $y_B[n] = x[n] - x[n-1]$

(a) What is the impulse response of each system?
(b) Is either of these systems stable? Causal?
(c) What is the impulse response of their parallel connection? Is the parallel system stable? Causal?
(d) What is the impulse response of their cascaded connection? Is the cascaded system stable? Causal?

D4.12 (Convolution in Practice)   Often, the convolution of a long sequence $x[n]$ and a short sequence $h[n]$ is performed by breaking the long signal into shorter pieces, finding the convolution of each short piece with $h[n]$, and gluing the results together. Let $x[n] = [1, 1, 2, 3, 5, 4, 3, 1]$, $h[n] = [4, 3, 2, 1]$.
(a) Split $x[n]$ into two equal sequences $x_1[n]$ and $x_2[n]$.
(b) Find the convolution $y_1[n] = h[n] \star x_1[n]$.
(c) Find the convolution $y_2[n] = h[n] \star x_2[n]$.
(d) Find the convolution $y[n] = h[n] \star x[n]$.
(e) How can you find $y[n]$ from $y_1[n]$ and $y_2[n]$?
(*Hint:* Shift $y_2[n]$ and use superposition. This forms the basis for the *overlap-add* method of convolution discussed in Chapter 12.)

D4.13 (Correlation)   Let $x[n] = \text{rect}[(n-4)/2]$ and $h[n] = \text{rect}[n/4]$.
(a) Find the autocorrelation R_{xx}.
(b) Find the autocorrelation R_{hh}.
(c) Find the cross correlation R_{xh}.
(d) Find the cross correlation R_{hx}.
(e) How are the results of parts (c) and (d) related?

D4.14 (Correlation) Find the correlation R_{xh} of the following signals:

(a) $x[n] = (a)^n u[n]$, $\quad h[n] = (a)^n u[n]$, $|a| < 1$

(b) $x[n] = n(a)^n u[n]$, $\quad h[n] = (a)^n u[n]$, $|a| < 1$

(c) $x[n] = \text{rect}[n/2N]$, $\quad h[n] = \text{rect}[n/2N]$

5 FOURIER SERIES

Our journey into the frequency domain begins with this chapter on Fourier series. The Fourier series describes *periodic signals* by combinations of harmonically related sinusoids or harmonic signals. This representation unfolds a perspective of periodic signals in the frequency domain in terms of their frequency content, or **spectrum**. It also reveals a reciprocity and duality in time and frequency. Operations on periodic signals in one domain have their duals in the other domain through their Fourier series description.

This chapter is divided into three parts:

Part 1   The development of the Fourier series, the concept of spectra, the effects of symmetry, and the properties of Fourier series.

Part 2   The applications of Fourier series to system analysis.

Part 3   Special topics in Fourier series including existence, convergence, the Gibbs effect, and smoothing windows.

Useful Background   The following topics provide a useful background:

1. Concepts related to periodic signals (Chapter 2).

2. The definition and properties of the sinc function (Chapter 2).

When we combine periodic signals with commensurate frequencies, we obtain another periodic signal $x_p(t)$. This is the process of **synthesis**. The **analysis,** or separation, of a periodic signal into its periodic components is not as easy. It requires that we address the following issues:

Which Periodic Components Best Describe $x_p(t)$?   The choice is not unique. A consistent approach demands that we pick the most elementary periodic signals, or **basis signals,** to describe $x(t)$. The choice falls squarely on sinusoids, because

1. They are among the simplest periodic signals and are easy to manipulate.
2. From a systems viewpoint, they are eigensignals of linear systems.
3. They provide a unique link with the frequency domain.

   We actually describe a **Fourier series** for $x_p(t)$ as a sum, in the right mix, of *harmonically related sinusoids* at the repetition frequency f_0 of $x_p(t)$ and its multiples (harmonics) kf_0. The choice of harmonic signals results in three additional advantages, as we shall discover subsequently:

1. The period of their combination also equals the period of $x_p(t)$.
2. It allows a simple, consistent, unique and independent scheme to find the coefficients (the right proportion) for each component. Only if we use this scheme do we obtain a *Fourier series* for $x_p(t)$ If not, we just have a plain old trigonometric series for $x_p(t)$
3. The Fourier series is *better* than any other. But better in what sense? Well, it yields the smallest *mean square error* in power between $x_p(t)$ and its series representation up to any number of terms or harmonics.

How Many Components Are Needed for an Exact Description of $x_p(t)$?   The actual number will naturally depend on the nature of $x_p(t)$. We may even require an infinite number of components for an exact representation.

Is a Fourier Series Possible for Any Arbitrary Periodic Signal?   The existence conditions for Fourier series (the so-called Dirichlet conditions, to be discussed later) are weak enough to assert that all periodic signals encountered in practice can be described by a Fourier series. Indeed, any exceptions are no more than mathematical curiosities.

5.1.1   *The Three Forms of a Fourier Series*
Due to the connection between sinusoids and complex exponentials, a Fourier series may be developed in three related forms—*trigonometric, polar* or *exponential.* The frequency may be expressed using ω_0 or f_0 with $\omega_0 = 2\pi f_0$.

5.1.2   *The Trigonometric Form*
To describe a periodic signal $x_p(t)$ with period T and *fundamental* frequency $f_0 = 1/T$, we choose sines and cosines at the frequencies $kf_0, k = 1, 2, 3, \ldots$. We include

a constant term to account for any dc offset in $x_p(t)$ and, with $\omega_0 = 2\pi f_0$, write its **trigonometric Fourier series** as:

$$x_p(t) = a_0 + a_1 \cos(\omega_0 t) + \cdots + a_k \cos(k\omega_0 t) + \cdots$$
$$+ b_1 \sin(\omega_0 t) + \cdots + b_k \sin(k\omega_0 t) + \cdots$$

This series may be expressed in summation notation as

$$x_p(t) = a_0 + \sum_{k=1}^{\infty} a_k \cos(k\omega_0 t) + b_k \sin(k\omega_0 t)$$

The weights a_0, a_k, and b_k are called the *trigonometric Fourier series coefficients,* and remain to be evaluated. The index starts at $k = 1$. The dc offset a_0 may be regarded as the cosine term $a_k \cos(k\omega_0 t)$ with $k = 0$.

5.1.3   The Polar Form

The **polar form** combines each pair of terms at the frequency $k\omega_0$ given by $a_k \cos(k\omega_0 t) + b_k \sin(k\omega_0 t)$ into the form $c_k \cos(k\omega_0 t + \theta_k)$.

$$x_p(t) = c_0 + \sum_{k=1}^{\infty} c_k \cos(k\omega_0 t + \theta_k)$$

With $c_0 = a_0$, the relation between c_k, a_k, and b_k is best found by resorting to the phasor representation of the time-domain terms as follows:

$$a_k \cos(k\omega_0 t) + b_k \sin(k\omega_0 t) \leftrightarrow a_k \angle 0° + b_k \angle -90° \quad c_k \cos(k\omega_0 t + \theta_k) \leftrightarrow c_k \angle \theta_k$$

This leads to

$$c_k \angle \theta_k = a_k - jb_k$$

Remark: The *magnitude* c_k is to be regarded as a *positive* quantity. A negative sign is absorbed into the angle θ_k (as a phase of $\pm 180°$ or $\pm \pi$ radians).

5.1.4   The Exponential Form

The **exponential form** is described by

$$x_p(t) = \sum_{k=-\infty}^{\infty} X_S[k] \exp(jk\omega_0 t)$$

Note here that the integer index k ranges from $-\infty$ to ∞. The coefficients are denoted by $X_S[k]$ to stress their discrete nature.

Each pair of terms $X_S[k] \exp(jk\omega_0 t) + X_S[-k] \exp(-jk\omega_0 t)$ at $\pm k\omega_0$ corresponds to the trigonometric pair $a_k \cos(k\omega_0 t) + b_k \sin(k\omega_0 t)$ at the frequency $k\omega_0$.

For $k = 0$, we have $X_S[0] = a_0$. For $k \neq 0$, we invoke Euler's relation to give

$$a_k \cos(k\omega_0 t) + b_k \sin(k\omega_0 t) = \tfrac{1}{2}a_k[e^{jk\omega_0 t} + e^{-jk\omega_0 t}] - j\tfrac{1}{2}b_k[e^{jk\omega_0 t} - e^{-jk\omega_0 t}]$$

Rearranging and comparing this with the exponential form gives

$$X_S[k]e^{jk\omega_0 t} + X_S[-k]e^{-jk\omega_0 t} = \tfrac{1}{2}(a_k - jb_k)e^{jk\omega_0 t} + \tfrac{1}{2}(a_k + jb_k)e^{-jk\omega_0 t}$$

The relation between the trigonometric and exponential forms is thus

$$X_S[k] = \tfrac{1}{2}(a_k - jb_k) \qquad X_S[-k] = \tfrac{1}{2}(a_k + jb_k) = X_S^*[k] \qquad k \geq 1$$

Remark: The index k for $X_S[k]$ can be negative, implying negative frequencies (which lead only to a phase reversal and provide no new information about $x_p(t)$).

5.1.5   *Coefficient Symmetry and Relationships*

For real signals, the $X_S[k]$ possess **conjugate symmetry** with $X_S[-k] = X_S^*[k]$. If we express $X_S[-k]$ as $X_S[-k] = \tfrac{1}{2}(a_{-k} - jb_{-k})$, the conjugate symmetry of $X_S[k]$ also suggests that $a_{-k} = a_k$ and $b_{-k} = -b_k$. This means that *the a_k show even symmetry, whereas the b_k display odd symmetry.* These relations are summarized by

$$X_S[k] = \tfrac{1}{2}(a_k - jb_k) = \tfrac{1}{2}c_k \angle \theta_k \qquad k \geq 1$$

The Argand diagram (Figure 5.1) is a useful way of showing the connections.

Figure 5.1   Relating the various Fourier series coefficients.

Relation between Fourier series coefficients

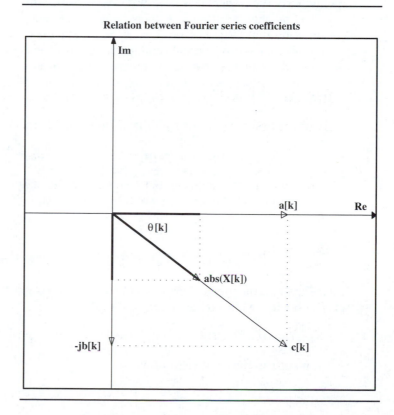

5.2
Fourier Analysis: Evaluating the Coefficients

The key to finding the Fourier series coefficients is *orthogonality*. Two signals are called **orthogonal** if the area under their product equals zero. *Harmonic signals* are orthogonal and obey the relations summarized in Table 5.1.

Table 5.1   Orthogonality of Sinusoids and Harmonic Signals.
($\int_T$ means integration over one period of our own choosing.)

Orthogonality Relation	Result
$\int_T \cos(m\omega_0 t)\sin(n\omega_0 t)\,dt$	0 (integer m, n)
$\int_T \cos(m\omega_0 t + \theta_1)\cos(n\omega_0 t + \theta_2)\,dt$	0 $(m \neq n)$
	$\frac{1}{2}T$ $(m = n)$
$\int_T \exp(m\omega_0 t)\exp(-n\omega_0 t)\,dt$	0 $(m \neq n)$
	T $(m = n)$

These relations suggest that over a one-period duration, the area under a sinusoid or under the product of two sinusoids at commensurate frequencies equals zero. The phase of the sinusoids is irrelevant. If the product of a sine and a cosine is involved, the frequencies need not even be different.

To find the coefficients, we need look at only one period of $x_p(t)$, since a representation that describes $x_p(t)$ over one period guarantees an identical representation over successive periods and, therefore, for the entire signal.

To invoke orthogonality, we integrate the Fourier series over T to give

$$\int_T x_p(t)\,dt = \int_T a_0\,dt + \cdots + \int_T a_k \cos(k\omega_0 t)\,dt + \cdots + \int_T b_k \sin(k\omega_0 t)\,dt + \cdots$$

All but the first term of the right-hand side equal zero, and we obtain

$$\int_T x_p(t)\,dt = \int_T a_0\,dt = a_0 T \qquad \text{or} \qquad a_0 = \frac{1}{T}\int_T x_p(t)\,dt$$

The dc offset a_0 thus represents the average value of $x_p(t)$.

If we multiply the series by $\cos(k\omega_0 t)$ and then integrate over T, we get

$$\int_T x_p(t)\cos(k\omega_0 t)\,dt = \int_T a_0 \cos(k\omega_0 t)\,dt$$
$$+ \cdots + \int_T a_k \cos(k\omega_0 t)\cos(k\omega_0 t)\,dt + \cdots + \int_T b_k \sin(k\omega_0 t)\cos(k\omega_0 t)\,dt + \cdots$$

Once again, orthogonality makes all the terms of the right-hand side of this equation zero except the one containing a_k and leads to

$$\int_T x_p(t)\cos(k\omega_0 t)\,dt = \int_T a_k \cos^2(k\omega_0 t)\,dt = a_k \int_T \tfrac{1}{2}[1 - \cos(2k\omega_0 t)]\,dt = \tfrac{1}{2}T a_k$$

This results in the coefficient a_k as

$$a_k = \frac{2}{T}\int_T x_p(t)\cos(k\omega_0 t)\,dt$$

Multiplying the series by $\sin(k\omega_0 t)$ and integrating, we similarly obtain

$$b_k = \frac{2}{T} \int_T x_p(t) \sin(k\omega_0 t)\, dt$$

The coefficients c_k and θ_k of the polar form, or $X_S[k]$ of the exponential form, may now be found using the appropriate transformations.

A formal expression for evaluating $X_S[k]$ directly is also found as follows:

$$X_S[k] = \tfrac{1}{2}a_k - j\tfrac{1}{2}b_k = \frac{1}{T} \int_T x_p(t) \cos(k\omega_0 t)\, dt - j\frac{1}{T} \int_T x_p(t) \sin(k\omega_0 t)\, dt$$

We now invoke Euler's relation, $\cos(k\omega_0 t) - j\sin(k\omega_0 t) = \exp(jk\omega_0 t)$, to obtain

$$X_S[k] = \frac{1}{T} \int_T x_p(t) \exp(-jk\omega_0 t)\, dt$$

Remarks: **1.** The coefficients show dependence on the harmonic index k.
2. Each coefficient is unique, is computed independently and does not affect any others. This concept describes the *finality of coefficients*.

Example 5.1
Fourier Series Coefficients
of a Rectangular
Pulse Train

Consider a rectangular pulse train (Figure 5.2) described by

$$x_p(t) = \begin{cases} A & 0 \le t \le t_0 \\ 0 & t_0 \le t \le T \end{cases}$$

Figure 5.2   A rectangular pulse train and its reconstructions.

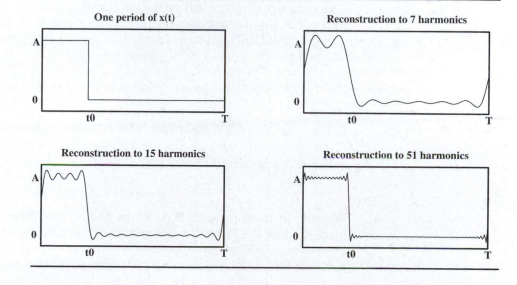

One period of x(t)

Reconstruction to 7 harmonics

Reconstruction to 15 harmonics

Reconstruction to 51 harmonics

We find the coefficients $X_S[k]$. The easy way to find $X_S[0]$ is graphically, as the average area. We have $X_S[0] = At_0/T$. The $X_S[k]$ are evaluated as

$$X_S[k] = \frac{1}{T} \int_0^T x_p(t) \exp(-jk\omega_0 t)\, dt$$

$$= \frac{1}{T} \int_0^{t_0} A \exp(-jk\omega_0 t)\, dt = \frac{-A}{jk\omega_0 T} \exp(-jk\omega_0 t)\Big|_0^{t_0}$$

Substituting for the limits, and absorbing the negative sign, we get

$$X_S[k] = \frac{A}{jk\omega_0 T}[1 - \exp(-jk\omega_0 t_0)]$$

To simplify this, we factor out $\exp(-jk\omega_0 t_0/2)$, and use Euler's relation:

$$X_S[k] = \frac{A}{jk\omega_0 T} \exp(-jk\omega_0 t_0/2)[\exp(jk\omega_0 t_0/2) - \exp(-jk\omega_0 t_0/2)]$$

$$= \frac{A}{jk\omega_0 T} \exp(-\tfrac{1}{2}jk\omega_0 t_0)[2j\sin(\tfrac{1}{2}k\omega_0 t_0)]$$

$$= \frac{2A}{k\omega_0 T} \sin(\tfrac{1}{2}k\omega_0 t_0) \exp(-\tfrac{1}{2}jk\omega_0 t_0)$$

This can also be rewritten in terms of a sinc function as

$$X_S[k] = \frac{At_0}{T}\left[\frac{\sin(\tfrac{1}{2}k\omega_0 t_0)}{\tfrac{1}{2}k\omega_0 t_0}\right] \exp(-\tfrac{1}{2}jk\omega_0 t_0) = \frac{At_0}{T} \operatorname{sinc}(kf_0 t_0) \exp(-j\pi k f_0 t_0)$$

Since $X_S[k] = \tfrac{1}{2}c_k \angle \theta_k$ for $k \geq 1$, we also have

$$c_k = 2|X_S[k]| = \frac{2At_0}{T} \operatorname{sinc}(kf_0 t_0) \qquad \theta_k = -\pi k f_0 t_0 \quad \text{for } k \geq 1$$

Figure 5.2 shows reconstructions of $x_p(t)$ from its harmonics. All display overshoot near the edges. The reason for this is discussed in Section 5.3.3.

Comment: The coefficients depend only on the duty ratio t_0/T. If t_0/T is a proper fraction m/n with $m < n$, the nth harmonic and its multiples vanish. Thus, if $t_0/T = 1/4$ or $2/4$ or $3/4$, the coefficients $X_S[k]$, $k = 4, 8, 12, \ldots$ (or the 4th harmonic and its multiples) all equal zero.

Note: This result cannot be extended to other periodic signals in general.   ▬▬

5.2.1   *Parseval's Relation*

Given the periodic signal $x_p(t)$, it is easy enough to evaluate the power contained in this signal in terms of the square of its rms value as

$$P_x = x_{\text{rms}}^2 = \frac{1}{T} \int_T x_p^2(t)\, dt$$

An equivalent formulation is based on the Fourier series of $x_p(t)$. When we replace $x_p^2(t)$ by its Fourier series and invoke orthogonality, products with different indices, such as $a_k a_m \cos(k\omega_0 t)\cos(m\omega_0 t), a_k b_m \cos(k\omega_0 t)\sin(m\omega_0 t)$ and $b_k b_m \sin(k\omega_0 t)\sin(m\omega_0 t)$, integrate to zero. Only the product terms with the same indices, of the form $a_k^2 \cos^2(k\omega_0 t)$ and $b_k^2 \sin^2(k\omega_0 t)$, survive. This leads to a familiar result. The total power in a combination of sinusoids at different frequencies is simply the sum of the power (the square of the rms value) in each sinusoid. We thus write

$$P_x = c_0^2 + \sum_{k=1}^{\infty} \left[\frac{c_k}{\sqrt{2}} \right]^2 = c_0^2 + \sum_{k=1}^{\infty} \tfrac{1}{2} c_k^2$$

Alternate formulations in terms of c_k or $X_S[k]$ are obtained by recognizing that $c_k^2 = a_k^2 + b_k^2 = 4|X_S[k]|^2$ and $|X_S[k]|^2 = X_S[k]X_S^*[k]$ to give

$$P_x = a_0^2 + \sum_{k=1}^{\infty} \tfrac{1}{2}[a_k^2 + b_k^2] = \sum_{k=-\infty}^{\infty} |X_S[k]|^2 = \sum_{k=-\infty}^{\infty} X_S[k]X_S^*[k]$$

To find the total power using the harmonics, we must sum an *infinte series*. It is far easier to obtain the *total* power in the time domain. But only the frequency-domain form allows us to find the power in a given harmonic, or up to a certain number of harmonics. The power P_N up to the Nth harmonic is simply

$$P_N = c_0^2 + \sum_{k=1}^{N} \tfrac{1}{2} c_k^2 = a_0^2 + \sum_{k=1}^{N} \tfrac{1}{2}[a_k^2 + b_k^2] = \sum_{k=-N}^{N} |X_S[k]|^2$$

This is a sum of positive quantities and is always less than the total signal power, which it approaches as more terms are added. It serves as a criterion to define the "goodness" of a given truncated representation.

The equivalence of the time-domain and frequency-domain expressions for the signal power forms the so-called **Parseval's relation**.

$$P_x = \frac{1}{T} \int_T x_p^2(t)\, dt = \sum_{k=-\infty}^{\infty} |X_S[k]|^2$$

Remark: Parseval's relation holds only if $x_p(t)$ is an energy signal over one period and, therefore, does not apply to functions such as the impulse.

Example 5.2
Using Parseval's Relation

For the $x_p(t)$ of Example 5.1, let $T = 2$, $t_0 = 1$. The power in $x_p(t)$ is

$$P_x = x_{\text{rms}}^2 = \frac{1}{T}\mathbb{A}[x_p^2(t)](0, T) = \frac{A^2 t_0}{T} = \tfrac{1}{2}A^2$$

The Fourier series coefficient $c_0 = A t_0 / T = \tfrac{1}{2}A$. With $\omega_0 = \pi$ and $\omega_0 t_0 = \tfrac{1}{2}\pi$,

$$c_k = \frac{2A t_0}{T} \left[\frac{\sin(k\omega_0 t_0/2)}{k\omega_0 t_0/2} \right] = A \frac{\sin(\tfrac{1}{2}k\pi)}{\tfrac{1}{2}k\pi} \qquad \text{for } k \geq 1$$

The c_k are zero for even k. The power in the odd harmonics is

$$P_k = \tfrac{1}{2} c_k^2 = \tfrac{1}{2}\left[\frac{A\sin(\tfrac{1}{2}k\pi)}{\tfrac{1}{2}k\pi} \right]^2 = 2A^2/k^2\pi^2 \quad (k \text{ odd})$$

(a) If we require the power up to the 2nd harmonic, we compute

$$P_0 + P_1 + P_2 = c_0^2 + \tfrac{1}{2}c_1^2 + \tfrac{1}{2}c_2^2 = \tfrac{1}{4}A^2 + 2A^2/\pi^2 = A^2[\tfrac{1}{4} + 2/\pi^2]$$

(b) The total power P_x can also be described by the infinite series

$$P_x = c_0^2 + \sum_{k=1}^{\infty} \tfrac{1}{2}c_k^2 = \frac{A^2}{4} + \frac{2A^2}{\pi^2} \sum_{k \text{ odd}}^{\infty} \frac{1}{k^2}$$

The infinite summation equals $\pi^2/8$ (Appendix A), and we indeed obtain

$$P_x = \frac{A^2}{4} + \frac{2A^2}{\pi^2}\left(\frac{\pi^2}{8}\right) = \frac{A^2}{2}$$

Comment: Naturally, this method of finding the total power is tedious!

5.2.2   *Notation and Terminology*

Tables 5.2 and 5.3 bring together the relations, notation and terminology used to describe the components of the various forms of a Fourier series.

Table 5.2   Fourier Series Forms and Relations.

Trigonometric	Polar	Exponential
$a_0 + \sum_{k=1}^{\infty} a_k \cos(k\omega_0 t) + b_k \sin(k\omega_0 t)$	$c_0 + \sum_{k=1}^{\infty} c_k \cos(k\omega_0 t + \theta_k)$	$\sum_{k=-\infty}^{\infty} X_S[k]\exp(jk\omega_0 t)$

Coefficients

Trigonometric	Polar	Exponential
$a_0 = \tfrac{1}{T}\int_T x_p(t)\,dt$	$a_k = \tfrac{2}{T}\int_T x_p(t)\cos(k\omega_0 t)\,dt$ $b_k = \tfrac{2}{T}\int_T x_p(t)\sin(k\omega_0 t)\,dt$	$X_S[k] = \tfrac{1}{T}\int_T x_p(t)\exp(-jk\omega_0 t)\,dt$

Relationships among the coefficients

$$a_0 = c_0 = X_S[0]$$
$$X_S[k] = \tfrac{1}{2}(a_k - jb_k) = \tfrac{1}{2}c_k \angle \theta_k \quad (k \geq 1)$$

Symmetry

$X_S[-k] = X_S^*[k]$ (conjugate)	$a_k = a_{-k}$ (even symmetry)	$b_k = -b_{-k}$ (odd symmetry)

Table 5.3   Fourier Series Terminology for $x_p(t)$.

Measure	Meaning
T, ω_0, f_0	Time period, fundamental frequency
$k, k\omega_0, kf_0$	Harmonic index, kth harmonic frequency
$c_0 = a_0 = X_S[0]$	The dc component (average value)
θ_k	The phase of the kth harmonic
$c_1 \cos(\omega_0 t + \theta_1)$	The fundamental component ($k = 1$)
$c_k \cos(k\omega_0 t + \theta_k)$	The kth harmonic component
$c_k = 2|X_S[k]| = (a_k^2 + b_k^2)^{1/2}$	Magnitude of the kth harmonic
$a_0^2 = c_0^2 = X_S^2[0]$	Power in the dc component
$\frac{1}{2}c_k^2 = 2|X_S[k]|^2 = \frac{1}{2}(a_k^2 + b_k^2)$	Power in the kth harmonic

---

Remarks:

1. In the Fourier series representation, the upper limit of the summation is $k = \infty$, since the exact representation for $x(t)$ may satisfy the equality only under this constraint.

2. An approximation to a finite number of harmonics describes a **truncated Fourier series.** Adding more terms improves the approximation. If the finite representation is exact, it is termed a **finite Fourier series.**

3. The Fourier series need not necessarily contain every harmonic up to the highest harmonic. Some (even the fundamental) could be absent.

4. In some cases, the Fourier series falls out as a direct application of trigonometric relations rather than the Fourier relations.

Example 5.3
A Finite Fourier Series

Using trigonometric identities, the signal $x_p(t) = 2 + 2\cos^2(2\pi t) - 2\sin(6\pi t)$ can be written as $x_p(t) = 3 + \cos(4\pi t) - 2\sin(6\pi t)$. This is a finite (and not truncated) Fourier series. To assign coefficients, we must first establish the fundamental frequency ω_0, which equals 2π (the GCD of 4π and 6π). By inspection, we have $a_0 = 3$, $a_2 = 1$, $b_3 = -2$, and all others zero. The period of $x_p(t)$ is given by $T = 2\pi/\omega_0 = 1$.

5.2.3   The Spectrum of Periodic Signals

The terms **spectral analysis** or **harmonic analysis** are often used to describe the analysis of a periodic signal $x_p(t)$ by Fourier series. The quantities a_k, b_k, c_k, θ_k or $X_S[k]$ describe the **spectral coefficients** of $x_p(t)$. Plots of these these coefficients, against the harmonic index k, or the frequency kf_0 (hertz) or $k\omega_0$ (radians per second), are called **spectra** or **spectral plots**. The two most commonly used spectral plots (Figure 5.3) are

1. The **magnitude spectrum,** a plot of harmonic magnitude c_k.
2. The **phase spectrum,** a plot of harmonic phase θ_k.

Figure 5.3   Various ways of plotting the Fourier series spectra. The spectra can also be plotted against f (at kf_0) or ω (at $k\omega_0$).

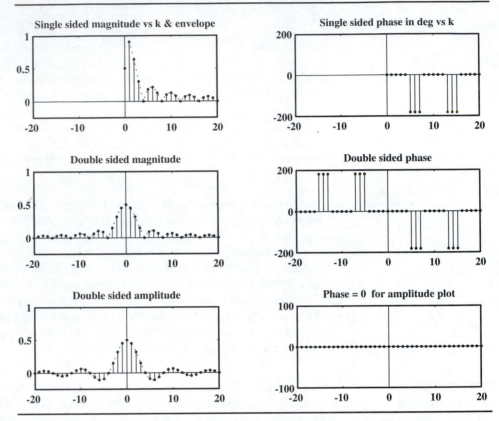

The magnitude and phase spectra yield information about the sinusoidal components of $x_p(t)$ in the *frequency domain*. Each pair of magnitude and phase spectra correspond, in the time domain, to a *unique* periodic signal.

The spectra describe *points* in the frequency domain since the spectral coefficients are defined only at *discrete* frequencies kf_0. For clarity, we often drop lines from these points on to the frequency axis, leading to the terminology **line spectra.** The lines themselves have no meaning.

The locus of the points in a spectral plot is called the **spectral envelope.** It is a continuous function which we can write as $X(f)$. Values of the discrete spectrum $X_S[k]$ equal $X(f)$ only at $f = kf_0$.

The Amplitude Spectrum   When the $X_S[k]$ are purely real or purely imaginary, we often sketch only the **amplitude spectrum** $X_S[k]$ (which could be negative) versus f, assuming a phase of zero for real $X_S[k]$ or $\pm 90°$ for imaginary $X_S[k]$.

Onesided and Twosided Spectra Onesided spectra refer to plots of c_0, c_k and θ_k for $k \geq 0$ or *positive frequencies*. Twosided spectra refer to plots of $|X_S[k]|$ and θ_k for all k, or all frequencies, *positive and negative*.

Symmetry of Twosided Spectra For real periodic signals, the $X_S[k]$ display conjugate symmetry, with $X_S[-k] = X_S^*[k]$. The twosided magnitude $|X_S[k]|$, or $\text{Re}\{X_S[k]\}$, show *even symmetry*. The twosided phase, or $\text{Im}\{X_S[k]\}$, display *odd symmetry* provided $X_S[0] \geq 0$ (since $X_S[0] < 0$ implies a phase of $180°$ or $-180°$, and destroys the odd symmetry).

Example 5.4
Spectra of a Rectangular Pulse Train

For the signal $x_p(t)$ of Figure 5.2 and Example 5.1, we choose $T = 1$, $t_0/T = \frac{1}{4}$, and $A = 2$ to obtain $X_S[0] = \frac{1}{2}$, and

$$X_S[k] = \frac{\frac{1}{2}\sin(\frac{1}{4}k\pi)}{\frac{1}{4}k\pi}\exp(-jk\pi/4) = \tfrac{1}{2}\text{sinc}(k/4)\exp(-jk\pi/4)$$

Figure 5.3 shows its onesided and twosided magnitude and phase spectra. Note the conjugate symmetry in the twosided spectra.

5.3
Simplifications Due to Signal Symmetry

The Fourier series of a periodic signal $x(t)$ devoid of symmetry must contain both odd (sine) and even (dc and cosine) components. For symmetric signals, simplifications through symmetry come about as follows (Figure 5.4):

1. An even signal must be made up of only even symmetric terms (dc and cosines). Hence $b_k = 0$, and the $X_S[k]$ must be purely real with $X_S[k] = a_k/2$.
2. An odd signal must be made up of only odd symmetric terms (sines). Hence $a_0 = a_k = 0$, and the $X_S[k]$ must be purely imaginary with $X_S[k] = -jb_k/2$.
3. A halfwave signal must likewise include only halfwave symmetric terms. *Only odd-indexed harmonics* (at f_0, $3f_0$, $5f_0$, ...) *display halfwave symmetry*. Even-indexed harmonics (at $2f_0$ $4f_0$, ...) complete an even number of cycles over the fundamental period T. Thus, each half-period of these is identical to the next half-period rather than an inverted replica (Figure 5.4) and *even harmonics cannot be halfwave symmetric*. As a result, a_0, a_k, b_k, c_k and $X_S[k]$ must vanish for even k. A half-wave symmetric signal thus contains only *odd-indexed harmonics*.

Remarks:

1. The terms *odd* and *even* are being used in two different contexts. For odd and even signals, they suggest symmetry. For odd and even k, they suggest an integer index or number. The meaning should, however, be clear from the context.
2. Halfwave symmetry is sometimes called *odd halfwave symmetry*. Yes, we also have *even halfwave symmetry* where only even-indexed harmonics survive and each half cycle is identical to the next. This is just a trick, because all we are really doing is using a period *twice* the true one to make all harmonics even multiples of *half* the fundamental frequency.

Figure 5.4   Illustrating the effects of symmetry.

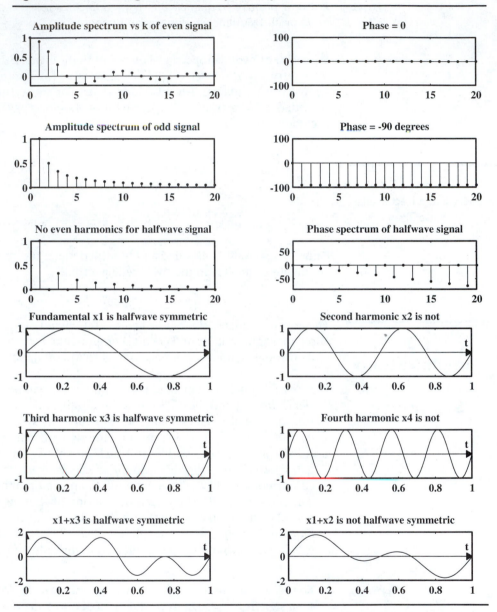

The nonzero coefficients may now be evaluated by using integration limits symmetric and invoking symmetry. The results, listed in Table 5.4, come about as follows:

Table 5.4   Fourier Series Coefficients of Symmetric Signals.

Symmetry	a_0	a_k	b_k
Even	$\dfrac{2}{T}\displaystyle\int_0^{T/2} x_e(t)\,dt$	$\dfrac{4}{T}\displaystyle\int_0^{T/2} x_e(t)\cos(k\omega_0 t)\,dt$	0
Odd	0	0	$\dfrac{4}{T}\displaystyle\int_0^{T/2} x_o(t)\sin(k\omega_0 t)\,dt$
Halfwave (k odd)	0	$\dfrac{4}{T}\displaystyle\int_0^{T/2} x_{\mathrm{hw}}(t)\cos(k\omega_0 t)\,dt$	$\dfrac{4}{T}\displaystyle\int_0^{T/2} x_{\mathrm{hw}}(t)\sin(k\omega_0 t)\,dt$
Even and halfwave (k odd)	0	$\dfrac{8}{T}\displaystyle\int_0^{T/4} x_{\mathrm{hwe}}(t)\cos(k\omega_0 t)\,dt$	0
Odd and halfwave (k odd)	0	0	$\dfrac{8}{T}\displaystyle\int_0^{T/4} x_{\mathrm{hwo}}(t)\sin(k\omega_0 t)\,dt$
Halfwave (k odd)	$X_S[k] = \dfrac{2}{T}\displaystyle\int_0^{T/2} x_{\mathrm{hw}}(t)\exp(-jk\omega_0 t)\,dt$		

1. If $x(t)$ is even, so is $x(t)\cos(k\omega_0 t)$. If $x(t)$ is odd, $x(t)\sin(k\omega_0 t)$ is even. We thus integrate only over $(0, \tfrac{1}{2}T)$ and scale the result by $4/T$.

2. If $x(t)$ is halfwave symmetric, we can also integrate over $(0, \tfrac{1}{2}T)$ but evaluate the resulting expressions only for odd k.

3. For *both even and halfwave* symmetry, only the a_k (k odd) are nonzero. For *both odd and halfwave symmetry*, only the b_k (k odd) are nonzero. The a_k (k odd) and b_k (k odd) can then be evaluated by integrating over $(0, \tfrac{1}{4}T)$ and scaling the result by $8/T$.

Remarks:

1. Note that the range of the integration limits is $(0, \tfrac{1}{2}T)$ or $(0, \tfrac{1}{4}T)$, *not any arbitrary half-period or quarter-period.*

2. The relation for $X_S[k]$ shows no simplifications except for halfwave symmetry, where we can use the limits $(0, \tfrac{1}{2}T)$ but only for odd k.

5.3.1   *Removing the DC Offset to Reveal Hidden Symmetry*

It is best to check for symmetry after removing the dc offset a_0. A signal that shows symmetry with its dc offset removed is said to possess *hidden symmetry*. If symmetry is present, we can use the symmetric version to evaluate all the Fourier coefficients except the dc offset (which must be computed from the original signal).

Remark: A time shift can also induce symmetry but it changes all coefficients. We develop relations between the coefficients in Section 5.4.

Example 5.5

Illustrating Concepts
Based on Symmetry

The spectra of $x(t)$ are shown in Figure 5.5. What can we say about $x(t)$? Harmonics at $f = 20, 60, 100, 140, \ldots$ suggest $f_0 = GCD(20, 60, 100, 140, \ldots) = 20$ Hz.

Figure 5.5   The spectra of the periodic signal for Example 5.5.

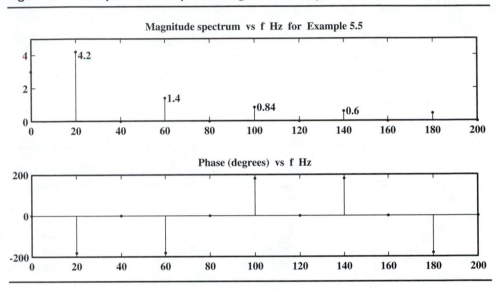

Thus $T = \frac{1}{20} = 0.05$ s. Since $k = 1, 3, 5, 7, \ldots$, and the dc component is nonzero, we have *hidden* halfwave symmetry. The phases of $\pm 180°$ imply cosine terms. We thus have even symmetry also.

The harmonic magnitudes $c_1 = 4.2$, $c_3 = 1.4$, $c_5 = 0.84$, $c_7 = 0.6, \ldots$ yield the ratios $c_3/c_1 = \frac{1}{3}$, $c_5/c_1 = \frac{1}{5}$, $c_7/c_1 = \frac{1}{7}, \ldots$. The convergence rate is thus $\propto 1/k$.

The power in $x(t)$ up to $k \leq 7$ is

$$P_7 = (3)^2 + \tfrac{1}{2}(4.2)^2[1 + (\tfrac{1}{3})^2 + (\tfrac{1}{5})^2 + (\tfrac{1}{7})^2] = 19.3329.$$

The total power P may be found by summing the infinite series

$$P_x = (3)^2 + \tfrac{1}{2}(4.2)^2[1 + (\tfrac{1}{3})^2 + (\tfrac{1}{5})^2 + (\tfrac{1}{7})^2 + \cdots]$$

This yields $P_x = 19.8812$. The rms value then equals $x_{\text{rms}} = \sqrt{P_x} = 4.4588$. Finally, $x(t)$ may be written as $x(t) = 3 - 4.2 \sum(1/k) \cos(40k\pi t)$ for odd k.

5.3.2   *Computing the Fourier Series Coefficients*

The Fourier coefficients of some periodic signals are listed in Table 5.5 and illustrated in Figure 5.6. We now illustrate the use of symmetry and other computational aspects.

Figure 5.6   Some periodic signals of Table 5.5 and their spectra.

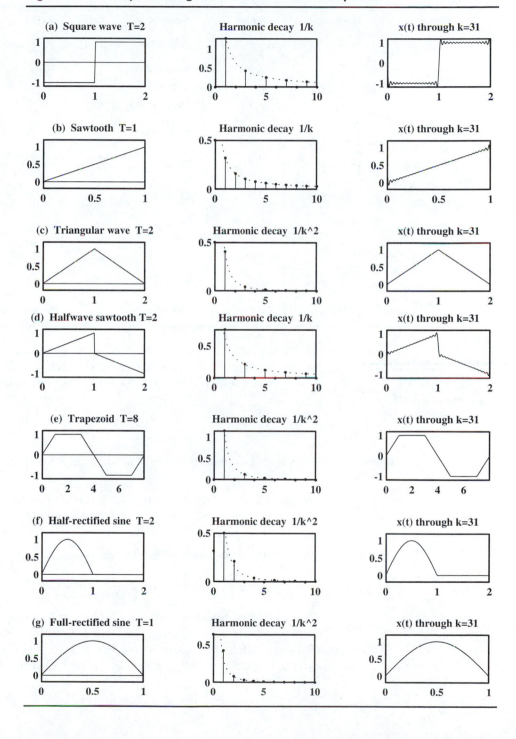

Table 5.5　Fourier Series Coefficients of Some Periodic Signals.

Signal	Expression	Range	Symmetry	$X_S[0]$	$X_S[k]$	Decay
Square wave 0 dc offset	1 -1	$0, \frac{1}{2}T$ $\frac{1}{2}T, T$	Odd & hw	0	$\dfrac{-j2}{k\pi}$ (k odd)	$1/k$
Triangle	$2t/T$ $2 - (2t/T)$	$0, \frac{1}{2}T$ $\frac{1}{2}T, T$	Even and hidden hw	$\dfrac{1}{2}$	$\dfrac{-2}{(k\pi)^2}$ (k odd)	$1/k^2$
Sawtooth	t/T	$0, T$	Hidden odd	$\dfrac{1}{2}$	$\dfrac{j}{2k\pi}$	$1/k$
Halfwave sawtooth	$2t/T$ $1 - (2t/T)$	$0, \frac{1}{2}T$ $\frac{1}{2}T, T$	Halfwave	0	$\dfrac{-2}{(k\pi)^2} + \dfrac{1}{jk\pi}$ (k odd)	$1/k$
Trapezoid	$2t/T$ 1	$0, \frac{1}{2}T$ $\frac{1}{2}T, T$	None	$\dfrac{3}{4}$	$\dfrac{\cos(k\pi) - 1}{2k^2\pi^2} + \dfrac{j}{2k\pi}$	$1/k$
Full rectified sine	$\sin(\pi t/T)$	$0, T$	Even	$\dfrac{2}{\pi}$	$\dfrac{2}{\pi(1 - 4k^2)}$	$1/k^2$
Parabola	$(t/T)^2$	$0, T$	None	$\dfrac{1}{3}$	$\dfrac{1}{2\pi^2 k^2} + \dfrac{j}{2\pi k}$	$1/k$
Even parabola	$(2t/T)^2$	$-\frac{1}{2}T, \frac{1}{2}T$	Even	$\dfrac{1}{6}$	$\dfrac{2\cos(k\pi)}{\pi^2 k^2}$	$1/k^2$
Even quadratic	$4t(1 - t/T)/T$	$0, T$	Even	$\dfrac{2}{3}$	$\dfrac{-2}{\pi^2 k^2}$	$1/k^2$
Impulse train	$A\delta(t - t_0)$	$0 < t_0 \le T$	None	$\dfrac{A}{T}$	$\dfrac{A}{T}\exp(jk\omega_0 t)$	Constant

---

Example 5.6
Computing the FS Coefficients

(a) *Sawtooth* (Figure 5.6b): $T = 1 (\omega_0 = 2\pi)$, $x_p(t) = t$, $0 \le t \le 1$. By inspection, $a_0 = \frac{1}{2}$. Assuming no symmetry, we find b_k $(k > 0)$ as follows:

$$b_k = 2\int_0^1 t\sin(2\pi kt)\,dt = \frac{2}{(2\pi k)^2}\left[\sin(2\pi kt) - 2\pi kt\cos(2\pi kt)\right]\Big|_0^1 = \frac{-1}{k\pi}$$

Note:　Since $x_p(t)$ has hidden odd symmetry, $a_k = 0, k \ne 0$.

(b) *Triangle* (Figure 5.6c): $T = 2 (\omega_0 = \pi)$, $x_p(t) = t, 0 \le t \le 1$. By inspection, $a_0 = \frac{1}{2}$. Since $x_p(t)$ is even, $b_k = 0, k \ne 0$. Using symmetry,

$$a_k = \frac{4}{T}\int_0^{T/2} x_p(t)\cos(k\omega_0 t)\,dt$$

$$= 2\int_0^1 t\cos(k\pi t)\,dt = \frac{2}{(\pi k)^2}\left[\cos(\pi kt) + \pi kt\sin(\pi kt)\right]\Big|_0^1$$

This simplifies to $a_k = 2[\cos(k\pi) - 1]/(k^2\pi^2), k > 0$. The a_k equal zero for even k and, since $a_0 \ne 0$, imply only hidden halfwave symmetry in $x_p(t)$.

(c) *Halfwave Sawtooth* (Figure 5.6d): $T = 2 (\omega_0 = \pi)$, $x_p(t) = t$, $0 \le t \le 1$. Since $x_p(t)$ is halfwave, $a_0 = 0$. We evaluate a_k and b_k *only for odd k* as

$$a_k = \frac{4}{T}\int_0^{T/2} x_p(t)\cos(k\omega_0 t)\,dt = 2\int_0^1 t\cos(k\pi t)\,dt$$

$$= \frac{2}{(\pi k)^2} \left[\cos(\pi k t) + \pi k t \sin(\pi k t) \right] \Big|_0^1$$

This simplifies to $a_k = 2[\cos(k\pi) - 1]/(k^2\pi^2) = -4/k^2\pi^2$ for odd k.

$$b_k = 2\int_0^1 t \sin(k\pi t)\, dt = \frac{2}{(\pi k)^2} \left[\sin(\pi k t) - k t \cos(\pi k t) \right] \Big|_0^1$$

$$= -2\frac{\cos(k\pi)}{k\pi} = 2/k\pi \quad \text{(odd } k\text{)}$$

(d) *Trapezoid* (Figure 5.6e): $T = 8, x_p(t) = t(0 \le t \le 1), x_p(t) = 1(1 \le t \le 2)$. It has both odd and hidden halfwave symmetry. Thus $a_0 = 0 = a_k$. Using symmetry,

$$b_k = \frac{8}{T} \int_0^{T/4} x_p(t) \sin(\tfrac{1}{4}k\pi t)\, dt = \int_0^1 t \sin(\tfrac{1}{4}k\pi t)\, dt + \int_1^2 \sin(\tfrac{1}{4}k\pi t)\, dt = I_1 + I_2$$

Now, simplifying I_1 and I_2 for odd k, we find

$$I_2 = \frac{-\cos(\tfrac{1}{4}k\pi t)}{\tfrac{1}{4}k\pi} \Big|_1^2 = \frac{4[-\cos(\tfrac{1}{2}k\pi) + \cos(\tfrac{1}{4}k\pi)]}{k\pi} = 4\frac{\cos(\tfrac{1}{4}k\pi)}{k\pi}$$

$$I_1 = \frac{1}{(\tfrac{1}{4}\pi k)^2} \left[\sin(\tfrac{1}{4}\pi k t) - \tfrac{1}{4}\pi k t \cos(\tfrac{1}{4}\pi k t) \right] \Big|_0^1 = \frac{16}{(\pi k)^2} \sin(\tfrac{1}{4}\pi k) - 4\frac{\cos(\tfrac{1}{4}\pi k)}{k\pi}$$

Thus $b_k = I_1 + I_2 = 16 \sin(\tfrac{1}{4}\pi k)/(k\pi)^2$ for odd k.

(e) *Half-Rectified Sine* (Figure 5.6f): $T = 2$, $x_p(t) = \sin(\pi t)$, $0 \le t \le 1$. The average value is found to be $a_0 = \mathbb{A}[x_p(t)]/T = 1/\pi$. There is no symmetry. We choose the range $(0, T)$ to find a_k and $b_k T$.

$$a_k = \int_0^1 \sin(\pi t) \cos(k\pi t)\, dt = \frac{1}{2} \int_0^1 \{\sin[(1 + k)\pi t] + \sin[(1 - k)\pi t\}\, dt$$

On simplification, using $\cos(1 \pm k)\pi = -\cos(k\pi)$, this yields

$$a_k = \frac{-1}{2\pi} \left[\frac{\cos[(1 + k)\pi t]}{1 + k} + \frac{\cos[(1 - k)\pi t]}{1 - k} \right] \Big|_0^1$$

$$= \frac{1}{\pi(1 - k^2)} [1 + \cos(k\pi)] = \frac{2}{\pi(1 - k^2)} \quad k \text{ even}$$

Similarly, b_k equals

$$b_k = \int_0^1 \sin(\pi t) \sin(k\pi t)\, dt = \frac{1}{2} \int_0^1 \{\cos[(1 - k)\pi t] - \cos[(1 + k)\pi t\}\, dt$$

$$= \frac{1}{2\pi} \left[\frac{\sin[(1 - k)\pi t]}{1 - k} + \frac{\sin[(1 + k)\pi t]}{1 + k} \right] \Big|_0^1 = 0$$

This may seem plausible to the unwary. But if $b_k = 0$, $x(t)$ must be even. It is not! Here is the catch: for $k = 1$, the first term has the indeterminate form $0/0$. Solution: evaluate b_1 separately, either by l'Hôpital's rule,

$$b_1 = \frac{1}{2\pi} \lim_{k \to 1} \left[\frac{\sin[(1 - k)\pi]}{1 - k} \right] = \frac{1}{2\pi} \lim_{k \to 1} \left[\frac{-\pi \cos[(1 - k)\pi]}{-1} \right] = \frac{1}{2}$$

or directly from the Fourier relation with $k = 1$,

$$b_1 = \int_0^1 \sin(\pi t) \sin(\pi t) \, dt = \int_0^1 \frac{1}{2}[1 - \cos(2\pi t)] \, dt = \frac{1}{2}$$

Thus $b_1 = \frac{1}{2}$ and $b_k = 0$ $(k > 1)$.

Comment: We must also evaluate a_1 separately. It actually works out to zero.

Beware: Indeterminate forms typically arise for signals with sinusoidal segments.

(f) *Impulse Train* (try sketching this): Period $T = 2$, $x(t) = A\delta(t)$.
          We have $X_S[0] = \mathbb{A}[x_p(t)]/T = A/2$. Using the sifting theorem, the $X_S[k]$ equal

$$X_S[k] = \frac{1}{T} \int_{-1}^{1} A\delta(t) \exp(-jk\omega_0 t) \, dt = A/2$$

Comment: The $X_S[k]$ are constant. This means we must sum an infinite number of cosines, all with peak value A, to reconstruct the impulse train.

Comment: If $x_p(t) = A_1\delta(t - t_1) + A_2\delta(t - t_2) + \cdots + A_N\delta(t - t_N)$, we obtain

$$X_S[k] = \frac{1}{T}[A_1 \exp(-jk\omega_0 t_1) + A_2 \exp(-jk\omega_0 t_2) + \cdots + A_N \exp(-jk\omega_0 t_N)]$$

In other words, for each impulse of strength A_n at t_n, we include the term $A_n \exp(-jk\omega_0 t_n)$ and divide the sum by T—just by inspection.

(g) *Impulse Pair Train* (sketch this too): Period $T = 1$, $x(t) = \delta(t - \frac{1}{8}) + \delta(t + \frac{1}{8})$.
Using the sifting property, superposition and Euler's relation, we obtain

$$X_S[k] = [\exp(j\tfrac{1}{8}k\omega_0) + \exp(-j\tfrac{1}{8}k\omega_0)] = 2\cos(\tfrac{1}{8}k\omega_0) = 2\cos(\pi k/4).$$

Note: The $X_S[k]$ show sinusoidal variation but no *decay* with k!   ───

5.3.3   *Signal Reconstruction*

The more harmonics we use to reconstruct a periodic signal, the better the reconstruction. The peaks in these harmonics occur at different times. As a result, if the number of harmonics N is not very large, the reconstruction shows oscillations (maxima and minima) or ripples. The number of peaks in the ripples provides a visual indication of the number of nonzero harmonics used in the reconstruction. Count the 20 peaks in the reconstruction of the sawtooth waveform by its 20 nonzero harmonics in Figure 5.7.

As N increases, we would expect the oscillations to die out and lead to perfect reconstruction for an infinite number of harmonics. Right? Not always! For signals with jumps, we can never get perfect reconstruction, since we simply cannot approximate a discontinuity by continuous functions (sinusoids). This intuitive

Figure 5.7   The Gibbs effect shows up in a reconstructed sawtooth wave with jumps but not in a reconstructed near-sawtooth wave (with no jumps).

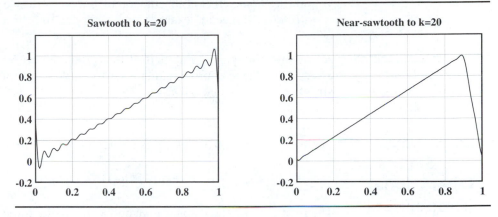

observation is confirmed by Figure 5.2 and Figure 5.6, which show reconstructions of various signals, including those with jumps.

5.3.4   The Gibbs Effect

The decay, or **convergence rate**, is intimately tied to reconstruction from a finite number of harmonics N. For signals with jumps, the harmonics decay as $1/k$ and we observe some remarkable anomalies as N is increased:

1. The oscillations and overshoot persist near the discontinuities, but become compressed (their frequency increases).
2. The peak overshoot moves closer to the discontinuity. As $N \to \infty$, this narrows into spikes, but its magnitude remains constant at about 9% of the magnitude of the discontinuity. This is independent of both the nature of the signal and N, as long as N is reasonably large.
3. At the discontinuity, the series converges (sums) to its midpoint.

The persistence of overshoot and imperfect reconstruction for signals with jumps describes the **Gibbs phenomenon** or **Gibbs effect**. Even when $N = \infty$, the spikes still remain, but as vertical lines that extend upward and downward by 9% of the discontinuity. The total power does, however, equal the power in the actual signal as predicted by Parseval's relation because the area under the spikes is zero and contributes nothing. The reconstruction is said to converge to the actual function in the *mean squared* sense.

Remarks: 1. The Gibbs effect is not just the appearance of overshoot for finite N. Rather, it is the lack of disappearance of overshoot as $N \to \infty$. If N is small, the overshoot may differ from 9% or be absent altogether.

2. The Gibbs effect occurs only for waveforms with jump discontinuities whose convergence is $1/k$. For all other waveforms, the convergence rate is $1/k^2$ or better, and the Gibbs effect is absent. For such waveforms we get better reconstruction with fewer terms and exact reconstruction with $N = \infty$.

<table>
<tr><td>

Example 5.7

Reconstruction and the Gibbs Effect

</td><td>

A true sawtooth pulse exhibits the Gibbs effect (Figure 5.7). A triangular pulse with a very steep slope approximating a sawtooth does not, and leads to perfect reconstruction for large enough k.

</td></tr>
</table>

5.4 Operational Properties

The Fourier series provides a link between time and frequency through

$$X_S[k] = \frac{1}{T} \int_{-T/2}^{T/2} x_p(t) \exp(-jk\omega_0 t)\, dt \qquad x_p(t) = \sum_{k=-\infty}^{\infty} X_S[k] \exp(jk\omega_0 t)$$

The relations may be displayed symbolically by the transform pair

$$x_p(t) \longleftrightarrow X_S[k]$$

The coefficients $X_S[k]$ are located at the index k or at $f = kf_0$ or $\omega = k\omega_0$. The properties that demonstrate how various operations affect $x_p(t)$ and its spectrum $X_S[k]$ are summarized in Table 5.6 (page 208) and Figure 5.8.

5.4.1   DC Shift, Amplitude Scaling and Superposition

A **dc shift** changes only the average value $X_S[0]$. **Amplitude scaling** scales only the magnitude of $X_S[0]$ and $X_S[k]$, not phase (Figure 5.8b). **Superposition** expresses the linearity of the Fourier series relations as

$$\alpha x_p(t) + \beta y_p(t) \longleftrightarrow \alpha X_S[k] + \beta Y_S[k]$$

5.4.2   Time Shift

A **time shift** changes $x_p(t)$ to $y_p(t) = x_p(t \pm \alpha)$. We may then write

$$y_p(t) = \sum_{k=-\infty}^{\infty} X_S[k] \exp[jk\omega_0(t \pm \alpha)] = \sum_{k=-\infty}^{\infty} X_S[k] \exp(\pm jk\omega_0 \alpha) \exp(jk\omega_0 t)$$

This represents a Fourier series with coefficients $X_S[k] \exp(\pm jk\omega_0 \alpha)$. Thus,

$$x_p(t - \alpha) \longleftrightarrow X_S[k] \exp(-jk2\pi f_0 \alpha)$$

The magnitude spectrum shows no change. The term $\exp(-jk2\pi f_0 \alpha)$ introduces an *additional linear phase* of $-2\pi k f_0 \alpha$. It changes the phase of each harmonic (Figure 5.8c) in proportion to its frequency kf_0. Since $X_S[k] = \frac{1}{2}(a_k - jb_k)$, both a_k and b_k change, but in a way that the new magnitude, $c_k(\text{new}) = [a_k^2(\text{new}) + b_k^2(\text{new})]^{1/2}$, equals $2|X_0[k](\text{old})|$. A time shift by an *integer* number of periods leaves all the coefficients unchanged.

Figure 5.8   How operations on a periodic signal affect its magnitude and phase spectra in accordance with the operational properties of Table 5.6.

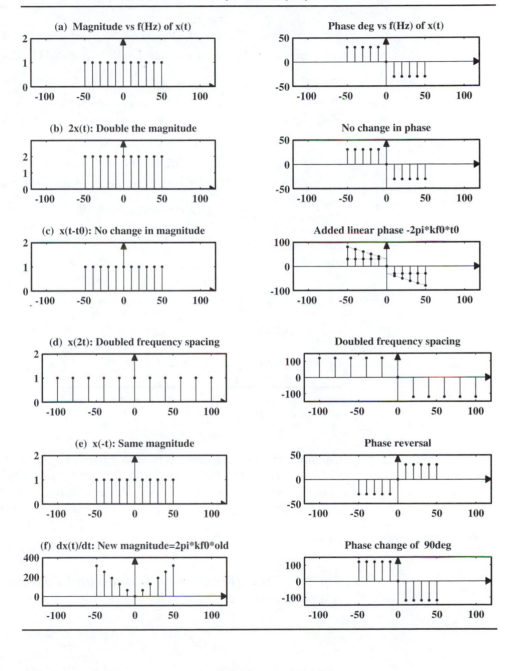

Table 5.6　Operational Properties of Fouier Series.
Period $= T$, $f_0 = 1/T$, $\omega_0 = 2\pi f_0$, $x_p(t) \longleftrightarrow X_S[k]$, $y_p(t) \longleftrightarrow Y_S[k]$

Property	Result
Superposition	$\alpha x_p(t) + \beta y_p(t) \longleftrightarrow \alpha X_S[k] + \beta Y_S[k]$
Derivative	$x'_p(t) \longleftrightarrow jk2\pi f_0 X_S[k] = jk\omega_0 X_S[k] \quad (k \neq 0)$
Integral	$\displaystyle\int_0^t x_p(t)\, dt \longleftrightarrow \dfrac{X_S[k]}{jk2\pi f_0}\ (k \neq 0) + C = \dfrac{X_S[k]}{jk\omega_0}\ (k \neq 0) + C$
Time delay	$x_p(t - \alpha) \longleftrightarrow X_S[k]\exp(-jk2\pi f_0\alpha) = X_S[k]\exp(-jk\omega_0\alpha)$
Time scaling	$x_p(\alpha t) \longleftrightarrow X_S[k]$ (harmonics located at $f = kf_0\alpha$)
Reversal	$x_p(-t) \longleftrightarrow X_S[-k] = X_S^*[k]$
Modulation	$\cos(m2\pi f_0 t)x_p(t) \longleftrightarrow \frac{1}{2}\{X_S[k - m] + X_S[k + m]\}$ (shifted by $\pm mf_0$)
Modulation	$\frac{1}{2}\{x_p(t + \alpha) + x_p(t - \alpha)\} \longleftrightarrow \cos(2\pi f_0\alpha)X_S[k]$
Discrete convolution	$x_p(t)y_p(t) \longleftrightarrow X_S[k] \star Y_S[k]$
Periodic convolution	$x_p(t) \bullet y_p(t) \longleftrightarrow X_S[k]Y_S[k]$
Symmetry	Real $x_p(t) \longleftrightarrow X_S[k] = X_S^*[-k]$
	Even $x_p(t) \longleftrightarrow$ Real $X_S[k]$
	Odd $x_p(t) \longleftrightarrow$ Imaginary $X_S[k]$
Central ordinates	$X_S[0] = \dfrac{1}{T}\displaystyle\int_T x_p(t)\, dt \qquad x_p(0) = \displaystyle\sum_{k=-\infty}^{\infty} X_S[k]$
Parseval's relation	$\dfrac{1}{T}\displaystyle\int_T x_p^2(t)\, dt = \displaystyle\sum_{k=-\infty}^{\infty} |X_S[k]|^2 = a_0^2 + \displaystyle\sum_{k=1}^{\infty} \tfrac{1}{2}(a_k^2 + b_k^2)$

NOTE: The spectral coefficients $X_S[k]$ and $Y_S[k]$ are located at $f = kf_0$ or $\omega = k\omega_0$ unless otherwise stated.

Example 5.8
Time Shift

Consider an odd sawtooth $x(t)$ with period T, $|X_S[k]| = A/k\pi$ and $\theta_k = -\frac{1}{2}\pi$. If $y(t) = x(t + \frac{1}{4}T)$, then $|Y_S[k]| = A/k\pi$, but its phase ϕ_k includes the additional linear term $k\omega_0 T/4 = \frac{1}{2}k\pi$. Its total phase is thus $\phi_k = \theta_k + \frac{1}{2}k\pi = \frac{1}{2}(k - 1)\pi$. This leads to alternating cosine and sine terms and a lack of symmetry in $y(t)$.

5.4.3　Time (Frequency) Scaling and Folding

A **time scaling** by α results in the series

$$x_p(\alpha t) = \sum_{k=-\infty}^{\infty} X_S[k]\exp(jk\omega_0\alpha t)$$

The spectral coefficients $X_S[k]$ are unaltered for any k, but located at $k\alpha f_0$ (or $k\alpha\omega_0$), because scaling to $x_p(\alpha t)$ changes the fundamental frequency of the scaled signal from f_0 to αf_0 (or $\alpha \omega_0$). In operational notation

$$x_p(\alpha t) \longleftrightarrow X_S[k] \quad \text{at} \quad \omega = k\omega_0\alpha \quad \text{or} \quad f = kf_0\alpha$$

The more compressed the time signal, the farther apart its harmonics in the frequency domain. (Figure 5.8d).

Reversal, or **folding**, is equivalent to a scale factor $\alpha = -1$, and results in

$$x_p(-t) \longleftrightarrow X_S[-k] = X_S^*[k]$$

The magnitude remains unchanged. The phase is reversed (Figure 5.8e).

Remark: We also have $x_p(t) + x_p(-t) \longleftrightarrow X_S^* + X_S = 2\text{Re}[X_S]$. This suggests that the Fourier coefficients of the even part of $x_p(t)$ equal $\text{Re}\{X_S[k]\}$. Similarly, the Fourier series coefficients of the odd part of $x_p(t)$ equal $\text{Im}\{X_S[k]\}$.

5.4.4 *Derivatives*

The derivative of $x_p(t)$ is found by differentiating each term in its series

$$\frac{dx_p(t)}{dt} = \sum_{k=-\infty}^{\infty} \frac{d}{dt}\left[X_S[k]\exp(jk\omega_0 t)\right] = \sum_{k=-\infty}^{\infty} jk\omega_0 X_S[k]\exp(jk\omega_0 t)$$

In operational notation,

$$x_p'(t) \longleftrightarrow jk\omega_0 X_S[k]$$

The dc term $X_S[0]$ vanishes, whereas all other $X_S[k]$ are multiplied by $jk\omega_0$. The magnitude of each harmonic is scaled by its frequency, $k\omega_0$, and the spectrum thus has significantly larger high-frequency content (Figure 5.8f).

Remark: The derivative operation enhances the sharp details and features in any time function. This suggests that the sharp details and features in a time signal contribute most to its high-frequency spectrum.

5.4.5 *Integration*

Upon integration, a periodic signal $x_p(t)$ with a dc offset $X_S[0]$ cannot stay periodic because the integral of $X_S[0]$ is a ramp that provides increasing contributions. For signals with $X_S[0] = 0$, however, we obtain

$$\int_0^t x_p(t)\,dt \longleftrightarrow \frac{X_S[k]}{jk\omega_0} \quad (k \neq 0) + \text{Constant}$$

The integrated signal is also periodic, but can be determined only to within a constant. Note that if we keep track of $X_S[0]$ separately, integration and differentiation may be thought of as inverse operations.

Remark: The coefficients $X_S[k]/jk\omega_0$ imply faster convergence and reduce the high-frequency components. Since integration is a smoothing operation, the smoother a signal, the smaller its high-frequency content.

5.4.6   *Fourier Series Coefficients by the Derivative Method*

The derivative property provides an elegant way of finding the Fourier series coefficients without the need for involved integration.

Let us denote the Fourier coefficients of successive derivatives of $x(t)$ by X_m, where m is the number of derivatives. We then have

$$X_1 = (jk\omega_0)X_S[k], \qquad X_2 = (jk\omega_0)^2 X_S[k], \ldots, \qquad X_m = (jk\omega_0)^m X_S[k]$$

Finding $X[k]$ is simply a matter of rearranging the above relations to

$$X_S[k] = \frac{X_S}{(jk\omega_0)} = \frac{X_S}{(jk\omega_0)^2} = \cdots = \frac{X_m}{(jk\omega_0)^m}$$

To obtain $X[k]$ from the mth derivative, we divide X_m by $(jk\omega_0)^m$.

The derivative method is based on this concept and two other key ideas:

1. Derivatives of piecewise signals sooner or later yield impulses.
2. The Fourier series coefficients of an impulse train are quite easy to obtain (by inspection or by the sifting property).

To find the coefficients $X[k]$ of a signal $x(t)$, here is what we do:

1. Take enough derivatives of $x(t)$ to obtain $x^{(m)}(t)$, which contains only impulses (and perhaps their derivatives).
2. Find the coefficients X_m of $x^{(m)}(t)$.
3. Divide X_m by $(jk\omega_0)^m$ to obtain $X[k]$.

This method obviates the need for tedious integration. In fact, we never need to work with higher-order derivatives of impulses if we are willing to do some bookkeeping. As impulses occur, we find their coefficients, remove them, and continue taking derivatives of the remaining waveform until done. To find $X_S[k]$, we simply divide each coefficient by $jk\omega_0$ as many times as we differentiated to obtain it, and sum the results. Note that a rectangular portion of width T contributes nothing to $X[k]$ (since it describes a constant) and may thus be ignored during the derivative process.

We illustrate the derivative method by some examples whose results and others are listed in Table 5.7. We have deliberately retained the factor $1/T$ while simplifying the coefficients.

Remark: $X_S[0]$ must be evaluated separately (since the dc component disappears after the very first derivative). So too must any special cases.

Table 5.7   Fourier Series Coefficients Using the Derivative Method.

Signal	Coefficient $X_S[k]$
Impulse train $x_p(t) = A\delta(t - t_0)$	$\dfrac{A}{T}\exp(-jk2\pi f_0 t_0)$
Rectangular pulse train $x_p(t) = A, |t| \le \frac{1}{2}t_0$	$\dfrac{At_0}{T}\operatorname{sinc}(kf_0 t_0)$
Triangular pulse train $x_p(t) = A(1 - 2|t|/t_0), |t| \le \frac{1}{2}t_0$	$\dfrac{At_0}{2T}\operatorname{sinc}^2(\frac{1}{2}kf_0 t_0)$
Exponential pulse train $x_p(t) = A\exp(-t/\tau), 0 \le t \le t_0$	$\dfrac{A}{T}\left[\dfrac{1 - \exp[-t_0(jk\omega_0 + 1/\tau)]}{jk\omega_0 + 1/\tau}\right]$
Sinusoidal pulse train $x_p(t) = A\cos(\pi t/t_0), |t| \le \frac{1}{2}t_0$	$\dfrac{1}{T}\left[\dfrac{2A\pi t_0 \cos(\frac{1}{2}k\omega_0 t_0)}{\pi^2 - (k\omega_0 t_0)^2}\right]$
Sawtooth pulse train $x_p(t) = At/t_0, 0 \le t \le t_0$	$\dfrac{1}{T}\left[\dfrac{A[\exp(-jk\omega_0 t_0) - 1]}{(k\omega_0)^2 t_0} - \dfrac{A\exp(-jk\omega_0 t_0)}{jk\omega_0}\right]$

Example 5.9

The Derivative Method

(a) For the rectangular pulse train of Figure 5.9a, $X_S[0] = 4/T$. Over one period $(-\frac{1}{2}T, \frac{1}{2}T)$, $x'(t)$ contains only impulses. The coefficients $X_1[k]$ of $x'(t)$ are, by inspection,

$$X_1[k] = \frac{1}{T}[2\exp(jk\omega_0) - 2\exp(-jk\omega_0)]$$

The coefficients $X_S[k]$ for $x(t)$ equal $X_1/jk\omega_0$, which can be simplified to

$$X_S[k] = \frac{1}{T}\left[\frac{2\exp(jk\omega_0) - 2\exp(-jk\omega_0)}{jk\omega_0}\right] = \frac{4}{T}\operatorname{sinc}(2kf_0)$$

(b) For $x(t) = 2\cos(\pi t/2)$, (Figure 5.9b), we obtain

$$x'(t) = -\pi\sin(\pi t/2) \qquad x''(t) = -\frac{1}{2}\pi^2\cos(\pi t/2) + \pi\delta(t+1) + \pi\delta(t-1)$$

The sum $y(t) = \frac{1}{4}\pi^2 x(t) + x''(t) = \pi\delta(t+1) + \pi\delta(t-1)$ contains only impulses and thus

$$Y_S[k] = \frac{1}{T}[\pi\exp(jk\omega_0) + \pi\exp(-jk\omega_0)] = \frac{2\pi}{T}\cos(k\omega_0)$$

We also have

$$Y_S[k] = \frac{1}{4}\pi^2 X_S[k] + (jk\omega_0)^2 X_S[k]$$

This leads to

$$X_S[k] = Y_S[k]/[\frac{1}{4}\pi^2 - k^2\omega_0^2]$$

or

$$X_S[k] = \frac{1}{T}\left[\frac{8\pi\cos(k\omega_0)}{\pi^2 - 4k^2\omega_0^2}\right]$$

Figure 5.9   Illustrating the derivative method for Example 5.9.

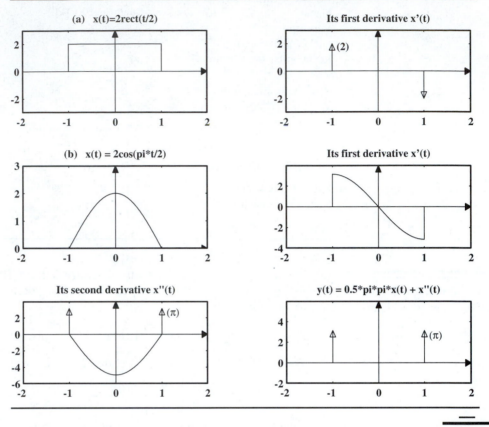

5.4.7   *Convergence Rate*

The Fourier series coefficients $X_S[k]$ of a periodic signal $x_p(t)$ usually decay with the harmonic index k as $1/k^n$. A larger n implies faster convergence, fewer terms in a truncated series for a given error, faster high-frequency decay, and a smoother signal. We can establish n without computing $X_S[k]$ by using the derivative property. Since the coefficients of an impulse train show no dependence on k, if $x_p(t)$ is differentiated n times before it first shows impulses, $X_S[k]$ must be proportional to $1/(jk\omega_0)^n$, and thus decay as $1/k^n$. A measure of smoothness may be ascribed to $x_p(t)$ in terms of the number of derivatives n it takes for the waveform to first result in impulses. In Figure 5.6, for example, the trapezoidal and half-rectified sine need two derivatives before yielding impulses, and show a convergence rate of $1/k^2$. Both are smoother than the sawtooth and square waves, which need only one derivative to produce impulses, and show a convergence rate of $1/k$. We stress that the convergence rate is a valid indicator only for large k.

Example 5.10
The Concept of
Convergence Rate

What is the convergence rate if $X_S[k] = A/k + jB/k^2$? For large k, B/k^2 is much smaller than A/k, and $X_S[k]$ thus decays as $1/k$ (and not $1/k^2$). For small k, of course, $X_S[k]$ will be affected by both terms.

5.4.8   Modulation in Time

If we multiply $x_p(t)$ by the harmonic signal $\exp(j\alpha t)$ we may write

$$x_p(t)\exp(j\alpha t) = \sum_{k=-\infty}^{\infty} X_S[k]\exp(j\alpha t)\exp(jk\omega_0 t) = \sum_{k=-\infty}^{\infty} X_S[k]\exp[j(k\omega_0 + \alpha)t]$$

The coefficients $X_S[k]$ are displaced from $k\omega_0$ to $k\omega_0 + \alpha$.

Similarly, the spectral coefficients of $x_p(t)\exp(-j\alpha t)$ are displaced to the frequencies $k\omega_0 - \alpha$. Since $\cos(\alpha t) = \frac{1}{2}[\exp(j\alpha t) + \exp(-j\alpha t)]$, the coefficients of $x_p(t)\cos(\alpha t)$ equal half the sum of the shifted spectral coefficients at the locations $k\omega_0 + \alpha$ and $k\omega_0 - \alpha$.

This result is called **modulation**. Multiplying a periodic signal $x_p(t)$ by $\cos(\alpha t)$ displaces the spectral coefficients $\pm\alpha$ and halves their magnitudes.

Example 5.11
Illustrating Modulation
in Time

Consider the signal $x_p(t) = \cos(k\omega_0 t)$. Its Fourier series coefficients are

$$\{X_S[k]\} = \{\tfrac{1}{2}, 0, \tfrac{1}{2},\} \text{ at } \{-k\omega_0, 0, k\omega_0\}.$$

If $x_p(t)$ is modulated by itself to give $y_0(t) = x_p(t)x_p(t) = x_p^2(t)$, we find $Y_S[k]$ by displacing the spectral locations by $\pm k\omega_0$ to $\{0, k\omega_0, 2k\omega_0\}$ and $\{-2k\omega_0, -k\omega_0, 0\}$. We halve the shifted coefficients and add

$$Y_S[k]_0 = \{\tfrac{1}{4}, 0, \tfrac{1}{4}\} + \{\tfrac{1}{4}, 0, \tfrac{1}{4}\} = \{\tfrac{1}{4}, 0, \tfrac{1}{2}, 0, \tfrac{1}{4}\}$$

The $Y_S[k]$ at the locations $\{-2k\omega_0, -k\omega_0, 0, k\omega_0, 2k\omega_0\}$ correspond to

$$y(t) = \tfrac{1}{2} + \tfrac{1}{2}\cos(2k\omega_0 t)$$

Check: Compare this with the result $x_p^2(t) = \cos^2(k\omega_0 t) = \tfrac{1}{2} + \tfrac{1}{2}\cos(2k\omega_0 t)$.

5.4.9   Frequency Modulation

If we shift $x_p(t)$ to the left and right by α and apply superposition, we obtain

$$x_p(t + \alpha) + x_p(t - \alpha) \longleftrightarrow X_S[k]\exp(-jk\omega_0\alpha) + X_S[k]\exp(jk\omega_0\alpha)$$

This simplifies to

$$x_p(t + \alpha) + x_p(t - \alpha) \longleftrightarrow 2X_S[k]\cos(k\omega_0\alpha)$$

and represents the dual of time-domain modulation.

Example 5.12

Illustrating Modulation and
Other Properties

Consider the signal $x_L(t) = \text{rect}(t/2t_C)$ of Figure 5.10 with period $T = 1$. Its Fourier series coefficients are $X_L[k] = 2t_C \text{sinc}(2kt_C)$. The other signals shown in Figure 5.10 may then be written as

$$x_H(t) = x_L(t - \tfrac{1}{2}) \qquad x_B(t) = x_L(t + t_0) + x_L(t - t_0) \qquad x_S(t) = \text{rect}(t) - x_B(t)$$

Using properties, the Fourier series coefficients of these signals are:

$$\begin{array}{ll} X_H[k] = X_L[k] \exp(-jk\pi) & \textbf{Using the shifting property.} \\ X_B[k] = 2\cos(2\pi kt_0)X_L[k] & \textbf{Using frequency modulation.} \\ X_S[k] = \text{sinc}(k) - X_B[k] & \textbf{Using superposition.} \end{array}$$

Since $\exp(-jk\pi) = (-1)^k$ and $\text{sinc}(k) = \delta[k]$, we can simplify these relations further. The results, summarized in Table 5.8, have important implications in digital filter design, which we discuss in Chapter 14.

Table 5.8   Fourier Series of Rectangular Pulse Trains.

Signal	Fourier series coefficient
$x_L(t) = \text{rect}(t/2t_C)$	$X_L[k] = 2t_C\text{sinc}(2kt_C)$
$x_H(t) = x_L(t - \tfrac{1}{2})$	$X_H[k] = X_L[k]\exp(-jk\pi) = (-1)^k 2t_C\text{sinc}(2kt_C)$
$x_B(t) = x_L(t + t_0) + x_L(t - t_0)$	$X_B[k] = 2\cos(2\pi kt_0)X_L[k] = 4t_C\cos(2\pi kt_0)\text{sinc}(2kt_C)$
$x_S(t) = \text{rect}(t) - x_B(t)$	$X_S[k] = \text{sinc}(k) - X_B[k] = \delta[k] - 4t_C\cos(2\pi kt_0)\text{sinc}(2kt_C)$

5.4.10   *Multiplication in Time and Frequency Convolution*

Discrete convolution is equivalent to polynomial multiplication. If we express $X_S[k]\exp(jk\omega_0t)$ as $X_S[k]W^k$, where $W = \exp(j\omega_0t)$, a finite Fourier series is just a polynomial in W. If $X_S[k]$ and $Y_S[k]$ are the coefficients of two finite Fourier series with identical periods, the coefficients $Z_S[k]$ of $x_p(t)y_p(t)$ will equal $X_S[k] \star Y_S[k]$. Thus,

$$x_p(t)y_p(t) \longleftrightarrow X_S[k] \star Y_S[k]$$

If the indices for X_S and Y_S span $\pm M$ and $\pm N$, we get

$$Z_S[k] = \sum_{n=-M-N}^{M+N} X_S[n]Y_S[n-k] \qquad k = -M-N, \ldots, -2, -1, 0, 1, 2, \ldots, M+N$$

If $x_p(t)$ or $y_p(t)$ is described by an infinite series, the $Z_S[k]$ represent an infinite summation not easily amenable to a closed-form solution.

Figure 5.10   Illustrating modulation and other properties for Example 5.12.

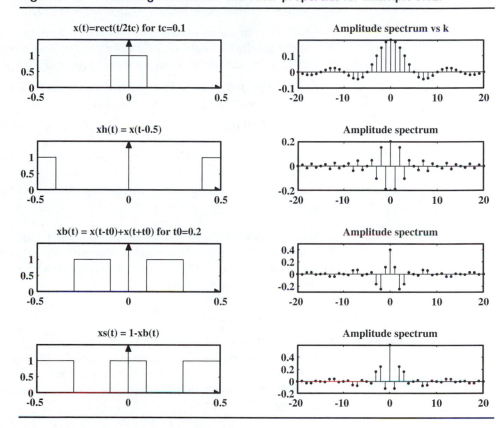

Remarks:

1. Remember that discrete convolution requires equispaced samples for both sequences. If $x_p(t)$ and $y_p(t)$ have different periods, the fundamental frequency f_0 of $x_p(t)y_p(t)$ will equal the GCD of the two frequencies. For correct results, both $X_S[k]$ and $Y_S[k]$ should have zeros inserted at the missing harmonics of f_0 before convolution.

2. The modulation property is a special case of the convolution property. In fact, multiplying $x(t)$ by $\cos(\omega_0 t)$ is equivalent to convolving the sequence $X_S[k]$ with the sequence $\{\frac{1}{2}, 0, \frac{1}{2}\}$.
   $\uparrow$

Example 5.13
Convolution in Frequency

(a) Let $x_p(t) = \cos(k\omega_0 t)$ with $\{X_S[k]\} = \{\frac{1}{2}, 0, \frac{1}{2}\}$ at $k = \{-1, 0, 1\}$. The product
$\uparrow$
$y_p(t) = x_p(t)x_p(t)$ corresponds to the convolution

$$Y_S[k] = \{\tfrac{1}{2}, 0, \tfrac{1}{2}\} \star \{\tfrac{1}{2}, 0, \tfrac{1}{2}\} = \{\tfrac{1}{4}, 0, \tfrac{1}{2}, 0, \tfrac{1}{4}\}$$

This describes $y_p(t) = \frac{1}{2} + \frac{1}{2}\cos(2k\omega_0 t) = \cos^2(k\omega_0 t)$ as expected.

(b) Let $x(t) = 2 + 4\sin(6\pi t)$ and $y(t) = 4\cos(2\pi t) + 6\sin(4\pi t)$. Their spectral coefficients are $X_S = \{j2, 2, -j2\}$ and $Y_S = \{j3, 2, 0, 2, -j3\}$. Since the fundamental frequencies of $x(t)$ and $y(t)$ are 6π and 2π, the fundamental frequency of the product is $\omega_0 = GCD(6\pi, 2\pi) = 2\pi$. We insert zeros at the missing harmonics 2π and 4π in X_S to get $X_S = \{j2, 0, 0, 2, 0, 0, -j2\}$. The sequence Y_S has no missing harmonics. The spectral coefficients Z_S of the product $x(t)y(t)$ are
$$Z_S = \{j2, 0, 0, 2, 0, 0, -j2\} \star \{j3, 2, 0, 2, -j3\}.$$

We find that $Z_S = \{-6, j4, 0, j10, 10, 0, 10, -j10, 0, -j4, -6\}$, which gives

$$z(t) = x(t)y(t) = 20\cos(2\pi t) + 20\sin(4\pi t) + 8\sin(8\pi t) - 12\cos(10\pi t)$$

Comment: You can check this result by direct multiplication of $x(t)$ and $y(t)$. ▬▬

5.4.11   *Frequency Multiplication and Convolution in Time*

The product of two periodic signals transforms into a discrete convolution. The dual of this property is that the product of two spectra corresponds to the convolution of their periodic time signals

$$x_p(t) \bullet y_p(t) \longleftrightarrow X_S[k]Y_S[k]$$

Remarks: **1.** Since $x_p(t)$ and $y_p(t)$ are periodic, it is important to realize that the convolution $x_p(t) \bullet y_p(t)$ represents their *periodic convolution*. The periods of $x_p(t)$ and $y_p(t)$ must also be commensurate. This makes sense because only then will the spectral coefficients be located at like harmonic frequencies and yield a nonzero product sequence.

2. The modulation property is the dual of the shifting property, whereas the scaling property is its own dual.

▬▬▬▬▬▬

Example 5.14
Convolution in Time

Consider two rectangular pulse trains $x_p(t)$ and $y_p(t)$ with identical periods T and identical heights 1, but different pulse widths t_1 and t_2. Their Fourier series coefficients are described by:

$$X_S[k] = (t_1/T)\text{sinc}(kf_0t_1)\exp(-jk\pi f_0t_1) \qquad Y_S[k] = (t_2/T)\text{sinc}(kf_0t_2)\exp(-jk\pi f_0t_2)$$

The Fourier coefficients of the signal $z_p(t) = x_p(t) \bullet y_p(t)$ are simply

$$Z_S[k] = X_S[k]Y_S[k] = (t_1t_2/T^2)\text{sinc}(kf_0t_1)\text{sinc}(kf_0t_2)\exp[-jk\pi f_0(t_1 + t_2)]$$

The signal $z_p(t)$ represents the periodic convolution of $x_p(t)$ and $y_p(t)$. We use the wraparound method (Sec. 4.5.5) to study several different cases:

(a) $t_1 = t_2 = 1$, $T = 2$. There is no wraparound and $z_p(t)$ describes a triangular waveform (Figure 5.11). The Fourier coefficients simplify to

$$Z_S[k] = \tfrac{1}{4}\text{sinc}^2(\tfrac{1}{2}k)\exp(-jk\pi) = \tfrac{1}{4}\text{sinc}^2(\tfrac{1}{2}k), \qquad k \text{ odd}$$

This real result reveals the halfwave and even nature of $z_p(t)$.

Figure 5.11 Illustrating the convolution property for Example 5.14.

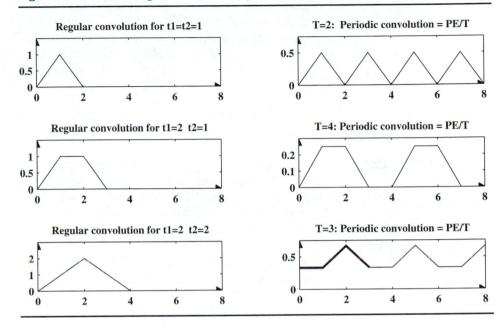

(b) $t_1 = 2$, $t_2 = 1$, $T = 4$. There is no wraparound and $z_p(t)$ describes a trapezoidal waveform (Figure 5.11). The Fourier coefficients simplify to

$$Z_S[k] = \tfrac{1}{8} \operatorname{sinc}(k/2)\operatorname{sinc}(k/4)\exp(-j\tfrac{3}{4}k\pi)$$

Now $Z_p[0] = \tfrac{1}{8}$. Since $\operatorname{sinc}(k/2) = 0$ for even k, $z_p(t)$ should possess only hidden halfwave symmetry. This is evident from Figure 5.11.

(c) $t_1 = 2$, $t_2 = 2$, $T = 3$. We have wraparound past 3 units and $z_p(t)$ describes the waveform of Figure 5.11. The Fourier coefficients simplify to

$$Z_S[k] = (4/9)\operatorname{sinc}^2(\tfrac{2}{3}k)\exp(-j4k\pi/3)$$

The signal is devoid of any symmetry, but since $\operatorname{sinc}(\tfrac{2}{3}k) = 0$ for $k = 3, 6, \ldots$, the third harmonic and its multiples are absent.

5.4.12 *The Central Ordinate Theorems*

The **central ordinate theorems** follow directly by setting $\omega_0 = 0$ or $t = 0$ in the defining relations for $x_p(t)$ and $X_S[k]$

$$X_S[0] = \frac{1}{T}\int_T x(t)\,dt \qquad x_p(0) = \sum_{k=-\infty}^{\infty} X_S[k]$$

For real signals, $x_p(0) = \sum \operatorname{Re}[X_S[k]]$, since $\operatorname{Im}[X[k]]$ is odd and sums to zero.

5.4.13   *Plancherel's Relation*

The power in $x_p(t)$ may be found in terms of either a time or frequency description based on Parseval's relation if $x_p(t)$ is a square integrable function over one period. A generalization of this is **Plancherel's relation** for the product of two arbitrary periodic signals $x_p(t)$ and $y_p(t)$ with identical periods. Denoting the trigonometric coefficients of $x_p(t)$ and $y_p(t)$ by (a_k, b_k) and (A_k, B_k), respectively, this relation reads

$$\frac{1}{T} \int_0^T x_p(t) y_p^*(t)\, dt = \sum_{k=-\infty}^{\infty} |\, X_S[k] Y_S^*[k] \,| = a_0 A_0 + \frac{1}{2} \sum_{k=1}^{\infty} a_k A_k + b_k B_k$$

5.5
System Response to Periodic Inputs

The Fourier series allows us to evaluate the *steady-state response* of an LTI system, due to a periodic input $x_p(t)$. Recall from Chapter 4 that the response to a harmonic $\exp(jk\omega_0 t)$ is $H(k\omega_0) \exp(jk\omega_0 t)$, where $H(k\omega_0)$ is the system function. This input-output relationship may be symbolically described by

$$\exp(jk\omega_0 t) \quad \longrightarrow \quad H(k\omega_0) \exp(jk\omega_0 t)$$

Since $x_p(t)$ is a weighted sum of the harmonics $X_S[k] \exp(jk\omega_0 t)$, superposition yields the response $y_p(t)$ to $x_p(t)$ and we can write

$$x_p(t) = \sum_{k=-\infty}^{\infty} X_S[k] \exp(jk\omega_0 t) \longrightarrow y_p(t) = \sum_{k=-\infty}^{\infty} H(k\omega_0) X_S[k] \exp(jk\omega_0 t)$$

With $H(k\omega_0) = H_k \angle \phi_k$ and $X_S[k] = \frac{1}{2} c_k \angle \theta_k$, we may express the coefficients of the output Fourier series in exponential or polar form as

$$Y_S[k] = H(k\omega_0) X_S[k] = \tfrac{1}{2} c_k H_k \angle (\phi_k + \theta_k)$$

Note that $y_p(t)$ is also periodic with period T and contains the same harmonic frequencies as the input, but with different magnitudes and phase. The convergence rate of $y_p(t)$, and its symmetry (or lack of it), depends on both $x_p(t)$ and the system function. For a halfwave symmetric input, the output is also halfwave symmetric.

Remark: The closed form for the response is very difficult to predict from its Fourier series. We could, of course, use convolution to find the exact time-domain result, but it lacks the power to untangle the frequency information for which the Fourier series is so well suited.

5.5.1   *Response of Electric Circuits to Periodic Inputs*

The response $y_p(t)$ of an electric circuit to a periodic input $x_p(t)$ is best found by phasor analysis. We replace sources and system variables by their phasor form, and the elements R, L and C by their impedances R, $j\omega L$ and $1/j\omega C$. We solve the mesh or node equations for this transformed circuit to obtain $H(\omega)$ as the ratio of the output and input. At each frequency $k\omega_0$, we compute $H(k\omega_0) = H_k \angle \phi_k$, $c_k H_k \angle \phi_k + \theta_k$ and the output $c_k H_k \cos(k\omega_0 t + \theta_k + \phi_k)$. The superposition of these outputs yields the response $y_p(t)$.

Example 5.15

System Response to Periodic Inputs

Consider a periodic sawtooth $v(t) = 4 - \sum (10/k) \sin(3kt)$ volts, applied to an RC circuit (Figure 5.12a) with $R = 1\,\Omega$ and $C = \frac{1}{3}F$. We find

$$H(\omega) = V/V_{in} = 1/(1 + j\omega/3)$$

With $H(0) = 1$, the dc response equals $c_0 H(0) = 4$ volts. With $\omega = k\omega_0$, $H(k\omega_0) = 1/(1 + jk\omega_0/3) = 1/(1 + jk)$. Since $c_k = 10/k$ and $\theta_k = 90°$, the phasor response to the kth harmonic is $V_k = H(k\omega_0)c_k \angle 90° = j10/k(1 + jk)$. For $k = 1, 2$ and 3, we successively compute

$$V_1 = 5\sqrt{2}\angle 45°, \qquad V_2 = \sqrt{5}\angle 26.6°, \qquad V_3 = \tfrac{1}{3}\sqrt{10}\angle 18.4°$$

The time-domain response through the third harmonic may then be written as

$$v_R(t) = 4 + 5\sqrt{2}\cos(3t + 45°) + \sqrt{5}\cos(6t + 26.6°) + \tfrac{1}{3}\sqrt{10}\cos(9t + 18.4°)$$

Comment: The response $V_k = j10/[k(1 + jk)]$ implies a convergence rate of $1/k^2$. Unlike the input, the output cannot contain jumps or exhibit Gibbs effect.

Figure 5.12 **(a) An RC lowpass filter. (b) A dc power supply and (c) its input and output showing ripple.**

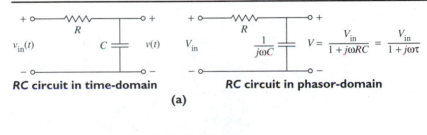

RC circuit in time-domain   RC circuit in phasor-domain

(a)

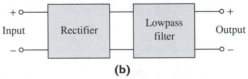

(b)

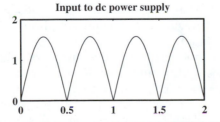

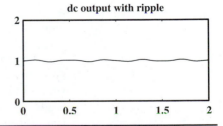

5.6
Applications of Fourier Series

Applications of the Fourier series typically depend upon a frequency-domain description of periodic signals and steady-state system analysis. We now describe several such applications.

5.6.1   DC Power Supplies

Conceptually a dc power supply involves feeding a periodic signal to a rectifier and lowpass filter (Figure 5.12b). The rectifier ensures that its output contains a dc component even if its input does not. Ideally, the filter should pass only this dc component in the rectified signal. In practice, the dc output is contaminated by a small amount of **ripple** due to the presence of harmonics (Figure 5.12c). The ripple is defined as the ratio of the magnitude of the largest nonzero harmonic output to the dc output. Since harmonic magnitudes often decay with frequency, the largest harmonic is typically the first nonzero harmonic. These concepts form the basis for the design of dc power supplies. Here is an example.

Example 5.16
Design of a
DC Power Supply

We design a dc power supply with a ripple R of less than 1%, using the input $A \sin(\omega_0 t)$ to a halfwave rectifier, followed by an RC lowpass filter (Figure 5.12) with $H(\omega) = 1/(1 + j\omega\tau)$. The Fourier series coefficients of the rectifier output are (Table 5.5)

$$c_0 = A/\pi \qquad c_1 = \tfrac{1}{2}A \qquad c_k = 2A/\pi(1 - k^2) \quad (k \text{ even})$$

The largest harmonic is c_1 at the fundamental frequency ω_0. The magnitude of the filter output due to the component at ω_0 equals

$$|V_1| = |c_1 H(\omega_0)| = |c_1/(1 + j\omega_0\tau)| \approx |c_1|/\omega_0\tau \qquad (\omega_0\tau \gg 1)$$

The dc output is simply c_1. For $R < 1\%$, we require $|c_0/V_1| > 100$ and

$$|c_0\omega_0\tau/c_1| > 100 \qquad \text{or} \qquad \tau > |100c_1/c_0\omega_0| = 25/f_0$$

The time constant τ can now be used to select the filter elements R and C.

5.6.2   Harmonic Distortion

Ideally, an amplifier with a gain A should amplify a pure cosine wave $\cos(\omega_0 t)$ to $A \cos(\omega_0 t)$. Distortion refers to any change in the output signal other than pure amplitude scaling or constant time delay. If the output suffers a nonlinear phase shift (with frequency), we get **phase distortion. Amplitude distortion** refers to nonuniform scaling of the input.

   Harmonic distortion results when a pure sinusoidal input is contaminated by spurious harmonics during amplification. This leads to **saturation**, or **clipping**, which may be symmetric or unsymmetric, as shown in Figure 5.13. The **clipping angle** ϕ in Figure 5.13 serves as a measure of the distortion. The larger its value, the more severe the distortion. The Fourier series of the clipped signal contains

Figure 5.13 Symmetric and unsymmetric clipping of a sine wave and the variation of harmonic distortion with clipping angle.

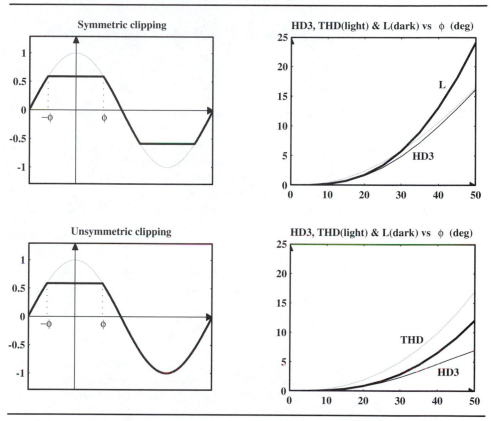

harmonics other than the desired fundamental and the magnitude of the fundamental is reduced from the ideal case. This reduction results in what is often called **signal loss**. It is the difference between the magnitude of an ideal amplifier A and the actual level of the fundamental component at the output, expressed as

$$L = |\frac{A - c_1}{A}|$$

Total harmonic distortion (THD) is a measure of the power contained in the unwanted harmonics as compared to the desired output. It is defined as the square root of the ratio of the power P_u in the unwanted harmonics and the power P_1 in the desired component (the fundamental). The square root operation is used because it has been customary to deal with rms values instead of power. Since the unwanted power P_u is simply the difference between the total ac power P_{AC} and the power P_1 in the fundamental, we have

$$\text{THD} = (P_u/P_1)^{1/2} = [(P_{AC} - P_1)/P_1]^{1/2}$$

Remark: Note that P_u describes only the power in the harmonics and excludes the dc component (even if present).

We can also find the distortion HD_k due only to the kth harmonic as

$$HD_k = (P_k/P_1)^{1/2} = |c_k/c_1|$$

Example 5.17

Harmonic Distortion

(a) Consider the input $\cos(10\pi t)$ to an amplifier with a gain of 10 leading to the response $y(t) = 8\cos(10\pi t) - 0.08\cos(30\pi t) + 0.06\cos(50\pi t)$.

    We note the absence of any second or even harmonic distortion. The third harmonic distortion HD_3 equals $|c_3/c_1| = 0.08/8 = 1\%$. The total harmonic distortion THD equals $\sqrt{P_U/P_1} = [(0.06)^2 + (0.08)^2/64]^{1/2} = 1.25\%$. The signal loss L equals $|(10 - 8)/10| = 20\%$.

(b) The Fourier coefficients of the symmetrically clipped cosine of Figure 5.13 are

$$c_1 = \frac{1}{\pi}\left[\sin 2\phi + \pi - 2\phi\right] \qquad c_k = \frac{2}{k\pi}\left[\frac{\sin(k+1)\phi}{k+1} - \frac{\sin(k-1)\phi}{k-1}\right] \quad (k \text{ odd})$$

No second harmonic distortion occurs, since all even harmonics are absent. The third harmonic distortion, HD_3, and the signal loss L, are given by

$$HD_3 = \left|\frac{c_3}{c_1}\right| = \left|\frac{\frac{1}{6}\sin 4\phi - \frac{1}{3}\sin 2\phi}{\sin 2\phi + \pi - 2\phi}\right| \qquad L = \left|\frac{A - c_1}{A}\right| = \frac{2\phi - \sin 2\phi}{\pi}$$

With $\phi = \frac{1}{8}\pi$, for example, we obtain $HD_3 \approx 2.25\%$ and $L \approx 2.49\%$. Figure 5.13 shows how HD_3 and L vary nonlinearly with the clipping angle ϕ (for both symmetric and unsymmetric clipping).

5.6.3   Tuned Circuits

The frequency response of the series RLC circuit of Figure 5.14 peaks at the **resonant frequency** ω_r, and its sharpness or selectivity is determined by a measure called the **quality factor**, or Q, which depends on its element values. At the resonant frequency, the magnitude of both the capacitor voltage and the inductor voltage are Q times the input magnitude. Related to Q is the **bandwidth** B_w, a measure of the frequency spread given by $B_w = \omega_r/Q$. The smaller the bandwidth, the larger the Q and the more selective the circuit. A circuit with high Q is called a **tuned circuit**.

Spectrum Analyzers   Tuned circuits play an important role in the design of **spectrum analyzers**, devices capable of separating a time signal into its frequency components. Conceptually, all we need is a parallel bank of highly selective circuits, each tuned to a different frequency. They act as a sieve to let through only those frequencies to which they are tuned.

Figure 5.14 **An *RLC* circuit and its frequency response.**

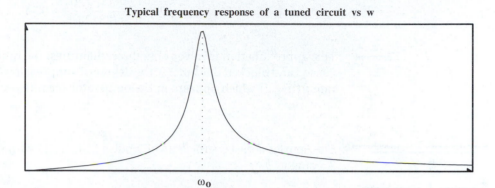

Typical frequency response of a tuned circuit vs w

ω_0

Example 5.18
Design of a Tuned Circuit

To extract the 25 krad/s component in a periodic signal with $\omega_0 = 1$ krad/s, let us feed it to a resonant circuit tuned to the 25th harmonic, and look at the capacitor voltage. The 25 krad/s component will be amplified by Q. Ideally, the tuned circuit should pass just the 25th harmonic, and block all others. In practice, we use high-Q circuits which provide high enough suppression.

Suppose the design calls for suppressing all frequencies other than 25 krad/s to less than 5% of the desired magnitude. It is then sufficient to satisfy this condition for only those harmonics adjacent to the 25th, since a high-Q circuit will attenuate the rest even more severely. In fact, we need consider only the 24th (and not the 26th) harmonic, because the harmonic magnitudes themselves decrease with frequency.

We can design such a circuit by specifying its Q and resonant frequency ω_r (which should obviously equal $25\omega_0$). To determine the Q, we find the magnitudes of the 25th and next-lower harmonic, and equate the ratio to the desired suppression. The capacitor voltage V at a frequency ω equals

$$V = IZ_C = \frac{V_{\text{in}}(1/j\omega C)}{R + j\omega L + 1/j\omega C} = \frac{V_{\text{in}}}{1 - \omega^2 LC + j\omega RC}$$

With $\omega_r^2 = 1/LC$ and $Q = 1/\omega_r RC$, this may be recast in terms of ω_r and Q as

$$V = \frac{V_{\text{in}}}{1 - (\omega/\omega_r)^2 + j(\omega/\omega_r)/Q} \qquad |V|^2 = \frac{|V_{\text{in}}|^2}{[1 - (\omega/\omega_r)^2]^2 + (\omega/\omega_r)^2/Q^2}$$

If the input is halfwave symmetric, with $c_k = A/k^2$ (k odd), the harmonic adjacent to the 25th is the 23rd (not the 24th). The squared magnitudes at $\omega = 25\omega_0 = \omega_r$ and $\omega = 23\omega_0$ are given by

$$|V_{25}| = Q|A/(25)^2| \qquad |V_{23}| = \frac{|A/(23)^2|}{\{[1 - (23/25)^2]^2 + (23/25)^2/Q^2\}^{1/2}}$$

Since the circuit is highly selective ($Q \gg 1$), $|V_{23}| \approx |A/(23)^2|/[1 - (23/25)^2]$. For the output at the 23th harmonic to equal 5% of the resonant output, we require

$|V_{25}/V_{23}| > 20$. This results in

$$Q > \frac{(25/23)^2(20)}{1 - (23/25)^2} \approx 154$$

Comment: This approximation involves only three quantities: the ratio k_p/k_b of the harmonics passed and blocked (25 and 23), the degree of suppression (20), and the convergence rate $n(n = 2)$ which appears in the numerator term $(k_p/k_r)^n$.

5.7
Some Limiting Representations

The Fourier series coefficients of a pulse train $x_p(t)$ with pulse width t_0 and period T equal

$$X_S[k] = \frac{1}{T} \int_0^T x_p(t) \exp(-jk\omega_0 t)\, dt = \frac{1}{T} \int_0^{t_0} x_p(t) \exp(-jk\omega_0 t)\, dt$$

The harmonic magnitudes typically decay with frequency.

Reducing the Pulse Width   If we keep T fixed and reduce the pulse width t_0, the harmonic locations remain unaltered, but their magnitudes change in an interesting way. Since $t_0/T \ll 1, \exp(-jk\omega_0 t) = \exp(-jk2\pi t/T) \approx 1$ over the interval $(0, t_0)$, and we obtain

$$X_S[k] = \frac{1}{T} \int_0^{t_0} x_p(t) \exp(-jk\omega_0 t)\, dt \approx \frac{1}{T} \int_0^{t_0} x_p(t)\, dt = X_S[0]$$

The $X_S[k]$ are constant. A periodic signal $x_p(t)$ with narrow pulses, regardless of the pulse shape, thus behaves much like an impulse train. The narrower the pulse, the less its high frequency spectrum decays with frequency (Figure 5.15).

Stretching the Period   Increasing T while holding the pulse width t_0 fixed also reveals some subtle but disturbing results (Figure 5.15).

1. The harmonics get closer. As $T \longrightarrow \infty$, the spectrum becomes continuous and may be represented by a curve rather than a set of discrete points (lines).
2. The pulses in $x_p(t)$ move further apart. As $T \longrightarrow \infty$, the periodic signal $x_p(t)$ reduces to a single pulse $x(t)$, and thus becomes nonperiodic.
3. The transition from a periodic to a nonperiodic signal also signifies a transition from a power signal to an energy signal.
4. The Fourier coefficients $X_S[k]$ (which depend on $1/T$) decrease. As $T \longrightarrow \infty$, the $X_S[k]$ become vanishingly small.

Clearly, as a periodic signal changes to an aperiodic one, its spectrum carries little meaning, since the $X_S[k]$ approach zero! How we resolve this forms the basis of the Fourier transform discussed in the next chapter.

Figure 5.15 **Effects of shrinking the pulse width and the period.**

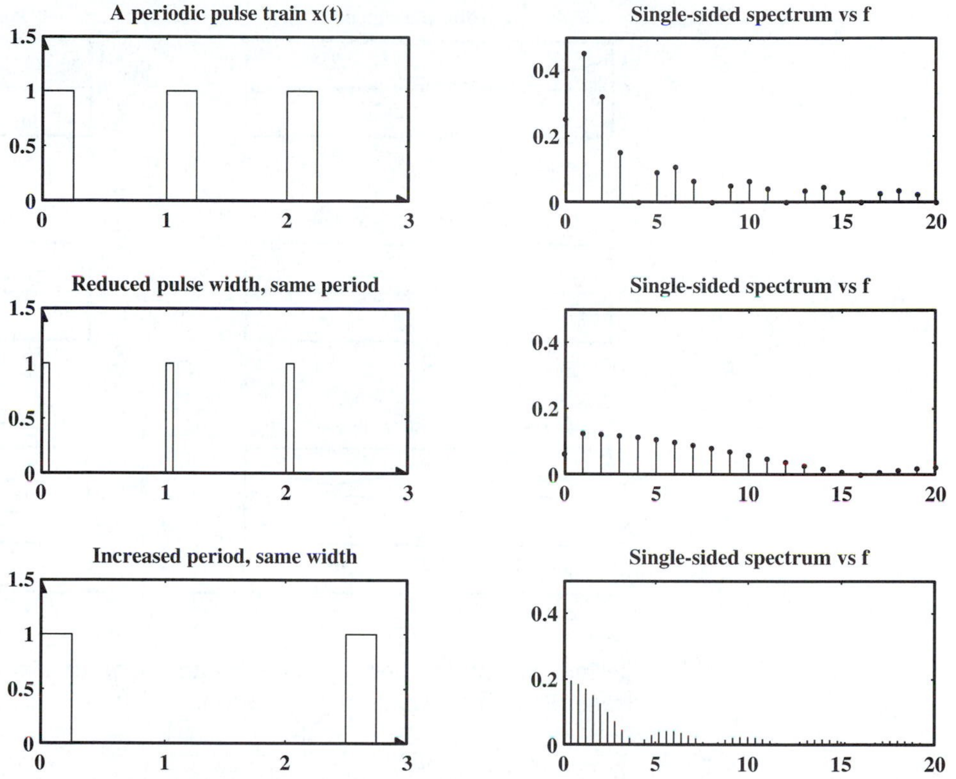

5.7.1 *Periodic Extension of Timelimited Signals*

The Fourier series actually describes a periodic signal $x_p(t)$ over only one period T. Its periodic extension is what corresponds to $x_p(t)$. We may describe any signal $x(t)$ over a finite interval (a, b) by a Fourier series $x_p(t)$, using a convenient period, provided it is understood that:

1. The series represents $x(t)$ only over (a, b) and
2. Outside (a, b), the series is a *periodic extension* of $x_p(t)$, (a, b).

This implies that the series will not converge to zero outside (a, b) but to its periodic extension. Many such Fourier series of $x(t)$ are possible, based on how we assign one period of the periodic signal, as illustrated in Figure 5.16. There is no unique representation.

Remark: No Fourier series can represent a timelimited function that is defined to be zero outside a given interval, since its periodic extension must have an infinite time

Figure 5.16   A signal x(t) and its several possible periodic continuations.

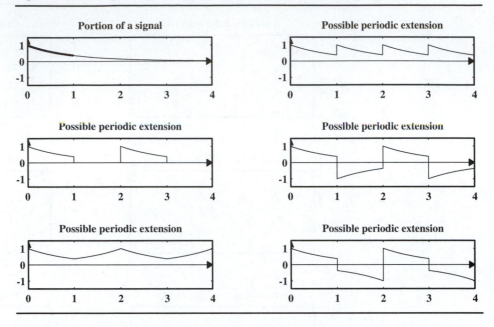

period. An infinite period results in an aperiodic signal, and requires the Fourier transform for its spectral description.

5.8
Special Topics: The Dirichlet Kernel and the Gibbs Effect

Reconstruction of a periodic signal by adding a finite number of harmonics is often called a **partial sum**. Truncation to N harmonics is equivalent to multiplying the Fourier series coefficients $X_S[k]$ pointwise by a rectangular **spectral window** or frequency-domain window $W_D[k] = \text{rect}(k/2N)$, which equals 1 for $-N \leq k \leq N$ and is zero elsewhere. The weighted coefficients $X_N[k]$ then equal

$$X_N[k] = X_S[k]W_D[k] = X_S[k]\text{rect}(k/2N)$$

The signal $w_D(t)$ corresponding to $W_D[k]$ is called the **Dirichlet kernel**. It represents the sum of harmonics of unit magnitude at multiples of f_0, and can be written as the summation

$$w_D(t) = \sum_{k=-N}^{N} \exp(-j2\pi k f_0 t)$$

From tables (see Appendix A), the closed form for $w_D(t)$ is

$$w_D(t) = \frac{\sin[(N + \tfrac{1}{2})\pi f_0 t]}{\sin(\pi f_0 t)} = (2N + 1)\frac{\text{sinc}[(2N + 1)f_0 t]}{\text{sinc}(f_0 t)} = \frac{M \, \text{sinc}(M f_0 t)}{\text{sinc}(f_0 t)}$$

where $M = 2N + 1$. This kernel is periodic with period $T = 1/f_0$ (Figure 5.17) and has some very interesting properties. Over one period

1. Its area equals T and it attains a maximum peak value of M and a minimum peak value of 1.
2. It shows N maxima, a positive main lobe of width $2T/M$, and decaying positive and negative sidelobes of width T/M, with $2N$ zeros at $kT/M, k = 1, 2, \ldots, 2N$.
3. The ratio R of the main lobe and peak sidelobe magnitudes stays nearly constant (≈ 4-4.7) for finite M, and $R \longrightarrow 1.5\pi \approx 4.71$ for very large M.
4. Increasing M increases the mainlobe height and compresses the sidelobes. As $M \longrightarrow \infty, w_D(t)$ approaches an impulse with strength T.

Since $X_N[k] = X_S[k]W_D[k]$, the reconstructed signal $x_N(t)$ equals the *periodic convolution* $w_D(t) \bullet x_p(t)$:

$$x_N(t) = w_D(t) \bullet x_p(t) = \frac{1}{T} \int_T w_D(\tau) x_p(t - \tau) \, d\tau$$

If $x_p(t)$ contains discontinuities, $x_N(t)$ exhibits an overshoot and oscillations near each discontinuity (Figure 5.17). With increasing N, the mainlobe and sidelobes become taller and narrower, and the overshoot (and associated undershoot) near the discontinuity becomes narrower, but more or less maintains its height.

Figure 5.17   One period of the Dirichlet kernel for $N = 5$ and $N = 10$ with $T = 1$. Its periodic convolution with a square wave (or any periodic signal with jump discontinuities) results in a reconstruction with overshoot.

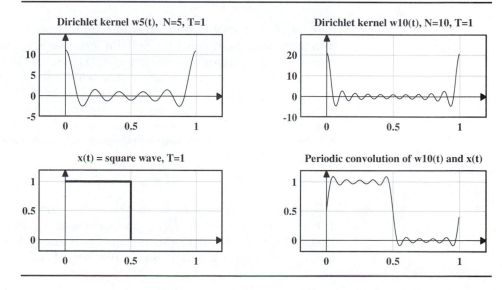

It can be shown that if a periodic function $x_p(t)$ jumps by J_k at times t_k, the Fourier reconstruction $x_p(t)$, as $N \longrightarrow \infty$, yields not only the function $x_p(t)$ but a pair of straight lines at each t_k, which extend equally above and below the discontinuity by an amount A_k, given by

$$A_k = \left[\frac{1}{2} - \frac{1}{\pi} \text{si}(\pi) \right] J_k = K J_k$$

where $\text{si}(x)$ is the **sine integral** defined as

$$\text{si}(x) = \int_0^x \frac{\sin(\xi)}{\xi} \, d\xi$$

and $K = \frac{1}{2} - \text{si}(\pi)/\pi = 0.0895$. The lines thus extend by about 9% on either side of the jump J_k. The fraction K is independent of both the nature of the periodic signal, and the number of terms used, as long as N is large. We are, of course, describing the Gibbs effect.

5.8.1   *Smoothing Spectral Windows*

The Gibbs effect is due to the *abrupt truncation* of the rectangular window. The use of a *tapered* spectral window yields better convergence and a much smoother reconstruction. The windowed signal $x_N(t)$ may be described by

$$x_N(t) = \sum_{k=-N}^{N} W_S[k] X_S[k] \exp(jk2\pi f_0 t)$$

Here, $W_S[k]$ is a spectral window whose coefficients or weights typically decrease with increasing $|k|$. It yields the periodic smoothing kernel $w_S(t)$. The signal $x_N(t)$ also equals the periodic convolution of $x_p(t)$ and $w_S(t)$.

Table 5.9 lists some useful spectral windows. The smoothed reconstruction corresponding to some of these is shown in Figure 5.18. *Every window (except the rectangular) provides a reduction, or elimination, of overshoot.* Which ones do we choose? Some of the criteria for window selection are discussed in Chapter 7 and require concepts based on Fourier transforms. At our level, though, it really makes little difference.

Remarks: 1. All windows display even symmetry and are positive, with $w[0] = 1$.
2. All windows (except the rectangular or boxcar window) are tapered.
3. For all (except the boxcar, Hamming, Kaiser and Bickmore), $w[\pm N] = 0$.
4. I_m represents the **modified Bessel function** of order m.

Table 5.9 Some Common Spectral Windows.

Window Type	Expression	Range
Boxcar	$\mathrm{rect}[k/2N] = 1$	$-N \leq k \leq N$
Fejer	$\mathrm{tri}(k/N) = 1 - |k|/N$	$-N \leq k \leq N$
vonHann	$\frac{1}{2} + \frac{1}{2}\cos(k\pi/N)$	$-N \leq k \leq N$
Hamming	$0.54 + 0.46\cos(k\pi/N)$	$-N \leq k \leq N$
Cosine	$\cos(k\pi/2N)$	$-N \leq k \leq N$
Riemann	$\mathrm{sinc}(k/N)$	$-N \leq k \leq N$
Parzen-2	$1 - k^2/N^2$	$-N \leq k \leq N$
Blackman	$0.42 + 0.5\cos(k\pi/N) + 0.08\cos(2k\pi/N)$	$-N \leq k \leq N$
Papoulis	$\frac{1}{\pi}|\sin(k\pi/N)| + (1 - |k|/N)\cos(k\pi/N)$	$-N \leq k \leq N$
Kaiser	$\dfrac{I_0(\pi\beta\sqrt{1 - (k/N)^2})}{I_0(\pi\beta)}$	$-N \leq k \leq N$
Modified Kaiser	$\dfrac{I_0(\pi\beta\sqrt{1 - (k/N)^2}) - 1}{I_0(\pi\beta) - 1}$	$-N \leq k \leq N$
Bickmore	$\left[\dfrac{\sqrt{1 - (k/N)^2}}{\pi\beta}\right]^{\nu-1/2} \left[\dfrac{I_{\nu-1/2}(\pi\beta\sqrt{1 - (k/N)^2})}{I_{\nu-1/2}(\pi\beta)}\right]$	$-N \leq k \leq N$
General Blackman	$a_0 + \sum_{n=1}^{L} 2a_n \cos(nk\pi/N)$	$-N \leq k \leq N$

Figure 5.18   Illustration of how the reconstruction of a periodic signal with discontinuities is smoothed by some of the windows of Table 5.9.

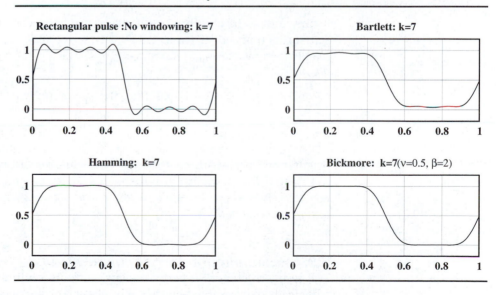

5.9
Special Topics: Truncation and Its Consequences

We now briefly address two aspects of truncated Fourier representations:

1. How good is a truncated approximation?
2. Could any other choice of coefficients lead to a better approximation?

If the finite term representation describes the function exactly, these questions become irrelevant. The terms *good* or *better* are used in the context of minimizing the *mean squared error* for several reasons. A pointwise comparison is time consuming and provides no consistent means of comparing different signals. The average error over one period is also a poor measure because cancellation of large positive and negative errors can give a false impression of a good approximation. Minimization of absolute error is an attractive possibility, but computationally cumbersome. The choice, by default, falls on the mean squared error. This measure embodies the advantage of the absolute error criterion in that positive and negative errors are prevented from canceling out due to the squaring operation. The squaring operation also provides a sensible weighting scheme that enhances the larger errors. Finally, it results in a measure which turns out to be *exactly* the difference between the average powers in the actual function and its approximation.

5.9.1   The Error in a Truncated Fourier Series

Given a periodic signal $x_p(t)$ and its Fourier series $x_N(t)$, truncated to N harmonics, the **mean squared error** ϵ_N is defined by

$$\epsilon_N = \frac{1}{T} \int_T [x_p(t) - x_N(t)]^2 \, dt$$

It turns out to be identical to the difference between the power contained in the actual function and its truncated Fourier series and may therefore be found quite readily using Parseval's relation:

$$\epsilon_N = \frac{1}{T} \int_T \left[\sum_{k=-\infty}^{\infty} X_S[k] \exp(jk\omega_0 t) - \sum_{k=-N}^{N} X_S[k] \exp(jk\omega_0 t) \right]^2 dt = P_x - P_N$$

If $\epsilon_N = 0$, $x_N(t)$ must be equal to $x_p(t)$ and not just an approximation.

An even more remarkable fact is that is it impossible to find a truncated representation $\tilde{x}_N(t)$, with coefficients $\tilde{X}_S[k]$ different from the Fourier coefficients $X_S[k]$, that has a smaller mean squared error as compared with the truncated Fourier series $x_N(t)$. In other words, *a Fourier series is the best least squares fit to a periodic signal*.

5.9.2   Summation by Arithmetic Means

The reconstruction of a periodic signal with discontinuities using partial sums is what leads to overshoot and the Gibbs effect. Certain operations on the partial sum result in a reduction or elimination of overshoot as summarized in Table 5.10.

Table 5.10   Smoothing by Operations on the Partial Sum. (Note: $M = 2N + 1$)

Operation	Result Spectral window	Result Convolution kernel	Reconstruction at discontinuities
Partial sum	$\text{rect}(k/2N)$	$\left[\dfrac{M\,\text{sinc}(Mf_0\tau)}{\text{sinc}(f_0\tau)} \right]$	9% overshoot (Gibbs effect)
Arithmetic mean of partial sum	$\text{tri}(k/N)$	$N\left[\dfrac{\text{sinc}(Nf_0 t)}{\text{sinc}(f_0 t)} \right]^2$	No overshoot Less steep slope
Integration of of partial sum	$\text{sinc}(k/N)$		1.2% overshoot Steeper slope

The Partial Sum   For a series represented by

$$x = r_0 + r_1 + r_2 + r_3 + \cdots + r_N + r_{N-1} + \cdots$$

The partial sum s_N to N terms is defined by

$$s_N = r_0 + r_1 + r_2 + r_3 + \cdots + r_{N-1}$$

Arithmetic Mean of Partial Sums   The **sum by arithmetic means** to N terms is defined as the average (arithmetic mean) of the intermediate partial sums:

$$\hat{x}_N = \frac{1}{N}[s_0 + s_1 + \cdots + s_{N-1} = \frac{1}{N}\sum_{k=0}^{N-1} s_k$$

This result is valid for any series, convergent or otherwise. It is particularly useful for finding the time signal from its Fourier series even when such a series diverges.

There are two important theorems relating to such a sum:

1. If the actual series converges to x, the sum by the method of arithmetic means also converges to the same value.
2. The Fourier series of an absolutely integrable signal $x(t)$ is summable by arithmetic means. The sum converges uniformly to $x(t)$ at every point of continuity and to the midpoint of discontinuities otherwise.

Writing out the partial sums, we have

$$s_1 = r_0$$
$$s_2 = r_0 + r_1$$
$$s_3 = r_0 + r_1 + r_2$$
$$\cdots$$
$$s_N = r_0 + r_1 + r_2 + \cdots + r_{N-1}$$

The sum of these partial sums may then be written

$$\hat{s}_N = s_1 + \cdots + s_N = Nr_0 + (N-1)r_1 + (N-2)r_2 + \cdots + r_{N-1}$$

The arithmetic mean of the partial sums then equals

$$\hat{x}_N = \frac{1}{N}[Nr_0 + (N-1)r_1 + (N-2)r_2 + \cdots + r_{N-1}] = \sum_{k=0}^{N-1} \frac{N-k}{N} r_k$$

This clearly reveals a *triangular* weighting on the individual terms r_k. For a periodic signal $x_p(t)$ with $r_k = X[k]\exp(jk\omega_0 t)$, the arithmetic mean $x_N(t)$ of the partial sums may be written, by analogy, as

$$x_N(t) = \sum_{k=-(N-1)}^{N-1} \frac{N-|k|}{N} X_S[k]\exp(jk\omega_0 t) = \sum_{k=-N}^{N} W_F[k]X_S[k]\exp(jk\omega_0 t)$$

In this result, $W_F[k]$ describes a tapered triangular window whose weights decrease linearly with $|k|$. With $W_F[\pm N] = 0$, we may write

$$W_F[k] = \text{tri}(k/N) = 1 - |k|/N \qquad -N \le k \le N$$

The reconstructed signal $x_N(t)$ is

$$x_N(t) = \frac{1}{T}\int_{-T/2}^{T/2} w_F(\tau)x_p(t-\tau)\,d\tau$$

where $w_F(t)$ can be shown to equal (Problem 5.26)

$$w_F(t) = \frac{1}{N}\left[\frac{\sin[(\frac{1}{2}N\omega_0 t)]}{\sin(\frac{1}{2}\omega_0 t)}\right]^2 = N\left[\frac{\text{sinc}(Nf_0 t)}{\text{sinc}(f_0 t)}\right]^2$$

This is the so-called **Fejer kernel**. It has a peak value of N and describes a periodic signal corresponding to the triangular spectral window $W_F[k]$. It is entirely positive, and, for a given N, its sidelobes are much smaller than those of a Dirichlet kernel (of the rectangular window) (Figure 5.19). These properties result in a reconstruction which is not only smoother but also *completely free of overshoot* for periodic signals with discontinuities (Figure 5.19). Even though partial sums lead to a poor reconstruction with overshoot for signals with jumps, the arithmetic means of these partial sums provide a smoother reconstruction without overshoot.

Integration of Partial Sums   The reconstruction based on integrating rather than averaging the partial sums is described by

$$x_N(t) = \frac{N}{T}\sum_{k=-N}^{N}\int_{t-\frac{T}{2N}}^{t+\frac{T}{2N}} X_S[k]\exp(jk\omega_0 t)\,dt$$

Carrying out the integration, and noting that $\omega_0 T = 2\pi$, we get

$$x_N(t) = \frac{N}{T}\sum_{k=-N}^{N} X_S[k]\exp(jk\omega_0 t)\left[\frac{\exp(jk\pi/N) - \exp(-jk\pi/N)}{jk\omega_0}\right]$$

Figure 5.19 **One period of the Fejer kernel for $N = 5$ and $N = 10$ with $T = 1$. Its periodic convolution with a square wave (or any periodic signal with or without jump discontinuities) yields a reconstruction free of overshoot.**

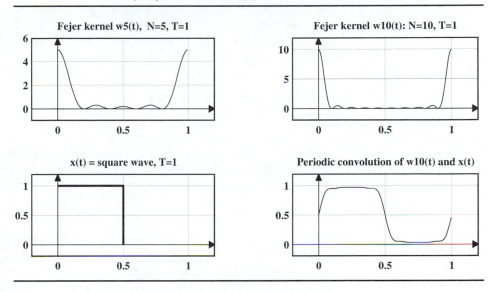

A final simplification, using Euler's relation, yields

$$x_N(t) = N \sum_{k=-N}^{N} X_S[k] \exp(jk\omega_0 t) \left[\frac{\sin(k\pi/N)}{k\pi} \right] = \sum_{k=-N}^{N} X_S[k] \text{sinc}(k/N) \exp(jk\omega_0 t)$$

The quantity $W_L[k] = \text{sinc}[k/N]$ describes the **Lanczos window**, which has a sinc taper. Its kernel, $w_L(t)$, reduces the reconstruction overshoot to less than 1.2% (as compared to 9% for the Dirichlet kernel) but does not eliminate it (as the Fejer kernel does). The reconstruction, however, shows a much steeper slope at the discontinuities as compared with the Fejer kernel.

5.10 Special Topics: Existence, Convergence and Uniqueness

To formalize the notion of existence, convergence and uniqueness of the Fourier series, we must address and answer the following questions:

1. Under what conditions will the Fourier series for $x_p(t)$ exist?
2. If it does, will the series converge?
3. If it does converge, will it converge for every t?
4. If it does converge for every t, will it converge to $x_p(t)$ for every t?

The first question deals with the existence of Fourier series, the next two with convergence, and the last with uniqueness.

The idea of existence may be justified by defining the coefficients of the series by the proposed Fourier relations and then showing that the series with these coefficients does indeed represent $x_p(t)$. We start from the opposite end and assume the coefficients can be found using the prescribed relations. Otherwise, the existence of Fourier series is very difficult to prove in a formal, rigorous sense. This approach requires that the Fourier series coefficients $|X_S[k]|$ be finite. Thus,

$$|X_S[k]| \le \int_T |x_p(t)| \, |\exp(jk\omega_0 t)| \, dt < \infty$$

Since $|\exp(jk\omega_0 t)| = 1$, we have

$$\int_T |x_p(t)| \, dt < \infty$$

Thus, $x_p(t)$ must be absolutely integrable over one period. A consequence of absolute integrability is the **Riemann-Lebesgue theorem** which states that the coefficients $X_S[k]$ approach zero as $k \longrightarrow \infty$:

$$\lim_{k \to \infty} \int_T x_p(t) \exp(jk\omega_0 t) \, dt = 0$$

This result also leads us toward the idea of convergence of the series with the chosen coefficients. Convergence to the actual function $x_p(t)$ for every value of t is known as **uniform convergence**. This requirement is satisfied for every continuous function $x_p(t)$ since we are trying to represent a function in terms of sines and cosines, which are themselves continuous. We face problems when reconstructing functions with jumps. If we require convergence to the midpoint of the jump at the discontinuity, we sidetrack such problems. In fact, this is exactly the condition that obtains for functions with jump discontinuities.

Even though our requirement calls for $x_p(t)$ to be absolutely integrable over one period, it includes square integrable (energy) signals for which

$$\int_T |x_p(t)|^2 \, dt < \infty$$

Obviously, every continuous or piecewise continuous signal, over one period, is square integrable. For such a signal, it turns out that the Fourier series also converges to $x_p(t)$ *in the mean* (or *in the mean squared sense*) in that, as more terms are added to the series $x_N(t)$ truncated to N terms, the mean squared error decreases, and approaches zero as $N \longrightarrow \infty$:

$$\int_T |x_p(t) - x_N(t)|^2 \, dt \longrightarrow 0 \qquad \text{as} \qquad N \longrightarrow \infty$$

A consequence of this result is that if $x_N(t)$ converges in the mean to $x_p(t)$, it cannot converge to any other function. In other words, $x_p(t)$ is *uniquely* represented by $x_N(t), N \longrightarrow \infty$ even though it may not converge pointwise to $x_p(t)$ at every value of t.

Uniqueness means that if two functions $x_p(t)$ and $y_p(t)$ possess the same Fourier series, then they are equal. Here, equality implies that the two functions could be different at a finite set of points (such as points of discontinuity) over one period. For signals that are square integrable over one period, convergence in the mean implies uniqueness. An important result in this regard is the converse of Parseval's relation, known as the **Riesz-Fischer theorem**, which, loosely stated, tells us that a series with finite total power ($P_x < \infty$) given in terms of the summation

$$P_x = \sum_{k=-\infty}^{\infty} |X_S[k]|^2 < \infty$$

must represent the Fourier series of a square integrable function.

5.10.1   *The Dirichlet Conditions*

The **Dirichlet conditions** describe *sufficient* conditions for the existence of Fourier series. A periodic function $x_p(t)$ will possess a unique Fourier series if, over one period:

1. $x_p(t)$ is absolutely integrable.
2. $x_p(t)$ has a finite number of maxima and minima and/or a finite number of finite discontinuities.

If these conditions are satisfied, the Fourier series will converge to

(a) $x_p(t)$ pointwise for every value of t where $x_p(t)$ is continuous and piecewise differentiable (convergence in the uniform sense) or
(b) the midpoint of the discontinuity at each jump discontinuity otherwise (nonuniform convergence).

If, in addition, $x_p(t)$ is square integrable over one period, the series will converge to $x_p(t)$ in a mean squared sense. The mean squared error will approach zero as the number of terms in the series approaches infinity.

We stress that the Dirichlet conditions are *sufficient* conditions that guarantee a Fourier series if satisfied. They do not preclude the existence of Fourier series for signals that violate some or all of these conditions, however. The periodic impulse train is such an example. Other classical examples of functions violating the Dirichlet conditions include

1. $x_p(t) = \sin(2\pi/t), 0 < t \le 1$ and $x_p(t) = \cos[\tan(\pi t)], 0 < t \le 1$, each of which possesses an infinite number of maxima and minima.
2. $x_p(t) = \tan(t), 0 \le t < \pi/2$, with an infinite discontinuity per period.
3. Dirichlet's own example, $x_p(t) = 1$ (rational t) or 0 (irrational t) over $(-\pi, \pi)$, which has an infinite number of finite discontinuities due to the infinite number of rational and irrational values t can attain over any finite interval.

Such functions, however, are mostly of theoretical interest.

Remark: More stringent formulations of sufficient conditions are based on how the operation of integration is interpreted. These involve concepts beyond the scope of this text, such as the theory of infinite sets, and lead to sufficient conditions that are no more useful in practice than the Dirichlet conditions themselves.

Since its inception, the theory of Fourier series has been an area of intense mathematical activity and research and many advances have resulted, most of little concern from a practical standpoint, and most well beyond our scope. But the convergence problem remains unsolved. No set of *necessary and sufficient* conditions has yet been formulated.

5.11 Fourier Series: A Synopsis

The Fourier series provides a unique representation of periodic signals in terms of harmonically related sinusoids or harmonic signals.

The concept	The idea behind Fourier series is as follows: Take a periodic signal $x(t)$ with period T. Choose harmonic signals at frequencies $f_0 = 1/T$ and its multiples kf_0. Combine in the right mix to generate an approximation to the signal $x(t)$.
The result	The result is a summation of sinusoids (or harmonic signals).

$$x(t) = a_0 + \sum_{k=1}^{\infty} a_k \cos(k\omega_0 t) + b_k \sin(k\omega_0 t)$$

$$= c_0 + \sum_{k=1}^{\infty} c_k \cos(k\omega_0 t + \theta_k) = \sum_{k=-\infty}^{\infty} X_S[k] e^{jk\omega_0 t}$$

The coefficients	The right mix (a_k, b_k or c_k, θ_k or $X_S[k]$) describes the Fourier coefficients. The coefficients X_S are complex with $X_S[-k] = X_S^*[k]$. The term a_0, c_0 or $X_S[0]$ is the dc offset. The terms $a_k \cos(k\omega_0 t) + b_k \sin(k\omega_0 t)$, $X_S[k] \exp(jk\omega_0 t) + X_S[-k] \exp(-jk\omega_0 t)$, and $c_k \cos(k\omega_0 t + \theta_k)$, are called harmonics.
Finality	If a new harmonic is added to the series, its coefficient neither depends on nor affects the coefficients of any other harmonics.
Symmetry	Even symmetric signals show only pure cosines. The $\|X_S\|$ are real. Odd symmetric signals show only pure sines. The $\|X_S\|$ are imaginary. Halfwave symmetric signals show only odd-indexed harmonics.
Magnitude and phase	At each frequency kf_0, the harmonic magnitude equals $(a_k^2 + b_k^2)^{1/2}$ or c_k or $2\|X_S[k]\|$ and the phase equals θ_k or $\angle X_S[k]$.
Spectra	Periodic signals have discrete spectra. A plot of c_k versus k and θ_k versus k for $k = 0, 1, 2, \ldots$ describes the onesided discrete magnitude and phase spectra of the periodic signal $x(t)$. A plot of $\|X_S[k]\|$ versus k and θ_k versus k for $k = 0, \pm 1, \ldots$ describes the twosided discrete magnitude and phase spectra of the periodic signal $x(t)$. For real signals, the twosided magnitude spectrum has even symmetry. For real signals with $a_0 > 0$, the twosided phase spectrum has odd symmetry.

Parseval's theorem	The power in $x(t)$ may be found either from $x(t)$ itself or from its spectral coefficients (a_k, b_k or c_k or $|X_S[k]|$).

$$P_x = \frac{1}{T} \int_T x^2(t)\, dt = a_0^2 + \sum_{k=1}^{\infty} \tfrac{1}{2}(a_k^2 + b_k^2) = c_0^2 + \sum_{k=1}^{\infty} \tfrac{1}{2}c_k^2 = \sum_{k=-\infty}^{\infty} |X_S[k]|^2$$

	Using Parseval's relation, we can also find the power in $x(t)$ up to a given frequency by including just the spectral power up to that frequency.
Reconstruction	The more harmonics we include, the better we approximate $x(t)$. Perfect reconstruction may involve an infinite number of harmonics.
Gibbs effect	A discontinuous periodic signal can never be reconstructed perfectly, no matter how many harmonics we include. The reconstructed signal always shows an overshoot ($\approx 9\%$ of jump) at the discontinuities.

Appendix 5A MATLAB Demonstrations and Routines

Getting Started on the Supplied MATLAB Routines

This appendix presents MATLAB routines related to concepts in this chapter and some examples of their use. See Appendix 1A for introductory remarks.

Note: The % sign and comments following it need not be typed.

Fourier Series

```
y=fseries('tri(t-1)',[4 0.2],[3 9 25],'hamming');
```

This generates the Fourier series coefficients for tri($t - 1$) with period $T = 4$ and time delay from the origin of 0.2 s. Spectra are plotted to 25 harmonics. Reconstructions are plotted to 3, 9 and 25 harmonics.

A smoothed reconstruction to $k = 25$ is also plotted using the hamming window. The output y has 7 columns: $k, a_k, b_k, c_k, \theta_k$, power, and cumulative power.

Other (optional) output arguments return the convergence rate (if found), the envelope and values of the reconstructed signals.

Demonstrations and Utility Functions

```
>>demofs                    %Interactive demonstration of Fourier series
>>fsbuild                   %Interactive FS reconstruction from coefficients
>>y=window('hamming',21)    %Returns values of the 21 point hamming window
```

Other window types include most of the windows listed in Table 5.9.

```
>>y=fskernel('d',11)        %One period of Dirichlet kernel, N=11 (400 points)
>>y=fskernel('f',11,200)    %One period of Fejer kernel, N=11 (200 points)
```

To plot the Dirichlet kernel with period $T = 1$, use the commands

```
>>t=(0:399)/400;   plot(t,y)
```

Appendix 5B
A Historical
Perspective

Jean Baptiste Joseph Fourier is dead. The man whose name adorns the two most celebrated techniques in spectral analysis was, in the words of his contemporary, Jacobi, a man who believed that the principal aim of mathematics was "public utility and the explanation of natural phenomena." Lest this description pass as an unqualified tribute to Fourier, Jacobi went on to add that Fourier did not "perceive the end of science as the honor of the human mind." What did he mean by that?

5B.1   *Prologue*

To read between Jacobi's lines, we must realize that Fourier was more an engineer or physicist rather than a mathematician. If we bear in mind the two key ingredients of the Fourier series, harmonically related sinusoids and the independence of coefficients, we find that in this specific context, three curious facts emerge:

1. Fourier's own memoirs indicate quite a lack of mathematical rigor in the manner in which he obtained his results concerning Fourier series.
2. Fourier certainly made no claim to originality in obtaining the formulas for calculating the coefficients and such formulas were, in fact, described by others years earlier.
3. Fourier was not the first to propose the use of harmonically related sinusoids to represent functions. This, too, was suggested well before his time.

What, then, is Fourier's claim to fame in the context of the series named after him? To be honest, his single contribution was to unify these ideas, look beyond the subtle details, and proclaim that any *arbitrary* function could be expressed as an *infinite* sum of harmonically related sinusoids. That Fourier realized the enormity of his hypothesis is evident from the fact that he spent his last years basking in the glory of his achievements, in the company of many a sycophant. But it is due more to the achievements of other celebrated mathematicians, both contemporaries and successors, who sought to instill mathematical rigor by filling in and clarifying the details of his grand vision and, in so doing, invented concepts that now form the very fabric of modern mathematical analysis, that Fourier's name is immortalized.

5B.2   *The Eighteenth Century*

Our historical journey into the development of Fourier series begins in the eighteenth century, a time of little rivalry and even lesser distinction between mathematicians and physicists. The concept of a function was a mathematical expression, a single mathematical formula that described its extent. A triangular waveform would not have been regarded as one function but two. And an arbitrary graph might not be represented by a function at all. The mid-eighteenth century saw the study of vibrating strings, which is subject to a partial differential equation, occupy the minds of three famous personalities of the day. The solution in functional form, assuming the initial string shape to be described by a single expression, was first proposed by d'Alembert in 1748 and extended a year later by Euler, arguably the most prolific mathematician in history, to include arbitrary shapes having different analytical expressions over different parts. It was Bernoulli who, in 1753, first suggested a trigonometric sine series as a general solution for string motion when it starts from rest. To put the problem in mathematical perspective, the one-dimensional differential equation governing the string motion y may be written as $\partial^2 f / \partial t^2 = c^2 \partial^2 f / \partial x^2$. The solution d'Alembert and Euler proposed was $f = \psi(x + ct) + \psi(x - ct)$ and involves traveling waves. Bernoulli's solution was of the form $f = K_1 \sin x \cos ct + K_2 \sin 2x \cos 2ct + \cdots$, when the string starts from rest. This solution, being the most general, he said, must include the solutions of Euler and d'Alembert. Euler objected on the grounds that such a series could represent an arbitrary function of a single variable in terms of such a series of sines only if the function were both odd and periodic.

In 1767, Lagrange, then only 23 but respected by both Euler and d'Alembert, obtained a solution tantalizingly close to the form with which we are now familiar, if only to transform it into the functional form of Euler. His only objective at the time was to defend Euler's claim against the critics. The solution that Lagrange transformed to fit the form given by Euler involved a time dependence that, when set to zero, would have led to a series of sines describing the string displacement, with coefficients given by the integral form for b_k as we now recognize it. Had Lagrange taken this one step more, the fame and glory would have been all his.

Clairaut in 1757, and Euler two decades later (on returning to the problem of vibrating strings in 1877), proposed evaluation of the coefficients of trigonometric series whose existence was already demonstrated by other means, more or less in the form we know them now. But they were still skeptical and believed that such results could be applied only to functions as then understood (a single analytical expression).

5B.3   *The Nineteenth Century*

The advent of the nineteenth century saw the emergence of a new and sometimes bitter rivalry between the proponents of mathematics and physics as the two fields became more well defined and the gulf between them widened. This was the time when Fourier burst on the scene. It was during the course of his studies in heat conduction, his first and lasting love (he even felt a sheer physical need for heat, always dressing heavily and living in excessively warm quarters), that he took the decisive step of suggesting a trigonometric series representation for an entirely arbitrary function, whether or not such a function could be described by mathematical expressions over its entire extent. He also realized that, given a function over a prescribed range, its value outside that range is in no way determined and this foreshadowed the concept of periodic extension central to the series representation.

These ideas were contained in Fourier's first paper on the theory of heat conduction, competed in draft form in 1804–1805 (while in Grenoble) and subsequently
presented to the Paris Academy in 1807. The referees encouraged Fourier to continue his study, even though his results generated as much skepticism and controversy as interest and excitement. In a letter rebutting his critics (written probably
to Lagrange), Fourier apologized for not acknowledging previous studies by Euler
and d'Alembert (due to his inability to refer to those works, he alleged). Yet, he
also showed his familiarity with such studies when he claimed that they were inad-
equate because they did not consider the range over which such expansions were
valid while denying their very existence for an arbitrary function. In any event, the
propagation of heat was made the subject of the *grand prix de mathematiques* for
1812. It is curious that there was only one other candidate besides Fourier! Whether
this topic was selected because of Fourier's goading and the implied challenge to
his critics or because of the genuine interest of the referees in seeing his work fur-
thered remains shrouded in mystery. Fourier submitted his prize paper late in 1811
and even though he won the prize, the judges, the famous trio of Laplace, Lagrange
and Legendre among them, were less than impressed with the rigor of Fourier's
methods and decided to withhold publication in the Academy's memoirs.

Remember that 1812 was also a time of great turmoil in France. Napoleon's
escape from Elba in 1815 and the second restoration led to a restructuring of
the Academy to which Fourier was finally nominated in 1816 amidst protests and
elected only a year later. And the year 1822 saw his election to the powerful post of
Secretary of the Academy and with it, the publication of his treatise on heat conduction which contained, without change, his first paper of 1807. That he had resented
criticism of his work all along became evident when, a couple of years later, he
caused publication in the Academy's memoirs of his prize-winning paper of 1811,
exactly as it was communicated. Amidst all the controversy, however, the importance of his ideas did no go unrecognized (Lord Kelvin, a physicist himself, called
Fourier's work a great mathematical poem) and soon led to a spate of activity by
Poisson, Cauchy and others. Cauchy, in particular, took the first step toward lend-
ing mathematical credibility to Fourier's work by proposing, in 1823, a definition
of an integral as the limit of a sum and giving it a geometric interpretation.

But it was Dirichlet's decade of dedicated work that truly patched the holes,
sanctified Fourier's method with rigorous proofs, and set the stage for further prog-
ress. He not only extended the notion of a function, giving a definition that went
beyond geometric visualization, but more important, he provided seminal results on
the sufficient conditions for the convergence of the series (1829) and the criterion
of absolute integrability (1837), which we encountered earlier in this chapter.

Motivated by Dirichlet, Riemann after completing his doctorate started work
on his probationary essay on trigonometric series in 1851 in the hope of gaining an
academic post at the University of Göttingen. This essay, completed in 1853, was
unfortunately published only in 1867, a year after he died. Meanwhile, his probationary lecture on *the hypotheses which lie at the foundations of geometry* on June
10, 1854, shook its very foundations and paved the way for his vision of a new

differential geometry (which was to prove crucial to the development of the theory of relativity). But even his brief excursion into Fourier series resulted in two major advances. First, as Dirichlet had done with functions, Riemann widened the concept of the integral of a continuous function beyond Cauchy's definition as the limit of a sum to include functions that were neither continuous nor contained a finite number of discontinuities. And more important, he proved that, for bounded integrable functions, the Fourier coefficients tend to zero as the harmonic index increases without limit, suggesting that convergence at a point depends only on the behavior of the function in the vicinity of that point.

The issues of convergence raised by Riemann in his essay—and the problems of convergence in general—motivated several personalities who made major contributions, Heine and Cantor among them. Heine, in 1870, established uniform convergence (first conceived in 1847) at all points except at discontinuities for functions subject to the Dirichlet conditions. Cantor's studies, on the other hand, probed the foundations of analysis and led him, in 1872, to propose the abstract theory of sets.

There were other developments in the nineteenth century, too. In 1875, du Bois-Raymond showed that if a trigonometric series converges to an integrable function $x(t)$, then that series must be the Fourier series for $x(t)$. And a few years later, in 1881, Jordan introduced the concept of functions of bounded variation leading to his own convergence conditions. Interestingly enough, Parseval, who proposed a theorem for summing series of products in 1805 that now carries his name, was not directly involved in the study of Fourier series and his theorem seems to have been first used in this context only a century later in 1893. The continuous form, to be encountered in the chapter on Fourier transforms and often referred to as the energy relation, was first used by Rayleigh in connection with black body radiation. Both forms were later (much later, in the twentieth century, in fact) generalized by Plancherel.

In 1899, Gibbs (who is remembered more for his work in thermodynamics leading to the phase rule and, lest we forget, for introducing the notation $\cdot$ and $\times$ to represent the dot product and cross product of vectors) reported the curious effect that now bears his name in a letter to *Nature*. The story goes that Michelson, who had developed a harmonic analyzer in 1898, observed this phenomenon for a square wave. Convinced that it was not an artifact of the device itself, he corresponded with Gibbs to seek an explanation. It is curious that the collected works of Gibbs, which include his letters to Michelson, make no reference to such correspondence. Gibbs' letter to *Nature*, which dealt with a sawtooth waveform, did not contain any proof and received scant attention at the time. Only in 1906 did Bôcher demonstrate how the Gibbs effect actually occurs for any waveform with discontinuities.

In his studies on Fourier series smoothing, Fejer was led to propose summation methods usually reserved for divergent series. In 1904, he showed that summation by the method of arithmetic means results in a smoothing effect at the discontinuities and the absence of the Gibbs phenomenon. This was to be the forerunner of the concept of spectral windows that are now so commonly used in spectral analysis.

5B.4   *The Twentieth Century*

As we move into the twentieth century, we cross another frontier in the development of Fourier methods, led this time by Lebesgue. Following in the footsteps of Borel, who in 1898 formulated the concept of measure of a set, Lebesgue in 1902 went on to reformulate the notion of the integral on this basis and, in 1903, extended Riemann's theorem to include unbounded functions provided integration was understood in the Lebesgue sense. In 1905, he presented a new sufficient condition for convergence that included all the previously known ones. And he settled the nagging problem of term-by-term integration of Fourier series, showing that it no longer depends on the uniform convergence of the series.

There have been many other developments, mostly of purely mathematical interest, as a result of Lebesgue's fundamental work: the conjecture of Lusin in 1915 that the Fourier series of a square integrable function converges except on a set of measure zero, proved finally in 1966 by Carleson; the Riesz-Fischer theorem, proving the converse of the Parseval's theorem; and others. All these developments have finally given Fourier series a respectability that mathematicians have craved for but engineers hardly cared about.

This, then, is where our saga ends. Even though there are certainly other aspects that deserve attention, we mention some of these only in passing. On the theoretical side are the Fourier integral (transform) itself, Schwartz's concept of generalized functions which we have been using so casually (in the form of impulse functions) in this book, and Wiener's exposition of generalized harmonic analysis. On the practical side, the advent of high speed computing in the fifties and the development of fast algorithms for Fourier analysis, starting with the work of Danielson and Lanczos in 1942 and of Cooley and Tukey in 1965, has led to a proliferation of Fourier-based methods in almost every conceivable discipline. And the end is nowhere in sight.

5B.5   *Epilogue*

In the words of Fourier, "the profound study of nature is the most fertile source of mathematical discoveries." Be that as it may, it surely would have been difficult, even for Fourier himself, to predict the household recognition of his name within the scientific world that would follow from his own study of a tiny aspect of that nature. Tiny or not, let us also remember that every time we describe a definite integral by the notation $\int_a^b$, we are using yet another one of his gifts. What else is there to say but *viva Fourier!*

P5.1 (Fourier Series Concepts) For each of the following periodic signals:

(1) $x_1(t) = 4 + 2\sin(4\pi t) + 3\sin(16\pi t) + 4\cos(16\pi t)$

(2) $x_2(t) = \sum_{\substack{k=-4 \\ k \neq 0}}^{4} \frac{6}{k}\sin(\frac{1}{2}k\pi)\exp(-jk\pi/3)\exp(jk6\pi t)$

(a) Write out its Fourier series (FS) in all forms.

(b) Which harmonics are present in its Fourier series?

(c) Compute the total signal power and rms value.

(d) Sketch its onesided and twosided magnitude and phase spectra.

P5.2 (Fourier Series Coefficients) Find the appropriate Fourier series coefficient for the following periodic signals described for one period:

(a) $X_S[k]$ for $x_p(t) = \exp(-t)$, $(0,1)$

(b) a_k for $x_p(t) = \text{rect}(t - \frac{1}{2})$, $(0,2)$

(c) b_k for $x_p(t) = (1 + t)$, $(0,1)$

P5.3 (Symmetry) The magnitude and phase spectra of two periodic signals $x_1(t)$ and $x_2(t)$ are sketched in Figure P5.3. For each signal:

(a) Which harmonics are present in its Fourier series?

(b) What symmetry (hidden or otherwise) is present, if any?

(c) Write out its Fourier series in polar form.

(d) What are the total signal power and rms value?

Figure P5.3

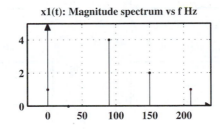
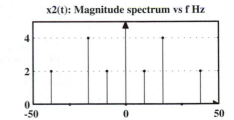
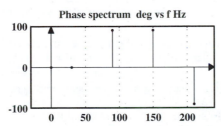
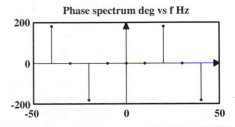

P5.4 (Symmetry) A portion of a periodic signal $x_p(t)$, whose time period is T, is described by $x_p(t) = t$, $(0, 1)$. Sketch $x_p(t)$ over $(-2T, 2T)$, and indicate what symmetry $x_p(t)$ possesses about $t = \frac{1}{2}T$ and $t = \frac{1}{4}T$ if

(a) $x_p(t)$ is even symmetric with $T = 2$.
(b) $x_p(t)$ is odd symmetric with $T = 2$.
(c) $x_p(t)$ is even and halfwave symmetric with $T = 4$.
(d) $x_p(t)$ is odd and halfwave symmetric with $T = 4$.

P5.5 (Coefficients) One period of various signals is shown in Figure P5.5:

(a) Use symmetry to set up an integral for computing a_k and b_k.
(b) Compute a_k, b_k and $X_S[k]$ and simplify for odd and even k. Evaluate special or indeterminate cases separately, if necessary.
(c) Sketch the magnitude and phase spectra for $|k| \le 5$.

Figure P5.5

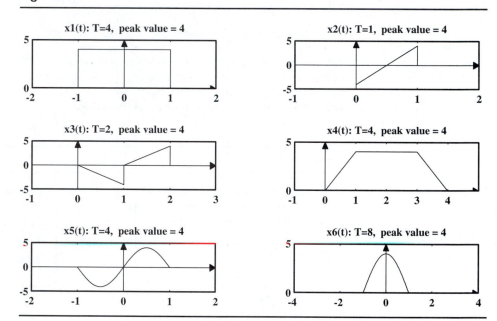

P5.6 (Properties) Let a_k and b_k be the trigonometric FS coefficients of $x_p(t)$. What are the trigonometric FS coefficients of $x_p(2t)$, $x_p(-t)$ and $x_p(-2t)$?

P5.7 (Properties) The magnitude and phase spectra of a periodic signal $x(t)$ are sketched in Figure P5.7.

(a) Write out the Fourier series for $x(t)$ and simplify.
(b) Sketch the magnitude and phase spectra of:
   (1) $x(2t)$ (2) $x(t - \frac{1}{6})$ (3) $x'(t)$

Figure P5.7

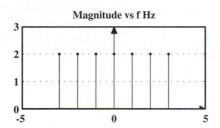
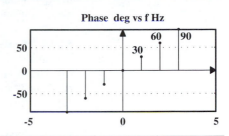

P5.8 (Derivative Method) Find the Fourier series coefficients of the following periodic signals with $T = 1$ using the derivative method.
(a) $\text{rect}(2t)$
(b) $\exp(t)\,\text{rect}(t - \frac{1}{2})$
(c) $t\,\text{rect}(t)$
(d) $\text{tri}(2t)$

P5.9 (Convergence and the Gibbs Effect) For each signal of Problem P5.5:
(a) Establish the convergence rate.
(b) Will the reconstruction show the Gibbs effect?
(c) What is the peak overshoot, if any, at each discontinuity?
(d) What value will its Fourier series converge to at $t = 0$?
(e) What value will its Fourier series converge to at $t = \frac{1}{2}T$?
(f) What value will its Fourier series converge to at $t = \frac{1}{4}T$?

P5.10 (Convergence) The Fourier series of a periodic signal is given by

$$x(t) = 2 + \sum_{k\,\text{odd}}^{\infty} \frac{4}{k\pi}\sin(k\pi t)$$

(a) What value does the series converge to at $t = 0$?
(b) What value does the series converge to at $t = \frac{1}{2}T$?
(c) What value does the series converge to at $t = \frac{1}{4}T$?

P5.11 (Gibbs Effect) A series RC circuit is excited by a periodic voltage with convergence rate $\propto 1/k$.
(a) Will the capacitor voltage exhibit the Gibbs effect?
(b) Will the resistor voltage exhibit the Gibbs effect?

P5.12 (Modulation) A periodic signal $x(t)$ is described by

$$x(t) = 2 + \sum_{k=1}^{4} (6/k)\sin^2(\tfrac{1}{2}k\pi)\cos(1600k\pi t)$$

(a) Sketch the spectrum of $x(t)$.
(b) Sketch the spectrum of the modulated signal $x(t)\cos(1600\pi t)$.
(c) Find the power in $x(t)$ and in the modulated signal.

P5.13 (Modulation) The signal $x_C(t) = \cos(2\pi f_C t)$ is modulated by an even symmetric periodic square wave with period T whose one period is described by $\mathrm{rect}(t/\tau)$. Find and sketch the spectrum of the modulated signal if $f_C = 1$ MHz, $T = 1$ ms and $\tau = 0.1$ ms.

P5.14 (System Analysis) A periodic signal $x(t) = |\sin(250\pi t)|$ is applied to an ideal filter. Sketch the filter output $y(t)$ if:
(a) The filter blocks all frequencies past 200 Hz.
(b) The filter passes only frequencies between 200 and 400 Hz.
(c) The filter blocks all frequencies past 400 Hz.

P5.15 (System Analysis) The signal $x(t) = 10|\sin(t)|$ volts is applied to a series RC circuit ($R = 1\ \Omega$, $C = 1$ F). The output is the capacitor voltage.
(a) Find and sketch the dc component of the output.
(b) Find and sketch the fundamental component of the output.
(c) Find the output power up to the second harmonic.

P5.16 (Application) The input to an amplifier is $x(t) = \cos(10\pi t)$ and the response is given by $y(t) = 10\cos(10\pi t) + 2\cos(30\pi t) + \cos(50\pi t)$.
(a) Compute the third harmonic distortion.
(b) Compute the total harmonic distortion.

P5.17 (Application) The signal $x(t) = \sin(2\pi f_0 t)$ is applied to the system whose output $y(t)$ is defined by
(1) $y(t) = \mathrm{abs}[x(t)]$ (2) $y(t) = \mathrm{sgn}[x(t)]$ (3) $y(t) = x^3(t)$

(a) Sketch the system output.
(b) What is the dc output?
(c) Which harmonics of the input are present in the output?
(d) What is the third harmonic distortion in the output, if any?
(e) What is the total harmonic distortion in the output?

P5.18 (Application) A square wave with zero dc offset and period T is applied to an RC circuit with time constant τ whose output is the capacitor voltage.
(a) Sketch the filter output if $\tau \approx 100\ T$.
(b) Sketch the filter output if $\tau \approx 0.001\ T$.

P5.19 (Application) Two periodic signals $x(t)$ and $y(t)$ described by

$$x(t) = 2 + \sum_{k=1}^{\infty} (4/k)\cos(1600k\pi t) \qquad y(t) = \sum_{k=1}^{\infty} (8/k^2)\sin(800k\pi t)$$

Each is passed through an ideal lowpass filter with $f_c = 1$ kHz to obtain the filtered signals $x_f(t)$ and $y_f(t)$. The filtered signals are then passed through a multiplier to obtain the signal $w_f(t) = x_f(t)y_f(t)$. The signal $w_f(t)$ is once again passed through an ideal lowpass filter with $f_c = 1$ kHz to obtain $z_f(t)$.
(a) Sketch the twosided spectra of $x_f(t)$, $y_f(t)$, $w_f(t)$ and $z_f(t)$.
(b) Sketch $z_f(t)$.

P5.20 **(Design)** Consider a square wave with zero dc offset and a period of 1ms as the input to an RC filter with $R = 1$ kΩ.

(a) Find C such that the output phase differs from the input phase by exactly $45°$ at 5 kHz.

(b) The half-power bandwidth is defined as the frequency range over which the output power exceeds half the input power. How is this related to the time constant of the RC filter? What is the half-power bandwidth of the RC filter designed in part (a)?

P5.21 **(Design)** We wish to design a dc power supply using a symmetric triangular wave with $T = 0.02$ s and zero dc offset fed to a halfwave rectifier followed by an RC filter with $R = 1$ kΩ. What value of C will ensure a ripple of less than 1%?

P5.22 **(Significance of Fourier Series)** To appreciate the significance of Fourier series, consider a periodic sawtooth signal $x(t) = 1$, $(0 < t < 1)$ for which the FS coefficients are $a_0 = \frac{1}{2}$, $b_k = -1/k\pi$. Its Fourier series up to the first harmonic is $x_1(t) \approx \frac{1}{2} - (1/\pi)\sin(2\pi t)$.

(a) Find the power P_T in $x(t)$, the power P_1 in $x_1(t)$ and the power error $P_T - P_1$.

(b) Approximate $x(t)$ by $y_1(t) = A_0 + B_1\sin(2\pi t)$ as follows. Pick two instants over $(0,1)$, say $t_1 = \frac{1}{4}$ and $t_2 = \frac{1}{2}$, and substitute to obtain:

$$y_1(t_1) = A_0 + B_1\sin(2\pi t_1) \qquad y_1(\tfrac{1}{2}) = \tfrac{1}{2}A_0 + B_1\sin(\pi) = A_0$$

$$y_1(t_2) = A_0 + B_1\sin(2\pi t_2) \qquad y_1(\tfrac{1}{4}) = A_0 + B_1\sin(\tfrac{1}{2}\pi) = A_0 + B_1$$

Solve for A_0 and B_1.

(c) Find the power P_y in $y_1(t)$ and the power error $P_T - P_y$. Does $x_p(t)$ or $y_1(t)$ have a smaller power error?

(d) Start with $t_1 = \frac{1}{6}$, $t_2 = \frac{1}{4}$ and recompute A_0 and B_1. Why are these different? Is the power error less than for part (c)? Is there a unique way to choose t_1 and t_2 to yield the smallest power error?

(e) If we want to extend this method to an approximation with many more harmonics and coefficients, what problems do you expect?

(f) Is the Fourier series method of computing the coefficients "better"?

P5.23 **(Reconstruction at a Discontinuity)** If t_- and t_+ denote instants on either side of a discontinuity in a signal $x(t)$, the mean square error in the Fourier series reconstruction $x_n(t)$ may be written as $\epsilon = [x_n(t) - x(t_-)]^2 + [x_n(t) - x(t_+)]^2$. Its minimum with respect to $x_n(t)$ occurs when $d\epsilon/dx_n = 0$. Show that this results in $x_n(t) = \frac{1}{2}[x(t_-) + x(t_+)]$, implying that the reconstructed value converges to the midpoint at a discontinuity.

P5.24 **(Closed Forms for Infinite Series)** Parseval's theorem provides an interesting approach to finding closed form solutions to infinite series. Generate such closed form results by using Parseval's theorem for the following signals:

(a) Sawtooth wave

(b) Triangular wave

(c) Full-rectified sine wave

P5.25 **(Spectral Bounds)** Starting with $x_p^{(n)}(t) \longleftrightarrow (j2\pi f_0)^n X_S[k]$ and the fundamental theorem of calculus $|\int x(\alpha)\,d\alpha| \le \int |x(\alpha)|\,d\alpha$, and noting that $|\exp(\pm j\theta)| = 1$, obtain

$$|X_S[k]| \le \frac{1}{T|2\pi f_0|^n}\int_T |x_p^{(n)}(t)|\,dt$$

This sets the nth bound on the spectrum in terms of the *absolute area* under the nth derivative of $x_p(t)$. Since the derivatives of an impulse possess zero area, the number of nonzero bounds that can be found equals the number of times we can differentiate before only derivatives of impulses occur.

Starting with $n = 0$, use this result to find all the nonzero spectral bounds for the following periodic signals defined over one period.

(a) $x_p(t) = \text{rect}(t), (T = 2)$ **(b)** $x_p(t) = \text{tri}(t), (T = 1)$ **(c)** $x_p(t) = |\sin(t)|, (T = \pi)$

P5.26 (Smoothing Kernels) The periodic signal (Dirichlet kernel) $d_p(t)$ corresponding to $W_S[k] = \text{rect}(k/2N)$ is $M \, \text{sinc}(Mf_0t)/\text{sinc}(f_0t)$, where $M = 2N + 1$. Use this result and the convolution property to find the periodic signal $f_p(t)$ (the Fejer kernel) corresponding to the triangular window $F_S[k] = \text{tri}(k/M)$.

P5.27 (Orthogonality) A set of signals ϕ_k that satisfy $\mathbb{A}[\phi_j\phi_k^*](a, b) = 0, j \neq k$ is said to be *orthogonal* over (a, b). If, in addition, the energy $E_k = \mathbb{A}[|\phi_k|^2](a, b) = 1$ in each ϕ_k, the set is called **orthonormal**. Sinusoids are examples of signals that are orthogonal over one composite period.

(a) Show that an even symmetric signal is always orthogonal to any odd symmetric signal over the symmetric duration $(-a, a)$.

(b) A set of functions $\{\phi_k\}$ is orthogonal only if $\mathbb{A}[\phi_j\phi_k^*] = 0, j \neq k$ for *every* pair of signals in the set. Of the following sets, which are orthogonal and which are also orthonormal?

1. $\{1, t, t^2\}(-1, 1)$

2. $\{\exp(-\frac{1}{2}t)u(t), (1 - t)\exp(-\frac{1}{2}t)u(t), (1 - 2t + \frac{1}{2}t^2)\exp(-\frac{1}{2}t)u(t)\}$

3. $\{\exp(-\frac{1}{2}t^2), t\exp(-\frac{1}{2}t^2)\}(-\infty, \infty)$

4. $\{\exp(-t), 2\exp(-t) - 3\exp(-2t), 3\exp(-t) - 12\exp(-2t) + 10\exp(-3t)\}(0, \infty)$

(c) An energy signal can be written as a sum of orthogonal energy signals $x(t) = \sum \alpha_k\phi_k$. This describes a generalized Fourier series. Show that the relation for computing α_k is $\alpha_k = \mathbb{A}[x(t)\phi_k^*]/E_k$.

(d) Set up the generalized Fourier series $x(t) \approx \alpha_0\phi_0 + \alpha_1\phi_1 + \alpha_2\phi_2$ for the following signals and compute the coefficients α_k:
$dx(t) = t\exp(-t/2)u(t)$ in terms of set (2) of part (b)
$x(t) = 12\exp(-3t)u(t)$ in terms of set (4) of part (b)

(e) Show that the energy in the generalized Fourier series equals $\sum |\alpha_k|^2 E_k$. Use this to compute the energy error in each of the series representations for $x(t)$ in part (d).

6 FOURIER TRANSFORMS

6.0
Scope and Objectives

The Fourier series provides us with a link between the time and frequency representation of periodic signals. The Fourier transform extends this link to arbitrary signals (aperiodic or otherwise). In this chapter, we look back to the consequences of transforming a periodic signal to an aperiodic one by the simple expedient of stretching the time period without limit. We show that such a transformation leads intuitively to the evolution of the direct and inverse Fourier transforms and provides a means of going back and forth between the series and transform results. This connection also forms the basis for the development of many properties of the Fourier transform which have their analogs in the Fourier series representation.

This chapter is divided into three parts:

Part 1 The Fourier transform and its properties.
Part 2 System analysis using the Fourier transform.
Part 3 Special topics, including existence and the central limit theorem.

Applications of the Fourier transform are covered in the next chapter.

Useful Background   The following topics provide a useful background:

1. The key concepts of Fourier series for periodic signals (Chapter 5).
2. The concept of periodic extension (Chapter 4).
3. The properties of the sinc and impulse functions (Chapter 2).

6.1
Introduction

The approach we adopt to develop the Fourier transform serves to unite the representations for periodic functions and their aperiodic counterparts, provided the ties that bind the two are also construed to be the very ones that separate them and are, therefore, understood in their proper context.

6.1.1   *The Fourier Series Revisited*

Recall that a periodic signal $x_p(t)$ with period T and its exponential Fourier series (FS) coefficients $X_S[k]$ are related by

$$x_p(t) = \sum_{k=-\infty}^{\infty} X_S[k] \exp(jk2\pi f_0 t) \qquad X_S[k] = \frac{1}{T} \int_{-T/2}^{T/2} x_p(t) \exp(-jk2\pi f_0 t)\, dt$$

The act of stretching the period T of a periodic signal $x_p(t)$ without limit is fraught with the following consequences:

1. The periodic signal $x_p(t)$ becomes aperiodic and represents a single pulse $x(t)$ corresponding to one period of $x_p(t)$.
2. The harmonic spacing f_0 approaches zero and the spectrum becomes a continuous curve rather than a set of discrete points. We can thus replace f_0 by an infinitesimally small quantity df such that $df \longrightarrow 0$ and the discrete frequency kf_0 by the continuous frequency variable f.
3. The transition from a periodic to an aperiodic signal also represents a transition from a power signal to an energy signal.
4. The spectral coefficients $X_S[k] \longrightarrow 0$ because the factor $1/T$ in the relation for finding $X_S[k]$ approaches zero. The $X_S[k]$ are no longer a useful indicator of the spectral content of the aperiodic signal $x(t)$.

6.1.2   *Toward the Fourier Transform*

Since the $X_S[k]$ no longer provide a useful representation of aperiodic signals, how do we find an alternative? One way is to redefine the coefficients $X_S[k]$ so as to eliminate its dependence on the offending factor $1/T$. We express $X_S[k]$ in the form

$$TX_S[k] = \int_{-T/2}^{T/2} x_p(t) \exp(-jk2\pi f_0 t)\, dt$$

Even though the product $TX_S[k]$ on the left-hand side becomes indeterminate as $T \longrightarrow \infty$ and $X_S[k] \longrightarrow 0$, the integral on the right-hand side often exists. Further, since $kf_0 \longrightarrow f$, the quantity $TX_S[k]$ describes a function of f, which we denote by $X(f)$, such that

$$X(f) = \lim_{T \to \infty} TX_S[k] = \int_{-\infty}^{\infty} x(t) \exp(-j2\pi f t)\, dt$$

This relation is also written in terms of the frequency ω as

$$X(\omega) = \int_{-\infty}^{\infty} x(t) \exp(-j\omega t)\, dt$$

The quantity $X(f)$ or $X(\omega)$ provides a frequency-domain representation of the aperiodic signal $x(t)$ and is called the **Fourier transform** (FT) of $x(t)$.

6.1.3 *Magnitude and Phase Spectra*

The Fourier transform of an aperiodic signal may be viewed as a consequence of removing the periodicity from its periodic counterpart. Yet there is a striking similarity between the nature of $X(f)$ for an aperiodic signal $x(t)$ and the discrete spectrum $X_S[k]$ of the corresponding periodic signal $x_p(t)$. The envelope of the scaled discrete magnitude spectrum $|TX_S[k]|$ of $x_p(t)$ is *identical* to the continuous magnitude spectrum $|X(f)|$ of $x(t)$, and the envelope of the phase spectrum of $x_p(t)$ is identical to the continuous phase spectrum of $x(t)$.

6.1.4 *Connections: Sampling and Periodic Extension*

Turning things around, the continuous spectrum of an aperiodic signal $x(t)$ sampled at the locations kf_0 describes a periodic signal $x_p(t)$ with period $T = 1/f_0$ which arises from replications of $x(t)$. Formally, *sampling in the frequency domain corresponds to periodic extension in the time domain. The sample spacing f_0 in the frequency domain is reciprocally related to the period T on the time domain through $T = 1/f_0$.*

This concept forms an important link between the two domains. The means of going back and forth between the transform $X(f)$ for an aperiodic $x(t)$ and the series coefficients $X_S[k]$ for its periodic extension $x_p(t)$ with period $T = 1/f_0$ are summarized in Table 6.1. The transforms of one period $x(t)$ of some of the periodic signals $x_p(t)$ of Table 5.7, using these transformations, are listed in Table 6.2.

Table 6.1 Relation Between Series $X_S[k]$ and Transform $X(f)$.

Direction	Relation
Series to transform	$X(f) = TX_S[k]\|_{kf_0=f}$
Transform to series	$X_S[k] = \frac{1}{T}X(f)\|_{f=kf_0}$

Table 6.2 Spectral Coefficients of Some Signals.

Periodic signal	Coefficient $X_S[k]$	$X(f)$ of one period
Impulse train $x_p(t) = A\delta(t - t_0), 0 < t_0 < T$	$\frac{A}{T}\exp(-j2\pi kf_0t_0)$	$A\exp(-j2\pi ft_0)$
Rectangular pulse train $x_p(t) = A, \|t\| \le \frac{1}{2}t_0$	$\frac{At_0}{T}\text{sinc}(kf_0t_0)$	$At_0\text{sinc}(ft_0)$
Triangular pulse train $x_p(t) = A(1 - 2\|t\|/t_0), \|t\| \le \frac{1}{2}t_0$	$\frac{At_0}{2T}\text{sinc}^2(\frac{1}{2}kf_0t_0)$	$\frac{1}{2}At_0\text{sinc}^2(\frac{1}{2}ft_0)$
Sinusoidal pulse train $x_p(t) = A\cos(\pi t/t_0), \|t\| \le \frac{1}{2}t_0$	$\frac{1}{T}\left[\frac{2A\pi t_0\cos(k\pi f_0t_0)}{\pi^2-(2k\pi f_0t_0)^2}\right]$	$\left[\frac{2A\pi t_0\cos(\pi ft_0)}{\pi^2-(2\pi ft_0)^2}\right]$

Remark: Only if $x(t)$ is timelimited with width $t_0 < T$ does the periodic extension correspond to a simple periodic *continuation* of $x(t)$.

The coefficients $X_S[k]$ of a train of rectangular pulses of height A and width t_0 centered about the origin (Figure 6.1) are

$$X_S[k] = \frac{1}{T} A t_0 \, \text{sinc}(k f_0 t_0)$$

The Fourier transform of the single pulse $x(t)$ that represents one period of $x_p(t)$ is then

$$X(f) = T X_S[k]_{k f_0 = f} = A t_0 \, \text{sinc}(f t_0)$$

Comment: Note the equivalence of the envelope (a sinc function), the removal of T, and the change to f (from $k f_0$), as we go from the series to the transform.   —

Figure 6.1   A triangular pulse and its periodic extension for various periods.

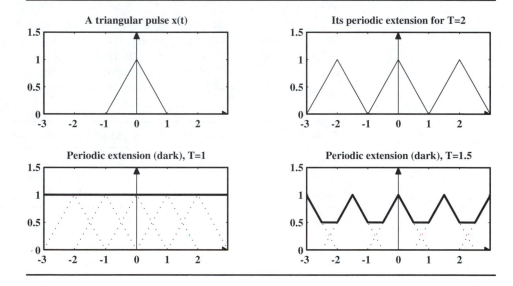

Suppose the Fourier transform of a triangular pulse $x(t)$ of unit height and width $t_0 = 2$ units, centered at the origin, is found to be $X(f) = \text{sinc}^2(f)$. The Fourier series coefficients of its corresponding periodic extension $x_p(t)$ with period T are

$$X_S[k] = X(k f_0)/T = \text{sinc}^2(k f_0)/T$$

The signals $x_p(t)$ for three choices of T are shown in Figure 6.1 and yield:

(a) $T = 2$: A triangular wave with $X_S[k] = \frac{1}{2}\text{sinc}^2(\frac{1}{2}k)$
(b) $T = 1$: A constant with $X_S[k] = \text{sinc}^2(k)$ or $X_S[0] = 1$ and $X_S[k] = 0, k \neq 0$
(c) $T = 1.5$: A triangular form with $X_S[k] = \frac{2}{3}\text{sinc}^2(\frac{2}{3}k)$ —

6.1.5   *Differences Between the Series and Transform*

Similarities aside, we also stress two subtle, but major, differences between periodic and aperiodic signals and their spectra.

1. The Fourier series always describes periodic signals (power signals).
2. Aperiodic signals lead to continuous frequency spectra.
3. Unlike $|X_S[k]|$, which indicates the magnitude of the components in $x_p(t)$ at the frequencies $f = kf_0$, $|X(f = kf_0)|$ does not represent magnitudes of the frequency components in $x(t)$. Since $X(f)$ equals $TX_S[k]$ in the limit as $T \longrightarrow \infty$, the actual magnitudes $X(f)/T$ approach zero at every frequency.

6.1.6   *The Inverse Fourier Transform*

Having overcome the problem of obtaining the spectrum $X(f)$ of $x(t)$, how do we recover $x(t)$ from its spectrum $X(f)$? Once again, we resort to the Fourier series. For a periodic signal $x_p(t)$, we have

$$x_p(t) = \sum_{k=-\infty}^{\infty} X_S[k]\exp(jk2\pi f_0 t)$$

The aperiodic signal $x(t)$ results when $T \longrightarrow \infty$, and requires $TX_S[k]$ instead of $X_S[k]$. We, therefore, rewrite the series as follows:

$$x_p(t) = \sum_{k=-\infty}^{\infty} TX_S[k]\exp(jk2\pi f_0 t)\frac{1}{T} = \sum_{k=-\infty}^{\infty} TX_S[k]\exp(jk2\pi f_0 t)f_0$$

As $T \longrightarrow \infty$, $x_p(t) \longrightarrow x(t)$, $kf_0 \longrightarrow f$, $TX_S[k] \longrightarrow X(f)$, and the summation tends to an integration over $(-\infty, \infty)$ with $f_0 \longrightarrow df \longrightarrow 0$. We obtain

$$x(t) = \int_{-\infty}^{\infty} X(f)\exp(j2\pi ft)\,df$$

If we use the variable ω instead, $d\omega = 2\pi\,df$ and

$$x(t) = \frac{1}{2\pi}\int_{-\infty}^{\infty} X(\omega)\exp(j\omega t)\,d\omega$$

This is the **inverse Fourier transform**, which allows us to obtain $x(t)$ from $X(f)$ or $X(\omega)$. The f-forms of the Fourier transform and its inverse are almost symmetric. The functions $x(t)$ and $X(f)$ form a unique transform pair, and their relationship is shown symbolically using a double arrow as

$$x(t) \longleftrightarrow X(f)$$

Notation: Unlike the Fourier series, which form we use, $X(f)$ or $X(\omega)$, does matter! We prefer the f-form. Some examples, and all properties, are also described in the ω-form.

6.1.7 *Symmetry Properties*

In analogy with the Fourier series coefficients $X_S[k]$ for periodic signals, the Fourier transform $X(f)$ is, in general, complex and may be represented in any of the following forms:

$$X(f) = \text{Re}\{X(f)\} + j\text{Im}\{X(f)\} = |X(f)| \angle \phi(f) = |X(f)|\exp\{j\phi(f)\}$$

For real signals, the magnitude $|X(f)|$ or $\text{Re}\{X(f)\}$ displays even symmetry, and the phase $\phi(f)$ or $\text{Im}\{X(f)\}$ displays odd symmetry. These and other symmetry properties are summarized in Table 6.3.

Table 6.3 Symmetry Properties of the Fourier Transform.

$x(t)$	$X(f)$
Real	Conjugate symmetric $X(f) = X^*(-f)$
Conjugate symmetric, $x(t) = x^*(-t)$	Real
Real and even	Real and even
Real and odd	Imaginary and odd

Spectrum for real signals

$x(t)$	$X(f)$	$|X(f)|$	$\phi(f)$	$\text{Re}[X(f)]$	$\text{Im}[X(f)]$
Real	Complex	Even	Odd	Even	Odd
Real, even	Real	Even	Odd	Even	Zero
Real, odd	Imaginary	Even	Odd	Zero	Odd

6.1.8 *Plotting Magnitude and Phase Spectra*

The magnitude and phase of $X(f)$ may be plotted in various ways (Figure 6.2).

Phase We may restrict the phase spectrum to values in the *principal range* $(-\pi, \pi)$. If the phase falls outside this range, we include phase jumps of $\pm 2n\pi$ to bring the phase within this range (Figure 6.2).

Real, Even Signals For even signals, $X(f)$ is always real. It is much more convenient to plot $X(f)$ as the **amplitude spectrum** and assume zero phase.

  If we do plot $|X(f)|$, the phase will be zero if $X(f) \geq 0$ and $\pm\pi$ if $X(f) < 0$. The phase spectrum accounts for sign changes in $X(f)$ as phase jumps of $\pm\pi$. The sign is chosen to bring the phase within the principal range.

Real, Odd Signals For odd signals, $X(f) = jA(f)$ is purely imaginary. We may thus plot the amplitude $A(f)$ and a phase of $\frac{1}{2}\pi$ for $f > 0$ and $-\frac{1}{2}\pi$ for $f < 0$. If we do plot $|X(f)| = |A(f)|$, the phase will equal $\frac{1}{2}\pi$ if $A(f) > 0$ and $\frac{1}{2}\pi \pm \pi$ if $A(f) < 0$ to account for sign changes in $A(f)$.

Figure 6.2 Various ways of plotting magnitude and phase spectra.

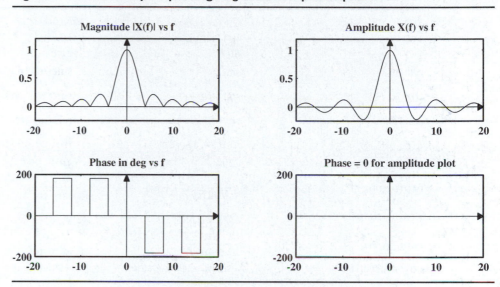

Phase Unwrapping   We may also *unwrap* the phase and plot it as a monotonic function. No jumps of 2π are included. The sign of jumps by $\pm\pi$ due to sign changes is selected to ensure that the phase spectrum is monotonic.

6.1.9   *A Catalog of Fourier Transform Pairs*

Table 6.4 (p. 256) and Figure 6.3 (pp. 257–258) illustrate the Fourier transforms of some useful signals. We remark that the transform of many power signals or signals that are not absolutely integrable (such as the constant, step and ramp) almost invariably includes impulses and/or their derivatives, whereas the transform of signals that grow exponentially or faster does not exist. The reason for this lies in the nature of convergence of the Fourier transform, which we discuss toward the end of this chapter. We illustrate the computation of Fourier transforms using the defining relation by some examples.

Example 6.3
Some Transform Pairs
from the Definition

(a) *The unit impulse*: The Fourier transform of $x(t) = \delta(t)$ is found using the sifting property of impulses to give

$$X(f) = \int_{-\infty}^{\infty} \delta(t)\exp(-j2\pi ft)\,dt = 1$$

The spectrum of an impulse is a constant for all frequencies (Figure 6.3a).

(b) *The decaying exponential*: The Fourier transform of $x(t) = \exp(-\alpha t)u(t)$ is

$$X(f) = \int_{0}^{\infty} \exp(-\alpha t)\exp(-j2\pi ft)\,dt = \int_{0}^{\infty} \exp[-(\alpha + j2\pi f)t]\,dt = \frac{1}{\alpha + j2\pi f}$$

Note how the magnitude decays for higher frequencies (Figure 6.3i).

Table 6.4   Some Useful Fourier Transform Pairs.

$x(t)$	$X(f)$	$X(\omega)$
$\delta(t)$	1	1
1	$\delta(f)$	$2\pi\delta(\omega)$
$\text{rect}(t)$	$\text{sinc}(f)$	$\text{sinc}(\omega/2\pi)$
$\text{tri}(t)$	$\text{sinc}^2(f)$	$\text{sinc}^2(\omega/2\pi)$
$\text{sinc}(t)$	$\text{rect}(f)$	$\text{rect}(\omega/2\pi)$
$u(t)$	$\frac{1}{2}\delta(f) + 1/(j2\pi f)$	$\pi\delta(\omega) + 1/j\omega$
$\delta'(t)$	$j2\pi f$	$j\omega$
$\exp(-\alpha t)u(t)$	$\dfrac{1}{\alpha + j2\pi f}$	$\dfrac{1}{\alpha + j\omega}$
$t\exp(-\alpha t)u(t)$	$\dfrac{1}{(\alpha + j2\pi f)^2}$	$\dfrac{1}{(\alpha + j\omega)^2}$
$\exp(-\alpha|t|)$	$\dfrac{2\alpha}{a^2 + 4\pi^2 f^2}$	$\dfrac{2\alpha}{\alpha^2 + \omega^2}$
$\cos(\pi t)\text{rect}(t)$	$\frac{1}{2}\text{sinc}(f + \frac{1}{2}) + \frac{1}{2}\text{sinc}(f - \frac{1}{2})$	$\frac{1}{2}\text{sinc}[(\omega + \pi)/2\pi] + \frac{1}{2}\text{sinc}[(\omega - \pi)/2\pi]$
$\exp(-\pi t^2)$	$\exp(-\pi f^2)$	$\exp\left(\dfrac{-\omega^2}{4\pi}\right)$
$\exp(-j2\pi\alpha t)$	$\delta(f + \alpha)$	$2\pi\delta(\omega + 2\pi\alpha)$
$\cos(2\pi\alpha t)$	$\frac{1}{2}[\delta(f + \alpha) + \delta(f - \alpha)]$	$\pi[\delta(\omega + 2\pi\alpha) + \delta(\omega - 2\pi\alpha)]$
$\sin(2\pi\alpha t)$	$\frac{1}{2}j[\delta(f + \alpha) - \delta(f - \alpha)]$	$j\pi[\delta(\omega + 2\pi\alpha) - \delta(\omega - 2\pi\alpha)]$
$\text{sgn}(t)$	$1/j\pi f$	$2/j\omega$
$\exp(-\alpha t)\cos(2\pi\beta t)u(t)$	$\dfrac{\alpha + j2\pi f}{(\alpha + j2\pi f)^2 + (2\pi\beta)^2}$	$\dfrac{\alpha + j\omega}{(\alpha + j\omega)^2 + (2\pi\beta)^2}$
$\exp(-\alpha t)\sin(2\pi\beta t)u(t)$	$\dfrac{2\pi\beta}{(\alpha + j2\pi f)^2 + (2\pi\beta)^2}$	$\dfrac{2\pi\beta}{(\alpha + j\omega)^2 + (2\pi\beta)^2}$
$tu(t)$	$\dfrac{j}{4\pi}\delta'(f) - \dfrac{1}{4\pi^2 f^2}$	$j\pi\delta'(\omega) - \dfrac{1}{\omega^2}$
$|t|$	$-\dfrac{1}{2\pi^2 f^2}$	$-\dfrac{2}{\omega^2}$
$\text{comb}(t) = \sum\delta(t - n)$	$\text{comb}(f) = \sum\delta(f - k)$	$(2\pi)\sum\delta(\omega - 2\pi k)$
$\sum\delta(t - nT)$	$\dfrac{1}{T}\sum\delta\left(f - \dfrac{k}{T}\right)$	$\dfrac{2\pi}{T}\sum\delta\left(\omega - \dfrac{2\pi k}{T}\right)$
$x_p(t) = \sum X_S[k]\exp(jk2\pi f_0 t)$	$\sum X_S[k]\delta(f - kf_0)$	$\sum 2\pi X_S[k]\delta(\omega - k\omega_0)$
$\dfrac{2}{1 + (2\pi t)^2}$	$\exp(-|f|)$	$\exp(-|\omega|/2\pi)$
$\dfrac{1}{\sqrt{|t|}}$	$\dfrac{1}{\sqrt{|f|}}$	$\sqrt{\dfrac{2\pi}{|\omega|}}$
$\cos[\pi(t^2 - \frac{1}{8})]$	$\cos[\pi(f^2 - \frac{1}{8})]$	$\cos\left(\dfrac{\omega^2 - \frac{1}{2}\pi^2}{4\pi}\right)$
$\text{sech}(\pi t)$	$\text{sech}(\pi f)$	$\text{sech}(\frac{1}{2}\omega)$
t^n	$\delta^{(n)}(f)/(-j2\pi)^n$	$2\pi(j)^n\delta^{(n)}(\omega)$
$\exp(\alpha t)u(t),\ \alpha > 0$	Not defined	Not defined
$\exp(\alpha t),\ \alpha > 0$	Not defined	Not defined
$\exp(\alpha t^2),\ \alpha > 0$	Not defined	Not defined

Figure 6.3   Spectra of some of the signals of Table 6.4.

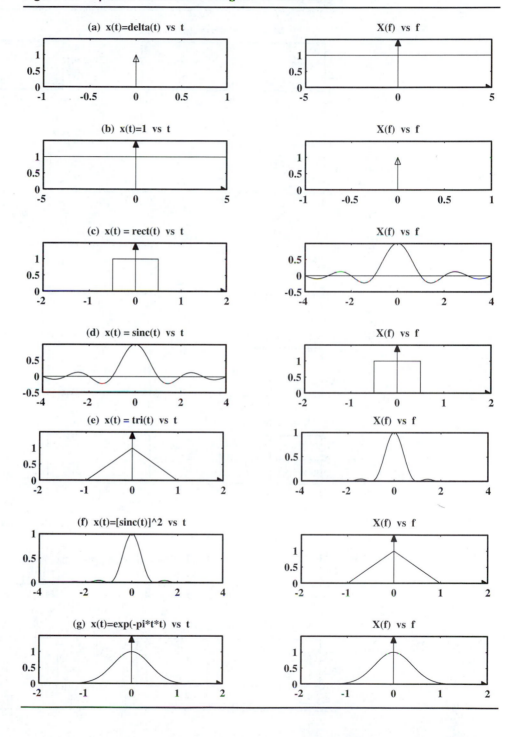

Figure 6.3   (continued)

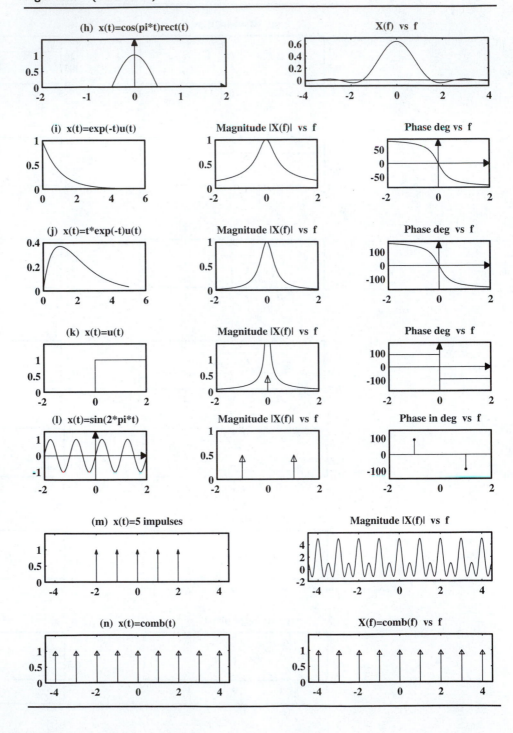

(c) *The rect function*: The signal $x(t) = \text{rect}(t)$ is unity over $(-\frac{1}{2}, \frac{1}{2})$ and zero else-where. We find $X(f)$ by writing the complex exponential term in Euler's form, and invoking symmetry, to give

$$X(f) = \int_{-1/2}^{1/2} \exp(-j2\pi ft)\,dt = \int_{-1/2}^{1/2} \cos(2\pi ft)\,dt$$

$$= \frac{\sin(2\pi ft)}{2\pi ft}\bigg|_{-1/2}^{1/2} = \frac{\sin(\pi f)}{\pi f} = \text{sinc}(f)$$

Note: Its amplitude spectrum is shown in Figure 6.3c. If we sketch the magnitude $|\text{sinc}(f)|$, the phase will equal $180°$ whenever $\text{sinc}(f) < 0$ (as in Figure 6.2). Note how the sidelobes decay as $1/f$.

Example 6.4
Fourier Transform of a
Cosine Pulse

Consider the half-cycle cosine $x(t) = \cos(\pi t)$, $-\frac{1}{2} \le t \le \frac{1}{2}$. We obtain $X(f)$ as

$$X(f) = \int_{-1/2}^{1/2} \cos(\pi t)\exp(-j2\pi ft)\,dt$$

$$= \int_{-1/2}^{1/2} \cos(\pi t)\cos(2\pi ft) - j\cos(\pi t)\sin(2\pi ft)\,dt$$

The second term is the integral of an odd function between symmetric limits and equals zero. The first term is even, and we invoke symmetry to get

$$X(f) = 2\int_{0}^{1/2} \cos(\pi t)\cos(2\pi ft)\,dt$$

$$= \int_{0}^{1/2} \cos\left[2\pi t(f + \frac{1}{2})\right]dt + \int_{0}^{1/2} \cos\left[2\pi t(f - \frac{1}{2})\right]dt$$

Upon integration, and substitution of limits, we obtain

$$X(f) = \frac{\sin[2\pi t(f + \frac{1}{2})]}{2\pi(f + \frac{1}{2})} + \frac{\sin[2\pi t(f - \frac{1}{2})]}{2\pi(f - \frac{1}{2})}\bigg|_{0}^{1/2}$$

$$= \frac{\sin[\pi(f + \frac{1}{2})]}{2\pi(f + \frac{1}{2})} + \frac{\sin[\pi(f - \frac{1}{2})]}{2\pi(f - \frac{1}{2})}$$

This may be expressed in terms of sinc functions, or in composite form, as

$$X(f) = \frac{1}{2}\text{sinc}\left(f + \frac{1}{2}\right) + \frac{1}{2}\text{sinc}\left(f - \frac{1}{2}\right) = \frac{2\cos(\pi f)}{\pi(1 - 4f^2)}$$

The signal and its amplitude spectrum are sketched in Figure 6.3h.

Note: The sinc form reveals a spectrum with $X(0) = \frac{1}{2}\text{sinc}(\frac{1}{2}) + \frac{1}{2}\text{sinc}(-\frac{1}{2}) = 2/\pi$ and $X(\frac{1}{2}) = \frac{1}{2}\text{sinc}(1) + \frac{1}{2}\text{sinc}(0) = 0 + \frac{1}{2} = \frac{1}{2}$. The cosine form reveals that $X(f)$ decays as $1/f^2$ and equals zero whenever $\cos(\pi f) = 0$ (for $|f| = \frac{1}{2}, \frac{3}{2}, \ldots$), suggesting a spectrum with sidelobes. In this form, $X(f)$ is indeterminate at $f = \frac{1}{2}$, but l'Hôpital's rule gives

$$X\left(\frac{1}{2}\right) = \lim_{f \to 1/2} \frac{-2\pi \sin(\pi f)}{-8\pi f} = \frac{1}{2}$$

as before. Note also that the width of the central lobe is *three* times the width of the sidelobes (unlike a sinc function, for which it is only twice).   ▬▬

6.2
Parseval's Relation for Energy Signals

The analogy with periodic functions is not yet complete. Since the Fourier series deals with power signals and the Fourier transform with energy signals, how do we find the energy in $x(t)$ from a knowledge of its spectrum $X(f)$, as we have done for the power in periodic signals? For periodic signals, we use the Parseval's relation

$$P_x = \frac{1}{T} \int_{-T/2}^{T/2} x_p^2(t)\, dt = \sum_{k=-\infty}^{\infty} |X_S[k]|^2$$

Since no averaging is required to find the energy in $x(t)$, and since $x(t)$ corresponds to one period of $x_p(t)$ as $T \longrightarrow \infty$, we have

$$E_x = \int_{-\infty}^{\infty} x^2(t)\, dt = \lim_{T \to \infty} \int_{-T/2}^{T/2} x_p^2(t)\, dt = \lim_{T \to \infty} T \sum_{k=-\infty}^{\infty} |X_S[k]|^2$$

As $T \longrightarrow \infty, f_0 \longrightarrow df$ and $|TX_S[k]| \longrightarrow |X(f)|$. We then obtain

$$E_x = \lim_{T \to \infty} \sum_{k=-\infty}^{\infty} |TX_S[k]|^2 f_0 = \int_{-\infty}^{\infty} |X(f)|^2\, df$$

Since $|X(f)|^2$ is always even, we may also write

$$E_x = \int_{-\infty}^{\infty} x^2(t)\, dt = \int_{-\infty}^{\infty} |X(f)|^2\, df = 2 \int_{0}^{\infty} |X(f)|^2\, df$$

Once again, if we use $\omega = 2\pi f$ instead of f, we obtain

$$E_x = \int_{-\infty}^{\infty} x^2(t)\, dt = \frac{1}{2\pi} \int_{-\infty}^{\infty} |X(\omega)|^2\, d\omega = \frac{1}{\pi} \int_{0}^{\infty} |X(\omega)|^2\, d\omega$$

This result is the counterpart of Parseval's relation for energy signals and is also called **Rayleigh's theorem**. It says that the energy in a signal may be found either from its time-domain representation $x(t)$ or from its frequency-domain representation in terms of the *magnitude spectrum* $|X(f)|$. The area under $x^2(t)$ or $|X(f)|^2$ equals the total signal energy. The energy in $x(t)$ over a frequency range $-f_1 \leq f \leq f_1$ or $-\omega_1 \leq \omega \leq \omega_1$ equals

$$E_1 = \int_{-f_1}^{f_1} |X(f)|^2\, df = 2 \int_{0}^{f_1} |X(f)|^2\, df = \frac{1}{\pi} \int_{0}^{\omega_0} |X(\omega)|^2\, d\omega$$

Remarks: 1. We stress that Parseval's relation applies only to energy signals for which $\mathbb{A}[x^2(t)]$ or $\mathbb{A}[|X(f)|^2]$ is finite.

2. An extension of Parseval's relation, called **Plancherel's relation**, applies to the product of two energy signals and reads

$$E_x = \int_{-\infty}^{\infty} x_1(t)x_2^*(t)\,dt = \int_{-\infty}^{\infty} X_1(f)X_2^*(f)\,df \quad \text{or} \quad \frac{1}{2\pi}\int_{-\infty}^{\infty} X_1(\omega)X_2^*(\omega)\,d\omega$$

Example 6.5

An Application of Parseval's Theorem

(a) ω-form: Consider $x(t) = \exp(-\alpha t)u(t)$. Its energy is $\mathbb{A}[x^2(t)] = 1/2\alpha$. To find the energy E_B in the band $(-\omega_B,\ \omega_B)$, we compute

$$|X(\omega)| = |1/(\alpha + j\omega)| = 1/(\alpha^2 + \omega^2)^{1/2}$$

$$E_B = \frac{1}{\pi}\int_0^{\omega_B} \frac{1}{\alpha^2 + \omega^2}\,d\omega = \left.\frac{\tan^{-1}(\omega/\alpha)}{\alpha\pi}\right|_0^{\omega_B} = \frac{\tan^{-1}(\omega_B/\alpha)}{\alpha\pi}$$

We can use this result in several ways:

1. To find total energy: With $\omega_B = \infty$, $E_x = \tan^{-1}(\infty)/\alpha\pi = \frac{1}{2}\pi/\alpha\pi = \frac{1}{2\alpha}$.
2. To find the energy E_B in a given frequency range: For example, the energy in the range $|\omega| \le \frac{1}{2}\alpha$ equals $E_B = (1/\pi\alpha)\tan^{-1}(\frac{1}{2})$.
3. To find the range $|\omega| \le \omega_B$ that contains a specified fraction of E_x: For what value of ω_B is 50% of the total energy contained in the range $|\omega| \le \omega_B$? We require $E_B = \frac{1}{2}E_x = 1/4\alpha$. This yields $1/4\alpha = (1/\alpha\pi)\tan^{-1}(\omega_B/\alpha)$. Solving for ω_B, we obtain $\omega_B = \alpha\tan(\pi/4) = \alpha$.

Comment: For the signal $\exp(-\alpha t)u(t)$, exactly half the total energy is contained in the frequency range $|\omega| \le \alpha$!

(b) What fraction of the total energy is contained in the central lobe of the spectrum of $x(t) = \text{rect}(t)$? The total energy in $x(t)$ is simply $E_x = \mathbb{A}[x^2(t)] = 1$.
Since $X(f) = \text{sinc}(f)$, whose central lobe extends over $|f| \le 1$, we seek

$$E_B = \int_{-1}^{1} \text{sinc}^2(f)\,df = 2\int_0^1 \text{sinc}^2(f)\,df$$

This integral can only be evaluated numerically and yields $E_B = 0.9028$.

Comment: About 90% of the energy is concentrated in the central lobe. This is true of any rectangular pulse with arbitrary height and width.

The operational properties of the Fourier transform are summarized in Figure 6.4 and Table 6.5 (p. 264). Many are analogous to their Fourier series counterparts.

1. For transform pairs, conversion from $X(f)$ to $X(\omega)$ is straightforward. We simply use the relation $\omega = 2\pi f$. For example:

$$\exp(-\alpha t)u(t) \longleftrightarrow \frac{1}{\alpha + j2\pi f} \quad \text{or} \quad \exp(-\alpha t)u(t) \longleftrightarrow \frac{1}{\alpha + j\omega}$$

$$\text{rect}(t) \longleftrightarrow \text{sinc}(f) \quad \text{or} \quad \text{rect}(t) \longleftrightarrow \text{sinc}(\omega/2\pi)$$

2. For properties, the conversion is not so obvious. In many, we simply require a notational change from f to ω, but in others, we must include (or omit) factors of 2π. Our suggestion: *Use one form consistently.* As mentioned earlier, we show examples of both, but prefer the f-form.

6.3.1   Superposition and Amplitude Scaling

Superposition of time signals yields a superposition of their transforms. Amplitude scaling by α scales both $x(t)$ and $X(\omega)$ by α (Figure 6.4b). These results follow from the linear nature of the defining relations.

6.3.2   Duality, or Symmetry or Similarity

The operations embodied in both the direct and inverse transforms involve finding the area under the product of a function and a complex exponential:

$$X(f) = \int_{-\infty}^{\infty} x(t)\exp(-j2\pi ft)\,dt \qquad x(t) = \int_{-\infty}^{\infty} X(f)\exp(j2\pi ft)\,df$$

If we reverse the roles of f and t and replace t by f (or f by t), we get

$$X(t) = \int_{-\infty}^{\infty} x(f)\exp(-j2\pi ft)\,df \qquad x(f) = \int_{-\infty}^{\infty} X(t)\exp(j2\pi ft)\,dt$$

In order to maintain the same form in the kernel, if we also replace t by $-t$ in the first relation and f by $-f$ in the second relation, we obtain

$$X(-t) = \int_{-\infty}^{\infty} x(f)\exp(j2\pi ft)\,df \qquad x(-f) = \int_{-\infty}^{\infty} X(t)\exp(-j2\pi ft)\,dt$$

These expressions are identical in form to the original forms for $X(f)$ and $x(t)$ and suggest that if $x(t)$ yields the transform $X(f)$, then a time signal $X(t)$ with the same shape as $X(f)$ results in a transform $x(-f)$ with the shape as $x(t)$, but folded. This also suggests the following mathematical form for the symmetry property, or *similarity theorem:*

$$\text{If} \quad x(t) \longleftrightarrow X(f) \quad \text{or} \quad X(\omega), \qquad \text{then} \quad X(t) \longleftrightarrow x(-f) \quad \text{or} \quad 2\pi x(-\omega)$$

The symmetry property clearly reveals the duality between the two domains. The change $t \longrightarrow -f$ accounts for the sign reversal in the exponential in the direct and inverse transforms. For even functions, we simply use $t \longrightarrow f$.

Figure 6.4 How operations on a time signal affect its magnitude and phase spectra based on the operational properties of the Fourier transform.

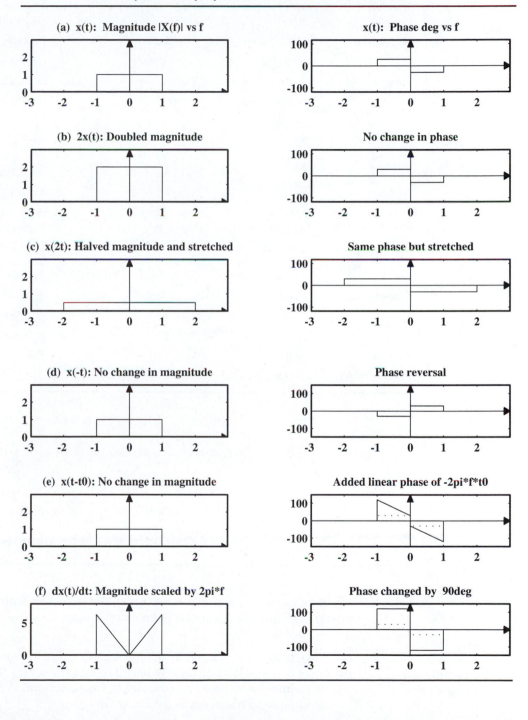

(a) x(t): Magnitude |X(f)| vs f

x(t): Phase deg vs f

(b) 2x(t): Doubled magnitude

No change in phase

(c) x(2t): Halved magnitude and stretched

Same phase but stretched

(d) x(-t): No change in magnitude

Phase reversal

(e) x(t-t0): No change in magnitude

Added linear phase of -2pi*f*t0

(f) dx(t)/dt: Magnitude scaled by 2pi*f

Phase changed by 90deg

Table 6.5   Summary of Operational Properties of the Fourier Transform.

Transform	$x(t)$	$X(f)$	$X(\omega)$
Symmetry	$X(t)$	$x(-f)$	$2\pi x(-\omega)$
Derivatives	$x'(t)$	$j2\pi f X(f)$	$j\omega X(\omega)$
	$x^{(n)}(t)$	$(j2\pi f)^n X(f)$	$(j\omega)^n X(\omega)$
Times-t	$-j2\pi t x(t)$	$X'(f)$	$2\pi X(\omega)$
	$(-j2\pi t)^n x(t)$	$X^{(n)}(f)$	$(2\pi)^n X^{(n)}(\omega)$
Integral	$\displaystyle\int_{-\infty}^{t} x(t)\,dt$	$\dfrac{1}{j2\pi f} X(f) + \frac{1}{2} X(0)\delta(f)$	$\dfrac{1}{j\omega} X(\omega) + \pi X(0)\delta(\omega)$
Scaling	$x(\alpha t)$	$\dfrac{1}{\lvert\alpha\rvert} X\!\left(\dfrac{f}{\alpha}\right)$	$\dfrac{1}{\lvert\alpha\rvert} X\!\left(\dfrac{\omega}{\alpha}\right)$
Folding	$x(-t)$	$X(-f)$	$X(-\omega)$
Shifting	$x(t-\alpha)$	$\exp(-j2\pi f\alpha)X(f)$	$\exp(-j\omega\alpha)X(\omega)$
	$\exp(j2\pi\alpha t)x(t)$	$X(f-\alpha)$	$X(\omega-2\pi\alpha)$
Modulation	$x(t)\cos(2\pi\alpha t)$	$\frac{1}{2}[X(f+\alpha)+X(f-\alpha)]$	$\frac{1}{2}[X(\omega+2\pi\alpha)+X(\omega-2\pi\alpha)]$
	$x(t)\sin(2\pi\alpha t)$	$\frac{1}{2}j[X(f+\alpha)-X(f-\alpha)]$	$\frac{1}{2}j[X(\omega+2\pi\alpha)-X(\omega-2\pi\alpha)]$
Convolution	$x_1(t) \star x_2(t)$	$X_1(f)X_2(f)$	$X_1(\omega)X_2(\omega)$
	$x_1(t)x_2(t)$	$X_1(f) \star X_2(f)$	$\dfrac{1}{2\pi}[X_1(\omega) \star X_2(\omega)]$
Conjugation	$x^*(t)$	$X^*(-f)$	$x^*(-\omega)$
Correlation	$x_1(t) \star\star x_2(t)$	$X_1(f)X_2^*(f)$	$X_1(\omega)X_2^*(\omega)$
Autocorrelation	$x(t) \star\star x(t)$	$X(f)X^*(f) = \lvert X(f)\rvert^2$	$X(\omega)X^*(\omega) = \lvert X(\omega)\rvert^2$

Initial value theorem	$x(0^+) = \displaystyle\lim_{\omega\to\infty}[j\omega X(\omega)]$
Central ordinates	$x(0) = \displaystyle\int_{-\infty}^{\infty} X(f)\,df = \dfrac{1}{2\pi}\int_{-\infty}^{\infty} X(\omega)\,d\omega \qquad X(0) = \int_{-\infty}^{\infty} x(t)\,dt$
Parseval's theorem	$\displaystyle\int_{-\infty}^{\infty} x^2(t)\,dt = \int_{-\infty}^{\infty} \lvert X(f)\rvert^2\,df = 2\int_{0}^{\infty} \lvert X(f)\rvert^2\,df$
	$\displaystyle\int_{-\infty}^{\infty} x^2(t)\,dt = \dfrac{1}{2\pi}\int_{-\infty}^{\infty} \lvert X(\omega)\rvert^2\,d\omega = \dfrac{1}{\pi}\int_{0}^{\infty} \lvert X(\omega)\rvert^2\,d\omega$
Plancherel's theorem	$\displaystyle\int_{-\infty}^{\infty} x(t)y^*(t)\,dt = \int_{-\infty}^{\infty} X(f)Y^*(f)\,df = \dfrac{1}{2\pi}\int_{-\infty}^{\infty} X(\omega)Y^*(\omega)\,d\omega$

Remark: When using the ω-form $X(\omega)$ (not $X(f)$), the factor 2π with $x(-\omega)$ (or $1/2\pi$ with $x(-t)$) also appears due to the defining relations.

Example 6.6

Using Duality

(a) Since $\mathrm{rect}(t) \longleftrightarrow \mathrm{sinc}(f)$, by duality $\mathrm{sinc}(t) \longleftrightarrow \mathrm{rect}(-f) = \mathrm{rect}(f)$.

(b) Since $\delta(t) \longleftrightarrow 1$, by duality we have $1 \longleftrightarrow \delta(-f) = \delta(f)$. The Fourier transform of an impulse in time is a constant in frequency and the Fourier transform of an impulse in frequency is a constant in time.

Comment: *It is worth memorizing these two Fourier transform pairs.*

(c) With $\exp(-\alpha t)u(t) \longleftrightarrow 1/(\alpha + j2\pi f)$, the duality property suggests that

$$1/(\alpha - j2\pi t) \longleftrightarrow \exp(-\alpha f)u(f)$$

Comment: Real signals have an even magnitude spectrum. The signal $1/(\alpha - j2\pi t)$ is complex precisely because the magnitude spectrum of $\exp(-\alpha f)u(f)$ is not even. ____

6.3.3   *Time Scaling and Frequency Scaling*

A scaling by α in one domain results in a scaling by $1/\alpha$ in the other. This follows from a change of variable. For $\alpha > 0$, for example, we have

$$\int_{-\infty}^{\infty} x(\alpha t)\exp(-j2\pi ft)\,dt \overset{\alpha t=\lambda}{\longrightarrow} \frac{1}{\alpha}\int_{-\infty}^{\infty} x(\lambda)\exp(-j2\pi f\lambda/a)\,d\lambda = \frac{1}{\alpha}X(f/\alpha)$$

Compression in one domain leads to a stretching and an amplitude reduction by $|\alpha|$ in the other, as illustrated in Figure 6.4c.

Remark: Unlike the analogous Fourier series property, the amplitude scaling factor $1/|\alpha|$ ensures that both the scaled signal and the scaled spectrum possess the same energy and thus satisfy Parseval's relation.

6.3.4   *Reversal or Folding*

A time-scale factor $\alpha = -1$ in the scaling property suggests that if $x(t)$ is folded, so is its transform. Thus

$$x(-t) \longleftrightarrow X(-f) = X^*(f) \quad \text{or} \quad X(-\omega) = X^*(\omega)$$

The magnitude spectrum remains unchanged, but the phase shows a sign reversal (Figure 6.4d). This property also suggests that

1. The transform of the *even extension* $x(t) + x(-t)$ equals $2\text{Re}[X(f)]$.
2. The transform of the *odd extension* $x(t) - x(-t)$ equals $2j\text{Im}[X(f)]$.
3. Clearly, the transform of the even part of $x(t)$ equals just $\text{Re}[X(f)]$.

Example 6.7
Using the Scaling Property

(a) The pedestal function $x(t)$ of Figure 6.5 may be described in terms of rect functions as $x(t) = \text{rect}(t/2) + 2\,\text{rect}(2t)$. Using the transform pair $\text{rect}(t) \longleftrightarrow \text{sinc}(f)$ and invoking superposition and scaling, we have

$$X(f) = (2)\text{sinc}(2f) + (2)(\tfrac{1}{2})\text{sinc}(\tfrac{1}{2}f) = 2\,\text{sinc}(2f) + \text{sinc}(\tfrac{1}{2}f)$$

(b) Consider the transform pair $\text{sinc}(t) \longleftrightarrow \text{rect}(f)$. Time scaling by α gives $\text{sinc}(\alpha t) \longleftrightarrow |\tfrac{1}{\alpha}|\text{rect}(f/\alpha)$. The smaller the value of α, the broader the sinc function becomes and the narrower the rect function. As $\alpha \longrightarrow 0$, the sinc function approaches a constant equal to unity, and $|\tfrac{1}{\alpha}|\text{rect}(f/\alpha)$ approaches the impulse

Figure 6.5   The signal $x(t)$ for Example 6.7.

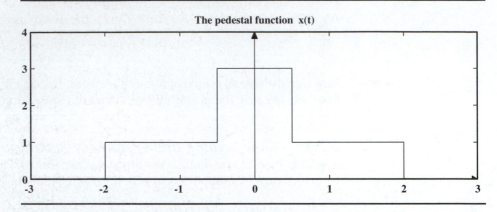

The pedestal function x(t)

$\delta(f)$. We then have the limiting result $1 \longleftrightarrow \delta(f)$. This result was also obtained previously using the duality property.

(c) ω-form: Consider the FT pair $x(t) = \exp(-t)u(t) \longleftrightarrow 1/1 + (j\omega)$ Scaling $x(t)$ to $x(\alpha t)$ gives

$$\exp(-\alpha t)u(\alpha t) = \exp(-\alpha t)u(t) \longleftrightarrow \left|\frac{1}{\alpha}\right| \frac{1}{1 + j(\omega/\alpha)} = \frac{1}{\alpha + j\omega}$$

Example 6.8
Using the Folding Property

(a) ω-form: Since $\exp(-\alpha t)u(t) \longleftrightarrow \dfrac{1}{\alpha + j\omega}$, we have $\exp(\alpha t)u(-t) \longleftrightarrow \dfrac{1}{\alpha - j\omega}$.

(b) ω-form: Since $\exp(-\alpha|t|)$ is the even extension of $\exp(-\alpha t)u(t)$,

$$\exp(-\alpha|t|) \longleftrightarrow 2\mathrm{Re}\left[\frac{1}{\alpha + j\omega}\right] = \frac{2\alpha}{\alpha^2 + \omega^2}$$

(c) ω-form: We may find the transform of $x(t) = \exp(-\alpha t)u(t) - \exp(\alpha t)u(-t)$ either from $2j\mathrm{Im}[1/(\alpha + j\omega)]$ or by using superposition as

$$\exp(-\alpha t)u(t) - \exp(\alpha t)u(-t) \longleftrightarrow \frac{1}{\alpha + j\omega} - \frac{1}{\alpha - j\omega} = \frac{-2j\omega}{\alpha^2 + \omega^2}$$

(d) ω-form: As $\alpha \longrightarrow 0, x(t) = \exp(-\alpha t)u(t) - \exp(\alpha t)u(-t) \longrightarrow \mathrm{sgn}(t)$ and

$$\mathrm{sgn}(t) \longleftrightarrow \lim_{\alpha \to 0}\left[\frac{-2j\omega}{\alpha^2 + \omega^2}\right] = \frac{2}{j\omega}$$

(e) The FT of $u(t)$: Since $u(t) = \frac{1}{2} + \frac{1}{2}\mathrm{sgn}(t)$, superposition yields

$$u(t) \longleftrightarrow \pi\delta(\omega) + 1/j\omega \quad \text{or} \quad u(t) \longleftrightarrow \tfrac{1}{2}\delta(f) + 1/j2\pi f$$

Comment: Since $\exp(-\alpha t)u(t) \longleftrightarrow 1/(\alpha + j\omega)$ and $\exp(-\alpha t)u(t) \longrightarrow u(t)$ as $\alpha \longrightarrow 0$, why does the FT of $u(t)$ not equal $1/j\omega$? This would imply that $u(t)$ is odd. The problem is that for complex quantities, all operations including limits must be invoked on the real and imaginary parts separately. If we do that, we get

$$\lim_{\alpha \to 0} \frac{1}{\alpha + j\omega} = \lim_{\alpha \to 0}\left[\frac{\alpha}{\alpha^2 + \omega^2}\right] + \lim_{\alpha \to 0}\left[\frac{-j\omega}{\alpha^2 + \omega^2}\right] = \lim_{\alpha \to 0}\left[\frac{\alpha}{\alpha^2 + \omega^2}\right] + \frac{1}{j\omega}$$

From part (b), the term in the limit corresponds to the FT of $\frac{1}{2}\exp(-\alpha|t|)$. As $\alpha \longrightarrow 0$, $\frac{1}{2}\exp(-\alpha|t|) \longrightarrow \frac{1}{2}$, whose FT is $\pi\delta(\omega)$.

Comment: The first limit may also be recognized as the limiting form of a Lorentzian signal (Section 2.8.1), which is an impulse.

6.3.5   *Time Shift*

A time shift of $x(t)$ to $x(t - \alpha)$ produces a linear change in the phase spectrum.

$$x(t - \alpha) \longleftrightarrow X(f)\exp(-j2\pi\alpha f)$$

To establish the FT of $x(t - \alpha)$, we let $\lambda = t - \alpha$ to give

$$\int_{-\infty}^{\infty} x(t - \alpha)\exp(-j2\pi ft)\,dt = \int_{-\infty}^{\infty} x(\lambda)\exp[-j2\pi f(\lambda + a)]\,d\lambda = \exp(-j2\pi fa)X(f)$$

The *phase difference* equals $-2\pi\alpha f$ and varies linearly with f (Figure 6.4e). The magnitude spectrum shows no change.

Remark: The quantity $\exp(-j2\pi f)$ may be regarded as a *unit shift operator* that shifts or delays the time function by 1 unit. Similarly, $\exp(-j2\pi f\alpha)$ contributes a delay of α units in the time signal.

Example 6.9
Using the Time-Shift Property

(a) The transform pair $\delta(t) \longleftrightarrow 1$ immediately suggests that

$$\delta(t - a) \longleftrightarrow \exp(-j2\pi fa)$$

Its magnitude spectrum is 1 and its phase equals $-2\pi fa$.

(b) Consider the pair $\mathrm{rect}(t) \longleftrightarrow \mathrm{sinc}(f)$. Shifting $\mathrm{rect}(t)$ to $\mathrm{rect}(t - \frac{1}{2})$ (which is a pulse between 0 and 1) results in the transform

$$\mathrm{rect}(t - \tfrac{1}{2}) \longleftrightarrow \mathrm{sinc}(f)\exp(-j\pi f)$$

Its amplitude spectrum is $\mathrm{sinc}(f)$ and its phase equals $-\pi f$.

(c) The even extension $x(t)$ and odd extension $y(t)$ of $\mathrm{rect}(t - \frac{1}{2})$ (Figure 6.6) have the transform

$$x(t) \longleftrightarrow 2\mathrm{Re}[\mathrm{sinc}(f)\exp(-j\pi f)] = 2\mathrm{sinc}(f)[\cos(\pi f)] = 2\mathrm{sinc}(2f)$$
$$y(t) \longleftrightarrow 2j\mathrm{Im}[\mathrm{sinc}(f)\exp(-j\pi f)] = 2j\mathrm{sinc}(f)[-\sin(\pi f)] = -j2\pi f\mathrm{sinc}^2(f)$$

Figure 6.6   The signal rect $(t - \frac{1}{2})$ and its odd and even extensions for Example 6.9.

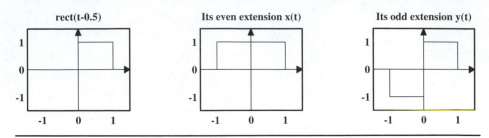

Comment: The result for $x(t)$ also follows by scaling sinc(t) to sinc$(\frac{1}{2}t)$.

(d) Consider the FT pair $\sin(2\pi\alpha t) \longleftrightarrow \frac{1}{2}j[\delta(f + \alpha) - \delta(f - \alpha)]$. We can find the transform of $\cos(2\pi\alpha t)$ by using the time-shift property as follows: Since $\cos(2\pi\alpha t) = \sin(2\pi\alpha t + \frac{1}{2}\pi) = \sin[2\pi\alpha(t + 1/4\alpha)]$, $t \longrightarrow t + 1/4\alpha$ and

$$\cos(2\pi\alpha t) \longleftrightarrow \frac{1}{2}j[\delta(f + \alpha) - \delta(f - \alpha)]\exp\left(\frac{j2\pi f}{4\alpha}\right)$$

Using the product property of impulses $f(x)\delta(x - \alpha) = f(\alpha)\delta(x - \alpha)$, we get

$$\cos(2\pi\alpha t) \longleftrightarrow \frac{1}{2}j[\delta(f + \alpha)]\exp(-j\tfrac{1}{2}\pi) - \frac{1}{2}j[\delta(f - \alpha)]\exp(j\tfrac{1}{2}\pi)$$

And, finally, since $\exp(\pm j\frac{1}{2}\pi) = \pm j$, we get the result

$$\cos(2\pi\alpha t) \longleftrightarrow \frac{1}{2}[\delta(f + \alpha) + \delta(f - \alpha)]$$

6.3.6   *Derivatives in Time*

The derivative of $x(t)$ results in its transform $X(f)$ being multiplied by $j2\pi f$. This follows from the definition if we use integration by parts (assuming that $x(t) \longrightarrow 0$ as $|t| \longrightarrow \infty$). Using the definition of $X(f)$, we have

$$\int_{-\infty}^{\infty} x'(t)\exp(-j2\pi ft)\, dt = x(t)\exp(-j2\pi ft)\Big|_{-\infty}^{\infty}$$

$$+ j2\pi f\int_{-\infty}^{\infty} x(t)\exp(-j2\pi ft)\, dt = j2\pi f X(f)$$

The magnitude spectrum increases in direct proportion to the frequency. The phase is augmented by 90° for all frequencies (Figure 6.4f). Since differentiation enhances sharp details and features in a time function, it follows that the sharp details in a signal are responsible for the high-frequency content of its spectrum.

Successive use of the derivative property leads to the general form

$$x^{(n)}(t) \longleftrightarrow (j2\pi f)^n X(f) \quad \text{or} \quad (j\omega)^n X(\omega)$$

Example 6.10

Using the Derivative Property

(a) Consider the Fourier transform pair $\text{rect}(t/2a) \longleftrightarrow 2a\,\text{sinc}(2af)$. Since the derivative of $\text{rect}(t/2\alpha)$ is $\delta(t+a) - \delta(t-a)$ (Figure 6.7), we get

$$\delta(t+a) - \delta(t-a) \longleftrightarrow j2\pi f[2a\,\text{sinc}(2af)] = j2\sin(2\pi af)$$

The spectrum is a sine wave whose phase equals 90° for $f > 0$. The spectrum may also be sketched as a full rectified sine whose phase alternates between 90° and −90° over each half-cycle (Figure 6.7).

Figure 6.7   Illustrating the derivative property for Example 6.10.

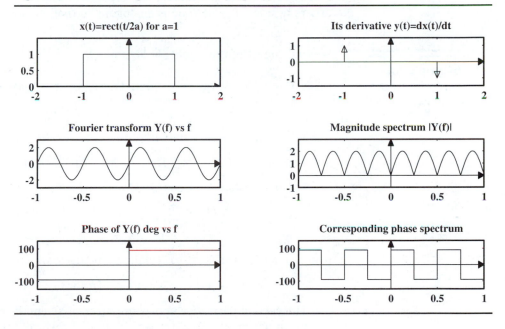

(b) We find the transform of $\sin(2\pi\alpha t)$ by using the result of part (a), and invoking the duality relation:

$$j2\sin(-2\pi\alpha t) \longleftrightarrow \delta(f+\alpha) - \delta(f-\alpha)$$
$$\sin(2\pi\alpha t) \longleftrightarrow \tfrac{1}{2}j[\delta(f+\alpha) - \delta(f-\alpha)]$$

The signal $\sin(2\alpha\pi t)$ is a sinusoid at the frequency $f = \alpha$. Its spectrum is a pair of impulses located at $f = \pm\alpha$, each with strength $\tfrac{1}{2}$. Its phase equals −90° at $f = \alpha$ and 90° at $f = -\alpha$.

(c) We find the FT of $\cos(2\pi\alpha t)$ from the result of part (b) and using the derivative property:

$$2\pi\alpha\cos(2\pi\alpha t) \longleftrightarrow (j2\pi f)(\tfrac{1}{2}j)[\delta(f+\alpha) - \delta(f-\alpha)]$$

Dividing by $2\pi\alpha$ and using the property $f(x)\delta(x - \alpha) = f(\alpha)\delta(x - \alpha)$, we get

$$\cos(2\pi\alpha t) \longleftrightarrow \tfrac{1}{2}[\delta(f + \alpha) + \delta(f - \alpha)]$$

Comment: The FT of $\cos(2\pi\alpha t)$ is a pair of impulses at $f = \pm\alpha$ with strength $\frac{1}{2}$. Its exponential Fourier series coefficients $X_S[k]$ also equal $\frac{1}{2}$ at $f = \pm\alpha$.

6.3.7   The Derivative Method of Computing Fourier Transforms

The derivative property, together with shifting and superposition, yields a powerful method of finding the transforms, not only of piecewise linear functions but many others besides. The method is analogous to that used with Fourier series.

Example 6.11
Transforms by the
Derivative Method

We find the FT of the signals of Figure 6.8.

(a) $x(t) = \text{tri}(t)$. Applying the derivative property twice, we obtain

$$x(t) \longleftrightarrow X(f) \qquad x'(t) \longleftrightarrow j2\pi f X(f) \qquad x''(t) \longleftrightarrow (j2\pi f)^2 X(f)$$

Since $x''(t)$ contains only impulses (Figure 6.8), we have

$$x''(t) \longleftrightarrow \exp(j2\pi f) - 2 + \exp(-j2\pi f) = -2 + 2\cos(2\pi f)$$

This suggests $-2 + 2\cos(2\pi f) = (j2\pi f)^2 X(f)$, and

$$X(f) = \frac{-2 + 2\cos(2\pi f)}{(j2\pi f)^2} = \frac{-2 + 2[1 - 2\sin^2(\pi f)]}{-4(\pi f)^2} = \frac{\sin^2(\pi f)}{(\pi f)^2} = \text{sinc}^2(f)$$

(b) $x(t) = \cos(\pi t), (-\frac{1}{2}, \frac{1}{2})$. Taking derivatives, we obtain

$$x(t) \longleftrightarrow X(f) \qquad x'(t) \longleftrightarrow j2\pi f X(f) \qquad x''(t) \longleftrightarrow (j2\pi f)^2 X(f)$$

The second derivative shows impulses of strength π at $\pm\frac{1}{2}$ and $-\pi^2\cos(\pi t)$. By adding $\pi^2 x(t)$ to $x''(t)$, we obtain just a pair of impulses (Figure 6.8), whose transform is

$$\pi^2 x(t) + x''(t) \longleftrightarrow \pi\exp(j2\pi f\tfrac{1}{2}) + \pi\exp(-j2\pi f\tfrac{1}{2}) = 2\pi\cos(\pi f)$$

Since $\pi^2 x(t) + x''(t) \longleftrightarrow \pi^2 X(f) + (j2\pi f)^2 X(f) = \pi^2(1 - 4f^2)X(f)$, we get

$$\pi^2(1 - 4f^2)X(f) = 2\pi\cos(\pi f) \text{ or } X(f) = \frac{2\cos(\pi f)}{\pi(1 - 4f^2)}$$

(c) $x(t) = \exp(-t), (0, 1)$. Taking its derivative, we obtain

$$x(t) \longleftrightarrow X(f) \qquad x'(t) \longleftrightarrow j2\pi f X(f)$$

Since $x'(t) = \delta(t) - \exp(-1)\delta(t - 1) - \exp(-t), (0, 1)$, the sum $x(t) + x'(t)$ rids us of the exponential part, and yields the transform

$$x(t) + x'(t) \longleftrightarrow 1 - \exp(-1)\exp(-j2\pi f)$$

Figure 6.8   The signals of Example 6.11 and their derivatives.

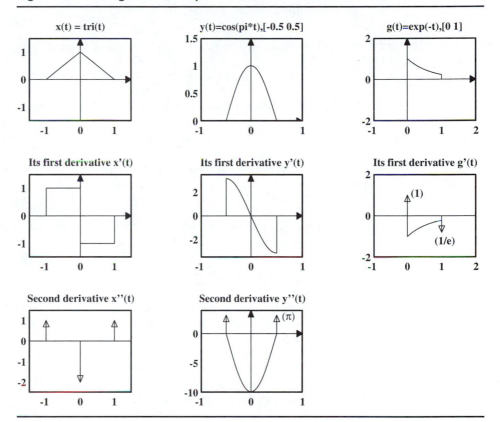

We also have $x(t) + x'(t) \longleftrightarrow X(f) + j2\pi f X(f) = (1 + j2\pi f)X(f)$, and thus

$$X(f) = \frac{1 - \exp[-(1 + j2\pi f)]}{1 + j2\pi f}$$

    —

6.3.8   *The Times-t Property: Derivatives in Frequency*

The transform pair $x'(t) \longleftrightarrow j2\pi f X(f)$ suggests the dual result

$$-j2\pi t x(t) \longleftrightarrow \frac{dX(f)}{df} \quad \text{or} \quad 2\pi \frac{dX(\omega)}{d\omega} \qquad tx(t) \longleftrightarrow \frac{j}{2\pi} \frac{dX(f)}{df} \quad \text{or} \quad j\frac{dX(\omega)}{d\omega}$$

For a formal proof, we use $t \exp(-j\omega t) = jd[\exp(-j\omega t)]/d\omega$, and interchange the order of differentiation and integration, as follows (in the ω-form):

$$\int_{-\infty}^{\infty} tx(t)e^{-j\omega t}\,dt = \int_{-\infty}^{\infty} x(t)^2\left[j\frac{de^{-j\omega t}}{d\omega}\,^2\right]dt = \left[j\frac{d}{d\omega}\int_{-\infty}^{\infty} x(t)e^{-j\omega t}\,dt\right] = j\frac{dX(\omega)}{d\omega}$$

Remark: The times-t and derivative properties are duals. Multiplication by the variable in one domain results in a derivative in the other.

Example 6.12
Using the Times-t
Property

(a) We find the transform of $r(t) = tu(t)$ from $u(t)$ as follows:

$$r(t) = tu(t) \longleftrightarrow (j/2\pi)\frac{d}{df}[\tfrac{1}{2}\delta(f) + 1/(j2\pi f)] = (j/4\pi)\delta'(f) - 1/(2\pi f)^2$$

(b) With $\exp(-\alpha t)u(t) \longleftrightarrow 1/(\alpha + j2\pi f)$, the times-$t$ property gives

$$t\exp(-\alpha t)u(t) \longleftrightarrow (j/2\pi)\frac{d}{df}\left[\frac{1}{\alpha + j2\pi f}\right] = \frac{1}{(\alpha + j2\pi f)^2}$$

(c) Applying this property successively to the result of part (b), we obtain

$$t^n\exp(-\alpha t)u(t) \longleftrightarrow \frac{n!}{(\alpha + j2\pi f)^{n+1}}$$

(d) ω-form: Using the times-t property on $\exp(-\alpha|t|)$, we obtain the transform of $t\exp(-\alpha|t|)$,

$$t\exp(-\alpha|t|) \longleftrightarrow j\frac{d}{d\omega}\left[\frac{2}{\alpha^2 + \omega^2}\right] = \frac{-4j\alpha\omega}{(\alpha^2 + \omega^2)^2}$$

Comment: This result may also be obtained if we start with $\exp(-\alpha|t|)$ and use the times-t and folding properties. Since $t\exp(-\alpha|t|)$ is the odd extension of $t\exp(-\alpha t)u(t)$, we have

$$t\exp(-\alpha|t|) \longleftrightarrow 2\text{Im}\left[\frac{1}{(\alpha + j\omega)^2}\right]$$

The transform simplifies to

$$2\text{Im}\left[\frac{1}{\alpha^2 - \omega^2 + 2j\alpha\omega}\right] = \frac{-4j\alpha\omega}{(\alpha^2 - \omega^2)^2 - 4\alpha^2\omega^2} = \frac{-4j\alpha\omega}{(\alpha^2 + \omega^2)^2}$$

6.3.9 High-Frequency Decay Rate

As with Fourier series coefficients, the Fourier transform of many signals decays with frequency, and the decay rate may be found (without evaluating $X(f)$) by taking successive derivatives of $x(t)$. If n is the number of derivatives it takes for $x(t)$ to first exhibit impulses (whose spectrum is constant with frequency), the decay rate is proportional to $1/f^n$.

Example 6.13
High-Frequency Decay

For the following signals, we list the number of derivatives for impulses to first appear and the high-frequency decay rate:

(a) $\text{rect}(t)$: One derivative; high-frequency decay rate: $1/f$.
(b) $\exp(-\alpha t)u(t)$: One derivative; high-frequency decay rate: $1/f$.

(c) $\text{tri}(t)$: Two derivatives; high-frequency decay rate: $1/f^2$.
(d) $[1 + \cos(2\pi t)]\text{rect}(t)$ (raised-cosine pulse): Three derivatives; high-frequency decay rate: $1/f^3$.

6.3.10   The Initial Value Theorem

For functions of the form $x(t)u(t)$, the behavior of $X(f)$ as $f \longrightarrow \infty$ is related to the behavior of $x(t)$ as $t \longrightarrow 0^+$ in terms of the **initial value theorem**, which reads

$$x(0^+) = \lim_{f \to \infty} [j2\pi f X(f)] \quad \text{or} \quad \lim_{\omega \to \infty} [j\omega X(\omega)]$$

Example 6.14
Using the Initial Value Theorem

(a) With $x(t) = \exp(-t)u(t) \longleftrightarrow 1/(1 + j2\pi f)$, the initial value theorem predicts

$$x(0^+) = \lim_{f \to \infty} j2\pi f[1/(1 + j2\pi f)] = \lim_{f \to \infty} \left[\frac{1}{1 + 1/j2\pi f} \right] = 1$$

(b) With $u(t) \longleftrightarrow \frac{1}{2}\delta(f) + 1/j2\pi f$, the initial value theorem predicts

$$u(0^+) = \lim_{f \to \infty} j2\pi f[\tfrac{1}{2}\delta(f) + 1/j2\pi f] = 1$$

Comment: Unlike $u(0)$ which is undefined, $u(0^+) = 1$ and is well defined.

6.3.11   Integration

Upon integration, the spectrum of a function is divided by $j\omega$ such that

$$\int_{-\infty}^{t} x(t)\,dt \longleftrightarrow \frac{X(f)}{j2\pi f} + \frac{1}{2}X(0)\delta(f) \quad \text{or} \quad \frac{X(\omega)}{j\omega} + \pi X(0)\delta(\omega)$$

Since $X(0)$ equals the area (not absolute area) under $x(t)$, this relation holds only if $\mathbb{A}[x(t)]$ is finite. The second term disappears if $\mathbb{A}[x(t)] = X(0) = 0$.

If $X(0) = 0$, integration and differentiation may be regarded as inverse operations.

The factor $1/f$ decreases the spectral content at higher frequencies. Since integration is a smoothing operation, the smoother a function, the less significant the high-frequency content in its spectrum.

Example 6.15
Applying the Integration Property

(a) We may use $\cos(2\pi\alpha t) \longleftrightarrow \frac{1}{2}[\delta(f + \alpha) + \delta(f - \alpha)]$ and the integration property to find the FT of $x(t) = \sin(2\pi\alpha t)$. Since $X[0] = 0$ and $x(t)$ equals the running integral of $2\pi\alpha\cos(2\pi\alpha t)$, the integration property (and simplification using the product property of impulses) yields

$$2\pi\alpha \int_{-\infty}^{t} \cos(2\pi\alpha t)\,dt \longleftrightarrow \frac{\pi\alpha[\delta(f + \alpha) + \delta(f - \alpha)]}{j2\pi f} = \frac{1}{2}j[\delta(f + \alpha) - \delta(f - \alpha)]$$

(b) Starting with $x(t) = \delta(t) \longleftrightarrow 1$ and noting that $X(0) = 1$ and $u(t)$ is just the running integral of $\delta(t)$, the transform of $u(t)$ may be written as

$$u(t) = \int_{-\infty}^{t} \delta(t)\, dt \longleftrightarrow \frac{X(f)}{j2\pi f} + \frac{1}{2}X(0)\delta(f) = \frac{1}{j2\pi f} + \frac{1}{2}\delta(f)$$

Comment: Could we extend this result to find the transform of $r(t)$ from $u(t)$? No, because $U(f) = \frac{1}{2}\delta(f) + 1/j2\pi f$ and thus $u(0)$ is infinite at $f = 0$.   ___

6.3.12   *Convolution in Time*

The transform of the convolution $x(t) \star y(t)$ equals the product $X(f)Y(f)$ or $X(\omega)Y(\omega)$. We interchange the order of integration and use a change of variables $(t - \lambda = \tau)$ in the defining integral (for the ω-form) as follows:

$$\int_{-\infty}^{\infty} \left[\int_{-\infty}^{\infty} x(t - \lambda)u(t - \lambda)y(\lambda)d\lambda \right] e^{-j\omega t}\, dt$$

$$= \int_{-\infty}^{\infty} \left[\int_{-\infty}^{\infty} x(t - \lambda)e^{-j\omega(t-\lambda)}\, dt \right] y(\lambda)e^{-j\omega\lambda}\, d\lambda = X(\omega)Y(\omega)$$

Thus, convolution in the time domain corresponds to multiplication in the frequency domain.

Remark: Unlike the Fourier series result, the convolution $x(t) \star y(t)$ represents the *regular (not periodic) convolution.*

___

Example 6.16
Using Convolution in Time

(a) We find the FT of $\text{tri}(t)$ using the result $\text{tri}(t) = \text{rect}(t) \star \text{rect}(t)$

$$\text{tri}(t) = \text{rect}(t) \star \text{rect}(t) \longleftrightarrow \text{sinc}(f)\text{sinc}(f) = \text{sinc}^2(f)$$

(b) ω-form: With $t\exp(-\alpha t)u(t) = \exp(-\alpha t)u(t) \star \exp(-\alpha t)u(t)$, and the transform pair $\exp(-\alpha t) \longleftrightarrow 1/(\alpha + j\omega)$, we obtain

$$t\exp(-\alpha t)u(t) = \exp(-\alpha t)u(t) \star \exp(-\alpha t)u(t) \longleftrightarrow \frac{1}{(\alpha + j\omega)^2}$$

(c) What is the convolution of two sinc functions? Starting with $\text{sinc}(t) \longleftrightarrow \text{rect}(f)$ and using scaling and convolution, we write

$$\text{sinc}(\alpha t) \star \text{sinc}(\beta t) \longleftrightarrow \left[\frac{1}{\alpha}\text{rect}\frac{f}{\alpha} \right] \left[\frac{1}{\beta}\text{rect}\frac{f}{\beta} \right]$$

The product of the two rect functions is just another rect function whose height equals $1/\alpha\beta$ and whose width is the smaller of the two. If $\beta > \alpha$, the product equals $[\text{rect}(f/\alpha)]/\alpha\beta$. The time function is then just a scaled sinc function that equals $(1/\beta)\text{sinc}(\alpha t)$, and we have the result

$$\text{sinc}(\alpha t) \star \text{sinc}(\beta t) = \frac{1}{\beta}\text{sinc}(\alpha t) \qquad \beta > \alpha$$

(d) Here is how we prove the integration property. The running integral of $x(t)$ is just the convolution of $x(t)$ with $u(t)$. With $u(t) \longrightarrow \frac{1}{2}\delta(f) + 1/j2\pi f$, we have

$$\int_{-\infty}^{t} x(t)\, dt = x(t) \star u(t) \longrightarrow X(f)\left[\frac{1}{2}\delta(f) + \frac{1}{j2\pi f}\right] = \frac{1}{2}X(0)\delta(f) + \frac{X(f)}{j2\pi f}$$

(e) To find the convolution $(1/\pi t) \star (1/\pi t)$, we start with $\mathrm{sgn}(t) \longrightarrow 1/j\pi f$. Duality gives $-1/j\pi t \longrightarrow \mathrm{sgn}(f)$. The convolution property then yields

$$(-1/j\pi t) \star (-1/j\pi t) \longrightarrow \mathrm{sgn}^2(f) = 1$$

Since $\delta(t) \longrightarrow 1$, we have $(-1/j\pi t) \star (-1/j\pi t) = \delta(t)$, and

$$(1/\pi t) \star (1/\pi t) = -\delta(t)$$

Comment: The convolved functions have odd symmetry, zero area, and infinite extent while their convolution has finite area and zero duration. Not quite what the duration and area properties of convolution would have suggested!

6.3.13   *Convolution in Frequency*

This property is a dual of the convolution-in-time property and reads

$$x(t)y(t) \longrightarrow X(f) \star Y(f) \qquad \text{or} \qquad \frac{1}{2\pi}X(\omega) \star Y(\omega)$$

Remarks: **1.** Unlike the Fourier series result, $X(f) \star Y(f)$ *represents a continuous (not discrete) convolution.*

   2. Since convolution is usually a stretching operation, the spectrum of the product will be "broader", and extend to higher frequencies.

Example 6.17
Using Frequency
Convolution

(a) The transform of $x(t) = 2\sin(2\pi\alpha t)\cos(2\pi\alpha t)$ is given by the convolution:

$$2\left(\frac{1}{2}\right)[\delta(f + \alpha) + \delta(f - \alpha)] \star \left(\frac{j}{2}\right)[\delta(f + \alpha) - \delta(f - \alpha)]$$

We simplify this result to

$$X(f) = \frac{j}{2}[\delta(f + 2\alpha) + \delta(f) - \delta(f - 2\alpha) - \delta(f)] = \frac{j}{2}[\delta(f + 2\alpha) - \delta(f - 2\alpha)]$$

This also describes the FT of $\sin(4\pi\alpha t)$ which $x(t)$ equals.

(b) Consider the tone burst signal $x(t) = \cos(2\pi f_0 t)\mathrm{rect}(t/t_0)$. Since this is the product of two signals, the convolution property gives

$$X(f) = \frac{1}{2}[\delta(f + f_0) + \delta(f - f_0)] \star t_0\mathrm{sinc}(ft_0)$$
$$= \frac{1}{2}t_0\mathrm{sinc}[t_0(f + f_0)] + \frac{1}{2}t_0\mathrm{sinc}[t_0(f - f_0)]$$

This is just the sum of two shifted sinc functions. The signal $x(t)$ and its spectrum are shown in Figure 6.9 for $t_0 = 2$ and $f_0 = 5$.

Figure 6.9   Illustrating frequency convolution for Example 6.17b.

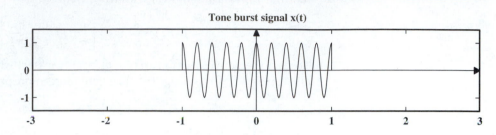

Tone burst signal x(t)

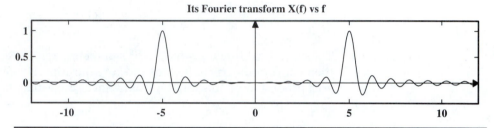

Its Fourier transform X(f) vs f

(c) With $t_0 = 1$ and $f_0 = \frac{1}{2}$, we get $x(t) = \cos(\pi t)\mathrm{rect}(t)$ and

$$X(f) = \tfrac{1}{2}[\delta(f + \tfrac{1}{2}) + \delta(f - \tfrac{1}{2})] \star \mathrm{sinc}(f) = \tfrac{1}{2}\,\mathrm{sinc}[(f + \tfrac{1}{2})] + \mathrm{sinc}[(f - \tfrac{1}{2})]$$

Comment: This result was obtained previously by several other means.   —

6.3.14   *The Timelimited-Bandlimited Theorem*

The convolution property of Fourier transforms may also be used to develop the **timelimited-bandlimited theorem**, which, in essence, says that *no signal can be both timelimited and bandlimited simultaneously.* Any timelimited function $x_L(t)$ may be regarded as the product of a function $x(t)$ extending for all time and a rect function that restricts its view over a finite duration. The spectrum $X_L(f)$ is just the convolution of $X(f)$ and a sinc function. Since a sinc function is of infinite extent, the duration property of convolution tells us that $X_L(f)$ must also be of infinite extent, regardless of how wide or narrow $X(f)$ itself is.

The convolution property can similarly be used to assert that a bandlimited spectrum must always correspond to a time signal of infinite extent.

Remarks: **I.** The theorem suggests that every signal must be of infinite extent:

(a) In only the time domain ($\mathrm{sinc}(t) \longleftrightarrow \mathrm{rect}(f)$, for example), or
(b) In only the frequency domain ($\mathrm{tri}(t) \longleftrightarrow \mathrm{sinc}^2(f)$, for example), or
(c) In both domains ($\exp(-|t|) \longleftrightarrow 2/(1 + \omega^2)$, for example).

2. In general, the spectrum of timelimited signals will show wiggles or decaying sidelobes because of the convolution with the sinc function.
3. Clearly, the derivatives of a bandlimited signal are also bandlimited. As a consequence, a bandlimited signal $x_B(t)$ must always be smooth and possess derivatives of all orders. If it didn't, we would get impulses sooner or later, leading to a spectrum of infinite extent. This does not mean that a function that possesses derivatives of all orders must be bandlimited. The Gaussian, for example, possesses derivatives of all orders, but its spectrum (also a Gaussian) is not bandlimited.

6.3.15 The Gibbs Effect and Signal Reconstruction

The timelimited-bandlimited theorem has important consequences. If we bandlimit the spectrum $|X(f)|$ of a signal $x(t)$ to a finite frequency f_B, the larger we make f_B, the better we approximate $x(t)$.

If $x(t)$ shows discontinuities, we can never reconstruct $x(t)$ perfectly from this spectrum no matter how large f_B is. The reconstructed signal shows an overshoot ($\approx$ 9% of jump) and converges to the midpoint at discontinuities. This is analogous to the Gibbs effect in the reconstruction of any periodic signal with discontinuities from a finite number of harmonics.

6.3.16 The Central Ordinate Theorems

The central ordinate theorems follow directly from the defining relations for $x(t)$ and $X(\omega)$ by setting $t = 0$ or $f = 0$.

$$X(0) = \int_{-\infty}^{\infty} x(t)\, dt \qquad x(0) = \int_{-\infty}^{\infty} X(f)\, df \quad \text{or} \quad \frac{1}{2\pi}\int_{-\infty}^{\infty} X(\omega)\, d\omega$$

Remarks:
1. For real signals, we need use only $\text{Re}[X(f)]$ to find $x(0)$. For such signals, $\text{Im}[X(f)]$ is odd and integrates to zero between $(-\infty, \infty)$.
2. If $x(t)$ possesses a discontinuity at the origin, we actually obtain $x(0)$ as the *average value* at the discontinuity. Thus, $x(0)$ does not necessarily correspond to the initial value $x(0^+)$.
3. $X(0)$ equals zero if $x(t)$ is odd or has the form $x(t) = y^{(n)}(t)$ because its transform $X(f) = (j2\pi f)^n Y(f)$ goes to zero at $f = 0$.

Example 6.18
Applying the Central Ordinate Theorems

(a) For the transform pair $\text{rect}(t) \longleftrightarrow \text{sinc}(f)$, $X(0) = \text{sinc}(0) = 1$ and is equal to $\mathbb{A}[\text{rect}(t)]$, whereas $x(0) = \text{rect}(0) = 1$ and equals $\mathbb{A}[\text{sinc}(f)]$.

(b) ω-form: For the transform pair $\exp(-\alpha t)u(t) \longleftrightarrow 1/(\alpha + j\omega)$, we find that $X(0) = A[\exp(-\alpha t)u(t)] = 1/\alpha$. This checks with $X(\omega)|_{\omega=0}$. With $\text{Re}[X(\omega)] = \alpha/[\alpha^2 + \omega^2]$, we find that $x(0) = \frac{1}{2}$ (and not 1) because

$$x(0) = \frac{1}{2\pi}\int_{-\infty}^{\infty} \frac{\alpha}{\alpha^2 + \omega^2}\, d\omega = \frac{1}{2\pi}\tan^{-1}\left(\frac{\omega}{\alpha}\right)\Bigg|_{-\infty}^{\infty} = \frac{1}{2\pi}\left(\frac{1}{2}\pi + \frac{1}{2}\pi\right) = \frac{1}{2}$$

6.3.17   The Time-Bandwidth Product

If we multiply the two central ordinate relations and rearrange, we get

$$\left[\frac{\int_{-\infty}^{\infty} x(t)\, dt}{x(0)}\right]\left[\frac{\int_{-\infty}^{\infty} X(f)\, df}{X(0)}\right] = 1$$

The left-hand side may be regarded as the product of measures defining time du-ration and bandwidth, respectively. This result implies that the time-bandwidth product is a constant. It holds only if $x(0)$ and $X(0)$ are finite. It also fails, for example, if $x(t)$ is odd or $X(f)$ has the form $f^n G(f)$.

Other measures of duration and bandwidth are discussed in the next chapter.

6.3.18   Modulation

The time-shift property $x(t - \alpha) \longleftrightarrow X(f)\exp(-j2\pi\alpha f)$ suggests the dual relation

$$x(t)\exp(j2\pi\alpha t) \longleftrightarrow X(f - \alpha) \quad \text{or} \quad X(\omega - 2\pi\alpha)$$

If we sum $X(f - \alpha)$ and $X(f + \alpha)$ and use Euler's relation, we obtain

$$x(t)\cos(2\pi\alpha t) \longleftrightarrow \tfrac{1}{2}[X(f + \alpha) + X(f - \alpha)]$$

The transform of the *modulated signal* $x(t)\cos(2\pi\alpha t)$ is half the sum of the shifted transforms $X(f + \alpha)$ and $X(f - \alpha)$. Its magnitude equals half the sum of the mag-nitudes $|X(f + \alpha)|$ and $|X(f - \alpha)|$ *only if* $x(t)$ *is bandlimited* to $|f| \le \alpha$. However, if $x(t)$ is even or odd (with $X(f)$ purely real or imaginary), the amplitude spectrum of $x(t)\cos(2\pi\alpha t)$ does equal half the sum of the shifted amplitude spectra, and its phase then equals the phase of $X(f)$. The modulation property is illustrated in Figure 6.10.

Remarks:
1. The modulation property is actually a special case of convolution in that if $x(t) \longleftrightarrow X(f)$ and $\cos(2\pi\alpha t) \longleftrightarrow \tfrac{1}{2}[\delta(f + \alpha) + \delta(f - \alpha)]$,

$$x(t)\cos(2\pi\alpha t) \longleftrightarrow X(f) \star \tfrac{1}{2}[\delta(f + \alpha) + \delta(f - \alpha)] = \tfrac{1}{2}[X(f + \alpha) + X(f - \alpha)]$$

2. Using $X(f + \alpha) - X(f - \alpha)$, we obtain the result for sine-wave modulation as

$$x(t)\sin(2\pi\alpha t) \longleftrightarrow \tfrac{1}{2}j[X(f + \alpha) - X(f - \alpha)]$$

Example 6.19

Using the Modulation Property

(a) Starting with the transform pair $1 \longleftrightarrow \delta(f)$ and using the modulation property $x(t)\exp(j2\pi f_0 t) \longleftrightarrow X(f - f_0)$, we get

$$\exp(j2\pi f_0 t) \longleftrightarrow \delta(f - f_0)$$

(b) ω-form: To find the transform of $\cos(\alpha t)$, we start with $1 \longleftrightarrow 2\pi\delta(\omega)$ and use the modulation property

$$\cos(\alpha t) = (1)[\cos(\alpha t)] \longleftrightarrow \pi[\delta(\omega + \alpha) + \delta(\omega - \alpha)]$$

Figure 6.10   Illustrating the modulation property.

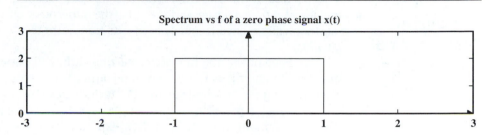

Spectrum vs f of a zero phase signal x(t)

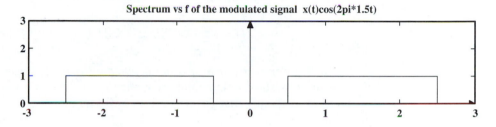

Spectrum vs f of the modulated signal x(t)cos(2pi*1.5t)

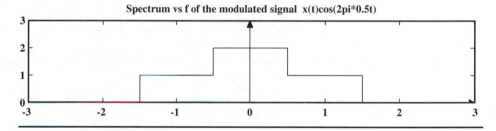

Spectrum vs f of the modulated signal x(t)cos(2pi*0.5t)

(c) ω-form: To find the FT of $\cos(\beta t)u(t)$, start with $u(t) \longleftrightarrow \pi\delta(\omega) + 1/j\omega$ and use modulation to give

$$\cos(\beta t)u(t) \longleftrightarrow \tfrac{1}{2}\pi[\delta(\omega+\beta) + \delta(\omega-\beta)] + \frac{1}{2}\left[\frac{1}{j(\omega+\beta)} + \frac{1}{j(\omega-\beta)}\right]$$

This simplifies to

$$\cos(\beta t)u(t) \longleftrightarrow \frac{j\omega}{\beta^2 + (j\omega)^2} + \frac{1}{2}\pi[\delta(\omega+\beta) + \delta(\omega-\beta)]$$

(d) ω-form: To find the transform of $\exp(-\alpha t)\sin(\beta t)u(t)$, we start with the pair $\exp(-\alpha t)u(t) \longleftrightarrow 1/(\alpha + j\omega)$ and get

$$\exp(-\alpha t)\sin(\beta t)u(t) \longleftrightarrow \frac{j}{2}\left[\frac{1}{\alpha+j(\omega+\beta)} - \frac{1}{\alpha+j(\omega-\beta)}\right] = \frac{\beta}{\beta^2 + (\alpha+j\omega)^2}$$

6.4 Fourier Transform of Periodic Signals

Yes, periodic signals can be described not only by a Fourier series, but also by a Fourier transform. The transform of a sinusoid $A\cos(2\pi f_0 t + \theta)$ is a pair of impulses at $\pm f_0$. A Fourier series is just the superposition of such sinusoids at the harmonically related frequencies kf_0. Its transform thus contains pairs of impulses at $\pm kf_0$.

We have now come full circle, starting with the Fourier series for periodic signals, stretching the period to develop Fourier transforms of aperiodic signals and, finally, using the Fourier transform to include periodic signals as well.

A formal expression for the Fourier transform of periodic signals is based on the exponential form $x_p(t) = \sum X_S[k]\exp(jk2\pi f_0 t)$. We invoke superposition and the FT pair $\exp(j2\pi f_0 t) \longleftrightarrow \delta(f - f_0)$ to obtain

$$x_p(t) = \sum_{k=-\infty}^{\infty} X_S[k]\exp(jk2\pi f_0 t) \longleftrightarrow \sum_{k=-\infty}^{\infty} X_S[k]\delta(f - kf_0)$$

The Fourier transform of a periodic signal is an impulse train. The impulses are located at the fundamental frequency f_0 and its harmonics kf_0. Their strengths equal the Fourier series coefficients $X_S[k]$ (Figure 6.11).

Figure 6.11　The Fourier transform of a periodic signal is an impulse train.

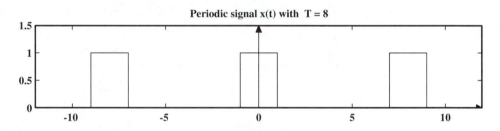

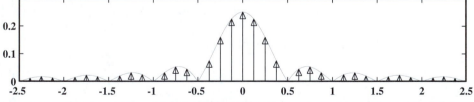

Remarks:
1. The impulse train is not periodic since the impulses are of different strengths $X_S[k]$. The only exception is the Fourier transform of a comb function, which is also a comb function.
2. Since a periodic signal is a power signal, its power is found by using Parseval's theorem for periodic signals.

Example 6.20

The Fourier Transform of Periodic Signals

(a) The Fourier series coefficients of $x(t) = \text{comb}(t)$ are $X_S[k] = 1$ (all k). The Fourier transform of this function is then simply (Figure 6.3n)

$$\text{comb}(t) = \sum_{k=-\infty}^{\infty} \delta(t-k) \longleftrightarrow \sum_{k=-\infty}^{\infty} \delta(f-k) = \text{comb}(f)$$

Comment: The transform of a comb signal is also a comb signal.

(b) The Fourier coefficients of the signal $x(t)$ of Figure 6.11 are $X_S[k] = \frac{1}{4}\text{sinc}(\frac{1}{4}k)$. The transform $X(f)$ thus equals

$$X(f) = \sum_{k=-\infty}^{\infty} \frac{1}{4}\text{sinc}\left(\frac{1}{4}k\right) \delta\left(f - \frac{1}{8}k\right)$$

Its magnitude spectrum is sketched in Figure 6.11.

6.5
System Analysis Using the Fourier Transform

Conceptually, the Fourier transform can be used for the analysis of relaxed LTI systems by resorting to the fact that the convolution operation in time corresponds to multiplication in frequency. The response $y(t)$ of a relaxed system with impulse response $h(t)$ and input $x(t)$ is $y(t) = x(t) \star h(t)$. In the frequency domain, this relation translates to

$$Y(\omega) = X(\omega)H(\omega)$$

The quantity $H(\omega) = Y(\omega)/X(\omega)$ defines the system **transfer function**. A relaxed LTI system is also described by a differential equation of the form

$$A_n\frac{d^n y}{dt^n} + \cdots + A_2\frac{d^2 y}{dt^2} + A_1\frac{dy}{dt} + A_0 y = B_m\frac{d^m x}{dt^m} + \cdots + B_2\frac{d^2 x}{dt^2} + B_1\frac{dx}{dt} + B_0 x$$

If $x(t) \longleftrightarrow X(\omega)$ and $y(t) \longleftrightarrow Y(\omega)$, it is a simple matter to transform this differential equation using the derivative property, and obtain

$$[A_n(j\omega)^n + \cdots + A_2(j\omega)^2 + A_1(j\omega) + A_0]Y(\omega)$$
$$= [B_m(j\omega)^m + \cdots + B_2(j\omega)^2 + B_1(j\omega) + B_m]X(\omega)$$

The transfer function $H(\omega)$ is then given by the ratio $Y(\omega)/X(\omega)$ as

$$H(\omega) = \frac{Y(\omega)}{X(\omega)} = \frac{B_m(j\omega)^m + \cdots + B_2(j\omega)^2 + B_1(j\omega) + B_0}{A_n(j\omega)^n + \cdots + A_2(j\omega)^2 + A_1(j\omega) + A_0}$$

For an LTI system, $H(\omega)$ is a ratio of polynomials in $j\omega$. It allows access to both the system impulse response and the differential equation.

Remarks: **1.** Since the convolution operation is defined for relaxed systems with zero initial conditions, $H(\omega)$ is also defined only for such systems.
   2. The order of the denominator polynomial equals the order of the system (the number of independent energy-storage elements present), provided there is no cancellation of common numerator and denominator factors.

6.5.1   *Finding the Transfer Function*

We can find the system transfer function $H(\omega)$ in one of three ways, depending on what information is available about the system.

1. Given the impulse response $h(t)$, its Fourier transform is $H(\omega)$.
2. Given the system input-output differential equation, we take its Fourier transform and extract the ratio $Y(\omega)/X(\omega)$ to obtain $H(\omega)$.
3. Given an electric circuit, we set up its differential equation using Kirchhoff's laws and transform it. A simpler way is to start with the transform of the constitutive relations for the circuit elements themselves. This leads to the concept of generalized impedance as follows:

$$
\begin{array}{lll}
v_R(t) = Ri_R(t) & V_R(\omega) = RI_R(\omega) & Z_R = V_R(\omega)/I_R(\omega) = R \\
v_L(t) = Ldi_L(t)/dt & V_L(\omega) = j\omega LI_L(\omega) & Z_L = V_L(\omega)/I_L(\omega) = j\omega L \\
i_C(t) = Cdv_C(t)/dt & I_C(\omega) = j\omega CV_C(\omega) & Z_C = V_C(\omega)/I_C(\omega) = 1/j\omega C
\end{array}
$$

The quantities Z_R, Z_C and Z_L have units of ohms and describe *generalized impedances.* We can transform a circuit directly to the frequency domain if we replace circuit elements by their impedances and sources and system variables by their transforms. Mesh and node equations based on Kirchhoff's laws in the frequency domain yield a much simpler algebraic form.

Example 6.21
Finding the Transfer Function

(a) If the impulse response of a system is $h(t) = e^{-t}u(t) + e^{-2t}u(t)$, then

$$
H(\omega) = \frac{1}{1 + j\omega} + \frac{1}{2 + j\omega} = \frac{3 + 2j\omega}{2 + 3j\omega + (j\omega)^2}
$$

(b) If the system is described by the differential equation

$$
y''(t) + 3y'(t) + 2y = 2x'(t) + 3x(t)
$$

we use the derivative property to get

$$
[(j\omega)^2 + 3j\omega + 2]Y(\omega) = [2j\omega + 3]X(\omega)
$$

Since $H(\omega) = Y(\omega)/X(\omega)$, we obtain the transfer function as

$$
H(\omega) = \frac{3 + 2j\omega}{2 + 3j\omega + (j\omega)^2}
$$

Note: Given $H(\omega)$, it is also possible to obtain the system differential equation. Cross multiplication gives $[(j\omega)^2 + 3j\omega + 2]Y(\omega) = [2j\omega + 3]X(\omega)$, and its inverse transform yields $y''(t) + 3y'(t) + 2y(t) = 2x'(t) + 3x(t)$.

(c) The *RLC* circuit shown in Figure 6.12a yields the transformed circuit of Figure 6.12b. Let $x(t) = \delta(t)$. Then $X(\omega) = 1$. If $i(t)$ is the output, the transformed circuit yields the transfer function $H(\omega) = I(\omega)/X(\omega) = I(\omega)$ as

$$
H(\omega) = I(\omega) = \frac{1}{2 + j\omega + 3/j\omega} = \frac{j\omega}{3 + 2j\omega + (j\omega)^2}
$$

Figure 6.12   An *RLC* circuit and its transformed version for Example 6.21.

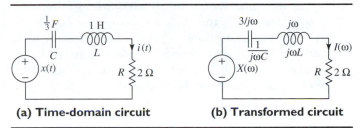

(a) Time-domain circuit   **(b) Transformed circuit**

6.5.2   *The Transfer Function and System Analysis*

The transfer function $H(\omega)$ allows evaluation of the response of a relaxed system (with zero initial conditions). It is also referred to as the **frequency response** of the system since a plot (or expression) of $|H(\omega)|$ and $\angle H(\omega)$ versus ω provides an indication of its magnitude and phase spectrum in the frequency domain.

Example 6.22

System Response to Arbitrary Inputs

(a) Let $H(\omega) = 4/(2 + j\omega)$ and let the input equal $x(t) = 3\exp(-2t)u(t)$. We evaluate the response $Y(\omega)$ as

$$Y(\omega) = H(\omega)X(\omega) = Y(\omega) = 12/(2 + j\omega)^2$$

Inverse transformation then yields $y(t) = 12t\exp(-2t)u(t)$.

(b) Let $H(\omega) = 4/(2 + j\omega)$ and let the input be $x(t) = 2u(t)$. Once again, we evaluate $Y(\omega) = H(\omega)X(\omega) = 4\left[\pi\delta(\omega) + \frac{1}{j\omega}\right]/(2 + j\omega)$. We separate terms, use the property $f(x)\delta(x) = f(0)\delta(x)$ and simplify

$$Y(\omega) = \frac{4\pi\delta(\omega)}{2 + j\omega} + \frac{4}{j\omega(2 + j\omega)} = 2\pi\delta(\omega) + \frac{4}{j\omega(2 + j\omega)}$$

The second term has no recognizable inverse, but it can be written as

$$\frac{4}{j\omega(2 + j\omega)} = \frac{2}{j\omega} - \frac{2}{2 + j\omega}$$

A formal method for obtaining this form is *partial fraction expansion* (see Chapter 8). The response $Y(\omega)$ now equals

$$Y(\omega) = 2\pi\delta(\omega) + \frac{2}{j\omega} - \frac{2}{2 + j\omega} = 2\left[\pi\delta(\omega) + \frac{1}{j\omega}\right] - \frac{2}{2 + j\omega}$$

Inverse transformation gives $y(t) = 2u(t) - 2\exp(-2t)u(t)$.

Comment: The inversion process is tedious due to the presence of impulses. We combined $2\delta(\omega)$ and $2/j\omega$ to obtain the inverse $2u(t)$. Term-by-term inversion gives $y(t) = 1 + \text{sgn}(t) - 2\exp(-2t)u(t)$, which is an equivalent result.

6.5.3   The Steady-State Response to Harmonic Inputs

The response of an LTI system to a harmonic input is also a harmonic at the input frequency and describes the *steady-state* response. The magnitude and phase of the output depend on the system transfer function at the applied frequency. To find the steady-state response $y_{ss}(t)$ to the harmonic input $x(t) = A\cos(\omega_0 t + \theta)$, we evaluate $H(\omega_0) = K\angle\phi$ and $y_{ss}(t) = KA\cos(\omega_0 t + \theta + \phi)$. In effect, the transfer function magnitude K multiplies the input magnitude A, and the transfer function phase ϕ adds to the input phase θ.

Remarks: **1.** For harmonic inputs at different frequencies, we evaluate $H(\omega)$ and the time-domain output at each input frequency and superpose the results.
2. The quantity $H(\omega)$ is also called the **steady-state transfer function**.

Example 6.23
System Response to
Harmonic Inputs

Let $H(\omega) = 4/(2 + j\omega)$ and let the input $x(t)$ be $x(t) = 3\sin(2t)$. The easiest way to find the output $y(t)$ for harmonic inputs is to treat $H(\omega)$ as the steady-state transfer function and use the following steps:

1. Evaluate $H(\omega) = |H(\omega)|\angle\theta$ at the frequency of the harmonic input. Since $\omega = 2$, we get $H(\omega) = 4/(2 + j2) = \sqrt{2}\angle -45°$.
2. Multiply the input magnitude by $|H(\omega)|$ and add θ to the input phase. Since $|H(\omega)| = \sqrt{2}$ and $\theta = -45°$, the output is $y(t) = 3\sqrt{2}\sin(2t - 45°)$.

Comment: Here are the steps involved in the (tedious) formal alternative:

$$X(\omega) = 3j\pi[\delta(\omega + 2) - \delta(\omega - 2)]$$
$$Y(\omega) = H(\omega)X(\omega) = 12j\pi[\delta(\omega + 2) - \delta(\omega - 2)]/(2 + j\omega)$$
$$y(t) = \text{IFT}[Y(\omega)]$$

By the way, simplification of $Y(\omega)$ will require the product property of impulses. We urge you not to use this approach unless you have time on your hands!

6.5.4   Concluding Remarks

The Fourier transform offers the distinct advantage of a frequency-domain view of the transfer function. Other than that, its use in general system analysis is either cumbersome or impossible because:

1. We are limited to relaxed systems (zero initial conditions).
2. Inputs that are not energy signals involve impulses.
3. Some signals (such as growing exponentials) cannot be handled at all.

The Laplace transform (discussed in Chapter 8), which is an extension of the Fourier transform, overcomes all of these limitations and is the method of choice for system analysis in the transformed domain. It would be fair to say that the *Fourier transform is better suited for signal analysis, whereas the Laplace transform is better suited for system analysis.*

6.6
Special Topics: Energy and Power Spectral Density

For an aperiodic signal $x(t)$, the quantity $|X(f)|^2$ describes the distribution of energy with frequency and is called the **energy spectral density**. In effect, the quantity $|X(f)|^2\Delta f$ equals the signal energy in the frequency band Δf. The area under $|X(f)|^2$ equals the total signal energy.

6.6.1   *Power Spectral Density*

The power spectrum $|X_S[k]|^2$ for a periodic signal $x_p(t)$ indicates the power in each component of its Fourier series. The **power spectral density** (PSD) $S_x(f)$, on the other hand, is a function that describes the *distribution* of power with frequency such that its area equals the total power in the signal. We thus write

$$P_x = \int_{-\infty}^{\infty} S_x(f)\, df$$

The units of $S_x(f)$ are WHz^{-1}. The power spectrum is plotted as a set of points or lines with heights $|X_S[k]|^2$. The area under a point or line has no meaning. We therefore describe $S_x(f)$ by a set of *impulses* of strength $|X_S[k]|^2$ at kf_0. The area $S_z(f)$ under this function equals the sum of the areas under each impulse and represents the power in the entire signal. The power spectral density $S(f)$ for a periodic signal is thus an *impulse train* described by

$$S(f) = \sum_{k=-\infty}^{\infty} |X_S[k]|^2 \delta(f - kf_0)$$

Integrating $S(f)$ and interchanging the order of summation and integration, we get the total power P_x in the periodic signal $x_p(t)$, as expected:

$$\int_{-\infty}^{\infty} S(f)\, df = \int_{-\infty}^{\infty} \left[\sum_{k=-\infty}^{\infty} |X_S|^2 \delta(f - kf_0) \right]$$

$$= \sum_{k=-\infty}^{\infty} \left[\int_{-\infty}^{\infty} |X_S|^2 \delta(f - kf_0)\, df \right] = \sum_{k=-\infty}^{\infty} |X_S|^2 = P_x$$

It is also possible to describe the PSD of a power signal $x(t)$, periodic or otherwise, in terms of a limiting form of its truncated version $x_T(t)$, as $T \longrightarrow \infty$. Truncation ensures that $x_T(t)$ represents an energy signal and its Fourier transform $X_T(f)$ yields the total energy via Parseval's energy relation

$$\int_{-\infty}^{\infty} x_T^2(t)\, dt = \int_{-\infty}^{\infty} |X_T(f)|^2\, df$$

Using the limits $(-T, T)$ on the left-hand side allows us to replace $x_T(t)$ by its original version $x(t)$ to give

$$\int_{-T}^{T} x^2(t)\, dt = \int_{-\infty}^{\infty} |X_T(f)|^2\, df$$

Dividing both sides by $2T$ and taking limits as $T \longrightarrow \infty$ results in

$$\lim_{T \to \infty} \frac{1}{2T} \int_{-T}^{T} x^2(t)\, dt = \lim_{T \to \infty} \frac{1}{2T} \int_{-\infty}^{\infty} |X_T(f)|^2\, df$$

The left-hand side is just the signal power in $x(t)$. Interchanging the integration and limiting operation on the right-hand side yields

$$P_x = \lim_{T \to \infty} \frac{1}{2T} \int_{-T}^{T} x^2(t)\, dt = \int_{-\infty}^{\infty} \left[\lim_{T \to \infty} \frac{1}{2T} |X_T(f)|^2 \right] df$$

Clearly, the integrand on the right-hand side must represent $S(f)$ if it is to yield the power when integrated. We thus express the PSD as

$$S(f) = \lim_{T \to \infty} \frac{|X_T(f)|^2}{2T}$$

The PSD is measured in watts per hertz and may be thought of as the average power associated with a 1-Hz frequency bin centered at f hertz. Note that $S(f)$ is always a real, nonnegative, and even function. If $x(t)$ is periodic, $S(f)$ is an impulse train.

6.6.2   The Wiener-Khintchine Theorem

Note that $|X_T(f)|^2$ is the transform of the autocorrelation $x_T(t) \star\star x_T(t)$. As $T \longrightarrow \infty$, $[x_T(t) \star\star x_T(t)]/2T$ approaches the autocorrelation $R_{xx}(t) = x(t) \star\star x(t)$, and $|X_T(f)|^2/2T$ approaches $S(f)$, and we have the transform pair

$$R_{xx}(t) \longleftrightarrow S(f)$$

This relation describes the celebrated **Wiener-Khintchine theorem.**

Remarks: 1. For power signals (periodic or otherwise), $R_{xx}(t)$ and $S(f)$ play a role analogous to the signal $x(t)$ and its spectrum $X(f)$.

2. Application of the central ordinate theorem yields

$$R_{xx}(0) = \int_{-\infty}^{\infty} S(f)\, df = P_x$$

and suggests that $R_{xx}(0)$ equals the average power in $x(t)$.

6.6.3   The Central Limit Theorem Revisited

The central limit theorem asserts that the repeated convolution of energy signals approaches a Gaussian. The Fourier transform of a Gaussian is also a Gaussian. Since convolution transforms to a product, we have an alternative interpretation of the central limit theorem. The product of the transforms of n energy signals must result in a Gaussian as $n \longrightarrow \infty$.

Example 6.24
The Central Limit Theorem

Consider a cascade of n systems, each having an identical impulse response given by $h(t) = \exp(-t)u(t)$. The impulse response $h_n(t)$ of the composite system is just the repeated convolution of $h(t)$ leading to the exact result and its Gaussian approximation for large n, reproduced here from Chapter 4.

$$h_n(t) = \exp(-t)u(t) \star \cdots \star \exp(-t)u(t) = \frac{t^2}{n!}\exp(-t)u(t) \approx \frac{1}{\sqrt{2\pi n}}\exp\left[-\frac{(t-n)^2}{2n}\right]$$

Since $h(t) = \exp(-t)u(t) \longleftrightarrow H(f) = 1/(1 + j2\pi f)$, the transform $H_n(f)$ of $h_n(t)$ is given by the repeated *product* of $H(f)$ with itself and equals

$$H_n(f) = [H(f)]^n = 1/(1 + j2\pi f)^n$$

We assert that as $n \longrightarrow \infty$, $H_n(f)$ also approaches a Gaussian. To see this, we take logarithms and use the expansion $\ln(1 + x) = x - \frac{1}{2}x^2 + \frac{1}{3}x^3 - \cdots$ to obtain

$$\ln[H_n(f)] = -n\ln[1 + j2\pi f] = -n[j2\pi f + \tfrac{1}{2}(2\pi f)^2 - \tfrac{1}{3}j(2\pi f)^3 + \cdots]$$

If we retain only terms up to the second order in f (which is equivalent to assuming that f decays at least as fast as $1/\sqrt{n}$), we have

$$\ln[H_n(f)] \approx -j2n\pi f - 2n\pi^2 f^2 \quad \text{or} \quad H_n(f) \approx \exp(-2n\pi^2 f^2)\exp(-j2n\pi f)$$

The magnitude spectrum $|H_n(f)| = \exp(-2n\pi^2 f^2)$ clearly has a Gaussian shape. The phase term $\exp(-j2n\pi f)$ simply accounts for the delay of n units.

Note: It is easily shown that $H_n(f)$, in fact, represents the exact Fourier transform of the Gaussian approximation to $h_n(t)$.

6.7
Special Topics: Existence, Convergence and Uniqueness

The *Dirichlet conditions* for Fourier series also establish sufficient conditions for the existence of Fourier transforms provided they are now applied to the entire extent of $x(t)$. These conditions may be restated as follows: A function $x(t)$ will possess a unique Fourier transform if

1. $x(t)$ is absolutely integrable.
2. $x(t)$ has a finite number of maxima and minima and/or a finite number of finite discontinuities.

   If the Dirichlet conditions are satisfied, then, as $f \longrightarrow \infty$, the Fourier transform of $x(t)$ will converge to

(a) $x(t)$ pointwise for every value of t where $x(t)$ is continuous and piecewise differentiable (convergence in the uniform sense) or
(b) $\frac{1}{2}[x(t^-) + x(t^+)]$ (the midpoint) at each discontinuity (nonuniform convergence).

If, in addition, $x(t)$ is an energy signal, the transform will converge to $x(t)$ in a mean squared sense and the mean squared error will approach zero as $f \longrightarrow \infty$ (as we include more frequencies to reconstruct $x(t)$).

Clearly, the Dirichlet conditions now impose much more severe restrictions for the existence of Fourier transforms because they apply to a function *over its entire extent*. For example, the step function, functions of polynomial and exponential growth, and periodic functions, clearly violate the finite area criterion and hence the Dirichlet conditions. If we allow generalized functions (impulses and their derivatives), the Fourier transform of *almost all* of these signals can be defined in what we call the *generalized sense*. The Dirichlet conditions are only *sufficient* conditions that guarantee a Fourier transform if satisfied. They do not preclude the existence of transforms for functions that violate some or all of them. To summarize, we can state that:

1. The Fourier transform of an absolutely integrable signal always exists.
2. The Fourier transform of signals that are not absolutely integrable (such as the step or sinusoid) almost always includes impulses.
3. The Fourier transform of signals that grow exponentially or faster (such as $\exp(\pm\alpha t)$ or $\exp(\pm\alpha t^2)$) does not exist.

6.8 The Fourier Transform: A Synopsis

The Fourier transform provides a unique representation of nonperiodic and periodic signals in the frequency domain.

The concept   Conceptually, we find the Fourier series coefficients for the periodic extension of $x(t)$ with arbitrary period T and find $TX_S[k]$ as $T \longrightarrow \infty$. As $T \longrightarrow \infty$, $X_S[k] \longrightarrow 0$ and provide no useful information, but the product $TX_S[k]$ often yields a meaningful spectrum called the Fourier transform, $X(f)$, of the signal $x(t)$. The signal $x(t)$ can also be reconstructed from its spectrum $X(f)$ using an inverse transformation.

$$X(f) = \int_{-\infty}^{\infty} x(t)\exp(-j2\pi ft)\,dt \qquad x(t) = \int_{-\infty}^{\infty} X(f)\exp(j2\pi ft)\,df$$

Symmetry   In general, $X(f)$ is complex with $X(-f) = X^*(f)$. For even symmetric signals, $X(f)$ is real. For odd symmetric signals, $X(f)$ is imaginary.

Spectra   A plot of the magnitude $|X(f)|$ versus f and phase $\angle X(f)$ versus f describes the continuous twosided magnitude and phase spectra of the signal $x(t)$.

For real $x(t)$, the twosided magnitude spectrum $|X(f)|$ has even symmetry. For real $x(t)$, the twosided phase spectrum $\angle X(f)$ has odd symmetry.

Parseval's theorem   The energy in $x(t)$ may be found either from $x(t)$ itself or from its spectral magnitude $|X(f)|$.

$$E_x = \int_{-\infty}^{\infty} x^2(t)\,dt = \int_{-\infty}^{\infty} |X(f)|^2\,df$$

Using Parseval's theorem, we can find the energy in $x(t)$ for a given frequency band by including just the spectral energy in that band.

Properties	There is a reciprocity between the time and frequency domains. The spectrum of a timelimited signal cannot be bandlimited, and vice versa. Compression in time results in stretching in frequency, and vice versa. Modulation in time results in a frequency shift, and vice versa. A time derivative results in the transform "times f," and vice versa. Convolution in time corresponds to multiplication in frequency and vice versa.
Periodic signals	The Fourier transform of periodic signals shows impulses located at kf_0 with strength $X_S[k]$. Since periodic signals are power signals, the power is now given by Parseval's relation for Fourier series.
Reconstruction	The larger the frequency band we include in the transform, the better we approximate $x(t)$. Perfect reconstruction requires using an infinite spectrum (if necessary).
Gibbs effect	A discontinuous signal can never be reconstructed perfectly, no matter how large a frequency band we include. The reconstructed signal always shows an overshoot ($\approx 9\%$ of jump) at the discontinuities.

Appendix 6A
MATLAB Demonstrations and Routines

Getting Started on the Supplied MATLAB Routines

This appendix presents MATLAB routines related to concepts in this chapter and some examples of their use. See Appendix 1A for introductory remarks.
Note: All polynomial coefficients must be entered in descending order.

Steady-State Response in Symbolic Form

```
yss=ssresp(ty,n,d,x)  %Computes steady-state response to sinusoids
```

Here $ty = $'s,' n and d are coefficients of $H(\omega) = N(\omega)/D(\omega)$ in descending powers of $j\omega$ and x is the array $[a\ w\ r]$ for an input of the form $x(t) = a\cos(w^*t + r)$.
To find the response of $H(\omega) = 1/(1 + j2\omega)$ to $x(t) = 4\cos(3t - 30°)$, use

```
>>yss=ssresp('s',[0 1],[2 1],[4 3 -pi/6])
```

MATLAB will respond with

```
yss=0.6576*cos(3*t-0.6141*pi)
```

yss is a string expression in t and may be plotted by first evaluating it. Try:

```
>>t=0:.05:10; ys=eval(yss);plot(t,ys)
```

Demonstrations and Utility Functions

`[m,p,w]=tfplot(ty,n,d,f,p1,p2);`

This returns the magnitude, phase (radians) and frequency (hertz) for $H(\omega) = N(\omega)/D(\omega)$ with coefficients of $N(\omega)$ and $D(\omega)$ in descending order. The argument $f = [Fl, Fh]$ defines the frequency limits (Hz), $p1 = 0$ for linear or 1 for log frequency, and $p2 = 0$ for no plots, 1 for magnitude, 2 for phase and $p2 > 2$ for both.

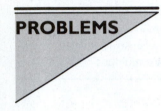

PROBLEMS

P6.1 (Fourier Transforms from the Fourier Series) Table 6.2 lists the Fourier series coefficients for several periodic signals. For each, choose $t_0 = 1$, $T = 4$, and
(a) Sketch the periodic signal $x_p(t)$ over $-2T \le t \le 2T$.
(b) Sketch the nonperiodic signal $x(t)$ that results as $T \longrightarrow \infty$.
(c) Use the listed $X_S[k]$ to find $X(f)$, the Fourier transform of $x(t)$.
(d) Sketch the magnitude and phase spectrum corresponding to $X(f)$.

P6.2 (FT From Definition) Sketch each signal $x(t)$ and, starting with the defining relation, find its Fourier transform $X(f)$. Use the integral tables in Appendix A.
(a) $x(t) = \text{rect}(t - \frac{1}{2})$ **(b)** $x(t) = 2t\,\text{rect}(t)$
(c) $x(t) = t\exp(-2t)u(t)$ **(d)** $x(t) = \exp(-2|t|)$

P6.3 (FT in Sinc Forms) Consider the signals $x(t)$ sketched in Figure P6.3.
(a) Express $x(t)$ as combinations of rect and/or tri functions.
(b) Evaluate $X(f)$ in terms of sinc functions.

P6.4 (Transforms) Sketch $x(t)$ and use properties to find and sketch $X(f)$.
(a) $x(t) = \exp(-2|t - 1|)$ **(b)** $x(t) = t\exp(-2|t|)$
(c) $x(t) = \exp(-2t)\cos(2\pi t)u(t)$ **(d)** $x(t) = u(1 - |t|)$
(e) $x(t) = u(1 - |t|)\text{sgn}(t)$ **(f)** $x(t) = \text{sinc}(t) \star \text{sinc}(2t)$
(g) $x(t) = \cos^2(2\pi t)\sin(2\pi t)$ **(h)** $x(t) = [1 - \exp(-t)]u(t)$

P6.5 (Properties) The Fourier transform of $x(t)$ is $X(f) = \text{rect}(f/2)$. Use properties to find and sketch the magnitude and phase of
(a) $x(t - 2)$ **(b)** $x'(t)$ **(c)** $x(-t)$ **(d)** $tx(t)$
(e) $x(2t)$ **(f)** $x(t)\cos(2\pi t)$ **(g)** $x^2(t)$ **(h)** $tx'(t)$
(i) $x(t) \star x(t)$

P6.6 (Properties) Consider the transform pair $x(t) \longleftrightarrow X(f)$ where $x(t) = t\exp(-2t)u(t)$. Without evaluating $X(f)$, find the time signals corresponding to
(a) $X(2f)$ **(b)** $X(f - 1) + X(f + 1)$ **(c)** $X'(f)$
(d) $fX'(f)$ **(e)** $j2\pi fX(2f)$ **(f)** $X(f/2)\cos(4\pi f)$
(g) $(1 - 4\pi^2 f^2)X(f)$

Figure P6.3

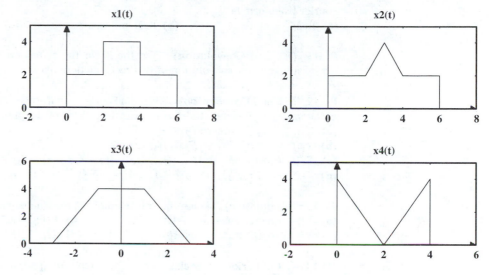

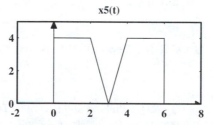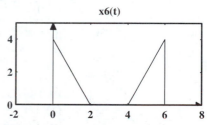

P6.7 (Properties) Find $x(0)$ from $x(t)$ directly and from the central ordinate theorem for the following signals.
(a) $x(t) = \text{rect}(t)$
(b) $x(t) = u(t)$
(c) $x(t) = \exp(-t)u(t)$
(d) $x(t) = \text{rect}(t - \frac{1}{2})$

P6.8 (Parseval's Theorem) A series RC circuit with $\tau = 1$ is excited by a voltage input $x(t)$. The output is the capacitor voltage. Find the total energy in the input and output and the output signal energy over $|f| \le 1/2\pi$ and over $|f| \le 1$ if
(a) $x(t) = \delta(t)$
(b) $x(t) = \exp(-t)u(t)$

P6.9 (Inverse Transforms) Using properties or otherwise, find the inverse Fourier transforms corresponding to each $X(f)$.
(a) $u(1 - |f|)$
(b) $j2\pi f/(1 + 4\pi^2 f^2)$
(c) $j2\pi f/(1 + 6j\pi f - 8\pi^2 f^2)$
(d) $\exp(-4j\pi f)/(1 + 4\pi^2 f^2)$
(e) $\text{sinc}(f/4)\cos(2\pi f)$
(f) $8\text{sinc}(2f)\text{sinc}(4f)$
(g) $4\text{sinc}(2f)/(1 + j2\pi f)$
(h) $\cos(\pi f)\exp(-j\pi f)/(1 + j2\pi f)$

P6.10 (Symmetry) Recall that conjugate symmetry of the Fourier transform applies only to real time signals. What is the real/complex nature and symmetry of the time signals whose Fourier transform is
(a) Real?   **(b)** Real and even?   **(c)** Real and odd?

P6.11 (FT and Convolution) Find the FT of the signal $x_3(t)$ shown in Figure P6.3 by expressing it as the convolution of two rect functions.

P6.12 (FT and Convolution) Given the Fourier transforms $X_1(f)$ and $X_2(f)$ of two signals $x_1(t)$ and $x_2(t)$, compute the Fourier transform $Y(f)$ of the product $x_1(t)x_2(t)$.
(a) $X_1(f) = X_2(f) = \text{rect}(f/4)$
(b) $X_1(f) = \text{rect}(f/2)$, $X_2(f) = X_1(f/2)$
(c) $X_1(f) = \delta(f) + \delta(f + 4) + \delta(f - 4)$, $X_2(f) = \delta(f + 2) + \delta(f - 2)$
(d) $X_1(f) = \text{tri}(f)$, $X_2(f) = \delta(f + 1) + \delta(f) + \delta(f - 1)$

P6.13 (Modulation) Sketch the spectrum of the modulated signal $y(t) = x(t)m(t)$ if
(a) $X(f) = \text{rect}(f), m(t) = \cos(\pi t)$ **(b)** $X(f) = \text{rect}(\frac{1}{4}f), m(t) = \cos(2\pi t)$
(c) $X(f) = \text{tri}(f), m(t) = \cos(10\pi t)$ **(d)** $X(f) = \text{tri}(f), m(t) = \text{comb}(2t)$

P6.14 (FT of Periodic Signals) Sketch the magnitude spectrum $|X(f)|$ and phase spectrum $\angle X(f)$ of the following periodic signals:
(a) $x(t) = 3 + 2\cos(10\pi t)$
(b) $x(t) = 4\,\text{comb}(t/2)$
(c) $x(t) = 3\cos(10\pi t) + 4\sin(10\pi t) + 6\cos(20\pi t + \frac{1}{4}\pi)$
(d) $x(t) = |\cos(\pi t)|$

P6.15 (Dirichlet Kernel)
(a) Let $x(t) = \delta(t + 1) + \delta(t) + \delta(t - 1)$. Find and sketch its Fourier transform $X(f)$.
(b) Consider the impulse sequence $x(t) = \sum \delta(t - k)$, where $k = -N, \ldots, N$. Show that $X(f) = M\,\text{sinc}(Mf)\,\text{sinc}(f)$ where $M = 2N + 1$. This result describes the Dirichlet kernel.
(c) What is the Fourier transform of $x(t) = \sum \delta(t - kt_0)$, $k = -N, \ldots, N$?

P6.16 (Application) A periodic signal $x(t)$ is passed through an ideal lowpass filter that passes frequencies below 5 Hz. Find and sketch the spectra of the filter input and output and the time-domain output if
(a) $x(t) = -4 + 2\sin(4\pi t) - 6\cos(12\pi t) + 3\sin(16\pi t) + 4\cos(16\pi t)$
(b) $x(t) = 8 - 8\cos^2(6\pi t)$ **(c)** $x(t) = |\sin(3\pi t)|$

P6.17 (Application) Each of the following signals is applied to an ideal lowpass filter that passes frequences below 2 Hz. Sketch the spectrum filtered signal.
(a) $x(t) = \exp(-t)u(t)$ **(b)** $x(t) = \exp(-|t|)$ **(c)** $x(t) = \text{sinc}(8t)$
(d) $x(t) = \text{rect}(2t)$ **(e)** $x(t) = |\sin(3\pi t/2)|$

P6.18 (System Response) The transfer function of a system is $H(\omega) = 16/(4 + j\omega)$. Find the time response for the following inputs:
(a) $x(t) = 4\cos(4t)$ **(b)** $x(t) = 4\cos(4t) - 4\sin(4t)$
(c) $x(t) = \exp(-4t)u(t)$ **(d)** $x(t) = 4\cos(4t) - 4\sin(2t)$

P6.19 (System Response) The transfer function $H(\omega)$ of a system is

$$H(\omega) = \frac{2 + 2j\omega}{4 + 4j\omega - \omega^2}$$

Find the time response and the spectrum for the following inputs.
(a) $x(t) = 4\cos(2t)$ **(b)** $x(t) = \exp(-t)u(t)$
(c) $x(t) = \delta(t)$ **(d)** $x(t) = 2\delta(t) + \delta'(t)$

P6.20 (System Formulation) Set up the system differential equation for each of the following systems.
(a) $H(\omega) = \dfrac{3}{2 + j\omega}$

(b) $H(\omega) = \dfrac{1 + j2\omega - \omega^2}{(1 - \omega^2)(4 - \omega^2)}$

(c) $H(\omega) = \dfrac{2}{1 + j\omega} - \dfrac{1}{2 + j\omega}$

(d) $H(\omega) = \dfrac{2j\omega}{1 + j\omega} - \dfrac{j\omega}{2 + j\omega}$

P6.21 (Frequency Response) Consider the following systems:

(1) $h(t) = \exp(-2t)u(t)$ (2) $h(t) = [1 - \exp(-2t)]u(t)$
(3) $h(t) = \text{sinc}(t)$ (4) $h(t) = t\exp(-t)u(t)$
(5) $h(t) = 0.5\delta(t)$ (6) $h(t) = [\exp(-t) - \exp(-2t)]u(t)$
(7) $h(t) = \delta(t) - \exp(-t)u(t)$

(a) Find the frequency response of each system.
(b) Which of these systems are stable and which are causal?
(c) Find the differential equation and order of each causal stable system.

P6.22 (Frequency Response) Find the frequency response and impulse response of the following systems.
(a) $y''(t) + 3y'(t) + 2y(t) = 2x'(t) + x(t)$
(b) $y''(t) + 4y'(t) + 4y(t) = 2x'(t) + x(t)$
(c) $y(t) = 0.2x(t)$

P6.23 (Power Spectral Density) Sketch the PSD (power spectral density) of
(a) $x(t) = 8\cos(10\pi t) + 4\sin(20\pi t) + 6\cos(30\pi t + \frac{1}{4}\pi)$
(b) $x(t) = 4 + 8\cos(10\pi t) + 4\sin(20\pi t)$

7 APPLICATIONS OF FOURIER TRANSFORMS

The Fourier transform finds widespread application in many diverse fields. In this chapter, we briefly examine those applications and extensions that are related directly to the area of system analysis and signal processing. This chapter is divided into three more or less independent parts:

Part 1 Ideal filters and their characteristics.
Part 2 Time-bandwidth relations.
Part 3 Modulation in communications systems.

Useful Background   Much of the material in this chapter relies on the concepts developed in the previous chapter and in Chapters 4 and 5.

We shall have a lot to say about filters in later chapters, but for now, we regard a filter as a device that passes a certain range of frequencies and blocks the rest. The range of frequencies passed defines the **passband** and the range of frequencies blocked defines the **stopband**. The band-edge frequencies are called the **cutoff frequencies**.

Ideally, we would like to have perfect transmission in the passband and perfect rejection in the stopband. Perfect transmission implies a transfer function with unity gain and zero phase over the passband or

$$H(f) = 1 \text{ over the passband}$$

In other words, the output $Y(f)$ equals the input $X(f)$ over the passband.

This concept leads to the **ideal filter** types listed in Table 7.1, whose magnitude spectra and impulse response are sketched in Figure 7.1.

Table 7.1   Classification of Common Filter Types.

Type	Passband	Stopband
Lowpass	0 to f_C	f_C to ∞
Highpass	f_C to ∞	0 to f_C
Bandpass	f_1 to f_2	0 to f_1 and f_2 to ∞
Bandstop	0 to f_1 and f_2 to ∞	f_1 to f_2

Figure 7.1   Ideal filters and their impulse response.

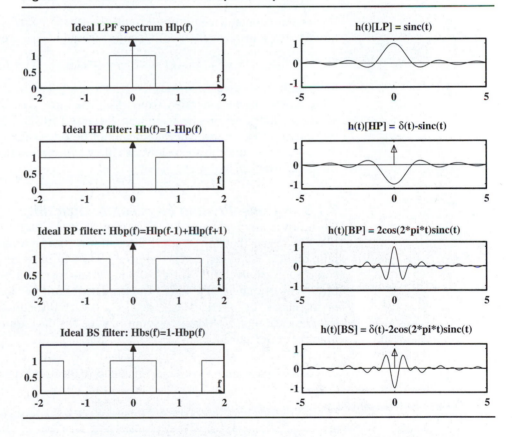

7.1.1   *Perfect and Ideal Filters*

The requirements of unity gain and zero phase in the passband turn out to be both impractical and unrealizable, and it is customary to relax these as follows. If the filter has *constant gain,* the output is an amplitude-scaled replica of the input. Similarly, if the output undergoes a *linear phase shift with frequency,* the output is a time-shifted replica of the input. A filter-transfer function with constant gain and linear phase, though not perfect, is referred to as **distortionless.**

   Amplitude distortion results if the gain is not constant over the required frequency range. **Phase distortion** results when the phase shift is not linear with frequency. The signal undergoes different delays at different instants, a situation termed **dispersion**.

7.1.2   *Distortionless Transmission*

Since a distortionless filter permits a linear phase shift and constant gain over the passband, we can write its transfer function over the passband as

$$H(f) = K \exp(-j2\pi f t_0) \quad \text{or} \quad Y(f) = KX(f)\exp(-j2\pi f t_0)$$

We have a constant magnitude K, and the phase $-2\pi f t_0$ varies linearly with frequency. For a lowpass filter with a cutoff frequency f_C, we have

$$H_{\mathrm{LP}}(f) = \exp(-j2\pi f t_0) \qquad |f| \le f_C$$

Its magnitude is given by $|H_{\mathrm{LP}}(f)| = \mathrm{rect}(f/2f_C)$. The transfer function of all other filter types can be generated from $H_{\mathrm{LP}}(f)$ (Figure 7.1). The impulse response of all these filters has a sinc form and extends forever (Figure 7.1). This suggests that *ideal filters are noncausal* and cannot be physically realized. In the following discussion, we concentrate on lowpass filters, whose impulse response has the form $h(t) = 2f_C\mathrm{sinc}[2f_C(t - t_0)]$.

7.1.3   *Causality and Physical Realizability*

For a physically realizable filter, we require $h(t) = 0$, $t < 0$. One might, of course, hope or argue that for an ideal filter, if we choose a large enough delay t_0, the tails of $h(t) = 2f_C\mathrm{sinc}[2f_C(t - t_0)]$ for $t \le 0$ will have very small magnitudes. We could then truncate $h(t)$ for $t < 0$ and come close to a physically realizable ideal filter. Unfortunately, this kind of argument cannot be sustained either in practice or in theory. In practice, large delays are very difficult to implement. In theory, the truncation of $h(t)$ leads to some curious results, summarized in Table 7.2 and illustrated in Figure 7.2.

Onesided Truncation   The causal onesided truncated $h(t)$ may be written as

$$\tilde{h}(t) = 2f_C\mathrm{sinc}[2f_C(t - t_0)]u(t)$$

Table 7.2   Effect of Truncating h(t) of an Ideal Filter.

Truncation	Causal?	Impulse response	Magnitude spectrum
None	No	$h(t) = 2f_C\mathrm{sinc}[2f_C(t - t_0)]$	Constant in passband
Onesided	Yes	$h(t)u(t)$	Ripples in passband and infinite ears at band-edges
Twosided	Yes	$h(t)[u(t) - u(t - T)]$	Ripples in passband and finite ears at band-edges

Figure 7.2   The impulse response of an ideal filter, its onesided and twosided truncation and the corresponding spectra.

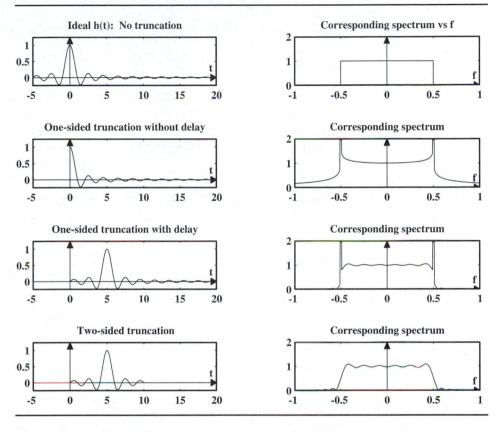

Its transfer function $\tilde{H}(f)$ may be described by the convolution

$$\tilde{H}(f) = \left[\frac{1}{2}\delta(f) + \frac{1}{j2\pi f} \right] \star \text{rect}(f/2f_C)\,\exp(j2\pi f t_0)$$

The convolution with $\delta(f)$ gives us $\text{rect}(f/2f_C)\exp(j2\pi f t_0)$, whose spectrum is identical to the ideal filter but with magnitude $\frac{1}{2}$. The convolution with $1/j2\pi f$ also yields a pulselike shape with a near-constant magnitude of $\frac{1}{2}$, to give a total magnitude of 1, as required. But this convolution becomes infinite at the pulse edges due to the term $1/f$. It also contains ripples due to the oscillations of $\exp(j2\pi f t_0)$. As the delay t_0 becomes large, the ripples decrease in magnitude, but the spikes never quite disappear.

Twosided Truncation   To avoid the infinite spikes due to the term $1/f$ which represents part of the transform of an *infinite* step, we also truncate $h(t)$ for positive time to make it *timelimited*. This is equivalent to multiplying $h(t)$ by a rectangular pulse and results in the causal impulse response $\tilde{h}(t)$ described by

$$\tilde{h}(t) = 2f_C \text{sinc}[2f_C(t - t_0)][u(t) - u(t - T)]$$

Its transfer function is described by the frequency-domain convolution

$$\tilde{H}(f) = \text{rect}(f/2f_C)\exp(j2\pi f t_0) \star T \text{sinc}(fT)\exp(-j\pi fT)$$

Since $\tilde{h}(t)$ is timelimited, $|\tilde{H}(f)|$ is of infinite extent and shows ripples that decay with frequency due to the decaying oscillations in $\text{sinc}(fT)$. It also shows *finite overshoot*, or ears, near the band edges. These ears are due to the Gibbs effect and persist as long as T remains finite. As $T \longrightarrow \infty$, $T\text{sinc}(fT)$ approaches an impulse, and $|\tilde{H}(f)|$ approaches the ideal filter response $\text{rect}(f/2f_C)$.

Step Response of Ideal Filters   The step response of an ideal lowpass filter is the convolution $u(t) \star h(t)$. Ignoring the time delay t_0, we have

$$y(t) = u(t) \star h(t) = u(t) \star 2f_C\text{sinc}(2f_Ct) = \int_{-\infty}^{t} 2f_C\text{sinc}(2f_C\lambda)\,d\lambda$$

This integral cannot be evaluated in closed form. Recognizing that the area under $\text{sinc}(2f_C\lambda)$ over $(-\infty, 0)$ equals $1/4f_C$, we can express $y(t)$ as

$$y(t) = \frac{1}{2} + \int_0^t 2f_C\text{sinc}(2f_C\lambda)d\lambda = \frac{1}{2} + \frac{1}{\pi}\int_0^t \frac{\sin(2\pi f_0\lambda)}{2\pi f_0\lambda}d(2\pi f_C\lambda) = \frac{1}{2} + \frac{1}{\pi}\text{si}(2\pi f_Ct)$$

where $\text{si}(x)$ describes the *sine integral* defined by

$$\text{si}(x) = \int_0^x \frac{\sin(\alpha)}{\alpha}\,d\alpha$$

The quantity $\text{si}(x)$ represents the area under $\sin(\alpha)/\alpha$ over $(0, x)$.

Intuitively, we expect $y(t)$ to follow the wiggles in the sinc variation of $h(t)$. The response $y(t)$, sketched against f_Ct in Figure 7.3, shows overshoot and decaying oscillations or *ringing* due to the Gibbs effect. Increasing the cutoff frequency f_C simply compresses $y(t)$, but the overshoot and ringing persist. As $f \longrightarrow \infty$, the response becomes identical to a unit step except for spikes at $t = 0$. For a *monotonic* step response $w(t)$, we require $w'(t) > 0$ which implies that $h(t) = \frac{dw(t)}{dt} > 0$. Clearly, an ideal filter can never have a monotonic step response because its impulse response (a sinc function) does not satisfy $h(t) > 0$.

Pulse Response   Using superposition, the response to $x(t) = \text{rect}(t)$ equals $y(t + \frac{1}{2}) - y(t - \frac{1}{2})$ and also shows overshoot and oscillations.

We summarize the main results of our discussion so far:

1. An ideal filter cannot possess a monotonic step response.
2. A physically realizable filter cannot display constant gain, or linear phase, over its passband.
3. Sharp cutoffs require large delays that are difficult to implement and produce undesirable ringing in the time response.

Figure 7.3 The impulse and step response of an ideal filter.

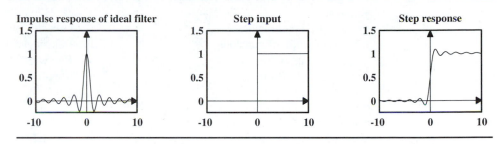

7.1.4 *Windows and Their Smoothing Effects*

Twosided truncation of $h(t)$ is equivalent to windowing it by a rectangular pulse. To reduce the spectral wiggles and ears due to abrupt truncation, we use smoothing windows for $h(t)$, much as we did for Fourier series smoothing. The difference is that they are now continuous time-domain windows.

Tables 7.3 and 7.4 list some commonly used time-domain windows (sometimes called **data windows**) and their spectra are sketched in Figure 7.4.

Table 7.3 Some Common Windows and Their Spectra.

Window	Expression $w(t)$ $(\lvert t \rvert \le \tfrac{1}{2})$	Spectrum $W(f)$
Boxcar	$\mathrm{rect}(t)$	$\mathrm{sinc}(f)$
Bartlett	$\mathrm{tri}(2t) = 1 - 2\lvert t \rvert$	$\tfrac{1}{2}\mathrm{sinc}^2(\tfrac{1}{2}f)$
vonHann	$\tfrac{1}{2} + \tfrac{1}{2}\cos(2\pi t)$	$\tfrac{1}{2}\mathrm{sinc}(f) + \tfrac{1}{4}\mathrm{sinc}(f+1) + \tfrac{1}{4}\mathrm{sinc}(f-1)$
Hamming	$0.54 + 0.46\cos(2\pi t)$	$0.54\mathrm{sinc}(f) + 0.23[\mathrm{sinc}(f+1) + \mathrm{sinc}(f-1)]$
Cosine	$\cos(\pi t)$	$\tfrac{1}{2}\mathrm{sinc}(f + \tfrac{1}{2}) + \tfrac{1}{2}\mathrm{sinc}(f - \tfrac{1}{2})$
Riemann	$\mathrm{sinc}(2t)$	$[\mathrm{si}(\pi f + \pi) - \mathrm{si}(\pi f - \pi)]/2\pi$
Parzen-1	$\begin{cases} 1 - 24\lvert t \rvert^2(1 - 2\lvert t \rvert) & \lvert t \rvert < \tfrac{1}{4} \\ 2(1 - 2\lvert t \rvert)^3 & \tfrac{1}{4} < \lvert t \rvert < \tfrac{1}{2} \end{cases}$	$\tfrac{3}{8}\mathrm{sinc}^4(\tfrac{1}{4}f)$
Parzen-2	$1 - 4t^2$	$2[\mathrm{sinc}(f) - \cos(\pi f)]/(\pi f)^2$ $[W(0) = 2/3]$
Blackman	$0.42 + 0.5\cos(2\pi t) + 0.08\cos(4\pi t)$	$0.42\mathrm{sinc}(f) + 0.25[\mathrm{sinc}(f+1) + \mathrm{sinc}(f-1)]$ $+0.04[\mathrm{sinc}(f+2) + \mathrm{sinc}(f-2)]$
Papoulis	$\frac{1}{\pi}\lvert \sin(2\pi t) \rvert + (1 - 2\lvert t \rvert)\cos(2\pi t)$	$\tfrac{1}{4}[\mathrm{sinc}(\tfrac{1}{2}f + \tfrac{1}{2}) + \mathrm{sinc}(\tfrac{1}{2}f - \tfrac{1}{2})]^2$
Kaiser	$\dfrac{I_0(\pi\beta\sqrt{1 - 4t^2})}{I_0(\pi\beta)},\quad 0 \le \pi\beta \le 10$	$\dfrac{\mathrm{sinc}[(f^2 - \beta^2)^{1/2}]}{I_0(\pi\beta)}$
Modified Kaiser	$\dfrac{I_0(\pi\beta\sqrt{1 - 4t^2}) - 1}{I_0(\pi\beta) - 1},\quad 1 \le \pi\beta \le 10$	$\dfrac{\mathrm{sinc}[(f^2 - \beta^2)^{1/2}] - \mathrm{sinc}(f)}{I_0(\pi\beta) - 1}$
Bickmore	$\left[\dfrac{\sqrt{1 - 4t^2}}{\pi\beta}\right]^{\nu-1/2}\left[\dfrac{I_{\nu-1/2}(\pi\beta\sqrt{1 - 4t^2})}{I_{\nu-1/2}(\pi\beta)}\right]$	$\dfrac{\sqrt{\tfrac{1}{2}\pi}J_\nu[\pi(f^2 - \beta^2)^{1/2}]}{[\pi(f^2 - \beta^2)^{1/2}]^\nu I_{\nu-1/2}(\pi\beta)}$
General Blackman	$a_0 + \sum_{k=1}^{N} 2a_k\cos(2\pi kt)$	$\sum_{k=-N}^{N} a_k\mathrm{sinc}(f - k)$ $(a_{-k} = a_k)$

Table 7.4   Special Characteristics of Window Functions.

Notation:

G_P: Peak gain of mainlobe G_S: Peak sidelobe gain

W_M: Half-width of mainlobe P_S: Peak sidelobe gain $\frac{G_S}{G_P}$ in dB

W_6: 6-dB half-width W_S: Half-width of main lobe to reach P_S

W_3: 3-dB half-width D_S: High-frequency decay rate (dB/oct)

Window	G_P	$\frac{G_S}{G_P}$	P_S dB	W_M	W_S	W_6	W_3	D_S dB/oct
Boxcar	1	0.2172	−13.3	1	0.81	0.6	0.44	−6
Parzen-2	0.6667	0.0862	−21.3	1.43	1.27	0.79	0.57	−12
Cosine	0.6366	0.0708	−23	1.5	1.35	0.81	0.59	−12
Riemann	0.5895	0.0478	−26.4	1.64	1.5	0.86	0.62	−12
Bartlett	0.5	0.0472	−26.5	2	1.62	0.88	0.63	−12
vonHann	0.5	0.0267	−31.5	2	1.87	1.0	0.72	−18
Hamming	0.54	0.0073	−42.7	2	1.91	0.9	0.65	−6
Papoulis	0.4053	0.0050	−46	3	2.7	1.18	0.85	−24
Parzen-1	0.375	0.0022	−53	4	3.25	1.27	0.91	−24
Blackman	0.42	0.0012	−58.1	3	2.82	1.14	0.82	−18
Kaiser ($\beta = 2.6$)	0.4314	0.0010	−59.9	2.8	2.72	1.11	0.80	−6
Bickmore								
$\beta = 2.6, \nu = \frac{1}{2}$	0.4314	0.0010	−59.9	2.8	2.72	1.11	0.80	−6
$\beta = 3, \nu = 3$	0.0013	0.0005	−65.7	3.63	3.54	1.36	0.97	−21

Remarks:

1. All windows are even symmetric and entirely positive.
2. All windows have been normalized to unit height and width for comparison.
3. All windows show a smooth taper. The boxcar, Hamming, Kaiser and Bickmore windows are abruptly truncated at $t = \pm\frac{1}{2}$.
4. The vonHann window is also known as the *Hanning* window.
5. The Riemann window $\text{sinc}^L(2t), L > 0$, is also known as the *Lanczos* window.
6. The Bartlett window is also called the *Fejer, Cesaro* or *Parzen* window.
7. The Kaiser window is also known as the *Taylor* window.
8. The Parzen-2 window is very similar to the so-called *Welch* window.
9. The coefficients of the general Blackman window are normalized to unit height using $a_0 + \Sigma 2a_k = 1$. The boxcar, vonHann, Hamming, Blackman and cosine windows fall out as special cases of this general window.
10. The magnitude in **decibels** (dB) is $X_{dB} = 20\log X$. An **octave** represents a *twofold* change (in frequency).
11. β controls the peak sidelobe gain.
12. For $\nu = \frac{1}{2}$, the Bickmore window reduces to the Kaiser window. Note that the parameter ν controls the sidelobe decay rate ($D_S = -6(\nu + \frac{1}{2})$ dB/oct).
13. Empirically determined relations for the Kaiser window are:

$$G_P = \frac{|\text{sinc}(j\beta)|}{I_0(\pi\beta)} \quad \frac{G_S}{G_P} = \frac{0.22\pi\beta}{\sinh(\pi\beta)} \quad W_M = 2(1 + \beta^2)^{1/2} \quad W_S = 2(0.661 + \beta^2)^{1/2}$$

Figure 7.4   Spectra of some of the data windows of Tables 7.3 and 7.4.

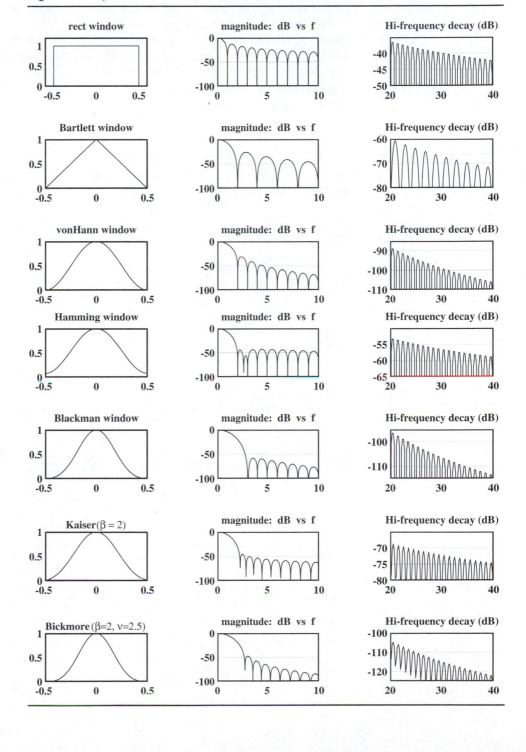

The main objectives in designing a window are to reduce both the mainlobe width (for a sharp transition) and the sidelobes (to minimize ripple). Unfortunately, these two objectives are incompatible. For all windows, a reduced sidelobe level can be achieved only at the expense of an increased mainlobe width. The compromise in using the various windows is between choosing the central lobe to be as narrow as possible and choosing the sidelobe level to be as small as possible. Any window except rectangular yields a marked improvement! The Hamming and Kaiser windows are among the most widely used.

Example 7.1
Windowing

Windowing the ideal impulse response by a triangular pulse (Figure 7.5) is equivalent to the convolution of the ideal spectrum and a sinc^2 spectrum. The smoothing effect is evident in Figure 7.5. A broader window would lead to an even sharper transition.

Figure 7.5   Windowing in time for Example 7.1.

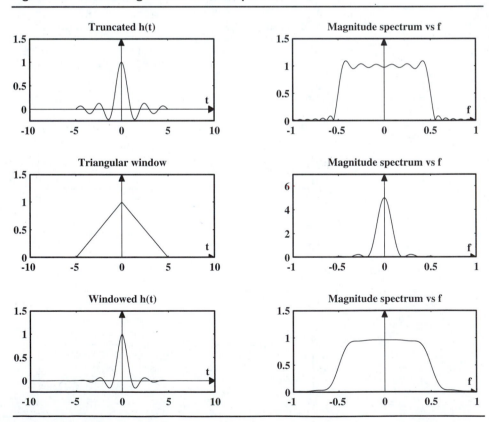

7.1.5   *Toward Real Filters*

Practical or real filters must be causal and stable. In the time domain, causality requires that $h(t) = 0, t < 0$. Stability requires that $|h(t)|$ be *absolutely integrable*. In the frequency domain, we are assured a causal, stable system only if

1. $|H(f)|$ is an energy signal (for stability), and
2. $H(f)$ satisfies the **Paley-Wiener criterion** (for causality)

$$\int_{-\infty}^{\infty} \frac{|\ln(|H(f)|)|}{1 + f^2} \, df < \infty$$

If $|H(f)| = 0$ over a finite frequency band, this integral becomes infinite. Note that $|H(f)|$ can be zero at some isolated frequencies and still satisfy the Paley-Wiener criterion. If satisfied, the Paley-Wiener criterion asserts that $|H(f)|\exp[j\theta(f)]$ describes a physically realizable system with magnitude $|H(f)|$ and phase $\theta(f)$. It does not tell us what $\theta(f)$ or $|H(f)|$ should be. In that sense, it has little practical appeal.

Example 7.2

Why an Ideal Filter is Physically Unrealizable

An ideal filter with $h(t) = \text{sinc}(\alpha t)$ is physically unrealizable because

1. Its impulse response $h(t)$ is noncausal and not absolutely integrable.
2. Even though $|H(f)| = (1/\alpha)\text{rect}(f/\alpha)$ is an energy signal, it violates the Paley-Wiener criterion (since $H(f)$ is zero past the cutoff frequency).

7.1.6   *Time-Domain Relationships*

Causality leads to a very powerful constraint in the time domain. Recall that the even and odd parts of $h(t)$ may be found as

$$h_e(t) = \tfrac{1}{2}[h(t) + h(-t)] \qquad h_o(t) = \tfrac{1}{2}[h(t) - h(-t)]$$

If $h(t) = 0$, $t < 0$, we have $h(-t) = 0, t > 0$. Thus, $h(t)$ and its folded version do not overlap, except at $t = 0$. The even part $\tfrac{1}{2}[h(t) + h(-t)]$ equals $\tfrac{1}{2}h(t)$ for $t > 0$. At $t = 0$, however, $h_e(t) = \tfrac{1}{2}[h(0) + h(0)] = h(0)$. We may, therefore, express $h(t)$ only in terms of $h_e(t)$ as

$$h(t) = h_e(t) + \text{sgn}(t)h_e(t) = \begin{cases} 2h_e(t) & t > 0 \\ h_e(0) & t = 0 \\ 0 & t < 0 \end{cases}$$

The odd part $h_o(t)$ also equals $\tfrac{1}{2}h(t)$ for $t > 0$. At $t = 0$, however, $h_o(0)$ is always zero regardless of the value of $h(0)$. We then express $h(t)$ in terms of $h_o(t)$ as

$$h(t) = \begin{cases} 2h_o(t) & t > 0 \\ 0 & t < 0 \end{cases}$$

Thus $h(t)$ is determined completely from $h_e(t)$, and determined everywhere except at $t = 0$, from $h_o(t)$. The odd and even parts are also related to each other.

$$h_o(t) = h_e(t)\text{sgn}(t)$$
$$h_e(t) = h_o(t)\text{sgn}(t) + h_e(0)$$

7.1.7 Frequency-Domain Relationships

Causality also imposes a powerful constraint in the frequency domain. Just as $h(t)$ can be found from its even part alone, so can $H(f)$. Since $h(t) = h_e(t) + \text{sgn}(t)h_e(t)$, $H(f)$ may be expressed as

$$H(f) = H_e(f) + \frac{1}{j\pi f} \star H_e(f)$$

We can also write the Fourier transform of $h(t)$ in complex form as

$$H(f) = H_R(f) + jH_I(f)$$

If $h(t)$ is real, $H_R(f)$ and $H_I(f)$ correspond to the Fourier transforms of $h_e(t)$ and $h_o(t)$, respectively. Comparison of the two forms for $H(f)$ leads to

$$h_e(t) \longleftrightarrow H_R(f) = H_e(f)$$
$$h_o(t) \longleftrightarrow H_I(f) = \frac{-1}{\pi f} \star H_e(f) = \frac{-1}{\pi f} \star H_R(f)$$

The convolution $H(x) \star 1/\pi x = \hat{H}(x)$ defines the **Hilbert transform** of $H(x)$. We may thus write

$$H_I(f) = \frac{-1}{\pi f} \star H_R(f) = -\hat{H}_R(f)$$

Since the real and imaginary parts of $H(f)$ are related (by the Hilbert transform), $H(f)$ can thus be found from its real part alone. This condition is referred to as **real-part sufficiency**. The transfer function $H(f)$ of a physically realizable system may also be found from its magnitude spectrum $|H(f)|$ alone. A transfer function derived from the magnitude spectrum alone is known as a **minimum-phase transfer function**. Its implications for filter design are discussed in Chapters 9 and 10.

7.2
Time-Bandwidth Relations

Time-bandwidth relations quantify the fundamental concept that measures of the impulse-response duration of a system are inversely related to measures of its bandwidth. After all, the narrower a signal, the wider its spectrum.

Since the step response and impulse response are related, the rise time of the step response also serves as a practical measure of impulse-response duration. The commonly used **10% to 90% rise time** T_r is defined as the time required for the step response $w(t)$ to rise from 10% to 90% of its final value

$$T_r = w(t_{90}) - w(t_{10})$$

A practical measure of bandwidth is the **half-power bandwidth**, the range of frequencies over which the magnitude exceeds $\sqrt{1/2}$ times its maximum.

Frequently used measures for duration and bandwidth are summarized in Table 7.5, and depend on the definitions used. Any of these measures can be used provided the appropriate quantities exist or are finite.

Table 7.5   Time-Bandwidth Measures.

Note:

For all systems:
$$t_0 = \frac{\mathbb{A}[th(t)]}{\mathbb{A}[h(t)]} \qquad\qquad T_0 = \frac{\mathbb{A}[t|h(t)|^2]}{\mathbb{A}[|h(t)|^2]}$$

For lowpass systems:
$$f_0 = 0 \qquad\qquad F_0 = 0$$

For bandpass systems:
$$f_0 = \frac{\mathbb{A}[f|H(f)|](0,\infty)}{\mathbb{A}[|H(f)|](0,\infty)} \qquad\qquad F_0 = \frac{\mathbb{A}[f|H(f)|^2](0,\infty)}{\mathbb{A}[|H(f)|^2](0,\infty)}$$

Entry	Duration T_r	Bandwidth B_f	Product $T_r B_f$
		Measures based on areas of $h(t)$ and $H(f)$	
1.	$\dfrac{\mathbb{A}[h(t)]}{h(0)}$	$\dfrac{\mathbb{A}[H(f)]}{H(0)}$	$T_r B_f = 1$
2.	$\dfrac{\mathbb{A}[|h(t)|]}{|h(t_0)|}$	$\dfrac{\mathbb{A}[|H(f)|]}{|H(f_0)|}$	$T_r B_f \geq 1$
		Measures based on energy	
3.	$\dfrac{\{\mathbb{A}[h(t)]\}^2}{\mathbb{A}[h^2(t)]}$	$\dfrac{\mathbb{A}[|H(f)|^2]}{|H(0)|^2}$	$T_r B_f = 1$
4.	$\dfrac{\{\mathbb{A}[|h(t)|]\}^2}{\mathbb{A}[|h(t)|^2]}$	$\dfrac{\mathbb{A}[|H(f)|^2]}{|H(f_0)|^2}$	$T_r B_f \geq 1$
		Measures based on moments of $h(t)$ and $H(f)$	
5.	$2\left[\dfrac{\mathbb{A}[t^2|h(t)|]}{\mathbb{A}[|h(t)|]}\right]^{1/2}$	$2\left[\dfrac{\mathbb{A}[f^2|H(f)|]}{\mathbb{A}[|H(f)|]}\right]^{1/2}$	$T_r B_f \geq 2/\pi$
6.	$2\left[\dfrac{\mathbb{A}[t^2|h(t)|]}{\mathbb{A}[|h(t)|]} - t_0^2\right]^{1/2}$	$2\left[\dfrac{\mathbb{A}[f^2|H(f)|](0,\infty)}{\mathbb{A}[|H(f)|](0,\infty)} - f_0^2\right]^{1/2}$	$T_r B_f \geq 2/\pi$
		Measures based on moments of $h^2(t)$ and $H^2(f)$	
7.	$2\left[\dfrac{\mathbb{A}[t^2|h(t)|^2]}{\mathbb{A}[|h(t)|^2]}\right]^{1/2}$	$2\left[\dfrac{\mathbb{A}[f^2|H(f)|^2]}{\mathbb{A}[|H(f)|^2]}\right]^{1/2}$	$T_r B_f \geq 1/\pi$
8.	$2\left[\dfrac{\mathbb{A}[t^2|h(t)|^2]}{\mathbb{A}[|h(t)|^2]} - T_0^2\right]^{1/2}$	$2\left[\dfrac{\mathbb{A}[f^2|H(f)|^2](0,\infty)}{\mathbb{A}[|H(f)|^2](0,\infty)} - F_0^2\right]^{1/2}$	$T_r B_f \geq 1/\pi$
		Measures based on practical values	
9.	10–90% rise time	Half-power bandwidth	

Remark: We consistently define the bandwidth for a *twosided* spectrum in Hz.

Entries 1 and 2 interpret the duration, or *width*, of $h(t)$ as a ratio of its area $\mathbb{A}[h(t)]$, and its *height* $h(0)$. A similar definition for the bandwidth of a lowpass filter gives $B_f = \mathbb{A}[H(f)]/H(0)$ and the equality $T_r B_f = 1$.

To accommodate sign changes or complex values, we use $|h(t)|$ and $|H(f)|$. All measures based on such absolute values satisfy the inequality $T_r B_f \geq C$.

To account for bandpass systems, we use $B_f = \mathbb{A}[H(f)]/H(f_0)$, where f_0 is a measure of the passband **center frequency**. Since the response of practical filters also shows a finite delay, we use $T_r = \mathbb{A}[h(t)]/h(t_0)$, where t_0 is a measure of **delay**. The delay is often computed as the mean of $h(t)$, or as the time at which the step response reaches 50% of its final value.

Entries 3 and 4 are energy measures based on **equivalent duration** and **equivalent bandwidth**. They lead to $T_r B_f \geq 1$, where the equality holds for $h(t) = \text{sinc}(\alpha t)$, the impulse response of an ideal filter.

Entries 5 and 6 rely on the moments of $h(t)$ and $H(f)$. The delay t_0 equals the mean, and T_r equals twice the normalized second central moment:

$$t_0 = \mathbb{A}[th(t)]/\mathbb{A}[h(t)] \qquad T_r = 2\mathbb{A}[(t-t_0)^2 h(t)]/\mathbb{A}[h(t)]$$

Frequency measures are similarly defined by

$$f_0 = \mathbb{A}[fH(f)]/\mathbb{A}[H(f)] \qquad B_f = 2\mathbb{A}[(f-f_0)^2 H(f)]/\mathbb{A}[H(f)]$$

They lead to the inequality $T_r B_f \geq 2/\pi$, with equality if $h(t)$ is a Gaussian.

Entries 7 and 8 define the **rms duration** and **rms bandwidth**, and are based on the moments of $|h(t)|^2$ and $|H(f)|^2$. They lead to the inequality $T_r B_f \geq 1/\pi$, with equality if $h(t)$ is a Gaussian.

Table 7.6 lists the time-bandwidth products for some common signals using the various definitions of Table 7.5. Of all functions, *the Gaussian has the smallest time-bandwidth product.*

Remarks:

1. In entries 1 and 3, Parseval's relation and the central ordinate theorems guarantee the equality $T_r B_f = 1$ for any transform pair $h(t) \longleftrightarrow H(f)$. For example, if $h(t) = \exp(-t/\tau)u(t)$, entry 1 yields $T_r = \tau$ and $B_f = 1/\tau$.
2. For $h(t) = \text{sinc}(t)$, entry 1 yields T_r as the half-width of its central lobe, and B_f as the frequency band outside of which $H(f)$ is zero.

Example 7.3
Time-Bandwidth Relations

(a) *A Practical First-Order Lowpass Filter*: Consider a lowpass filter with $h(t) = (1/\tau)\exp(-t/\tau)u(t), H(f) = 1/(1 + j2\pi f\tau), w(t) = [1 - \exp(-t/\tau)]u(t)$. Using entry 9 of Table 7.5, the rise time T_r is computed by finding $w(t_{10}) = 1 - \exp(-t_{10}/\tau) = 0.1, w(t_{90}) = 1 - \exp(-t_{90}/\tau) = 0.9$. Thus $t_{10} = \tau \ln(10/9)$, $t_{90} = \tau \ln(10)$, and $T_r = t_{90} - t_{10} = \tau \ln(9)$. The half-power bandwidth B_f is the range of frequencies for which $|H(f)| \geq \sqrt{1/2}$. This corresponds to the range $-1/2\pi\tau \leq f \leq 1/2\pi\tau$. Thus, $B_f = 1/\pi\tau$ and $T_r B_f = \ln(9)/\pi = 0.6994$.

Table 7.6   Time-Bandwidth Product of Common Signals.

Signal $h(t)$	Transform $H(f)$	Entry used from Table 7.5	Duration T_r	Bandwidth B_f	Product $T_r B_f$
$\text{sinc}(t)$	$\text{rect}(f)$	#1	1	1	1
		#3	1	1	1
$\exp(-\alpha t)u(t)$	$\dfrac{1}{\alpha + j2\pi f}$	#1	$1/\alpha$	α	1
		#3	$2/\alpha$	$\alpha/2$	1
$t\exp(-\alpha t)u(t)$	$\dfrac{1}{(\alpha + j2\pi f)^2}$	#2 $(t_0 = 2/\alpha)$	$e^2/2\alpha$	$\alpha/2$	$e^2/4$
		#3	$4/\alpha$	$\alpha/4$	1
		#7	$2\sqrt{3}/\alpha$	$2/\alpha$	$4\sqrt{3}$
		#8 $(T_0 = \frac{3}{2\alpha})$	$\sqrt{3}/\alpha$	2α	$2\sqrt{3}$
$\exp(-\pi t^2)$	$\exp(-\pi f^2)$	#1	1	1	1
		#3	$\sqrt{2}$	$1/\sqrt{2}$	1
		#5	$\sqrt{2/\pi}$	$\sqrt{2/\pi}$	$2/\pi$
		#7	$\sqrt{1/\pi}$	$\sqrt{1/\pi}$	$1/\pi$

Comment: This result is often expressed using the *onesided* bandwidth as $T_r B = 0.35$ (with B in hertz) or $T_r B_w \approx 2.2$ (with B_w in radians per second). It is also used as an approximation for higher-order systems.

(b) *A Distortionless Ideal Lowpass Filter:* Consider an ideal filter with $H(f) = \text{rect}(f)\exp(-j2\pi f t_0)$, $h(t) = \text{sinc}(t - t_0)$, where t_0 is the delay, in an effort to make $h(t)$ causal. Its step response is

$$w(t) = h(t) \star u(t) = \frac{1}{2} + \frac{1}{\pi}\text{si}[2\pi(t - t_0)]$$

Both $h(t)$ and $w(t)$ are sketched in Figure 7.6 (p. 308). Define $w_f = 1$ as the final value of $w(t)$, T_f as the time to first reach w_f, T_{lz} as the time of the last zero crossing (at $t = 0$ in Figure 7.6), and T_{10} and T_{90} as the times to reach 10% and 90% of w_f after the last zero. Since $|H(f)| = \text{rect}(f)$, $B_f = 1$. We compute the following numerically estimated measures for duration and time-bandwidth product:

   1. $T_r = T_f - T_{lz} = 1.234,$ $T_r B_f = 1.234$
   2. $T_r = T_{90} - T_{lz} = 1.064,$ $T_r B_f = 1.064$
   3. $T_r = T_{90} - T_{10} = 0.892,$ $T_r B_f = 0.892$

Comment: The last result is often listed as $T_r B = 0.446$ (using the *onesided* bandwidth).

Figure 7.6   Finding the time-bandwidth product for an ideal filter.

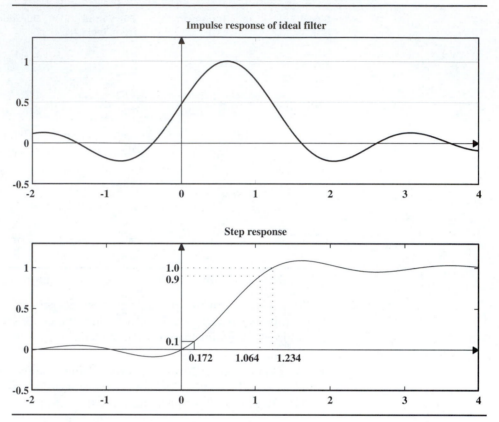

7.2.1   *A Final Thought*

From these examples, an important observation that can be generalized to any system is that regardless of the actual definition used (as long as it is reasonable) or the actual numbers obtained, *the time-bandwidth product is a constant* that satisfies the inequality

$$T_r B_f \geq C$$

where C depends on the definitions of time and bandwidth. The time-bandwidth product is an important concept that represents the equivalent of the celebrated *uncertainty principle* of quantum mechanics. We cannot simultaneously make both time and frequency measures arbitrarily small. The more localized a signal is in time, the more spread out it is in frequency.

Remark: We also have an analogy for *DT* signals. A finer resolution in the spectrum of a sampled signal requires that we sample the signal for a longer duration. We cannot simultaneously make the frequency resolution and sampling duration arbitrarily small.

7.3
Modulation in Communication Systems

Communication of information often requires transmission of signals, or *messages*, through space by converting electrical signals to electromagnetic waves using antennas. Efficient radiation requires antenna dimensions comparable to the wavelength of the signal. The wavelength λ of a radiated signal equals c/f, where c is the speed of light (at which electromagnetic waves travel) and f is the signal frequency. Transmission of low-frequency messages, such as speech and music, would require huge antenna sizes measured in kilometers. A practical approach is to shift low-frequency messages to a much higher frequency using **modulation** and transmit the modulated signal instead. This requires a means of recovering the original message at the receiving end, a process called **demodulation**.

Modulation Schemes   Several modulation methods are used in practice. In **amplitude modulation** (AM), the message modulates the amplitude of a high-frequency carrier. In **phase modulation** (PM) and **frequency modulation**, the message modulates the carrier phase or frequency.

7.3.1   *Schemes for Amplitude Modulation (AM)*

We examine several schemes based on modulating the amplitude of a high-frequency carrier $x_C(t) = \cos(2\pi f_C t)$ by a message signal $x_S(t)$ bandlimited to a frequency $f_B \ll f_C$. These schemes are illustrated in Figures 7.7–7.8.

Standard AM   In this scheme, the modulated signal includes the carrier as a separate component. The message signal $x_S(t)$ modulates the carrier $x_C(t)$ to give a modulated signal of the form

$$x_M(t) = [A_C + x_S(t)]\cos(2\pi f_C t)$$

With $A_S = |x_S(t)|_{\max}$ and $\beta = A_S/A_C$, we may rewrite this as

$$x_M(t) = A_C \cos(2\pi f_C t) + \frac{\beta}{A_S} x_S(t) A_C \cos(2\pi f_C t)$$

Here, β is called the **modulation index**. If $x_S(t)$ varies slowly compared to the carrier, the envelope of $x_M(t)$ follows $x_S(t)$ provided $0 < \beta < 1$ (or $|x_S(t)| < A_C$). For $|\beta| > 1$, the envelope no longer resembles $x_S(t)$ and we have **overmodulation** (Figure 7.7). The spectrum of the modulated signal $x_M(t)$ equals

$$X_M(f) = \frac{A_C}{2}[\delta(f + f_C) + \delta(f - f_C)] + \frac{\beta A_C}{2A_S}[X_S(f + f_C) + X_S(f - f_C)]$$

It exhibits symmetry about $\pm f_C$. The portion for $f > |f_C|$ is called the **upper sideband** and that for $f < |f_C|$, the **lower sideband**. Its bandwidth equals $2f_B$ (centered about $\pm f_C$), which is twice the signal bandwidth.

The lowpass signal $x_S(t)$ is bandlimited to f_B and called a **baseband signal**. The signal $x_M(t)$ is also bandlimited to $f_B + f_C$, but it is called a **bandpass signal** , since its spectrum does not include the origin.

Figure 7.7   Schemes for amplitude modulation.

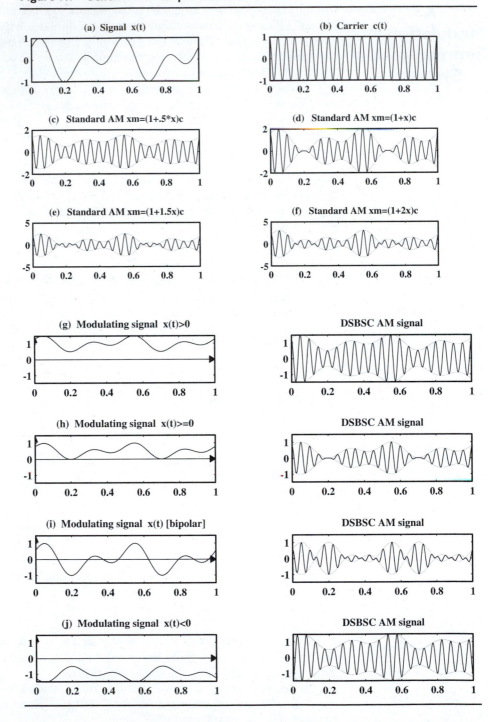

Figure 7.8 The spectra of AM signals and synchronous demodulation.

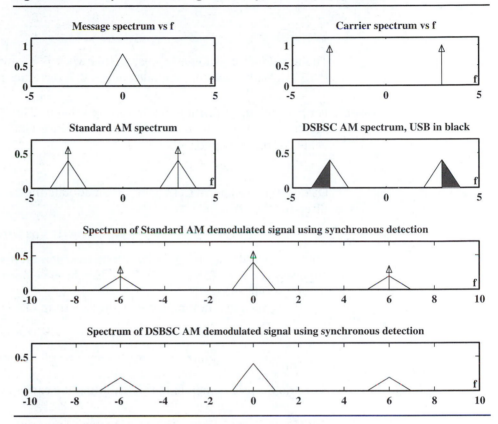

Efficiency   The **efficiency** of modulation is a measure of the power required to transmit just the message component. This power is only a fraction of the total transmitted power since the modulated signal also includes the carrier. We define the efficiency η as the ratio of the power P_S in the two sidebands and the total power P_M.

Example 7.4

Single-Tone AM

For a single-tone message $x_S(t) = A_S \cos(2\pi f_0 t)$ with $f_0 \ll f_C$, we may write

$$x_M(t) = [A_C + x_S(t)]\cos(2\pi f_C t) = [1 + \beta\cos(2\pi f_0 t)]A_C\cos(2\pi f_C t)$$

where $\beta = A_S/A_C$. Expanding this result, we get

$$x_M(t) = A_C\cos(2\pi f_C t) + \tfrac{1}{2}A_C\beta\cos[2\pi(f_C + f_0)t] + \tfrac{1}{2}A_C\beta\cos[2\pi(f_C - f_0)t]$$

The spectrum of $x_M(t)$ consists of the carrier component with strength A_C, and two sidebands at $f_C \pm f_0$ with strengths $\tfrac{1}{2}A_C$ and zero phase.

The power P_M in the modulated signal equals

$$P_M = \tfrac{1}{2}A_C^2 + \tfrac{1}{2}[\tfrac{1}{4}A_C^2\beta^2 + \tfrac{1}{4}A_C^2\beta^2] = \tfrac{1}{4}A_C^2[2 + \beta^2]$$

The power in each sideband is $\frac{1}{4}A_C^2\beta^2$. The efficiency η is thus

$$\eta = \frac{\frac{1}{4}A_C^2\beta^2}{\frac{1}{4}A_C^2[2+\beta^2]} = \frac{\beta^2}{2+\beta^2}$$

The maximum possible efficiency occurs when $\beta = 1$ (with $A_S = A_C$) and equals $\frac{1}{3}$ (33.33%). With a modulation index of $\beta = 0.5$, the efficiency drops to $\frac{1}{9}$ (11.11%).

Comment: For periodic signals with $\beta < 1$, it turns out that $\eta < 1/(1 + p_f^2)$ where p_f is the **peak factor**, the ratio of the peak and rms value. Since $x_{rms} \le x_p$, we have $p_f \ge 1$. Large values of p_f imply poor efficiency.

DSBSC AM   To improve efficiency, we modulate the carrier $x_C(t)$ by multiplying it directly by $x_S(t)$

$$x_M(t) = x_S(t)x_C(t) = x_S(t)\cos(2\pi f_c t)$$

This is a special case of standard AM with *no separate carrier* called double-sideband suppressed-carrier (DSBSC) AM. Note that the envelope of $x_M(t)$ follows $x_S(t)$ only if $x_S(t) \ge 0$ (Figure 7.7g–j).

Using the modulation property, the spectrum of $x_M(t)$ is given by

$$X_M(f) = \frac{1}{2}[X_S(f+f_C) + X_S(f-f_C)]$$

Since $x_S(t)$ is bandlimited to f_B, the spectrum still occupies a band of $2f_B$ (centered about $\pm f_C$), as with standard AM. The efficiency of DSBSC AM, however, is 100%, since no power is expended in transmitting the carrier.

SSB (Single-Sideband) AM   This scheme makes use of the symmetry in the spectrum of the modulated signal to reduce the transmission bandwidth. Conceptually, we can transmit just the upper sideband (by using a bandpass or highpass filter) or the lower sideband (by using a bandpass or lowpass filter). This process requires filters with sharp cutoffs. A more practical method is based on the idea of the Hilbert transform as a phase-shifting operation, which we briefly introduce here. Recall that the Hilbert transform of a signal $x(t)$ is defined by the convolution

$$\hat{x}(t) = \frac{1}{\pi t} \star x(t)$$

The spectrum $\hat{X}(f)$ of the Hilbert transformed signal may be regarded as the product of $X(f)$ with the transform of $1/\pi t$. In other words,

$$\hat{X}(f) = -j\,\text{sgn}(f)X(f)$$

This implies that the Hilbert transform of any signal $x(t)$ may be obtained by passing it through a system with a transfer function $H(f)$ given by

$$H(f) = -j\,\text{sgn}(f) = \begin{cases} -j & f > 0 \\ j & f < 0 \end{cases}$$

The Hilbert transform may, therefore, also be viewed as an operation that shifts the phase of $x(t)$ by $-\frac{1}{2}\pi$. A system that shifts the phase of a signal by $\pm\frac{1}{2}\pi$ (or $\pm 90°$) is called a **Hilbert transformer** or a **quadrature filter**.

To generate an SSB AM signal, we proceed as follows:

1. First, $x_S(t)$ is modulated by $x_C(t)$ to give the **in-phase component**

$$x_{\text{MI}}(t) = x_C(t)x_S(t) = \cos(2\pi f_C t)x_S(t)$$

2. Next, both $x_S(t)$ and $x_C(t)$ are shifted in phase by $-90°$ to yield their Hilbert transforms $\hat{x}_S(t)$ and $\hat{x}_C(t)$. Note that shifting the carrier $x_C(t) = \cos(2\pi f_C t)$ by $-90°$ simply yields $\hat{x}_C(t) = \sin(2\pi f_C t)$.

3. Then, $\hat{x}_C(t)$ is modulated by $\hat{x}_S(t)$ to give the **quadrature component**

$$x_{\text{MQ}}(t) = \sin(2\pi f_C t)\hat{x}_S(t)$$

4. The lower and upper sideband are obtained from $x_{\text{MI}}(t)$ and $x_{\text{MQ}}(t)$ as

$$x_{\text{ML}}(t) = x_{\text{MI}}(t) + x_{\text{MQ}}(t) = x_S(t)\cos(2\pi f_C t) + \hat{x}_S(t)\sin(2\pi f_C t)$$
$$x_{\text{MU}}(t) = x_{\text{MI}}(t) - x_{\text{MQ}}(t) = x_S(t)\cos(2\pi f_C t) - \hat{x}_S(t)\sin(2\pi f_C t)$$

Remark: Since the transmitted modulated signal contains only one sideband, its envelope does not correspond to the message signal.

Example 7.5
Single-Tone SSB
Modulation

For a single-tone message $x_S(t) = \cos(2\pi f_0 t)$, which modulates the carrier $\cos(2\pi f_C t)$ with $f_0 \ll f_C$, the in-phase component equals

$$x_{\text{MI}}(t) = \cos(2\pi f_0 t)\cos(2\pi f_C t) = \tfrac{1}{2}\cos[2\pi t(f_0 - f_C)] + \tfrac{1}{2}\cos[2\pi t(f_0 + f_C)]$$

The phase-shifted (by $-90°$) signals $\sin(2\pi f_0 t)$ and $\sin(2\pi f_C t)$ yield the quadrature component

$$x_{\text{MQ}}(t) = \sin(2\pi f_0 t)\sin(2\pi f_C t) = \tfrac{1}{2}\cos[2\pi t(f_0 - f_C)] - \tfrac{1}{2}\cos[2\pi t(f_0 + f_C)]$$

The sum and difference of $x_{\text{MI}}(t)$ and $x_{\text{MQ}}(t)$ generates the SSB signals:

$$x_{\text{ML}}(t) = x_{\text{MI}}(t) + x_{\text{MQ}}(t) = \cos[2\pi t(f_0 - f_C)] = \text{LSB signal}$$
$$x_{\text{MU}}(t) = x_{\text{MI}}(t) - x_{\text{MQ}}(t) = \cos[2\pi t(f_0 + f_C)] = \text{USB signal}$$

7.3.2   *Detection of AM Signals*

At the receiver, the message $x_S(t)$ may be recovered using **demodulation** or **detection**. We discuss two detection schemes.

Synchronous (Coherent) Detection Synchronous detection (Figure 7.8) entails modulation of the modulated signal by $\cos(2\pi f_C t)$, followed by lowpass filtering. For standard AM, with $x_M(t) = [A_C + x_S(t)]\cos(2\pi f_C t)$, the demodulated signal equals

$$x_D(t) = x_M(t)\cos(2\pi f_C t) = [A_C + x_S(t)]\cos(2\pi f_C t)\cos(2\pi f_C t)$$

Using the identity $\cos^2(\alpha) = \frac{1}{2}(1 + \cos 2\alpha)$, a little algebra yields

$$x_D(t) = \frac{1}{2}[A_C + x_S(t)][1 + \cos(4\pi f_C t)]$$
$$= \frac{1}{2}[A_C + x_S(t)] + \frac{1}{2}[A_C + x_S(t)]\cos(4\pi f_C t)$$

Its spectrum $X_D(f)$ reveals a component at $2f_C$ which can be eliminated by passing $x_D(t)$ through a lowpass filter whose bandwidth exceeds f_B. The filtered signal equals $\frac{1}{2}[A_C + x_S(t)]$. If $x_S(t)$ contains no dc component, we use ac coupling to eliminate $\frac{1}{2}A_C$ and recover $\frac{1}{2}x_S(t)$.

For DSBSC AM, no carrier is present, and we just recover $\frac{1}{2}x_S(t)$.

For SSB AM, the demodulated signal may be written as

$$x_D(t) = [x_S(t)\cos(2\pi f_C t) \mp \hat{x}_S(t)\sin(2\pi f_C t)]\cos(2\pi f_C t)$$

With $\cos^2(\alpha) = \frac{1}{2}(1 + \cos 2\alpha)$ and $\sin(\alpha)\cos(\alpha) = \frac{1}{2}\sin(2\alpha)$, this simplifies to

$$x_D(t) = \frac{1}{2}x_S(t)[1 + \cos(4\pi f_C t)] \mp \frac{1}{2}\hat{x}_S(t)\sin(4\pi f_C t)$$

Once again, passing $x_D(t)$ through a lowpass filter removes the two frequency components at $2f_C$ and yields the message signal $\frac{1}{2}x_S(t)$.

This method is called **synchronous** or **coherent detection**, since it requires a signal whose frequency and phase (but not amplitude) is synchronized or matched to the carrier signal at the transmitter. The coherent carrier may be obtained using **carrier extraction** at the receiver or transmitting a small fraction of the carrier (a *pilot signal*) along with $x_M(t)$. Synchronous detection is costly in practice.

Asynchronous (Envelope) Detection For standard AM (with $\beta \leq 1$) or DSBSC AM (with $x_s(t) \geq 0$), we can also recover the message $x_s(t)$ as the envelope of the demodulated signal using the circuit of Figure 7.9. The diode turns on only when the input $x_D(t)$ exceeds the output and the RC lowpass filter yields the envelope $x_S(t)$. The detector output is a faithful replica of $x_S(t)$ only if $f_C \gg f_B$. We choose $\tau > 1/f_C$ to ensure that the output is maintained between peaks, and $\tau < 1/f_B$ to ensure that the capacitor discharges rapidly enough to follow downward excursions of the message. This scheme is quite inexpensive and finds widespread use in commercial AM receivers.

7.3.3   Multiplexing

Conceptually, we can transmit many messages simultaneously simply by using carriers at different frequencies for each message. This even allows the messages to be demodulated separately. The problem is that this scheme is wasteful of equipment, requiring many local oscillators, and even more wasteful of transmission bandwidth and power. An alternative is to use *multiplexing*, a method that allows

Figure 7.9   Envelope detection of AM signals.

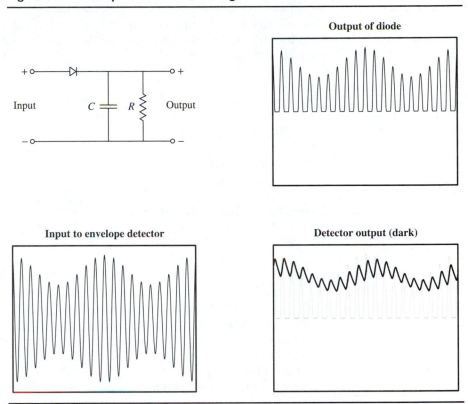

both transmission and reception of several messages simultaneously using a single carrier and thus makes efficient use of the transmission bandwidth.

Frequency Multiplexing   In **frequency multiplexing**, we allocate portions of the transmission bandwidth to different messages in a way that their spectra do not overlap. If the bandwidth of all the messages $x_1, x_2, \ldots, x_N$ is less than f_B, we can implement frequency multiplexing as follows:

1. Choose modulating frequencies $f_M, 2f_M, \ldots, Nf_M$, with $f_M > f_B$.
2. Modulate $x_1, x_2, \ldots, x_N$ using SSB modulation at $f_M, 2f_M, \ldots, Nf_M$.
3. Combine the modulated signals into a single message signal $x_S(t)$.
4. Transmit $x_S(t)$, using any modulation scheme and a carrier at $f_C \gg Nf_M$.
5. At the receiver, demodulate $x_D(t)$ using the carrier at f_C.
6. Extract $x_1, x_2, \ldots, x_N$ by demodulation with carriers at $f_M, 2f_M, \ldots, Nf_M$.

Time Multiplexing In **time multiplexing**, we sample each message and allocate different channels, or time slots, to different messages in a way that the messages themselves do not overlap. For channels with equal bandwidth f_B, the sampler must switch at a rate exceeding $2f_B$. This method thus requires the same bandwidth as frequency multiplexing.

Example 7.6
Frequency Multiplexing

The bandwidth of speech signals is usually confined to 3.4 kHz. Let us choose $f_M = 4$ kHz to avoid spectral overlap. If there are five messages, the spectrum of the composite message $x_S(t)$ will show the messages separated by 0.6 kHz in the bands 0.6–4.0, 4.6–8, 8.6–12, 12.6–16 and 16.6–20 kHz. The bandwidth of the composite signal is thus 20 kHz.

Heterodyning **Heterodyning** or **mixing** is a practical scheme for demodulation of signals. The modulated signal $x_M(t)$ is multiplied by another carrier generated by an oscillator whose frequency f_M can be varied. The resulting *heterodyned signal* $x_{IF}(t)$ equals $x_M(t)\cos(2\pi f_M t)$ and has a spectrum centered at $f_{IF} = |f_C - f_M|$. The carrier, in effect, shifts the spectrum of $x_M(t)$ to the **intermediate frequency** f_{IF}. The oscillator frequency f_M is adjusted to make f_{IF} a fixed value, typically 455 kHz, and $x_{IF}(t)$ is passed through a bandpass filter with a center frequency of f_{IF}. The filtered output is finally demodulated to recover $x_S(t)$. By adjusting the oscillator frequency f_M, this scheme can be used to tune in messages modulated by different frequencies, or to receive a particular message from frequency multiplexed signals. Since both the bandpass filter and the demodulator operate at the fixed frequency f_{IF}, this simplifies their design. Receivers using the principle of heterodyning are called **superheterodyne receivers**.

Example 7.7
Heterodyning

If the carrier frequency f_C is between 550 kHz–1600 kHz, the oscillator must be made adjustable between 95 kHz and 1145 kHz, in order to shift incoming signals to 455 kHz.

7.4
Angle
Modulation

Angle modulation collectively refers to schemes that vary the phase, or frequency, of a carrier signal in proportion to the message signal. Consider a sinusoidal carrier with variable phase

$$x(t) = A_C \cos[2\pi f_C t + \theta_C(t)]$$

The argument of the sinusoid may itself be regarded as an **instantaneous phase angle** $\phi_C(t)$ with

$$\phi_C(t) = 2\pi f_C t + \theta_C(t)$$

The carrier completes one full cycle as $\phi_C(t)$ changes by 2π. The **peak phase deviation** is defined as $|\theta(t)|_{\max}$.

The derivative $\frac{d\phi_C}{dt}$ describes the rate of change of $\phi_C(t)$ and yields the **instantaneous frequency** as

$$\omega_M(t) = \frac{d\phi_C}{dt} = 2\pi f_C + \frac{d\theta_C}{dt} \quad \text{or} \quad f_M(t) = f_C + \frac{1}{2\pi}\frac{d\theta_C}{dt}$$

The **peak frequency deviation** equals $|\theta_C'(t)|_{max}$ rad/s or $\frac{1}{2\pi}|\theta_C'(t)|_{max}$ Hz.

Phase modulation (PM) refers to a scheme that varies $\phi_C(t)$ by varying the phase $\theta_C(t)$, whereas **frequency modulation (FM)** describes a scheme that varies $\phi_C(t)$ by varying the instantaneous frequency (Figure 7.10).

Figure 7.10   PM and FM signals. Note the reversal in the PM signal at t = 1.

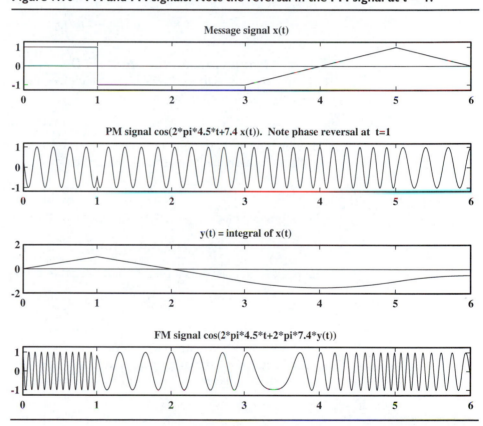

7.4.1   *Phase Modulation (PM)*

In phase modulation, the phase $\theta_C(t)$ of the carrier is varied in proportion to $x_S(t)$. If $x_S(t)$ is constant, $\theta_C(t) = C$ and $f_M(t) = f_C$. If $x_S(t)$ shows a linear time dependence, $\theta_C(t) = A + Bt$ and $f_M(t) = f_C + B/2\pi$. In either case, the instantaneous frequency

$f_M(t)$ is constant. Formally, a phase modulated signal $x_{PM}(t)$ may be written

$$x_{PM}(t) = A_C \cos[2\pi f_C t + \theta_C(t)] = A_C \cos[2\pi f_C t + k_P x_S(t)]$$

where $\theta_C(t) = k_P x_S(t)$. The constant k_P (in radians/units of x_S) is called the **phase sensitivity**. The instantaneous phase and frequency equal

$$\phi_{PM}(t) = 2\pi f_C t + k_P x_S(t) \qquad f_{PM}(t) = f_C + \frac{k_P}{2\pi}\frac{dx_S}{dt}$$

The peak frequency deviation equals

$$\Delta f_P = \frac{k_P}{2\pi}|x_S'(t)|_{\max}$$

In analogy with AM, we also define a modulation index β_P, called the **deviation ratio**, in terms of the highest frequency f_B in the message as

$$\beta_P = \Delta f_P/f_B$$

7.4.2 Frequency Modulation (FM)

In frequency modulation, the instantaneous frequency of the carrier is varied linearly with $x_S(t)$, and may be written as

$$f_{FM}(t) = f_C + k_F x_S(t)$$

The constant k_F (in Hz/units of x_S) is called the **frequency sensitivity**.

Upon integrating $2\pi f_{FM}(t)$, we obtain the modulated angle $\phi_{FM}(t)$ as

$$\phi_{FM}(t) = 2\pi f_C t + 2\pi k_F \int x_S(t)\,dt$$

The frequency modulated signal $x_{FM}(t)$ may then be expressed as

$$x_{FM}(t) = A_C \cos\left[2\pi f_C t + 2\pi k_F \int x_S(t)\,dt\right]$$

The peak frequency deviation Δf_F, and deviation ratio β_F, simply equal

$$\Delta f_F = k_F|x_S(t)|_{\max} \qquad \beta_F = \Delta f_F/f_B$$

Remarks:
1. A PM signal may be viewed as a special case of an FM signal, with the modulating quantity as $\frac{dx_S(t)}{dt}$ (instead of $x_S(t)$).
2. Similarly, an FM signal may be viewed as a special case of a PM signal, with the modulating quantity as $\int x_S(t)\,dt$ (instead of $x_S(t)$).

Example 7.8
Concepts in Angle Modulation

Consider the angle-modulated signal $x_M(t) = \cos[200\pi t + 0.4\sin(10\pi t)]$.

(a) The carrier frequency and message frequency are $f_C = 100$ Hz and $f_0 = 5$ Hz. Its instantaneous phase is $\phi_C(t) = 200\pi t + 0.4\sin(10\pi t)$. The peak phase deviation thus equals 0.4 rad/s. Its instantaneous frequency equals $f_M(t) = (1/2\pi)\frac{d\phi_C}{dt} = 100 + 2\cos(10\pi t)$. The peak frequency deviation Δf thus equals 2 Hz. The deviation ratio β then equals $= \Delta f/f_0 = 2/5 = 0.4$.

(b) If $x_M(t)$ is regarded as a PM signal then $k_P x_S(t) = 0.4 \sin(10\pi t)$. Suppose $k_P = 5$. Then $x_S(t) = 0.08 \sin(10\pi t)$. As a check, we evaluate $\Delta f_P = \frac{k_P}{2\pi} |x'_S(t)|_{\max} = 5(0.8\pi)/2\pi = 2$ Hz.

(c) If $x_M(t)$ is regarded as an FM signal, then $2\pi k_F \int x_S(t)\, dt = 0.4 \sin(10\pi t)$. Suppose $k_F = 4$. Taking derivatives, we get $2\pi k_F x_S(t) = 8\pi x_S(t) = 4\pi \cos(10\pi t)$ or $x_S(t) = 0.5 \cos(10\pi t)$. As a check, we evaluate $\Delta f_F = k_F |x_S(t)|_{\max} = 4(0.5) = 2$ Hz.

7.4.3 Differences Between AM, FM and PM

There are several major differences between amplitude-modulated signals and angle-modulated signals.

1. In contrast to the varying envelope of AM signals, the envelope of FM or PM signals is always constant.
2. The zero crossings of AM signals are always equispaced. For PM and FM signals, the angle varies with time, and so does the spacing between the zero crossings.
3. Unlike AM, an FM signal is a *nonlinear* function of $x_S(t)$. As a result, the spectrum of FM signals is governed by a much more complex form.
4. As a consequence of this nonlinearity, the transmission bandwidth of FM signals is much larger than that of a corresponding AM signal.

7.4.4 Narrowband Angle Modulation

If $\theta_C(t) \ll 1$, we get a rather simple result for the modulated signal $x_M(t)$ and its spectrum $X_M(f)$. The signal $x_M(t) = A_C \cos[2\pi f_c t + \theta_C(t)]$ may be expanded as

$$x_M(t) = A_C \cos(2\pi f_c t)\cos[\theta_C(t)] - A_C \sin(2\pi f_c t)\sin[\theta_C(t)]$$

For small x, $\cos(x) \approx 1$ and $\sin(x) \approx x$, and we can approximate $x_M(t)$ by

$$x_M(t) \approx A_C \cos(2\pi f_c t) - A_C \theta_C(t)\sin(2\pi f_c t)$$

If $\Theta_C(f)$ represents the Fourier transform of $\theta_C(t)$, the spectrum $X_M(f)$ of the modulated signal can be described by

$$X_M(f) = \tfrac{1}{2}A_C[\delta(f + f_C) + \delta(f - f_C) - j\Theta_C(f + f_C) + j\Theta_C(f - f_C)]$$

For PM, $\theta_C(t)$ equals $k_P x_S(t)$ and $\Theta_C(f) = k_P X_S(f)$, leading to

$$X_{\mathrm{PM}}(f) = \tfrac{1}{2}A_C[\delta(f + f_C) + \delta(f - f_C) - jk_P X_S(f + f_C) + jk_P X_S(f - f_C)]$$

For FM, $\theta_C(t) = k_F \int x_S(t)\, dt$ and $\Theta_C(f) = k_F X_S(f)/j2\pi f$ (if we assume that $x_S(t)$ has no dc component, such that $X_S(0) = 0$), and we get

$$X_{\mathrm{FM}}(f) = \tfrac{1}{2}A_C\left[\delta(f + f_C) + \delta(f - f_C) - k_F \frac{X_S(f + f_C)}{(f + f_C)} + k_F \frac{X_S(f - f_C)}{(f - f_C)}\right]$$

The spectrum of both PM and FM signals comprises two impulses at $f \pm f_C$, with strength $\tfrac{1}{2}A_C$, and two sidebands (Figure 7.11).

Figure 7.11   Spectra of FM and PM signals.

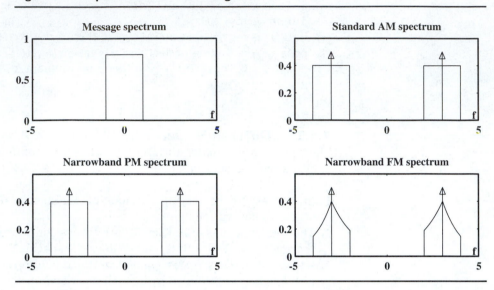

The sidebands for PM retain the same shape as the message spectrum $X_S(f)$, but for FM, the factors $1/(f \pm f_C)$ make $X_S(f)$ droop as we move away from f_C. This makes FM more susceptible to noise. To counter the unwanted effects of noise, we use **preemphasis** to boost high frequencies in the message *before modulation* (using a differentiating filter with $|H(f)| \propto f$). At the receiver, demodulation is followed by **deemphasis** (using a lowpass filter) to recover the original message.

Example 7.9
Single-Tone Angle
Modulation

Let us choose $\theta_C(t) = \beta \sin(2\pi f_0 t)$. Then $f_M(t) = f_C + \beta f_0 \cos(2\pi f_0 t)$.

Since $|\cos(2\pi f_0 t)| \leq 1$, $f_M(t)$ varies between $f_C + \beta f_0$ and $f_C - \beta f_0$, with a *peak frequency excursion* $\Delta f = \beta f_0$. For narrowband FM, $\theta_C(t) \ll 1$, and thus $\beta \ll 1$. If we substitute for $\theta_C(t)$, the modulated signal $x_M(t)$ equals

$$x_M(t) = A_C \cos[2\pi f_C t + \theta_C(t)] = A_C \cos[2\pi f_C t + \beta \sin(2\pi f_0 t)]$$

Using trigonometric identities, we expand this to

$$x_M(t) = A_C \cos(2\pi f_C t)\cos[\beta \sin(2\pi f_0 t)] - A_C \sin(2\pi f_C t)\sin[\beta \sin(2\pi f_0 t)]$$

For $x \ll 1$, $\cos(x) \approx 1$, $\sin(x) \approx x$, and we have the narrowband approximation

$$x_M(t) \approx A_C \cos(2\pi f_C t) - A_C \beta \sin(2\pi f_C t)\sin(2\pi f_0 t)$$

Using trigonometric identities once again, this can be rewritten as

$$x_M(t) = A_C \cos(2\pi f_C t) + \tfrac{1}{2}A_C\beta \cos[2\pi(f_C + f_0)t] - \tfrac{1}{2}A_C\beta \cos[2\pi(f_C - f_0)t]$$

The spectrum of $x_M(t)$ comprises a carrier at f_C of strength A_C and two components at $f_C \pm f_0$ (both of strength $\tfrac{1}{2}A_C\beta$), whose phase equals zero at $f = f_C + f_0$ and 180°

at $f = f_C - f_0$. Thus a narrowband angle-modulated signal is similar to an AM signal except for a phase difference of $180°$ in the component at $f = f_C - f_0$.

Comment: Like AM, the bandwidth of the angle-modulated signal $x_M(t)$ also equals $2f_0$.

7.5
Special Topics: Wideband Angle Modulation

The analysis of wideband angle modulation is often quite cumbersome. Some useful results obtain if $\theta_C(t)$ is assumed to be periodic. In such a case, we express $x_M(t)$, using Euler's identity, as

$$x_M(t) = A_C \cos[2\pi f_C t + \theta_C(t)] = \text{Re}\,[A_C \exp[j2\pi f_C t + j\theta_C(t)]]$$

Since $\theta_C(t)$ is assumed periodic, $\exp[j\theta_C(t)]$ is also periodic and can be expressed in terms of its Fourier series:

$$\exp[j\theta_C(t)] = \sum_{k=-\infty}^{\infty} X_S[k]\exp(j2k\pi f_0 t) \qquad X_S[k] = \frac{1}{T}\int_T \exp[j(\theta_C - 2k\pi f_0 t)]\,dt$$

Substituting for $\exp(j\theta_C)$ into the expression for $x_M(t)$, we obtain

$$x_M(t) = \text{Re}\left[A_C \sum_{k=-\infty}^{\infty} X_S[k] \exp[j2\pi(f_C + kf_0)t] \right]$$

The spectrum of $x_M(t)$ shows components at the frequencies $f_C \pm kf_0$.

7.5.1   *Bandwidth of Wideband FM*

Theoretically, the spectrum suggests an infinite bandwidth for FM. In practice, most of the power is concentrated in a few harmonics about f_C. The bandwidth may thus be regarded as finite, but much larger than that for AM. There are two ways to estimate the bandwidth B of wideband FM.

The first is **Carson's Rule**, an empirical relation, which says

$$B = 2f_B(1 + \beta) = 2(\Delta f + f_B) \approx 2f_B\beta \qquad (\beta > 100)$$

For *narrowband* FM ($\beta < 0.25$ or so), $B \approx 2f_B$. More than 98% of the power lies in the band B for single-tone modulation. Carson's rule underestimates the bandwidth and, for $\beta > 2$, it is sometimes modified to $B = 2f_B(2 + \beta)$.

The second method is a numerical approximation based on applying Parseval's theorem to the signal $\exp[j\theta_C(t)]$ and its spectral coefficients $X_S[k]$.

$$P = \frac{1}{T}\int_T |\exp(j\theta_C)|^2 dt = \sum_{k=-\infty}^{\infty} |X_S[k]|^2 = 1$$

Since $|\exp(j\theta_C)| = 1$, the total power equals 1. The sum of $|X_S[k]|^2$ through a finite number of harmonics $k = N$ equals the fraction of the total power over the frequency band $f_C \pm Nf_0$. If we define B as the range over which a large fraction, say 98%, of the total power is concentrated, we can estimate N, and compute the bandwidth as $B \approx 2Nf_0$.

Table 7.7 lists some signals used for frequency modulation. The coefficients $X_S[k]$ for each modulated signal in this table are given by:

Square Wave $X_S[k] = \frac{1}{2}\text{sinc}[\frac{1}{2}(\beta - k] + \frac{1}{2}(-1)^{-k}\text{sinc}[\frac{1}{2}(\beta + k)]$

Sinusoid $X_S[k] = J_k(\beta) = \frac{1}{2\pi}\int_0^{2\pi} \exp[j(\beta \sin\tau - k\tau)]\, d\tau$. The $J_k(\beta)$ represent **Bessel functions** of integer order k and argument β.

Table 7.7 Signals for Wideband FM.

Signal	Instaneous frequency ($T = 1/f_0$)		Form of $X_S[k]$
Square wave	$\begin{cases} f_C + \beta f_0 \\ f_C + \beta f_0 \end{cases}$	$\begin{array}{l} -\frac{1}{4}T \le t \le \frac{1}{4}T \\ \frac{1}{4}T \le t \le \frac{3}{4}T \end{array}$	Sinc functions
Sinusoid	$f_C + \beta f_0 \cos(2\pi t/T)$	$0 \le t \le T$	Bessel functions
Sawtooth	$f_C - \beta f_0(1 - 2t/T)$	$0 \le t \le T$	Fresnel integrals
Triangular	$\begin{cases} f_C + \beta f_0(1 + 4t/T) \\ f_C + \beta f_0(1 - 4t/T) \end{cases}$	$\begin{array}{l} -\frac{1}{2}T \le t \le 0 \\ 0 \le t \le \frac{1}{2}T \end{array}$	Fresnel integrals

---

Sawtooth Wave $X_S[k] = \frac{1}{\alpha}\exp(-j2\pi\lambda_k^2)\left[C(\lambda_k) - C(\gamma_k) + jS(\lambda_k) - jS(\gamma_k)\right]$. Here $\alpha = \sqrt{2\beta}$, $\lambda = 2k/\alpha + \frac{1}{2}\alpha$, $\gamma_k = 2k/\alpha - \frac{1}{2}\alpha$, and the quantities $C(x)$ and $S(x)$ are **Fresnel integrals** of the *first* and *second kind*, defined by

$$C(x) = \int_0^x \cos(\tfrac{1}{2}\pi\tau^2)\, d\tau \qquad S(x) = \int_0^x \sin(\tfrac{1}{2}\pi\tau^2)\, d\tau$$

Triangular Wave $X_S[k] = \frac{1}{\alpha}[C(\lambda_k) - C(\gamma_k)]\cos(\frac{1}{2}\pi\lambda_k^2) + \frac{1}{\alpha}[S(\lambda_k) - S(\gamma_k)]\sin(\frac{1}{2}\pi\lambda_k^2)$. Here $\alpha = \sqrt{2\beta}$, $\lambda_k = k/\alpha + \frac{1}{2}\alpha$ and $\gamma_k = k/\alpha - \frac{1}{2}\alpha$.

Figure 7.12 shows the spectra of the resulting FM signals for each case (for $\beta = 10$), centered about f_C. Except for square-wave modulation, the $|X_S[k]|$ are seen to be negligible for $|k| > \beta$, and most of the power is concentrated in the components in the range $f_C \pm \beta f_0$.

7.6
Applications of Fourier Transforms: A Synopsis

Ideal filters A perfect filter requires unity gain and zero phase in the passband. For a distortionless filter, the response is a scaled, shifted version of the input. In either case, the impulse response $h(t)$ has a sinc form and extends for all time. An ideal filter is thus noncausal.

Physically realizable filters Such filters cannot possess a monotonic step response or a constant magnitude spectrum and linear phase. If we choose a large delay and truncate $h(t)$ for $t < 0$ to make an ideal filter causal, its spectrum shows infinite spikes. Twosided truncation produces finite ears due to the Gibbs effect. To reduce these we use smoothing windows.

Figure 7.12 Signals for wideband FM and spectra of the modulated signal ($\beta = 10$).

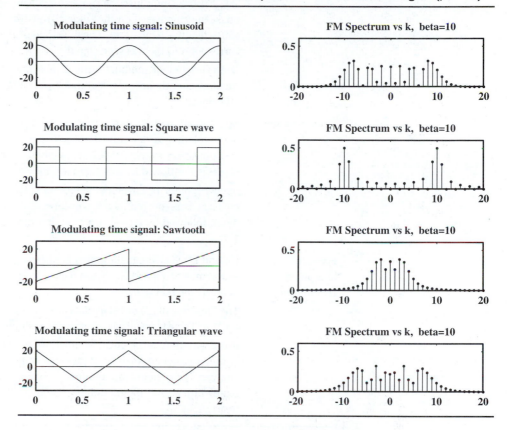

For a causal system, $h(t)$ is determined completely from its even part, and $H(f)$ can be completely determined from its real part. The transfer function of a physically realizable system may also be found from its magnitude spectrum $|H(f)|$ and is called a minimum-phase transfer function.

Time-bandwidth relations

Measures of the impulse response duration (or rise time) are inversely related to measures of bandwidth. Commonly used definitions for both are based on areas and moments of $h(t)$ and $H(f)$. The time-bandwidth product is a constant.

Modulation

Modulation is used in the long-distance transmission of low-frequency messages. The message modulates the amplitude, phase or frequency of a high-frequency carrier to yield the modulated signal.

In standard AM, the modulated signal includes the carrier; in DSBSC AM, it does not. For standard AM, the modulation index determines whether the envelope of the modulated signal follows the message.

SSB AM transmits just the upper (or lower sidebands) using filters or phase shifting based on the Hilbert transform (which is an operation that shifts the phase by $-\frac{1}{2}\pi$). The envelope of the modulated signal does not correspond to the message signal.

The efficiency of modulation is a measure of the power required to transmit just the message component. DSBSC AM and SSB AM yield 100% efficiency.

Demodulation

Demodulation recovers the message from a modulated signal. Synchronous demodulation works with all AM schemes. It uses a signal whose frequency and phase are synchronized or matched to the carrier. The modulated signal is multiplied by this signal and lowpass filtered.

Asynchronous, or envelope, detection is used for AM or DSBSC AM. The message is recovered as the envelope of the demodulated signal using a diode followed by an RC circuit with a time constant between $1/f_C$ and $1/f_B$.

Frequency and phase modulation

FM and PM vary the instantaneous phase or frequency of the carrier. A PM signal may be viewed as a special case of an FM signal with the modulating quantity as $dx_S(t)/dt$.

The envelope of FM or PM signals is always constant but the zero crossings are not. Their transmission bandwidth is much larger than that for AM.

If the variation in the angle is small, we have narrowband FM or PM which is much easier to analyze. Wideband angle modulation is difficult to analyze. For periodic modulation, the bandwidth depends on the message bandwidth f_B and the modulation index β. For large β, the bandwidth is nearly $2f_B\beta$.

Appendix 7A
MATLAB
Demonstrations
and Routines

Getting Started on the MATLAB Demonstrations

This appendix presents MATLAB routines related to concepts in this chapter and some examples of their use. See Appendix 1A for introductory remarks.

Spectral Measures of Windows

```
>>y=winspec('hamming')
```

plots the Hamming window and its spectrum and returns spectral measures in the column vector y whose seven columns are

```
y=[Peak;normalized peak sidelobe; pSL(dB); lobewidth; width to pSL;w6dB;w3dB]
```

The routine can be used for most of the windows listed in Table 7.4.

Modulation

```
[xm,xp,xf]=modsig(ty,x,c,k)
```

generates a modulated signal of type ty = 'am', 'pm', or 'fm' using x as the message. Here $c = [f_C, ts]$ is a two-element array of the carrier frequency and the sampling interval, and $k = \beta$ for AM, $k = k_P$ for PM and $k = k_F$ for FM.

To plot an FM signal with $x = \text{tri}(t-1)$, $f_C = 2$, $k_F = 5$ use

```
>>t=0:0.01:2;x=tri(t-1);modsig('fm',x,[2 0.01],5);
```

```
>>y=wbfm(ty,beta,n)
```

This computes the power in n harmonics for a wideband FM signal with modulation index beta. The string argument ty is 'si' for sine, 'st' for sawtooth, 'tr' for triangular or 'sq' for square-wave modulation.

If no output arguments are given, wbfm plots the magnitude spectrum. For $n < 1$, y returns the number of harmonics containing the specified fraction of the total power.

For sine-wave modulation with beta=6, the power in 4 harmonics is

```
>>y=wbfm('si',6,4)
```

MATLAB responds with

```
y=0.5759
```

To find the number of harmonics that contain 80% of the total power, use

```
>>y=wbfm('si',6,0.8)
```

MATLAB responds with

```
y=5
```

To see the spectrum, use

```
wbfm('si',6,0.8)
```

Utility Functions

```
>>si(x,a)        %Sine integral [of sin (ax)/ax]
>>si2(x,a)       %Sine squared integral
                 [of sin^2(alpha x)/(alpha x)^2]
>>besjn(n,x)     %Bessel function of integer order n
>>cx(x)          %Fresnel integral of the first kind C(x)
>>sx(x)          %Fresnel integral of the second kind S(x)
```

To plot the integral of sinc(t) from 0 to 5, use

```
>>t=0:.02:5;y=si(t,pi);plot(t,y)
```

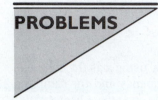

PROBLEMS

P7.1 (Ideal Filters) Relate the frequency response and impulse response of each of the following filter types in terms of the frequency response $H_{LP}(f)$ and impulse response $h_{LP}(t)$ of an ideal lowpass filter with cutoff frequency f_C.
(a) A lowpass filter with cutoff frequency $\frac{1}{4}f_C$.
(b) A highpass filter with cutoff frequency $2f_C$.
(c) A bandpass filter with center frequency $8f_C$ and passband $2f_C$.
(d) A bandstop filter with center frequency $8f_C$ and stopband $2f_C$.

P7.2 (Frequency Response of Filters) Sketch the magnitude and phase of each of the following filters and find their response to the input $\cos(\pi t)$.
(a) $h(t) = 4\mathrm{sinc}[4(t-1)]$ **(b)** $H(f) = \mathrm{tri}(f/6)\exp(j\pi f)$

P7.3 (Time-Bandwidth Product) Consider a lowpass filter whose impulse response is given by $h(t) = (1/\tau)\exp(-t/\tau)u(t)$.
(a) Compute the time-bandwidth product using entry 1 of Table 7.5.
(b) Compute the time-bandwidth product using entry 3 of Table 7.5.
(c) Compute the time-bandwidth product using $T_r = \mathbb{A}[h(t)]/h(t_0)$, where t_0 is the time at which the step response reaches 50% of its final value, and B_f is as defined in entry 1 of Table 7.5.

P7.4 (Time-Bandwidth Product) Compute the time-bandwidth product of an ideal lowpass filter assuming T_r as the width of the central lobe of its impulse response $h(t)$.

P7.5 (Time-Bandwidth Product) Compute the time-bandwidth product for the systems whose impulse response is:
(a) $h(t) = \mathrm{tri}(t)$ using entry 3 of Table 7.5.
(b) $h(t) = \exp(-\pi t^2)$ using entry 5 of Table 7.5.

P7.6 (Standard AM) Consider the message $x(t) = 2\cos(2\pi f_1 t) + \cos(2\pi f_2 t)$ used to modulate the carrier $\cos(2\pi f_C t)$ to generate $x_M(t) = [A_C + x(t)]\cos(2\pi f_C t)$.
(a) What value of A_C ensures a modulation index $\beta = 0.5$?
(b) What is the efficiency η when $\beta = 0.5$?
(c) Sketch the spectrum of $x(t)$ and $x_M(t)$ if $f_1 = 10$ Hz, $f_2 = 20$ Hz and $f_C = 100$ Hz.

P7.7 (Standard AM) Consider the modulated signal

$$x_M(t) = 1000\{1 + \sum(1/2k)\cos(2000k\pi t)\}\cos(20000\pi t)\}, \qquad k = 1,2,5$$

(a) What frequencies are present in the modulated signal?
(b) What are the magnitudes of the components at each frequency?
(c) What is the total transmitted power?
(d) What is the modulation index?
(e) What is the efficiency?

P7.8 (Standard AM) An AM station operates with a moulation index of 0.8 and transmits a total power of 50 kW.

(a) What is the power in the transmitted carrier?

(b) What fraction of the total power resides in the message?

P7.9 (Standard AM) Find the efficiency of a single-tone AM modulator with a modulation index of 0.5 if the carrier power equals 32 W.

P7.10 (Synchronous Demodulation) The signal $x(t) = 2\cos(2\pi f_1 t) + \cos(2\pi f_2 t)$ is used to modulate the carrier $\cos(2\pi f_C t)$, to generate $x_M(t) = [A_C + x(t)]\cos(2\pi f_C t)$. The modulated signal $x_M(t)$ is synchronously demodulated.

(a) Sketch the spectrum of the demodulated signal.

(b) Sketch the spectrum of an ideal lowpass filter that can be used to recover $x(t)$.

P7.11 (DSBSC AM) Consider the message $x(t) = 2\cos(2\pi f_1 t) + \cos(2\pi f_2 t)$ used to modulate the carrier $\cos(2\pi f_C t)$ to generate a DSBSC AM signal $x_M(t)$.

(a) Write an expression for $x_M(t)$ and sketch its spectrum.

(b) What fraction of the total signal power is contained in the sidebands?

(c) What is the modulation efficiency?

P7.12 (DSBSC AM) Consider the message $x(t) = 2\cos(2\pi f_1 t)$ used to modulate the carrier $\cos(2\pi f_C t)$, to generate a DSBSC AM signal $x_M(t)$.

(a) Write an expression for $x_M(t)$ and sketch its spectrum.

(b) Sketch the spectrum of the demodulated signal if demodulation is achieved by the signal $\cos(2\pi f_C t)$.

P7.13 (DSBSC AM) A 1.5 MHz carrier is modulated by a music signal with frequencies from 50 Hz to 15 kHz. What is the range of frequencies over which the upper and lower sidebands extend?

P7.14 (Synchronous Detection) Synchronous detection requires both phase and frequency coherence. Let $x(t) = 2\cos(2\pi f_0 t)$ be used to modulate the carrier $\cos(2\pi f_C t)$ to generate a DSBSC AM signal $x_M(t)$.

(a) Find and sketch the spectrum of the demodulated signal if the demodulating carrier has the form $\cos(2\pi f_C t + \theta)$.

(b) Are there any values of θ for which it is impossible to recover the message?

(c) Let the demodulating carrier have a frequency offset to become $\cos[2\pi(f_C + \Delta f)t]$. Find and sketch the spectrum of the demodulated signal. Do we recover the message signal in this case?

P7.15 (Envelope Detection) The following signals are used to generate DSBSC AM signals. Which signals can be recovered using envelope detection?

(a) $x(t) = \cos(2\pi f_1 t)$ (b) $x(t) = 2 + \cos(2\pi f_1 t)$

(c) $x(t) = 2\cos(2\pi f_1 t) + \cos(2\pi f_2 t)$ (d) $x(t) = 2 + 2\cos(2\pi f_1 t) + \cos(2\pi f_2 t)$

(e) $x(t) = 4 + 2\cos(2\pi f_1 t) + \cos(2\pi f_2 t)$

P7.16 (Envelope Detection) A standard AM signal $x_M(t) = [A_C + x(t)]\cos(2\pi f t)$ is to be demodulated by an envelope detector. The message $x(t)$ is bandlimited to 10 kHz, and the carrier frequency is 100 kHz. What are the allowable limits on the detector time constant to avoid distortion?

P7.17 (SSB AM) The message $x(t) = 2\cos(2\pi f_1 t) + \cos(2\pi f_2 t)$ modulates the carrier $\cos(2\pi f_C t)$, to generate $x_M(t) = x(t)\cos(2\pi f_C t)$. Identify the upper and lower sideband components of $x_M(t)$.

P7.18 (SSB AM) The message $x(t) = \text{sinc}^2(t)$ modulates the carrier $\cos(10\pi t)$ to generate the modulated signal $x_M(t) = x(t)\cos(2\pi f_C t)$.
(a) Sketch the spectrum of $x(t)$, the carrier, and $x_M(t)$.
(b) Sketch the spectrum of the LSB SSB signal.
(c) If the signal in part (b) is synchronously demodulated, sketch the spectrum of the demodulated signal.

P7.19 (Instantaneous Frequency) Find the instantaneous frequency of
(a) $x(t) = \cos(10\pi t + \frac{1}{4}\pi)$ $\qquad\qquad$ **(b)** $x(t) = \cos(10\pi t + 2\pi t)$
(c) $x(t) = \cos(10\pi t + 2\pi t^2)$ $\qquad\qquad$ **(d)** $x(t) = \cos[10\pi t + 2\sin(2\pi t)]$

P7.20 (FM) An FM signal is described by $x(t) = A\cos[2\pi 10^7 t + 50\sin(2\pi 10^4 t)]$.
(a) Identify the carrier frequency.
(b) Identify the frequency of the modulating signal.
(c) Find the peak frequency deviation and modulation index.

P7.21 (FM) The peak deviation in commercial FM is 75 kHz. The frequency of the modulating signal varies between 50 Hz and 15 kHz. What permissible range of modulation index can we work with?

P7.22 (FM) A signal, bandlimited to 15 kHz, is to be transmitted using FM. The peak deviation is 30 kHz. What bandwidth is required? You may use Carson's rule or its modification as appropriate.

P7.23 (Wideband FM) Consider an FM signal with $\beta = 5$. Find the power contained in the harmonics $f_C \pm k f_0$, $k = 0, 1, 2, 3$ for
(a) Single-tone sinusoidal modulation.
(b) Square-wave modulation.

P7.24 (The Hilbert Transform) Unlike most other transforms, the Hilbert transform belongs to the same domain as the signal transformed. The Hilbert transform shifts the phase of $x(t)$ by $-\frac{1}{2}\pi$. Find the Hilbert transforms of
(a) $x(t) = \cos(2\pi f t)$ $\qquad\qquad$ **(b)** $x(t) = \sin(2\pi f t)$
(c) $x(t) = \cos(2\pi f t) + \sin(2\pi f t)$ $\qquad\qquad$ **(d)** $x(t) = \exp(-j2\pi f t)$
(e) $x(t) = \delta(t)$ $\qquad\qquad$ **(f)** $x(t) = 1/(\pi t)$

P7.25 (Properties of the Hilbert Transform) Using some of the results of the preceding problem as examples, or otherwise, verify the following properties of the Hilbert transform.
(a) The magnitude spectra of $x(t)$ and $\hat{x}(t)$ are identical.
(b) The Hilbert transform of $x(t)$ taken twice returns $-x(t)$.
(c) The Hilbert transform of an even function is odd, and vice versa.
(d) The Hilbert transform of $x(at)$ is $\text{sgn}(a)\hat{x}(at)$.
(e) The Hilbert transform of $x(t) \star y(t)$ equals $x(t) \star \hat{y}(t)$ or $\hat{x}(t) \star y(t)$.
(f) The Hilbert transform of a real signal is also real.

P7.26 (Scaling) The Hilbert transform of $\text{sinc}(t)$ is $\frac{1}{2}\pi t \text{sinc}^2(t)$. Use the scaling property to find the Hilbert transform of $\text{sinc}(\alpha t)$.

P7.27 (Modulation) A signal $x_B(t)$ bandlimited to f_B modulates a high-frequency carrier $x_C(t) = \cos(2\pi f_C t)$. If $f_C > f_B$, show that the Hilbert transform of the modulated signal $x_M(t) = x_B(t)\cos(2\pi f_C t)$ is given by $\hat{x}_M(t) = x_B(t)\sin(2\pi f_C t)$.

8 LAPLACE TRANSFORMS

Broadly speaking, the Fourier transform is a very useful tool for *signal analysis*, whereas the Laplace transform is more useful for *system analysis*. The genesis of the Laplace transform may be viewed at several levels:

1. A relation arising from the response (using convolution) of a linear system to the complex exponentials $\exp(st)$ that are its eigensignals. We described this connection in Chapter 4.
2. A modified version of the Fourier transform.
3. An independent transformation method.

In order to keep the discussion self-contained, we start with the last approach and explore the other viewpoints and connections between the various transform methods at the end of this chapter.

This chapter is divided into four parts:

Part 1 An introduction to the Laplace transform and its properties.
Part 2 The inverse Laplace transform.
Part 3 System analysis using the Laplace transform.
Part 4 Connections between various transform methods.

Useful Background   The following topics provide a useful background:

1. The Fourier transform and its properties (Chapter 6)
2. Convolution and transformed-domain methods (Chapter 4).

8.1
The Laplace Transform

The **Laplace transform** $X(s)$ of a signal $x(t)$ is defined as

$$X(s) = \int_{0-}^{\infty} x(t)\exp[-(\sigma + j\omega)t]\,dt = \int_{0-}^{\infty} x(t)\exp(-st)\,dt$$

The complex quantity $s = \sigma + j\omega$ generalizes the concept of frequency to the complex domain. It is often referred to as the **complex frequency**, with ω measured in radians/second, and σ in nepers/second. The relation between $x(t)$ and $X(s)$ is denoted symbolically by the **transform pair**

$$x(t) \longleftrightarrow X(s)$$

The double arrow suggests a one-to-one correspondence between the signal $x(t)$ and its Laplace transform $X(s)$.

Remarks: 1. *Uniqueness*: The Laplace transform provides unique correspondence only for causal signals, since the Laplace transform of $x(t)$ and its causal version $x(t)u(t)$ are clearly identical due to the lower limit of 0.
2. *Lower limit*: To include signals such as $\delta(t)$, which are discontinuous at the origin, the lower limit is chosen as $0-$ whenever appropriate.

Twosided Laplace Transform   We also define the twosided Laplace transform by allowing the lower limit to be $-\infty$. This form finds limited use and is not pursued here.

8.1.1   Convergence

The Laplace transform is an *integral operation*. It exists if $x(t)\exp(-\sigma t)$ is absolutely integrable. Clearly, only certain choices of σ will make this happen. The range of σ which ensures existence defines the **region of convergence** (ROC) of the Laplace transform. For example, the Laplace transform $X(s)$ of $x(t) = \exp(2t)u(t)$ exists only if $\sigma > 2$, which ensures that the product is a decaying exponential. Thus, $\sigma > 2$ defines the region of convergence of the transform $X(s)$. Except in the first example that follows, we avoid explicit mention of the ROC.

8.1.2   Some Laplace Transform Pairs

We use the defining relation to develop a stock list of Laplace transforms. These and other useful transform pairs are listed in Table 8.1 (p. 332).

8.1.3   Some Transforms from the Defining Relation

We provide some examples that illustrate the use of the defining relation to find Laplace transforms and establish the region of convergence.

Example 8.1
Laplace Transforms from
the Defining Relation

(a) *Impulse:* Using the sifting property, the transform of $x(t) = \delta(t)$ equals

$$X(s) = \int_{0-}^{\infty} \delta(t)\exp(-st)\,dt = 1 \quad \text{and thus} \quad \delta(t) \longleftrightarrow 1$$

Table 8.1   Laplace Transforms of Some Signals.

$x(t)$	$X(s)$	$x(t)$	$X(s)$
$\delta(t)$	1	$t\cos(\beta t)u(t)$	$\dfrac{s^2 - \beta^2}{(s^2 + \beta^2)^2}$
$u(t)$	$1/s$	$t\sin(\beta t)u(t)$	$\dfrac{2s\beta}{(s^2 + \beta^2)^2}$
$tu(t)$	$1/s^2$	$\cos^2(\beta t)u(t)$	$\dfrac{s^2 + 2\beta^2}{s(s^2 + 4\beta^2)}$
$t^2 u(t)$	$2/s^3$	$\sin^2(\beta t)u(t)$	$\dfrac{2\beta^2}{s(s^2 + 4\beta^2)}$
$t^n u(t)$	$n!/s^{n+1}$	$\exp(-\alpha t)\cos(\beta t)u(t)$	$\dfrac{s + \alpha}{(s + \alpha)^2 + \beta^2}$
$\exp(-\alpha t)u(t)$	$\dfrac{1}{s + \alpha}$	$\exp(-\alpha t)\sin(\beta t)u(t)$	$\dfrac{\beta}{(s + \alpha)^2 + \beta^2}$
$t\exp(-\alpha t)u(t)$	$\dfrac{1}{(s + \alpha)^2}$	$t\exp(-\alpha t)\cos(\beta t)u(t)$	$\dfrac{(s + \alpha)^2 - \beta^2}{\{(s + \alpha)^2 + \beta^2\}^2}$
$t^n \exp(-\alpha t)u(t)$	$\dfrac{n!}{(s + \alpha)^{n+1}}$	$t\exp(-\alpha t)\sin(\beta t)u(t)$	$\dfrac{2\beta(s + \alpha)}{\{(s + \alpha)^2 + \beta^2\}^2}$
$\cos(\beta t)u(t)$	$\dfrac{s}{s^2 + \beta^2}$	$\dfrac{1}{\sqrt{\pi t}}u(t)$	$\dfrac{1}{\sqrt{s}}$
$\sin(\beta t)u(t)$	$\dfrac{\beta}{s^2 + \beta^2}$	$\exp(\alpha t)u(t) \quad \alpha > 0$	$1/(s - \alpha)$
$\mathrm{sinc}(\alpha t)u(t)$	$\dfrac{1}{\pi\alpha}\tan^{-1}(\pi\alpha/s)$	$\exp(\alpha t^2)u(t) \quad \alpha > 0$	Not defined

---

Note: The choice $0-$ for the lower limit is due to the fact that $\delta(t)$ is discontinuous at $t = 0$. The ROC is the entire s plane.

(b) *Step:* From the defining relation, the transform of $x(t) = u(t)$ equals

$$X(s) = \int_{0-}^{\infty} u(t)\exp(-st)\, dt = \int_{0-}^{\infty} \exp(-st)\, dt = 1/s \quad \text{and thus} \quad u(t) \longleftrightarrow 1/s$$

Note: $\exp(-st) = \exp[-(\sigma + j\omega)t]$ equals 0 at the upper limit only if $\sigma > 0$. Thus, $\sigma > 0$ defines its region of convergence.

(c) *Ramp:* The transform of $x(t) = r(t)$ equals

$$X(s) = \int_{0}^{\infty} r(t)\exp(-st)\, dt = \int_{0}^{\infty} t\exp(-st)\, dt = 1/s^2 \quad \text{and thus} \quad r(t) \longleftrightarrow 1/s^2$$

Note: The region of convergence for this result is also $\sigma > 0$.

(d) *Decaying exponential:* The transform of $x(t) = \exp(-\alpha t)u(t)$ equals

$$X(s) = \int_0^\infty \exp(-\alpha t)\exp(-st)\,dt = \int_0^\infty \exp[-(s+\alpha)t]\,dt = 1/(s+\alpha)$$

The region of convergence for this result is $\sigma > -\alpha$.

(e) *Switched cosine:* With $x(t) = \cos(\alpha t)u(t)$, we use Euler's relation to get

$$X(s) = \int_0^\infty \cos(\alpha t)\exp(-st)\,dt$$

$$= \frac{1}{2}\int_0^\infty \{\exp[-(s+j\alpha)t] + \exp[-(s-j\alpha)t]\}\,dt = \frac{s}{s^2 + \alpha^2}$$

8.2
Properties of the Laplace Transform

The operational properties of the Laplace transform are summarized in Table 8.2 (p. 334). Most of these properties follow from the defining integral. The **linearity** property comes about because

$$\int_0^\infty [\alpha x_1 + \beta x_2]\exp(-st)\,dt = \alpha \int_0^\infty x_1 \exp(-st)\,dt + \beta \int_0^\infty x_2 \exp(-st)\,dt$$

$$= \alpha X_1(s) + \beta X_2(s)$$

The **times-exp** property also falls out directly from the defining integral.

$$\int_0^\infty x(t)\exp(-\alpha t)\exp(-st)\,dt = \int_0^\infty x(t)\exp[-(s+\alpha)t]\,dt = X(s+\alpha)$$

The **times-sin** and **times-cos** properties form a direct extension if the sines and cosines are expressed as exponentials using Euler's relation.

$$\int_0^\infty x(t)\cos(\alpha t)e^{-st}\,dt = \frac{1}{2}\int_0^\infty x(t)[e^{-j\alpha} + e^{j\alpha}]e^{-st}\,dt = \frac{1}{2}[X(s+j\alpha) + X(s-j\alpha)]$$

The **time-scaling** property results from a change of variable.

$$\int_0^\infty x(\alpha t)\exp(-st)\,dt \xrightarrow{\alpha t=\lambda} = \frac{1}{\alpha}\int_0^\infty x(\lambda)\exp(-s\lambda/\alpha)\,d\lambda = \frac{1}{\alpha}X(s/\alpha)$$

Thus, compression in one domain results in a stretching in the other.

The **time-shift** property also results from a change of variable.

$$\int_\alpha^\infty x(t-\alpha)\exp(-st)\,dt \xrightarrow{t-\alpha=\lambda} = \int_0^\infty x(\lambda)\exp[-s(\lambda+\alpha)]\,d\lambda = \exp(-\alpha s)X(s)$$

Remark: The quantity $\exp(-s)$ may be regarded as a *unit shift operator* that shifts or delays the time function by 1 unit. Similarly, $\exp(-s\alpha)$ contributes a delay of α units to the time signal.

The **derivative** property follows from the defining relation.

$$\int_{0-}^\infty x'(t)\exp(-st)\,dt = x(t)\exp(-st)\Big|_{0-}^\infty + s\int_{0-}^\infty x(t)\exp(-st)\,dt = -x(0-) + sX(s)$$

Table 8.2   Operational Properties of the Laplace Transform.

Note: $x(t)$ is to be regarded as the causal sign $x(t)u(t)$.

Property	x(t)	X(s)
Superposition	$\alpha x_1(t) + \beta x_2(t)$	$\alpha X_1(s) + \beta X_2(s)$
Times-exp	$\exp(-\alpha t)x(t)$	$X(s + \alpha)$
Times-cos	$\cos(\alpha t)x(t)$	$\frac{1}{2}[X(s + j\alpha) + X(s - j\alpha)]$
Times-sin	$\sin(\alpha t)x(t)$	$j\frac{1}{2}[X(s + j\alpha) - X(s - j\alpha)]$
Time Scaling	$x(\alpha t), \ \alpha > 0$	$\frac{1}{\alpha}X(s/\alpha)$
Time Shift	$x(t - \alpha)u(t - \alpha), \ \alpha > 0$	$\exp(-s\alpha)X(s)$
Times-t	$tx(t)$	$-dX(s)/ds$
	$t^n x(t)$	$(-1)^n d^n X(s)/ds^n$
Derivative	$x'(t)$	$sX(s) - x(0-)$
	$x''(t)$	$s^2 X(s) - sx(0-) - x'(0-)$
	$x^n(t)$	$s^n X(s) - s^{n-1}x(0-) - \cdots - x^{n-1}(0-)$
Integral	$\int_{0-}^{t} x(t)\,dt$	$X(s)/s$
	$\int_{-\infty}^{t} x(t)\,dt$	$\dfrac{X(s)}{s} + \dfrac{\mathbb{A}[x(t)](-\infty, 0)}{s}$
Convolution	$x_1(t) \star x_2(t)$	$X_1(s)X_2(s)$
Multiplication	$x_1(t)x_2(t)$	$\dfrac{1}{2\pi j}X_1(s) \star X_2(s)$
Switched periodic, $x_1(t)$ = one period	$x_p(t)u(t)$	$X_1(s)/[1 - \exp(-sT)]$

Initial value theorem	$x(0+) = \lim\limits_{s \to \infty}[sX(s)]$ If $X(s)$ is strictly proper
Final value theorem	$x(t)|_{t \to \infty} = \lim\limits_{s \to 0}[sX(s)]$ If poles of $X(s)$ lie in LHP

The **times-t** property involves expressing $t\exp(-st)$ as $-\frac{d}{ds}[\exp(-st)]$ and interchanging the order of differentiation and integration.

$$\int_{0-}^{\infty} tx(t)e^{-st}\,dt = \int_{0-}^{\infty} x(t)\left[-\frac{d}{ds}(e^{-st})\right]dt = -\frac{d}{ds}\int_{0-}^{\infty} x(t)e^{-st}\,dt = -\frac{dX(s)}{ds}$$

Remark: The times-t and derivative properties are duals. Multiplication by the variable (t or s) in one domain results in a derivative in the other.

The **convolution** property involves changing the order of integration and a change of variable $(t - \lambda = \tau)$.

$$\int_0^\infty \left[\int_0^\infty x(t - \lambda)u(t - \lambda)y(\lambda)d\lambda \right] e^{-st}\, dt = \int_{0-}^\infty \left[\int_\lambda^\infty x(t - \lambda)e^{-s(t-\lambda)}\, dt \right] y(\lambda)e^{-s\lambda}\, d\lambda$$

$$= X(s)Y(s)$$

8.2.1   Using the Properties

To develop the Laplace transform of unfamiliar or complicated signals, it is often much easier to start with known transform pairs and then use properties. We illustrate this approach in the following examples.

Example 8.2

An Exponentially Damped Ramp

Let us find the Laplace transform of $x(t) = t \exp(-3t)u(t)$.

(a) *Easiest way:* Start with the known pair $r(t) = tu(t) \longleftrightarrow 1/s^2$ and use the times-exp property to give $\exp(-3t)tu(t) \longleftrightarrow 1/(s + 3)^2$.

(b) *Still easy:* Start with the known pair $\exp(-3t)u(t) \longleftrightarrow 1/(s + 3)$ and use the times-t property to give $t \exp(-3t)u(t) \longleftrightarrow -d/ds[1/(s + 3)] = 1/(s + 3)^2$.

(c) *Another way:* Start with $\exp(-t)tu(t) \longleftrightarrow 1/(s + 1)^2$ as in part (a). Rewrite $x(t)$ as $\frac{1}{3}(3t)\exp(-3t)u(t)$, and use scaling to get $X(s) = \frac{1}{3}[\frac{1}{3}/(\frac{1}{3}s + 1)^2] = 1/(s + 3)^2$.

Example 8.3

An Arbitrary Sinusoid

To find the Laplace transform of $x(t) = \cos(3t + \frac{1}{4}\pi)u(t)$, use trigonometric identities to write $x(t) = [(1/\sqrt{2})\cos(3t) - (1/\sqrt{2})\sin(3t)]u(t)$. From known transform pairs, and superposition, we obtain $X(s) = (1/\sqrt{2})(s - 3)/(s^2 + 9)$.

Example 8.4

Using the Time-Shift Property

Here are two ways to find the Laplace transform of $x(t) = t^2 u(t - 1)$:

(a) Start with $u(t) \longleftrightarrow 1/s$, shift to give $u(t - 1) \longleftrightarrow \exp(-s)/s$, and use the times-$t$ property twice to give $t^2 u(t - 1) \longleftrightarrow d^2/ds^2[\exp(-s)/s]$. This results in $t^2 u(t - 1) \longleftrightarrow \exp(-s)[2/s^3 - 2/s^2 + 1/s]$.

(b) Rewrite $x(t)$ as

$$(t - 1 + 1)^2 u(t - 1) = (t - 1)^2 u(t - 1) - 2(t - 1)u(t - 1) + u(t - 1)$$

Now, use the time-shift property to obtain $X(s) = \exp(-s)[2/s^3 - 2/s^2 + 1/s]$.

Comment: In part (b) we rewrote the expression in terms of the argument $t - 1$ of the *shifted* step. This ensures correct use of the time-shift property.

Two ways to find the transform of $x(t) = \exp(-3t)\cos(2t)u(t)$ are:

Example 8.5
Using the Times-cos
Property

(a) Start with the known pair $y(t) = \exp(-3t)u(t) \longleftrightarrow Y(s) = 1/(s+3)$, and use the times-cos property $\cos(\alpha t)y(t) \longleftrightarrow \frac{1}{2}[Y(s+j\alpha) + Y(s-j\alpha)]$ to give

$$X(s) = \tfrac{1}{2}[1/(s+j2+3) + 1/(s-j2+3)] = (s+3)/[(s+3)^2 + 4]$$

(b) Start with the known pair $\cos(2t)u(t) \longleftrightarrow s/s^2 + 4$, and use the times-exp property to give $\exp(-3t)\cos(2t)u(t) \longleftrightarrow (s+3)/[(s+3)^2 + 4]$.

Example 8.6
Signals Described
Graphically

We find the transforms of the signals shown in Figure 8.1. With $w(t) = u(t) - u(t-1)$, the time-shift property gives $W(s) = 1/s - \exp(-s)/s$. The signal $x(t)$ may be described as $tw(t)$. The times-t property gives

$$X(s) = -\frac{d}{ds}[1 - \exp(-s)/s] = \frac{1}{s} - \frac{1}{s^2} + \exp(-s)/s^2.$$

We can also write $x(t) = u(t) - r(t) + r(t-1)$ to get $X(s) = 1/s - 1/s^2 + \exp(-s)/s^2$.

Figure 8.1   The signals whose Laplace transform is found in Example 8.6.

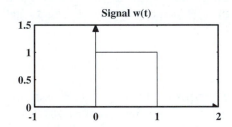

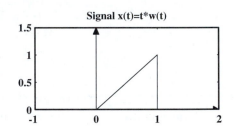

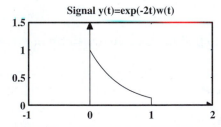

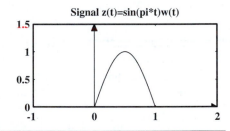

The signal $y(t)$ may be described as $\exp(-2t)w(t)$. Using the times-exp property, we obtain $Y(s) = \{1 - \exp[-(s+2)]\}/(s+2)$.

The signal $z(t)$ may be described as $\sin(\pi t)w(t)$. Using the times-sin property, we obtain

$$Z(s) = \frac{1}{2}j\left[\frac{1 + \exp[-(s+j\pi)]}{s+j\pi} - \frac{1 + \exp[-(s-j\pi)]}{s-j\pi}\right]$$

With $\exp(\pm j\pi) = -1$, this result simplifies to $Z(s) = \pi[1 + \exp(-s)]/(s^2 + \pi^2)$.

Comment: An easier way is to write $z(t) = \sin(\pi t)u(t) + \sin[\pi(t-1)]u(t-1)$ and use the shift property to get

$$Z(s) = \pi/(s^2 + \pi^2) + \pi\exp(-s)/(s^2 + \pi^2) = \pi[1 + \exp(-s)]/(s^2 + \pi^2)$$

Example 8.7
Using the Derivative
Property

The onesided Laplace transforms of $\sin(\alpha t)$ and $\sin(\alpha t)u(t)$ are identical. We use the derivative property to show this as follows.

(a) Start with the pair $x(t) = \cos(\alpha t) \longleftrightarrow s/(s^2 + \alpha^2)$ and use the derivative property to give $-\alpha\sin(\alpha t) \longleftrightarrow s^2/(s^2 + \alpha^2) - x(0-)$. Since $x(0-) = 1$, we get $-\alpha\sin(\alpha t) \longleftrightarrow s^2/(s^2 + \alpha^2) - 1 = -\alpha^2/(s^2 + \alpha^2)$. Thus, $\sin(\alpha t) \longleftrightarrow \alpha/(s^2 + \alpha^2)$.

(b) Start with $y(t) = \cos(\alpha t)u(t) \longleftrightarrow s/(s^2 + \alpha^2)$. The derivative property gives $\cos(\alpha t)\delta(t) - \alpha\sin(\alpha t)u(t) \longleftrightarrow s^2/(s^2 + \alpha^2) - y(0-)$. With $\cos(\alpha t)\delta(t) = \delta(t)$, and $y(0-) = 0$, we get $\delta(t) - \alpha\sin(\alpha t)u(t) \longleftrightarrow s^2/(s^2 + \alpha^2)$. Since $\delta(t) \longleftrightarrow 1$, superposition gives $-\alpha\sin(\alpha t)u(t) \longleftrightarrow s^2/(s^2 + \alpha^2) - 1$. This leads to the result $\sin(\alpha t)u(t) \longleftrightarrow \alpha/(s^2 + \alpha^2)$, as before.

Comment: The initial value $x(0-)$ plays an important role, especially when we deal with noncausal signals.

8.2.2   *Laplace Transform of Switched Periodic Signals*

Consider a periodic signal $x_1(t)$ turned on at $t = 0$. It may be expressed as $x(t) = x_p(t)u(t)$. If the period of $x_p(t)$ equals T and $x_1(t)$ describes $x(t)$ over $(0-T)$, the signal $x(t)$ may be represented by $x_1(t)$ and its superposed, shifted versions as

$$x(t) = x_p(t)u(t) = x_1(t) + x_1(t - T) + x_1(t - 2T) + \cdots$$

With $X_1(s)$ as the Laplace transform of $x_1(t)$, the time-shift property gives the Laplace transform $X(s)$ of $x(t)$ as

$$X(s) = X_1(s) + e^{-sT}X_1(s) + e^{-2sT}X_1(s) + \cdots = X_1(s)[1 + e^{-sT} + e^{-2sT} + \cdots]$$

Using the closed-form result for the infinite series (Appendix A) gives

$$X(s) = X_1(s)/(1 - e^{-sT})$$

Example 8.8
A Switched Sawtooth
Pulse Train

One period $x_1(t)$, $(0, 1)$ of the switched sawtooth $x(t)$ of Figure 8.2 (p. 338) has the Laplace transform $X_1(s) = 1/s - 1/s^2 + \exp(-s)/s^2$ (see Example 8.6). The Laplace transform $X(s)$ of the signal $x(t)$ is then $X_1(s)/(1 - e^{-s})$ and equals

$$X(s) = \frac{1/s - 1/s^2 + \exp(-s)/s^2}{1 - e^{-s}} = \frac{s - 1 + e^{-s}}{s^2(1 - e^{-s})}$$

Figure 8.2   The periodic signal for Example 8.8.

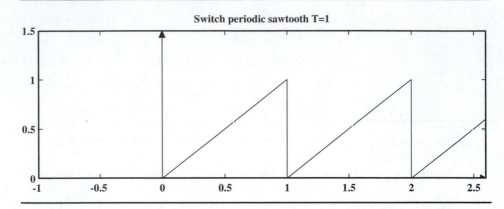

8.3
Poles, Zeros and the s-Plane

The Laplace transform of many signals is a rational function (a ratio of polynomials) in s, whose general form can be expressed as

$$X(s) = \frac{P(s)}{Q(s)} = \frac{b_0 s^M + b_1 s^{M-1} + \cdots + b_M}{s^N + a_1 s^{N-1} + \cdots + a_N}$$

It is customary to normalize the coefficient of s^N to unity.

Recall that $X(s)$ is a *proper rational function* if the degree N of $Q(s)$ exceeds or equals the degree M of $P(s)$, or $N - M \geq 0$. Similarly, $X(s)$ is a *strictly proper rational function* if $N - M \geq 1$, or the degree of N of $Q(s)$ exceeds (not just equals) the degree M of $P(s)$.

An Nth-order polynomial has N roots, which may be real or complex. If the polynomial coefficients are real, the complex roots will always occur in conjugate pairs. Some of the roots (or complex conjugate pairs of roots) may even be repeated. If we denote the roots of $P(s)$ by z_i, $i = 1, 2, \ldots, M$ and the roots of $Q(s)$ by p_k, $k = 1, 2, \ldots, N$, then $X(s)$ may be expressed in factored form as

$$X(s) = \frac{P(s)}{Q(s)} = K \frac{(s - z_1)(s - z_2) \cdots (s - z_M)}{(s - p_1)(s - p_2) \cdots (s - p_N)}$$

The M roots of $P(s)$ are termed the **zeros** or **finite zeros** of $X(s)$, and the N roots of $Q(s)$ are called the **poles** or **finite poles** of $X(s)$. Here, we assume that *factors common to $P(s)$ and $Q(s)$ have been canceled.*

If $P(s)$ and $Q(s)$ are of unequal degree, it is also customary to "equalize" the number of poles and zeros by introducing poles and zeros at infinity. A pole occurs at $s = \infty$ if $X(\infty) = \infty$. A zero occurs at $s = \infty$ if $X(\infty) = 0$. If $N > M$, the quantity $r_d = N - M$ defines the number of **zeros at infinity**. For $M > N$, $r_n = M - N$ defines the number of **poles at infinity**. If we include poles and zeros at infinity, $X(s)$ has the same number of poles and zeros.

8.3.1 *Pole-Zero Plots*

A plot of the poles (denoted by crosses) and zeros (denoted by circles) of a rational function $X(s)$ in the *s*-plane constitutes the **pole-zero plot** and provides a visual picture of the root locations. For repeated roots, we show the multiplicity alongside the roots.

Clearly, we can find $X(s)$ directly from a pole-zero plot of the root locations but only to within the multiplicative constant K. If the value of K is also shown on the plot, $X(s)$ is then known in its entirety.

Example 8.9
A Pole-Zero Plot

Consider $H(s) = 2s(s + 1)/[(s + 3)(s^2 + 4)(s^2 + 4s + 5)]$. The numerator degree is 2. The denominator degree is 5. There are thus three zeros at infinity. The two finite zeros are $s = 0$ and $s = -1$. The locations of the five finite poles are $s = -3$, $s = \pm j2$, $s = -2 \pm j$. There are no poles at infinity (Figure 8.3).

Figure 8.3 Pole-zero plot for the rational function
$$H(s) = 2s(s + 1)/[(s + 3)(s^2 + 4)(s^2 + 4s + 5)].$$

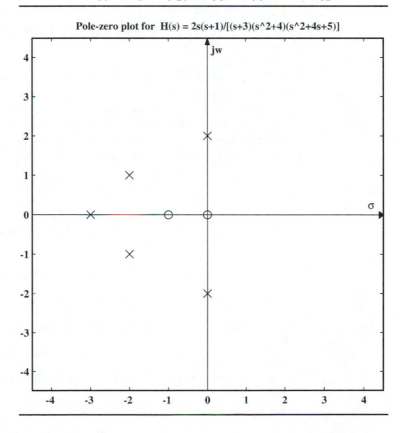

Pole-zero plot for $H(s) = 2s(s+1)/[(s+3)(s^2+4)(s^2+4s+5)]$

8.3.2   *The Initial and Final Value Theorems*

Initial Value Theorem   The **initial value theorem** predicts the initial value $x(0+)$ of the time signal $x(t)$ *from its strictly proper transform* $X(s)$ as

$$x(0+) = \lim_{s\to\infty}[sX(s)]$$

This theorem predicts the initial value at $t = 0+$ and not $t = 0$.

Remarks: **1.** If $X(s)$ is not strictly proper, we get correct results by using the *strictly proper* part of $X(s)$ obtained after long division since the remaining part corresponds to impulses and their derivatives which occur only at $t = 0$ and are zero at $t = 0+$.
2. If $X(s)$ or its strictly proper part equals $P(s)/Q(s)$ and $P(s)$ is of degree M and $Q(s)$ is of degree N, $x(0+) = 0$ if $N - M > 1$. If $N - M = 1$, $x(0+)$ equals the ratio of the leading coefficients of $P(s)$ and $Q(s)$.

Final Value Theorem   The final value theorem predicts the value of the time signal $x(t)$ as $t \to \infty$ from its transform $X(s)$, and reads

$$x(\infty) = \lim_{s\to 0}[sX(s)]$$

It applies only if the poles of $X(s) = P(s)/Q(s)$, with common factors in $P(s)$ and $Q(s)$ canceled, lie in the left half of the s-plane. The only $j\omega$-axis pole permitted is a simple pole at $s = 0$. Here is what to expect:

1. $x(\infty) = 0$ if all poles of $X(s)$ lie to the left of the $j\omega$-axis (because $x(t)$ comprises only damped exponentials).
2. $x(\infty)$ is constant if there is a single pole at $s = 0$ (because $x(t)$ includes a step).
3. $x(\infty)$ is indeterminate if there are conjugate pole pairs on the $j\omega$-axis (because $x(t)$ includes sinusoids whose final value is indeterminate). *The final value theorem yields erroneous results if used in this case.*

Example 8.10
Initial and Final Value Theorems

(a) With $X(s) = 12(s + 1)/[s(s^2 + 4)]$, we find
Initial value: $x(0+) = \lim_{s\to\infty} sX(s) = 0$ since $N - M = 2$ for $X(s)$.
Final value: Indeterminate ($X(s)$ has conjugate poles $s = \pm j2$ on $j\omega$-axis).
(b) With $X(s) = (2s + 6)/[s(4s + 2)]$, we find

$$x(0+) = \lim_{s\to\infty} sX(s) = \lim_{s\to\infty}(2s + 6)/(4s + 2) = \lim_{s\to\infty}(2 + 6/s)/(4 + 2/s) = \frac{1}{2}$$
$$x(\infty) = \lim_{s\to 0} sX(s) = \lim_{s\to 0}(s + 6)/(s + 2) = 3$$

(c) With $X(s) = (4s + 5)/(2s + 1)$, we use long division to rewrite it as

$$X(s) = 2 + 3/(2s + 1) = 2 + Y(s)$$

The strictly proper part $Y(s)$ gives the initial value of $x(t)$ as

$$x(0+) = \lim_{s\to\infty} sY(s) = \lim_{s\to\infty}(3s)/(2s + 1) = \frac{3}{2}$$

For the final value we use $X(s)$ directly, to obtain

$$x(\infty) = \lim_{s \to 0} sX(s) = \lim_{s \to 0}(4s^2 + 5s)/(2s + 1) = 0$$

8.3.3   The Transfer Function

The concept of the transfer function lies at the heart of system analysis using transform methods. *The transfer function is defined only for relaxed, LTI systems* in one of two equivalent ways:

1. The ratio of the transformed output $Y(s)$ and transformed input $X(s)$.
2. The transform of the system impulse response $h(t)$.

The convolution $y(t) = x(t) \star h(t)$ describes the response of a relaxed system. Since the convolution operation transforms to a product, we have

$$Y(s) = X(s)H(s) \quad \text{or} \quad H(s) = Y(s)/X(s)$$

An LTI system may also be described by the differential equation

$$\frac{d^N y}{dt^N} + a_1 \frac{d^{N-1}y}{dt^{N-1}} + \cdots + a_{N-1}\frac{dy}{dt} + a_N y = b_0 \frac{d^M x}{dt^M} + b_1 \frac{d^{M-1}x}{dt^{M-1}} + \cdots + b_{M-1}\frac{dx}{dt} + b_M x$$

Its Laplace transform results in the transfer function $H(s) = Y(s)/X(s)$ as

$$H(s) = \frac{Y(s)}{X(s)} = \frac{b_0 s^M + b_1 s^{M-1} + \cdots + b_M}{s^N + a_1 s^{N-1} + \cdots + a_N}$$

The transfer function is thus a ratio of polynomials in s. It allows access to both the impulse response and the differential equation of the relaxed system in the time domain.

Example 8.11
H(s) and h(t) from a Differential Equation

(a) Consider a system governed by $y''(t) + 3y'(t) + 2y(t) = 2x'(t) + 3x(t)$. Using the derivative property, we obtain $s^2 Y(s) + 3sY(s) + 2Y(s) = 2sX(s) + X(s)$ or $H(s) = Y(s)/X(s) = (2s + 3)/(s^2 + 3s + 2)$.

(b) We can write $H(s) = 1/(s + 1) + 1/(s + 2)$. The impulse response thus equals $h(t) = [\exp(-t) + \exp(-2t)]u(t)$.

(c) Since $H(s) = Y(s)/X(s)$, we can also write $[s^2 + 3s + 2]Y(s) = [2s + 3]X(s)$. This leads to $y''(t) + 3y'(t) + 2y(t) = 2x'(t) + 3x(t)$.

Comment: $H(s), h(t)$, and the differential equation describe the same system.

8.3.4   Poles and Zeros of the Transfer Function

The transfer function may be also be expressed in factored form as

$$H(s) = \frac{Y(s)}{X(s)} = \frac{b_0 s^M + b_1 s^{M-1} + \cdots + b_M}{s^N + a_1 s^{N-1} + \cdots + a_N} = K\frac{(s - z_1) \cdots (s - z_M)}{(s - p_1) \cdots (s - p_N)}$$

The quantity $G = b_M/a_N$ is called the **dc gain**, since $H(s) \to b_M/a_N$ as $s \to 0$.

Remarks:
1. The poles of $H(s)$ are also called **natural modes**, or **natural frequencies**. The poles of $Y(s) = H(s)X(s)$ determine the form of the system response. The natural frequencies of $H(s)$ always appear in the system response unless they are canceled by any corresponding zeros in $X(s)$.
2. The zeros of $H(s)$ are often called **transmission zeros**. Since $H(s) = 0$ at its zeros, $Y(s) = X(s)H(s)$ is also zero. The zeros of $H(s)$ may thus be regarded as the frequencies that are blocked by the system.

8.4
The Inverse Laplace Transform

The inverse Laplace transform is also governed by an integral operation but with respect to the variable s. For a particular value of σ in the region of convergence, the signal $x(t)$ is given by

$$x(t) = \int_{-\infty}^{\infty} X(\sigma + j2\pi f)\exp[(\sigma + j2\pi f)t]\,df$$

$$= \exp(\sigma t)\int_{-\infty}^{\infty} X(\sigma + j2\pi f)\exp(j2\pi ft)\,df$$

The change of variable $s = \sigma + j2\pi f$ gives $df = ds/2\pi$, and we obtain

$$x(t) = \frac{1}{2\pi j}\int_{\sigma - j\infty}^{\sigma + j\infty} X(s)\exp(-st)\,ds$$

This relation holds for both onesided and twosided transforms. Since s is complex, the solution requires a knowledge of complex variables. A simpler alternative, and the only one we discuss in this text, is to express $X(s)$ as a sum of terms whose inverse transforms can be recognized from a stock list of transform pairs. We form such a sum by resorting to a *partial fraction expansion* of $X(s)$.

8.4.1   *Partial Fraction Expansion*

The denominator $Q(s)$ of an Nth-order rational function can be factored into linear (perhaps repeated) factors. A **partial fraction expansion** (PFE) allows a *strictly proper rational fraction* $P(s)/Q(s)$ to be expressed as a sum of N terms, whose numerators are constant, and whose denominators correspond to the linear and/or repeated factors of $Q(s)$. This, in turn, allows us to relate such terms to their corresponding inverse transform.

The partial fraction method requires not only a strictly proper $X(s)$ but also a denominator $Q(s)$ in factored form. The form of the expansion depends on the nature of the factors in $Q(s)$. The partial fraction expansions for the two most important cases are summarized in Table 8.3 and are based on normalizing the leading coefficient of $Q(s)$ to unity. The constants in the partial fraction expansion are often called **residues**.

Distinct Linear Factors   If $Q(s)$ contains only distinct linear factors, we write $X(s)$ in the form:

$$X(s) = \frac{P(s)}{(s + p_1)(s + p_2)\cdots(s + p_m)\cdots(s + p_N)}$$

Table 8.3   Partial Fraction Expansions.

Factors	$X(s) = P(s)/Q(s)$	PFE	Constants (residues)
Distinct linear	$\displaystyle\prod_{m=1}^{N} \frac{P(s)}{(s+p_m)}$	$\displaystyle\sum_{m=1}^{N} \frac{K_m}{(s+p_m)}$	$K_m = (s+p_m)X(s)\Big\|_{s=-p_m}$
Repeated linear	$\displaystyle\frac{1}{(s+r)^k}\prod_{m=1}^{N}\frac{P(s)}{(s+p_m)}$	$\displaystyle\sum_{m=1}^{N}\frac{K_m}{(s+p_m)} + \sum_{n=0}^{k-1}\frac{A_n}{(s+r)^{k-n}}$	$A_n = \dfrac{1}{n!}\dfrac{d^n}{ds^n}[(s+r)^k X(s)]\Big\|_{s=-r}$

$$= \frac{K_1}{(s+p_1)} + \frac{K_2}{(s+p_2)} + \cdots + \frac{K_m}{(s+p_m)} + \cdots + \frac{K_N}{(s+p_N)}$$

To find the mth coefficient K_m, we multiply both sides by $(s+p_m)$ to get

$$(s+p_m)X(s) = K_1\frac{(s+p_m)}{(s+p_1)}$$
$$+ K_2\frac{(s+p_m)}{(s+p_2)} + \cdots + K_m + \cdots + K_N\frac{(s+p_m)}{(s+p_N)}$$

With both sides evaluated at $s = -p_m$, we obtain K_m as

$$K_m = (s+p_m)X(s)|_{s=-p_m}$$

$X(s)$ will, in general, contain terms with real constants and pairs of terms with complex conjugate residues, and can be written as

$$X(s) = \frac{K_1}{(s+p_1)} + \frac{K_2}{(s+p_2)} + \cdots$$
$$+ \frac{A_1}{(s+r_1)} + \frac{A_1^*}{(s+r_1^*)} + \frac{A_2}{(s+r_2)} + \frac{A_2^*}{(s+r_2^*)} + \cdots$$

We make two remarks:

1. For a real root, the residue (coefficient) will also be real.
2. For each pair of complex conjugate roots, the residues will also be complex conjugates, and we thus need compute only one of these.

Repeated Factors   The preceding results require modification if $Q(s)$ contains repeated factors. If a factor is repeated k times, it is said to have a *multiplicity* of k. If $Q(s)$ contains the repeated term $(s+r)^k$ and one other linear factor, the PFE of $X(s) = P(s)/Q(s)$ may be set up as

$$X(s) = \frac{P(s)}{(s+p_1)(s+r)^k}$$
$$= \frac{K_1}{(s+p_1)} + \frac{A_0}{(s+r)^k} + \frac{A_1}{(s+r)^{k-1}} + \frac{A_2}{(s+r)^{k-2}} + \cdots + \frac{A_{k-1}}{(s+r)}$$

Observe that the constants A_j ascend in index j from 0 to $k-1$, whereas the denominators $(s+r)^m$ descend in power m from k to 1.

The coefficient K_n for a nonrepeated term $K_n/(s + p_n)$ is found by evaluating $(s + p_n)X(s)$ at $s = -p_n$. For our case,

$$K_1 = (s + p_1)X(s)|_{s=-p_1}$$

The coefficients A_j associated with the repeated root require $(s + r)^k X(s)$, or its derivatives, for their evaluation. We successively find

1. A_0 by evaluating $(s + r)^k X(s)$ at $s = -r$.

2. A_1 by evaluating $\dfrac{d}{ds}[(s + r)^k X(s)]$ at $s = -r$.

3. A_2 by evaluating $\dfrac{1}{2!}\dfrac{d^2}{ds^2}[(s + r)^k X(s)]$ at $s = -r$.

4. A_m by evaluating $\dfrac{1}{m!}\dfrac{d^m}{ds^m}[(s + r)^k X(s)]$ at $s = -r$.

Even though this process allows us to find the coefficients independently of each other, the algebra in finding the derivatives can become tedious if the multiplicity k of the roots exceeds 2 or 3.

The Inverse Transform   Once the partial fraction expansion is established, the inverse transform for each term can be found with the help of a transform-pair table. A summary of the various forms appears in Table 8.4. The entries in this table are based on the following observations:

1. All terms show an exponential decay or damped exponential form.
2. Terms corresponding to real factors will have the form $K \exp(-pt)u(t)$.
3. Terms corresponding to each complex conjugate pair of roots will be of the form $A \exp(-rt) + A^* \exp(-r^*t)$. Using Euler's relation, this can be reduced to a real term in one of the forms listed in Table 8.4.

Example 8.12
Nonrepeated Poles

Let $X(s) = (2s^3 + 8s^2 + 4s + 8)/[(s)(s + 1)(s^2 + 4s + 8)]$. This can be factored as

$$X(s) = \frac{K_1}{s} + \frac{K_2}{s + 1} + \frac{A}{s + 2 + j2} + \frac{A^*}{s + 2 - j2}$$

We successively evaluate

$$K_1 = sX(s)|_{s=0} = (2s^3 + 8s^2 + 4s + 8)/[(s + 1)(s^2 + 4s + 8)]|_{s=0} = \frac{8}{8} = 1$$

$$K_2 = (s + 1)X(s)|_{s=-1} = (2s^3 + 8s^2 + 4s + 8)/[s(s^2 + 4s + 8)]|_{s=-1} = \frac{10}{-5} = -2$$

$$A = (s + 2 + j2)X(s)|_{s=-2-j2}$$

$$= (2s^3 + 8s^2 + 4s + 8)/[s(s + 1)(s + 2 - j2)]|_{s=-2-j2} = \frac{3}{2} + j\frac{1}{2}$$

The partial fraction expansion thus becomes

$$X(s) = \frac{1}{s} - \frac{2}{s+1} + \frac{1.5 + j\frac{1}{2}}{s+2+j2} + \frac{1.5 - j\frac{1}{2}}{s+2-j2}$$

With $\frac{3}{2} + j\frac{1}{2} = 1.581\angle 18.4° = 1.581\angle 0.1024\pi$, Table 8.4 yields $x(t)$ either as

$$x(t) = u(t) - 2\exp(-t)u(t) + 2\exp(-2t)[1.5\cos(2t) + \tfrac{1}{2}\sin(2t)]u(t)$$

or

$$x(t) = u(t) - 2\exp(-t)u(t) + 3.162\exp(-2t)\cos(2t - 0.1024\pi)u(t)$$

Comment: As a check, confirm that the LT of $x(t)$ does match $X(s)$. Try it!

___

Example 8.13
Repeated Poles

Let $X(s) = 4/[(s+1)(s+2)^3]$. Its partial fraction expansion is

$$X(s) = \frac{K_1}{s+1} + \frac{A_0}{(s+2)^3} + \frac{A_1}{(s+2)^2} + \frac{A_2}{(s+2)}$$

We compute $K_1 = (s+1)X(s)|_{s=-1} = 4/(s+2)^3|_{s=-1} = 4$. Since $(s+2)^3X(s) = 4/(s+1)$, we also successively compute

$$A_0 = 4/(s+1)|_{s=-2} = -4$$

$$A_1 = \frac{d}{ds}[4/(s+1)]|_{s=-2} = -4/(s+1)^2|_{s=-2} = -4$$

$$A_2 = \left(\frac{1}{2}\right)\frac{d^2}{ds^2}[4/(s+1)]|_{s=-2} = \left(\frac{1}{2}\right)(8)/(s+1)^3|_{s=-2} = -4$$

Table 8.4   Inverse Transforms of Partial Fraction Terms.

Partial fraction term	Inverse transform
$\dfrac{K}{s+\alpha}$	$K\exp(-\alpha t)u(t)$
$\dfrac{K}{(s+\alpha)^n}$	$\dfrac{K}{(n-1)!}t^{n-1}\exp(-\alpha t)u(t)$
$\dfrac{Cs+D}{(s+\alpha)^2+\beta^2}$	$\exp(-\alpha t)[C\cos(\beta t) + \dfrac{D-\alpha C}{\beta}\sin(\beta t)]u(t)$
$\dfrac{A+jB}{s+\alpha+j\beta} + \dfrac{A-jB}{s+\alpha-j\beta}$	$2\exp(-\alpha t)[A\cos(\beta t) + B\sin(\beta t)]u(t)$
$\dfrac{A+jB}{(s+\alpha+j\beta)^n} + \dfrac{A-jB}{(s+\alpha-j\beta)^n}$	$\dfrac{2}{(n-1)!}t^{n-1}\exp(-\alpha t)[A\cos(\beta t) + B\sin(\beta t)]u(t)$
$\dfrac{M\angle\theta}{s+\alpha+j\beta} + \dfrac{M\angle-\theta}{s+\alpha-j\beta}$	$2M\exp(-\alpha t)\cos(\beta t - \theta)u(t)$
$\dfrac{M\angle\theta}{(s+\alpha+j\beta)^n} + \dfrac{M\angle-\theta}{(s+\alpha-j\beta)^n}$	$\dfrac{2M}{(n-1)!}t^{n-1}\exp(-\alpha t)\cos(\beta t - \theta)u(t)$

This gives the result

$$X(s) = \frac{4}{s+1} - \frac{4}{(s+2)^3} - \frac{4}{(s+2)^2} - \frac{4}{s+2}$$

From Table 8.4, $x(t) = [4\exp(-t) - 2t^2\exp(-2t) - 4t\exp(-2t) - 4\exp(-2t)]u(t)$.

8.4.2   *Some Nonstandard Forms*

Certain exceptions can also be handled by using the partial fraction method.

1. If $X(s)$ is a rational function but not strictly proper, we use long division to express $X(s)$ as the sum of a quotient $G(s)$ and a strictly proper function $X_1(s)$. The inverse transform then equals $x_1(t) + g(t)$.
2. If the numerator of $X(s)$ contains exponentials of the form $\exp(-\alpha s)$, this reflects a time delay. First, we find the partial fractions and the inverse without this factor. Then, we replace t by $t - \alpha$ in this inverse to account for the time delay it produces.

Example 8.14
An Improper Rational Function

Consider $X(s) = (s^2 + 4s + 2)/[(s+1)(s+2)]$. Since $X(s)$ is not strictly proper, we use long division to obtain $X(s) = 1 + s/[(s+1)(s+2)]$. A partial fraction expansion of the second term leads to $X(s)$ and $x(t)$ as

$$X(s) = 1 + 2/(s+2) - 1/(s+1) \qquad x(t) = \delta(t) + 2\exp(-2t)u(t) - \exp(-t)u(t)$$

Example 8.15
The Effect of Delay

Let $X(s) = (se^{-2s} + 1)/[(s+1)(s+2)]$. We split this into $X(s) = se^{-2s}/[(s+1) \cdot (s+2)] + 1/[(s+1)(s+2)] = e^{-2s}X_1(s) + X_2(s)$. Here, $X_1(s) = s/[(s+1)(s+2)] = 2/(s+2) - 1/(s+1)$ and $x_1(t) = 2\exp(-2t)u(t) - \exp(-t)u(t)$. Also, $X_2(s) = 1/(s+1)(s+2) = 1/(s+1) - 1/(s+2)$, and we have $x_2(t) = \exp(-t)u(t) - \exp(-2t)u(t)$. From the time-shift property, the inverse transform of $e^{-2s}X_1(s)$ equals $x_1(t-2)$. Combining this with $x_2(t)$, we obtain the inverse transform $x(t)$ as $x(t) = \{2\exp[-2(t-2)] - \exp[-(t-2)]\}u(t-2) + \{\exp(-t) - \exp(-2t)\}u(t)$.

8.5
The s-Plane and BIBO Stability

In the time domain, BIBO stability of an LTI system requires a differential equation in which the highest derivative of the input never exceeds the highest derivative of the output and a characteristic equation whose roots have negative real parts. Equivalently, we require the impulse response $h(t)$ to be absolutely integrable. In the s-domain, this translates into requiring a *proper* transfer function $H(s)$ (with common factors canceled) whose poles have negative real parts or lie in the left-half of the s-plane (excluding the $j\omega$-axis). Here is why:

1. A pole in the right half-plane leads to exponential growth in $h(t)$. We thus have an unbounded response for any input, bounded or otherwise. For example, $H(s) = 1/(s - 2)$ results in the growing exponential $\exp(2t)$.

2. Multiple poles on the $j\omega$-axis lead to terms of the form $1/(s^2 + \alpha^2)^2$ or $1/s^2$ in $H(s)$ and result in polynomial growth of the form $t\cos(\alpha t + \beta)$. For example, $H(s) = 1/s^2$ produces a growing ramp, and $H(s) = 1/(s^2 + 1)^2$ produces the growing term $t\cos(t)$.

3. Simple poles on the $j\omega$-axis lead to factors of the form $1/(s^2 + \alpha^2)$ or $1/s$ in $H(s)$. If the applied input also has the same form, the response will have factors of the form $1/s^2$ or $1/(s^2 + \alpha^2)^2$ and show polynomial growth. This excludes the $j\omega$-axis from the region of stability.

None of the terms corresponding to these poles is absolutely integrable. *For BIBO stability, $H(s)$ (with common factors canceled) must be proper, and its poles must lie entirely in the left-half plane (LHP) and exclude the $j\omega$-axis.* This is both a necessary and sufficient condition for stability.

Remarks:
1. BIBO stability requires a system differential equation in which the highest derivative of the input never exceeds the highest derivative of the output. This condition is equivalent to a proper transfer function $H(s) = P(s)/Q(s)$ where the degree M of $P(s)$ never exceeds the degree N of $Q(s)$. If $M > N$, the system will be unstable due to $j\omega$-axis poles at $s = j\omega = \infty$. Alternatively, $h(t)$ will contain derivatives of $\delta(t)$ that are not absolutely integrable.

2. If a linear system has no right half-plane zeros or poles, the system is called a **minimum-phase system**. We discuss minimum-phase systems in Chapters 9 and 10.

3. A system with poles in the left half-plane and simple finite poles on the $j\omega$-axis is also called **wide sense stable**, or **marginally stable**.

4. A polynomial with real coefficients whose roots lie entirely in the left half of the s-plane is called **strictly Hurwitz**. If it also has simple $j\omega$-axis roots, it is called **Hurwitz**. The coefficients of Hurwitz polynomials are all nonzero and of the same sign. A formal method called the **Routh test** allows us to check if a polynomial is Hurwitz (without having to find its roots) and also gives us the number (but not the location) of right half-plane roots.

Example 8.16
Stability

Here are some examples of stable and unstable systems:

$H(s) = (s + 2)/(s^2 - 2s - 3)$. Unstable due to an RHP pole at $s = 3$.

$H(s) = (s + 1)/(s^2 + 4)$. Marginally stable due to simple poles at $s = \pm j2$.

$H(s) = (s^2 + 2)/(s^2 + 3s + 2)$. Stable, since $M = N$ and there are no RHP poles.

$H(s) = (s^3 + 2)/(s^2 + 3s + 2)$. Unstable, since $M > N$ (or due to a pole at ∞).

8.5.1   *Other Forms of Stability*

BIBO stability relates to external inputs. Other definitions of stability rely only on the specifics of a system, such as its structure or state.

Asymptotic Stability   If the zero-input response of a system decays to zero with time, we classify the system as being **asymptotically stable**. Such a system has a transfer function whose poles lie entirely within the LHP and this also implies BIBO stability. BIBO stability and asymptotic stability are often used interchangeably.

Any passive linear system that includes one or more resistive elements is always asymptotically stable. The resistive elements dissipate energy and allow the system to relax to zero state no matter what the initial state.

Liapunov Stability   If the zero-input response always remains bounded (it may or may not approach zero), we classify the system as being **stable in the sense of Liapunov**. In addition to poles in the LHP, the transfer function of such a system may also contain *simple* (not repeated) poles on the $j\omega$-axis. Asymptotic stability is a special case of Liapunov stability.

A passive system with resistors is both asymptotically and Liapunov stable. A passive system with only lossless elements (e.g., L and C) is Liapunov stable but not asymptotically stable. The energy in such a system due to a nonzero initial state remains constant. It cannot increase, because the system is passive. It cannot be dissipated, because the system is lossless.

8.6
The Laplace Transform and System Analysis

The Laplace transform is a useful tool for the analysis of LTI systems. The convolution, derivative, and integration properties provide the key to simplifying analysis using Laplace transforms because

1. Convolution transforms to the much simpler multiplication operation.
2. Transformed differential equations assume an algebraic form.

For a relaxed LTI system, the response $Y(s)$ to an input $X(s)$ is $H(s)X(s)$, where the transfer function $H(s)$ may be found in several ways:

1. From the Laplace transform of the impulse response $h(t)$.
2. From the transformed differential equation.
3. From an s-domain model of an electric circuit.

If the system is not relaxed, the effect of initial conditions is easy to include. In any case, the response is obtained in the transformed domain. A time-domain solution requires inverse transformation. This additional step is a penalty exacted by all transformed-domain methods.

8.6.1   *Systems Described by Differential Equations*

For a system governed by a differential equation and subject to arbitrary initial conditions, we transform the differential equation (using the derivative property), incorporate the effect of initial conditions, and invoke partial fractions to find the time-domain result. Recall that for a linear system with nonzero initial conditions, the total response equals the sum of two components:

1. The zero-state response (due only to the input).
2. The zero-input response (due only to the initial conditions).

This separation is easily accomplished using Laplace transforms. We can also obtain the natural and forced components if required. Here is an example.

Example 8.17
Solving Differential
Equations

Consider a system governed by $y''(t) + 3y'(t) + 2y(t) = 4\exp(-2t)$ with initial conditions $y(0) = 3, y'(0) = 4$. Transformation to the s-domain, using the derivative property, yields

$$s^2 Y(s) - sy(0) - y'(0) + 3[sY(s) - y(0)] + 2Y(s) = 4/(s+2)$$

(a) *Total response:* Substitute for the initial conditions and rearrange:

$$[s^2 + 3s + 2]Y(s) = 3s + 4 + 9 + 4/(s+2) = 3s + 13 + 4/(s+2)$$
$$= (3s^2 + 19s + 30)/(s+2)$$

Upon simplification, we obtain $Y(s)$ and its partial fraction form as

$$Y(s) = (3s^2 + 19s + 30)/[(s+1)(s+2)^2]$$
$$= K_1/(s+1) + A_0/(s+2)^2 + A_1'/(s+2)$$

Solving for the constants, we obtain

$$K_1 = (3s^2 + 19s + 30)/(s+2)^2|_{s=-1} = 14$$
$$A_0 = (3s^2 + 19s + 30)/(s+1)|_{s=-2} = -4$$
$$A_1 = \frac{d}{ds}[(3s^2 + 19s + 30)/(s+1)]|_{s=-2} = -11$$

Upon inverse transformation, $y(t) = 14\exp(-t) - 4t\exp(-2t) - 11\exp(-2t)$.

Comment: As a check, we confirm that $y(0) = 3$ and $y'(0) = -14 + 22 - 4 = 4$.

The forced response is zero, so the natural response equals $y(t)$ itself.

(b) *Zero-input and zero-state response:* For the zero-state response, we ignore initial conditions to obtain $[s^2 + 3s + 2]Y_{zs}(s) = 4/(s+2)$. This yields

$$Y_{zs}(s) = 4/[(s+2)(s^2 + 3s + 2)] = 4/(s+1) - 4/(s+2)^2 - 4/(s+2)$$
$$y_{zs}(t) = [4\exp(-t) - 4t\exp(-2t) - 4\exp(-2t)]u(t)$$

For the zero-state response, we ignore the input to obtain

$$[s^2 + 3s + 2]Y_{zi}(s) = 3s + 13$$
$$Y_{zi}(s) = (3s + 13)/(s^2 + 3s + 2) = 10/(s + 1) - 7/(s + 2)$$
$$y_{zi}(t) = [10\exp(-t) - 7\exp(-2t)]u(t)$$

We confirm that

$$y(t) = y_{zs}(t) + y_{zi}(t) = [14\exp(-t) - 4t\exp(-2t) - 11\exp(-2t)]u(t)$$

8.6.2   The Transfer Function and Circuit Analysis

For circuit analysis, we write Kirchhoff's laws and transform these to establish $H(s)$ or the system response. This process is entirely equivalent to transforming the circuit itself to the s-domain using constitutive relations for circuit elements and then writing the (much simpler) algebraic equations for the transformed circuit. For a relaxed RLC circuit, we obtain the transformations

$$v_R(t) = Ri_R(t) \longleftrightarrow V_R(s) = RI_R(s)$$
$$v_L(t) = L\frac{di_L(t)}{dt} \longleftrightarrow V_L(s) = sLI_L(s)$$
$$i_C(t) = C\frac{dv_C(t)}{dt} \longleftrightarrow I_C(s) = sCV_C(s)$$

These relations describe a general form of Ohm's law $V(s) = Z(s)I(s)$. The quantity $Z(s) = V(s)/I(s)$ defines an *impedance* with units of ohms. The impedance of the various elements may be described as (Figure 8.4)

$$\frac{V_R(s)}{I_R(s)} = Z_R = R \qquad \frac{V_L(s)}{I_L(s)} = Z_L = sL \qquad \frac{V_C(s)}{I_C(s)} = Z_C = \frac{1}{sC}$$

We transform a circuit by replacing the elements R, L and C by their impedances Z_R, Z_L and Z_C, and replacing sources by their Laplace transform. Node or mesh equations for this **transformed circuit** now assume a much simpler algebraic form.

Remark: It is important to keep in mind that the definitions of both impedance and the transfer function are based on *relaxed* LTI systems.

8.6.3   Circuits with Nonzero Initial Conditions

For a circuit with nonzero initial conditions, we can include the effect of the initial conditions $i_L(0-)$ and $v_C(0-)$ in the constitutive equations for L and C to give

$$v_L(t) = L\frac{di_L(t)}{dt} \longleftrightarrow V_L(s) = L[sI_L(s) - i_L(0-)] = Z_LI_L(s) - Li_L(0-)$$
$$i_C(t) = C\frac{dv_C(t)}{dt} \longleftrightarrow I_C(s) = C[sV_C(s) - v_C(0-)] = V_C(s)/Z_C - Cv_C(0-)$$

This allows the initial conditions to be included by modeling (Figure 8.4):

Figure 8.4   Modeling elements with zero and nonzero initial conditions.

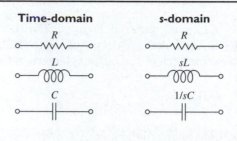

1. A capacitor C by the impedance $Z_C = 1/sC$ *in parallel with a current source* $-Cv_C(0-)$.

2. An inductor L by the impedance $Z_L = sL$ *in series with a voltage source* $-Li_L(0-)$.

Remark: The *constants* $Li_L(0-)$ and $Cv_C(0-)$ represent the Laplace transform of the *impulses* $Li_L(0-)\delta(t)$ and $Cv_C(0-)\delta(t)$, which are responsible for the sudden changes in state leading to the nonzero initial conditions $i_L(0-)$ and $v_C(0-)$. The Laplace transform thus automatically accounts for any initial conditions. A great advantage indeed!

The inductor current and capacitor voltage may also be written as

$$I_L(s) = \frac{V_L(s)}{Z_L} + \frac{i_L(0-)}{s} \qquad V_C(s) = I_L(s)Z_C + \frac{v(0-)}{s}$$

This provides an alternative means of including the effect of initial conditions in the transformed system by modeling

1. A capacitor by $Z_C = 1/sC$ *in series with a voltage source* $v_C(0-)/s$.

2. An inductor by $Z_L = sL$ *in parallel with a current source* $i_L(0-)/s$.

The two forms are equivalent. We may easily go back and forth between them using Thevenin's and Norton's theorems (source transformations).

Example 8.18
Response of a Relaxed System

For the RL circuit shown in Figure 8.5a with zero initial conditions, transformation results in the s-domain circuit of Figure 8.5b. The transfer function and impulse response for each of the outputs labeled is found as

$$H_R(s) = \frac{V_R(s)}{V(s)} = \frac{2}{4s+2} = \frac{\frac{1}{2}}{s+\frac{1}{2}} \qquad h_R(t) = \frac{1}{2}\exp(-\frac{1}{2}t)u(t)$$

$$H_L(s) = \frac{V_L(s)}{V(s)} = \frac{4s}{4s+2} = 1 - \frac{\frac{1}{2}}{s+\frac{1}{2}} \qquad h_L(t) = \delta(t) - \frac{1}{2}\exp(-\frac{1}{2}t)u(t)$$

$$H_I(s) = I(s)/V(s) = \frac{1}{4s+2} = \frac{\frac{1}{4}}{s+\frac{1}{2}} \qquad h_I(t) = \frac{1}{4}\exp(-\frac{1}{2}t)u(t)$$

Figure 8.5   A circuit and its transformed versions for Examples 8.18 and 8.19.

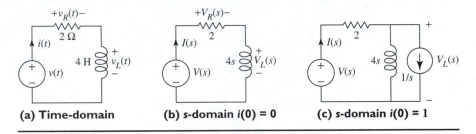

(a) Time-domain   **(b) s-domain i(0) = 0**   **(c) s-domain i(0) = 1**

Note how the same system has different transfer functions, depending on what is specified as the output. If the output is specified as $v_L(t)$ and the input is $v(t) = \exp(-t)u(t)$, we find

$$V_L(s) = V(s)H_L(s) = \frac{s}{(s+1)(s+\frac{1}{2})} = \frac{2}{s+1} - \frac{1}{s+\frac{1}{2}}$$

R The response then equals $v_L(t) = 2\exp(-t)u(t) - \exp(-\frac{1}{2}t)u(t)$.

Example 8.19
Response of a System with Nonzero Initial State

For the RL circuit of Example 8.18 with $v(t) = \exp(-t)u(t)$, assume that the initial current $i(0) = 1$ A and the response is $v_L(t)$. The transformed s-domain circuit is shown in Figure 8.5c. A node equation gives

$$\frac{V_L(s) - V(s)}{2} + \frac{V_L(s)}{4s} + \frac{1}{s} = 0 \quad \text{or} \quad \left[\frac{1}{2} + \frac{1}{4s}\right]V_L(s) = \frac{1}{2}V(s) - \frac{1}{s}$$

With $V(s) = 1/(s+1)$, we solve for $V_L(s)$ to get

$$V_L(s) = -\frac{(s+2)}{(s+1)(s+\frac{1}{2})} = \frac{2}{s+1} - \frac{3}{s+\frac{1}{2}}$$

Taking inverse transforms, we finally obtain

$$v_L(t) = 2\exp(-t)u(t) - 3\exp(-\tfrac{1}{2}t)u(t)$$

Example 8.20
Zero-State and
Zero-Input Response

From Example 8.18, the response $v_L(t) = 2\exp(-t)u(t) - \exp(-\tfrac{1}{2}t)u(t)$ is the zero-state response. The initial condition in Example 8.19 is responsible for the additional term $-2\exp(-\tfrac{1}{2}t)u(t)$. This corresponds to the zero-input response $v_{zi}(t)$. We could also compute $v_{zi}(t)$ separately by short circuiting the input and finding $V_{zi}(s)$ from the node equation $[V_{zi}(s)/2] + V_{zi}(s)/4s + 1/s = 0$. Upon solving this, we get $V_{zi}(s) = -2/(s + \tfrac{1}{2})$ and $v_{zi}(t) = -2\exp(-\tfrac{1}{2}t)u(t)$.

8.6.4   Zero-Input Response from the Transfer Function

For an LTI system with nonzero initial conditions, the response $Y(s)$ in the s-domain is the sum of

1. The zero-state response (due only to the *transformed input*) and
2. The zero-input response (due to the *transformed initial conditions*).

With $H(s) = N(s)/D(s)$, the response $Y(s)$ of a relaxed system to an input $X(s)$ may be expressed as $Y(s) = X(s)H(s) = X(s)N(s)/D(s)$. This corresponds to the zero-state response. If the system is not relaxed, the initial conditions result in an additional contribution, the zero-input response $Y_{zi}(s)$, which may be expressed in the form $Y_{zi}(s) = N_{zi}(s)/D(s)$. To evaluate $Y_{zi}(s)$, we must find and transform the system differential equation using the initial conditions.

Example 8.21
Zero-Input Response from
the Transfer Function

Consider the transfer function of a relaxed third-order system given by

$$H(s) = Y(s)/X(s) = \frac{(s+1)}{s^3 + 4s^2 + 2s + 3}$$

Let $x(t) = \exp(-t)u(t), y(0) = 6, y'(0) = -8$, and $y''(0) = 2$.

The zero-state response $y_{zs}(t)$ is found as the inverse Laplace transform of

$$Y_{zs}(s) = H(s)X(s) = \frac{1}{s^3 + 4s^2 + 2s + 3}$$

To find the zero-input response, we note that the operational form of the homogeneous differential equation is $[s^3 + 4s^2 + 2s + 3]y = 0$. Using the derivative property and initial conditions, this transforms to

$$s^3 Y_{zi}(s) - s^2 y(0) - sy'(0) - y''(0) + 4[s^2 Y_{zi}(s) - sy(0) - y'(0)]$$
$$+ 2[sY_{zi}(s) - y(0)] + 3Y_{zi}(s) = 0$$

We then find $Y_{zi}(s) = (6s^2 + 16s - 18)/(s^3 + 4s^2 + 2s + 3)$. The zero-input response $y_{zi}(t)$ is the inverse transform of $Y_{zi}(s)$.

8.6.5   The Frequency Response

Harmonic signals are eigensignals of LTI systems. The system response to a harmonic input is a harmonic at the same frequency. For a harmonic input of the form $x(t) = \exp(j\omega t)u(t)$, applied to an LTI system with transfer function $H(s)$, the response in the s-domain may be written as

$$Y(s) = H(s)X(s) = H(s)/(s - j\omega)$$

The total response requires partial fraction expansion and subsequent inverse transformation. The steady-state component has the form $K\exp(j\omega t)$, since all other terms decay with time. To find the constant K, we simply use the partial fraction approach to obtain

$$K = (s - j\omega)Y(s)|_{s=j\omega} = H(s)|_{s=j\omega} = H(\omega)$$

Thus, the steady-state response to the input $\exp(j\omega t)$ has the form $H(\omega)\exp(j\omega t)$. Since $H(\omega)$ is, in general, complex, it is often expressed as

$$H(\omega) = |H(\omega)|\angle\phi(\omega)$$

This provides a direct indication of the magnitude and phase at any frequency and is called the **frequency response**, or the **steady-state transfer function**, of the system. It represents the transfer function for harmonic (and dc) inputs and is found from $H(s)$ by the substitution $s \rightarrow j\omega$ (with $\omega = 0$ for dc). Note that $H(-\omega) = H^*(\omega)$ and possesses *conjugate symmetry*. This means that $|H(\omega)|$ is an even function, and $\theta(\omega)$ is an odd function.

Remark: The frequency response makes sense primarily for stable systems for which the natural response does indeed decay with time (even though it is frequently used to study the stability of control systems).

Example 8.22

Frequency Response of a First-Order Filter

Consider the transfer function $H(s) = 4s/(s + 2)$. We have $H(\omega) = 4j\omega/(j\omega + 2)$. The magnitude and phase of the frequency response are given by

$$|H(\omega)| = 4\omega/\sqrt{4 + \omega^2} \qquad \phi(\omega) = 90° - \tan^{-1}(\omega/2)$$

We note that $|H(\omega)|$ increases monotonically from $|H(0)| = 0$ to a maximum of $H_{max} = |H(\infty)| = 4$. The frequency $\omega = 2$ when $|H(2)| = 4/\sqrt{2} = H_{max}/\sqrt{2}$ defines the half-power frequency. The phase decreases from a maximum of 90° at $\omega = 0$ to 0° as $\omega \rightarrow \infty$. The phase at $\omega = 2$ is $\phi(2) = 90° - \tan^{-1}(1) = 45°$. Figure 8.6 shows a plot of $|H(\omega)|$ and $\phi(\omega)$. The system attenuates low frequencies and passes high frequencies. It describes a *highpass filter*.

8.6.6   The Steady-State Response to Harmonic Inputs

The steady-state response of an LTI system to a harmonic is also a harmonic at the same frequency. The magnitude and phase of the input are modified by the magnitude and phase of the system transfer function $H(\omega)$ at the applied frequency. To find the steady-state response $y_{ss}(t)$ to the harmonic input $x(t) = A\cos(\omega_0 t + \theta)$,

Figure 8.6   Frequency response of $H(s) = 2s/(s + 2)$ for Example 8.22.

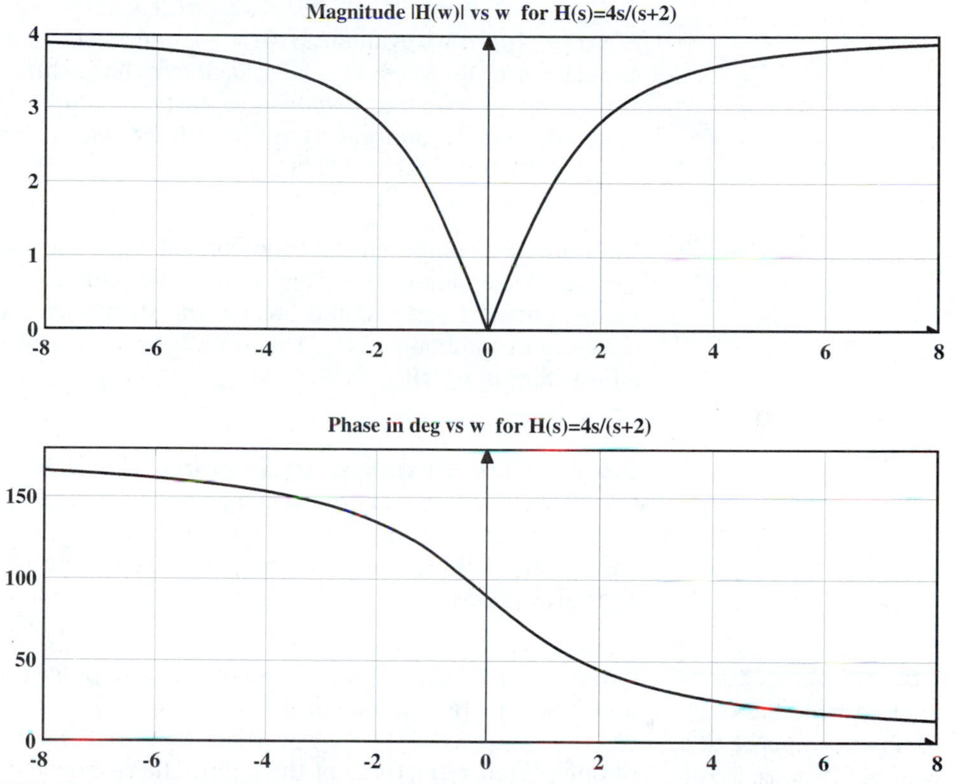

Magnitude |H(w)| vs w   for H(s)=4s/(s+2)

Phase in deg vs w   for H(s)=4s/(s+2)

we evaluate $H(\omega_0) = K\angle\phi$ and $y_{ss}(t) = KA\cos(\omega_0 t + \theta + \phi)$. In effect, we multiply the input magnitude A by the transfer function magnitude K, and add the transfer function phase ϕ to the input phase θ.

Remark: If the input comprises harmonics at different frequencies, we must evaluate both $H(\omega)$ and the individual *time-domain response* at each frequency before using superposition to find the total response.

Example 8.23
Steady-State Response to
Sinusoidal Inputs

Consider a system with $H(s) = Y(s)/X(s) = (s - 2)/(s^2 + 4s + 4)$. We find the steady-state response $y_{ss}(t)$ due to each of the following inputs:

1. $x(t) = 8\cos(2t)$:   Since $\omega = 2$,   we   have $H(2) = (2j - 2)/(-4 + j8 + 4) = \sqrt{2}/4\angle 45°$. This changes the input $x(t) = 8\cos(2t)$ to yield the steady-state response

$$y_{ss}(t) = 8(\sqrt{2}/4)\cos(2t + 45°) = 2\sqrt{2}\cos(2t + 45°)$$

2. $x(t) = 8\cos(2t) + 16\sin(2t)$: Both terms are at the same frequency, $\omega = 2$. Since $H(2) = \sqrt{2}/4\angle 45°$, each input term is affected by this transfer function to give $y_{ss}(t) = 2\sqrt{2}\cos(2t + 45°) + 4\sqrt{2}\sin(2t - 45°)$.

3. $x(t) = 4u(t)$: We evaluate $H(0) = -\frac{1}{2}$ and $y_{ss}(t) = (4)(-\frac{1}{2})u(t) = -2u(t)$.

4. $x(t) = 4u(t) + 8\cos(2t + 15°)$: Each term has a different frequency. With $H(0) = -\frac{1}{2}$, the steady-state response to $4u(t)$ is $-2u(t)$. With $H(2) = \sqrt{2}/4\angle 45°$, the steady-state response to $8\cos(2t)$ is $2\sqrt{2}\cos(2t + 60°)$. By superposition $y_{ss}(t) = -2u(t) + 2\sqrt{2}\cos(2t + 60°)$.

Comment: Note how we computed the time-domain response *before using superposition.* Of course, we can find $y_{ss}(t)$ by first finding the *total response* $y(t)$ using Laplace transforms, partial fractions, and inverse transforms, and then discarding the natural component to obtain $y_{ss}(t)$. This roundabout method is hardly worth the effort if all we want is $y_{ss}(t)$!

8.6.7   The Response to Switched Periodic Inputs

For periodic inputs $x_p(t)u(t)$ switched on at $t = 0$, we can use the Laplace transform to find the response $y_1(t)$ for the first period and then invoke time invariance and superposition to find the total response as the sum of $y_1(t)$ and its successively shifted versions.

Example 8.24
Complete Response to a
Switched Periodic Signal

Consider an RC circuit excited by a rectangular pulse train that starts at $t = 0$ (Figure 8.7). The transfer function of the circuit is $H(s) = 1/(s + 1)$. The Laplace transform of $x(t)$ equals $X_1(s)/(1 - e^{-2s})$, where $X_1(s) = 1/s - e^{-s}/s$ is the Laplace transform of one period $x_1(t)$, $(0, 2)$ of the input. The response equals

$$V_C(s) = \frac{1 - e^{-s}}{s(s + 1)(1 - e^{-2s})} = \left[\frac{1}{s} - \frac{1}{s+1}\right]\frac{1 - e^{-s}}{1 - e^{-2s}}$$

If the factor $1/(1 - e^{-2s})$ were not present, the response $v_1(t)$ would equal

$$v_1(t) = [1 - e^{-t}]u(t) - [1 - e^{-(t-1)}]u(t - 1)$$

Including the factor $1/(1 - e^{-2s})$ results in the superposition of the shifted versions of $v_1(t)$, and leads to the total response $v_C(t)$ as

$$v_C(t) = v_1(t) + v_1(t - 2)u(t - 2) + v_2(t - 4)u(t - 4) + \cdots$$

It is not easy to simplify the response for the nth period unless closed-form results for the appropriate polynomial sums are available.

8.6.8   The Steady-State Response to Switched Periodic Inputs

The response to a switched periodic input $x_p(t)u(t)$ comprises the natural response that decays with time and a periodic steady-state component. Subtracting the natural from the total response $y_1(t)$ for one period of the input can yield the steady-state response, but this requires some effort.

Figure 8.7   The *RC* circuit for Example 8.24, its switched periodic input and the resulting response.

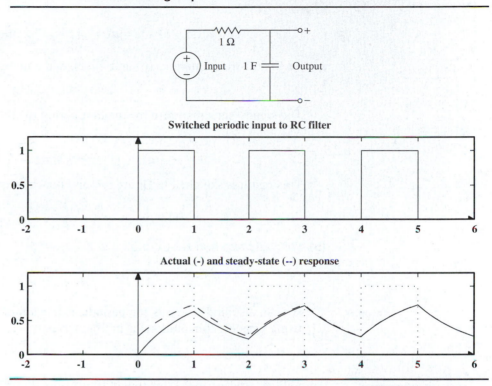

Formally, the form of the natural response may be expressed as

$$y_N(t) = \sum_k A_k \exp(p_k t)$$

Here, p_k are the poles of the transfer function $H(s)$ and A_k are constants. The steady-state response $y_{ss}(t)$ for the first period $(0, T)$ is then

$$y_{ss}(t) = y_1(t) - y_N(t) \qquad 0 \le t \le T$$

The constants A_k may be evaluated in one of two ways:

1. By equating $y_{ss}(0) = y_{ss}(T)$, which ensures periodicity of $y_{ss}(t)$.
2. By evaluating $A_k = (s - p_k)X(s)H(s)|_{s=p_k}$, where $X(s)$ is the Laplace transform of $x_p(t)u(t)$.

Example 8.25
Steady-State Response to a Switched Periodic Signal

Consider the previous example of an *RC* circuit with $H(s) = 1/(s + 1)$ excited by a rectangular pulse train $x_p(t)u(t)$ with period $T = 2$ and $X(s) = X_1(s)/(1 - e^{-2s})$, where $X_1(s) = 1/s - (e^{-s}/s)$ is the Laplace transform of the first period $(0, 2)$ of the input. The natural response equals $v_N(t) = K\exp(-t)u(t)$. To find the steady-state response, we subtract $v_N(t)$ from $v_1(t)$ (the response to the first period of the input) to give

$$v_{ss}(t) = v_1(t) - v_N(t) = [1 - e^{-t}]u(t) - [1 - e^{-(t-1)}]u(t-1) - K\exp(-t)u(t)$$

(a) To find K, we evaluate $v_{ss}(0)$ and $v_{ss}(2)$, and equate the two, to give

$$0 - K = (1 - e^{-2}) - (1 - e^{-1}) - Ke^{-2} \qquad \text{or} \qquad K = -1/(1 + e)$$

The steady-state response for the first period $(0, 2)$ then equals

$$v_{ss}(t) = v_1(t) - K\exp(-t)u(t)$$
$$= [1 - e^{-t}]u(t) - [1 - e^{-(t-1)}]u(t-1) + [e^{-t}/(1 + e)]u(t)$$

This result is sketched in Figure 8.7, and may be expressed by intervals as

$$v_{ss}(t) = \begin{cases} 1 - [e^{-(t-1)}/(1 + e)] & 0 \le t \le 1 \\ [e^{-(t-2)}/(1 + e)] & 1 \le t \le 2 \end{cases}$$

(b) We could also find K by evaluating

$$K = (s + 1)H(s)X(s)\Big|_{s=-1} = \frac{1 - e^{-s}}{s(1 - e^{-2s})}\Big|_{s=-1} = -\frac{1}{1 + e}$$

Comment: We can also find $v_{ss}(t)$ as the periodic extension of $v_1(t)$ with period 2, or by using periodic convolution (as in Chapter 4).

8.7
Connections

The Laplace transform is related to both the time-domain operation of convolution and to the Fourier transform as we now describe.

8.7.1   *The Laplace Transform and Convolution*

The response of a causal LTI system to a complex exponential input described by $x(t) = \exp(-st)$ may be written in convolution form as

$$y(t) = x(t) \star h(t) = \int_0^\infty A\exp[s(t - \lambda)]h(\lambda)\,d\lambda = A\exp(st)\int_0^\infty h(\lambda)\exp(-s\lambda)\,d\lambda$$

The output thus equals the input $A\exp(st)$ multiplied by

$$H(s) = \int_0^\infty h(\lambda)\exp(-s\lambda)\,d\lambda$$

The quantity $H(s)$ is the **system transfer function** and is described by a weighted sum of complex exponentials. This description in terms of complex exponentials may be extended to any causal signal $x(t)u(t)$ to give

$$X(s) = \int_0^\infty x(\lambda)\exp(-s\lambda)\,d\lambda$$

This relation transforms the causal signal $x(t)$ to the function $X(s)$, and defines the *onesided Laplace transform* of $x(t)$.

The response $y(t) = H(s)x(t)$ may be written as

$$Y(s) = \int_0^\infty y(t)\exp(-st)\,dt = \int_0^\infty H(s)x(t)\exp(-s)\,dt = H(s)X(s)$$

Since $y(t) = x(t) \star h(t)$, the operation of convolution transforms to the operation of multiplication in the s-domain, and thus

$$x(t) \star h(t) \longleftrightarrow X(s)H(s)$$

8.7.2 The Laplace Transform and Fourier Transform

The Fourier transform describes a signal as a sum (integral) of weighted harmonics, or complex exponentials. Its major drawbacks are that

1. It cannot handle exponentially growing signals.
2. It cannot handle initial conditions in system analysis.

In addition, the Fourier transform of signals that are not absolutely integrable usually includes impulses.

   The Laplace transform attempts to overcome the shortcomings inherent in the Fourier transform by modifying the Fourier transform as follows:

1. We redefine it as the sum of exponentially weighted harmonics.

$$X(s) = \int_{-\infty}^\infty x(t)\exp[-(\sigma + j2\pi f)t]\,dt = \int_{-\infty}^\infty x(t)\exp(-st)\,dt$$

This defines the *twosided Laplace transform*. The exponential weighting has the form $\exp(\mp\sigma t)$. It serves as a **convergence factor** that allows transformation of exponentially growing signals of the form $\exp(\pm\alpha t)$.
2. We change the lower limit in the integration to zero.

$$X(s) = \int_0^\infty x(t)\exp(-st)\,dt$$

This defines the *onesided Laplace transform*. It allows application to causal signals and permits the analysis of systems with arbitrary initial conditions to arbitrary inputs.

8.7.3 The Convergence Factor

For signals that are not absolutely integrable, an appropriate choice of σ can make $x(t)\exp(-\sigma t)$ absolutely integrable and ensure a Laplace transform that requires no impulses. The choice of σ that makes convergence possible defines the region of convergence.

   We point out that even though the convergence factor $\exp(-\sigma t)$ can handle exponentially growing signals of the form $\exp(\mp\alpha t)$, neither the Laplace transform nor the Fourier transform can handle signals of faster growth, such as $\exp(\pm\alpha t^2)$ and $\exp(\alpha t^2)u(t)$.

Remark: Due to the convergence factor $\exp(-\sigma t)$, the Laplace transform $X(s)$ of a signal $x(t)u(t)$ no longer displays the kind of symmetry present in its Fourier transform $X(f)$.

8.7.4   *Correspondence Between LT and FT Pairs*

The Fourier transform of a signal $x(t)$ corresponds to its Laplace transform $X(s)$, with s replaced by $j\omega$, but only if

1. The signal is causal, ensuring identical limits for both.
2. The signal is absolutely integrable. This ensures that the value of σ includes $\sigma = 0$ (the $j\omega$-axis) to allow this replacement.

This relationship allows us to relate the Laplace transform and Fourier transform of causal signals, as listed in Table 8.5. This table also includes transforms of causal signals such as $u(t)$ and $\cos(\alpha t)u(t)$, which are not absolutely integrable and show no apparent correspondence. We note, however, that the Laplace transform of these signals is just the nonimpulsive portion of the Fourier transform $X(\omega)$ with $j\omega \to s$! We can thus always find the Laplace transform of causal signals from their Fourier transform, but not the other way around.

Remarks:
1. Most operational properties of the Laplace and Fourier transforms also show correspondence. These include superposition, shifting, scaling, convolution and products. The derivative property must be modified to account for initial values. Other properties, such as symmetry, do not translate to the Laplace transform.
2. Since $s = \sigma + j\omega$ is complex, the Laplace transform may be plotted as a surface in the s-plane. The Fourier transform (with $s \to j\omega$) is just the cross section of this plot along the $j\omega$-axis.

Example 8.26

Absolutely Integrable Causal Signals

(a) The signal $x(t) = \exp(-\alpha t)u(t)$ is absolutely integrable. Its Fourier transform is $X(f) = 1/(j2\pi f + \alpha)$. We can thus find the Laplace transform from $X(f)$ as $X(s) = 1/(s + \alpha)$.

Comment: The signal $\exp(-\sigma t)\exp(-\alpha t)u(t)$ is absolutely integrable if $\sigma > -\alpha$. For positive α, this includes $\sigma = 0$ or the $j\omega$-axis.

(b) The signal $x(t) = \delta(t)$ is also absolutely integrable. Since $X(f) = 1$, we have $X(s) = 1$. ▭

Example 8.27

A Causal Signal That Is Not Absolutely Integrable

The signal $x(t) = u(t)$ is not absolutely integrable, but $\exp(-\sigma t)x(t)$ is absolutely integrable for $\sigma > 0$. Since this excludes $\sigma = 0$ (the $j\omega$-axis), we find the Laplace of $u(t)$ by dropping the impulsive part of $X(\omega) = \pi\delta(\omega) + 1/j\omega$ and replacing $j\omega$ by s, to give $X(s) = 1/s$. ▭

Table 8.5   Laplace and Fourier Transform Pairs Compared.

$X(t)$	$X(s)$	$X(\omega)$
	LT and FT of causal absolutely integrable signals	
$\delta(t)$	1	1
$\delta'(t)$	s	$j\omega$
$\exp(-\alpha t)u(t)$	$\dfrac{1}{s+\alpha}$	$\dfrac{1}{j\omega+\alpha}$
$t\exp(-\alpha t)u(t)$	$\dfrac{1}{(s+\alpha)^2}$	$\dfrac{1}{(j\omega+\alpha)^2}$
$\exp(-\alpha t)\cos(\beta t)u(t)$	$\dfrac{s+\alpha}{(s+\alpha)^2+\beta^2}$	$\dfrac{j\omega+\alpha}{(\alpha+j\omega)^2+\beta^2}$
$\exp(-\alpha t)\sin(\beta t)u(t)$	$\dfrac{\beta}{(s+\alpha)^2+\beta^2}$	$\dfrac{\beta}{(\alpha+j\omega)^2+\beta^2}$
	Causal signals for which the LT and FT show no correspondence	
$u(t)$	$\dfrac{1}{s}$	$\dfrac{1}{j\omega}+\pi\delta(\omega)$
$tu(t)$	$\dfrac{1}{s^2}$	$\dfrac{1}{(j\omega)^2}+j\pi\delta'(\omega)$
$\cos(\beta t)u(t)$	$\dfrac{s}{s^2+\beta^2}$	$\dfrac{j\omega}{(j\omega)^2+\beta^2}+\pi[\delta(\omega+\beta)+\delta(\omega-\beta)]$
$\sin(\beta t)u(t)$	$\dfrac{\beta}{s^2+\beta^2}$	$\dfrac{\beta}{(j\omega)^2+\beta^2}+j\pi[\delta(\omega+\beta)-\delta(\omega-\beta)]$
$\exp(\alpha t)u(t)\quad \alpha>0$	$\dfrac{1}{s-\alpha}$	Not defined
$\exp(\alpha t^2)u(t)\quad \alpha>0$	Not defined	Not defined

8.8
The Laplace Transform: A Synopsis

The Laplace transform

The Laplace transform is an integral operation defined by

$$X(s) = \int_{0-}^{\infty} x(t)\exp(-st)\,dt$$

It allows transformation of time signals and systems to the s-domain or complex frequency domain. The onesided Laplace transform integrates $x(t)\exp(-st)$ from $0-$ to ∞ and is used for causal signals of the form $x(t)u(t)$. The LT is a linear operation. The LT of complicated signals is found from simpler forms by using properties.

Properties

The quantity $\exp(-s\alpha)$ may be regarded as a shift operator. The transform $X(s)\exp(-s\alpha)$ describes the delayed signal $x(t-\alpha)u(t-\alpha)$. Convolution in time corresponds to multiplication in the s-domain. The Laplace transform of many signals is a ratio of polynomials in s whose roots may be plotted in the s-plane on a pole-zero plot.

The inverse Laplace transform	This is also governed by an integral form but is typically evaluated using partial fraction expansions.
The transfer function	$H(s)$ is defined for a relaxed LTI system as the ratio of the output $Y(s)$ and input $X(s)$, or the transform of the system impulse response $h(t)$. For LTI systems $H(s)$ is a ratio of polynomials in s.
System stability	The poles (denominator roots) of $H(s)$ determine the nature of the system response and system stability. For BIBO stability, the transfer function must be proper and all the poles of $H(s)$, with common factors in $H(s)$ canceled, must have negative real parts (lie in the LHP).
System response	The Laplace transform is useful for system analysis. Laplace transforms convert the convolution integral to a product. For relaxed systems, the response $Y(s)$ equals $X(s)H(s)$.
	Laplace transforms convert differential equations to algebraic equations. Initial conditions can be accounted for during the transformation. We can analyze circuits by transforming them using generalized impedances and including the effect of initial conditions.
Frequency response	The frequency response $H(\omega)$ is found from $H(s)$ by the substitution $s \longrightarrow j\omega$ (with $\omega = 0$ for dc). It represents the transfer function for harmonic inputs and possesses conjugate symmetry. The steady-state response to a sinusoid $A\cos(\omega_0 t + \theta)$ is $A|H|\cos(\omega_0 t + \theta + \phi)$ where $|H|$ and ϕ are the magnitude and phase of the steady-state transfer function $H(\omega_0)$.
Connections	The Laplace transform $X(s)$ of a causal, absolutely integrable signal equals its Fourier transform $X(\omega)$ with $j\omega$ replaced by s.

Appendix 8A MATLAB Demonstrations and Routines

Getting Started on the MATLAB Demonstrations

This appendix presents MATLAB routines related to concepts in this chapter and some examples of their use. See Appendix 1A for introductory remarks.

Note: All polynomial arrays must be entered in descending powers.

Laplace Transform and Its Symbolic Inverse

```
y=ltr([A b c w r])      Finds the LT of x(t) = A exp(-bt)t^C cos(wt + r)u(t).
x=ilt('tf',p,q)         Returns ILT of X(s) = P(s)/Q(s) in symbolic form.
```

Example

To find the Laplace transform of $4t\exp(-3t)\cos(t)$, use

```
>>y=ltr([4 3 1 1 0])
```

MATLAB returns:

y=[0 0 4 24 32; 1 12 56 120 100]

To find the ILT, use

>>x=ilt('tf',[0 0 4 24 32],[1 12 56 120 100])

MATLAB returns

x=4*t.*exp(-3*t).*cos(t)

Analytical Response in Symbolic Form

[yt,yzs,yzi]=sysresp2('ty',n,d,p,q,ic)   Computes system response.

Here, $ty = $'s,' n, d are the coefficient arrays of $N(s)$ and $D(s)$ of $H(s) = N(s)/D(s)$ and p, q are the coefficient arrays of $P(s)$ and $Q(s)$ of $X(s) = P(s)/Q(s)$.

Example

Let $H(s) = (s + 2)/(s^2 + 4s + 3)$, $x = 2t \exp(-2t)u(t)$, $y(0) = 3$, and $y'(0) = -1$. Then $X(s) = 2/(s + 2)^2 = 2/(s^2 + 4s + 4)$. To find the response, use the command

>>[yt,yzs,yzi]=sysresp2('s',[1 2],[1 4 3],2,[1 4 4],[3 -1])

MATLAB returns

```
yt=-2*exp(-2*t)+5*exp(-t)                    %Total response
yzs=1*exp(-3*t)-2*exp(-2*t)+1*exp(-t)        %Zero-state response
yzi=-1*exp(-3*t)+4*exp(-t)                    %Zero-input response
```

These *string functions* can be evaluated using the 'eval' command.

Partial Fractions

[r,p,k]=tf2pf(n,d)   Returns PFE as constants, poles and impulsive terms.
[n,d]=pf2tf(r,p,k)   Returns the TF from the PF expansion.

Example

To find the PF expansion of $H(s) = (s + 2)/(s^2 + 4s + 3)$, use

>>[r,p,k]=tf2pf([1 2],[1 4 3])

MATLAB returns

r=[0.5000;0.5000], p=[-3;-1], k=[]

To convert to TF, use

>>[n,d]=pf2tf([0.5;0.5],[-3;-1],[])

MATLAB returns

```
n=[0 1 2], d=[1 4 3]
```

Demonstrations and Utility Functions

```
plotpz(p,q,'s')
```                                 Plots poles and zeros of $X(s) = P(s)/Q(s)$.
```
tfplot('s',n,d,[fl fh],0,3)
```           Plots TF magnitude and phase over [fl fh] Hz.

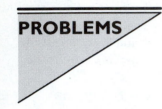

**PROBLEMS**

**P8.1  (The Laplace Transform and its ROC)**  Use the defining relation to find the Laplace transform and its region of convergence for
**(a)** $x(t) = \exp(-3t)u(t)$                    **(b)** $x(t) = \exp(3t)u(t)$

**P8.2  (Laplace Transforms)**  Find the Laplace transforms of
**(a)** $\exp(-2t + 4)u(t)$         **(b)** $t\exp(-2t + 4)u(t)$         **(c)** $\exp(-2t + 4)u(t - 1)$
**(d)** $[1 - \exp(-t)]u(t)$         **(e)** $(t - 2)u(t - 1)$         **(f)** $(t - 2)^2 u(t - 1)$
**(g)** $\cos(t - \frac{1}{4}\pi)u(t)$         **(h)** $\cos(t - \frac{1}{4}\pi)u(t - \frac{1}{4}\pi)$         **(i)** $\cos(t)u(t - \frac{1}{4}\pi)$
**(j)** $|\sin(\pi t)|u(t)$         **(k)** $u[\sin(\pi t)]u(t)$

**P8.3  (Properties)**  Given the Laplace transform pair $x(t) \longleftrightarrow 4/(s + 2)^2$, find the Laplace transform of the following without computing $x(t)$.
**(a)** $x(t - 2)$         **(b)** $x(2t)$         **(c)** $x(2t - 2)$
**(d)** $x'(t)$         **(e)** $x'(t - 2)$         **(f)** $x'(2t)$
**(g)** $x'(2t - 2)$         **(h)** $\exp(-2t)x(t)$         **(i)** $\exp(-2t)x(2t)$
**(j)** $\exp(-2t - 2)x(2t - 2)$         **(k)** $tx'(t)$         **(l)** $2t\exp(-2t)x'(2t - 2)$

**P8.4  (Properties)**  Given the Laplace transform pair $\exp(-2t)u(t) \longleftrightarrow X(s)$, find the time signal corresponding to the following.
**(a)** $X(2s)$         **(b)** $X'(s)$         **(c)** $sX(s)$
**(d)** $sX'(s)$         **(e)** $sX'(2s)$         **(f)** $\exp(-2s)X(s)$
**(g)** $s\exp(-2s)X(2s)$

**P8.5  (Pole-Zero Patterns)**  Sketch the pole-zero patterns of the following systems. Which of these describe stable systems?
**(a)** $H(s) = (s + 1)^2/(s^2 + 1)$                    **(b)** $H(s) = s^2/[(s + 2)(s^2 + 2s - 3)]$
**(c)** $H(s) = (s - 2)/[s(s + 1)]$                    **(d)** $H(s) = 2(s^2 + 4)/[s(s^2 + 1)(s + 2)]$
**(e)** $H(s) = 16/[s^2(s + 4)]$                    **(f)** $H(s) = 2(s + 1)/[s(s^2 + 1)^2]$

**P8.6  (Initial and Final Value Theorems)**  Find the initial and final values for each $X(s)$ for which the theorems apply.
**(a)** $X(s) = s/(s + 1)$                    **(b)** $X(s) = (s + 2)^2/(s + 1)$
**(c)** $X(s) = (s + 1)^2/(s^2 + 1)$                    **(d)** $X(s) = s^2/[(s + 2)(s^2 + 2s + 2)]$
**(e)** $X(s) = 2/[s(s + 1)]$                    **(f)** $X(s) = 2(s^2 + 1)/[s(s + 2)(s + 5)]$
**(g)** $X(s) = 16/(s^2 + 4)^2$                    **(h)** $X(s) = 2(s + 1)/[s(s^2 + 1)^2]$

**P8.7** **(Initial Value Theorem)** If $H(s) = N(s)/D(s)$ is a ratio of polynomials with $N(s)$ of degree $N$ and $D(s)$ of degree $D$, in which of the following cases does the initial value theorem apply, and if it does, what can we say about $h(0+)$?

**(a)** $D \leq N$          **(b)** $D > N + 1$          **(c)** $D = N + 1$

**P8.8** **(Inverse Transforms from Partial Fractions)** Find the time signal corresponding to the following Laplace transforms.

**(a)** $H(s) = 2/[s(s + 1)]$

**(b)** $H(s) = 2s/[(s + 1)(s + 2)(s + 3)]$

**(c)** $H(s) = 4s/[(s + 3)(s + 1)^2]$

**(d)** $H(s) = 4(s + 2)/[(s + 3)(s + 1)^2]$

**(e)** $H(s) = 2/[(s + 2)(s^2 + 4s + 5)]$

**(f)** $H(s) = 4(s + 1)/[(s + 2)(s^2 + 2s + 2)]$

**(g)** $H(s) = 2(s^2 + 2)/[(s + 2)(s^2 + 4s + 5)]$

**(h)** $H(s) = 2/[(s + 2)(s + 1)^3]$

**(i)** $H(s) = 4/(s^2 + 2s + 2)^2$

**(j)** $H(s) = 4s/(s^2 + 2s + 2)^2$

**(k)** $H(s) = 32/(s^2 + 4)^2$

**(l)** $H(s) = 4/[s(s^2 + 1)^2]$

**P8.9** **(Inverse Transforms for Nonstandard Forms)** Find the time signal corresponding to the following Laplace transforms.

**(a)** $H(s) = s/(s + 1)$

**(b)** $H(s) = (s + 2)^2/(s + 1)$

**(c)** $H(s) = (s + 1)^2/(s^2 + 1)$

**(d)** $H(s) = s^3/[(s + 2)(s^2 + 2s + 2)]$

**(e)** $H(s) = 4(s - 2e^{-s})/[(s + 1)(s + 2)]$

**(f)** $H(s) = 4(s^2 - e^{-s})/[(s + 1)(s + 2)]$

**(g)** $H(s) = (1 - e^{-s})/[s(1 - e^{-3s})]$

**(h)** $H(s) = (1 - e^{-s})/[s^2(1 + e^{-s})]$

**P8.10** **(Inversion and Partial Fractions)** There are several ways to set up a partial fraction expansion (PFE) and find the inverse transform. For a quadratic denominator $[(s + a)^2 + b^2]$ with complex roots, we use a linear term $As + B$ in the numerator. For $H(s) = (As + B)/[(s + a)^2 + b^2]$, we find $h(t) = \exp(-at)[K_1 \cos(bt) + K_2 \sin(bt)]$.

**(a)** Find $K_1$ and $K_2$ in terms of $A, B, a$ and $b$.

**(b)** Let $H(s) = 2(s^2 + 2)/[(s + 2)(s^2 + 2s + 2)] = C/(s + 2) + (As + B)/[(s + a)^2 + b^2]$. Find $A, B$ and $C$ by comparing the numerator of $H(s)$ with the assumed form and find $h(t)$.

**(c)** Extend these results to find $x(t)$ if $X(s) = 40/[(s^2 + 4)(s^2 + 2s + 2)]$.

**P8.11** **(Inversion and Partial Fractions)** Let $H(s) = (s + 2)/(s + 3)(s + 4)^3$. Its PFE has the form $H(s) = K/(s + 3) + A_0/(s + 4)^3 + A_1/(s + 4)^2 + A_2/(s + 4)$. Finding $K$ and $A_0$ is easy but $A_1$ and $A_2$ require derivatives of $(s + 1)^3 H(s)$. An alternative is to recognize that $H(s)$ and its PFE are valid for any value of $s$, excluding the poles. All we need is to evaluate the PFE at two values of $s$ to yield two equations in two unknowns, assuming $K$ and $A_0$ are already known.

**(a)** Try this approach using $s = -2$ and $-5$ to find the PFE constants $A_1$ and $A_2$, assuming you have already evaluated $K$ and $A_0$.

**(b)** Repeat part (a) choosing $s = 0$ and $s = -6$. Is there a "best choice"?

**P8.12** **(Transfer Function)** Find the transfer function and impulse response of the systems shown in Figure P8.12 (p. 366).

**P8.13** **(Transfer Function)** Find the transfer function, differential equation and order of the following systems and check for their stability.

**(a)** $h(t) = \exp(-2t)u(t)$

**(b)** $h(t) = [1 - \exp(-2t)]u(t)$

**(c)** $h(t) = t\exp(-t)u(t)$

**(d)** $h(t) = 0.5\delta(t)$

**(e)** $h(t) = \delta(t) - \exp(-t)u(t)$

**(f)** $h(t) = [\exp(-t) + \exp(-2t)]u(t)$

**Figure P8.12**

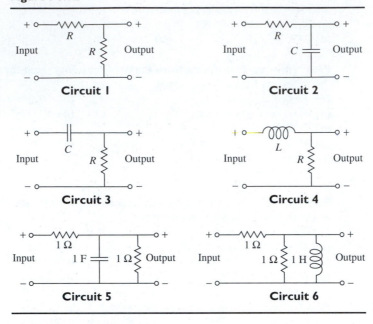

Circuit 1

Circuit 2

Circuit 3

Circuit 4

Circuit 5

Circuit 6

**P8.14 (Transfer Function)** Find the transfer function and impulse response of the following systems.
**(a)** $y''(t) + 3y'(t) + 2y(t) = 2x'(t) + x(t)$
**(b)** $y''(t) + 4y'(t) + 4y(t) = 2x'(t) + x(t)$
**(c)** $y(t) = 0.2x(t)$

**P8.15 (System Formulation)** Set up the system differential equations from the following transfer functions.
**(a)** $H(s) = 3/(s + 2)$                       **(b)** $H(s) = (1 + 2s + s^2)/[(1 + s^2)(4 + s^2)]$

**(c)** $H(s) = \dfrac{2}{1+s} - \dfrac{1}{2+s}$            **(d)** $H(s) = \dfrac{2s}{1+s} - \dfrac{s}{2+s}$

**P8.16 (System Response)** The transfer function $H(s)$ of a system is

$$H(s) = \frac{2 + 2s}{4 + 4s + s^2}$$

Find the response for the following inputs.
**(a)** $x(t) = \delta(t)$                     **(b)** $x(t) = 2\delta(t) + \delta'(t)$
**(c)** $x(t) = \exp(-t)u(t)$            **(d)** $x(t) = t\exp(-t)u(t)$
**(e)** $x(t) = 4\cos(2t)u(t)$          **(f)** $x(t) = [4\cos(2t) + 4\sin(2t)]u(t)$

**P8.17 (System Analysis)** Find the zero-state, zero-input and total response of the following systems, assuming $x(t) = \exp(-2t)u(t), y(0) = 1$ and $y'(0) = 2$.
**(a)** $y''(t) + 4y'(t) + 3y(t) = 2x'(t) + x(t)$
**(b)** $y''(t) + 4y'(t) + 4y(t) = 2x'(t) + x(t)$
**(c)** $y''(t) + 4y'(t) + 5y(t) = 2x'(t) + x(t)$

**P8.18 (System Analysis)** For each circuit shown in Figure P8.18:
**(a)** Find the transfer function and impulse response.
**(b)** Find the response to $x(t) = \exp(-t)u(t)$, assuming $v_C(0) = 0$, $i_L(0) = 0$.
**(c)** Find the response to $x(t) = u(t)$, assuming $v_C(0) = 1$, $i_L(0) = 2$.

**Figure P8.18**

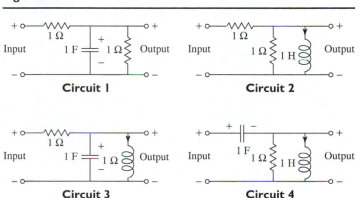

**Circuit I**      **Circuit 2**

**Circuit 3**      **Circuit 4**

**P8.19 (Steady-State Response)** The transfer function $H(s)$ of a system is

$$H(s) = \frac{2 + 2s}{4 + 4s + s^2}$$

Find the steady-state response for the following inputs.
**(a)** $x(t) = 4u(t)$ **(b)** $x(t) = 4\cos(2t)u(t)$
**(c)** $x(t) = [\cos(2t) + \sin(2t)]u(t)$ **(d)** $x(t) = [4\cos(t) + 4\sin(2t)]u(t)$

**P8.20 (Response to Periodic Inputs)** Find the steady-state response and the total response of an $RC$ lowpass filter with time constant $\tau = 2$ for the switched periodic inputs $x(t)$ whose one period $x_1(t)$ is given by
**(a)** $x_1(t) = u(t) - u(t-1), T = 2$ **(b)** $x_1(t) = t[u(t) - u(t-1)], T = 1$

**P8.21 (Response to Periodic Inputs)** Find the steady-state response and the total response of the following systems to the switched periodic inputs $x(t)$ whose one period equals $x_1(t)$.
**(a)** $H(s) = (s+3)/(s^2 + 3s + 2)$,   $x_1(t) = [u(t) - u(t-1)], T = 2$
**(b)** $H(s) = (s+2)/(s^2 + 4s + 3)$,   $x_1(t) = \text{tri}(t-1), T = 2$

**P8.22 (Stability)** A perfect differentiator is described by $y(t) = dx/dt$.
**(a)** Find its transfer function $H(s)$ and use the condition for BIBO stability to show that the system is unstable.
**(b)** Verify your conclusion by finding the impulse response $h(t)$ and applying the condition for BIBO stability in the time domain.
**(c)** Repeat parts (a) and (b) for a perfect integrator.

**P8.23  (Model Order Reduction)**  For a stable system, the effect of poles much farther from the $j\omega$-axis than the rest is negligible after some time and the behavior of the system may be approximated by a lower-order model from the remaining or **dominant poles** as long as the new system is also stable.

**(a)** Let $H(s) = 100/[(s + 1)(s + 20)]$. Find $h(t)$. Discard the term that makes the smallest contribution to approximate $h(t)$ by $h_R(t)$ and establish the reduced model $H_R(s)$.

**(b)** If the poles $(s + \alpha_k)$ of $H(s)$ to be neglected are written in the form $\alpha(1 + s/\alpha_k)$, $H_R(s)$ may also be computed directly from $H(s)$ if we discard just the factors $(1 + s/\alpha)$ from $H(s)$. Obtain $H_R(s)$ from $H(s)$ using this method and explain any differences from the results of part (a).

**(c)** As a rule of thumb, poles with magnitudes 10 times larger than the rest may be neglected. Use this idea to find the reduced model $H_R(s)$ and its order if
$H(s) = 400/[(s^2 + 2s + 200)(s + 20)(s + 2)]$.

# 9 APPLICATIONS OF LAPLACE TRANSFORMS

*9*

---

## 9.0
## Scope and Objectives

The Laplace transform is widely used in many diverse fields. In this chapter, we briefly examine those applications and extensions that are related directly to the area of system analysis and signal processing. This chapter is divided into three parts.

**Part 1** Sketching the frequency response using Bode plots.
**Part 2** The analysis and characteristics of real filters.
**Part 3** Scaling and frequency transformations.

Useful Background   Much of the material in this chapter relies on the concepts developed in the previous chapter and Chapters 5 and 6.

---

## 9.1
## Frequency Response and Bode Plots

For LTI systems, the frequency response describes plots of the magnitude and phase of $H(\omega)$ versus frequency. Bode plots allow us to plot the frequency response over wide ranges by using logarithmic *scale compression*. For a *dimensionless* $H(\omega)$, the magnitude is plotted in **decibels**, $H_{dB}$, as $20 \log(|H(\omega)|)$ against $\log(\omega)$. Since any phase variation can be brought into the range $\pm\pi$, the phase $\phi(\omega)$ is typically plotted on a *linear* scale versus $\log(\omega)$.

For LTI systems whose transfer function is a ratio of polynomials, a rough sketch can be quickly generated using straight-line approximations called **asymptotes** over different frequency ranges to obtain **asymptotic Bode plots**.

**Remark:**   Since $\log(0) = -\infty$, and the logarithm of a negative number is not real, log and Bode plots use only positive frequencies, and exclude dc.

The log operator provides nonlinear compression. A **decade** (ten-fold) change in $|H(\omega)|$ results in exactly a 20-dB change in its decibel magnitude, because

$$20 \log |10H(\omega)| = 20 \log |H(\omega)| + 20 \log 10 = H_{dB} + 20$$

$$20 \log |H(\omega)/10| = 20 \log |H(\omega)| - 20 \log 10 = H_{dB} - 20$$

Similarly, a twofold (an octave) change in $|H(\omega)|$ results in approximately a 6-dB change in its decibel magnitude, because

$$20 \log |2H(\omega)| = 20 \log |H(\omega)| + 20 \log 2 = H_{dB} + 6$$

$$20 \log |\tfrac{1}{2}H(\omega)| = 20 \log |H(\omega)| - 20 \log 2 = H_{dB} - 6$$

Thus, if $|H(\omega)| = A\omega^k$, the slope of $H_{dB}$ versus $\omega$ is $20k$ dB/decade or $6k$ dB/octave.

**The Transfer Function**    The steady-state transfer function, or frequency response, of an LTI system is a ratio of polynomials in $\omega$ and may be expressed as

$$H(\omega) = \frac{B_m(j\omega)^m + \cdots + B_2(j\omega)^2 + B_1(j\omega) + B_0}{A_n(j\omega)^n + \cdots + A_2(j\omega)^2 + A_1(j\omega) + A_0}$$

We make three important observations about such a transfer function:

1. As $\omega \longrightarrow \infty$, $H(\omega)$ is proportional to $(j\omega)^k$, where $k = m - n$. At high frequencies, its decibel magnitude $H_{dB}$ shows a slope of $-20k$ dB/decade, and its phase approaches $(j)^{m-k}$ or $90(m - k)°$.
2. As $\omega \longrightarrow 0$, $H(\omega) \longrightarrow H(0)$ and is also proportional to some power of $j\omega$. For nonzero $A_0$ and $B_0$, $H(0)$ is a constant, with $|H(0)| = B_0/A_0$, and a phase that equals zero if $H(0)$ is positive, or $\pm180°$ otherwise.
3. Being a ratio of polynomials in $j\omega$, the numerator and denominator can be factored into linear and quadratic factors with real coefficients. A standard representation is obtained by setting the real part of each factored term to unity to obtain the form

$$H(\omega) = \frac{K_1(j\omega)^k(1 + j\omega/a_1)(1 + j\omega/a_2) \cdots [1 + C_1(j\omega/c_1) + (j\omega/c_1^2)] \cdots}{(1 + j\omega/d_1)(1 + j\omega/d_2) \cdots [1 + E_1(j\omega/e_1) + (j\omega/e_1^2)] \cdots}$$

The factored form offers two advantages:

1. Upon taking logarithms, products transform to a sum. This allows us to sketch the decibel magnitude of the simpler individual factors separately and then use superposition to plot the composite response.
2. The phase is the algebraic sum of the phase due to each factor. We can thus plot the composite phase as their superposition.

### 9.1.1 *Bode Plots of Linear and Repeated Linear Factors*

There are actually only four types of terms that we must consider in order to plot the spectrum for any rational transfer function: constants, terms of the form $(j\omega)^{\pm k}$, $(1 + j\omega/a)^{\pm k}$, and quadratic terms. The magnitude and phase approximations of these are listed in Table 9.1 and plotted in Figure 9.1 (p. 372).

---

**Table 9.1   Magnitude and Phase Approximations.**

| Term | Magnitude in dB |
|------|-----------------|
| $K$ (constant) | Constant for all $\omega$ for both positive and negative $K$ |
| $(j\omega)^{\pm k}$ | Linear slope of $\pm 20k$ dB/decade for all $\omega$ |
| $(1 + j\omega/a)^{\pm k}$ | 0 dB for $\omega \le a$, linear slope of $\pm 20k$ dB/decade for $\omega \ge a$ |

| Term | Phase |
|------|-------|
| $K$ (constant) | Zero phase for all $\omega$ for $K > 0$, $180°$ for $K < 0$ |
| $(j\omega)^{\pm k}$ | Constant phase of $\pm 90k°$ for all $\omega$ |
| $(1 + j\omega/a)^{\pm k}$ | Zero for $\omega \le a/10$, $\pm 90k°$ for $\omega \ge 10a$, linear in between |

---

**The Constant**   The decibel value of a constant $K$ is $H_{dB} = 20 \log |K|$ dB for all $\omega$. Its phase is zero if $K > 0$, and $\pm 180°$ if $K < 0$.

**The Term $j\omega$**   Its decibel magnitude is simply $H_{dB} = 20 \log(\omega)$ dB, which describes a straight line with a slope of 20 dB/decade for all $\omega$. A convenient reference for plotting this term is $\omega = 1$ when $H_{dB} = 0$. The phase equals $90°$ for all $\omega$.

**The Repeated Term $(1 + j\omega/a)^k$**   Its decibel value is $H_{dB} = 20 \log |1 + j\omega/a|$ dB, and yields separate linear approximations for low and high frequencies. For $\omega \ll a$, $H_{dB} \approx 20 \log(1) = 0$ dB, which is a straight line with zero slope. For $\omega \gg a$, $H_{dB} \approx 20k \log(\omega/a)$ dB, which is a straight line with a slope of $20k$ dB/decade. Negative values of $k$ correspond to terms in the denominator, and suggest negative slopes.

The two straight lines intersect at $\omega = \omega_B = a$ where

$$H_{dB} = 20k \log |1 + j\omega_B/a| = 20k \log |1 + j| = 3k \text{ dB}$$

**Break Frequency**   The frequency $\omega_B$ is called the **break frequency**, or **corner frequency**. For a linear factor of the form $1/(1 + j\omega/\omega_B)$, $\omega_B$ it is also called the **half-power frequency**, or the **3-dB frequency**.

The difference between the asymptotic and true magnitude equals

1. $3k$ dB at $\omega_B = a$, and decreases for all other frequencies.
2. $k$ dB at frequencies an octave above and below $\omega_B$.
3. Nearly 0 dB at frequencies a decade above and below $\omega_B$.

**Figure 9.1    Bode magnitude and phase plots of linear and repeated factors.**

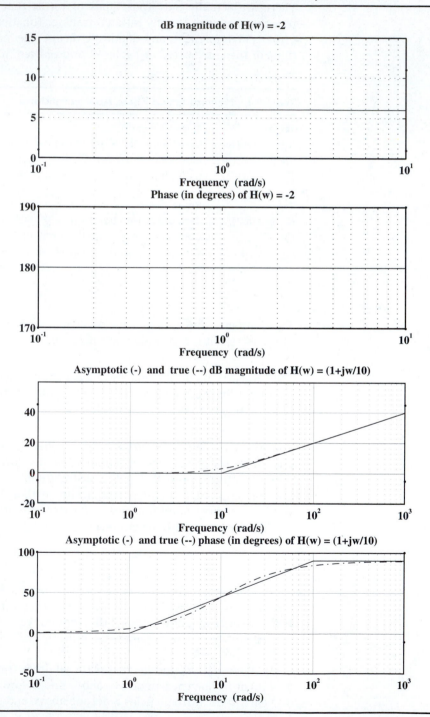

**Figure 9.1**   **(continued)**

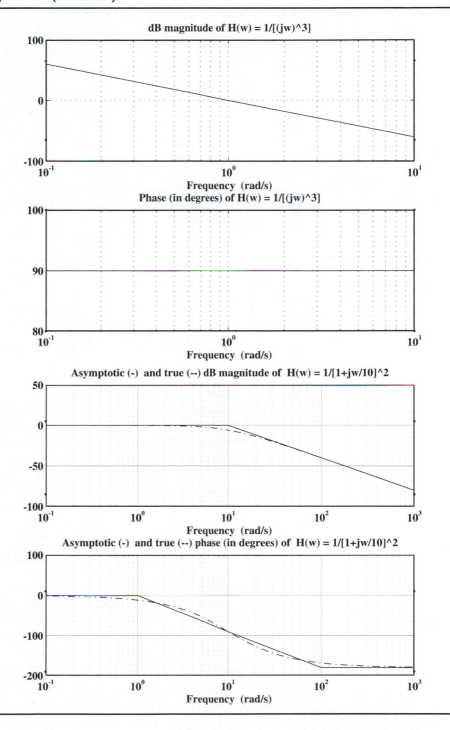

The phase of $(1 + j\omega/a)^k$ equals $\phi(\omega) = k\tan^{-1}(\omega/a)$. At low and high frequencies, this may be approximated as

$$\omega \ll a: \qquad \phi(\omega) = \tan^{-1}(\omega/a) \approx 0°$$

$$\omega \gg a: \qquad \phi(\omega) = \tan^{-1}(\omega/a) \approx 90k°$$

At $\omega_B = a$, we have the *exact* result $\phi(\omega) = k\tan^{-1}(1) = \frac{1}{4}k\pi = 45k°$.

To generate an asymptotic phase plot, we need a linear approximation in the intermediate frequency range. The break frequency provides a convenient reference. Since the difference between the true and asymptotic magnitude almost disappears within a decade of the break frequency, it is customary to start the high and low frequency ranges a decade above and below $\omega_B$. We approximate the phase by a straight line over this two-decade range. Since the phase increases by $90k°$ over this range, the slope of this straight line is $45k°$/decade, and it passes through $45k°$ at $\omega = \omega_B$ (Figure 9.1).

**Example 9.1**

Asymptotic Plots

(a) The asymptotic plot of $1/(1 + j\omega/10)^2$ (Figure 9.1) is 0 dB until $\omega = 10$ and has an asymptote with a slope of $-40$ dB/decade past $\omega = 10$. The true magnitude at $\omega = 10$ is $-6$ dB. The phase plot reveals a slope of $-90°$/decade starting at $\omega = 1$, an initial phase of $0°$, and a final phase of $180°$.

(b) The term $(j\omega)^{-3}$ has a decibel magnitude with a slope of $-60$ dB/decade, and a phase of $-270°$ or $90°$ for all frequencies.

## 9.1.2    *Bode Magnitude Plots*

The Bode magnitude plot for a transfer function with several terms is the sum of similar plots for each of its individual terms. Since the asymptotic plots are linear, so is the composite plot. It can actually be sketched directly if we keep track of the following:

1. A constant term shifts the entire plot. We include this as the last step.
2. Linear terms of the form $(1 + j\omega/\omega_B)^{\pm k}$ start out with zero slope. The starting slope is thus zero if terms of the form $(j\omega)^{\pm k}$ are absent.
3. The term $(j\omega)^{\pm k}$, if present, always dictates the slope of the initial low-frequency asymptote, which equals $\pm 20k$ dB. The asymptote always passes through 0 dB at $\omega = 1$ *if we ignore the constant term*.
4. If the break frequencies are arranged in ascending order, the slope changes at each successive break frequency. The change in slope equals $\pm 20k$ dB/decade for $(1 + j\omega/\omega_B)^{\pm k}$. The slope increases by $20k$ dB/decade if $\omega_B$ corresponds to a numerator term, and decreases by $20k$ dB/decade otherwise.

A rough sketch of the true plot may also be generated by adding approximate **correction factors** at the break frequencies. The correction factor at the break frequency $\omega_B$, corresponding to a linear factor $(1 + j\omega/\omega_B)$, equals

**1.** 3 dB if the surrounding break frequencies are at least a decade apart.
**2.** 2 dB if an adjacent break frequency is an octave away.

For repeated factors with multiplicity $M$, these values are multiplied by $M$. At all other frequencies, the true value can only be computed from $H(\omega)$.

---

**Example 9.2**
Composite Magnitude Plots

**(a)** *Linear Factors:*

Let
$$H(\omega) = \frac{40(\frac{1}{4} + j\omega)(10 + j\omega)}{j\omega(20 + j\omega)}$$

In standard form,

$$H(\omega) = \frac{5(1 + j\omega/\frac{1}{4})(1 + j\omega/10)}{j\omega(1 + j\omega/20)}$$

The break frequencies in ascending order are:

$$\omega_1 = \tfrac{1}{4} \text{ (numerator)}, \quad \omega_2 = 10 \text{ (numerator)}, \quad \omega_3 = 20 \text{ (denominator)}$$

The term $1/j\omega$ provides a starting asymptote $-20$ dB/decade and 0 dB at $\omega = 1$. We can now sketch a composite plot by including the other terms:

At $\omega_1 = \frac{1}{4}$ (numerator), the slope changes by $+20$ dB/decade to 0 dB/decade.
At $\omega_2 = 10$ (numerator), the slope changes by $+20$ dB/decade to $+20$ dB/decade.
At $\omega_3 = 20$ (denominator), the slope changes by $-20$ dB/decade to 0 dB/decade.

We find the asymptotic magnitudes at the break frequencies:
$\omega_1 = \frac{1}{4}$ (numerator): 12 dB (2 octaves below $\omega = 1$)
$\omega_2 = 10$ (numerator): 12 dB (zero slope from $\omega = \frac{1}{2}$ to $\omega = 10$)
$\omega_3 = 20$ (denominator): 18 dB (1 octave above $\omega = 10$, with slope 6 dB/octave)

Finally, the constant 5 shifts the plot by $20\log(5) = 14$ dB (Figure 9.2, p. 376). For a true plot, we apply correction factors at the break frequencies:

| $\omega_B$ | Surrounding frequency | Correction | True dB magnitude |
|---|---|---|---|
| $\frac{1}{4}$ (numerator) | 10 (> 1 decade away) | $+3$ dB | $14 + 12 + 3 = 29$ dB |
| 10 (numerator) | 20 (1 octave away) | $+2$ dB | $14 + 12 + 2 = 28$ dB |
| 20 (denominator) | 10 (1 octave away) | $-2$ dB | $14 + 18 - 2 = 30$ dB |

For the exact decibel magnitude, say at $\omega = 20$, we use $H(\omega)$ to compute

$$H_{\text{dB}} = 20\log \left| 5\frac{(1 + j\omega/\frac{1}{4})(1 + j\omega/10)}{j\omega(1 + j\omega/20)} \right|_{\omega=20}$$

$$= 20\log \left| 5\frac{\sqrt{(6401)(5)}}{20\sqrt{2}} \right| = 30.0007 \text{ dB}$$

**(b)** *Repeated Factors:*

Consider
$$H(\omega) = \frac{(1 + j\omega)(1 + j\omega/100)}{(1 + j\omega/10)^2(1 + j\omega/300)}$$

Its Bode plot is sketched in Figure 9.2. We make the following remarks:

The starting slope is 0 dB/decade since a term of the form $(j\omega)^k$ is absent.
The slope changes by $-40$ dB/decade at $\omega_B = 10$ due to the repeated factor.
For a true plot, the correction factor at $\omega_B = 10$ (denominator) is $-6$ dB.
The true magnitudes at $\omega = 100$ and $\omega = 300$, which are not an octave apart, must be computed from $H(\omega)$.

**Figure 9.2    Bode magnitude plots for the transfer functions of Example 9.2.**

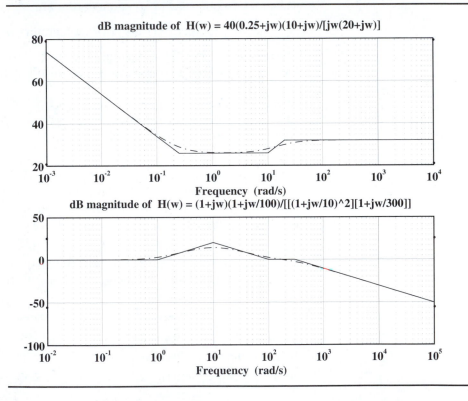

### 9.1.3    *Bode Phase Plots*

The Bode phase plot for a transfer function with several terms is just the sum of similar plots for each of its individual terms. Since asymptotic plots are linear, the composite plot can be sketched directly without sketching the individual terms if we note the following:

1. A positive constant does not affect the phase plot.
2. A negative constant contributes 180° (or −180°) to the phase.
3. The term $(j\omega)^{\pm k}$ shifts the plot by $\pm 90k°$ and may be used in the end.
4. The initial and final slopes are zero.
5. The initial phase is due only to terms like $(j\omega)^{\pm k}$ and equals $\pm 90k°$.
6. The final phase equals $90n°$, where $n$ is the difference between the order of the numerator and denominator polynomial. These values serve as a consistency check.
7. If the break frequencies are arranged in ascending order, the slope changes at frequencies one decade on either side of each successive break frequency. The slope increases by $\pm 45k°$/decade for a factor of the form $(1 + j\omega/\omega_B)^{\pm k}$.

**Example 9.3**
Composite Phase Plot for Linear Terms

Consider $H(\omega) = (\frac{1}{2} + j\omega)(10 + j\omega)/[j\omega(20 + j\omega)^2]$. To sketch its Bode phase plot (Figure 9.3), we first check the initial and final phase: $\phi_i = -90°$ and $\phi_f = -90°$. We arrange break frequencies and their multiplicity: $\frac{1}{2}$ (numerator), 10 (numerator), 20(denominator, $k = 2$).

**Figure 9.3    Bode phase plot for the transfer function of Example 9.3.**

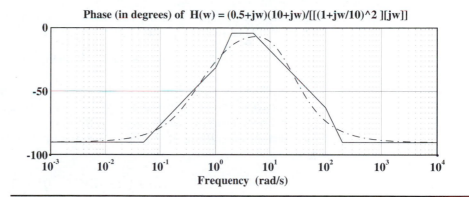

We find the slopes for the composite plot due to terms like $(1 + j\omega/\omega_B)^k$.

$\omega_B = 0.5$ (numerator):   $+45°$/decade between 0.05 and 5
$\omega_B = 10$ (numerator):   $+45°$/decade between 1 and 100
$\omega_B = 20$ (denominator, $k = 2$):   $-90°$/decade between 2 and 200

The frequencies at which the slopes change are, 0.05, 1, 2, 5, 100, 200. The slopes for the composite plot are found as

$0°$/decade before 0.05 and past 200
$45°$/decade between 0.05 and 1 (due to the first term only)

90°/decade between 1 and 2 (due to the first two terms)
0°/decade between 2 and 5 (due to all three terms)
−45°/decade between 5 and 100 (due to the last two terms)
−90°/decade between 100 and 200 (due to the last term only)

We sketch the asymptotic plot and shift it by −90° due to the term $1/j\omega$. As a check, both the initial and final values of the phase equal −90°. ⎯⎯

### 9.1.4    *More on the Quadratic Term*

A quadratic term in the denominator may be written in the form

$$H(\omega) = 1/[1 + j2\zeta(\omega/a) + (j\omega/a)^2]$$

The quantity $\zeta$ is referred to as the **damping factor.** For $\zeta = 1$, $H(\omega)$ is a perfect square and can be handled as a repeated linear factor. For $\zeta > 1$, $H(\omega)$ can always be factored into the form $(1 + j\omega/b)(1 + j\omega/c)$, and handled as a product of linear factors. For $0 < \zeta < 1$, we get complex roots, and this is the *only* case we must deal with separately for both magnitude and phase.

**Magnitude Plot**   For low frequencies ($\omega \ll a$) we have $H_{dB} \approx 20\log(1) = 0$ dB. For high frequencies ($\omega \gg a$), since $(\omega/a)^2 \gg (\omega/a)$, we also have

$$H_{dB} \approx -20\log|(j\omega/a)^2| = -40\log(\omega/a) \qquad \omega \gg a$$

This describes a straight line with a slope of −40 dB/decade.

The two intersect at $\omega = a$ and yield an asymptotic Bode plot identical to a squared linear term. So where is the difference? In the true plot, of course! The *exact* decibel magnitude at the break frequency depends on the quantity $\zeta$, and equals

$$H_{dB} = 20\log|1 + j2\zeta(\omega_B/a) + (j\omega_B/a)^2| = 20\log|1 + j2\zeta + j^2| = 20\log(2\zeta)$$

The true magnitude plot (Figure 9.4) for $0 < \zeta < 1$ reveals that:

**1.** For $\zeta = 1$, $H_{dB}$ equals −6 dB, just as it does for $(1 + j\omega/a)^2$.
**2.** For $\zeta = \frac{1}{2}$, $H_{dB}$ equals 0 dB. The true value equals the asymptotic value at the break frequency. The true plot shows some peaking elsewhere.
**3.** For $\zeta > \frac{1}{2}$, $H_{dB}$ is always negative, and no peaking occurs.
**4.** For $\zeta < \frac{1}{2}$, $H_{dB}$ is always positive and the response is always peaked. The smaller the value of $\zeta$, the more peaked the response.

For a peaked response, the break frequency $\omega_B = a$ may not correspond to the frequency $\omega_M$ where $|H(\omega)|$ is a maximum. The frequency $\omega_M$ must actually be found by setting $d|H(\omega)|/d\omega$ equal to zero.

**Phase Plot**   For $\zeta = 1$, the asymptotic phase plot is identical to that for a squared linear factor. For $\zeta < 1$, it depends on $\zeta$. The transition between 0° and 180° occurs in less than two decades (Figure 9.4).

For very small values of $\zeta$, say 0.2 or less, the phase change from 0° to $-180°$ occurs in an almost steplike fashion and the asymptotes extend almost up to the break frequency itself.

**Figure 9.4   Bode plots for a quadratic factor for various values of $\zeta$.**

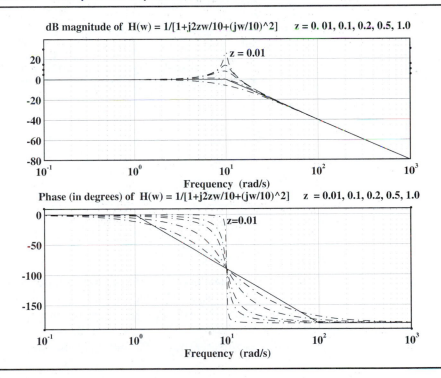

dB magnitude of $H(w) = 1/[1+j2zw/10+(jw/10)^2]$      z = 0. 01, 0.1, 0.2, 0.5, 1.0

Phase (in degrees) of $H(w) = 1/[1+j2zw/10+(jw/10)^2]$      z = 0.01, 0.1, 0.2, 0.5, 1.0

**Example 9.4**

Composite Plot for Quadratic Terms

Consider the transfer function $H(\omega) = 50(2 + j\omega)/(-\omega^2 + 2j\omega + 100)$.

With $-\omega^2 = (j\omega)^2$, we write this in the standard form

$$H(\omega) = \frac{(1 + j\omega/2)}{1 + j2(0.1)(\omega/10) + (j\omega/10)^2}$$

The Bode magnitude plot (Figure 9.5, p. 380) is 0 dB up to $\omega = 2$ with a 20 dB/decade asymptote starting at $\omega = 2$ and another of $-20$ dB/decade starting at $\omega = 10$. The asymptotic value at $\omega = 10$ equals 14 dB. The true value at $\omega = 2$ equals 3 dB. No correction factor is required. With $\zeta = 0.1$, $-20\log(2\zeta) = -20\log(0.2) = 14$ dB, and the true value at $\omega = 10$ differs from the asymptotic value by 14 dB, and equals $14 + 14 = 28$ dB.

**Figure 9.5** **Bode plots for the transfer function of Example 9.4.**

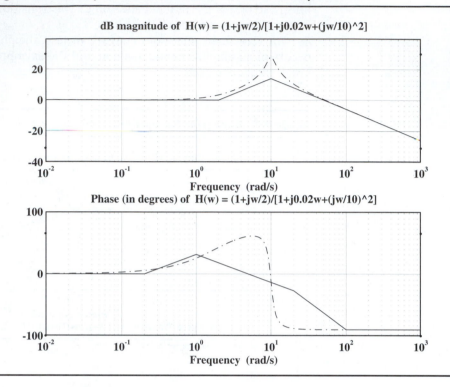

The phase plot (Figure 9.5) shows zero phase until $\omega = 0.2$, followed by asymptotes with slopes of 45°/decade until $\omega = 1$, −45°/decade until $\omega = 20$ and −90°/decade until $\omega = 100$. Past $\omega = 100$, the phase stays constant at −90°.

*Comment:* Note how different the asymptotic plot is compared with the actual. —

## 9.2
## Analysis of Analog Filters

For real filters, it is impossible to achieve constant gain and linear phase in the passband. The frequency response of such filters is specified by the magnitude $|H(\omega)|$ and phase $\angle\theta(\omega)$ of the transfer function $H(\omega)$.

### 9.2.1 *Phase, Delay and Minimum-Phase Systems*

With $H(\omega) = |H(\omega)|\angle\theta(\omega)$, we define the **phase delay** $t_p$ as $-\theta(\omega)/\omega$, the negative phase normalized by $\omega$, and the **group delay** $t_g$ as the negative slope of the phase $\theta(\omega)$ (Figure 9.6). Formally,

$$t_p = -\frac{\theta(\omega)}{\omega} \qquad t_g = -\frac{d\theta(\omega)}{d\omega}$$

If $\theta(\omega)$ varies linearly with frequency, $t_p$ and $t_g$ are not only constant, but are equal. For rational transfer functions, $\theta(\omega)$ is a transcendental function, but *the group delay is always a rational function* of $\omega^2$ and is much easier to work with in many filter applications.

---

**Figure 9.6   The phase delay at a frequency** $\omega_0$ **equals** $-\theta/\omega_0$. **The group delay equals the negative slope** $-d\theta/d\omega$ **at** $\omega_0$.

---

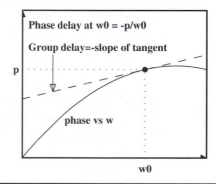

Phase delay at w0 = -p/w0

Group delay=-slope of tangent

p

phase vs w

w0

---

Consider a system with its transfer function $H(s) = K\Pi(s - z_k)/\Pi(s - p_k)$ in factored form, where $K > 0$. Replacing $K$ by $-K$ or a factor $s - a$ by $s + a$ in this factored form changes only the phase of its frequency response $H(\omega)$. If $K > 0$, and all the poles and zeros of $H(s)$ lie in the LHP, it defines a **minimum-phase system**. Such a system has the smallest group delay, and smallest deviation from zero phase, at every frequency among all transfer functions with the same magnitude spectrum. A *stable* system is called **mixed phase** if some of its zeros lie in the RHP and **maximum phase** if all its zeros lie in the RHP. Of all possible stable systems with the same magnitude response, there is only one minimum-phase system.

**Example 9.5**
Illustrating the
Minimum-Phase Concept

Consider the transfer functions of the following stable systems:

$$H_1(s) = \frac{(s + 1)(s + 2)}{(s + 3)(s + 4)(s + 5)}$$

$$H_2(s) = \frac{(s - 1)(s + 2)}{(s + 3)(s + 4)(s + 5)}$$

$$H_3(s) = \frac{(s - 1)(s - 2)}{(s + 3)(s + 4)(s + 5)}$$

All have the same magnitude but different phase and delay. $H_1(s)$ is minimum phase (no zeros in the RHP), $H_2(s)$ is mixed phase (one zero outside the RHP), and $H_3(s)$ is maximum phase (all zeros in the RHP).

### 9.2.2   *Transfer Function from the Magnitude Spectrum*

The transfer function $H(s)$ of a physically realizable system found from its magnitude spectrum $|H(\omega)|$ alone is also a minimum-phase transfer function. A simple method for finding $H(s)$ starts with the magnitude-squared function $|H(\omega)|^2$, which may be written as

$$|H(\omega)|^2 = H(\omega)H^*(\omega) = H(\omega)H(-\omega)$$

Since $H(\omega)$ possesses conjugate symmetry, $|H(\omega)|^2$ describes a nonnegative, even polynomial in $\omega^2$. We can, therefore, express $|H(\omega)|^2$ as

$$|H(\omega)|^2 = H(\omega)H^*(\omega) = H(\omega)H(-\omega) = N(\omega^2)/D(\omega^2)$$

Since $H(s) = P(s)/Q(s)$, where $P(s)$ and $Q(s)$ are polynomials in $s$ with real coefficients, we write $|H(\omega)|^2$ as

$$|H(\omega)|^2 = H(s)H(-s)|_{s=j\omega} \qquad \text{or} \qquad H(s)H(-s) = N(\omega^2)/D(\omega^2)|_{\omega^2=-s^2}$$

This form suggests that both the poles and zeros of $H(s)H(-s)$ representing $|H(\omega)|^2$ must possess **quadrantal symmetry** (which describes both conjugate symmetry and mirror symmetry about the $j\omega$-axis). The system transfer function $H(s) = KP(s)/Q(s)$ is then obtained by selecting $P(s), Q(s)$ and $K$ as follows:

1. To ensure a *stable* system, we choose the left-half-plane (LHP) roots of $Q(\omega^2)$ with $\omega = s/j$. This is the only acceptable choice for $Q(s)$.
2. To ensure a *minimum-phase* system, we choose only left-half-plane (LHP) roots of $P(\omega^2)$ with $\omega = s/j$. Other choices, though possible, do not yield a minimum-phase system.
3. We select a value of $K > 0$. Typically, $K$ is chosen to match the magnitudes of $H(s)$ and $H(\omega)$ at a convenient frequency (dc, say).

---

**Example 9.6**
Finding a Transfer Function from $|H(\omega)|$

Consider a system for which $|H(\omega)|^2 = 4(9 + \omega^2)/(4 + 5\omega^2 + \omega^4)$. To find the minimum-phase transfer function of this system, we replace $j\omega$ by $s$ (or $\omega^2$ by $-s^2$) to give $|H(s)|^2 = H(s)H(-s) = 4(9 - s^2)/(4 - 5s^2 + s^4)$. We factor this to get $H(s)H(-s) = 4(3 - s)(3 + s)/[(2 - s)(2 + s)(1 - s)(1 + s)]$.

The required $H(s)$ is found by choosing only the LHP poles and zeros as

$$H(s) = K(s + 3)/[(s + 1)(s + 2)]$$

Since $|H(\omega)|$ has a dc magnitude of $\sqrt{9} = 3$, and $H(s) = 3K/2$ at $s = 0$, we choose $K = 2$ to match the dc gains. Thus $H(s) = 2(s + 3)/[(s + 1)(s + 2)]$.   ▬

### 9.2.3   *The Frequency Response: A Graphical Interpretation*

The factored form or pole-zero plot of $H(s)$ is useful for a qualitative picture of the frequency response. Consider the transfer function

$$H(s) = \frac{8s}{s^2 + 2s + 26} = \frac{8s}{(s + 1 - j5)(s + 1 + j5)}$$

Its pole-zero plot is shown in Figure 9.7.

At $\omega = 3$, for example, the frequency response $H(3)$ may be written

$$H(3) = \frac{8(j3)}{(1 - j_2)(1 + j8)}$$

Analytically, the magnitude is the ratio of the magnitudes of each term.

**Figure 9.7   Pole-zero plot, frequency response evaluation and the 3-D s-plane magnitude for $H(s) = 8s/(s^2 + 2s + 26)$.**

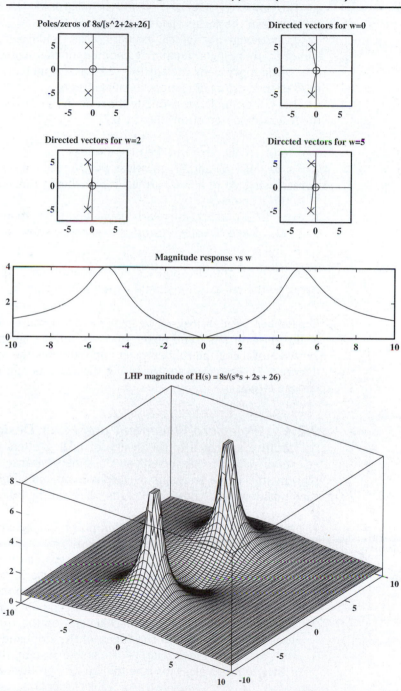

Graphically, the complex terms $j3, 1 - j2$, and $1 + j8$ may be viewed in the $s$-plane as vectors $\vec{N}_1$, $\vec{D}_1$ and $\vec{D}_2$ directed from each pole or zero location to $s = j3$. The ratio $|N_1|/(|D_1||D_2|)$ yields the magnitude $|H(3)|$, as before. Similarly, the difference in the angles yields the phase of $H(3)$.

Even though a graphical evaluation can yield exact results, it is so much more suited to obtaining a qualitative estimate of the magnitude response. We start at $\omega = 0$ and move along the positive $j\omega$-axis (for increasing $\omega$), and observe how the ratio of the vectors influences the magnitude.

For our example, we estimate the ratio $|N_1|/(|D_1||D_2|)$ as $\omega$ is varied and make the following observations (Figure 9.7):

1. $\omega = 0$:  $N_1$ is zero, and the magnitude is zero.
2. $0 < \omega < 5$:  $N_1$ and $D_2$ increase, but $D_1$ decreases. The response is small but increasing. At $\omega = 5, D_1$ attains its smallest value, and we expect to see a peak in the response.
3. $\omega \gg 5$:  $N_1$ and $D_1$ are nearly equal, and the response may be approximated by $1/|D_2|$. Since $D_2$ is increasing, the response shows a decrease.

This is typical of the response of a bandpass filter. If the pole locations are moved closer to the $j\omega$-axis, we should expect a larger peak at $\omega = 5$.

**The Rubber Sheet Analogy**   If we imagine the $s$-plane as a rubber sheet, tacked down at the zeros of $H(s)$, and poked up to an infinite height at the pole locations, the curved surface of the rubber sheet approximates the magnitude of $H(s)$ for any $s$ (Figure 9.7). The cross section along the $j\omega$-axis approximates the frequency response $H(\omega)$.

### 9.2.4    Pole-Zero Placement and Filter Design

An intuitive approach to the design of analog filters is based on the qualitative effects of pole-zero placement on the filter response. The strategy for pole-zero placement is based on mapping the passband and stopband along the $j\omega$-axis and then positioning the poles and zeros based on the following reasoning:

1. *Conjugate symmetry*: All complex poles and zeros must be paired with their complex conjugates.
2. *Zero placement and minimum phase*: Zeros can be placed anywhere in the $s$-plane, and result in decreased response. For a filter with minimum phase, the zeros must lie to the left of the $j\omega$-axis. Zeros on the $j\omega$-axis result in a null in the response. Thus, the stopband should contain zeros on, or near, the $j\omega$-axis. A good starting choice is to place a zero (on the $j\omega$-axis) in the middle, or two zeros (on the $j\omega$-axis) at the edges, of the stopband.
3. *Pole placement and stability*: For a stable system, the poles must lie to the left of the $j\omega$-axis. Poles closer to the $j\omega$-axis produce a large gain over a narrower

bandwidth. Clearly, the passband should contain poles near the $j\omega$-axis for large passband gains. A good starting choice is to place a pole (near the $j\omega$-axis) in the middle, or two poles at the edges, of the passband.

4. *Transition region*: To achieve a steep transition from the stopband to the passband, a good choice is to pair each stopband zero with a pole close to it and to the $j\omega$-axis.

5. *Pole-zero interaction*: Poles and zeros interact to produce a composite response that may not match qualitative predictions. Their placement may have to be changed, or other poles and zeros may need to be added, to tweak the response. Poles closer to the $j\omega$-axis, or farther away from each other, produce small interaction. Poles closer to each other, or farther from the $j\omega$-axis, produce more interaction. The closer we wish to approximate a given response, the more the number of poles and zeros we require, and the higher the filter order.

Filter design using pole-zero placement involves trial and error. The pole-zero patterns of some simple filter configurations as suggested by the above guidelines and their magnitude response are sketched in Figure 9.8 (p. 386). Note that bandstop and bandpass filters must have an order of two or higher.

A formal approach for the design of causal, physically realizable filters from magnitude specifications is discussed in the next chapter.

## 9.3 Analysis of First-Order and Second-Order Filters

For a filter with a rational transfer function given by $H(s) = P(s)/Q(s)$, the degree of the denominator $Q(s)$ defines the filter order.

The general first-order transfer function may be described by

$$H(s) = P(s)/Q(s) = K(a_0 + a_1 s)/(1 + b_1 s)$$

For this filter to be stable, we require $b_1 > 0$.

Various choices for $a_k$ result in the filter types whose transfer function $H(s)$ and frequency response $H(\omega)$ are summarized in Table 9.2.

**Table 9.2    Transfer Function of First-Order Filters.**

$$H(s) = \frac{P(s)}{Q(s)} = K\frac{a_0 + a_1 s}{1 + b_1 s} = K\frac{a_0 + a_1 s}{1 + s/\omega_C}$$

| Type | Constraints | $H(s)$ | $H(\omega)$ | $t_p$ | $t_g$ |
|---|---|---|---|---|---|
| Lowpass | $a_1 = 0$ | $\dfrac{K}{1 + b_1 s}$ | $\dfrac{K}{1 + j(\omega/\omega_C)}$ | $\dfrac{1}{\omega}\tan^{-1}(\omega/\omega_C)$ | $\omega_C/(\omega^2 + \omega_C^2)$ |
| Highpass | $a_0 = 0$ | $\dfrac{Ks}{1 + b_1 s}$ | $\dfrac{K(j\omega/\omega_p)}{1 + j(\omega/\omega_C)}$ | $\dfrac{1}{\omega}[\tan^{-1}(\omega/\omega_C) - \tfrac{1}{2}\pi]$ | $\omega_C/(\omega^2 + \omega_C^2)$ |
| Allpass | $a_1 = -b_1$ | $K\dfrac{1 - b_1 s}{1 + b_1 s}$ | $K\dfrac{1 - j(\omega/\omega_C)}{1 + j(\omega/\omega_C)}$ | $\dfrac{2}{\omega}\tan^{-1}(\omega/\omega_C)$ | $2\omega_C/(\omega^2 + \omega_C^2)$ |

**Figure 9.8** Pole-zero patterns and their corresponding frequency responses.

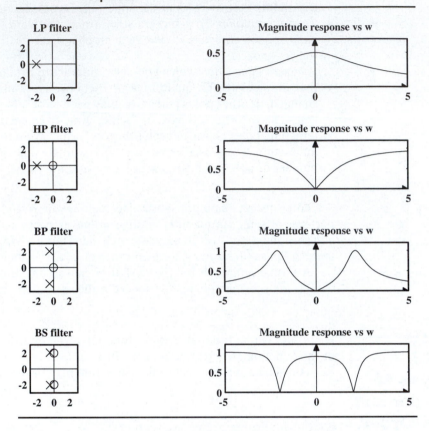

The denominator of $H(\omega)$ is often expressed as $1 + j(\omega/\omega_C)$, where $\omega_C = 1/b_1$ is called the **cutoff frequency**. For lowpass and highpass filters, $\omega_C$ equals the half-power frequency at which $|H(\omega)|$ equals $1/\sqrt{2}$ times $H_{max}$ its maximum value. The cutoff frequency is related to the time constant $\tau$ by $\omega_C = 1/\tau$.

For a first-order lowpass filter, the phase delay and group delay equal

$$t_p = -\frac{\theta(\omega)}{\omega} = \frac{1}{\omega}\tan^{-1}(\omega/\omega_C) \qquad t_g = -\frac{d\theta(\omega)}{d\omega} = \omega_C/(\omega^2 + \omega_C^2)$$

The magnitude $|H(\omega)|$ of an **allpass filter** (Table 9.2) is constant for all frequencies. Its phase $\theta(\omega)$, and delay $t_p$, are frequency dependent and described by

$$\theta(\omega) = \tan^{-1}(-\omega/\omega_C) - \tan^{-1}(\omega/\omega_C) = -2\tan^{-1}(\omega/\omega_C)$$

$$t_p = -\frac{\theta(\omega)}{\omega} = \frac{2}{\omega}\tan^{-1}(\omega/\omega_C) \qquad t_g = -\frac{d\theta(\omega)}{d\omega} = 2\omega_C/(\omega^2 + \omega_C^2)$$

### 9.3.1 Second-Order Filters

A second-order filter has the general form

$$H(s) = P(s)/Q(s) = K(a_0 + a_1 s + a_2 s^2)/(1 + b_1 s + b_2 s^2)$$

Stability requires that $b_1$ and $b_2$ be nonzero and positive. Various choices for the coefficients $a_k$ result in the filter types listed in Table 9.3.

---

**Table 9.3    Transfer Function of Second-Order Filters.**

$$H(s) = \frac{P(s)}{Q(s)} = K\frac{a_0 + a_1 s + a_2 s^2}{1 + b_1 s + b_2 s^2} = K\frac{K(a_0 + a_1 s + a_2 s^2)}{1 + s/\omega_p Q + (s/\omega_p)^2}$$

| Type | Constraints | $H(s)$ | $H(\omega)$ |
|---|---|---|---|
| Lowpass | $a_1 = a_2 = 0$ | $\dfrac{K}{1 + b_1 s + b_2 s^2}$ | $\dfrac{K}{1 + j(\omega/\omega_p)/Q + (j\omega/\omega_p)^2}$ |
| Highpass | $a_0 = a_1 = 0, a_2 = 1$ | $\dfrac{Ks^2}{1 + b_1 s + b_2 s^2}$ | $\dfrac{K(j\omega/\omega_p)^2}{1 + j(\omega/\omega_p)/Q + (j\omega/\omega_p)^2}$ |
| Bandpass | $a_0 = a_2 = 0, a_1 = 1$ | $\dfrac{Ks}{1 + b_1 s + b_2 s^2}$ | $\dfrac{K(j\omega/\omega_p)}{1 + j(\omega/\omega_p)/Q + (j\omega/\omega_p)^2}$ |
| Bandstop | $a_1 = 0, a_0 = 1, a_2 = b_2$ | $\dfrac{K(1 + b_2 s^2)}{1 + b_1 s + b_2 s^2}$ | $\dfrac{K[1 + (j\omega/\omega_p)^2]}{1 + j(\omega/\omega_p)/Q + (j\omega/\omega_p)^2}$ |
| Allpass | $a_k = (-1)^k b_k$ | $K\dfrac{1 - b_1 s + b_2 s^2}{1 + b_1 s + b_2 s^2}$ | $K\dfrac{1 - j(\omega/\omega_p)/Q + (j\omega/\omega_p)^2}{1 + j(\omega/\omega_p)/Q + (j\omega/\omega_p)^2}$ |

---

The denominator $Q(s)$ of the transfer function $H(s)$ is often expressed as

$$Q(s) = 1 + s/\omega_p Q + (s/\omega_p)^2$$

Here $\omega_p = \sqrt{1/b_2}$ is called the **undamped natural frequency**, or **pole frequency**, and $Q = 1/b_1\omega_p$ represents the **quality factor** and is a measure of the losses in the circuit. For highpass and lowpass circuits, $Q = K/|H(\omega_p)|$ at $\omega = \omega_p$. For bandpass filters, $Q$ is typically defined in terms of the half-power bandwidth $\Delta\omega$ as $Q = \omega_p/\Delta\omega$. In practice, $\omega_p$ and $\Delta\omega$ are evaluated from $\theta(\omega)$ and $Q$ is estimated from $|H(\omega)|$, as described in Table 9.4.

---

**Table 9.4    Parameter Measurements for Second-Order Filters.**
Note: $\omega_p$ = pole frequency, $\omega_1, \omega_2$ = half-power frequencies

| Filter | $\omega_p$ found when | $Q$ found as |
|---|---|---|
| Lowpass | $\theta(\omega_p) = -90°$ | $|H(\omega_p)|/|H(0)|$ |
| Highpass | $\theta(\omega_p) = 90°$ | $|H(\omega_p)|/|H(\omega \to \infty)|$ |
| Bandpass | $\theta(\omega_p) = 0$ | $(\omega_2 - \omega_1)/\omega_p$ |

### 9.3.2     *Relating Time and Frequency Measures*

Since a sharp magnitude response $|H(\omega)|$ can be achieved only at the expense of a poorer delay performance, the time response of filters depends on how the frequency response is optimized. Time measures are often based on the step response and are summarized in Table 9.5. These measures are applicable to a much wider class of excitations, because the impulse response, which is an indicator of system performance, is, in fact, related to the step response.

---

**Table 9.5     Step Response Measures for Real Filters.**

| Measure | Explanation |
|---|---|
| Time delay | Time between application of input and appearance of response; typical measure: Time to reach 50% of final value. |
| Rise time | Measure of the steepness of initial slope of response; typical measure: Time to rise from 10% to 90% of final value. |
| Overshoot | Deviation (if any) beyond the final value; typical measure: Peak overshoot. |
| Settling time | Time for oscillations to settle to within a specified value; typical measure: Time to settle to within 5% of final value. |
| Damping | Rate of decay toward final value; typical measure: Damping factor $\alpha$ in $\exp(-\alpha t)\cos(\beta t + \theta)$. |

---

### 9.3.3     *Second-Order Lowpass Filters*

For lowpass filters, $Q$ relates both the magnitude $|H(\omega)|$ of the frequency response and the step response, as summarized in Table 9.6.

---

**Table 9.6     Response of Second-Order Lowpass Filters.**

| Value of $Q$ | Magnitude $|H(\omega)|$ | Step response $w(t)$ |
|---|---|---|
| $Q < \frac{1}{2}$ | Monotonic | Monotonic |
| $Q > \sqrt{\frac{1}{2}}$ | Overshoot | Overshoot |
| $\frac{1}{2} < Q < \sqrt{\frac{1}{2}}$ | Monotonic | Overshoot |

---

**Frequency Response**     With $H(\omega) = 1/[1 + j(\omega/\omega_p)/Q + (j\omega/\omega_p)^2]$, the magnitude $|H(\omega)|$ is monotonic and shows no peaking if $Q < 1\sqrt{2}$ and $Q = 1\sqrt{2}$ describes the boundary between monotonicity and peaking. For $Q > 1\sqrt{2}$, $|H(\omega)|$ peaks near $\omega_p$, and the peaking increases with $Q$. The frequency $\omega_{pk}$ and peak magnitude $H_{pk}$ at

$\omega_{pk}$ are found by setting $d|H(\omega)|/d\omega$ to zero. We find that

$$H_{pk} = \frac{A_{LP}Q}{\omega_p[1 - (1/4Q^2)]^{\frac{1}{2}}} \qquad \omega_{pk} = \omega_p[1 - (1/2Q^2)]^{\frac{1}{2}} \qquad Q > 1/\sqrt{2}$$

For $Q \gg 1$, $\omega_{pk}$ and $H_{pk}$ can be approximated by

$$\omega_{pk} \approx \omega_p(Q \gg 1) \qquad H_{pk} = A_{LP}Q/\omega_p \quad (Q \gg 1)$$

The highest $Q$ we can use for a monotonic $|H(\omega)|$ is $Q = \sqrt{\frac{1}{2}}$. The transfer function $H(s)$ for such a filter may be written

$$H(s) = \frac{K}{s^2 + s(\omega_p\sqrt{2}) + \omega_p^2}$$

This describes a *second-order Butterworth lowpass filter* (see Chapter 10).

**Step Response** The step response shows either a smooth rise to the final value (**overdamped**) or overshoot and decaying oscillations about the final value (**underdamped**). The nature of the step response depends on the poles of $H(s)$, which may be written as

$$p_{1,2} = -\sigma_k \pm j\omega_k = -\omega_p/2Q \pm j\omega_p\{1 - (1/4Q^2)\}^{\frac{1}{2}}$$

For $Q > \frac{1}{2}$, the poles are complex conjugates and the response shows overshoot and oscillations (ringing) and results in a small rise time but a large settling time. The frequency of oscillations increases with $Q$.

For $Q < \frac{1}{2}$, the poles are real and distinct. The response approaches the steady-state value monotonically but with a larger rise time.

For $Q = \frac{1}{2}$, the poles are real and equal. The response is **critically damped**. It yields the fastest monotonic approach to the steady-state value with no overshoot. The value $Q = \frac{1}{2}$ represents the transition between monotonicity and peaking in the step response.

In general, for a monotonic $|H(\omega)|$ as well as a small rise time, we must tolerate some overshoot in the step response. The overshoot depends largely on how sharp the transition between the passband and stopband is. A sharper transition also means a higher filter order, resulting in larger delay times and more nonlinearity in the phase characteristics. For filters designed with flat delay in mind, the transition is made more gradual and leads to a smoother step response largely free of overshoot. This is typical of the compromises one makes in designing practical filters.

---

**Example 9.7**
**A Second-Order Lowpass Butterworth Filter**

Consider a second-order lowpass Butterworth filter with unit pole frequency and $H(s) = 1/(s^2 + s\sqrt{2} + 1)$. Rewriting $H(s)$ as $H(s) = 1/[(s + 1/\sqrt{2})^2 + (1/\sqrt{2})^2]$, we find the impulse response as $h(t) = \sqrt{2}\exp(-t/\sqrt{2})\sin(t/\sqrt{2})$.

The step response $y_s(t)$ corresponds to the inverse transform of

$$Y_s(s) = H(s)/s = 1/[s(s^2 + s\sqrt{2} + 1)]$$

Using partial fractions, we find

$$X_s(s) = \frac{1}{s} + \frac{-\frac{1}{2} - j\frac{1}{2}}{s + 1/\sqrt{2} - j/\sqrt{2}} + \frac{-\frac{1}{2} + j\frac{1}{2}}{s + 1/\sqrt{2} + j/\sqrt{2}}$$

$$y_s(t) = [1 + \sqrt{2}\exp(-t/\sqrt{2})\cos(t/\sqrt{2} + 135°)]u(t)$$

The magnitude $|H(\omega)|$ and step response are plotted in Figure 9.9. While $|H(\omega)|$ is monotonic, the step response is underdamped and shows overshoot.

**Figure 9.9    Frequency and step response for the filter of Example 9.7.**

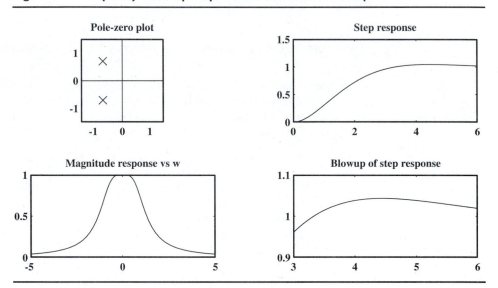

**Stability**    The poles of a second-order filter may be described as

$$p_k = -\sigma_k \pm j\omega_k = |p_k|\angle\theta_k \qquad |p_k| = [\sigma_k^2 + \omega_k^2]^{1/2} = \omega_p \qquad \theta_k = \pm\tan^{-1}(\omega_k/\sigma_k)$$

The magnitude of the poles always equals $\omega_p$. The poles thus lie on a circle of radius $\omega_p$ in the $s$-plane. Their orientation $\theta_k$ is related to $Q$. For $Q > \frac{1}{2}$, the poles are complex conjugates. The *degree* of stability depends on the pole locations. The larger the $Q$, the closer the poles move to the $j\omega$-axis, and the less stable the system. From a practical standpoint, a high $Q$ makes the circuit more sensitive to component variations. In the extreme case, $Q = \infty$ corresponds to poles on the $j\omega$-axis, and describes an ideal lossless resonant system.

The denominator of $H(s)$ may also be expressed in terms of its roots as

$$s^2 + s(\omega_p/Q) + \omega_p^2 = (s - p_1)(s - p_2) = (s + \sigma_k - j\omega_k)(s + \sigma_k + j\omega_k)$$

leading to the result

$$s^2 + s(\omega_p/Q) + \omega_p^2 = s^2 + 2\sigma_k s + \sigma_k^2 + \omega_k^2 \quad \text{with} \quad \omega_p^2 = \sigma_k^2 + \omega_k^2$$

The $Q$ may also be expressed in terms of $\sigma_k$ and $\omega_k$ as

$$Q = \tfrac{1}{2}[1 + (\omega_k^2/\sigma_k^2)]^{1/2} \qquad Q \approx \tfrac{1}{2}(\omega_k/\sigma_k) \quad \text{for} \quad Q \gg 1$$

**Phase Response**  The phase and delay of second-order filters depends on the $Q$ and the type of filter considered. The phase shows nonlinearity in the passband, especially toward the band edges.

For lowpass filters, the phase equals

$$\theta_{\text{LP}}(\omega) = -\tan^{-1}\left|\frac{(\omega/\omega_p)/Q}{1 - (\omega/\omega_p)^2}\right|$$

The phase varies from 0 at $\omega/\omega_p = 0$ to $-\pi$ as $\omega/\omega \to \infty$, and depends on $Q$. For $Q < 1$, the phase is monotonic, but for large $Q$, an inflection occurs at $\omega/\omega_p \approx 1$.

For an allpass filter, the phase and group delay equal

$$\theta_{\text{AP}}(\omega) = -2\tan^{-1}\left|\frac{(\omega/\omega_p)/Q}{1 - (\omega/\omega_p)^2}\right|$$

$$t_g(\omega) = -\frac{d\theta_{\text{AP}}(\omega)}{d\omega} = \frac{2(\omega_p/Q)(\omega_p^2 + \omega^2)}{(\omega_p^2 - \omega^2)^2 + (\omega_p^2/Q^2)\omega^2}$$

Allpass filters are often used in cascade with other systems to linearize their phase. They are designed to provide the required *phase lag* that makes the overall phase as linear as possible. Naturally, this additional phase lag results in an *increased system delay*.

### 9.3.4  *Some Examples*

The characteristics of some common passive filters are summarized in Table 9.7 (p. 394) and Figure 9.10 (pp. 392–393). The design of some of these is studied in Chapter 10. Here, we concentrate on their analysis by considering a few examples.

---

**Example 9.8**
A Linear Phase Bessel Filter

---

The second-order Bessel filter of Figure 9.10 yields $H(s)$ and $H(\omega)$ as

$$H(s) = 3/(s^2 + 3s + 3) \qquad H(\omega) = 3/[(3 - \omega^2) + j3\omega]$$

Comparison of the denominator with $s^2 + s(\omega_p/Q) + \omega_p^2$ suggests that $\omega_p = \sqrt{3}$ and $Q = 1/\sqrt{3}$. Since $Q < 1/\sqrt{2}$ we should expect an underdamped step response. We find the half-power frequency by setting $|H(\omega)| = \sqrt{\tfrac{1}{2}}$ or $|H(\omega)|^2 = \tfrac{1}{2}$ to get

$$9/[(3 - \omega^2)^2 + 9\omega^2] = \tfrac{1}{2} \qquad \text{or} \qquad \omega^4 + 3\omega^2 - 9 = 0$$

The solution for positive $\omega$ yields $\omega^2 = -\tfrac{3}{2} + \tfrac{1}{2}\sqrt{90} = 1.8541$ and $\omega = 1.3617$. The phase of $H(\omega)$ is given by $\theta(\omega) = -\tan^{-1}[3\omega/(3 - \omega^2)]$. The group delay is

$$t_g = -\frac{d\theta(\omega)}{d\omega} = \frac{3\omega^2 + 9\omega^4 + 3}{\omega^2 + 9}$$

The magnitude, phase, group delay and step response of this filter are sketched in Figure 9.10. Note the monotonic magnitude and step response, and an almost linear phase (or constant delay) in the passband.

---

**Figure 9.10    The analog filters of Table 9.7 and their responses.**

---

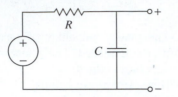

**a. First-order lowpass filter**

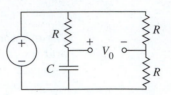

**b. First-order allpass filter**

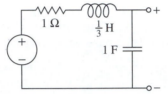

**c. Second-order lowpass (Bessel) filter**

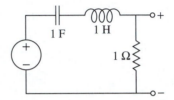

**d. Second-order bandpass filter**

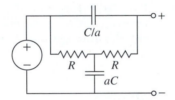

**e. Second-order bridged-*T*
bandstop filter**

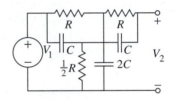

**f. Second-order twin-*T*
bandstop filter**

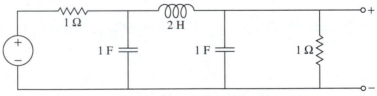

**g. Third-order lowpass (Butterworth) filter**

**Figure 9.10** **(continued)**

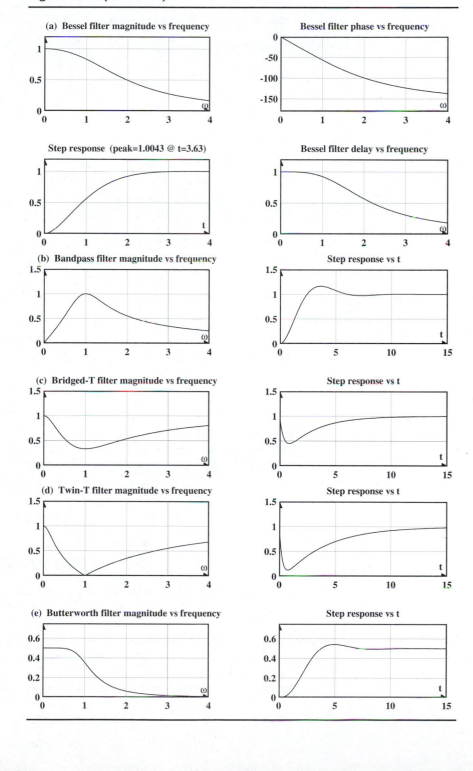

(a) Bessel filter magnitude vs frequency

Bessel filter phase vs frequency

Step response  (peak=1.0043 @ t=3.63)

Bessel filter delay vs frequency

(b) Bandpass filter magnitude vs frequency

Step response vs t

(c) Bridged-T filter magnitude vs frequency

Step response vs t

(d) Twin-T filter magnitude vs frequency

Step response vs t

(e) Butterworth filter magnitude vs frequency

Step response vs t

**Table 9.7   Examples of Passive Filters.**

| Order | Configuration | Transfer function | Frequency characteristics |
|---|---|---|---|
| First | Lowpass | $\dfrac{1}{1 + sRC}$ | Monotonic magnitude |
| First | Allpass | $0.5\dfrac{1 - sRC}{1 + sRC}$ | Constant magnitude |
| Second | Lowpass | $\dfrac{3}{s^2 + 3s + 3}$ | Constant delay (Bessel filter) |
| Second | Bandpass | $\dfrac{s}{s^2 + s + 1}$ | Peaked magnitude |
| Second | Bandstop | $\dfrac{s^2 + 2s/aRC + (1/RC)^2}{s^2 + (2 + a^2)s/aRC + (1/RC)^2}$ | Bridged-$T$ filter |
| Second | Bandstop | $\dfrac{s^2 + (1/RC)^2}{s^2 + 4s/RC + (1/RC)^2}$ | Twin-$T$ filter |
| Third | Lowpass | $\dfrac{0.5}{(s + 1)(s^2 + s + 1)}$ | Monotonic magnitude (Butterworth filter) |

---

**Example 9.9**
A Twin-$T$ Bandstop or Notch Filter

For the twin-$T$ bandstop filter of Figure 9.10, we find

$$H(s) = \frac{V_2(s)}{V_1(s)} = \frac{s^2 + (1/RC)^2}{s^2 + 4s/RC + (1/RC)^2}$$

The frequency response $H(\omega)$ is found by the substitution $s \longrightarrow j\omega$ to give

$$H(\omega) = \frac{1/R^2C^2 - \omega^2}{-\omega^2 + 1/R^2C^2 + j4\omega/RC}$$

Note that $H(0) = 1 = H(\infty)$. Also, $H(\omega) = 0$ when $\omega = 1/RC$. These results confirm the bandstop nature of this filter.

The frequency $\omega_0 = 1/RC$ is called the **notch frequency** or **null frequency**.

The stopband is the region between the half-power frequencies $\omega_1$ and $\omega_2$. At the half-power frequencies, $|H(\omega)| = \sqrt{\tfrac{1}{2}}$ or $|H(\omega)|^2 = \tfrac{1}{2}$ and we have

$$|H(\omega)|^2 = \frac{(1/R^2C^2 - \omega^2)^2}{[(1/R^2C^2 - \omega^2)^2 + 16\omega^2/R^2C^2]} = \frac{1}{2}$$

Rearranging and simplifying this equation, we obtain

$$(1/R^2C^2 - \omega^2)^2 = 16\omega^2/R^2C^2 \quad \text{or} \quad (1/R^2C^2 - \omega^2) = \pm 4\omega/RC$$

The solution of the two quadratics yields the two positive frequencies

$$\omega_1 = \omega_0(\sqrt{5} - 2) \qquad \omega_2 = \omega_0(\sqrt{5} + 2)$$

The width of the stopband thus equals $\Delta\omega = \omega_2 - \omega_1 = 4\omega_0$.

The phase of $H(\omega)$ equals $0°$ at $\omega = 0$, $\omega = \omega_0$ and $\omega = \infty$. The phase equals $-45°$ at $\omega_1$ and $+45°$ at $\omega_2$.

The magnitude and phase plots of this filter are sketched in Figure 9.10.

**Example 9.10**
A Bandpass Filter

For the bandpass filter of Figure 9.10, we find $H(s)$ and $H(\omega)$ as

$$H(s) = \frac{s}{s^2 + s + 1} \qquad H(\omega) = \frac{j\omega}{(1 - \omega^2) + j\omega}$$

Comparison of the denominator with $s^2 + s(\omega_p/Q) + \omega_p^2$ suggests that $\omega_p = 1$ and $Q = 1$. Since $Q > 1/\sqrt{2}$, we should expect an oscillatory step response.

The step response of this filter is given by

$$W(s) = H(s)/s = \frac{1}{s^2 + s + 1} = \frac{1}{(s + \frac{1}{2})^2 + (\frac{1}{2}\sqrt{3})^2}$$

Inverse transformation yields $w(t) = (2/\sqrt{3}) \exp(-\frac{1}{2}t) \sin(\frac{1}{2}\sqrt{3}t)$.

This damped sinusoid nearly dies out in about 10 s (5 time constants), or less than three half-cycles. The frequency response $H(\omega)$ and step response of this filter are plotted in Figure 9.10.

## 9.4
## Scaling and Frequency Transformations

The design of filters based on normalized frequencies uses frequency scaling to establish the actual transfer function. **Impedance scaling** is used to achieve better impedance matching, or to change design values to practically available values without changing the filter transfer function. These scaling operations are summarized in Table 9.8.

**Table 9.8   Frequency and Impedance Scaling.**
Note: Frequency scaling is from the frequency $\omega_0$ to the frequency $\omega_n$.

| Scaling | $\omega_x =$ | $R \longrightarrow$ | $L \longrightarrow$ | $C \longrightarrow$ | $H(s) \longrightarrow$ |
|---|---|---|---|---|---|
| Impedance | | $K_z R$ | $K_z L$ | $C/K_z$ | $H(s)$ |
| Frequency | $\omega_n/\omega_0$ | $R$ | $L\omega_x$ | $C/\omega_x$ | $H(s/\omega_x)$ |
| Combined | $\omega_n/\omega_0$ | $K_z R$ | $K_z L/\omega_x$ | $C/(K_z\omega_x)$ | $H(s/\omega_x)$ |

### 9.4.1   *Impedance Scaling*

If $H(s)$ is the ratio of voltages or currents, an impedance scaling by $K_z$ scales the impedances $R, sL$ and $1/sC$ to $K_z R, K_z sL$ and $K_z/sC$, respectively. Thus resistor and inductor values are both scaled by $K_z$. The capacitor values are scaled by $1/K_z$. The transfer function $H(s)$ remains unchanged.

### 9.4.2   *Frequency Scaling*

**Frequency scaling** from a frequency $\omega_0$ to a new frequency $\omega_n$ involves the transformation $\omega \longrightarrow \omega/\omega_x$, where $\omega_x = \omega_n/\omega_0$. The impedances $R, j\omega L$ and $1/j\omega C$ transform to $R, j(\omega/\omega_x)L$ and $1/[j(\omega/\omega_x)C]$, and result in the scaled element values $R, L/\omega_x$ and $C/\omega_x$. The transfer function $H(s)$ is scaled to $H(s/\omega_x)$.

Frequency and impedance scaling may also be used in combination, as listed in Table 9.8.

### 9.4.3 *Prototype Transformations*

Most strategies for the design of filters concentrate on a **lowpass prototype** using a **normalized frequency** $\nu$. Frequency transformations are required to convert to the lowpass prototype, and back to the required filter type. The transformations from the lowpass prototype are summarized in Table 9.9.

**Table 9.9    Transformations from Lowpass Prototypes.**

| Transformation | Rule | Comments |
|---|---|---|
| LP2LP | $s \longrightarrow s/\omega_x$ | $\omega_x = \omega_{new}/\omega_{old}$ |
| LP2HP | $s \longrightarrow \omega_x/s$ | $\omega_x = \omega_{new}/\omega_{old}$ |
| LP2BP | $s \longrightarrow \dfrac{s^2 + \omega_x^2}{sB_w}$ | $\omega_x = $ center frequency <br> $B_w = \omega_2 - \omega_1 = $ bandwidth, $\omega_1\omega_2 = \omega_x^2$ |
| LP2BS | $s \longrightarrow \dfrac{sB_w}{s^2 + \omega_x^2}$ | $\omega_x = $ center frequency <br> $B_w = \omega_2 - \omega_1 = $ bandwidth, $\omega_1\omega_2 = \omega_x^2$ |

### 9.4.4    *The Lowpass to Lowpass (LP2LP) Transformation*

The lowpass-to-lowpass (LP2LP) transformation $s \longrightarrow s/\omega_x$ is just frequency scaling by another name. It is a *linear transformation* since every frequency is scaled by the same factor $\omega_x$. Using this scaling, the quadratic factor $s^2 + As + B$ transforms to $(s/\omega_x)^2 + A(s/\omega_x) + B$.

### 9.4.5    *The Lowpass to Highpass (LP2HP) Transformation*

The lowpass-to-highpass (LP2HP) transformation $s \longrightarrow 1/s$ is the reciprocal of the lowpass-to-lowpass transformation. It is *nonlinear* but invariant to the frequency $\omega = 1$. If the unit cutoff frequency is to transform to $\omega_x$, we must also use the LP2LP scaling $s \longrightarrow s/\omega_x$. The general LP2HP transformation is thus

$$s \longrightarrow 1/(s/\omega_x) = \omega_x/s.$$

The second-order lowpass transfer function $H_{LP}(s)$ with $\omega_p = 1$ is given by

$$H_{LP}(s) = \frac{1}{1 + s/Q_p + s^2}$$

The transformation $s \longrightarrow \omega_x/s$, results in the highpass form

$$H_{HP}(s) = H_{LP}(\omega_x/s) = \frac{1}{1 + \omega_x/(sQ_p) + (\omega_x/s)^2} = \frac{s^2}{s^2 + s\omega_x/Q_p + \omega_x^2}$$

Its pole frequency is $\omega_p = \omega_x$, and its $Q$ equals that of $H_{LP}(s)$.

### 9.4.6 The Lowpass to Bandpass (LP2BP) Transformation

Since a bandpass characteristic is essentially a combination of lowpass and a high-pass behavior, a lowpass-to-bandpass (LP2BP) transformation has the generic form $s \longrightarrow s/\alpha + \beta/s$. The general transformation is governed by

$$s \longrightarrow \frac{\omega_x}{B_w}\left[\frac{s}{\omega_x} + \frac{\omega_x}{s}\right] = \frac{s^2 + \omega_x^2}{sB_w} \qquad \omega_{LP} \longrightarrow \frac{\omega_{BP}^2 - \omega_x^2}{\omega_{BP}B_w}$$

Here, $B_w$ is the bandwidth $\omega_H - \omega_L$, and $\omega_x$ is the *geometric* center frequency with $\omega_H \omega_L = \omega_x^2$. This *nonlinear* transformation maps the lowpass origin into the band-pass frequency $\omega_x$, and the unit lowpass frequency to two bandpass frequencies $\omega_L$ and $\omega_H$ with $B_w = \omega_H - \omega_L$. Any pair of geometrically symmetric bandpass frequencies $\omega_a$ and $\omega_b$, with $\omega_b > \omega_x > \omega_a$ and $\omega_a \omega_b = \omega_x^2$, corresponds to the *normalized* lowpass frequency $(\omega_b - \omega_a)/B_w$. The lowpass frequency at infinity is mapped to the bandpass origin.

For a first-order lowpass transfer function $H_{LP}(s) = 1/(s+1)$, the LP2BP trans-formation results in (Figure 9.11, p. 398)

$$H_{BP}(s) = \frac{1}{(\omega_x/B_w)[s/\omega_x + \omega_x/s] + 1} = \frac{sB_w}{s^2 + sB_w + \omega_x^2}$$

This has the required bandpass form with

$$\omega_x^2 = \omega_L \omega_H \qquad B_w = \omega_H - \omega_L = \frac{\omega_x}{Q}$$

In general, each lowpass pole transforms into a pair of bandpass poles, and yields a transfer function with *twice* the order of the lowpass filter. Each quadratic factor also transforms to two second-order terms with identical $Q$ (but larger than the original) but different pole frequencies (different from each other and the original). The poles must, in general, be found using numerical methods. Even though the LP2BP transformation may not yield the simplest circuit, it offers a simpler alternative to designing the bandpass filter directly.

### 9.4.7 The Lowpass to Bandstop (LP2BS) Transformation

A bandstop filter may be viewed as a bandpass filter with its passband and stop-band reversed. The lowpass-to-bandstop (LP2BS) transformation shows exactly this reversal and is given by

$$s \longrightarrow \frac{sB_w}{s^2 + \omega_x^2} \qquad \omega_{LP} \longrightarrow \frac{\omega_{BS}B_w}{\omega_x^2 - \omega_{BS}^2}$$

Here $B_w = \omega_H - \omega_L$ and $\omega_x^2 = \omega_H \omega_L$. The lowpass origin maps to the bandstop frequency $\omega_x$, and the unit lowpass frequency maps to the pair of bandstop fre-quencies $\omega_a$ and $\omega_b$. Since the roles of the passband and the stopband are now reversed, a pair of geometrically symmetric bandstop frequencies $\omega_a$ and $\omega_b$, with $\omega_b > \omega_x > \omega_a$ and $\omega_a \omega_b = \omega_x^2$, translates to the *normalized* lowpass frequency $\omega_{LP} = B_w/(\omega_b - \omega_a)$.

**Figure 9.11    Illustrating the LP2BP transformation.**

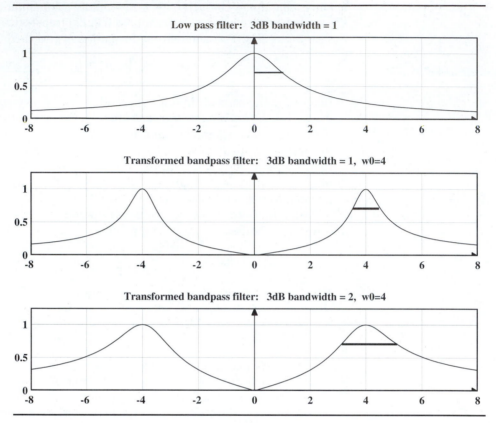

Low pass filter:   3dB bandwidth = 1

Transformed bandpass filter:   3dB bandwidth = 1,  w0=4

Transformed bandpass filter:   3dB bandwidth = 2,  w0=4

**Remark:**   For highpass and bandstop forms, we could also use the seemingly simpler transformations $H_{\text{HP}}(f) = 1 - H_{\text{LP}}(f)$ and $H_{\text{BS}}(f) = 1 - H_{\text{BP}}(f)$. These are not suitable when the passband or stopband contains ripples (instead of being monotonic), because the ripples are relocated from the passband to stopband and vice versa. These forms are seldom used in practice.

### 9.4.8    *Developing the Lowpass Prototype*

Filter design also requires transformations to convert the given frequency specifications to those that apply to a lowpass prototype (LPP). For lowpass filters, we normalize specifications with respect to the passband edge. For highpass filters, we use $\omega \longrightarrow 1/\omega$ followed by normalization.

A bandpass-to-lowpass or bandstop-to-lowpass transformation maps the center frequency to the lowpass origin. It also maps geometrically symmetric frequency pairs to single lowpass frequencies. If the band-edge frequencies are given in increasing order as $[\omega_1, \omega_2, \omega_3, \omega_4]$, *geometric symmetry* implies that $\omega_2 \omega_3 = \omega_1 \omega_4 = \omega_x^2$. It also implies that the **transition widths** $\omega_2 - \omega_1$ and $\omega_4 - \omega_3$ are not equal. For bandpass filters, we find $B_w = \omega_3 - \omega_2, \omega_p = B_w$, and $\omega_s = \omega_4 - \omega_1$. For bandstop filters, on the other hand, we find $B_w = \omega_4 - \omega_1, \omega_p = 1/(\omega_3 - \omega_2)$, and $\omega_s = 1/B_w$. In either case, normalization by $\omega_p$ yields the LPP specifications $\nu_p = 1$, and $\nu_s = \omega_s/\omega_p$. Subsequent LP2BP or LP2BS transformation of the LPP uses the computed values of $\omega_x$ and $B_w$.

If the given specifications $[\omega_1, \omega_2, \omega_3, \omega_4]$ are not geometrically symmetric, the band-edges (or center frequency) must be relocated to induce geometric symmetry, but in such a way that the new transition widths do not exceed the specified values. Table 9.10 formalizes the selection of band-edges for various choices. The first entry in this table is based on fixed passband edges, the second on fixed stopband edges, the third on a fixed center frequency, and the last on a compromise. *The fixed passband choice is very common.*

**Table 9.10   Transformation to Lowpass Prototype for BP and BS Filters.**

$[\omega_1\ \omega_2\ \omega_3\ \omega_4]$ = band-edge frequencies with $\omega_1 < \omega_2 < \omega_3 < \omega_4$ For bandpass filters, the passband edges are $\omega_2$ and $\omega_3$. For bandstop filters, the passband edges are $\omega_1$ and $\omega_4$.

| Requirement | Center frequency | Choices for band-edge frequencies |
|---|---|---|
| Fixed $\omega_2, \omega_3$ | $\omega_x^2 = \omega_2\omega_3$ | If $\omega_1\omega_4 < \omega_x^2$, $\omega_1 = \omega_x^2/\omega_4$, else $\omega_4 = \omega_x^2/\omega_1$ |
| Fixed $\omega_1, \omega_4$ | $\omega_x^2 = \omega_1\omega_4$ | If $\omega_2\omega_3 < \omega_x^2$, $\omega_3 = \omega_x^2/\omega_2$, else $\omega_2 = \omega_x^2/\omega_3$ |
| Fixed $\omega_x$ | $\omega_x$ | If $\omega_2\omega_3 < \omega_x^2$, $\omega_3 = \omega_x^2/\omega_2$, else $\omega_2 = \omega_x^2/\omega_3$ |
| Fixed $\omega_x$ | $\omega_x$ | If $\omega_1\omega_4 < \omega_x^2$, $\omega_1 = \omega_x^2/\omega_4$, else $\omega_4 = \omega_x^2/\omega_1$ |
| Compromise | $\omega_x^2 = (\omega_1\omega_2\omega_3\omega_4)^{1/2}$ | If $\omega_1\omega_4 > \omega_2\omega_3$, $\omega_3 = \omega_x^2/\omega_2$ and $\omega_4 = \omega_x^2/\omega_1$, else $\omega_2 = \omega_x^2/\omega_3$ and $\omega_1 = \omega_x^2/\omega_4$ |

**Example 9.11**
Developing Prototype
Specifications

(a) For a lowpass filter with $\omega_p = 200$ and $\omega_s = 500$, we normalize by $\omega_p$ to get the LPP with $\nu_p = 1$ and $\nu_s = 2.5$. The actual filter is designed from the LPP using the LP2LP transformation $s \to s/\omega_x$ with $\omega_x = 200$.

(b) For a highpass filter with $\omega_p = 500$ and $\omega_s = 200$, we obtain the reciprocals $[1/500, 1/200]$ and normalize by $1/500$ to get the LPP $\nu_p = 1$ and $\nu_s = 2.5$. The highpass filter is designed from the LPP using the LP2HP transformation $s \to \omega_x/s$ with $\omega_x = 500$.

(c) Consider a bandpass filter with $\omega = [\omega_1, \omega_2, \omega_3, \omega_4] = [16, 18, 32, 48]$. For a fixed passband $B_w = \omega_3 - \omega_2$, we choose $\omega_x^2 = \omega_2\omega_3 = 576$. From Table 9.10, since $\omega_1\omega_4 > \omega_x^2$, we recompute $\omega_4 = \omega_x^2/\omega_1 = 36$, which ensures both geometric symmetry, and a smaller transition width (as compared to changing $\omega_1$). We thus have $\omega = [16, 18, 32, 36]$, $B_w = \omega_3 - \omega_2 = 32 - 18 = 14$ and stopband $= \omega_4 - \omega_1 = 36 - 16 = 20$. The LPP specifications become

$$\nu_p = 1, \nu_s = (\omega_4 - \omega_1)/B_w = 20/14 = 1.4286$$

The bandpass filter is designed from the LPP using the LP2BP transformation

$$s \to (s^2 + \omega_x^2)/sB_w = (s^2 + 576)/14s$$

(d) Consider a bandstop filter with $\omega = [\omega_1, \omega_2, \omega_3, \omega_4] = [16, 18, 32, 48]$. For a fixed passband $B_w = \omega_4 - \omega_1$, we compute $\omega_x^2 = (16)(48) = 768$. From Table 9.10, since $\omega_2\omega_3 < \omega_x^2$, we recompute $\omega_3 = \omega_x^2\omega_2 = 42.667$ to give $\omega = [16, 18, 42.6667, 48], B_w = \omega_4 - \omega_1 = 48 - 16 = 32, \quad \omega_3 - \omega_2 = 24.6667$. The LPP specifications become

$$\nu_p = 1, \nu_s = B_w/(\omega_3 - \omega_2) = 32/24.6667 = 1.2973$$

The bandstop filter is designed from the LPP using the LP2BS transformation $s \to sB_w/(s^2 + \omega_x^2) = 32s/(s^2 + 768)$.

---

## 9.5
## Application of Laplace Transforms: A Synopsis

**Frequency response**

For LTI systems, the frequency response is a measure of the magnitude and phase of $H(\omega)$ versus frequency.

**Bode plots**

These plots use a logarithmic scale for magnitude and frequency but not phase. For a rational $H(s)$, we can generate linear approximations to the frequency response. These are a fair approximation for magnitude but less so for phase, especially for quadratic factors with complex roots.

**Real filters**

Real filters are described by their magnitude, phase and delay. The phase delay is the phase normalized by frequency. The group delay is the negative slope of the phase. For rational transfer functions, the phase is transcendental but the group delay is rational. A stable minimum-phase transfer function can be found from its magnitude spectrum and has the smallest phase delay. All its poles and zeros lie in the LHP.

**Filter design**

The ratio of vectors directed from each zero and pole to a given $s$ provides magnitude information which describes a surface in the $s$-plane. If $s = j\omega$, we obtain the frequency response. An intuitive approach to filter design is based on pole-zero placement. Complex poles and zeros must have conjugates and lie in the LHP for stability and minimum phase. Zeros on the $j\omega$-axis result in a null in the response. Filter design using pole-zero placement involves trial and error.

**First-order filters**

These are described by their half-power frequency, which is proportional to the reciprocal of the time constant.

**Second-order filters**

These are described in terms of $Q$, a measure of the losses in a circuit, which relates both the magnitude $|H(\omega)|$ and the step response. The magnitude $|H(\omega)|$ is monotonic, with no peaking if $Q < 1\sqrt{2}$. The step response is underdamped for $Q > \frac{1}{2}$ and results in a small rise time but large settling time. The larger the $Q$, the less stable the filter. The phase and delay is nonlinear in the passband, especially near the band-edge.

In general, for a monotonic $|H(\omega)|$ and small rise time, we must tolerate some overshoot in the step response.

Transformations    Impedance scaling is used for impedance matching. Most filter design is based on lowpass prototypes and frequency transformations.

The LP2LP transformation $s \longrightarrow s/\omega_x$ is linear. The LP2HP transformation $s \longrightarrow \omega_x/s$ is nonlinear. The LP2BP transformation $s \longrightarrow (s^2 + \omega_x^2)/Bs$ involves the geometric center frequency $\omega_x$ and the bandwidth $B$. It maps a lowpass frequency to two geometrically symmetric frequencies. Its reciprocal is the LP2BS transformation.

Prototype specifications    Transformations to convert the given frequency specifications to a lowpass prototype rely on normalization by the passband edge for HP and LP filters, or on ensuring geometrical symmetry for BP and BS filters.

## Appendix 9A
## MATLAB
## Demonstrations
## and Routines

### Getting Started on the Supplied Matlab Routines

This appendix presents MATLAB routines related to concepts in this chapter and some examples of their use. See Appendix 1A for introductory remarks.

**Note:**   All polynomial arrays must be entered in descending powers.

### Asymptotic Bode Plots

```
bodelin(n,d,ty)
```

Plots asymptotic and actual Bode plots for $H(s) = N(s)/D(s)$. Use $ty = 0$ for magnitude only, $ty = 1$ for phase only, $ty >= 2$ for both.

Example

For a Bode plot of $H(s) = (s + 1)/(s^2 + 5s + 100)$, try

```
>>bodelin([1 1],[1 5 100],3)
```

Minimum-Phase Transfer Function

```
y=tfmin(n,d)
```

Returns the minimum-phase transfer function from $|H(\omega)|^2 = N(\omega)/D(\omega)$.

Example

To find the minimum phase $H(s)$ from $|H(\omega)|^2 = (4\omega^2 + 16)/(\omega^6 + 1)$, try

```
y=tfmin([4 0 16],[1 0 0 0 0 0 1])
```

MATLAB returns $y = [0\ 0\ 2\ 4;\ 1\ 2\ 2\ 1]$, meaning $H(s) = (2s + 4)/(s^3 + 2s^2 + 2s + 1)$.

Lowpass Prototypes and Frequencey Transformations

```
[w,w0,bw]=conv2lpp(ty,pb,sb)
```

Converts ty = 'lp','hp','bp' or 'bs' specs to normalized LPP specs $w = [wp, ws]$. Also returns center frequency ($w0$), and bandwidth (bw).

```
[n,d]=lp2af(ty,p,q,w0,bw)
```

Converts LP filter $H_1(s) = P(s)/Q(s)$ to filter types ty='lp', 'hp', 'bp' or 'bs' with scaling frequency w0 and (for bp and bs) bandwidth bw.

Example

To transform a BS filter with pb=[10 40]rad/s and sb=[20 30]rad/s, use

```
[w,w0,B]=conv2lpp('bp',[10 40],[20 30])
```

MATLAB returns

```
w=[1.0000 1.8000], w0=20, B=30
```

To transform the designed LP, say $H(s) = 4/(s + 4)$, to the BS filter, use

```
>>[n,d]=lp2af('bs',[0 4],[1 4],20,30)
```

MATLAB will return

```
n = [1 0 400], d=[1 7.5 400]
```

Thus, $H_{BS}(s) = (s^2 + 400)/(s^2 + 7.5s + 400)$. To plot its magnitude, use

```
tfplot(n,d,[0 8],0,1)
```

Note that f = 8 Hz corresponds nearly to $w \approx 50$ rad/s.

Demonstrations and Utility Functions

```
splane3d(sig,w,n,d)
```

Plots a 3D complex frequency plot of $X(s) = N(s)/D(s)$. The range of σ and ω is specified as

```
sig=[siglo sighi] and w=[wlo whi]
```

PROBLEMS

P9.1 (Bode Plots) Sketch the Bode magnitude plot for each $H(s)$ and the Bode phase plot for the $H(s)$ of parts (a)-(c):

(a) $H(\omega) = \dfrac{10 + j\omega}{(1 + j\omega)(100 + j\omega)}$

(b) $H(\omega) = \dfrac{10j\omega(10 + j\omega)}{(1 + j\omega)(100 + j\omega)}$

(c) $H(\omega) = \dfrac{100 + j\omega}{j\omega(10 + j\omega)^2}$

(d) $H(\omega) = \dfrac{1 + j\omega}{(10 + j\omega)(5 + j\omega)}$

(e) $H(\omega) = \dfrac{10j\omega(10 + j\omega)}{(1 + j\omega)(5 + j\omega)^2}$

(f) $H(\omega) = \dfrac{10(1 + j\omega)}{j\omega(100 + 10j\omega - \omega^2)}$

P9.2 (Bode Plots) The transfer function $H(\omega)$ of a system is given by

$$H(\omega) = \frac{10 + 10j\omega}{(10 + j\omega)(2 + j\omega)}$$

Compute the following in decibels (dB) at $\omega = 1$, 2 and 10 rad/s:
(a) The asymptotic magnitude.
(b) The corrected magnitude.
(c) The exact magnitude.

P9.3 (Bode Plots) The transfer function $H(\omega)$ of a system is given by

$$H(\omega) = \frac{1 + j\omega}{(0.1 + j\omega)^2}$$

(a) Find the asymptotic phase in degrees at $\omega = 0.01$, 0.1, 1 and 10 rad/s.
(b) What is the exact phase in degrees at $\omega = 0.01$, 0.1, 1 and 10 rad/s?

P9.4 (Minimum-Phase Systems) Find the stable minimum phase transfer function from the following $|H(\omega)|^2$ and classify each filter by type. Also find a stable transfer function that is not minimum phase, if possible.

(a) $\dfrac{\omega^2}{\omega^2 + 4}$

(b) $\dfrac{4\omega^2}{\omega^4 + 17\omega^2 + 16}$

(c) $\dfrac{\omega^4 + 8\omega^2 + 16}{\omega^4 + 17\omega^2 + 16}$

P9.5 (Minimum-Phase Systems) Find the minimum-phase transfer function corresponding to each Bode magnitude plot sketched in Figure P9.5.

P9.6 (Phase Delay and Delay) Find the phase and group delay for each of the following filters:
(a) $H(s) = (s - 1)/(s + 1)$
(b) $H(s) = 1/(s^2 + \sqrt{2}s + 1)$
(c) $H(s) = 3/(s^2 + 3s + 3)$

P9.7 (Allpass Filters) Using the ideas of pole-zero placement, argue that for a stable all-pass filter, the location of each LHP pole must be matched by a RHP zero. Does an allpass filter represent a minimum-phase, mixed-phase or maximum-phase system?

Figure P9.5

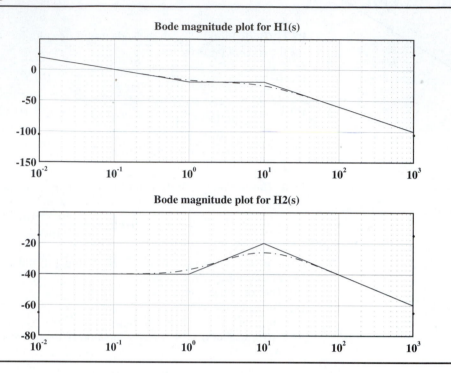

Bode magnitude plot for H1(s)

Bode magnitude plot for H2(s)

P9.8 (Allpass Filters)

(a) Argue that the group delay t_g of a stable allpass filter is always greater than or equal to zero.

(b) The overall delay of cascaded systems is simply the sum of the individual delays. Argue that a minimum-phase system has the smallest group delay from among all filters with the same magnitude by cascading it with an allpass filter.

P9.9 (Filter Design)

To design a second-order notch filter with a notch frequency of ω_0, we select conjugate zeros at $s = \pm j\omega_0$. The poles are then placed close to the zeros but to the left of the $j\omega$-axis to ensure a stable filter. We thus locate the poles at $s = -\alpha\omega_0 \pm j\omega_0$, where $\alpha \ll 1$ determines the sharpness of the notch.

(a) Find $H(s)$ for a notch filter with $\omega_0 = 10$ rad/s and α as a variable.

(b) Find the bandwidth of the notch filter for $\alpha = 0.1$.

P9.10 (Filter Design)

Let us design a bandpass filter with lower and upper cutoff ω_1 and ω_2 by locating conjugate poles at $s = -\alpha\omega_1 \pm j\omega_1$ and $s = -\alpha\omega_2 \pm j\omega_2$ (where $\alpha < 1$) and a pair of zeros at $s = 0$.

(a) Find $H(s)$ for a bandpass filter with $\omega_1 = 40$ rad/s, $\omega_2 = 50$ rad/s and $\alpha = 0.1$.

(b) What is the filter order?

(c) Estimate the center frequency and bandwidth.

(d) What is the effect of changing α on the frequency response?

P9.11 **(Filter Analysis)** Find the impulse response and step response of each of the filters listed in Table 9.7.

P9.12 **(Scaling)** For the lowpass RC filter (Circuit #1) of Table 9.7, use impedance and frequency scaling to transform it to $\omega_c = 1000$ rad/s by replacing the capacitor of 1 F by a capacitor of 1 μF.

P9.13 **(Prototype Transformations)** Let $H(s) = 1/(s^2 + s + 1)$. Assume a passband of unity and use frequency transformations to find the transfer function of:
(a) A bandpass filter with passband = 1 rad/s and center frequency = 1 rad/s
(b) A bandpass filter with passband = 10 rad/s and center frequency = 100 rad/s
(c) A bandstop filter with stopband = 1 rad/s and center frequency = 1 rad/s
(d) A bandstop filter with stopband = 2 rad/s and center frequency = 10 rad/s

P9.14 **(Lowpass Prototype)** For each set of filter specifications, find the normalized lowpass prototype frequencies. Assume a fixed passband, where necessary.
(a) Lowpass: Passband < 2 kHz, Stopband > 3 kHz
(b) Highpass: Passband > 3 kHz, Stopband < 2 kHz
(c) Bandpass: Passband 10-15 kHz, Stopband < 5 kHz and > 20 kHz
(d) Bandstop: Passband < 20 kHz and > 40 kHz, Stopband 30-34 kHz

P9.15 **(Lowpass Prototype)** For each set of bandpass filter specifications, find the normalized lowpass prototype frequencies. Assume a fixed center frequency f_0 (if given) or a fixed passband (if given) or fixed stopband, in that order.
(a) Passband 10-14 Hz, Stopband: < 5 Hz and > 20 Hz, $f_0 = 12$ Hz
(b) Passband 10-40 Hz, Stopband: < 5 Hz
(c) Passband 10-40 Hz, Stopband: < 5 Hz, $f_0 = 25$ Hz
(d) Stopband: < 5 Hz and > 50 Hz, Lower passband edge: 15 Hz, $f_0 = 25$ Hz

P9.16 **(Lowpass Prototype)** For each set of bandstop filter specifications, find the normalized lowpass prototype frequencies. Assume a fixed center frequency f_0 (if given) or a fixed passband (if given) or fixed stopband, in that order.
(a) Bandstop: Passband < 20 Hz and > 40 Hz, Stopband: 26-36 Hz, $f_0 = 30$ Hz
(b) Passband < 20 Hz and > 80 Hz, Lower stopband edge = 25 Hz
(c) Stopband 40-80 Hz, Lower passband edge = 25 Hz
(d) Passband: < 20 Hz and > 60 Hz, Lower stopband edge: 30 Hz, $f_0 = 40$ Hz

10 ANALOG FILTERS

In Chapter 7, we described why ideal filters cannot be physically realized; and in Chapter 9, we used Laplace transforms to analyze some practical filters. This chapter details the synthesis and design of practical analog filters subject to given magnitude and phase specifications. The design is based on approximating these specifications using polynomial and rational functions. We cover the following five classical filter approximations:

1. Butterworth (maximally flat) filters.
2. Chebyshev, or Chebyshev I, (equiripple) filters.
3. Inverse Chebyshev, or Chebyshev II, filters.
4. Elliptic (or Cauer) filters.
5. Bessel (linear phase) filters.

The discussion of each filter type is supplemented by a tabulated design summary and includes its major advantages and shortcomings.

Useful Background   The following topics provide a useful background:

1. Ideal filters (Chapter 7).
2. Real filters and their magnitude and phase response (Chapter 9).
3. Finding a stable causal transfer function $H(s)$ from its magnitude spectrum $|H(\omega)|$ (Chapter 9).
4. Frequency transformations (Chapter 9).

10.1 Introduction

In the context of electrical engineering, a **filter** is often described as a *device* or a *physical process* that discriminates against or modifies certain features of a signal in some prescribed fashion. What exactly does a filter discriminate against? If you guessed *noise*, you are quite right. In the time domain, noise reduction may be viewed as a smoothing operation and is achieved by *smoothing filters* or *interpolating filters*. Often, we require the response to be tailored to meet specifications in the frequency domain. The unwanted components or "noise," might, for example, represent a range of frequencies that needs to be suppressed. In this context, a filter describes a *frequency-selective device* that allows us to shape the frequency response (magnitude or phase or both) in a prescribed manner. Yet another, less restrictive viewpoint is to think of a filter as a model that represents the dynamics of a system, the goal being to design a filter that best represents a given set of input-output data.

10.1.1 *Analog Filters*

This chapter deals with analog filters and their design. Analog filters are indispensable in many situations, including digital signal processing. The front end of most digital signal-processing devices is an analog filter, called an **antialiasing filter**, which limits the frequency of the input signal to a range the digital filter can handle, based on sampling requirements. A popular method for the design of digital filters also starts with an analog filter that meets the required specifications, which is then mapped into an appropriate digital realization.

Analog filters using only passive elements are capable of covering a broad frequency range. **Active filters**, which utilize operational amplifiers and other active devices, are almost universally used in analog signal-processing applications, except at extremely high frequencies.

10.2 Filter Terminology

Since the response of a real filter can only approximate that of an ideal filter, we choose parameters best suited to assess filter performance from both the theoretical and design standpoint. Such specifications, called **transmission specifications**, are usually based on the frequency response for *positive frequencies* and summarized in Table 10.1 and Figure 10.1 (p. 408).

The **magnitude**, $|H(\omega)|$, is called **gain**. Its reciprocal, $1/|H(\omega)|$, is called **attenuation**. Both are typically described in dB in terms of $|H(\omega)|^2$ by

$$G(\omega) = 10 \log |H(\omega)|^2 \text{ dB} \qquad A(\omega) = 10 \log |1/H(\omega)|^2 \text{ dB}$$

Note that $A(\omega) = -G(\omega)$ in decibels. If $|H(\omega)|$ is rational, $|H(\omega)|^2$ is an even, rational function of ω^2. The roots ω_k of $|H(\omega)|^2$ show both conjugate and mirror symmetry. Such symmetry is called **quadrantal symmetry** and forms the basis for finding a causal stable transfer function $H(s)$ from $|H(\omega)|^2$ (as described in Chapter 9).

Table 10.1   Filter Terminology.

Specification	Notation	Characteristics
Frequency response	$\lvert H(\omega)\rvert \angle \theta(\omega)$	Magnitude and phase characteristics
Order		Highest power of ω in denominator of $H(\omega)$
Gain	$\lvert H(\omega)\rvert$	or $20\log\lvert H(\omega)\rvert$ dB or $10\log\lvert H(\omega)\rvert^2$ dB
Attenuation (loss)	$\lvert 1/H(\omega)\rvert$	or $20\log\lvert 1/H(\omega)\rvert$ dB or $10\log\lvert 1/H(\omega)\rvert^2$ dB
Passband edge	ω_p	Region of almost constant gain
Stopband edge	ω_s	Region of almost zero gain
Transition band	ω_p to ω_s	Region between passband and stopband
Passband attenuation	A_p (dB)	The maximum attenuation in the passband
Stopband attenuation	A_s (dB)	The minimum attenuation in the stopband
Passband ripple	ϵ	Deviation from maximum passband gain
Stopband ripple	δ	Deviation from minimum stopband gain
Roll-off rate	dB/decade	Slope of H_{dB} at high frequencies
Phase	$\theta(\omega)$	The phase at various frequencies
Phase delay	$-\theta(\omega)/\omega$	The phase $-\theta(\omega)$ normalized by ω
Group delay	$-d\theta(\omega)/d\omega$	The negative slope of $\theta(\omega)$

Figure 10.1   Illustrating the filter terminology of Table 10.1.

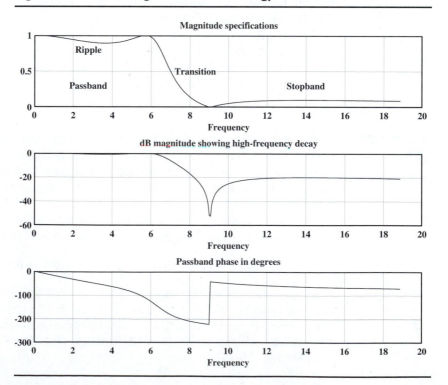

The highest power of ω^2 in the denominator $D(\omega^2)$ of $|H(\omega)|^2 = N(\omega^2)/D(\omega^2)$ (or of s in the denominator of $H(s)$) determines the **filter order**.

The **passband** is the region of almost constant gain extending to the passband edge ω_p. The extent of the passband is usually termed the **bandwidth**.

The **stopband** is the region of very small (ideally zero) gain.

The **transition band** is sandwiched between ω_p and ω_s and made as narrow as possible (ideally zero). Its extent defines the **transition width**.

Deviations from the average passband gain are defined by **passband ripple**.

Deviations from the average stopband gain are defined by **stopband ripple**.

The maximum attenuation in the passband is denoted by A_p, and is measured in decibels (dB). The minimum attenuation in the stopband is denoted A_s, and is measured in decibels (dB).

The **high frequency attenuation rate**, or **roll off**, is the slope of $|H(\omega)|$ at high frequencies in decibels per decade (dB/dec.).

With $H(\omega) = |H(\omega)|\angle\theta(\omega)$, the **phase delay**, t_p, is defined as

$$t_p = \frac{-\theta(\omega)}{\omega}$$

The **group delay**, t_g, is defined as the negative slope of the phase function

$$t_g = \frac{-d\theta(\omega)}{d\omega}$$

If $\theta(\omega)$ varies linearly with frequency, t_p and t_g are not only constant but are equal. For rational transfer functions, $\theta(\omega)$ is a transcendental function, but the *group delay is always a rational function of ω^2*, as is $|H(\omega)|^2$, and is thus much easier to work in many filter applications.

10.3
Classical Lowpass Filter Approximations

The conventional approach to the design of analog filters starts with a *normalized lowpass prototype* (LPP) using a normalized frequency variable ν, and a subsequent frequency transformation to the required form.

For a normalized lowpass filter, $|H(\nu)|^2$ may be expressed as

$$|H(\nu)|^2 = \frac{1}{1 + L_n^2(\nu)}$$

Here, $L_n(\nu)$ is an nth-order polynomial or rational function, making $L_n^2(\nu)$ an even function. Since we require $|H(\nu)|^2$ to be nearly unity in the passband $|\nu| < \nu_p$, and almost zero in the stopband $|\nu| > \nu_s$, we must satisfy

$$L_n(\nu) \approx 0 \quad |\nu| < \nu_p \qquad L_n(\nu) \longrightarrow \infty \quad |\nu| > \nu_s$$

The various classical filters differ primarily in the choice for $L_n(\nu)$ to best meet desired specifications, as summarized in Table 10.2 and Figure 10.2 (p. 410).

Butterworth, Chebyshev I and Bessel filters use polynomial approximations. An nth-order polynomial can be differentiated any number of times, integrated over any interval, and completely specified by only $n + 1$ coefficients. Chebyshev II and elliptic filters use approximations based on *rational functions* (a ratio of polynomials).

Table 10.2   Classical Lowpass Filter Approximations.

Type	Denominator of $|H(\nu)|^2$	Characteristics
Butterworth (maximally flat)	$1 + \epsilon^2 \nu^{2n}$	Maximally flat at $\nu = 0$. Monotonic in both the passband and stopband.
Chebyshev I	$1 + \epsilon^2 T_n^2(\nu)$	Equiripple response in the passband. Monotonic response in the stopband.
Chebyshev II (inverse)	$1 + [1/\epsilon^2 T_n^2(1/\nu)]$	Monotonic response in the passband. Equiripple response in the stopband.
Elliptic	$1 + \epsilon^2 R_n^2(\nu, \delta)$	Equiripple response in passband and stopband.
Bessel	$H(s) = 1/B(s)$	Linear phase in the passband.

Figure 10.2   The various classical lowpass filters listed in Table 10.2.

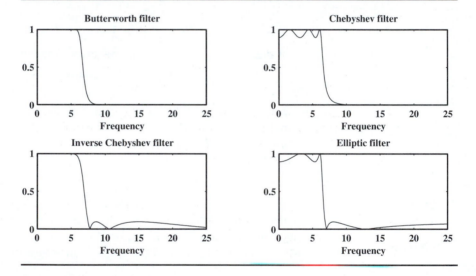

10.3.1   *The Design Process*

Filter design is essentially a three-step process:

1. Establishing the filter specifications for a given performance.
2. Determining the transfer function that meets the specifications.
3. Realizing the transfer function in hardware.

The fewer, or less stringent, the specifications, the better the possibility of achieving both design objectives and circuit implementation. The design based on any set of performance specifications is, at best, a compromise at all three steps. At the first, the actual filter may never meet performance specifications if they are too stringent; at the second, the same set of specifications may lead to several possible different realizations; and at the third, inherent component tolerances could render a circuit useless, if based on too critical a set of design values.

Filter design techniques differ primarily in how performance is specified. A trivial specification might involve just a range of frequencies to be processed (passed or rejected). A more stringent constraint might involve a specific amount of passband ripple and/or stopband attenuation and/or transition width. Tighter specifications might additionally demand linear phase and/or a minimum filter order and/or a specific filter type.

Typically, the specifications include the passband and stopband frequencies and attenuations and, optionally, the half-power (3-dB) frequency and the dc gain.

Design Steps   The design process entails the following steps:

1. Convert frequency specifications to apply for a lowpass prototype.
2. Find the filter order N and $|H_N(v)|^2$ for the lowpass prototype.
3. Find the normalized lowpass transfer function $H_N(s)$ from $|H_N(v)|^2$.
4. Use frequency transformations to denormalize $H_N(s)$ to $H_A(s)$.

Normalizing Frequency   It is common practice to normalize the frequency specifications by a normalizing frequency (such as the passband edge). We design the LPP to *meet* attenuation requirements at unit normalized frequency and *exceed* attenuation requirements at all the other frequencies. Since denormalization involves just a frequency scaling, the denormalizing frequency may then be chosen to meet specifications at any frequency of our choice. A common choice for denormalization is the passband edge.

Design Equations   The attenuation equations based on the required form for $|H(v)|$ or $L_n(v)$ are all we need to design the required filter.

The attenuation $A(v)$ may be described in terms of $L_n(v)$ as

$$A(v) = 10\log(1/|H(v)|^2) = 10\log[1 + L_n^2(v)]\text{ dB}$$

For a maximum attenuation A_p dB in the passband $|v| \leq v_p$ and a minimum attenuation A_s dB in the stopband $|v| \geq v_s$, we require

$$A_p \leq 10\log|1 + L_n^2(v_p)| \qquad A_s \geq 10\log|1 + L_n^2(v_s)|$$

10.4
The
Butterworth
Approximation

The Butterworth filter is based on a *monotonic* polynomial $L(v)$ of the form

$$L(v) = a_0 + a_1v + \cdots + a_nv^n$$

The idea is to choose the coefficients a_k to ensure that $L(v)$ is as close to zero as possible in the passband. A comparison of the polynomials v^k in Figure 10.3 suggests that *no combination can be better than the single term v^n*. The best approximation is thus $L(v) = a_nv^n$, with all the other $a_k = 0$.

Figure 10.3   **The monotonic polynomial $f(v)=Bv^n$ has the smallest deviation from zero over $(0 \leq v \leq 1)$ and leads to the lowpass Butterworth filter.**

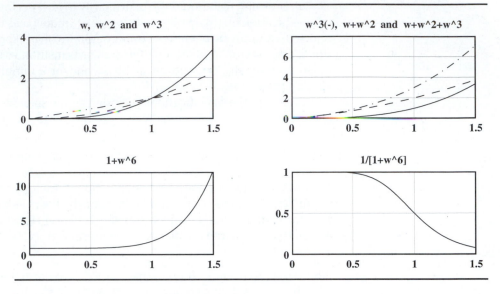

Formally, the flatter $L(v)$ is at $v = 0$, the less it will deviate from zero at other frequencies. We ensure *maximal* flatness by forcing $L(v)$ and as many of its derivatives as possible to zero at $v = 0$. For example,

$$L(v) = 0 = a_0 + a_1 v + a_2 v^2 + \cdots + a_n v^n \longrightarrow a_0 = 0$$
$$L'(v) = 0 = a_1 + a_2 v + \cdots + a_n v^{n-1} \longrightarrow a_1 = 0$$
$$L''(v) = 0 = a_3 + a_3 v + \cdots + a_n v^{n-2} \longrightarrow a_2 = 0$$

With the successive coefficients a_k equaling zero, we must finally retain the last term to avoid a trivial result, leaving us with $L(v) = a_n v^n$ where a_n is simply a measure of the deviation from zero. We, in effect, use up the free parameters a_k (*the degrees of freedom*) to achieve maximal flatness at the origin. This is why a Butterworth filter is often called a **maximally flat** approximation.

Remark: For a *maximally flat rational function*, the denominator is the sum of the numerator $N(v)$ and the term Av^n (with the numerator order less than n)

$$L(v) = \frac{N(v)}{N(v) + Av^n}$$

For the Butterworth approximation, obviously $N(v) = 1$.

10.4.1   *The Butterworth Filter*

Using the maximally flat polynomial with $a_n = \epsilon$, we may express $|H(\nu)|^2$ for an nth-order **Butterworth lowpass filter** as

$$|H(\nu)|^2 = \frac{1}{1 + L_n^2(\nu)} = \frac{1}{1 + \epsilon^2 \nu^{2n}}$$

The quantity ϵ^2 is a measure of the deviation from unit passband magnitude.

10.4.2   *Design Recipe for Butterworth Filters*

The design recipe for Butterworth filters is summarized in Table 10.3.

Table 10.3   Lowpass Butterworth Filter Design. Note: $\omega_3 =$ half-power frequency (3-dB frequency).

Design step	If ω_3 is not specified	If ω_3 is also specified
Specifications	$A_p, A_s, \omega_p, \omega_s$	$A_p, A_s, \omega_p, \omega_s$ and ω_3
Normalization	$\omega_N = \omega_p$ (passband edge)	$\omega_N = \omega_3$ (3-dB frequency)
Finding ϵ^2	$\epsilon^2 = 10^{0.1A_p} - 1$	$\epsilon^2 = 1$
Finding n	$n = \dfrac{\log[(10^{0.1A_s} - 1)/\epsilon^2]^{1/2}}{\log(\omega_s/\omega_p)}$	Find n_s and n_p using ν_s and ν_p Choose $n = \max(n_s, n_p)$
3-dB frequency	$\nu_3 = (1/\epsilon)^{1/n}$	$\nu_3 = 1$
Stopband attenuation	$A_s = 10\log[1 + \epsilon^2(\omega_s/\omega_p)^{2n}]$	$A_s = 10\log[1 + (\omega_s/\omega_3)^{2n}]$
Normalized poles	$p_k = -\sin\theta_k + j\cos\theta_k, \quad \theta_k = (2k-1)\pi/2n\,\text{(rad)}, \quad k = 1, 2, \ldots, n$	
Factored form	$H_N(s) = \dfrac{K}{Q_N(s)}, \quad Q_N(s) = \begin{cases} (s+1) \displaystyle\prod_{k=1}^{(n-1)/2} [s^2 + s(2\sin\theta_k) + 1] & n \text{ odd} \\[2ex] \displaystyle\prod_{k=1}^{n/2}[s^2 + s(2\sin\theta_k) + 1] & n \text{ even} \end{cases}$	
Polynomial form	$Q_N(s) = q_0 + q_1 s + q_2 s^2 + \cdots + q_n s^n, \quad q_0 = 1, \quad q_k = \dfrac{\cos[(k-1)\pi/2n]}{\sin(k\pi/2n)} q_{k-1}$	
K for unit gain (dc or peak)	$K = Q_N(0) = 1$	$K = Q_N(0) = 1$
Denormalization	$\omega_D = \begin{matrix} \omega_p/\tilde\nu_p & \text{(passband edge)} \\ \omega_s/\tilde\nu_s & \text{(stopband edge)} \end{matrix}$	$\omega_D = \omega_3$ (3-dB frequency)
Denormalized $H_A(s)$	$H_A(s) = H_N(s/\omega_D)$	$H_A(s) = H_N(s/\omega_D)$

We first normalize the frequencies by the passband edge ω_p, to obtain the *normalized frequencies* $\nu_p = \omega_p/\omega_p = 1$ and $\nu_s = \omega_s/\omega_p$. The form of $|H(\nu)|^2$ requires only ϵ and the filter order n, which we find from

$$A(\nu) = 10\log|1/H(\nu)|^2 = 10\log(1 + \epsilon^2\nu^{2n})$$

We find ϵ by evaluating the attenuation at $\nu = \nu_p = 1$

$$A_p = A(\nu_p) = 10\log(1 + \epsilon^2) \quad \text{or} \quad \epsilon^2 = 10^{0.1A_p} - 1$$

Note that if ω_p is the half-power frequency (with $A_p = 3.01$ dB), then $\epsilon^2 = 1$. At the stopband edge $\nu = \nu_s$, the attenuation equals

$$A_s = A(\nu_s) = 10\log(1 + \epsilon^2 \nu_s^{2n})$$

Substituting for ϵ^2 and simplifying, we obtain the filter order n as

$$n = \frac{\log[(10^{0.1A_s} - 1)/\epsilon^2]^{1/2}}{\log(\nu_s)} = \frac{\log[(10^{0.1A_s} - 1)/(10^{0.1A_p} - 1)]^{1/2}}{\log(\omega_s/\omega_p)}$$

The normalized half-power frequency ν_3 is found by setting $|H(\nu_3)|^2 = \frac{1}{2}$ or

$$\epsilon^2 \nu_3^{2n} = 1 \qquad \text{or} \qquad \nu_3 = (1/\epsilon)^{1/n}$$

Remarks:

1. The filter order must be *rounded up to an integer*. If the attenuation is specified at other frequencies, n must be found for each such frequency, and the *largest* value chosen for the order.
2. A smaller ϵ^2 (a flatter passband) or a larger stopband attenuation A_s (a sharper transition) requires a filter of higher order.

Example 10.1

Butterworth Lowpass Filter Attenuation

Consider the design of a Butterworth lowpass filter to meet the following specifications: $A_p \leq 1$ dB for $\omega \leq 4$ rad/s, $A_s \geq 20$ dB for $\omega \geq 8$ rad/s.

We normalize with respect to ω_p to get $\nu_p = \omega_p/\omega_p = 1$ and $\nu_s = \omega_s/\omega_p = 2$. Following the design procedure, we successively compute:

$$\epsilon^2 = 10^{0.1A_p} - 1 = 0.2589 \quad n = 4.289 \longrightarrow n = 5 \quad \nu_3 = 1.145$$

Using the attenuation equation with $\epsilon^2 = 0.2589$, we find

$$A_p = A(\nu_p) = 1 \text{ dB} \quad A_3 = A(\nu_3) = 3.01 \text{ dB} \quad A_s = A(\nu_s) = 24.25 \text{ dB}$$

We thus meet the specs at ν_p and ν_3, and exceed the specs at ν_s.

Example 10.2

Butterworth Lowpass Filter Order

We find the order of a Butterworth lowpass filter specified by the attenuations of 2 dB at 5, 10 dB at 8, 20 dB at 10 and 40 dB at 30 rad/s. Using $\omega_N = 5$ such that $\nu_p = 1$, we first compute $\epsilon^2 = 10^{0.1A_p} - 1 = 0.5849$. We compute $n = \{\log[(10^{0.1A_x} - 1)/\epsilon^2]^{1/2}\}/\log(\nu_x)$ for each attenuation to get

$$A_1 = 10 \text{ dB} \qquad \nu = 8/5 = 1.6 \qquad n = 2.91$$
$$A_2 = 20 \text{ dB} \qquad \nu = 10/5 = 2 \qquad n = 3.7$$
$$A_3 = 40 \text{ dB} \qquad \nu = 30/5 = 6 \qquad n = 2.72$$

The largest value of n is rounded up to give the filter order as $n = 4$.

10.4.3   The Normalized Butterworth Filter

A **normalized Butterworth filter** refers to a filter *normalized with respect to the half-power frequency* (with $\epsilon^2 = 1$). For such a filter

$$|H_N(\nu)|^2 = |H(\nu/\nu_3)|^2 = \frac{1}{1 + (\nu/\nu_3)^{2n}} = \frac{1}{1 + \nu_N^{2n}}$$

It is customary to find $H_N(s)$ from $|H_N(\nu)|^2$, and then denormalize to $H(s)$.

10.4.4 *The Normalized Transfer Function*

To establish $H_N(s)$ from $|H_N(v)|^2$, recall from Chapter 9 that we must

1. Replace v_N^2 by $-s^2$ (or v_N by s/j).
2. Relate the resulting function to $H_N(s)H_N(-s)$.
3. Extract $H_N(s)$ using the left half-plane (LHP) roots of $H_N(s)H_N(-s)$.

Following these steps, we replace v_N by s/j in the normalized $|H(v)|^2$.

$$H(s)H(-s) = |H_N(v)|^2\big|_{v_N=s/j} = \frac{1}{1+(-s^2)^n} = \frac{1}{Q(s)}$$

The poles of $H(s)$ are found from

$$1 + (-s^2)^n = 0 \quad \text{or} \quad (-s^2)^n = -1 \quad \text{or} \quad (-1)^n p_k^{2n} = -1$$

We may write this relation in terms of the poles p_k as

$$\exp(-j\pi n)p_k^{2n} = \exp[j(2k-1)\pi], \qquad k = 1, 2, \ldots, 2n$$

This yields the $2n$ **normalized poles** (Figure 10.4)

$$p_k = \exp[j(\theta_k + \tfrac{1}{2}\pi)], \qquad \theta_k = (2k-1)\pi/2n, \qquad k = 1, 2, \ldots, 2n$$

Figure 10.4   Pole locations of a normalized Butterworth lowpass filter.

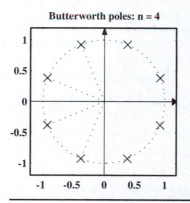

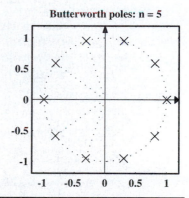

The angle θ_k (*radians*) describes the angular orientation with respect to the positive $j\omega$-axis. With $p_k = \sigma_k + j\omega_k$, the real and imaginary parts of the poles equal

$$\sigma_k = -\sin(\theta_k), \qquad \omega_k = \cos(\theta_k), \qquad k = 1, 2, \ldots, 2n$$

Using the identity $\sin^2(\theta) + \cos^2(\theta) = 1$, the pole magnitudes $|p_k|$ are given by

$$|p_k| = (\sigma_k^2 + \omega_k^2)^{1/2} = 1$$

These results show that:

1. The *normalized poles lie on a unit circle* in the s-plane.
2. The poles are equispaced $\pi/2n$ radians apart with $\theta_k = (2k-1)\pi/2n$, where θ_k is measured with respect to the positive $j\omega$-axis.
3. The poles can never lie on the $j\omega$-axis, since $2k-1$ can never be even.
4. For odd n, there is always one pair of real poles located at $s = \pm 1$.

The LHP poles used to form $Q_N(s)$ are p_k, $k = 1, 2, \ldots, n$. Each conjugate pole pair yields the real quadratic factor

$$(s + p_k)(s + p_k^*) = s^2 + 2(\sin\theta_k)s + \sin^2\theta_k + \cos^2\theta_k = s^2 + s(2\sin\theta_k) + 1$$

Its pole frequency is $\omega_{pk} = 1$, and its pole quality factor is $Q_k = 1/[2\sin(\theta_k)]$. For odd n, we also have the linear factor $(s + 1)$. The factored form for $H_N(s)$ may then be written

$$H_N(s) = \frac{K}{Q_N(s)}, \quad Q_N(s) = \begin{cases} (s+1)\prod_{k=1}^{(n-1)/2}[s^2 + s(2\sin\theta_k) + 1] & n \text{ odd} \\ \prod_{k=1}^{n/2}[s^2 + s(2\sin\theta_k) + 1] & n \text{ even} \end{cases}$$

The constant K is chosen as $Q_N(0)$ for unit dc gain (or unit peak gain). This form is useful for realizing higher-order filters by cascading second-order sections (with an additional first-order section if n is odd).

We can also express the denominator $Q_N(s)$ in polynomial (unfactored) form

$$Q_N(s) = q_0 + q_1 s + q_2 s^2 + \cdots + q_n s^n$$

Since each linear or quadratic factor is symmetric (with its leading and trailing coefficients equal to 1), so is their product (polynomial multiplication is equivalent to convolution and the convolution of symmetric sequences yields a symmetric result). This implies that $q_k = q_{n-k}$. To find q_k, we use the recursion relation

$$q_k = \frac{\cos[(k-1)\pi/2n]}{\sin(k\pi/2n)} q_{k-1} \qquad q_0 = 1$$

We need evaluate just half the number of coefficients to establish $Q_N(s)$. Table 10.4 lists both the factored and polynomial forms for $Q_N(s)$ for various n.

10.4.5   *Denormalization*

Since $|H(\nu)|^2 = |H_N(\nu/\nu_3)|^2$, we have $H(s) = H(s/\nu_3)$. The poles of the *denormalized* $H(s)$ thus lie on a circle of radius $\nu_3 = (1/\epsilon)^{1/n}$. If we denormalize $H(s)$ to $H_A(s) = H(s/\omega_p)$, $H_A(s)$ *exactly* meets the given passband specification. This is equivalent to denormalizing $H_N(s)$ directly to obtain $H_A(s) = H_N(s/\omega_p\nu_3)$. To *exactly* meet the attenuation A_x at any other frequency ω_x, we use $H_N(\nu)$ to find the actual frequency $\tilde{\nu}_x$ that corresponds exactly to A_x. From the attenuation expression for $A(\nu)$ with $\nu = \tilde{\nu}_x$, we obtain

$$\tilde{\nu}_x = [(10^{0.1A_x} - 1)]^{1/2n} = (10^{0.1A_x} - 1)^{1/2n}$$

Table 10.4   Butterworth Polynomials.

Order n	Denominator polynomial $Q_N(s)$ in polynomial form
1	$1 + s$
2	$1 + \sqrt{2}s + s^2$
3	$1 + 2s + 2s^2 + s^3$
4	$1 + 2.613s + 3.414s^2 + 2.613s^3 + s^4$
5	$1 + 3.236s + 5.236s^2 + 5.236s^3 + 3.236s^4 + s^5$
6	$1 + 3.864s + 7.464s^2 + 9.141s^3 + 7.464s^4 + 3.864s^5 + s^6$
7	$1 + 4.494s + 10.103s^2 + 14.606s^3 + 14.606s^4 + 10.103s^5 + 4.494s^6 + s^7$
8	$1 + 5.126s + 13.138s^2 + 21.848s^3 + 25.691s^4 + 21.848s^5 + 13.138s^6 + 5.126s^7 + s^8$

Order n	Denominator polynomial $Q_N(s)$ in factored form
1	$1 + s$
2	$1 + \sqrt{2}s + s^2$
3	$(1 + s)(1 + s + s^2)$
4	$(1 + 0.76536s + s^2)(1 + 1.84776s + s^2)$
5	$(1 + s)(1 + 0.6180s + s^2)(1 + 1.6180s + s^2)$
6	$(1 + 0.5176s + s^2)(1 + \sqrt{2}s + s^2)(1 + 1.9318s + s^2)$
7	$(1 + s)(1 + 0.4450s + s^2)(1 + 1.2456s + s^2)(1 + 1.8022s + s^2)$
8	$(1 + 0.3986s + s^2)(1 + 1.1110s + s^2)(1 + 0.6630s + s^2)(1 + 1.9622s + s^2)$

We then choose the denormalizing frequency ω_D to ensure that $\omega_D \tilde{v}_x = \omega_x$. This gives $\omega_D = \omega_x / \tilde{v}_x$ and the denormalized filter

$$H_A(s) = H_N(s/\omega_D)$$

To meet passband specifications, for example, $\tilde{v}_x = \tilde{v}_p$ and we must compute $\tilde{v}_p$ and $\omega_D = \omega_p / \tilde{v}_p$. It turns out that $\tilde{v}_p = 1/\nu_3$ and $\omega_D = \omega_p \nu_3$, as before.

Example 10.3
Butterworth Lowpass
Filter Poles

For the 5th-order Butterworth lowpass filter of Example 10.1, the normalized LHP poles are located around a unit circle with orientations from the positive $j\omega$-axis given by $\theta_k = (2k - 1)\pi/2n = (2k - 1)18°, k = 1, 2, \ldots, 5$.

Since n is odd, one pole ($k = 3$) is located on the real axis and corresponds to $(s + 1)$. The quadratic terms corresponding to the two remaining pole pairs are

$$s^2 + s(2\sin\theta_1) + 1 = s^2 + 0.618s + 1 \quad \text{and} \quad s^2 + s(2\sin\theta_2) + 1 = s^2 + 1.618s + 1$$

Thus, $H_N(s) = K/Q_N(s)$, where $Q_N(s)$ may be written in factored form as

$$Q_N(s) = (s + 1)(s^2 + 0.618s + 1)(s^2 + 1.618s + 1)$$

We choose $K = 1$ for unit dc gain. To find the polynomial form of Q_N, we use the recursion relation to find its coefficients q_k. With $q_0 = 1$, we obtain

$$q_1 = \frac{q_0}{\sin(\pi/10)} = 3.236 \quad \text{and} \quad q_2 = \frac{q_1 \cos(\pi/10)}{\sin(\pi/5)} = 5.236$$

Using symmetry, we have $q_3 = q_2$, $q_4 = q_1$ and $q_5 = q_0$. Thus,

$$Q_N(s) = 1 + 3.236s + 5.236s^2 + 5.236s^3 + 3.236s^4 + s^5$$

We exactly meet specs at the 3-dB frequency if we denormalize $H_N(s)$ by $\omega_p = 4$. To exactly meet passband specs at ω_p, we use $\omega_D = \omega_p/\tilde{\nu}_p$ where

$$\tilde{\nu}_p = (10^{0.1A_p} - 1)^{1/2n} = 0.8736 \qquad \omega_D = \omega_p/\tilde{\nu}_p = 4.5787$$

This gives the denormalized transfer function $H_A(s) = H_N(s/4.5787)$ as

$$H_A(s) = \frac{2012.4}{2012.4 + 1422.3s + 502.6s^2 + 109.8s^3 + 14.82s^4 + s^5}$$

The frequency response of this filter is sketched in Figure 10.5.

Comment: To exactly meet stopband specs at ω_s, we use $\omega_D = \omega_s/\tilde{\nu}_s$ where

$$\tilde{\nu}_s = (10^{0.1A_s} - 1)^{1/2n} = 1.5833 \quad \text{and} \quad \omega_D = \omega_s/\tilde{\nu}_s = 5.0527$$

Figure 10.5   Response of the Butterworth filter designed in Example 10.3.

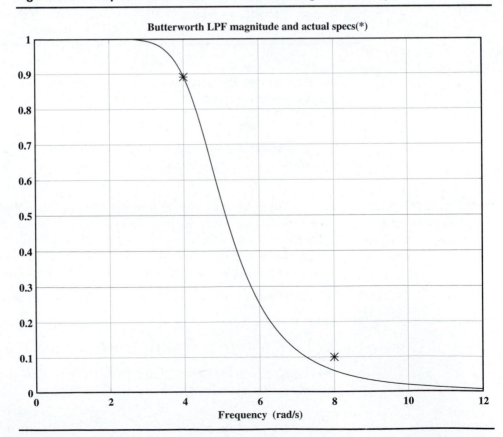

The denormalized transfer function now equals $H_A(s) = H_N(s/5.0527)$, and meets specifications exactly at the stopband edge. We can easily confirm this by computing the attenuation. With $\epsilon^2 = 0.2589$, the attenuation equation gives

$$A_p = A(\omega_p/\omega_D) = 0.4 \text{ dB} \quad A_3 = A(\omega_3/\omega_D) = 1.38 \text{ dB} \quad A_s = A(\omega_s/\omega_D) = 20 \text{ dB}$$

10.4.6   A Design That Meets Additional Specifications

Often, the design also specifies a half-power frequency ω_3. In this case, we normalize the frequencies by ω_3. This also implies that $\epsilon^2 = 1$. To find the filter order, we solve the attenuation relation for n at both $v_p = \omega_p/\omega_3$ and $v_s = \omega_s/\omega_3$, and choose n as the larger value. Successive design steps are identical to the standard design (Table 10.3). Final denormalization uses the frequency $\omega_D = \omega_3$ and exactly meets specifications at the half-power frequency.

Example 10.4
Butterworth Lowpass
Filter Design

Consider the design of a Butterworth lowpass filter to meet the following specifications: $A_p \leq 1$ dB for $\omega \leq 4$, $A_s \geq 20$ dB for $\omega \geq 8$ and $\omega_3 = 6$.

Since ω_3 is given, we normalize the band-edge frequencies to $v_p = \omega_p/\omega_3 = 4/6$ and $v_s = \omega_s/\omega_3 = 8/6$. With $\epsilon^2 = 1$ in the attenuation relation, we find

$$A_p = 10 \log(1 + v_p^{2n}) \qquad n = 0.5 \log(10^{0.1 A_p} - 1)/\log(v_p) = 1.6663$$
$$A_s = 10 \log(1 + v_s^{2n}) \qquad n = 0.5 \log(10^{0.1 A_s} - 1)/\log(v_s) = 7.9865$$

Thus $n = 8$. We now exactly meet specs at v_3. As a check, we find that

$$A_p = A(v_p) = 0.0066 \text{ dB} \qquad A_3 = A(v_3) = 3.01 \text{ dB} \qquad A_s = A(v_s) = 20.03 \text{ dB}$$

The denormalized transfer function is given by $H_A(s) = H_N(s/\omega_3) = H_N(s/6)$.

10.4.7   High-Frequency Response

For an nth-order Butterworth filter, we expect an attenuation of $20n$ dB/decade for high frequencies. In fact, for $v \gg 1$, $A(v)$ may be approximated by

$$A(v) = 10 \log(1 + \epsilon^2 v^{2n}) \approx 10 \log(\epsilon^2 v^{2n}) = 20 \log \epsilon + 20n \log v$$

10.4.8   Phase Characteristics

The phase is best evaluated from $H_N(s)$ in factored form. The contribution due to a linear factor, if present, is simply

$$H_N(s) = \frac{1}{s+1} \qquad H_N(v) = \frac{1}{1+jv} \qquad \phi(v) = -\tan^{-1}(v)$$

For a quadratic factor, conversion from $H_N(s)$ to $H_N(v)$ leads to

$$H_N(s) = \frac{1}{s^2 + 2\sin\theta_k s + 1}$$

$$H_N(v) = \frac{1}{(1 - v^2) + j2v\sin\theta_k} \qquad \phi_N(v) = -\tan^{-1}\left[\frac{2v\sin\theta_k}{1 - v^2}\right]$$

The total phase $\phi(v)$ is just the sum of contributions due to each quadratic factor and the linear factor (if present), and may be written as

$$\phi(v) = -\sum_{k=1}^{\text{int}(n/2)} \tan^{-1}\left[\frac{2v\sin\theta_k}{1 - v}\right] - \begin{cases} \tan^{-1}(v) & n \text{ odd} \\ 0 & n \text{ even} \end{cases}$$

An expression for the group delay may also be found by evaluating $-d\phi/dv$. Using the relations

$$\frac{d}{dx}[\tan^{-1}(x)] = \frac{1}{1 + x^2} \quad \text{and} \quad \frac{d}{dx}[\tan^{-1}f(x)] = \frac{1}{1 + f^2(x)}\frac{d}{dx}[f(x)]$$

and subsequent simplification, we obtain the two equivalent forms

$$t_g(v) = \sum_{k=1}^{\text{int}(n/2)} \frac{2(1 + v^2)\sin\theta_k}{1 - 2v^2\cos 2\theta_k + v^4} + \left.\begin{cases} \frac{1}{1 + v^2} & n \text{ odd} \\ 0 & n \text{ even} \end{cases}\right\} = \frac{1}{1 + v^{2n}}\sum_{k=1}^{n}\frac{v^{2(k-1)}}{\sin\theta_k}$$

Remarks: 1. The phase at high frequencies approaches $-n\pi/2$ radians.
2. The group delay is fairly constant for smaller frequencies, but shows peaking near the passband edge. As the filter order increases, the delay becomes less flat, and the peaking is more pronounced.
3. Peaking distortion can be minimized by using delay equalizers (allpass filters) in cascade with the filter.
4. Denormalization of the phase or group delay can be carried out in much the same way as for the normalized transfer function.

10.4.9   The Butterworth Highpass Form

The transformation $v \to 1/v$ yields the highpass form $|H_{HP}(v)|^2 = |H_{LP}(1/v)|^2$. For Butterworth filters, finding $1 - |H_{LP}(v)|^2$ yields an identical result because

$$1 - |H_{LP}(v)|^2 = 1 - \frac{1}{1 + v^{2n}} = \frac{v^{2n}}{1 + v^{2n}} = \frac{1}{1 + (1/v)^{2n}} = |H_{LP}(1/v)|^2$$

10.4.10   Concluding Remarks

Poor Transition Characteristics   All the degrees of freedom are spent in making the Butterworth magnitude response maximally flat at the origin. This yields a monotonic frequency response free of overshoot and ripples but also results in a slow transition from the passband to the stopband. Sharper transitions require higher filter orders, which means increased complexity and expense.

Increased Pole Q for Large Order   A high order also means increased pole Q. The pole Q for the first root location is given by

$$Q_1 = \frac{1}{2\sin\theta_1} = \frac{1}{\sin(\pi/2n)}$$

As n increases, so does Q_1, because $\sin(\pi/2n)$ decreases. The problem of increased pole Q for large order is, in fact, typical of all filters and results in increased susceptibility to element variation.

10.5
The Chebyshev
Approximation

The Chebyshev approximation is also based on a polynomial fit. Instead of selecting the coefficients (degrees of freedom) of the polynomial $L_n(v)$ to make the response maximally flat at the origin alone, we now try to make the response uniformly good by ensuring zero loss at several frequencies spread out over the passband and minimizing the error (deviation from zero loss) at all other frequencies in the passband. Such an approximation is called a **uniform approximation**.

10.5.1   *The Equiripple Concept*

A uniform approximation requires the maxima and minima of the error to have equal magnitudes or be **equiripple** (Figure 10.6, p. 423). The number of points where the error is zero decides the order of the approximating polynomial. The equiripple polynomial with the smallest maximum absolute error is called a **minmax polynomial**. Minmax approximations usually require iterative methods. The good news is that the Chebyshev polynomials are themselves minmax, and approximations based on Chebyshev polynomials are easy to obtain.

Remark: Note that least squares methods, which only minimize the mean squared error, provide no guarantee that a small mean squared error will also result in a small absolute error *throughout the approximation interval*.

10.5.2   *The Chebyshev Polynomials*

The **Chebyshev polynomial** (*of the first kind*) of degree n is described by a relation that looks trigonometric in nature:

$$T_n(x) = \cos(n\cos^{-1}x)$$

$T_n(x)$ is actually a polynomial of degree n. The polynomial form shows up if we use the transformation $x = \cos\theta$ to give $T_n(x) = T_n(\cos\theta) = \cos(n\theta)$. Using trigonometric identities for $\cos(n\theta)$, we easily obtain the first few Chebyshev polynomials (Figure 10.7, p. 423) as

$\cos 0 = 1$	$\cos\theta = \cos\theta$	$\cos 2\theta = 2\cos^2\theta - 1$	$\cos 3\theta = 4\cos^3\theta - 3\cos\theta$
$T_0(x) = 1$	$T_1(x) = x$	$T_2(x) = 2x^2 - 1$	$T_3(x) = 4x^3 - 3x$

With $T_0(x) = 1$, the polynomials can also be obtained from the recursion

$$T_{n+1}(x) = 2xT_n(x) - T_{n-1}(x) \qquad n \geq 1 \qquad (T_0(x) = 1)$$

For example, $T_4(x) = 2xT_3(x) - T_2(x) = 8x^4 - 8x^2 + 1$.

Table 10.5 (p. 423) lists the first few Chebyshev polynomials, Table 10.6 (p. 424) lists some of their properties and Figure 10.7 illustrates their characteristics.

The leading coefficient of an nth-order Chebyshev polynomial always equals 2^{n-1}. The constant coefficient is always ± 1 for even n, and 0 for odd n. The nonzero

---

Figure 10.6   **A function and its approximation. If the error is larger in one direction, we can reduce it by shifting the approximation in the other. Ultimately the best fit will show equiripple behavior.**

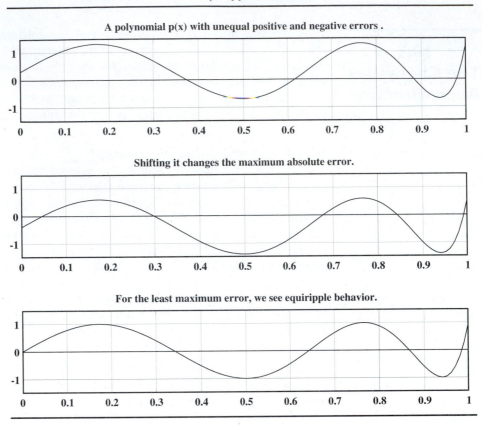

A polynomial p(x) with unequal positive and negative errors .

Shifting it changes the maximum absolute error.

For the least maximum error, we see equiripple behavior.

---

coefficients alternate in sign and point to the oscillatory nature of the polynomial. The form $T_n(x) = \cos(n\theta)$, which oscillates between 1 and -1, is clearly suggestive of the equiripple nature of these polynomials. The transformation $x = \cos\theta$ provides a link between the polynomial form in x and the trigonometric form in θ.

Chebyshev polynomials possess two remarkable characteristics in the context of both polynomial approximation and filter design.

1. *The equiripple property:* $T_n(x)$ oscillates about zero in the interval $(-1, 1)$ with $n + 1$ maxima and minima equal to ± 1. These are located at

$$x_m = \cos(k\pi/n) \qquad k = 0, 1, \ldots, n$$

It also possesses n zeros in the interval $(-1, 1)$ located at

$$x_z = \cos(\theta_k) \qquad \theta_k = (2k - 1)\pi/2n \qquad k = 1, 2, \ldots, n$$

Outside the interval $(-1, 1)$, $|T_n(x)|$ shows a very steep increase.

Figure 10.7   The Chebyshev polynomials of orders 2 to 5.

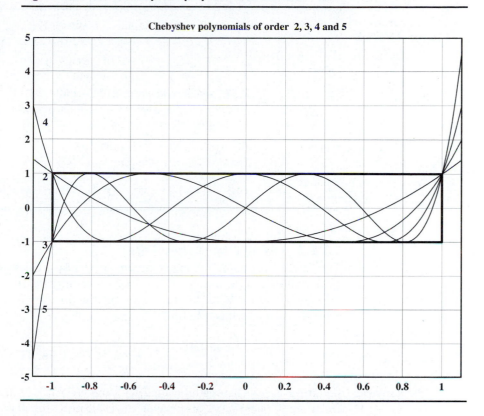

Chebyshev polynomials of order 2, 3, 4 and 5

Table 10.5   Chebyshev Polynomials.

Order n	$T_n(x)$
0	1
1	x
2	$2x^2 - 1$
3	$4x^3 - 3x$
4	$8x^4 - 8x^2 + 1$
5	$16x^5 - 20x^3 + 5x$
6	$32x^6 - 48x^4 + 18x^2 - 1$
7	$64x^7 - 112x^5 + 56x^3 - 7x$
8	$128x^8 - 256x^6 + 160x^4 - 32x^2 + 1$

---

2. *Smallest maximum absolute value:* Over $(-1, 1)$, the **normalized Chebyshev polynomial** $\tilde{T}_n(x) = T_n(x)/2^{n-1}$ has the smallest *maximum* absolute value (which equals $1/2^{n-1}$) among all normalized nth-degree polynomials. This is the celebrated **Chebyshev theorem.**

Table 10.6   Properties of Chebyshev Polynomials.

| Symmetry | $T_n(x) = T_n(-x)$ n even $T_n(x) = -T_n(-x)$ n odd |
| | $T_{2n}(x) = 2T_n^2(x) - 1$ (since $\cos(2\alpha) = 2\cos^2(\alpha) - 1$) |

| Values | $T_n(1) = 0$ $T_n(0) = \begin{cases} 1 & n \text{ even} \\ 0 & n \text{ odd} \end{cases}$ |

$-1 < x < 1$	$T_n(x)$ oscillates between ± 1 in the interval $-1 < x < 1$.
$x > 1$	For $|x| > 1, T_n(x)$ shows a monotonic increase with x.
$x \gg 1$	For large x, $T_n(x) \approx 2^{n-1}x^n$. As $|x| \longrightarrow \infty, |T_n(x)| \longrightarrow \infty$

| Evaluation | For $x < 1, T_n(x) = \cos[n(\cos^{-1}x)]$ |
| | For $x > 1, T_n(x) = \cosh[n(\cosh^{-1}x)]$ |

| Computation for $x \geq 1$ | $\cosh^{-1}(x) = \ln[x + (x^2 - 1)^{1/2}], \quad \sinh^{-1}(x) = \ln[x + (x^2 + 1)^{1/2}]$ |

10.5.3   The Chebyshev Filter

The Chebyshev approach approximates $L_n(v)$ by a Chebyshev polynomial. A peak deviation of ϵ in the passband $|v| \leq v_p = 1$ requires $L_n(v_p) = \epsilon$. Since the peak variation of $T_n(v)$ for $|v| \leq 1$ is unity, we must choose $L_n(v) = \epsilon T_n(v)$. The magnitude $|H(v)|^2$ then assumes the form

$$|H(v)|^2 = \frac{1}{1 + L_n^2(v)} = \frac{1}{1 + \epsilon^2 T_n^2(v)}$$

Remark: Squaring $T_n(v)$ ensures an even $T_n^2(v)$ and $|H(v)|^2$ for any order n. Note that the form $L_n^2(v) = \epsilon^2 T_{2n}(v)$ also yields an even polynomial for any n, since $T_{2n}(x) = 2T_n^2(x) - 1$. This choice, however, is much less common.

Figure 10.8 shows the genesis of a Chebyshev I filter characteristic from $T_n(v)$. The magnitude of the passband maxima and minima is governed only by ϵ^2, but a larger n yields more of them and results in a sharper transition due to the higher polynomial order. In fact, the filter order equals the number of maxima and minima within the passband excluding the passband edge (Figure 10.8).

Example 10.5
Chebyshev Filter Order

The 2nd-order Chebyshev filter of Figure 10.8 shows a total of 2 maxima and minima in the passband. Since $T_2(0) = 0$, and $|H(0)|$ equals 0.9 (Figure 10.8), we find $|H(0)|^2 = 1/(1 + \epsilon^2) = (0.9)^2$. This yields $\epsilon^2 = 0.2346$. Since $T_2(v) = 2v^2 - 1$, we have $T_2^2(v) = 4v^4 - 4v^2 + 1$. Thus,

$$|H(v)|^2 = \frac{1}{1 + \epsilon^2 T_2^2(v)} = \frac{1}{1 + \epsilon^2(4v^4 - 4v^2 + 1)} = \frac{1}{1.2346 - 0.9383v^2 + 0.9383v^4}$$

Figure 10.8   Genesis of the lowpass Chebyshev filter.

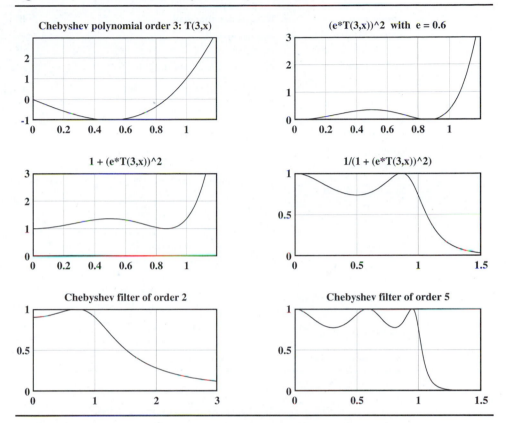

10.5.4   *Design Recipe for Chebyshev I Filters*

The design procedure for Chebyshev I filters is summarized in Table 10.7 (p. 426).

We start by normalizing the frequencies by the passband edge ω_p to obtain $\nu_p = \omega_p/\omega_p = 1$ and $\nu_s = \omega_s/\omega_p$. The design requires n and ϵ.

The design equation is the attenuation relation given by

$$A(\nu) = 10\log|1/H(\nu)|^2 = 10\log[1 + \epsilon^2 T_n^2(\nu)]$$

Since $\nu_p = 1$ and $T_n(1) = 1$, ϵ^2 is found by evaluating $A(\nu)$ at $\nu = \nu_p = 1$ to give

$$A_p = A(\nu_p) = A(1) = 10\log(1 + \epsilon^2) \text{ or } \epsilon^2 = 10^{0.1A_p} - 1$$

This relation is identical to the normalized Butterworth case. Once again, *if ω_p corresponds to the half-power frequency (with $A_p \approx 3$ dB), then $\epsilon^2 = 1$.*

The Chebyshev I filter order n is established by evaluating $A(\nu)$ at $\nu = \nu_s$,

$$A_s = A(\nu_s) = 10\log[1 + \epsilon^2 T_n^2(\nu_s)]$$

Table 10.7   Lowpass Chebyshev Filter Design. Note: ω_3 = half-power frequency (3-dB frequency).

Design step	If ω_3 is not specified	If ω_3 is also specified
Specifications	$A_p, A_s, \omega_p, \omega_s$	$A_p, A_s, \omega_p, \omega_s$ and ω_3
Finding ϵ^2	$\epsilon^2 = 10^{0.1A_p} - 1$	$\epsilon^2 = 10^{0.1A_p} - 1$
Normalization	$\omega_N = \omega_p$ (passband edge)	$\omega_N = \omega_3/\nu_3$ (3-dB frequency)
Finding n	$n = \dfrac{\cosh^{-1}[(10^{0.1A_s} - 1)/\epsilon^2]^{1/2}}{\cosh^{-1}(\omega_s/\omega_p)}$	Find ω_N and compute $\tilde{A}_s$. If $\tilde{A}_s < A_s$, let $n = n + 1$, and loop.
3-dB frequency	$\nu_3 = \cosh\left[\dfrac{1}{n}\cosh^{-1}\left(\dfrac{1}{\epsilon}\right)\right]$	$\nu_3 = \cosh\left[\dfrac{1}{n}\cosh^{-1}\left(\dfrac{1}{\epsilon}\right)\right]$
Stopband attenuation	$A_s = 10\log[1 + \epsilon^2 T_n^2(\omega_s/\omega_p)]$	$\tilde{A}_s = 10\log[1 + \epsilon^2 T_n^2(\omega_s\nu_3/\omega_3)]$
Normalized poles	$p_k = -\sin(\theta_k)\sinh(\alpha) + j\cos(\theta_k)\cosh(\alpha)$ $\qquad k = 1, 2, \ldots, n$	$\alpha = (1/n)\sinh^{-1}(1/\epsilon)$ $\theta_k = (2k - 1)\pi/2n$
Factored form	$H_N(s) = \dfrac{K}{Q_N(s)}, Q_N(s) = \begin{cases} [s + \sinh(\alpha)] \displaystyle\prod_{k=1}^{(n-1)/2} (s - p_k)(s - p_k^*) & n \text{ odd} \\ \displaystyle\prod_{k=1}^{n/2}(s - p_k)(s - p_k^*) & n \text{ even} \end{cases}$	
K for unit peak gain	$K = \begin{cases} Q_N(0)/\sqrt{1 + \epsilon^2} & n \text{ even} \\ Q_N(0) & n \text{ odd} \end{cases}$	
Denormalization	$\omega_D = \begin{matrix} \omega_p & \text{(passband edge)} \\ \omega_s/\tilde{\nu}_s & \text{(stopband edge)} \end{matrix}$	$\omega_D = \omega_3/\nu_3$ (3-dB frequency)
Denormalized $H_A(s)$	$H(s) = H_N(s/\omega_D)$	$H(s) = H_N(s/\omega_D)$

Since $\nu_s > 1$, the computation of $T_n(\nu_s)$ poses problems if the trigonometric form is used, since $\cos^{-1}(\nu_s)$ is imaginary for $\nu_s > 1$. We therefore resort to the hyperbolic form, $T_n(\nu_s) = \cosh(n\cosh^{-1}\nu_s)$, and write

$$A_s = 10\log[1 + \{\epsilon\cosh[n(\cosh^{-1}\nu_s)]\}^2]$$

The filter order is obtained by solving this equation, and yields

$$n = \frac{\cosh^{-1}[(10^{0.1A_s} - 1)/\epsilon^2]^{1/2}}{\cosh^{-1}(\nu_s)} = \frac{\cosh^{-1}[(10^{0.1A_s} - 1)/(10^{0.1A_p} - 1)]^{1/2}}{\cosh^{-1}(\omega_s/\omega_p)}$$

Note the similarity between this result and the analogous relation for finding the Butterworth filter order.

The half-power frequency ν_3 at which $|H(\nu)|^2 = \frac{1}{2}$ or $\epsilon^2 T_n^2(\nu) = 1$ can be evaluated in terms of ϵ and the filter order n from $\epsilon T_n(\nu_{3\text{dB}}) = 1$, yielding

$$\nu_3 = \cosh[(1/n)\cosh^{-1}(1/\epsilon)] \qquad \epsilon \leq 1$$

We can also evaluate the order n in terms of the half-power frequency as

$$n = \cosh^{-1}[(1/\epsilon)/\cosh^{-1}(\nu_3)]$$

Example 10.6

Chebyshev Lowpass Filter Attenuation

Consider the design of a Chebyshev lowpass filter to meet the following specifications: $A_p \leq 1$ dB for $\omega \leq 4$, $A_s \geq 20$ dB for $\omega \geq 8$. Following the design procedure, we successively compute: $\nu_p = \omega_p/\omega_p = 1$, $\nu_s = \omega_s/\omega_p = 2$ and

$$\epsilon^2 = 10^{0.1A_p} - 1 = 0.2589, \quad \text{and} \quad n = 2.783 \longrightarrow n = 3$$

We also find the half-power frequency as $\nu_3 = \cosh[\frac{1}{3}\cosh^{-1}(1/\epsilon)] = 1.0949$. Using the attenuation equation, we compute

$$A_p = A(\nu_p) = 1 \text{ dB} \qquad A_s = A(\nu_s) = 22.456 \text{ dB}$$

We thus meet the specifications at ν_p and exceed them at ν_s.

10.5.5   The Passband Ripple and Filter Order

Since $T_n(\nu)$ oscillates between 0 and 1, $|H(\nu)|$ in turn varies between 1 and $\sqrt{1/(1+\epsilon^2)}$ within the passband. The maximum ripple magnitude corresponding to the attenuation A_p can thus be described by

$$\text{Ripple magnitude} = 1 - \sqrt{1/(1+\epsilon^2)}$$

The magnitude at the passband edge $\nu_p = 1$ is given by

$$|H(\nu_p)|^2 = |H(1)|^2 = 1/(1+\epsilon^2)$$

Since $T_n(0)$ is 1 for even n, and 0 for odd n, the dc gain $|H(0)|$ and dc attenuation $A(0)$ depend on the filter order and equal

$$\begin{aligned} |H(0)| &= \frac{1}{\sqrt{1+\epsilon^2}} & A(0) &= A_p & n \text{ even} \\ |H(0)| &= 1 & A(0) &= 0 & n \text{ odd} \end{aligned}$$

The *peak* gain, however, equals 1 for both odd n and even n.

10.5.6   The Normalized Transfer Function

To find the normalized transfer function $H_N(s)$, we substitute $\nu = s/j$ into $|H(\nu)|^2$, and obtain a relation for $H_N(s)H_N(-s)$ as

$$H_N(s)H_N(-s) = |H(\nu)|^2_{\nu=s/j} = \frac{1}{1+\epsilon^2 T_n^2(s/j)}$$

We obtain the poles of $H_N(s)$ from

$$1 + \epsilon^2 T_n^2(s/j) = 0 \quad \text{or} \quad T_n(s/j) = \cos[n\cos^{-1}(s/j)] = \pm j/\epsilon$$

It is convenient to define a new complex variable z in the form

$$z = \theta + j\alpha = \cos^{-1}(s/j) \quad \text{or} \quad s/j = \cos(z)$$

This results in

$$\cos(nz) = \cos(n\theta + jn\alpha) = \cos(n\theta)\cosh(n\alpha) - j\sin(n\theta)\sinh(n\alpha) = \pm j/\epsilon$$

Equating the real and imaginary parts, we obtain the two relations

$$\cos(n\theta)\cosh(n\alpha) = 0 \qquad \sin(n\theta)\sinh(n\alpha) = \pm 1/\epsilon$$

Since $\cosh(n\alpha) \geq 1$ for all $n\alpha$, the first relation gives

$$\cos(n\theta) = 0 \quad \text{or} \quad \theta_k = (2k-1)\pi/2n \quad k = 1, 2, \ldots, 2n$$

Note that θ_k (*in radians*) has the same form as for the Butterworth roots. The sign for θ_k establishes the sign for the real part of the roots. For positive θ, $\sin(n\theta) = 1$, and the second relation yields

$$\sinh(n\alpha) = 1/\epsilon \quad \text{or} \quad \alpha = (1/n)\sinh^{-1}(1/\epsilon)$$

Since $s/j = \cos(z)$, we have

$$s = j\cos(z) = j\cos(\theta_k + j\alpha) = \sin(\theta_k)\sinh(\alpha) + j\cos(\theta_k)\cosh(\alpha)$$

The poles of $H_N(s)$ are the LHP roots with negative real parts and are

$$p_k = -\sin(\theta_k)\sinh(\alpha) + j\cos(\theta_k)\cosh(\alpha) \qquad k = 1, 2, \ldots, n$$

With $p_k = \sigma_k + j\omega_k$, the real and imaginary parts of the poles are given by

$$\sigma_k = -\sin(\theta_k)\sinh(\alpha) = -\sinh[(1/n)\sinh^{-1}(1/\epsilon)]\sin(\theta_k)$$

$$\omega_k = \cos(\theta_k)\cosh(\alpha) = \cosh[(1/n)\sinh^{-1}(1/\epsilon)]\cos(\theta_k)$$

Using the identity $\sin^2\alpha + \cos^2\alpha = 1$, we also obtain

$$\frac{\sigma_k^2}{\sinh^2(\alpha)} + \frac{\omega_k^2}{\cosh^2(\alpha)} = 1$$

This suggests that *the Chebyshev poles lie on an ellipse* in the s-plane, with a major semiaxis (along the $j\omega$-axis) that equals $\cosh(\alpha)$ and a minor semiaxis (along the σ-axis) that equals $\sinh(\alpha)$ (Figure 10.9).

Comparing Chebyshev and Butterworth Poles   The poles of the normalized Butterworth filter transfer function are described by

$$\sigma_{BN} = -\sin(\theta_k) \qquad \omega_{BN} = \cos(\theta_k)$$

Comparison reveals that the real part of the Chebyshev pole is scaled by $\sinh(\alpha)$, and the imaginary part by $\cosh(\alpha)$. Since Q depends on the ratio of the imaginary and real parts, the Q of the Chebyshev poles is scaled by $\coth(\alpha)$, and is larger than that for a Butterworth filter of the same order. This is a consequence of passband ripples.

Geometric Construction for Pole Locations   The Chebyshev poles may be found directly by evaluating the n normalized Butterworth poles (with $\epsilon = 1$) that lie on a unit circle and then contracting this circle to an ellipse. This gives the Chebyshev poles as

$$p_C = \sigma_{BN}\sinh(\alpha) + j\omega_{BN}\cosh(\alpha) \qquad \alpha = (1/n)\sinh^{-1}(1/\epsilon)$$

and leads to the following geometric evaluation (Figure 10.9):

Figure 10.9   **The lowpass Chebyshev poles lie on an ellipse and can be found from the Butterworth poles using a geometric construction.**

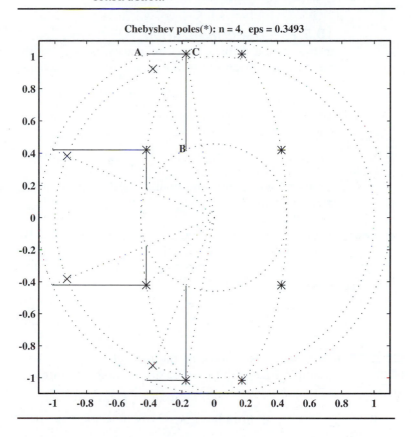

Chebyshev poles(*): n = 4, eps = 0.3493

1. Draw concentric circles of radii $\cosh(\alpha)$ and $\sinh(\alpha)$ with radial lines along θ_k, the angular orientation of the Butterworth poles.
2. On each radial line, label the point of intersection with the larger circle (of radius $\cosh(\alpha)$) as A_K, and with the smaller circle as B_k.
3. Draw a horizontal line through A_k, and a vertical line through B_k. Their point of intersection C_k represents a Chebyshev pole location.

The LPP transfer function $H_N(s) = K/Q_N(s)$ may be readily obtained in factored form by expressing the denominator $Q_N(s)$ as

$$Q_N(s) = (s - p_1)(s - p_2) \cdots (s - p_k)$$

Each conjugate pole pair will yield a quadratic factor. For odd n, $Q_N(s)$ will also contain the linear factor $s + \sinh(\alpha)$.

The constant K in $H_N(s) = K/Q_N(s)$ is chosen for a desired gain. For a peak gain of unity, $K = Q_N(0)$ for odd n, and $K = Q_N(0)/\sqrt{1 + \epsilon^2}$ for even n. For unit dc gain, $K = Q_N(0)$ for any filter order.

10.5.7   *Denormalization*

In the standard form, denormalization is achieved by using $\omega_D = \omega_p$ to exactly meet the specifications at the passband edge ω_p. This gives

$$H_A(s) = H_N(s/\omega_D) = H_N(s/\omega_p)$$

To exactly meet requirements at any other frequency ω_x, we find the actual frequency $\tilde{v}_x$ corresponding to A_x from the expression for $A(v)$ with $v = \tilde{v}_x$, as

$$\tilde{v}_x = \cosh\left\{ \frac{1}{n} \cosh^{-1}\left[\frac{1}{\epsilon}(10^{0.1A_x} - 1)^{1/2} \right] \right\}$$

We then choose the denormalizing frequency ω_D to ensure that $\omega_D \tilde{v}_x = \omega_x$. This gives $\omega_D = \omega_x/\tilde{v}_x$ and $H(s) = H_N(s/\omega_D)$.

Remarks: 1. For $n = 1$, $T_1(v) = v$ and $|H(v)|^2 = 1/(1 + \epsilon^2 v^2)$. This is identical to the form for a first-order Butterworth filter.

2. The following relations may sometimes prove helpful in computations involving Chebyshev filter design:

$$\cosh^{-1} x = \ln[x + (x^2 - 1)^{1/2}] \qquad \sinh^{-1} x = \ln[x + (x^2 + 1)^{1/2}] \qquad x \geq 1$$
$$\cos^{-1} x = \tan^{-1}[(1 - x^2)^{1/2}/x] \qquad \sin^{-1} x = \tan^{-1}[x/(1 - x^2)^{1/2}] \qquad x < 1$$

Example 10.7
Chebyshev Lowpass
Filter Poles

To find the LHP poles of the normalized 3rd-order Chebyshev lowpass filter of Example 10.6 (with $\epsilon^2 = 0.2589$), we first compute

$$\alpha = (1/n)\sinh^{-1}(1/\epsilon) = 0.4760$$
$$\theta_k(\text{rad}) = (2k - 1)\pi/2n = (2k - 1)\pi/6° \qquad (k = 1, 2, 3)$$

The LHP poles $s_k = -\sinh(\alpha)\sin(\theta_k) + j\cosh(\alpha)\cos(\theta_k)$ then yield

$$s_1 = -\sinh(0.476)\sin(\pi/6) + j\cosh(0.476)\cos(\pi/6) = -0.2471 + j0.966$$
$$s_2 = -\sinh(0.476)\sin(\pi/2) + j\cosh(0.476)\cos(\pi/2) = -0.4942$$
$$s_3 = -\sinh(0.476)\sin(5\pi/6) + j\cosh(0.476)\cos(5\pi/6) = -0.2471 - j0.966$$

The two conjugate poles, s_1 and s_3, correspond to the quadratic term

$$(s + 0.2471 - j0.966)(s + 0.2471 + j0.966) = (s^2 + 0.4942s + 0.9492)$$

With $H_N(s) = K/Q_N(s)$, the factored and polynomial forms for $Q_N(s)$ are thus

$$Q_N(s) = (s + 0.4942)(s^2 + 0.4942s + 0.9492) = s^3 + 0.9883s^2 + 1.2384s + 0.4913$$

For *unit dc gain*, we choose $K = Q_N(0) = 0.4913$. To satisfy the specifications of Example 10.6 at the passband edge, we must denormalize using $\omega_D = 4$ to obtain

$$H_A(s) = H_N(s/4) = 31.4436/(31.4436 + 19.8145s + 3.9534s^2 + s^3)$$

The magnitude response of this filter is shown in Figure 10.10a.

Figure 10.10   Magnitude response of the lowpass and bandpass Chebyshev filters designed in Examples 10.7–10.8.

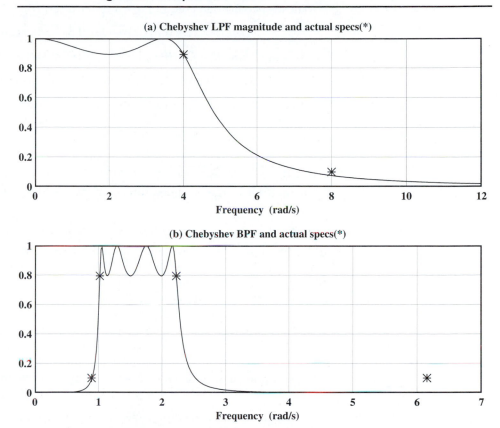

(a) Chebyshev LPF magnitude and actual specs(*)

(b) Chebyshev BPF and actual specs(*)

Comment: To exactly meet specs at ω_s, we choose $\omega_D = \omega_s / \tilde{\nu}_s$ where

$$\tilde{\nu}_s = \cosh\left\{\frac{1}{n}\cosh^{-1}\left[\frac{1}{\epsilon}(10^{0.1A_s}-1)^{1/2}\right]\right\}$$

We compute $\tilde{\nu}_s = 1.8441$ and $\omega_D = \omega_s / \tilde{\nu}_s = 4.3381$ and $H_A(s) = H_N(s/4.3381)$. We now meet specs at ω_s and exceed them at ω_p. As a check, we compute

$$A_p = A(\omega_p / \omega_D) = 0.151 \text{ dB} \qquad A_s = A(\omega_s / \omega_D) = 20 \text{ dB} \qquad \underline{}$$

Example 10.8

Chebyshev Bandpass Filter Design

Let us design a Chebyshev bandpass filter for which we are given:
Band-edges: $[\omega_1, \omega_2, \omega_3, \omega_4] = [0.89, 1.019, 2.221, 6.155]$ rad/s
Maximum Passband Attenuation: $A_p = 2$ dB
Minimum Stopband Attenuation: $A_s = 20$ dB
The frequencies are not geometrically symmetric. Following Section 9.4.8, we assume fixed passband edges, and compute $\omega_x^2 = \omega_2 \omega_3$. Since $\omega_1 \omega_4 > \omega_x^2$, we use Table 9.10 to recompute $\omega_4 = \omega_x^2 / \omega_1 = (1.019)(2.221)/0.89 = 2.54$. Finally, with

$B_w = \omega_3 - \omega_2 = 2.221 - 1.019 = 1.202$, we find the LPP specifications $\omega_{Ap} = 1$, $\omega_{As} = (\omega_4 - \omega_1)/B_w = (2.54 - 0.89)/1.202 = 1.3738$.

These specifications require a 4th-order filter, $H_N(s) = K/Q_N(s)$, where

$$Q_N(s) = s^4 + 0.7162s^3 + 1.2565s^2 + 0.5168s + 0.2058$$

For *peak unit gain*, $K = Q_N(0)/\sqrt{1 + \epsilon^2} = 0.1634$ (since n is even), and we get

$$H_N(s) = 0.1634/(s^4 + 0.7162s^3 + 1.2565s^2 + 0.5168s + 0.2058)$$

We transform this using the LP2BP transformation $s \longrightarrow (s^2 + \omega_x^2)/sB_w$, where $\omega_x = (\omega_2\omega_3)^{1/2} = [(1.019)(2.221)]^{1/2} = 1.5045$, $B_w = 1.202$. This leads to an 8th-order analog bandpass filter with $H_{BP}(s)$ described by

$$\frac{0.34s^4}{s^8 + 0.86s^7 + 10.87s^6 + 6.75s^5 + 39.39s^4 + 15.27s^3 + 55.69s^2 + 9.99s + 26.25}$$

The magnitude of the designed filter is shown in Figure 10.10b.   ▬▬

10.5.8   *A Design That Meets Additional Specifications*

Often, the design also specifies a half-power frequency ω_3. In this case, we start by normalizing the frequencies with respect to the passband edge and computing ϵ^2 and n as before. The filter order that meets specifications at the half-power frequency is then found iteratively as follows.

1. We find ν_3 and choose a normalizing frequency ω_N which ensures $\omega_3/\omega_N = \nu_3$ ($\omega_N = \omega_3/\nu_3$). We use this to compute $\tilde{\nu}_s = \omega_s/\omega_N$ and $A(\tilde{\nu}_s)$.
2. If $A(\tilde{\nu}_s) \geq A_s$, we are done since we now meet or exceed the stopband specifications. If not, we increase n by 1 and repeat the previous step.

Subsequent steps for finding $H_N(s)$ are identical to the standard design. Final denormalization now uses the denormalizing frequency $\omega_D = \omega_3/\nu_3$.

▬▬▬▬▬▬

Example 10.9

Chebyshev Lowpass Filter Design

We design a Chebyshev lowpass filter to meet the specifications: $A_p \leq 1$ dB for $\omega \leq 4$ rad/s, $A_s \geq 20$ dB for $\omega \geq 8$ rad/s, and $\omega_3 = 6$ rad/s. We normalize by ω_p to get $\nu_p = 1$ and $\nu_s = \omega_s/\omega_p = 8/6$. Next, we compute

$$\epsilon^2 = 10^{0.1A_p} - 1 = 0.2589 \quad n = 2.783 \longrightarrow n = 3 \quad \nu = 1.0949$$

Since $\omega_3 = 6$, we pick $\omega_n = \omega_3/\tilde{\nu}_3$. We then successively increase n and compute $\tilde{\nu}_3, \omega_N = \omega_3/\tilde{\nu}_3, \tilde{\nu}_s = \omega_s/\omega_N$ and $\tilde{A}_s = A(\tilde{\nu}_s)$ until $\tilde{A}_s \geq A_s$. We find

n	$\tilde{\nu}_3$	ω_N	$\tilde{\nu}_s = \omega_s/\omega_N$	$\tilde{A}_s = A(\tilde{\nu}_s)$	A_s (given)	$\tilde{A}_s \geq A_s$?
3	1.0949	5.4801	1.4598	12.5139	20	No
4	1.0530	5.6980	1.4040	18.4466	20	No
5	1.0338	5.8037	1.3784	24.8093	20	Yes

Thus, the required filter order is $n = 5$, and $\omega_N = \omega_3/\tilde{\nu}_3 = 5.8037$. We find $Q_N(s)$ as before, choose $K = Q_N(0)$ for peak unit gain, and obtain the normalized transfer function as $H_N(s) = K/Q_N(s)$. Finally, we denormalize using $\omega_D = \omega_N$ to obtain $H_A(s) = H_N(s/\omega_N) = H_N(s/5.8037)$. This design exactly meets specs at ω_3 and exceeds them at both ω_p and ω_s. The attenuation equation confirms that

$$A_p = A(\omega_p/\omega_D) = 0.4 \text{ dB}, \quad A_3 = A(\omega_3/\omega_D) = 3.01 \text{ dB}, \quad A_s = A(\omega_s/\omega_D) = 24.81 \text{ dB}$$

The magnitude of the designed filter is shown in Figure 10.11.

Figure 10.11   **Magnitude response of the lowpass and bandpass Chebyshev filter designed in Example 10.9.**

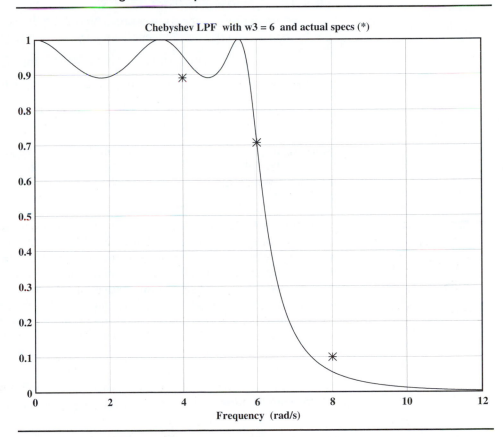

Chebyshev LPF with w3 = 6 and actual specs (*)

Frequency (rad/s)

10.5.9   *The High-Frequency Response*

Since $|T_n(\nu)|$ increases monotonically (and rapidly) for $\nu > 1$, the magnitude response exhibits a fairly steep monotonic decrease outside the passband. The attenuation at high frequencies may be approximated by

$$A(\nu) = 10\log[1 + \epsilon^2 T_n^2(\nu)] \approx 20\log \epsilon T_n(\nu)$$

Since $T_n(v) \longrightarrow 2^{n-1}v^n$ for large v, this approximation yields

$$A(v) \approx 20 \log \epsilon 2^{n-1}v^n = 20 \log \epsilon + 20(n-1) \log 2 + 20n \log v$$

The high-frequency response shows an attenuation rate of $20n$ dB/decade.

Remark: The Chebyshev filter provides an additional $6(n-1)$ dB of attenuation in the stopband compared with the Butterworth filter. This results in a much sharper transition for the same filter order. This improvement is made at the expense of introducing ripples in the passband, however.

10.5.10 *Phase Characteristics*

A closed-form rational function expression for the time delay involves the so-called *Chebyshev polynomial of the second kind*, $U_n(x)$, defined by

$$U_n(x) = \sin[(n+1)\cos^{-1}(x)] / \sin[\cos^{-1}(x)]$$

In terms of this function, the group delay $t_g(v)$ is given by

$$t_g(v) = \frac{1}{1 + \epsilon^2 T_n^2(v)} \sum_{k=1}^{n} \frac{U_{2k-2}(v)\sinh[(2n - 2k + 1)v]}{\epsilon^2 \sin \theta_k}$$

Since $T_n(v) \longrightarrow 2^{n-1}v^n$ for large v, the phase at high frequencies approaches $-n\pi/2$ radians. The phase response of the Chebyshev filter is not nearly as linear as that of a Butterworth filter of the same order.

10.6
The Inverse Chebyshev Approximation

The motivation for the inverse Chebyshev approximation stems from the need to develop a filter magnitude characteristic with a steep transition like the Chebyshev filter and a phase response with good delay properties like the Butterworth filter. We therefore seek a magnitude response with a maximally flat passband to improve the delay performance and somehow retain equiripple behavior to ensure a steep transition. The equiripple property clearly dictates the use of Chebyshev polynomials. To improve delay properties, we need, somehow, to transfer the ripples to a region outside the passband. This is achieved by a frequency transformation that reverses the characteristics of the normal Chebyshev response. This results in equiripple behavior in the stopband and a monotonic response in the passband, which, it turns out, is also maximally flat. This is the basis for the **inverse Chebyshev filter**, or **Chebyshev II filter**.

10.6.1 *The Inverse Chebyshev Filter*

Consider the normalized nth-order Chebyshev filter described by

$$|H_C(v)|^2 = \frac{1}{1 + L(v^2)} = \frac{1}{1 + \epsilon^2 T_n^2(v)}$$

The highpass transformation $v \longrightarrow 1/v$ results in

$$|H_C(1/v)|^2 = \frac{1}{1 + \epsilon^2 T_n^2(1/v)}$$

It translates the ripples to a band extending from $v = 1$ to $v = \infty$ and the monotonic response to the region $v \leq 1$ as required (Figure 10.12). What we need next is a means to convert this to a lowpass form. We cannot use the inverse transformation $v \longrightarrow 1/v$ again because it would relocate the ripples in the passband. Subtracting $|H_C(v)|^2$ from unity, however, results in

$$|H(v)|^2 = 1 - |H_C(1/v)|^2 = 1 - \frac{1}{1 + \epsilon^2 T_n^2(1/v)} = \frac{\epsilon^2 T_n^2(1/v)}{1 + \epsilon^2 T_n^2(1/v)}$$

Figure 10.12   Genesis of the lowpass inverse Chebyshev filter.

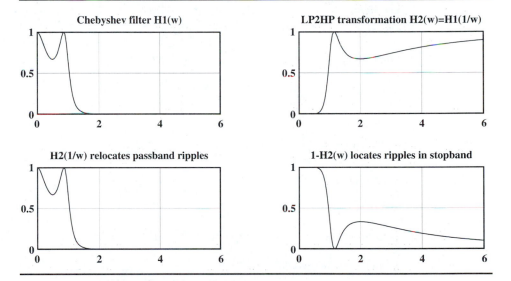

The function $|H(v)|^2$ now possesses a lowpass form that is monotonic in the passband and rippled in the stopband, just as required (Figure 10.12). The ripples in $|H(v)|^2$ start at $v = 1$, which defines the start of the stopband, the frequency where the response *first* reaches the maximum stopband magnitude. This suggests that we normalize the frequency with respect to ω_s, the stopband edge. In keeping with previous forms, we may express the inverse Chebyshev filter characteristic as

$$|H(v)|^2 = \frac{\epsilon^2 T_n^2(1/v)}{1 + \epsilon^2 T_n^2(1/v)} = \frac{1}{1 + [1/\epsilon^2 T_n^2(1/v)]} = \frac{1}{1 + L_n^2(v)}$$

The function $L_n^2(v)$ is now a rational function rather than a polynomial. To show that $|H(v)|^2$ yields a maximally flat response at $v = 0$, we start with $T_n(1/v)$ (using $n = 2$ and $n = 3$ as examples)

$$T_2(1/v) = 2(1/v)^2 - 1 = \frac{2 - v^2}{v^2} \qquad T_3(1/v) = 4(1/v)^3 - 1/v = \frac{4 - 3v^2}{v^3}$$

Substituting into $|H(\nu)|^2$, we obtain

$$|H_2(\nu)|^2 = \frac{\epsilon^2(4 - 4\nu^2 + \nu^4)}{\epsilon^2(4 - 4\nu^2 + \nu^4) + \nu^4} \qquad |H_3(\nu)|^2 = \frac{\epsilon^2(16 - 24\nu^2 + 9\nu^4)}{\epsilon^2(16 - 24\nu^2 + 9\nu^4) + \nu^6}$$

Both have the required form $N(\nu)/[N(\nu) + A\nu^{2n}]$ for maximal flatness.

10.6.2   *Design Recipe for Inverse Chebyshev Filters*

The design steps for inverse Chebyshev filters are listed in Table 10.8.

Table 10.8   Lowpass Inverse Chebyshev Filter Design. Note: ω_3 = half-power frequency (3-dB frequency).

Design step	If ω_3 is not specified	If ω_3 is also specified
Specifications	$A_p, A_s, \omega_p, \omega_s$	$A_p, A_s, \omega_p, \omega_s$ and ω_3
Finding ϵ^2	$1/\epsilon^2 = 10^{0.1A_s} - 1$	$1/\epsilon^2 = 10^{0.1A_s} - 1$
Normalization	$\omega_N = \omega_s$ (stopband edge)	$\omega_N = \omega_3/\nu_3$ (3-dB frequency)
Finding n	$n = \dfrac{\cosh^{-1}[1/(10^{0.1A_p} - 1)/\epsilon^2]^{1/2}}{\cosh^{-1}(\omega_s/\omega_p)}$	Find ω_N and compute $\tilde{A}_s$. If $\tilde{A}_s < A_s$, let $n = n + 1$, and loop
3-dB frequency	$1/\nu_3 = \cosh\left[\dfrac{1}{n}\cosh^{-1}\left(\dfrac{1}{\epsilon}\right)\right]$	$1/\nu_3 = \cosh\left[\dfrac{1}{n}\cosh^{-1}\left(\dfrac{1}{\epsilon}\right)\right]$
Passband/Stopband attenuation	$10\log\{1 + [\epsilon^2 T_n^2(\omega_s/\omega_p)]^{-1}\}$	$10\log\{[1 + [\epsilon^2 T_n^2(\omega_3/\nu_3\omega_s)]^{-1}\}$
Normalized poles	$p_N = \dfrac{1}{-\sin(\theta_k)\sinh(\alpha) + j\cos(\theta_k)\cosh(\alpha)}$	$\alpha = (1/n)\sinh^{-1}(1/\epsilon)$ $\theta_k = (2k - 1)\pi/2n$
Normalized zeros	$z_k = j\omega_k = j\sec(\theta_k) \qquad k = 1, 2, \ldots, \text{int}(n/2)$	
Transfer function	$H_N(s) = K\dfrac{P_N(s)}{Q_N(s)} = K\dfrac{(s^2 + \omega_1^2)(s^2 + \omega_2^2)\cdots(s^2 + \omega_m^2)}{(s - p_1)(s - p_2)\cdots(s - p_n)}$	$m = \text{int}(n/2)$
K for unit gain (dc or peak)	$K = Q_N(0)/P_N(0)$	
Denormalization	$\omega_D = \begin{array}{l} \omega_s \quad \text{(stopband edge)} \\ \omega_p/\tilde{\nu}_p \quad \text{(passband edge)} \end{array}$	$\omega_D = \omega_3/\nu_3$ (3-dB frequency)
Denormalized $H_A(s)$	$H_A(s) = H_N(s/\omega_D)$	$H_A(s) = H_N(s/\omega_D)$

For the inverse Chebyshev filter, we normalize by the stopband frequency ω_s such that $\nu_s = 1$ and $\nu_p = \omega_p/\omega_s$. The design attenuation equation is

$$A(\nu) = 10\log[1/|H(\nu)|^2] = 10\log(1 + [1/\epsilon^2 T_n^2(1/\nu)])$$

We find ϵ^2 by evaluating the attenuation at the stopband $\nu = \nu_s = 1$ to give

$$A_s = A(\nu_s) = 10\log(1 + 1/\epsilon^2) \quad \text{or} \quad 1/\epsilon^2 = 10^{0.1A_s} - 1$$

If ω_s equals the half-power frequency, we find that $\epsilon^2 = 1$.

The order n is established by evaluating the attenuation at $\nu = \nu_p$:

$$A_p = A(\nu_p) = 10\log(1 + [\epsilon^2 T_n^2(1/\nu_p)]^{-1})$$

Since $v_p < 1$ and $1/v_p > 1$, evaluation of $T_n(1/v_p)$ requires the hyperbolic form and yields

$$A_p = 10 \log \left[1 + \frac{1}{\{\epsilon \cosh[n(\cosh^{-1} 1/v_p)]\}^2} \right]$$

The order n is then obtained from this relation as

$$n = \frac{\cosh^{-1}[1/(10^{0.1A_p} - 1)\epsilon^2]^{1/2}}{\cosh^{-1}(1/v_p)} = \frac{\cosh^{-1}[(10^{0.1A_s} - 1)/(10^{0.1A_p} - 1)]^{1/2}}{\cosh^{-1}(\omega_s/\omega_p)}$$

The second form for n is identical to that for a Chebyshev I filter.

The half-power frequency v_3, at which $|H(v)|^2 = \frac{1}{2}$, can be expressed in terms of ϵ and n as

$$1/v_3 = \cosh[(1/n)\cosh^{-1}(1/\epsilon)] \qquad \epsilon \le 1$$

This form is analogous to the *reciprocal* of the relation for finding v_3 for the Chebyshev I filter.

Example 10.10
Inverse Chebyshev Filter Attenuation

We design an inverse Chebyshev lowpass filter to meet the specifications $A_p \le 1$ dB for $\omega \le 4$, and $A_s \ge 20$ dB for $\omega \ge 8$. Using the design equations, we compute:

$$v_p = \omega_p/\omega_s = \tfrac{1}{2}, \quad v_s = 1, \quad \epsilon^2 = 1/(10^{0.1A_s} - 1) = 0.0101 \qquad n = 2.783 \longrightarrow n = 3$$

From $1/v_3 = \cosh[(1/n)\cosh^{-1}(1/\epsilon)]$, the half-power frequency is $v_3 = 0.65$. From the attenuation equation, we compute

$$A_p = A(v_p) = 0.5936 \text{ dB} \qquad A_s = A(v_s) = 20 \text{ dB}$$

We thus *exactly* meet the specs at ω_s, and exceed the specs at ω_p.

10.6.3   The Stopband Ripple

The quantity $1/\epsilon^2$ is related to the *stopband ripple* just as ϵ^2 was related to the passband ripple in the Chebyshev I filter. Note, however, that *the parameter ϵ^2* (which is a measure of the ripple in both cases) *is not the same* for both.

The magnitude at the stopband edge $v_s = 1$ is given by

$$|H(1)|^2 = \frac{1}{1 + (1/\epsilon^2)} = \frac{\epsilon^2}{1 + \epsilon^2}$$

In contrast to the Chebyshev I filter, the dc gain $|H(0)|$ always equals 1 regardless of filter order, since $T_n(1/v) \longrightarrow \infty$ for any n.

10.6.4   *The Normalized Transfer Function*

To find $H(s)$, we start with the substitution $v = s/j$ into $|H(v)|^2$ to obtain the relation for $H(s)H(-s)$

$$H(s)H(-s) = |H(v)|^2_{v=s/j} = \frac{\epsilon^2 T_n^2(1/v)}{1 + \epsilon^2 T_n^2(1/v)} \Big|_{v=s/j}$$

The poles of the transfer function $H(s)$ represent the LHP roots p_k of

$$[1 + \epsilon^2 T_n^2(1/v)]|_{v=s/j} = 0$$

This is similar to the corresponding relation for the Chebyshev I filter. The argument of T_n here is $1/v$ instead of v. This means that we can actually apply the Chebyshev I relations for evaluating the inverse Chebyshev roots provided we use the appropriate value for ϵ that applies here and invert the resulting relations to obtain the actual roots.

The LHP poles p_k may then be written explicitly as $p_k = 1/(\tilde{\sigma}_k + j\tilde{\omega}_k)$, with

$$\tilde{\sigma}_k = -\sinh[(1/n)\sinh^{-1}(1/\epsilon)]\sin(\theta_k) \quad \tilde{\omega}_k = \cosh[(1/n)\sinh(1/\epsilon)]\cos(\theta_k)$$

where $\theta_k = (2k - 1)\pi/2n, k = 1, 2, \ldots, n$. Note that ϵ now refers to the inverse Chebyshev parameter.

For odd n, we have a real root $s = -1/\sinh[(1/n)\sinh^{-1}(1/\epsilon)]$. The poles of the inverse Chebyshev filter are not simply reciprocals of the Chebyshev I poles, because of the different ϵ in the two cases. Nor do they lie on an ellipse, even though their locus does resemble an ellipse that is elongated, much like a dumbbell.

The zeros of the inverse Chebyshev representation are just the roots of

$$T_n^2(1/v)|_{v=s/j} = 0$$

Recall that the zeros v_z of a Chebyshev polynomial are given by

$$v_z = \cos(\theta_k) \qquad k = 1, 2, \ldots, n$$

Replacing v_z by s/j, ignoring the root at infinity for odd n (corresponding to $\theta_k = \frac{1}{2}\pi$), and invoking conjugate symmetry, the zeros are given by

$$z_k = \pm j\sec(\theta_k) = \pm j\omega_k \qquad k = 1, 2, \ldots, \text{int}(\tfrac{1}{2}n)$$

These conjugate terms yield the factors $s^2 + \omega_k^2$, and we may express $H_N(s)$ as

$$H_N(s) = K\frac{P_n(s)}{Q_N(s)} = K\frac{(s^2 + \omega_1^2)(s^2 + \omega_2^2)\cdots(s^2 + \omega_m^2)}{(s - p_1)(s - p_2)\cdots(s - p_n)} \quad m = \text{int}(\tfrac{1}{2}n)$$

To ensure unit dc gain (or unit peak gain), we choose $K = Q_N(0)/P_N(0)$.

Remark: For the same values of n and ϵ, the coefficients of $Q_N(s)$ are identical to those of a Chebyshev I transfer function denominator $Q_C(s)$ in reversed order. The two are thus related by $Q_N(s) = s^n Q_C(1/s)$.

10.6.5   *Denormalization*

To exactly meet attenuation specifications at the stopband edge requires denormalization of $H_N(s)$ using $\omega_D = \omega_s$ to give $H_A(s) = H_N(s/\omega_s)$.

To exactly meet requirements at any other frequency ω_x, we find the actual frequency $\tilde{v}_x$ corresponding to $A_x = A(\tilde{v}_x)$ from the attenuation equation as

$$1/\tilde{v}_x = \cosh\left\{\frac{1}{n}\cosh^{-1}\left[\frac{1}{\epsilon(10^{0.1A_x} - 1)^{1/2}}\right]\right\}$$

We then choose the denormalizing frequency ω_D to ensure that $\omega_D \tilde{v}_x = \omega_x$. This gives $\omega_D = \omega_x/\tilde{v}_x$ and $H_A(s) = H_N(s/\omega_D)$.

Example 10.11

Inverse Chebyshev Filter Poles

To find $H_N(s)$ for the normalized 3rd-order inverse Chebyshev lowpass filter of Example 10.8 with $\epsilon^2 = 0.0101$, we successively compute

$$\alpha = (1/n)\sinh^{-1}(1/\epsilon) = 0.9977 \quad \theta_k = (2k-1)\pi/2n = (2k-1)\pi/6° \quad (k = 1, 2, 3)$$

Then $1/s_k = -\sinh(\alpha)\sin(\theta_k) + j\cosh(\alpha)\cos(\theta_k)$ yields

$$1/s_1 = -\sinh(0.9977)\sin(\pi/6) + j\cosh(0.9977)\cos(\pi/6) = -0.5859 + j1.3341$$

$$1/s_2 = -\sinh(0.9977)\sin(\pi/2) + j\cosh(0.9977)\cos(\pi/2) = -1.1717$$

$$1/s_3 = -\sinh(0.9977)\sin(5\pi/6) + j\cosh(0.9977)\cos(5\pi/6) = -0.5859 - j1.3341$$

The LHP poles are $[s1, s2, s3] = [-0.276 - j0.6284, -0.8534, -0.276 + j0.6284]$. The two conjugate poles correspond to the quadratic term

$$(s + 0.276 + j0.6284)(s + 0.276 - j0.6284) = (s^3 + 0.552s + 0.471)$$

The factored and polynomial forms for $Q_N(s)$ are thus

$$Q_N(s) = (s + 0.853)(s^2 + 0.552s + 0.471) = s^3 + 1.405s^2 + 0.942s + 0.402$$

Since $n = 3$, we have only two LHP zeros given by $\pm j\sec(\theta_1) = \pm j1.1547$. The numerator $P_N(s)$ corresponding to this is $(s^2 + 1.1547^2) = (s^2 + 1.3333)$. Thus, $H_N(s) = KP_N(s)/Q_N(s) = K(s^2 + 1.3333)/(s^3 + 1.4054s^2 + 0.9421s + 0.402)$. For unit dc gain, we choose $K = Q_N(0)/P_N(0) = 0.402/1.3333 = 0.3015$. To satisfy the specifications of Example 10.8 at the stopband edge, we denormalize using $\omega_D = \omega_s = 8$ to obtain $H_A(s) = H_N(s/8)$ as

$$H_A(s) = (205.832 + 2.412s^2)/(205.832 + 60.294s + 11.243s^2 + s^3)$$

Comment: To exactly meet the passband specs, we choose $\omega_D = \omega_p/\tilde{v}_p$, where

$$1/\tilde{v}_p = \cosh\left\{\frac{1}{n}\cosh^{-1}\left[\frac{1}{\epsilon(10^{0.1A_p} - 1)^{1/2}}\right]\right\}$$

We compute $\tilde{v}_p = 0.5423$, $\omega_D = \omega_p/\tilde{v}_p = 7.3765$ and $H_A(s) = H_N(s/7.3765)$. The design exactly meets specs at ω_p. In fact, the attenuation confirms that

$$A_p = A(\omega_p/\omega_D) = 1 \text{ dB} \qquad A_s = A(\omega_s/\omega_D) = 28.6083 \text{ dB}$$

The magnitude of this filter is shown in Figure 10.13a (p. 440).

Figure 10.13   Magnitude response of the inverse Chebyshev filters designed in Examples 10.11–10.12.

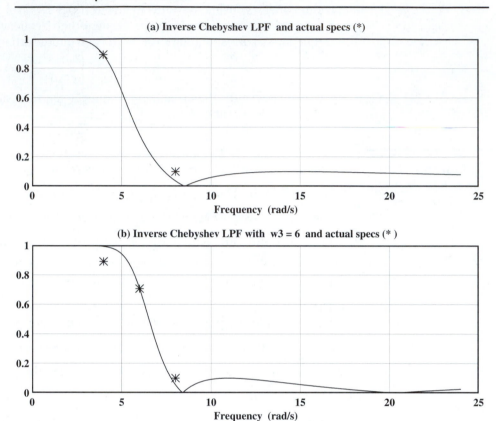

(a) Inverse Chebyshev LPF and actual specs (*)

(b) Inverse Chebyshev LPF with w3 = 6 and actual specs (*)

10.6.6   *A Design That Meets Additional Specifications*

Often, the design also specifies the half-power frequency ω_3. In this case, we start by normalizing the frequencies with respect to the stopband edge, and find ϵ^2 and n as before. The filter order that meets the half-power specifications is then found iteratively as follows.

1. We find ν_3 and choose a normalizing frequency ω_N that ensures $\omega_3/\omega_N = \nu_3$ or $\omega_N = \omega_3/\nu_3$. We use this to compute $\tilde{\nu}_s = \omega_s/\omega_N$ and $A(\tilde{\nu}_s)$.

2. If $A(\tilde{\nu}_s) \geq A_s$, we are done, since we now meet or exceed the stopband specifications. If not, we increase n by 1 and repeat the previous step.

Knowing n and ϵ, we find $H_N(s)$ as in the standard design. Final denormalization uses $\omega_D = \omega_N = \omega_3/\nu_3$ and yields $H_A(s) = H_N(s/\omega_N)$.

Example 10.12
Inverse Chebyshev Filter
Design

Consider the design of a Chebyshev II lowpass filter that meets the specifications $A_p \leq 1$ dB for $\omega \leq 4$, $A_s \geq 20$ dB for $\omega \geq 8$ and $\omega_3 = 6$.

From the Example 10.11, we have $\epsilon^2 = 0.0101$, $n = 3$, and $\nu_3 = 0.65$. Since $\omega_3 = 6$, we pick $\omega_N = \omega_3/\tilde{\nu}_3$. We successively increase n and compute $\tilde{\nu}_3$, ω_N, $\tilde{\nu}_s = \omega_s/\omega_N$ and $\tilde{A}_s = A(\tilde{\nu}_s)$ until $\tilde{A}_s \geq A_s$. We find

n	$\tilde{\nu}_3$	ω_N	$\tilde{\nu}_s = \omega_s/\omega_N$	$\tilde{A}_s = A(\tilde{\nu}_s)$	A_s (given)	$\tilde{A}_s \geq A_s$?
3	0.65	9.231	0.8667	11.6882	20	No
4	0.7738	7.7535	1.0318	25.2554	20	Yes

Thus the required filter order is $n = 4$ and $\omega_N = \omega_3/\tilde{\nu}_3 = 7.7535$.

We now compute $H_N(s)$ and denormalize it using $\omega_D = \omega_N = 7.7535$ to get $H_A(s) = H_N(s/7.7535)$. This design exactly meets specs at ω_3. We use the attenuation equation, with $\nu = \omega/\omega_D$, to confirm that

$$A_p = A(\omega_p/\omega_D) = 0.061 \text{ dB}, \quad A_3 = A(\omega_3/\omega_D) = 3.01 \text{ dB}$$
$$A_s = A(\omega_s/\omega_D) = 25.25 \text{ dB}$$

The filter magnitude is shown in Figure 10.13b.

Note: With ω_3 not specified, both Chebyshev filter types (I and II) yield $n = 3$ for identical specifications. With ω_3 also given, the Chebyshev I filter yields $n = 5$, but the Chebyshev II filter yields $n = 4$ (a lower order).

10.6.7   Concluding Remarks

The inverse Chebyshev filter inherits the improved delay of the maximally flat Butterworth filter and the steeper transition of the Chebyshev I filter by permitting ripples in the stopband. The poles have a smaller Q compared with a Chebyshev I filter of the same order, making it less sensitive to component variations. The finite transmission zeros, however, result in a more complex realization (with more elements).

10.7
The Elliptic
Approximation

The elliptic approximation provides ripples in both bands by using rational functions whose numerator and denominator both display the equiripple property. Such an approximation can be described by

$$|H(\nu)|^2 = \frac{1}{1 + L_n^2(\nu)} = \frac{1}{1 + \epsilon^2 R_n^2(\nu, \delta)}$$

Here, $R_n(\nu, \delta)$ is the so-called **Chebyshev rational function**, ϵ^2 describes the passband ripple and the additional parameter δ provides a measure of the stopband ripple magnitude. Clearly, $R_n(\nu, \delta)$ must be sought in the rational function form $A(\nu^2)/B(\nu^2)$, with both the numerator and denominator satisfying near optimal constraints and possessing properties analogous to Chebyshev polynomials. This implies that, for a given order n,

1. $R_n(\nu, \delta)$ should exhibit oscillatory behavior with equal extrema in the passband ($|\nu| < 1$), with all its n zeros within the passband.
2. $R_n(\nu, \delta)$ should exhibit oscillatory behavior, with equal extrema in the stopband ($|\nu| > 1$), with all its n poles within the stopband.
3. $R_n(\nu, \delta)$ should be even if n is even and odd if n is odd.

We also impose the additional simplifying constraint

$$R_n(1/\nu, \delta) \propto 1/R_n(\nu, \delta)$$

This provides a reciprocal relation between the poles and zeros of R_n and suggests that if we can find a function of this form with equiripple behavior in the passband $0 < \nu < 1$, it will automatically result in equiripple behavior in the reciprocal range $1 < \nu < \infty$ representing the stopband. The functional form for $R_n(\nu, \delta)$ that meets these criteria may be described in terms of its root locations ν_k by

$$R_n(\nu, \delta) = C\nu^N \prod_{k=1}^{\text{int}(n/2)} \frac{\nu^2 - \nu_k^2}{\nu^2 - B/\nu_k^2}$$

where $\textbf{int}(x)$ denotes the integer part of x, ν_k is a root of the numerator, B and C represent constants, and $N = 0$ for even n and $N = 1$ for odd n.

The Elliptic Approximation   An alternative to the trial-and-error method for establishing $R_n(\nu, \delta)$ is to find a transformation $\nu = g(\phi)$ that transforms R_n, a polynomial in ν, to a periodic function in ϕ much as $\nu = \cos \phi$ transformed the Chebyshev polynomial $T_n(\nu)$ into the periodic function $\cos(n\phi)$ with its attendant equiripple character. It turns out that such a transformation is indeed possible but involves *elliptic integrals* and *elliptic functions*. Before examining this transformation, we introduce the concept of elliptic functions and integrals.

10.7.1   *Elliptic Integrals and Elliptic Functions*

The **elliptic integral of the first kind**, u, is actually a function of two arguments, the *amplitude*, ϕ, and the *parameter*, m, and is defined by

$$u(\phi, m) = \int_0^{\phi} (1 - m \sin^2 x)^{-1/2}\, dx$$

The dependence of u on ϕ is better visualized by taking the derivative

$$\frac{du}{d\phi} = (1 - m \sin^2 \phi)^{-1/2}$$

With $m = 0$, we have the trivial result

$$\left.\frac{du}{d\phi}\right|_{m=0} = 1 \qquad u(\phi, 0) = \phi$$

If the amplitude ϕ equals $\pi/2$, we get the **complete elliptic integral of the first kind**, $u(\pi/2, m)$, denoted by $K(m)$, which is now a function only of m:

$$K(m) = u(\pi/2, m) = \int_0^{\pi/2} (1 - m \sin^2 x)^{-1/2} \, dx$$

The **modulus** k is related to the parameter m by $m = k^2$. The **complementary complete elliptic integral of the first kind** is denoted by $K'(m)$. With $k'^2 = 1 - k^2$ and $m' = 1 - m$, it is defined by

$$K'(m) = K(m') = u(\pi/2, m') = \int_0^{\pi/2} (1 - m' \sin^2 x)^{-1/2} \, dx$$

We thus have $K'(m) = K(1 - m)$. The quantity m' is called the **complementary parameter**, and k' is called the **complementary modulus**.

The **Jacobian elliptic functions** also involve two arguments, one being the elliptic integral $u(\phi, m)$, and the other the parameter m. Some of these functions are

Jacobian elliptic sine	$\text{sn}(u, m) = \sin \phi$
Jacobian elliptic cosine	$\text{cn}(u, m) = \cos \phi$
Jacobian elliptic difference	$\text{dn}(u, m) = d\phi/du$

All display periodicity, resemble trigonometric functions, and even satisfy some of the same identities. For example,

$$\text{sn}^2 + \text{cn}^2 = 1 \quad \text{cn}^2 = 1 - \text{sn}^2 \quad \text{dn} = d\phi/du = (1 - m \sin^2 \phi)^{1/2} = (1 - m\,\text{sn}^2)^{1/2}$$

The Jacobian elliptic sine, $\text{sn}(u, m)$ (Figure 10.14, p. 444), behaves much like the trigonometric sine, except that it is flattened and elongated. The degree of elongation (called the *period*) and flatness depend on the parameter m. For $m = 0$, $\text{sn}(u, m)$ is identical to $\sin(\phi)$. For small u, $\text{sn}(u, m)$ closely follows the sine and with increasing m, becomes more flattened and elongated, reaching an infinite period when $m = 1$. This ability of $\text{sn}(u, m)$ to change shape with changing m is what provides us with the means to characterize the function R_n.

The complete elliptic integral K and its complement K' are related to the Jacobian sine function by the relations

$$K(m) = u(\pi/2, m) = \int_0^1 [(1 - z^2)(1 - mz^2)]^{-1/2} \, dz \qquad z = \text{sn}(au + b, m)$$
$$K'(m) = \int_1^{1/k} [(z^2 - 1)(1 - mz^2)]^{-1/2} \, dz \qquad z = \text{sn}(au + b, m)$$

The Jacobian elliptic functions are actually *doubly periodic* for complex arguments. In particular, $\text{sn}(u, m)$, where u is complex, has a real period of $4K$ and an imaginary period of $2K'$.

10.7.2 *The Chebyshev Rational Function*

The transformation for ν in the rational function $R_n(\nu, \delta)$ that achieves the desired equiripple behavior is of the form $\nu = \text{sn}(\alpha K, m)$, where K is itself the complete

---

Figure 10.14   The Jacobian elliptic sine sn(u, m).

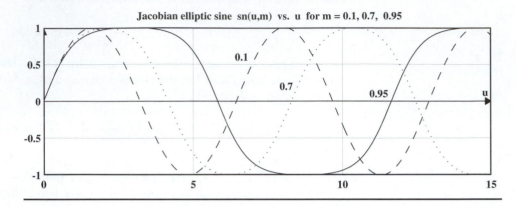

elliptic integral of the first kind. Without going into the involved details of its development, the transformation $v = \text{sn}(\alpha K, m)$ results in a Jacobian elliptic function of the form

$$R_n(v, \delta) = Cv^N \prod_{k=1}^{\text{int}(n/2)} \frac{v^2 - z_k^2}{v^2 - p_k^2} \qquad N = \begin{cases} 0 & n \text{ even} \\ 1 & n \text{ odd} \end{cases}$$

Here, $N = 0$ for even n and $N = 1$ for odd n. The constant C is found by recognizing that $R_n(v, \delta) = 1$ at $v = 1$, leading to

$$C = \prod_{k=1}^{\text{int}(n/2)} \frac{1 - p_k^2}{1 - z_k^2} \qquad k = 1, 2, \ldots, n$$

The poles p_k and zeros z_k are imaginary, and their locations depend on the order n. They are described in terms of elliptic functions and integrals as

$$R_n(v, \delta) = Cv^N \prod_{k=1}^{\text{int}(n/2)} \frac{v^2 - [\text{sn}(u_k, m)]^2}{v^2 - [v_s/\text{sn}(u_k, m)]^2} \qquad \begin{array}{ll} u_k = (2k-1)K(m)/n, & n \text{ even} \\ u_k = 2kK(m)/n, & n \text{ odd} \\ m = 1/v_s^2 \end{array}$$

Here, v_s is the first value of v at which $R_n(v, \delta) = \delta$, sn is the Jacobian elliptic sine function, $K(m) = K(1/v_s^2)$ is the complete elliptic integral of the first kind and δ is the maximum magnitude of oscillations for $|v| > v_s$. The zeros of $R_n(v, \delta)$, with $K(m) = K(1/v_s^2)$, are given by

$$z_k = \begin{cases} \text{sn}[2kK(m)/n], & n \text{ odd} \\ \text{sn}[(2k-1)K(m)/n], & n \text{ even} \end{cases} \qquad k = 1, 2, \ldots, \text{int}(\tfrac{1}{2}n)$$

and the pole locations are related to the zeros by

$$p_k = v_s/z_k \qquad k = 1, 2, \ldots, \text{int}(\tfrac{1}{2}n)$$

The frequencies v_{pass} of the passband maxima are given by

$$v_{\text{pass}} = \begin{cases} \text{sn}[(2k+1)K(m)/n], & n \text{ odd} \\ \text{sn}[2kK(m)/n], & n \text{ even} \end{cases} \qquad k = 1, 2, \ldots, \text{int}(\tfrac{1}{2}n)$$

The frequencies v_{stop} of the stopband maxima are related to v_{pass} by

$$v_{\text{stop}} = v_s/v_{\text{pass}}$$

At these frequencies, $R_n(v, \delta)$ is given by

$$R_n(v, \delta) = \begin{cases} \pm 1 & v = v_{\text{pass}} \\ \pm \delta & v = v_{\text{stop}} \end{cases}$$

It also obeys the reciprocal relation $R_n(v, \delta) = \delta/R_n(v_s/v, \delta)$.

Filter Order   Like the Jacobian elliptic functions, $R_n(v, \delta)$ is also doubly periodic. The periods are not independent but are related through the order n. With $m = 1/v_s^2$ (as before), and $p = 1/\delta_s^2$, this relationship is given by

$$n = \frac{K(m)K'(p)}{K'(m)K(p)} = \frac{K(1/v_s^2)K'(1/\delta^2)}{K'(1/v_s^2)K(1/\delta^2)} = \frac{K(1/v_s^2)K(1-1/\delta^2)}{K(1-1/v_s^2)K(1/\delta^2)}$$

10.7.3   *Design Recipe for Elliptic Filters*

The design recipe for elliptic filters is summarized in Table 10.9 (p. 446).

We normalize by the passband edge to obtain $v_p = 1$ and $v_s = \omega_s/\omega_p$. The parameters ϵ and δ are found from the attenuation relation

$$A(v) = 10\log[1 + \epsilon^2 R_n^2(v, \delta)]$$

At the passband edge $v = v_p = 1$, $R_n(v_p, \delta) = 1$, and we find

$$A_p = A(v_p) = 10\log[1 + \epsilon^2] \qquad \text{or} \qquad \epsilon^2 = 10^{0.1 A_p} - 1$$

If ω_p equals the half-power frequency $(A_p = 3.01 \text{ dB})$, *then $\epsilon^2 = 1$.* At the stopband edge $v = v_s$, $R_n(v_s, \delta) = \delta$, and

$$A_s = A(v_s) = 10\log[1 + \epsilon^2 \delta^2] \qquad \text{or} \qquad \delta^2 = (10^{0.1 A_s} - 1)/\epsilon^2$$

The order n is next evaluated from the relation

$$n = \frac{K(1/v_s^2)K'(1/\delta^2)}{K'(1/v_s^2)K(1/\delta^2)} = \frac{K(1/v_s^2)K(1-1/\delta^2)}{K(1-1/v_s^2)K(1/\delta^2)}$$

If n does not work out to an integer, we must (iteratively) find a value of v_s and $K(1/v_s^2)$ that satisfies the above relation for the chosen integer n. This ensures that the design *exactly* meets passband specifications.

Table 10.9   Lowpass Elliptic Filter Design.

Design step	Design relation
Specifications	$G_{dB}, A_p, A_s, \omega_p, \omega_s$
Normalization	$\omega_N = \omega_p$ (passband edge)
Finding ϵ^2 and δ^2	$\epsilon^2 = 10^{0.1A_p} - 1$ and $\delta^2 = (10^{0.1A_s} - 1)/\epsilon^2$
Constants m and p	$m = 1/v_s^2 = (\omega_p/\omega_s)^2$ $p = 1/\delta^2$
Finding n	$n = \dfrac{K(m)K'(p)}{K'(m)K(p)} = \dfrac{K(1/v_s^2)K'(1/\delta^2)}{K'(1/v_s^2)K(1/\delta^2)}$
Normalized zeros and poles	$z_N = \pm jp_k$ $$p_N = \text{LHP roots of } \prod^{\text{int}(n/2)} (v^2 - p_k^2)^2 - (\epsilon Cv^N)^2 \prod^{\text{int}(n/2)} (v^2 - z_k^2)^2 = 0$$ where $$z_k = \begin{cases} sn[2kK(m)/n] & n \text{ odd} \\ sn[(2k-1)K(m)/n & n \text{ even} \end{cases} \quad \begin{matrix} N = 0 & (n \text{ even}) \\ N = 1 & (n \text{ odd}) \end{matrix} \quad C = \prod_{k=1}^{\text{int}(n/2)} \dfrac{1 - p_k^2}{1 - z_k^2}$$ $p_k = v_s/z_k$
Normalized $H_N(s)$	$H_N(s) = K\dfrac{P(s)}{Q(s)} = K\dfrac{\prod^{\text{int}(n/2)}(s^2 + |z_N|^2)}{\prod^n(s - p_N)}$
K for unit peak gain	$K = \begin{cases} Q(0)/[P(0)\sqrt{1 + \epsilon^2}] & n \text{ even} \\ Q(0)/P(0) & n \text{ odd} \end{cases}$
Denormalization	$\omega_D = \omega_p$ (passband edge)
Denormalized $H_A(s)$	$H_N(s/\omega_D)$

To find the transfer function $H(s) = KP(s)/Q(s)$, we start with $|H(v)|^2$:

$$|H(v)|^2 = \frac{1}{1 + \epsilon^2 R_n^2(v, \delta)}$$

$$= \frac{\prod^{\text{int}(n/2)} (v^2 - p_k^2)^2}{\prod^{\text{int}(n/2)} (v^2 - p_k^2)^2 - (\epsilon Cv^N)^2 \prod^{\text{int}(n/2)} (v^2 - z_k^2)^2}$$

The zeros, z_N, of $H(s)$ are the imaginary poles p_k of $R_n(v, \delta)$ with $v \longrightarrow s/j$ and thus $z_{Nk} = \pm jp_k$. These conjugate symmetric zeros yield the numerator of $H(s)$ as a function of s^2. The denominator of $H(s)$ requires the LHP roots of

$$\prod^{\text{int}(n/2)} (v^2 - p_k^2)^2 - (\epsilon Cv^N)^2 \prod^{\text{int}(n/2)} (v^2 - z_k^2)^2 = 0 \qquad N = \begin{cases} 0 & n \text{ even} \\ 1 & n \text{ odd} \end{cases}$$

The constant C is defined as before, and equals

$$C = \prod_{k=1}^{\text{int}(n/2)} \frac{1 - p_k^2}{1 - z_k^2} \qquad z_k = \begin{cases} sn[2kK(m)/n, & n \text{ odd} \\ sn[(2k-1)K(m)/n], & n \text{ even} \end{cases} \qquad k = 1, 2, \ldots, \text{int}(\tfrac{1}{2}n)$$

For $n > 2$, finding the denominator roots often requires numerical methods. For unit dc gain, we choose $K = Q(0)/P(0)$. For a peak gain of unity, we choose $K = Q(0)/P(0)$ for odd n, and $K = Q(0)/[P(0)\sqrt{1 + \epsilon^2}]$ for even n (since $H(0) = 1$ for odd n and $H(0) = 1/\sqrt{1 + \epsilon^2}$ for even n). Final denormalization uses $\omega_D = \omega_p$ to yield $H_A(s) = H_N(s/\omega_D) = H_N(s/\omega_p)$.

Example 10.13
Elliptic Filter Design

Consider the design of an elliptic lowpass filter to meet the following specifications: $A_p = 2$ dB, $A_s = 20$ dB, $\omega_p = 4$ rad/s, $\omega_s = 8$ rad/s.

Following the design procedure, we normalize by ω_p to get $\nu_p = 1$, $\nu_s = \omega_s/\omega_p = 2$. We compute

$$\epsilon^2 = 10^{0.1A_p} - 1 = 0.5849 \text{ and } \delta^2 = (10^{0.1A_s} - 1)/\epsilon^2 = 169.2617$$

We let $m = 1/\nu_s^2 = 0.25$ and $p = 1/\delta^2 = 0.0059$. Then, $K(m) = 1.6858$, $K'(m) = K(1 - m) = 2.1565$, $K(p) = 1.5731$ and $K'(p) = K(1 - p) = 3.9564$. The filter order is given by

$$n = \frac{K(m)K'(p)}{K'(m)K(p)} = \frac{(1.6858)(3.9564)}{(2.1565)(1.5731)} = 1.97$$

We choose $n = 2$. Note that the actual ν_s that yields $n = 2$ is $\nu_s = 1.942$, but we develop the following design using $\nu_s = 2$. With $n = 2$,

$$\text{sn}[(2k - 1)K(m)/n, m] = \text{sn}[K(m)/n, m] = \text{sn}(0.8429, 0.25) = 0.7321$$

The zeros and poles of $R_n(\nu, \delta)$ are then $z_k = 0.7321$ and $p_k = \nu_s/z_k = 2.7321$. The zeros of $H(s)$ are at the poles of R_n. Thus $z_n = \pm j2.7321$, and with $H(s) = KP(s)/Q(s)$, we get $P(s) = s^2 + 2.7321^2 = s^2 + 7.4641$. To find $Q(s)$, we first compute the constant C as

$$C = \prod_{k=1}^{\text{int}(n/2)} \frac{1 - p_k^2}{1 - z_k^2} = \frac{1 - 7.4641}{1 - 0.5359} = -13.9282$$

and, with $N = 0$, we evaluate the LHP roots of

$$\prod^{\text{int}(n/2)} (\nu^2 - p_k^2)^2 - (\epsilon C \nu^N)^2 \prod^{\text{int}(n/2)} (\nu^2 - z_k^2)^2 = (\nu^2 - p^2)^2 - \epsilon^2 C^2 (\nu^2 - z^2)^2 = 0$$

Upon simplification, we obtain $114.47\nu^4 - 136.54\nu^2 + 88.299 = 0$. The two LHP roots are $p_N = -0.3754 \pm j0.8587$, and $Q(s)$ thus equals

$$Q(s) = (s + 0.3754 + j0.8587)(s + 0.3754 - j0.8587) = s^2 + 0.7508s + 0.8783$$

The normalized transfer function may now be written as

$$H_N(s) = K\frac{s^2 + 7.4641}{s^2 + 0.7508s + 0.8783}$$

For a *peak* gain of unity, we choose $K = 0.8783/(7.4641\sqrt{1 + \epsilon^2}) = 0.09347$. With $\omega_D = \omega_p = 4$, the denormalized transfer function $H_A(s)$ becomes

$$H_A(s) = H_N(s/\omega_p) = H_N(s/4) = \frac{0.09347s^2 + 11.1624}{s^2 + 3.0033s + 14.0527}$$

Figure 10.15 | $H(f)$| for the elliptic filter designed in Example 10.13.

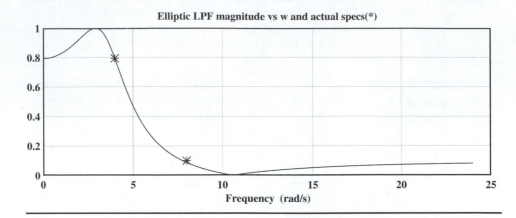

The magnitude spectrum of this filter is shown in Figure 10.15.

10.7.4 Concluding Remarks

Of all the approximations that we have discussed, elliptic filters yield the lowest filter order for given specifications by permitting ripples in both the passband and stopband. For a given order, they exhibit the steepest transition region, but the very presence of the ripples also makes for the most nonlinear phase and the worst delay characteristics.

10.8
The Bessel
Approximation

The Bessel approximation is used in filter design as a means of obtaining linear phase or flat delay in the passband. Unlike other filter forms, it is much easier to start with a transfer function $H(s)$ directly rather than derive it from a parametric model. Like other filters, however, we work with a normalized *unit delay*. Denormalization to a desired delay t_g is then accomplished by the transformation $s \longrightarrow t_g s$ in the transfer function $H(s)$. We start with the nth-order lowpass transfer function $H(s)$ given by

$$H(s) = \frac{K}{q_0 + q_1 s + q_2 s^2 + \cdots + q_n s^n}$$

The steady-state transfer function $H(\omega)$ may then be written as

$$H(\omega) = \frac{K}{(q_0 - q_2\omega^2 + q_4\omega^4 + \cdots) + j(q_1\omega - q_3\omega^3 + q_5\omega^5 + \cdots)}$$

And the phase $\phi(\omega)$ readily follows as

$$\phi(\omega) = -\tan^{-1}\left[\frac{q_1\omega - q_3\omega^3 + q_5\omega^5 + \cdots}{q_0 - q_2\omega^2 + q_4\omega^4 + \cdots}\right]$$

Recall that even though the phase function is transcendental in nature, its derivative, which represents the time delay, is a rational function. It is also an even function (of ω^2) with no pole at the origin.

We can use several approaches to find the coefficients q_k that result in a constant delay.

The Maximally Flat Approach   One method is to invoke maximal flatness and find the required polynomial coefficients by setting the first derivative of $\phi(\omega)$ to 1 (for unit delay) and all higher-order derivatives (except the nth) to zero. To illustrate this approach, consider a first-order transfer function $H(s)$ and its magnitude and phase described by

$$H(s) = K/(q_0 + s) \qquad H(\omega) = K/(q_0 + j\omega) \qquad \phi(\omega) = -\tan^{-1}(\omega/q_0)$$

To establish q_0, we seek a linear phase function $\tilde{\phi}(\omega) \approx -\omega$ for unit delay. With $\tan^{-1}(x) \approx x - \frac{1}{3}x^3 + \frac{1}{5}x^5 - \frac{1}{7}x^7 + \cdots$, $\phi(\omega)$ may be approximated by

$$\phi(\omega) = -\tan^{-1}(\omega q_1/q_0) \approx -[\omega/q_0 - \tfrac{1}{3}(\omega/q_0)^3 + \cdots]$$

Since only q_0 is involved, comparison with $\tilde{\phi}(\omega)$ gives $q_0 = 1$ and

$$H(s) = K/(1 + s)$$

For $K = 1$, the dc gain equals 1. The delay $-d\phi(\omega)/d\omega$ results in

$$t_g = -\frac{d\phi}{d\omega} = \frac{d}{d\omega}[\tan^{-1}(\omega)] = \frac{1}{1 + \omega^2}$$

and clearly represents a maximally flat function.

Consider a second-order transfer function with magnitude and phase

$$H(s) = \frac{K}{q_0 + q_1 s + s^2} \qquad H(\omega) = \frac{K}{(q_0 - \omega^2) + j\omega q_1}$$

$$\phi(\omega) = -\tan^{-1}\left[\frac{q_1 \omega}{q_0 - \omega^2}\right] = -\tan^{-1}\left[\frac{\omega q_1/q_0}{1 - \omega^2/q_0}\right]$$

Using $\tan^{-1}(x) \approx x - \frac{1}{3}x^3 + \frac{1}{5}x^5 - \frac{1}{7}x^7 + \cdots$, $\phi(\omega)$ can be approximated by

$$\phi(\omega) \approx -\left[\frac{\omega q_1/q_0}{1 - \omega^2/q_0}\right] + \frac{1}{3}\left[\frac{\omega q_1/q_0}{1 - \omega^2/q_0}\right]^3 - \cdots$$

If the denominators of each term are expressed in terms of the first-order binomial approximation $(1 - x)^{-n} \approx (1 + nx)$ for $x \ll 1$, we obtain

$$\phi(\omega) \approx -\omega q_1/q_0[1 + \omega^2/q_0] + \tfrac{1}{3}(\omega q_1/q_0)^3[1 + 3\omega^2/q_0] - \cdots$$

Simplifying and retaining terms to ω^3, we get

$$\phi(\omega) \approx -\omega q_1/q_0 - \omega^3[q_1/q_0^2 - \tfrac{1}{3}(q_1/q_0)^3]$$

Comparison with the required phase $\tilde{\phi}(\omega) \approx -\omega$ yields the *nonlinear* equations

$$q_1/q_0 = 1 \qquad \text{and} \qquad [q_1/q_0^2 - \tfrac{1}{3}(q_1/q_0)^3] = 0$$

Substitution of the first equation into the second gives $1/q_0 - 1/3 = 0$. We thus have $q_0 = 3$ and $q_1 = 3$. With these values, $H(s)$ becomes

$$H(s) = \frac{K}{3 + 3s + s^2}$$

We choose $K = 3$ for dc gain. Using the result

$$\frac{d}{d\omega} \tan^{-1}[h(\omega)] = \frac{1}{1 + h^2(\omega)} \frac{d}{d\omega}[h(\omega)]$$

the delay $-d\phi(\omega)/d\omega$ equals

$$t_g = -\frac{d\phi}{d\omega} = \frac{d}{d\omega}(\tan^{-1}[3\omega/(3 - \omega^2)]) = \frac{(9 + 3\omega^2)}{(9 + 3\omega^2) + \omega^4}$$

This is again a maximally flat function.

This approach becomes quite tedious for higher-order filters due to the nonlinear equations involved (even for the second-order case).

The Continued Fraction Approach   This approach is based on the fact that ideally, the filter output $x_0(t)$ should represent a replica of the input $x_i(t)$ but delayed or shifted by a normalized delay of 1 such that

$$x_0(t) = x_i(t - 1)$$

The transfer function of the filter that achieves this must be given by

$$H(s) = \frac{X_0(s)}{X_i(s)} = \frac{X_i(s)e^{-s}}{X_i(s)} = \frac{1}{e^s}$$

We approximate e^s to obtain $H(s) = P(s)/Q(s)$ in polynomial form. Now, for a stable, physically realizable filter, it turns out that if $Q(s)$ can be written as the sum of an even polynomial $E(s)$ and an odd polynomial $O(s)$, the ratio $E(s)/O(s)$ results in positive coefficients. We thus seek a form for e^s in terms of odd and even functions. Such a form is given by

$$e^s = \cosh(s) + \sinh(s)$$

Now, $\cosh(s)$ and $\sinh(s)$ may be approximated by the even and odd series

$$\cosh(s) = 1 + \frac{s^2}{2!} + \frac{s^4}{4!} + \cdots \qquad \sinh(s) = s + \frac{s^3}{3!} + \frac{s^5}{5!} + \cdots$$

Their ratio, written as a continued fraction, yields

$$\frac{\cosh(s)}{\sinh(s)} = \frac{E(s)}{O(s)} = \frac{1}{s} + \cfrac{1}{\frac{3}{s} + \cfrac{1}{\frac{5}{s} + \cfrac{1}{\frac{7}{s} + \cfrac{1}{\ddots \quad \frac{2n-1}{s} + \cfrac{1}{\ddots}}}}}$$

We see that all the terms in the continued fraction are positive.

We truncate this expansion to the required order, reassemble $E(s)/O(s)$ and establish the polynomial form $Q(s) = E(s) + O(s)$, and $H(s) = K/Q(s)$. Finally, we choose K so as to normalize the dc gain to unity.

Example 10.14
The Continued Fraction
Method

(a) For a second-order polynomial approximation to unit delay, we have

$$\frac{E(s)}{O(s)} = \frac{1}{s} + \frac{1}{\dfrac{3}{s}} = \frac{1}{s} + \frac{s}{3} = \frac{3 + s^2}{3s}$$

This leads to the polynomial representation

$$Q(s) = E(s) + O(s) = (3 + s^2) + (3s) = 3 + 3s + s^2$$

and the transfer function (normalized to unit dc gain)

$$H(s) = \frac{3}{3 + 3s + s^2}$$

This is identical to the $H(s)$ based on maximal flatness.

(b) A third-order polynomial would require the truncation

$$\frac{E(s)}{O(s)} = \frac{1}{s} + \frac{1}{\dfrac{3}{s} + \dfrac{1}{\dfrac{5}{s}}} = \frac{1}{s} + \frac{1}{\dfrac{3}{s} + \dfrac{s}{5}} = \frac{1}{s} + \frac{5s}{15 + s^2} = \frac{15 + 6s^2}{15s + s^3}$$

leading to the polynomial approximation

$$Q(s) = E(s) + O(s) = (15 + 6s^2) + (15s + s^3) = 15 + 15s + 6s^2 + s^3$$

The transfer function $H(s)$, normalized to unit dc gain, is then

$$H(s) = \frac{15}{15 + 15s + 6s^2 + s^3}$$

10.8.1   *Bessel Polynomials*

The continued fraction approach can actually be developed into systematic procedures for evaluating either $Q(s)$ or its coefficients and results in the family of so-called **Bessel polynomials** or **Thomson polynomials**, all of which result in linear phase over some range and in delays that satisfy the constraint of maximal flatness at the origin.

Table 10.10 (p. 452) shows the first few Bessel polynomials $Q(s)$ in both unfactored and factored form. The numerator K of the transfer function $H(s) = K/Q(s)$ simply equals the constant term in $Q(s)$ for unit dc gain. Bessel polynomials may be generated in one of two ways.

1. Starting with $Q_0(s) = 1$ and $Q_1(s) = s + 1$, we use recursion to generate

$$Q_n(s) = (2n - 1)Q_{n-1}(s) + s^2 Q_{n-2}(s) \qquad Q_0(s) = 1, \quad Q_1(s) = s + 1$$

Table 10.10   Bessel Polynomials.

Order n	Denominator polynomial $Q_n(s)$ in unfactored form
0	1
1	$s + 1$
2	$s^2 + 3s + 3$
3	$s^3 + 6s^2 + 15s + 15$
4	$s^4 + 10s^3 + 45s^2 + 105s + 105$
5	$s^5 + 15s^4 + 105s^3 + 420s^2 + 945s + 945$
6	$s^6 + 21s^5 + 210s^4 + 1{,}260s^3 + 4{,}725s^2 + 10{,}395s + 10{,}395$
7	$s^7 + 28s^6 + 378s^5 + 3{,}150s^4 + 17{,}325s^3 + 62{,}370s^2 + 135{,}135s + 135{,}135$

Order n	Denominator polynomial $Q_n(s)$ in factored form
0	1
1	$s + 1$
2	$s^2 + 3s + 3$
3	$(s + 2.32219)(s^2 + 3.67782s + 6.45944)$
4	$(s^2 + 5.79242s + 9.14013)(s^2 + 4.20578s + 11.4878)$
5	$(s + 3.64674)(s^2 + 6.70391s + 14.2725)(s^2 + 4.64934s + 18.15631)$

---

2. Starting with $Q_n(s) = q_0 + q_1 s + q_2 s^2 + \cdots + q_n s^n$, we find the q_k from

$$q_n = 1 \qquad q_k = \frac{(2n - k)!}{2^{n-k} \, k!(n - k)!} \qquad k = 0, 1, 2, \ldots, n - 1$$

As a check, the value of q_0 equals the product $(1)(3)(5) \cdots (2n - 1)$.

10.8.2   *Magnitude and Delay of Bessel Filters*

The First Approach   One approach for computing the magnitude and delay of Bessel filters is to start with the transfer function

$$H(s) = \frac{Q_n(0)}{Q_n(s)} = \frac{Q_n(0)}{q_0 + q_1 s + q_2 s^2 + \cdots + q_n s^n}$$

The denominator of the steady-state transfer function $H(\omega)$ may then be written as the sum of an even part $E(\omega)$ and an odd part $O(\omega)$ as

$$
\begin{aligned}
H_n(\omega) &= \frac{Q_n(0)}{(q_0 - q_2 \omega^2 + q_4 \omega^4 + \cdots) + j(q_1 \omega - q_3 \omega^3 + q_5 \omega^5 + \cdots)} \\
&= \frac{Q_n(0)}{E_n(\omega) + jO_n(\omega)}
\end{aligned}
$$

leading to

$$|H_N(\omega)|^2 = \frac{Q_n(0)}{E_n^2(\omega) + O_n^2(\omega)}$$

The phase $\phi(\omega)$ and delay t_g correspond to

$$\phi(\omega) = -\tan^{-1}\left[\frac{O(\omega)}{E(\omega)}\right]$$

$$t_g = -\frac{d\phi}{d\omega} = \left[\frac{E(\omega)\dfrac{dO(\omega)}{d\omega} - O(\omega)\dfrac{dE(\omega)}{d\omega}}{E^2(\omega) + O^2(\omega)} \right]$$

A simpler result emerges if we rewrite the expression for $|H_n(\omega)|^2$ as

$$E_n^2(\omega) + E_o^2(\omega) = Q_n(0)/|H_n(\omega)|^2$$

The delay for first- and second-order filters can then be described by

$$t_{g1} = \frac{1}{1+\omega^2} = 1 - \frac{\omega^2}{1+\omega^2} = 1 - \frac{\omega^2}{E_1^2 + O_1^2} = 1 - \omega^2|H(\omega)|^2/Q_1^2(0)$$

$$t_{g2} = \frac{(9+3\omega^2)}{(9+3\omega^2)+\omega^4} = 1 - \frac{\omega^4}{(9+3\omega^2)+\omega^4} = 1 - \frac{\omega^4}{E_2^2 + O_2^2}$$
$$= 1 - \omega^4|H(\omega)|^2/Q_2^2(0)$$

By induction, the delay of an nth-order transfer function equals

$$t_{gn} = 1 - \omega^{2n}/(E_n^2 + O_n^2) = 1 - \omega^{2n}|H(\omega)|^2/Q_n^2(0)$$

The Second Approach   A second approach to finding magnitude and delay is based on the relationship between *Bessel polynomials* and the more familiar *Bessel functions*. In fact, Bessel polynomials are so named because of this relationship. The **spherical Bessel function of the first kind**, $J_n(x)$, is related to the integer-order Bessel function $J_n(x)$ by

$$J(n,x) = J_n(x) = \sqrt{\pi/2x}J_{n+1/2}(x)$$

It forms the basis for relating Bessel polynomials to Bessel functions. In particular, the Bessel polynomial $Q_n(1/s)$ may be expressed as

$$Q_n(1/j\omega) = j^{-n}[\cos(n\pi)J_{-n-1}(\omega) - jJ_n(\omega)]\,\omega\exp(j\omega)$$

The expressions for the magnitude, phase and delay of the Bessel filter in terms of Bessel functions derive from this relationship and yield the following results:

$$|H(\omega)|^2 = \frac{Q_n^2(0)}{\omega^{2n+2}[J_n^2(\omega) + J_{-n-1}^2(\omega)]}$$

$$\phi(\omega) = -\omega + \tan^{-1}[\cos(n\pi)J_n(\omega)/J_{-n-1}(\omega)]$$

$$t_g = -\frac{d\phi}{d\omega} = \left[1 - \frac{1}{\omega^2[J_n^2(\omega) + J_{-n-1}^2(\omega)]}\right] = 1 - \omega^{2n}|H(\omega)|^2/Q_n^2(0)$$

Magnitude Response   As the filter order increases, the magnitude response of Bessel filters tends to a Gaussian shape (Figure 10.16, p. 454). On this basis, it turns out that the attenuation A_x at any frequency ω_x for reasonably large filter order ($n > 3$, say) can be approximated by

$$A_x \approx \frac{10\omega_x^2}{(2n-1)\ln(10)}$$

The frequency ω_p corresponding to a specified passband attenuation A_p equals

---

Figure 10.16 **Magnitude of Bessel lowpass filters of various orders. With increasing order, the magnitude approximates a Gaussian shape.**

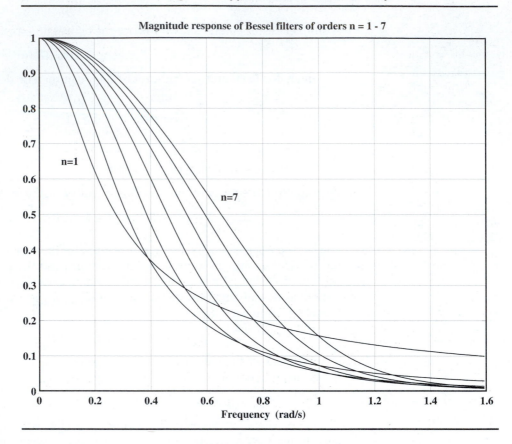

Magnitude response of Bessel filters of orders n = 1 - 7

$$\omega_p \approx \sqrt{0.1 A_p (2n - 1)\ln(10)} \qquad n \geq 3$$

In particular, the half-power frequency ω_3 is given by

$$\omega_3 \approx \sqrt{(2n - 1)\ln(2)} \qquad n \geq 3$$

10.8.3   *Design Specifications*

The design of Bessel filters actually involves finding just the filter order n for a set of specifications, which can include

1. A constant delay t_g (to within a prescribed tolerance), and a maximum attenuation A_p in the passband.

2. A constant delay t_g (to within a prescribed tolerance), and a rise time smaller than T_r. Specifying T_r is equivalent to specifying the half-power frequency ω_3, since the time-bandwidth product $T_r\omega_3$ is a constant, with $T_r\omega_3 \approx 2.2$.

10.8.4 Design Recipe for Bessel Filters

A formal design procedure involves the following steps:

1. We normalize the passband frequency ω_p to $\nu = \omega_p t_g$ for unit delay.
2. We either assume a low enough filter order n, or estimate n from

$$A_p \approx \frac{10\nu^2}{(2n-1)\ln(10)} \qquad n \approx \frac{5\nu^2}{A_p\ln(10)}$$

3. We find the normalized transfer function $H_N(s)$ and calculate $A(\nu)$ from either of the following:

$$|H(\nu)|^2 = \frac{Q_n^2(0)}{\nu^{2n+2}[J_n^2(\nu) + J_{-n-1}^2(\nu)]} \qquad A(\nu) = -10\log|H(\nu)|^2$$

$$H(\nu) = \frac{Q_n(0)}{E(\nu) + jO(\nu)} \quad |H(\nu)|^2 = \frac{Q_n(0)}{E^2(\nu) + O^2(\nu)} \qquad A(\nu) = -10\log|H(\nu)|^2$$

4. The delay is then computed from

$$t_n = 1 - \nu^{2n}|H(\nu)|^2/Q_n^2(0)$$

5. If $1 - t_n$ exceeds the given tolerance, or $A(\nu)$ exceeds A_p, we increase the filter order n by 1 and repeat steps 3 and 4.
6. Finally, we set up $H_N(s)$ and denormalize using $s \longrightarrow st_g$ to obtain $H_A(s) = H_N(st_g)$, which provides the required delay t_g.

Remark: Another popular, but much less flexible approach, resorts to nomograms of delay and attenuation for various filter orders to establish the smallest order that best matches specifications.

Example 10.15
Design of Bessel Filters

We wish to design a Bessel filter with a constant delay of $t_0 = 0.02s$ to within 0.1% up to $\omega = 100$ rad/s and an attenuation of no more than 2.5 dB.

The normalized frequency for unit delay is $\nu = \omega t_0 = (100)(0.02) = 2$. A first estimate of the filter order is given by $n \approx 5\nu^2/A_p\ln(10) = 5(4)/2.5\ln(10) \approx 3.47$. Thus we choose $n = 4$.

Next, we compute the actual attenuation and normalized delay. We illustrate by using the Bessel function approach and successively computing:

Coefficients of $Q_4(s) = [1, 18, 45, 105, 105]$; $Q_4(0) = 105$
$J(n, \omega_n) = J(4, 2) = 0.0141$, $J(-n-1, \omega_n) = J(-5, 2) = 4.4613$
$|H(\nu)|^2 = (105)^2/(2^{10})[(0.0141)^2 + (4.4613)^2] = 0.541$
$A_x = -10\log(0.541) = 2.67$ dB and $t_n = 1 - (2^8)(0.541)/(105)^2 = 0.987$

Since $1 - t_n > 0.001$ and $A(v) > A_p$, neither the attenuation nor the normalized delay meet requirements. We therefore increase n to 5 and recompute:

Coefficients of $Q_5(s) = [1, 15, 105, 420, 945, 945]$; $Q_5(0) = 945$
$J(n, \omega_n) = J(5, 2) = 0.0026$, $J(-n - 1, \omega_n) = J(-6, 2) = -18.5914$
$|H(v)|^2 = (945)^2/(2^{14})[(0.0026)^2 + (18.59)^2] = 0.631$
$A_x = -10\log(0.631) = 2.001$ dB and $t_n = 1 - (2^{10})(0.631)/(945)^2 = 0.9993$

Since t_n is within 0.1% of unity and $A_x \approx 2$, we are done and choose $n = 5$. The normalized transfer function $H_N(s)$ equals

$$H_N(s) = \frac{945}{s^5 + 15s^4 + 105s^3 + 420s^2 + 945s + 945}$$

With $t_0 = 0.02$, the actual transfer function is $H_A(s) = H_N(st_0) = H_N(s/50)$.
The magnitude, phase and delay of this filter is shown in Figure 10.17.

Figure 10.17   Characteristics of the Bessel filter of Example 10.15.

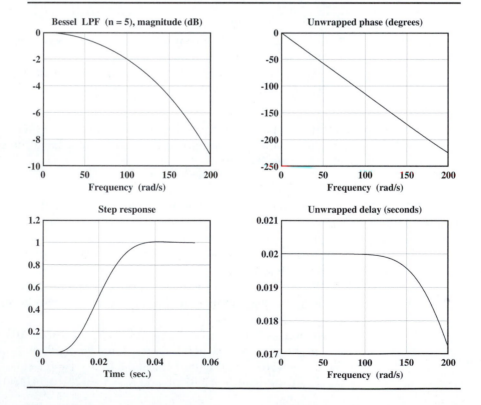

10.9
Analog Filters:
A Synopsis

Analog filter design	Analog filter design is based on polynomial or rational function approximations for the magnitude squared function to meet magnitude or phase specifications. Typically, these include the passband and stopband frequencies and attenuations and perhaps the 3-dB frequency.
Design approach	The conventional approach starts with a normalized lowpass prototype and subsequent transformation to the required form. The prototype is often designed with unit passband.
Design equation	The design equation is the attenuation of the required polynomial or rational function form. This yields the order n and ripple ϵ. The transfer function is then found from the magnitude specifications.
Butterworth filters	Butterworth filters are based on a polynomial approximation to achieve the flattest response at the origin. They are often normalized with respect to the half-power frequency. Their response is monotonic in both the passband and stopband and shows a fairly constant delay, except near the passband edge. The poles of the transfer function lie on a circle. They exhibit the slowest transition from the passband to the stopband for a given order and require the highest filter order for given specifications.
Chebyshev filters	Chebyshev filters are also based on a polynomial (Chebyshev) fit. Their response is monotonic in the stopband and rippled in the passband. The filter order equals the number of maxima and minima within the passband. The transfer function poles lie on an ellipse whose axes are related to the radius of the Butterworth filter poles. They provide a sharper transition and more attenuation in the stopband compared with Butterworth filters at the expense of ripples in the passband. The phase response of Chebyshev filters is not as linear as a Butterworth filter of the same order.
Inverse Chebyshev filters	Inverse Chebyshev filters are based on a rational function approximation involving Chebyshev polynomials. Their response is monotonic in the passband and rippled in the stopband. They inherit the improved delay of the Butterworth filter and the steeper transition of the Chebyshev filter by permitting ripples in the stopband.
Elliptic filters	Elliptic filters are based on rational function approximations involving elliptic functions. Their response is rippled in both bands. By permitting this, they yield the sharpest transition but also the worst delay for a given order. For given specifications, they require the smallest order.
Bessel filters	Bessel filters are based on approximations for linear phase. Their design specifications require a given delay and maximum attenuation over a desired frequency range. The design starts directly with a transfer function with unit normalized delay whose coefficients are based on a maximally flat approximation to unit delay. The magnitude response of Bessel filters tends to a Gaussian shape for large orders.

Appendix 10A MATLAB Demonstrations and Routines

Getting Started on the MATLAB Demonstrations

This appendix presents MATLAB routines related to concepts in this chapter and some examples of their use. See Appendix 1A for introductory remarks.

Note: All polynomial arrays must be entered in descending powers.

Classical Analog Filter Design

```
[tf,tfn]=afd(na,ty,a,fp,fs,f3,pl)
```

Designs classical analog filters. na(name) = 'bw', 'c1', 'c2' or 'el'. ty(type) = 'lp','hp','bp' or 'bs', pl='p' for plot or 'o' for omit plots, a = attenuation $[A_p \ A_s]$ in dB, fp & fs = passband & stopband edge(s) in Hz, f3 = OPTIONAL 3-dB edge(s) or center frequency (for bp and bs).
Returns the actual TF and the normalized LPP TF.

Example

To design a Butterworth bandpass filter with $A_p = 2$ dB, $A_s = 20$ dB, passband edges 200 Hz and 300 Hz and stopband edges 100 Hz and 400 Hz, use

```
>> [tf,tfn]=afd('bw','bp',[3 20],[200 300],[100 400],'p')
```

MATLAB plots the poles, magnitude, phase and delay and returns with

```
>> tfn=[0 0 0 1.3076; 1.0000 2.1870 2.3915 1.3076]
```

Bessel Filter Design

```
h=afdbess(d,fp,a)
```

Designs lowpass Bessel filters with delay d, passband edge fp and maximum passband attenuation a, and plots their magnitude, phase and delay. Returns h as denominator of actual and normalized TF. The numerator in in either case equals the constant (last) term of the denominator.

Example

To design a Bessel filter with a delay of 0.01 s up to 25 Hz and maximum passband attenuation of 1.5 dB, use

```
>>h=afdbess(0.01,25,1.5)
```

Demonstrations and Utility Functions

`delay('ty',n)`	Rational function for delay of filter type 'ty' and order n.
`buttpole(n)`	Plots the poles of a Butterworth LPP of order n.
`chebpole(n,r)`	Geometric construction of Chebyshev poles with ripple r (dB).
`chebyt(n,x)`	Returns the Chebyshev polynomial of order n.
`chcoeff(n)`	Computes the coefficients of nth-order Chebyshev polynomial.
`besstf(n)`	Computes coefficients of Bessel filter of order n.
`bessord(d,f,a)`	Computes order of Bessel LPF for given delay d,f (Hz) and A_p.
`lpp(ty,n,r)`	LPP for filter types ty, order n and ripple r dB.

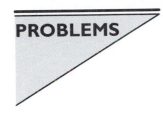

PROBLEMS

P10.1 (Butterworth Poles) Sketch the pole locations of a 3rd-order and 4th-order Butterworth lowpass filter assuming:
(a) Unit half-power frequency v_3.
(b) $\epsilon = 0.707$ and a passband of 1 rad/s.
(c) $\epsilon = 0.707$ and a passband edge of 100 Hz.

P10.2 (Butterworth Filters) Set up the form of $|H(v)|^2$ for a 3rd-order Butterworth lowpass filter with $\epsilon = 0.7$ and find:
(a) The attenuation at $v = 0.5$ and $v = 2$.
(b) The half-power frequency v_3.
(c) The pole orientations and magnitudes.
(d) The normalized transfer function $H(s)$.
(e) The high-frequency decay rate.

P10.3 (Butterworth Filters) Consider a 5th-order lowpass Butterworth filter with a passband of 1 kHz, and a maximum passband attenuation of 1 dB. What is the attenuation of this filter at 2 kHz?

P10.4 (Butterworth Filters) We wish to design a Butterworth lowpass filter with a 2-dB bandwidth of 1 kHz and an attenuation of at least 25 dB beyond 2 kHz. What is the actual attenuation of the designed filter at $f = 1$ kHz and $f = 2$ kHz?

P10.5 (Butterworth Filter Design) Design a Butterworth filter with a peak gain of 2 that meets the following sets of specifications:
(a) 3-dB passband of 0–4 rad/s and high frequency decay = 60 dB/decade.
(b) 1-dB passband of 0–10 Hz and filter order $n = 2$.
(c) Minimum gain = 1.8 over 0–5 Hz and filter order $n = 2$.

P10.6 (Butterworth Filter Design) Design a Butterworth filter that meets the following sets of specifications. Assume $A_p \leq 2$ dB and $A_s \geq 40$ dB. Assume a fixed passband, where necessary.
(a) Lowpass: passband < 2 kHz, stopband > 6 kHz.
(b) Highpass: passband > 4 kHz, stopband < 1 kHz.

(c) Lowpass: passband < 10 Hz, stopband: > 50 Hz, f_{3dB} = 15 Hz.
(d) Bandpass: passband 20–30 Hz, stopband: < 10 Hz and > 50 Hz.
(e) Bandstop: passband < 10 Hz and > 60 Hz, stopband: 20–25 Hz.

P10.7 (Butterworth Filter Design) Design a Butterworth *bandpass* filter to meet the following sets of specifications. Assume $A_p \le 2$ dB and $A_s \ge 40$ dB. Assume a fixed center frequency (if given) or a fixed passband (if given) or fixed stopband, in that order.
(a) Passband 30–50 Hz, stopband: < 5 Hz and > 400 Hz, f_0 = 40 Hz.
(b) Passband 20–40 Hz, stopband: < 5 Hz.
(c) Stopband: < 5 Hz and > 50 Hz, lower passband edge: 15 Hz, f_0 = 20 Hz.

P10.8 (Butterworth Filter Design) Design a Butterworth *bandstop* filter to meet the following sets of specifications. Assume $A_p \le 2$ dB and $A_s \ge 40$ dB. Assume a fixed center frequency (if given) or a fixed passband (if given) or fixed stopband, in that order.
(a) Passband < 20 Hz and > 50 Hz, stopband: 26–36 Hz, f_0 = 30 Hz.
(b) Stopband 30–100 Hz, lower passband edge= 50 Hz.
(c) Stopband: 50–200 Hz, lower passband edge: 80 Hz, f_0 = 90 Hz.

P10.9 (Chebyshev Poles) Analytically evaluate the poles of:
(a) A 3rd-order Chebyshev lowpass filter with $\epsilon = 0.5$.
(b) A 4th-order Chebyshev lowpass filter with passband ripple= 2 dB.

P10.10 (Chebyshev Poles) Use a geometric construction to locate the poles of a Chebyshev lowpass filter with:
(a) $\epsilon = 0.4$ and order $n = 3$.
(b) A passband ripple of 1 dB and order $n = 4$.

P10.11 (Chebyshev Filters) Set up $|H(\nu)|^2$ for a 3rd-order Chebyshev lowpass filter with $\epsilon = 0.7$ and find:
(a) The attenuation at $\nu = 0.5$ and $\nu = 2$.
(b) The half-power frequency ν_3.
(c) The pole orientations and magnitudes.
(d) The transfer function $H(s)$.
(e) The high-frequency decay rate.

P10.12 (Chebyshev Filters) Consider a 5th-order lowpass Chebyshev filter with a passband of 1 kHz and a passband ripple of 1 dB. What is the attenuation of this filter in decibels (dB) at $f = 1$ kHz and $f = 2$ kHz?

P10.13 (Chebyshev Filters) The minimum passband gain of a Chebyshev filter equals 9.55 when the maximum gain is 10. What is the value of the ripple factor ϵ?

P10.14 (Chebyshev Filters) A 4th-order Chebyshev filter shows a passband ripple of 1 dB up to 2 kHz. What is the normalized and actual half-power frequency of this filter?

P10.15 (Chebyshev Filter Design) Design a Chebyshev filter with unit peak gain that meets the following sets of specifications:
(a) 3-dB passband of 0–4 rad/s and high-frequency decay = 60 dB/decade.

(b) 1-dB passband of 0–10 Hz and filter order $n = 2$.

(c) Minimum gain = 0.9 over 0–5 Hz and filter order $n = 2$.

P10.16 (Chebyshev Filter Design) Design a Chebyshev filter to meet the following sets of specifications. Assume $A_p \leq 2$ dB and $A_s \geq 40$ dB. Assume a fixed passband, where necessary.

(a) Lowpass: passband < 2 kHz, stopband > 6 kHz.

(b) Highpass: passband > 4 kHz, stopband < 1 kHz.

(c) Lowpass: passband < 10 Hz, stopband: > 50 Hz, $f_{3dB} = 15$ Hz.

(d) Bandpass: passband 20–30 Hz, stopband: < 10 Hz and > 50 Hz.

(e) Bandstop: passband < 10 Hz and > 60 Hz, stopband: 20–25 Hz.

P10.17 (Chebyshev Filter Design) Design a Chebyshev *bandpass* filter that meets the following sets of specifications. Assume $A_p \leq 2$ dB and $A_s \geq 40$ dB. Assume a fixed center frequency (if given) or a fixed passband (if given) or fixed stopband, in that order.

(a) Passband 30–50 Hz, stopband: < 5 Hz and > 400 Hz, $f_0 = 40$ Hz.

(b) Passband 20–40 Hz, stopband: < 5 Hz.

(c) Stopband: < 5 Hz and > 50 Hz, lower passband edge: 15 Hz, $f_0 = 20$ Hz.

P10.18 (Chebyshev Filter Design) Design a Chebyshev *bandstop* filter that meets the following sets of specifications. Assume $A_p \leq 2$ dB and $A_s \geq 40$ dB. Assume a fixed center frequency (if given) or a fixed passband (if given) or fixed stopband, in that order.

(a) Passband < 20 Hz and > 50 Hz, stopband: 26–36 Hz, $f_0 = 30$ Hz.

(b) Stopband 30–100 Hz, lower passband edge= 50 Hz.

(c) Stopband: 50–200 Hz, lower passband edge: 80 Hz, $f_0 = 90$ Hz.

P10.19 (Inverse Chebyshev Poles) Find the poles and pole locations of a 3rd-order inverse Chebyshev lowpass filter with $\epsilon = 0.1$.

P10.20 (Inverse Chebyshev Filters) Set up $|H(\nu)|^2$ for a 3rd-order inverse Chebyshev lowpass filter with $\epsilon = 0.1$ and find

(a) The attenuation at $\nu = 0.5$ and $\nu = 2$.

(b) The half-power frequency ν_3.

(c) The pole orientations and magnitudes.

(d) The transfer function $H(s)$.

(e) The high-frequency decay rate.

P10.21 (Inverse Chebyshev Filters) Consider a 4th-order inverse Chebyshev lowpass filter with a stopband past 2 kHz and a stopband ripple of 40 dB. What is the attenuation of this filter in decibels (dB) at $f = 1$ kHz and $f = 2$ kHz?

P10.22 (Inverse Chebyshev Filter Design) Design an inverse Chebyshev filter to meet the following sets of specifications. Assume $A_p \leq 2$ dB and $A_s \geq 40$ dB. Assume a fixed passband, where necessary.

(a) Lowpass: passband < 2 kHz, stopband > 6 kHz.

(b) Highpass: passband > 4 kHz, stopband < 1 kHz.

(c) Lowpass: passband < 10 Hz, stopband: > 50 Hz, $f_{3dB} = 15$ Hz.

(d) Bandpass: passband 20–30 Hz, stopband: < 10 Hz and > 50 Hz.

(e) Bandstop: passband < 10 Hz and > 60 Hz, stopband: 20–25 Hz.

P10.23 (Inverse Chebyshev Filter Design)   Design an inverse Chebyshev *bandpass* filter that meets the following sets of specifications. Assume $A_p \leq 2$ dB and $A_s \geq 40$ dB. Assume a fixed center frequency (if given) or a fixed passband (if given) or fixed stopband, in that order.
(a) Passband 30–50 Hz, stopband: < 5 Hz and > 400 Hz, $f_0 = 40$ Hz.
(b) Passband 20–40 Hz, stopband: < 5 Hz.

P10.24 (Inverse Chebyshev Filter Design)   Design an inverse Chebyshev *bandstop* filter that meets the following sets of specifications. Assume $A_p \leq 2$ dB and $A_s \geq 40$ dB. Assume a fixed center frequency (if given) or a fixed passband (if given) or fixed stopband, in that order.
(a) Passband < 20 Hz and > 50 Hz, stopband: 26–36 Hz, $f_0 = 30$ Hz.
(b) Stopband 30–100 Hz, lower passband edge $= 50$ Hz.

P10.25 (Elliptic Filter Design)   Design a filter that meets the following sets of specifications. Assume $A_p \leq 2$ dB and $A_s \geq 40$ dB. Assume a fixed passband, where necessary.
(a) Lowpass: passband < 2 kHz, stopband > 6 kHz.
(b) Highpass: passband > 6 kHz, stopband: < 1 kHz.

P10.26 (Bessel Filters)   Design a Bessel filter with a passband edge of 100 Hz and a passband delay of 1 ms with a 1% tolerance for each of the following cases:
(a) The passband edge corresponds to the half-power bandwidth.
(b) The maximum attenuation in the passband is 0.3 dB.

P10.27 (The Maximally Flat Concept)   The general form of a maximally flat rational function is $R(\nu) = N(\nu)/[N(\nu) + A\nu^n]$, where $N(\nu)$ is a polynomial of order less than n.
(a) What is $N(\nu)$ for a Butterworth lowpass filter?
(b) Show that $R(\nu)$ is maximally flat for $n = 2$ using $N(\nu) = b_0 + b_1\nu$.
(c) Show that $R(\nu)$ is maximally flat for $n = 3$ using $N(\nu) = b_1\nu$.

P10.28 (Delay)
(a) Find a general expression for the delay of a 2nd-order lowpass filter described by the transfer function $H(s) = K/(s^2 + As + B)$.
(b) Use the results of part (a) to compute the delay of the following filters and explain if any of the delays has a maximally flat form.
   1. A 2nd-order normalized Butterworth lowpass filter.
   2. A 2nd-order normalized Chebyshev lowpass filter with $A_p = 1$ dB.
   3. A 2nd-order normalized Bessel lowpass filter.

11 SAMPLING AND THE DTFT

11.0 Scope and Objectives

A sampled or *DT* signal is defined only at discrete times. The very act of sampling represents a potential loss of information except for bandlimited signals (sampled at a high enough rate). This is the celebrated *sampling theorem*, which allows us to recover a bandlimited analog signal exactly from its sampled version with no loss of information.

It turns out that the spectrum of a sampled or *DT* signal is not only continuous but periodic. It is, in fact, a periodic extension of the spectrum of its underlying *CT* signal. This periodicity is a consequence of the duality and reciprocity between time and frequency and leads to the formulation of the *discrete-time Fourier transform* (DTFT).

This chapter is divided into five parts:

Part 1 The sampling operation and the sampling theorem.
Part 2 Schemes for the recovery of sampled signals using interpolation.
Part 3 Sundry aspects including sampling errors and quantization.
Part 4 The spectrum of *DT* signals and the DTFT.
Part 5 Properties and applications of the DTFT.

Useful Background
1. Concepts dealing with *DT* signals (Chapter 2).
2. The properties of the Fourier transform and the connection between the Fourier series and Fourier transform.

11.1 Sampling

Conceptually, the operation of sampling a *CT* signal $x_C(t)$ can be described by several different schemes. We describe three such schemes: *ideal sampling (impulse sampling), natural sampling* and *flat-top sampling*. Their essential features are summarized in Table 11.1 and Figure 11.1.

Table 11.1   Sampling Operations.

Type	Sampling signal	Sampled signal	Spectrum of sampled signal
Ideal	$s_I(t) = \sum \delta(t - nt_s)$	$x_C(t)s_I(t)$	$\sum(1/t_s)X_C(f - kS_F)$
Natural	$s_N(t) = s_p(t)$	$x_C(t)s_p(t)$	$\sum S[k]X_C(f - kS_F)$
Flat-top	$s_A(t) = s_I(t)$	$[x_C(t)s_I(t)] \star h_f(t)$	$\mathrm{sinc}(f/S_F)\sum X_C(f - kS_F)$

Figure 11.1   Illustrating ideal, natural and flat-top sampling.

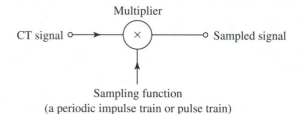

Multiplier

CT signal $\circ\!\!\longrightarrow$ $\times$ $\longrightarrow\!\!\circ$ Sampled signal

Sampling function
(a periodic impulse train or pulse train)

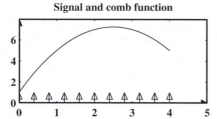

Signal and comb function

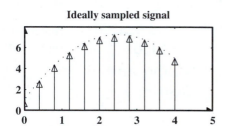

Ideally sampled signal

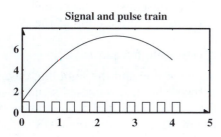

Signal and pulse train

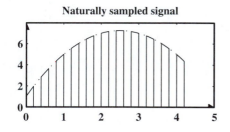

Naturally sampled signal

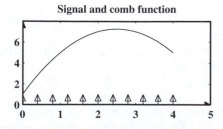

Signal and comb function

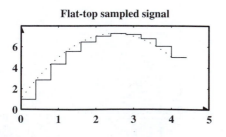

Flat-top sampled signal

11.1.1   *Ideal Sampling*

Ideal sampling describes a sampled signal $x_S(t)$ as a weighted sum of impulses, the weights being equal to the values of the *CT* signal $x_S(t)$ at the impulse locations nt_s. The sampled signal may be regarded as a product of $x_C(t)$ and a periodic impulse train $s_I(t)$ (Figure 11.1). Thus,

$$x_S(t) = x_C(t)s_I(t) = x_C(t) \sum_{n=-\infty}^{\infty} \delta(t - nt_s) = \sum_{n=-\infty}^{\infty} x_C(nt_s)\delta(t - nt_s)$$

The *DT* signal $x_S[n]$ is just the sequence of sample values $\{x_C(nt_s)\}$.

11.1.2   *Natural Sampling*

Natural sampling is based on passing $x_C(t)$ through a switch that opens and closes every t_s seconds (Figure 11.1). The action of the switch can be modeled as a periodic pulse train $s_N(t)$ of unit height, with period t_s and pulse duration t_d. The sampled signal $x_S(t)$ equals $x_C(t)$ for the t_d seconds that the switch remains closed and is zero when the switch is open. It may be described as the product

$$x_S(t) = x_C(t)s_N(t)$$

11.1.3   *Flat-Top (Zero-Order-Hold) Sampling*

In **flat-top sampling**, the signal $x_C(t)$ is also sampled periodically but the sampled value is held constant for the duration t_s (Figure 11.1) using a practical **zero-order-hold** circuit. It is equivalent to using an impulse train to sample $x_C(t)$, and passing the resulting signal through a system whose impulse response is a pulse of height 1 and duration t_s described by $h_f(t) = \text{rect}[(t - \frac{1}{2}t_s)/t_s]$. The sampled signal $x_S(t)$ can be regarded as the convolution of $h_f(t)$ and the signal $x_C(t)s_I(t)$:

$$x_S(t) = [x_C(t)s_I(t)] \star h_f(t) = \left[\sum_{n=-\infty}^{\infty} x_C(nt_s)\delta(t - nt_s) \right] \star h_f(t)$$

11.1.4   *What the Sampling Operation Reveals*

In each of these three schemes, the sampling operation leads to an apparent loss of information in the sampled signal $x_S(t)$, when compared with its underlying *CT* counterpart $x_C(t)$.

We now study the sampling operation from two related perspectives. Analysis in the frequency domain leads to the celebrated *sampling theorem* which establishes conditions for sampling with no loss of information. Analysis in the time domain leads to methods of recovering analog signals from their sampled versions.

11.2
The Sampling Theorem

The smaller the sampling interval, the less the loss of information we expect in the sampled signal $x_S(t)$. Intuitively, there must always be some loss of information, no matter how small an interval we use. Fortunately, our intuition notwithstanding, it is indeed possible to sample signals without any loss of information. The catch is that the signal $x_S(t)$ must be bandlimited to some frequency f_B. If it is, we obtain a sampled signal $x_S(t)$ with the same information content as the *CT* signal $x_C(t)$ provided the $x_C(t)$ signal is sampled at a rate exceeding $2f_B$ samples/second.

The **sampling theorem** simply formalizes this remarkable fact: *A signal $x_C(t)$ bandlimited to a frequency f_B can be sampled without loss of information if the sampling rate S_F exceeds $2f_B$ (or the sampling interval t_s is smaller than $\frac{1}{2f_B}$).* Put another way, it is possible to exactly recover a bandlimited analog signal from its samples if the sampling rate is suitably chosen.

The *critical* sampling rate $S_N = 2f_B$ is often called the **Nyquist rate** or **Nyquist frequency**. The critical sampling interval $t_N = 1/S_N = 1/2f_B$ is called the **Nyquist interval**.

Example 11.1

Let the signals $x_1(t)$ and $x_2(t)$ be bandlimited to 2 kHz and 3 kHz, respectively. If you sketch their spectra (as pulses, say) you can find the Nyquist rate for the signals below as follows:

Signal	S_N	Remarks
$2x_1(t)$	4 kHz	The spectral width does not change.
$x_1(2t)$	8 kHz	Compression in time means spectral expansion.
$x_2(t-3)$	6 kHz	A time shift only affects the phase spectrum.
$x_1(t) + x_2(t)$	6 kHz	The spectrum is the sum of the individual spectra.
$x_1(t)x_2(t)$	10 kHz	Frequency convolution. Use the duration property.
$x_1(t) \star x_2(t)$	4 kHz	The spectrum is the product of the two spectra.

To understand the sampling theorem, we describe each of the three sampling operations from a frequency domain viewpoint. Consider a signal $x_C(t)$ bandlimited to f_B whose spectrum $X_C(f)$ is shown in Figure 11.2. What is important about the spectrum is not its shape, but its bandlimited nature.

11.2.1   *Spectrum of Ideally Sampled Signals*

The impulse sampling function $s_I(t)$ is a periodic signal with period t_s. Its Fourier series coefficients $S[k]$ equal $1/t_s$. Its Fourier transform is a train of impulses with strength $S[k]$ and separation $S_F = 1/t_s$. Thus,

$$S_I(f) = \sum_{k=-\infty}^{\infty} (1/t_s)\delta(f - kS_F)$$

The sampled signal $x_S(t)$ is the product of $x_C(t)$ and $s_I(t)$. Its spectrum $X_p(f)$ may, therefore, be described as the convolution

$$X_p(f) = X_C(f) \star S_I(f) = X_C(f) \star \frac{1}{t_s} \sum_{k=-\infty}^{\infty} \delta(f - kS_F) = \frac{1}{t_s} \sum_{k=-\infty}^{\infty} X_C(f - kS_F)$$

In words, $X_p(f)$ describes replicated versions of $X_C(f)$. It is actually the *periodic extension* of $X_C(f)$, with a principal period between $-\frac{1}{2}S_F$ and $\frac{1}{2}S_F$ (Figure 11.2). The shifted replicas of $(1/t_s)X_C(f)$ are called **spectral orders** or **spectral images**.

Figure 11.2   Spectrum of a bandlimited *CT* signal and its ideally sampled versions for oversampling, critical sampling, and undersampling.

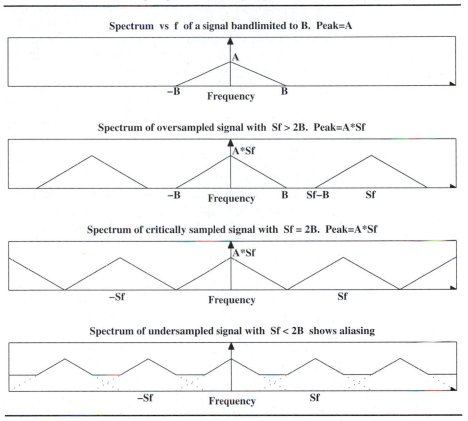

As long as the images do not overlap, the principal period equals $(1/t_s)X_C(f)$ (Figure 11.2). We can recover $X_C(f)$ (and hence $x_C(t)$) by passing the sampled signal through an ideal lowpass filter with a cutoff frequency of $\frac{1}{2}S_F$ (or f_B) and a gain of t_s.

Since the spectral image at the origin extends over $(-f_B, f_B)$, and the next image extends over $(S_F - f_B, S_F + f_B)$ (Figure 11.2), for no overlap we require

$$S_F - f_B > f_B \qquad \text{or} \qquad S_F > 2f_B$$

This sets the smallest possible sampling rate to guarantee perfect recovery of $x_C(t)$ from its sampled version $x_S(t)$ as $S_N = 2f_B$, the Nyquist rate. This corresponds to a maximum sampling interval of $t_N = 1/2f_B$. If the sampling rate S_F is smaller than S_N, the spectral images overlap and the principal period $(-\frac{1}{2}S_F, \frac{1}{2}S_F)$ of the periodic extension $X_p(f)$ is no longer a replica of $X_C(f)$ (Figure 11.2). In this case, we cannot recover $x_C(t)$. There is a loss of information due to **undersampling**.

These observations suggest several important results:

1. *Periodicity*: The spectrum of the ideally sampled signal is periodic, with a period that equals the sampling frequency S_F.
2. *Bandlimiting*: Sampling is a bandlimiting operation in the sense that the principal period of the spectrum is bandlimited to the frequency range $|f| \leq \frac{1}{2}S_F$. The highest frequency we can observe in the spectrum is $\frac{1}{2}S_F$ and depends only on the sampling rate S_F, not on the highest frequency f_B in $x_C(t)$. To ensure that the highest frequency in the spectrum of the sampled signal exceeds f_B, we must choose $S_F \geq 2f_B$.
3. *Aliasing*: Undersampling results in spectral overlap. Components beyond $\frac{1}{2}S_F$ fold back to frequencies lower than $\frac{1}{2}S_F$. This is aliasing. The frequency $\frac{1}{2}S_F$ is called the **folding frequency**. Aliasing is more pronounced for spectral components closer to the folding frequency.

11.2.2   Spectrum of Naturally Sampled Signals

For natural sampling, the sampling function $s_N(t) = x_p(t)$ is a periodic pulse (not impulse) train. For a rectangular pulse train $x_N(t)$ with period t_s and Fourier series coefficients $S[k]$, the Fourier transform $X_N(f)$ is a train of impulses with strength $S[k]$ and separation $S_F = 1/t_s$ is given by

$$X_N(f) = \sum_{k=-\infty}^{\infty} S[k]\delta(f - kS_F)$$

The sampled signal $x_S(t)$ is the product of $x_C(t)$ and $s_N(t)$. Its spectrum $X_S(f)$ is thus described by the convolution

$$X_S(f) = X_C(f) \star X_N(f) = X_C(f) \star \sum_{k=-\infty}^{\infty} S[k]\delta(f - kS_F) = \sum_{k=-\infty}^{\infty} S[k]X_C(f - kS_F)$$

Again, $X_S(f)$ is a superposition of $X_C(f)$ and its shifted replicas. Since the $S[k]$ are not equal, $X_S(f)$ is not periodic (Figure 11.3). If there is no spectral overlap, the image about the origin equals $S[0]X_C(f)$. We can then recover $X_C(f)$ (and hence $x_C(t)$) by passing the sampled signal through an ideal lowpass filter with a cutoff frequency of $\frac{1}{2}S_F$ and a gain of $1/S[0]$.

We make the following remarks:

1. *Lack of Periodicity*: The spectrum of the sampled signal using natural sampling is not periodic. But since the spectral images appear at multiples of S_F, the highest frequency we can observe in the spectrum is still $\frac{1}{2}S_F$. It depends only on the sampling rate and not on the manner in which the sampling is performed.
2. *Aliasing*: Aliasing occurs for sampling rates S_F less than $2f_B$.
3. In theory, natural sampling can be performed by any periodic signal with nonzero dc offset $S[0]$. To recover $x_C(t)$, we use an ideal lowpass filter with a gain of $1/S[0]$, the reciprocal of the dc offset.

Figure 11.3 Spectrum of a bandlimited *CT* signal, and the spectra of its naturally sampled and flat-top sampled versions.

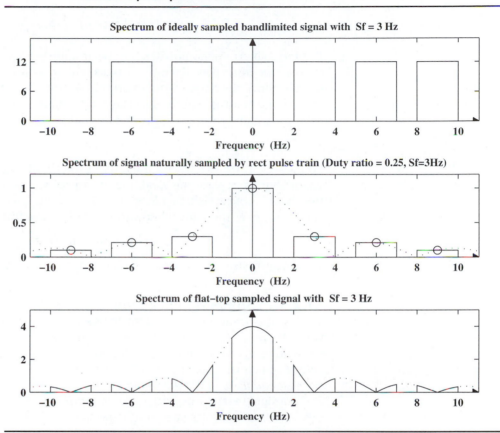

11.2.3 *Spectrum of Flat-Top Sampled Signals*

In flat-top sampling, ideal sampling is followed by a zero-order-hold to keep the signal levels constant between samples. This is equivalent to passing the ideally sampled signal through a system whose impulse response, $h_f(t) = \text{rect}([t - \frac{1}{2}t_s]/t_s)$, is a rectangular pulse of width t_s. The sampled signal $x_S(t)$ is then described by the convolution

$$x_S(t) = h_f(t) \star [x_C(t)s_I(t)] = h_f(t) \star \left[\sum_{n=-\infty}^{\infty} x_C(nt_s)\delta(t - nt_s) \right]$$

The Fourier transform $H_f(f)$ of $h_f(t) = \text{rect}([t - \frac{1}{2}t_s]/t_s)$ is the sinc function

$$H_f(f) = t_s \text{sinc}(ft_s) \exp(-j\pi ft_s) = t_s \text{sinc}(f/S_F) \exp(-j\pi f/S_F)$$

The transform of $x_C(t)s_I(t)$ equals $(1/t_s) \sum X_C(f - kS_F)$. The transform of the sampled signal $x_S(t)$ is given by the product

$$X_S(f) = \text{sinc}(f/S_F) \exp(-j\pi f/S_F) \sum_{k=-\infty}^{\infty} X_C(f - kS_F)$$

The term $\text{sinc}(f/S_F)$ attenuates the spectral images $\Sigma X_C(f - kS_F)$ (Figure 11.3) and causes *sinc distortion*. Since the images still show up at multiples of S_F, we must sample above the Nyquist rate to prevent overlap. The higher the sampling rate S_F, the less the distortion in the spectral image about the origin. An ideal lowpass filter with unity gain recovers the *distorted* signal

$$\tilde{X}(f) = \text{sinc}(f/S_F) \exp(-j\pi f/S_F) X_C(f)$$

To recover $X_C(f)$ exactly, we must use a filter that negates the effects of the sinc distortion. We replace the ideal filter by a practical **equalization** or **compensation filter** with a concave-shaped magnitude spectrum $|H_r(f)| = 1/|\text{sinc}(f/S_F)|$ (and linear phase) over the signal bandwidth $|f| \leq \frac{1}{2}f_B$. The filter transfer function is thus

$$H_r(f) = \exp(-j2\pi f t_r)/\text{sinc}(f/S_F), \qquad |f| \leq f_B$$

The linear phase $-2\pi f t_r$ produces a constant delay t_r in the filter output.

11.2.4   *Sampling Pure Sinusoids*

The Nyquist frequency for a sinusoid $\cos(2\pi f_0 t + \theta)$ equals $2f_0$. The Nyquist interval is $1/2f_0$, or $\frac{1}{2}T$. This amounts to taking *more than* two samples per period. If we acquire just two samples per period, starting at a zero crossing, all the sample values will equal zero. The moral is that we must *choose a sampling rate exceeding, not just equaling, the Nyquist rate*.

If a signal $x(t) = \cos(2\pi f_0 t + \theta)$ is sampled at S_F, the sampled signal is $x_s[n] = \cos(2\pi n f_0/S_F + \theta)$. Its spectrum is periodic, with principal period $(-\frac{1}{2}S_F, \frac{1}{2}S_F)$. If $f_0 < \frac{1}{2}S_F$, there is no aliasing and the principal period shows a pair of impulses at $\pm f_0$ (with strength $\frac{1}{2}$). If $f_0 > \frac{1}{2}S_F$, we have aliasing. The components at $\pm f_0$ are aliased to a lower frequency $\pm f_a$ in the principal range.

11.2.5   *Finding Aliased Frequencies*

To find the aliased frequency $|f_a|$, we select an integer $k = N$ which places $f_a = f_0 - NS_F$ in the principal range $(-\frac{1}{2}S_F, \frac{1}{2}S_F)$. The spectrum describes a sampled version of $\cos(2\pi f_a t + \theta)$ (if $f_a > 0$) or $\cos(-2\pi f_a t + \theta) = \cos(2\pi f_a t - \theta)$ (if $f_a < 0$). This phase reversal is due to conjugate symmetry of components at $\pm f_a$. Naturally, if $S_F > 2f_0$, there is no aliasing.

Example 11.2
Aliasing and Sampled Sinusoids

Consider the sinusoid $x(t) = A\cos(2\pi f_0 t + \theta)$ with $f_0 = 100$ Hz.

(a) If $x(t)$ is sampled at $S_F = 300$ Hz, no aliasing occurs, since $S_F > 2f_0$.
(b) If $x(t)$ is sampled at $S_F = 80$ Hz, we obtain $f_a = f_0 - S_F = 20$ Hz, which is in the range $(-\frac{1}{2}S_F, \frac{1}{2}S_F)$. This describes a sampled version of $A\cos[2\pi(20)t + \theta]$.

(c) If $S_F = 160$ Hz, we obtain $f_0 - S_F = -60$ Hz. The aliased sinusoid is a sampled version of $A\cos[2\pi(-60)t + \theta] = A\cos[2\pi(60)t - \theta]$. Note the phase reversal.

(d) If $S_F = 60$ Hz, we obtain $f_0 - 2S_F = -20$ Hz. The aliased sinusoid is a sampled version of $A\cos[2\pi(-20)t + \theta] = A\cos[2\pi(20)t - \theta]$. Note the phase reversal.

11.2.6   Sampling Periodic Signals

A periodic signal $x_p(t)$ with period T can be described by a sum of sinusoids at the frequency $f_0 = 1/T$ and its harmonics kf_0 (Chapter 5). If the highest frequency in such a representation is f_H, we choose $S_F > 2f_H$ to identify all frequencies uniquely without aliasing. If all possible harmonics are present, we must decide on the highest frequency we wish to identify—say, Nf_0—select $S_F > 2Nf_0 = 2N/T$ and accept some aliasing due to frequencies higher than Nf_0 (unless we filter them out).

Example 11.3
Sampling Periodic Combinations

Let $x_p(t) = 8\cos(2\pi t) + 6\cos(8\pi t) + 4\cos(22\pi t) + 6\sin(32\pi t) + \cos(58\pi t) + \sin(66\pi t)$.

   If it is sampled at $S_F = 10$ Hz, the last 4 terms will be aliased. The sampled signal will describe an analog signal whose first two terms are identical to $x_p(t)$, but whose other components are at the aliased frequencies. The following table shows what to expect:

f_0	Aliasing?	Aliased frequency f_a	Analog equivalent
1	No ($f_0 < \frac{1}{2}S_F$)		$8\cos(2\pi t)$
4	No ($f_0 < \frac{1}{2}S_F$)		$6\cos(8\pi t)$
11	Yes	$11 - S_F = 1$	$4\cos(2\pi t)$
16	Yes	$16 - 2S_F = -4$	$6\sin(-8\pi t) = -6\sin(8\pi t)$
29	Yes	$29 - 3S_F = -1$	$\cos(-2\pi t) = \cos(2\pi t)$
33	Yes	$33 - 3S_F = 3$	$\sin(6\pi t)$

Comment: Since $6\cos(8\pi t) - 6\sin(8\pi t) = 6\sqrt{2}\cos(8\pi t + 45°)$, we recover the analog signal $x_S(t) = 13\cos(2\pi t) + \sin(6\pi t) + 6\sqrt{2}\cos(8\pi t + 45°)$, which cannot be distinguished from $x_p(t)$ at the instants $t = nt_s$, where $t_s = 0.1$ s.

   To avoid aliasing of $x_p(t)$, we must actually sample $x_p(t)$ at $S_F > 2f_H = 66$ Hz.

Example 11.4
Why Aliasing Causes Problems

Suppose we sample a sinusoid $x(t)$ at 30 Hz and obtain the periodic spectrum of Figure 11.4, with components at ±10 Hz, ±20 Hz, ±40 Hz, ±50 Hz, ...

   What can we say about $x(t)$? We can certainly identify the period as 30 Hz, and thus $S_F = 30$ Hz. But we simply cannot uniquely identify $x(t)$ because it could be

1. A sinusoid at 10 Hz showing no aliasing
2. A sinusoid at 20 Hz or 50 Hz or 80 Hz, ..., causing aliasing
3. A combination of the above!

Figure 11.4   Spectrum of a sinusoid sampled at 30 Hz for Example 11.4.

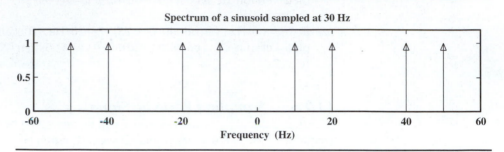

Comment: If we *knew* that $x(t)$ was a 50-Hz sinusoid, aliasing would *not* be a problem, since we could recover $x(t)$ by using a bandpass filter. It is precisely because we have no a priori information about $x(t)$ that we use the *principal period* as a means to identify $x(t)$ for better or worse.

11.2.7   Sampling Bandpass Signals

The spectrum of **bandpass signals** occupies a range of frequencies between f_L and f_H, where f_L is greater than zero. The quantity $B = f_H - f_L$ is a measure of the bandwidth of the signal. Even though a Nyquist rate of $2f_H$ can be used to recover such signals, we can often get by with a lower sampling rate. To retain all the information in $x_C(t)$, we actually need a sampling rate that aliases the entire spectrum to a lower frequency range without overlap. The smallest such frequency is $S_F = 2f_H/N$, where N is the integer part of f_H/B. For recovery, we must sample at exactly S_F. Other choices result in bounds on the sampling frequency given by

$$2f_H/k < S_F < 2f_L/(k-1) \qquad k = 1, 2, \ldots, N$$

Here, the integer k can range from 1 to N, with $k = N$ yielding the smallest value $S_F = 2f_H/N$, and $k = 1$ corresponding to the Nyquist rate $2f_H$.

Example 11.5
Sampling of
Bandpass Signals

Consider a bandpass signal $x_C(t)$ with $f_L = 4$ kHz and $f_H = 6$ kHz. Then $B = 2$ kHz and $N = \text{int}(f_H/B) = 3$. We require 12 kHz$/k < S_F < 8$ kHz $/(k-1)$. This leads to the following ranges for S_F, and the spectra of Figure 11.5.

k	Bounds on S_F	Remarks
3	4 kHz–4 kHz	The smallest S_F; requires bandpass filter for recovery.
2	6 kHz–8 kHz	Requires bandpass filter for recovery.
1	12 kHz–∞	No aliasing; we can use a lowpass filter for recovery.

Figure 11.5   Illustrating the sampling of bandpass signals.

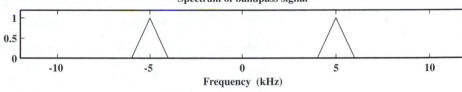

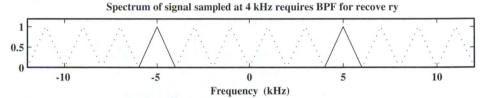

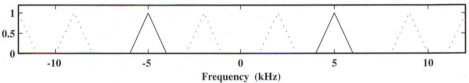

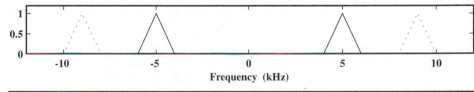

11.3 Sampling, Interpolation and Signal Recovery

For a sampled sequence obtained from a *CT* signal $x_C(t)$, an important aspect is the recovery of the original signal from its samples. This requires "filling in" the missing details, or interpolating between the sampled values. The nature of the interpolation that recovers $x_C(t)$ may be discerned by considering the sampling operation in the time domain. Of the three sampling operations considered earlier, only ideal (or impulse) sampling leads to a truly *DT* signal $x_S[n]$ whose samples equal the strengths of the impulses $x_C(nt_s)\delta(t - nt_s)$ at the sampling instants nt_s. This is the only case we pursue.

11.3.1   *Ideal Recovery and the Sinc Interpolating Function*

The ideally sampled signal $x_S(t)$ is the product of $x_C(t)$ and the impulse sampling function $s_I(t) = \sum \delta(t - nt_s)$ and may be written as

$$x_S(t) = x_C(t) \sum_{n=-\infty}^{\infty} \delta(t - nt_s) = \sum_{n=-\infty}^{\infty} x_C(nt_s)\delta(t - nt_s) = \sum_{n=-\infty}^{\infty} x_S[n]\delta(t - nt_s)$$

The *DT* signal $x_S[n]$ is just the sequence of samples $\{x_C(nt_s)\}$. We can recover $x_C(t)$ by passing $x_S(t)$ through a lowpass filter with a gain of t_s and a cutoff frequency of f_B. The impulse response of such a filter is a sinc function given by

$$h_f(t) = 2t_s f_B \operatorname{sinc}(2f_B t)$$

The recovered signal $x_C(t)$ may therefore be described as the convolution

$$x_C(t) = x_S(t) \star h_f(t) = \left[\sum_{n=-\infty}^{\infty} x_C(nt_s)\delta(t - nt_s) \right] \star h_f(t) = \sum_{n=-\infty}^{\infty} x_S[n]h_f(t - nt_s)$$

This describes the superposition of shifted versions of $h_f(t)$ weighted by the sample values $x_S[n]$. Substituting for $h_f(t)$, we obtain the following result that allows us to recover $x_C(t)$ exactly from its samples $x_S[n]$ as a sum of scaled shifted versions of sinc functions:

$$x_C(t) = \sum_{n=-\infty}^{\infty} 2t_s f_B x_S[n]\operatorname{sinc}[2f_B(t - nt_s)]$$

This result is valid for *any oversampled signal* with $t_s \le 1/2f_B$.

If we choose $S_F = 2f_B$ (critical Nyquist rate) and $t_s = 1/2f_B$, we obtain

$$x_C(t) = \sum_{n=-\infty}^{\infty} x_S[n]\operatorname{sinc}[(t - nt_s)/t_s]$$

Here is what this result says. At each sampling instant, replace the sample value $x_S[n]$ by a sinc function whose peak value equals $x_S[n]$ and whose zero crossings occur at all the other sampling instants. The sum of these sinc functions yields the analog signal $x_C(t)$ as illustrated in Figure 11.6.

Sinc interpolation is rather unrealistic from a practical viewpoint. The infinite extent of the sinc means that it cannot be implemented on line and perfect reconstruction requires all its past and future values. We could truncate it on either side after its magnitude becomes small enough. But unfortunately, it decays very slowly and must be preserved for a fairly large duration (covering many past and future sampling instants) in order to provide a reasonably accurate reconstruction. Since the function is also smoothly varying, it cannot properly reconstruct a discontinuous signal at the discontinuities even with a large number of values.

Remark: Sinc interpolation forms the yardstick by which all other schemes are measured in their ability to reconstruct bandlimited signals. This is why $\sin(x)/x$ is also called the **cardinal interpolating function**.

11.3.2   *Interpolation*

Since the sinc interpolating function is a poor choice in practice, we must look to other interpolating signals. If a *CT* signal $x_C(t)$ is to be recovered from its sampled version $x_S[n]$ using an interpolating function $i(t)$, what must we require of $i(t)$ to best approximate $x_C(t)$? At the very least, the interpolated approximation $x_i(t)$ should match $x_C(t)$ at the sampling instants nt_s. This suggests that $i(t)$ should

Figure 11.6 The signal recovered from its sampled version $x[n] = \{1, 2, 3, 4\}$ using sinc interpolation.

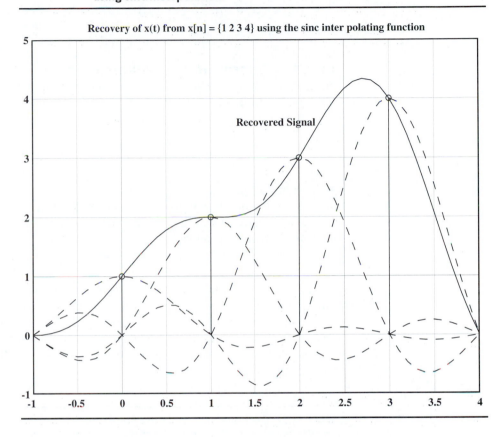

Recovery of x(t) from x[n] = {1 2 3 4} using the sinc inter polating function

equal zero at all sampling instants except the origin where it must equal unity such that

$$i(t) = \begin{cases} 1 & t = 0 \\ 0 & t = nt_s \end{cases} \qquad n = \pm 1, \pm 2, \pm 3, \dots$$

In addition, we also require $i(t)$ to be absolutely integrable to ensure that it stays finite between the sampling instants.

11.3.3 *Interpolation as Convolution*

The interpolated signal may be regarded as the output of a system whose impulse response is $h_i(t) = i(t)$, and whose input is the ideally sampled version of $x_C(t)$ described by the impulse train

$$x_S(t) = \sum_{n=-\infty}^{\infty} x_C(nt_s)\delta(t - nt_s)$$

The interpolated signal $x_i(t)$ is simply the convolution of $i(t)$ with $x_C(t)$

$$i(t) \star x_S(t) = i(t) \star \sum_{n=-\infty}^{\infty} x_C(nt_s)\delta(t - nt_s) = \sum_{n=-\infty}^{\infty} x_S[n]i(t - nt_s) = x_i(t)$$

11.3.4   Interpolation as Summation

Interpolation may also be regarded as a summation of shifted versions of the interpolating function. At each instant nt_s, we erect the interpolating function $i(t - nt_s)$, scale it by $x_S[n]$ and sum to obtain $x_i(t)$ as

$$x_i(t) = \sum_{n=-\infty}^{\infty} x_S[n]i(t - nt_s)$$

At a sampling instant $t = kt_s$, the interpolating function $i(kt_s - nt_s)$ equals zero unless $n = k$ when it equals unity. As a result, $x_i(t)$ *exactly* equals $x_C(t)$ at each sampling instant. At all other times, the interpolated signal $x_i(t)$ is only an approximation to the actual signal $x_C(t)$.

11.3.5   Practical Interpolating Functions

The nature of $i(t)$ dictates the nature of the interpolating system both in terms of its causality, stability and physical realizability. It also determines how good the reconstructed approximation is. The characteristics of some interpolating functions commonly encountered in practice are summarized in Table 11.2 and illustrated in Figure 11.7.

Table 11.2   Common Interpolating Functions.

Type	Form of $i(t)$	Interpolated Signal $x_i(t)$
Constant	$\mathrm{rect}[(t - \tfrac{1}{2}t_s)/t_s]$	$\sum_n x_S[n]\mathrm{rect}[(t - nt_s - \tfrac{1}{2}t_s)/t_s]$
Linear	$\mathrm{tri}(t/t_s)$	$\sum_n x_S[n]\mathrm{tri}[(t - nt_s)/t_s]$
Sinc	$\mathrm{sinc}(t/t_s)$	$\sum_n x_S[n]\mathrm{sinc}[(t - nt_s)/t_s]$
Raised cosine	$\mathrm{sinc}(t/t_s)\dfrac{\cos(\pi Rt/t_s)}{1 - (2Rt/t_s)^2}$	$\sum_n x_S[n]\dfrac{\mathrm{sinc}[(t - nt_s)/t_s]\cos[\pi R(t - nt_s)/t_s]}{1 - [2R(t - t_s)/t_s]^2}$

There is no "best" interpolating signal. Some are better in terms of their accuracy, others are better in terms of their cost effectiveness, still others are better in terms of their numerical implementation.

Step Interpolating Function   The widely used step interpolating function yields a stepwise or staircase approximation to $x_C(t)$. Even though it appears crude, there are several aspects that make it useful from a practical standpoint:

Figure 11.7   **Interpolating functions for signal recovery and the recovered signals for the sampled sequence $x[n] = \{1\ 2\ 3\ 4\}$.**

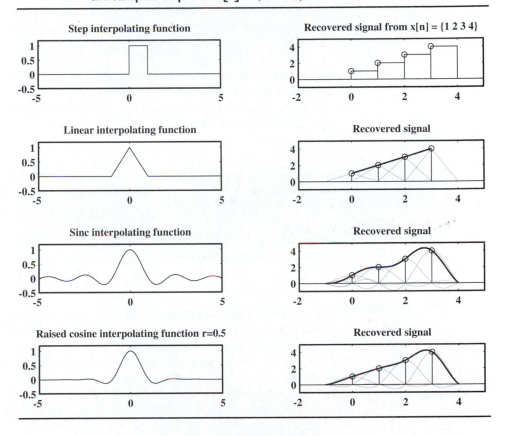

1. At any instant t between two sampling instants nt_s and $(n+1)t_s$, the reconstructed signal is given just by the sample value $x_S[n]$. Thus,

$$x_i(t) = x_S[n] \qquad nt_s \le t < (n+1)t_s$$

The operation does not depend on any future values and is well suited for *on-line* or *real-time* processing where the output is produced at the same rate as the incoming data.

2. Step interpolation results in exact reconstruction of signals that are piecewise constant.

A system that performs step interpolation is called a **zero-order-hold**. A practical digital-to-analog converter (DAC) for sampled signals uses a zero-order-hold for a staircase approximation (step interpolation) followed by a lowpass (anti-imaging) filter (for smoothing the steps).

Linear Interpolating Function   The linear interpolating function generates a linear approximation between sampled values. At any instant t between adjacent sampling instants nt_s and $(n+1)t_s$, the reconstructed signal equals $x_S[n]$ plus an increment that depends on the slope of the line joining $x_S[n]$ and $x_S[n+1]$. We have

$$x_i(t) = x_S[n] + (t - nt_s)\{x_S[n+1] - x_S[n]\}/t_s \qquad nt_s \leq t < (n+1)t_s$$

This operation requires one future value of the input and cannot actually be implemented on-line. It can, however, be realized with a delay of one sampling interval t_s, which is tolerable in many situations. Systems performing linear interpolation are also called **first-order hold** circuits. They yield exact reconstructions of piecewise linear signals.

Raised Cosine Interpolating Function   The sinc interpolating function forms the basis for several others described by the generic relation

$$i(t) = g(t)\operatorname{sinc}(t/t_s)$$

where $g(0) = 1$. One of the more commonly used of these is the **raised cosine interpolating function** described by

$$i_{\mathrm{rc}}(t) = \frac{\cos(\pi R t/t_s)}{1 - (2Rt/t_s)^2}\operatorname{sinc}(t/t_s) \qquad 0 \leq R \leq 1$$

Here R is called the **rolloff factor**. Like the sinc interpolating function, $i_{\mathrm{rc}}(t)$ equals 1 at $t = 0$ and 0 at the other sampling instants. It exhibits faster decaying oscillations on either side of the origin for $R > 0$ as compared to the sinc function. This faster decay results in improved reconstruction if the samples are not acquired at exactly the sampling instants (in the presence of jitter, that is). It also allows fewer past and future values to be used in the reconstruction as compared to the sinc interpolating function. The terminology *raised cosine* is actually based on the shape of its spectrum. For $R = 0$, the raised cosine reduces to the sinc interpolating function.

11.4
Some Consequences of the Sampling Theorem

The crux of the sampling theorem is not just the choice of an appropriate sampling rate. More important, the processing of a *CT* signal is equivalent to the processing of its Nyquist sampled version, because it retains the same information content as the original. This is how the sampling theorem is used in practice. It forms the link between analog and digital signal processing and allows us to use digital techniques to manipulate analog signals. When we sample a signal $x_C(t)$ at the instants nt_s, we imply that the spectrum of the sampled signal is periodic with period $P = S_F$ and bandlimited to a highest frequency $f_B = \frac{1}{2}S_F = 1/2t_s$.

11.4.1   *How Useful is the Sampling Theorem?*

Here is a dilemma. A bandlimited signal can never be timelimited. For perfect recovery, it must be sampled at the bandlimited or Nyquist rate, but forever! If we take only a finite number of its samples, we effectively have a timelimited signal whose spectrum can never be bandlimited and that cannot be perfectly recovered by sampling at any rate. So, how useful is the sampling theorem? Quite useful, actually.

In practice, many timelimited signals are indeed very nearly bandlimited. The sampling theorem serves as a very useful measure of how to sample such signals. We can use the sampling theorem even when a signal is not quite bandlimited. How we do this is based on the idea of *prefiltering*.

11.4.2   *Prefiltering*

Low sampling rates can cause aliasing. Components aliased to some frequency f_x will then add to other aliased or unaliased components already present at f_x and change the entire character of the signal. For most signals, the magnitude of the components decreases at higher frequencies, and the effects of aliasing are thus more pronounced closer to the folding frequency $\frac{1}{2}S_F$.

To prevent aliasing, it is common practice to prefilter noisy signals or signals for which only certain frequencies are of interest *before sampling*. Such bandlimiting filters are called **antialiasing filters**. A practical analog-to-digital converter (ADC) uses an antialiasing filter and a zero-order-hold (to generate the sampled signal) followed by a quantizer and encoder (to make it suitable for any digital signal processing).

11.4.3   *Oversampling*

The highest frequency f_H we can identify without aliasing depends only on S_F. In theory, **oversampling** (at a rate much higher than the Nyquist rate S_N) may appear wasteful, because it can lead to the manipulation of a large number of samples. In practice, however, it offers several advantages. First, it provides adequate separation between spectral images. Second, it minimizes the effects of the sinc distortion caused by zero-order-hold sampling. Finally, it allows us to relax the sharp cutoff requirements of the recovery filters that recover the analog signal from its samples. How much the sampling frequency S_F exceeds S_N depends on how well a signal can be bandlimited (using practically designed filters).

11.4.4   *Choice of Sampling Instants*

The sampling theorem tells us nothing about how to choose the sampling *locations*. A shift in the sampling instants changes only the phase spectrum of the sampled signal. All we need is a set of uniformly spaced instants.

It turns out that perfect recovery is possible even without uniform sampling as long as we satisfy the Nyquist rate and take S_N samples *on the average*. While this is a comforting result, it has little practical appeal.

11.5
Special Topics:
Sampling Errors

Sampling errors can arise from various sources such as

1. Undersampling, which leads to aliasing.
2. Roundoff error in sample values.
3. Truncation of the sinc interpolating function.
4. Jitter when samples are not acquired at exactly the sampling instants.

11.5.1   *Truncation Error*

If the truncated approximation to the reconstructed signal $x_C(t)$ is denoted by $x_R(t)$, and we let $i_n(t) = \text{sinc}([t - nt_s]/t_s)$, the reconstruction error may be written as

$$\epsilon_N(t) = x_C(t) - x_R(t) = \sum_{n=-\infty}^{\infty} x_C(nt_s)i_n(t) - \sum_{n=-N}^{N} x_S[n]i_n(t) = \sum_{|n|>N} x_S[n]i_n(t)$$

We now invoke a property of sinc functions based on their orthogonality:

$$\int_{-\infty}^{\infty} i_m(t)i_n(t)\, dt = \begin{cases} t_s & m = n \\ 0 & m \neq n \end{cases}$$

This property allows us to describe the energy E_C in $x_C(t)$ and the energy E_R in the truncated representation as

$$E_C = t_s \sum_{n=-\infty}^{\infty} x_S^2[n] \qquad E_R = t_s \sum_{n=-N}^{N} x_S^2[n]$$

The difference $E_N = E_C - E_R$ is a measure of the energy in $\epsilon_N(t)$. It actually equals the energy in $\epsilon_N(t)$ and also describes the mean squared error. The fractional energy error $(E_C - E_R)/E_N$ may now be readily evaluated in terms of the known sample values. In some cases, it may be easier to obtain the total energy E_C in $x_C(t)$ by resorting to Parseval's theorem.

If $i_n(t)$ is bandlimited to f_B, so is $\epsilon_N(t)$. It is then possible to obtain a bound on the absolute error $|\epsilon_N(t)|$ in terms of its energy E_N as

$$|\epsilon_N(t)| \leq (2f_B E_N)^{1/2}$$

11.5.2   *Roundoff Error*

If the difference between the sampled value $\tilde{x}_S[n]$ and its true value $x_S[n] = x_C(nt_s)$ at the sampling instant nt_s is denoted $v[n]$, the error $v(t)$ in the reconstructed signal equals

$$v(t) = \sum_{k=-\infty}^{\infty} v[n]i_n(t)$$

As before, since $i_n(t)$ is bandlimited, so is $v(t)$. The roundoff error can then be shown to be bounded in terms of the energy E_v by

$$|v(t)| \leq (2f_B E_v)^{1/2} \qquad \text{where} \qquad E_v = t_s \sum_{k=-\infty}^{\infty} v^2[n]$$

11.5.3   *Timing Jitter*

In practice, the phase noise on the sampling clock can result in errors, or **jitter**, in the time of occurrence of the true sampling instant by some amount τ called the *jitter tolerance*. This leads to a difference between the exact and actual sample values. Its effects can be studied only in statistical terms.

11.6
Special Topics:
Quantization

The importance of *DT* signals stems from the proliferation of high-speed digital computers for signal processing. Due to the finite memory limitations of such machines, we can process only finite data sequences. This requires not only sampling *CT* signals but also *quantizing* (rounding or truncating) signal amplitudes to a finite set of values. Since quantization affects only the signal amplitude, both *CT* and *DT* signals can be quantized. For digital signal processing, however, we need quantized *DT* signals, or *digital* signals.

Each quantized sample is represented as a group (word) of zeroes and ones (*bits*) that can be processed digitally. The finer the quantization, the longer the word. Like sampling, improper quantization leads to loss of information. But unlike sampling, no matter how fine the quantization, its effects are irreversible, since word lengths must necessarily be finite. The systematic treatment of quantization theory is very difficult because

1. Finite word lengths appear as nonlinear effects.
2. Quantization always introduces some noise, whose effects can be described only in statistical terms.

Quantization is usually considered only in the final stages of any design, and many of its effects (such as overflow and limit cycles) are beyond the realm of this text.

11.6.1   *Uniform Quantizers*

Quantizers are devices that operate on a signal to produce a finite number of amplitude levels or quantization levels. It is common practice to use **uniform quantizers** with equal quantization levels. A quantizer with an even number of levels between the symmetric extremes A and $-A$ is called a **midriser**, whereas one with an odd number of levels is called a **midtread**. The two forms, shown in Figure 11.8, reveal that midriser quantizers do not have a quantization level corresponding to zero.

Figure 11.8   Midriser and midtread quantizers.

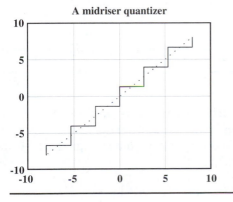
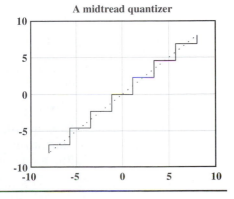

The number of levels in most quantizers used in an ADC is invariably a power of 2. If $L = 2^B$, each of the L levels is coded to a binary number and each signal value is represented in binary form as a B-bit word corresponding to its quantized value. A 4-bit quantizer is thus capable of 2^4 (or 16) levels and a 12-bit quantizer yields 2^{12} (or 4096) levels. If we need a zero quantization level, we can include an extra negative level.

A signal may be quantized by **rounding** to the nearest quantization level, **truncation** to a level smaller than the next higher one, or **sign-magnitude truncation**, which is rather like truncating absolute values and then using the appropriate sign. These operations are illustrated in Figure 11.9.

Figure 11.9   Illustration of quantization by rounding, truncation and sign-magnitude truncation.

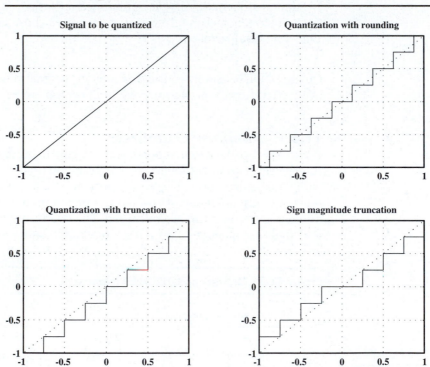

11.6.2   Quantization Error and Quantization Noise

The **quantization error**, naturally enough, depends on the number of levels. If the quantized signal corresponding to a *DT* signal $x[n]$ is denoted by $x_Q[n]$, the quantization error $\epsilon[n]$ equals

$$\epsilon[n] = x[n] - x_Q[n]$$

It is customary to define the **quantization signal-to-noise ratio** (SNR_Q) as the ratio of the power P_S in the signal and the power P_N in the error $\epsilon[n]$ (or noise). This is usually measured in decibels, and we thus obtain

$$P_S = \frac{1}{N}\sum_{n=0}^{N-1} x^2[n] \quad P_N = \frac{1}{N}\sum_{n=0}^{N-1}\epsilon^2[n] \quad \text{SNR}_Q(\text{dB}) = 10\log\frac{P_S}{P_N} = 10\log\frac{\sum x^2[n]}{\sum \epsilon^2[n]}$$

The effect of quantization errors due to rounding or truncation is quite difficult to quantify analytically unless statistical estimates are used.

Consider a signal $x(t)$ whose maximum variation, or **dynamic range**, D is given by $D = x_{\max} - x_{\min}$. If $x(t)$ is quantized to L levels, the **quantization step size**, or *resolution*, Δ is defined as

$$\Delta = D/L$$

For a signal $x[n]$ quantized and rounded to $x_Q[n]$, the maximum value of the quantization error must lie between $-\frac{1}{2}\Delta$ and $\frac{1}{2}\Delta$. If L is large, the error is equally likely to take on any value between $-\frac{1}{2}\Delta$ and $\frac{1}{2}\Delta$. The quantization error is said to be *uniformly distributed*. It turns out that a *statistical estimate* of SNR in decibels, denoted by SNR_S, is provided by

$$\text{SNR}_S(\text{dB}) = 10\log P_S + 10.8 + 20\log(L) - 20\log(D)$$

For a *B*-bit quantizer with $L = 2^B$ levels (and $\Delta = D/2^B$), we obtain

$$\text{SNR}_S(\text{dB}) = 10\log(P_S) + 10.8 + 6B - 20\log(D)$$

This result suggests a 6-dB improvement in the SNR for each additional bit. It also suggests a reduction in the SNR if the dynamic range D is chosen to exceed the signal limits.

Remark: In practice, signal levels do not often reach the extreme limits of their dynamic range. This allows us to increase the SNR by choosing a smaller value of D but at the expense of some distortion (at very high signal levels).

Example 11.6
Quantization Effects

(a) Consider the ramp $x(t) = 2t$ over $(0, 1)$. For a sampling interval of 0.1 s and $L = 4$, we obtain the sampled, quantized (rounded) and error signals as

$$x[n] = \{0, 0.2, 0.4, 0.6, 0.8, 1.0, 1.2, 1.4, 1.6, 1.8, 2.0\}$$
$$x_Q[n] = \{0, 0.0, 0.5, 0.5, 1.0, 1.0, 1.0, 1.5, 1.5, 2.0, 2.0\}$$
$$e[n] = \{0, 0.2, -0.1, 0.1, -0.2, 0.0, 0.2, -0.1, 0.1, -0.2, 0.0\}$$

We can now compute the SNR in one of two ways:

$$\text{SNR}_Q = 10\log(P_S/P_N) = 10\log\{\textstyle\sum x^2[n]/\sum e^2[n]\}$$
$$= 10\log(15.4/0.2) = 18.9\ \text{dB}$$
$$\text{SNR}_Q = 10\log P_S + 10.8 + 20\log(L) - 20\log(D)$$

With $D = 2$ and $N = 11$, we obtain

$$\text{SNR}_S = 10 \log \left[\frac{1}{N} \sum_{n=0}^{N-1} x^2[n] \right] + 10.8 + 20 \log(4) - 20 \log(2) = 18.7 \text{ dB}$$

Comment: Why the difference? Because SNR_S is a *statistical estimate*. The larger the number of samples N, the less SNR_Q and SNR_S differ. For $N = 500$, for example, we find $\text{SNR}_Q = 18.0751$ dB and $\text{SNR}_S = 18.0748$ dB.

Comment: If we assume $x(t)$ to be periodic, we find $P_S = \mathbb{A}[x^2(t)]/T = 4/3$ and

$$\text{SNR}_S = 10 \log(4/3) + 10 \log(12) + 20 \log(4) - 20 \log(2) = 18.062 \text{ dB}$$

(b) Consider the sinusoid $x(t) = A \cos(2\pi f t)$. The power in $x(t)$ is $P_S = \frac{1}{2} A^2$. The dynamic range equals $D = 2A$. For a B-bit quantizer, we obtain SNR_S as

$$\text{SNR}_S = 10 \log(P_S) + 10.8 + 6B - 20 \log(D)$$
$$= 10 \log(\tfrac{1}{2} A^2) + 10 \log(12) + 6B - 20 \log(2A)$$
$$\text{SNR}_S = -10 \log(2) + 20 \log(A) + 10 \log(12) + 6B - 20 \log(A) - 20 \log(2)$$

This simplifies to the widely used result

$$\text{SNR}_S = 6B + 10 \log(1.5) = (6B + 1.76) \text{ dB}$$

11.6.3   Coefficient Quantization

During the realization of digital filters, the filter coefficients cannot be specified to arbitrary precision and must be truncated or rounded. The effects of such **coefficient quantization** are also difficult to quantify. They can range in severity from minor changes in the frequency response to instability (an example of which appears in Chapter 14).

11.7
The Discrete-Time Fourier Transform (DTFT)

Ideal sampling forms the link between *CT* and *DT* signals. A *DT* signal may be regarded as a sequence whose values correspond to the strengths of the *CT* impulses in an ideally sampled analog signal. The **discrete-time Fourier transform** (DTFT) describes the spectrum of *DT* signals. It simply formalizes the idea that ideal sampling imposes bandlimited periodicity on the spectrum of a *DT* signal. In Chapter 4, we described how the DTFT arises from convolution by considering discrete eigensignals of linear systems. Here, we provide a different interpretation based on the sampling operation and Fourier transform. This also serves to highlight the periodic nature of the DTFT.

Neither the time signal nor its Fourier transform are, in general, either discrete or periodic. The Fourier transform of periodic signals does result in a discrete spectrum described by an impulse train. In fact, the Fourier series becomes a special case of the Fourier transform if we allow

I. The impulses in the discrete spectrum to be replaced by a discrete sequence whose sample values equal the impulse strengths $X_S[k]$.

2. The time signal to remain periodic.

This leads to the Fourier series relations for a discrete spectrum and its corresponding periodic time signal:

$$X_S[k] = \frac{1}{T} \int_T x_p(t) \exp(-j2\pi k f_0 t)\, dt \qquad x_p(t) = \sum_{k=-\infty}^{\infty} X_S[k] \exp(j2\pi k f_0 t)$$

Here, the frequency separation f_0 is the reciprocal of the period $T = 1/f_0$.

The dual nature of the Fourier transform suggests that a periodic spectrum must, in turn, describe an impulse train in the time domain. Just as we did for the Fourier series, we can now relate such a periodic spectrum and its corresponding sampled sequence in the time domain if we allow:

I. The impulses in the time signal to be replaced by a discrete sequence whose sample values equal the impulse strengths $x_S[n]$.

2. The spectrum to remain periodic.

We are, in effect, interchanging time and frequency and can write a dual of the Fourier series relations for a sampled signal $x_S[n]$ and its periodic spectrum $X_p(f)$ with period S_F using an appropriate change of notation:

$$X_p(f) = \sum_{n=-\infty}^{\infty} x_S[n] \exp(-j2\pi n f t_s) \qquad x_S[n] = \frac{1}{S_F} \int_{S_F} X_p(f) \exp(j2\pi n f t_s)\, df$$

The time separation (sampling interval) t_s equals the reciprocal of S_F, and $X_p(f)$ is periodic with principal period $(-\frac{1}{2}S_F, \frac{1}{2}S_F)$. For a *DT* sequence, $t_s = 1/S_F = 1$ and the *digital* frequency $F = f/S_F$, and we obtain:

$$X_p(F) = \sum_{n=-\infty}^{\infty} x[n] \exp(-j2\pi n F) \qquad x[n] = \int_1 X_p(F) \exp(j2\pi n F)\, dF$$

The quantity $X_p(F)$ defines the DTFT of the *DT* signal $x[n]$. It is periodic in F with principal period $(-\frac{1}{2}, \frac{1}{2})$.

Remarks: **I.** Since the sampled signal $x_S[n]$ has the same sample values as $x[n]$, but spaced t_s apart, $X_p(f)$ is simply a frequency scaled (stretched by S_F) version of $X_p(F)$ with principal period $(-\frac{1}{2}S_F, \frac{1}{2}S_F)$, and plotted against the analog frequency f.

2. If $x_S[n]$ is derived from a *bandlimited CT* signal $x_C(t)$, *sampled above the Nyquist rate*, $X_p(f)$ is related to the Fourier transform of $x_C(t)$ by $X_p(f) = (1/t_s)X_C(f)$ over the *principal period* $(-\frac{1}{2}S_F, \frac{1}{2}S_F)$.

11.7.1   *Fourier Series, the DTFT and Duality*

The Fourier series and DTFT are duals. Given a periodic signal $x(t)$ with period $T = 1$, and its spectral coefficients $X_S[k]$, the **folded periodic signal** $X_p(-F)$ describes the DTFT of the discrete sequence $x[n]$. The folding is due to sign reversal of the exponents in the exponential terms in the defining relations. If $x(t)$ possesses even symmetry, we don't even need the folding.

11.8
Properties of the DTFT

Since the DTFT is the dual of the Fourier series, we can carry over the entire arsenal of operational properties of the Fourier series virtually unchanged to the DTFT. Other transform pairs can also be developed from the analogy between the Fourier transform and the DTFT if we replace time samples by impulses. Since the time signal is a sequence, some relations such as the derivative property do not apply to the DTFT, while others are unique to the DTFT and find no counterpart in the Fourier series. The various properties are summarized in Table 11.3.

Table 11.3   Operational Properties of the DTFT.

Symmetry properties for real *DT* signals

$x[n]$	$X_p(F)$
Any real signal	$X_p(F) = X_p^*(-F) = X_p(1 - F)$ (Even magnitude and odd phase)
	OR Even real part and odd imaginary part
Real and Even	Real and Even
Real and Odd	Imaginary and Odd

Properties arising from duality with Fourier series

Property	DTFT result	FS result (with $T=1$)
Superposition	$x[n] + y[n] \longleftrightarrow X_p(F) + Y_p(F)$	$x_p(t) + y_p(t) \longleftrightarrow X_S[k] + Y_S[k]$
Times scalar	$\alpha x[n] \longleftrightarrow \alpha X_p(F)$	$\alpha x_p(t) \longleftrightarrow \alpha X_S[k]$
Time delay	$x[n - \alpha] \longleftrightarrow X_p(F) \exp(-j2\pi F\alpha)$	$x_p(t - \alpha) \longleftrightarrow X_S[k] \exp(-jk\omega_0\alpha)$
Reversal	$x[-n] \longleftrightarrow X_p(-F) = X_p^*(F)$	$x_p(-t) \longleftrightarrow X_S[-k] = X_S^*[k]$
Modulation	$x[n] \exp(j2\pi n\alpha) \longleftrightarrow X_p(F - \alpha)$	$x_p(t) \exp(jm\omega_0 t) \longleftrightarrow X_S[k - m]$
Product	$x[n]y[n] \longleftrightarrow X_p(F) \bullet Y_p(F)$	$x_p(t)y_p(t) \longleftrightarrow X_S[k] \star Y_S[k]$
Convolution	$x[n] \star y[n] \longleftrightarrow X_p(F)Y_p(F)$	$x_p(t) \bullet y_p(t) \longleftrightarrow X_S[k]Y_S[k]$
Central ordinates	$X_p[0] = \sum_n x[n]$	$X_S[0] = \int_1 x_p(t)\, dt$
	$x[0] = \int_1 X_p(F)\, dF$	$x_p(0) = \sum_n X_S[k]$
Parseval's relation	$\sum_n |x[n]|^2 = \int_1 |X_p(F)|^2\, dF$	$\int_1 x_p^2(t)\, dt = \sum_k |X_S[k]|^2$

Additional properties of the DTFT

Property	Relation
Difference	$x[n] - x[n - 1] \longleftrightarrow [1 - \exp(-j2\pi F)]X_p(F)$
Modulation	$x[n] \cos(2\pi n\alpha) \longleftrightarrow \frac{1}{2}[X_p(F + \alpha) + X_p(F - \alpha)]$
Times-n	$nx[n] \longleftrightarrow \dfrac{-1}{j2\pi} \dfrac{dX_p(F)}{dF}$
Summation	$\sum_{k=-\infty}^{n} x[k] \longleftrightarrow \dfrac{X_p(F)}{1 - \exp(-j2\pi F)} + \frac{1}{2}X_p(0)\delta(F)$ (one period)

Symmetry   The symmetry properties of the DTFT follow from the defining relations. For a *real DT* signal $x[n]$:

1. The DTFT possesses conjugate symmetry about the origin with $X_p(-F) = X_p^*(F)$. This results in an *even magnitude spectrum* and an *odd phase spectrum* or an even real part and an odd imaginary part. Such conjugate symmetry characterizes every frequency domain transform for real signals.
2. Since $X_p(F)$ is periodic with principal period $(-\frac{1}{2}, \frac{1}{2})$, this periodicity leads to $X_p(-F) = X_p(1-F) = X_p^*(F)$. This means that we also have conjugate symmetry about $F = \frac{1}{2}$.
3. For a real, even sequence, $X_p(F)$ is always real, and for a real, odd sequence, $X_p(F)$ is always imaginary.

   The DTFT for real *DT* signals can therefore be plotted in one of two ways as illustrated in Figure 11.10:

Figure 11.10   Illustration of the symmetry of the DTFT for real signals.

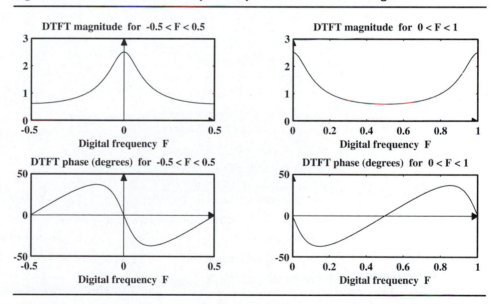

1. Over the principal period $(-\frac{1}{2}, \frac{1}{2})$ with conjugate symmetry about $F = 0$, which shows an even magnitude spectrum and an odd phase spectrum.
2. Over the period $(0, 1)$ with conjugate symmetry about $F = \frac{1}{2}$.

11.8.1   *Some DTFT Pairs*
A list of DTFT pairs appears in Table 11.4 and Figure 11.11. These can be developed in various ways:

1. We can use the defining relations and properties directly.
2. We can use duality between the Fourier series and the DTFT.
3. We can sample a *CT* signal $x(t)$ to get $x[n]$, and obtain $X_p(F)$ as a *periodic extension* of $X(f)$ with period 1. For signals whose Fourier transforms $X(f)$ are not bandlimited, the results are often unwieldy.

We provide some examples illustrating each approach.

Table 11.4   DTFT of Some Useful Signals.

DTFT pairs from defining relation

$x[n]$	DTFT $X_p(F)$ for one period $(-\frac{1}{2}, \frac{1}{2})$
$\delta[n]$	1
$\alpha^n u[n], \quad \lvert\alpha\rvert < 1$	$\dfrac{1}{1 - \alpha \exp(-j2\pi F)}$
$n\alpha^n u[n], \quad \lvert\alpha\rvert < 1$	$\dfrac{\alpha \exp(-j2\pi F)}{[1 - \alpha \exp(-j2\pi F)]^2}$
$(n+1)\alpha^n u[n], \quad \lvert\alpha\rvert < 1$	$\dfrac{1}{[1 - \alpha \exp(-j2\pi F)]^2}$
$\alpha^{\lvert n\rvert}, \quad \lvert\alpha\rvert < 1$	$\dfrac{1 - \alpha^2}{1 - 2\alpha \cos(2\pi F) + \alpha^2}$
$u[n]$	$\dfrac{1}{1 - \exp(-j2\pi F)} + \dfrac{1}{2}\delta(F)$
$\mathrm{rect}[n/2N]$	$\dfrac{\sin(M\pi F)}{\sin(\pi F)} = \dfrac{M\,\mathrm{sinc}(MF)}{\mathrm{sinc}(F)} \quad (M = 2N + 1)$
$\mathrm{tri}[n/N]$	$\dfrac{\sin^2(N\pi F)}{N \sin^2(\pi F)} = \dfrac{N\,\mathrm{sinc}^2(NF)}{\mathrm{sinc}^2(F)}$
$\cos(n\pi/2N)\mathrm{rect}(n/2N)$	$\dfrac{\sin(\pi/2N)\cos(2\pi NF)}{\cos(2\pi F) - \cos(\pi/2N)}$

DTFT pairs from the Fourier series

DTFT pair		Fourier series pair $(T = 1)$	
$x[n]$	$X_p\,(F)$ (one period)	$x_p\,(t)$ (one period)	$X_S[k]$
$2\alpha\,\mathrm{sinc}[2n\alpha]$	$\mathrm{rect}(F/2\alpha)$	$\mathrm{rect}(t/2\alpha)$	$2\alpha\,\mathrm{sinc}(2k\alpha)$
$\alpha\,\mathrm{sinc}^2[n\alpha]$	$\mathrm{tri}(F/\alpha)$	$\mathrm{tri}(t/\alpha)$	$\alpha\,\mathrm{sinc}^2(k\alpha)$
$\dfrac{2\pi\alpha\cos(n\pi\alpha)}{\pi^2 - (2n\pi\alpha)^2}$	Sinusoidal pulse $\cos(\pi F/\alpha) \quad \lvert F\rvert \le \frac{1}{2}\alpha$	Sinusoidal pulse $\cos(\pi t/\alpha) \quad \lvert t\rvert \le \frac{1}{2}\alpha$	$\dfrac{2\pi\alpha\cos(k\pi\alpha)}{\pi^2 - (2k\pi\alpha)^2}$

DTFT pairs from *FT* of bandlimited signals

DTFT pair (one period)		Fourier transform pair	
$x[n]$	$X_p(F)$ (one period)	$x_p(t)$	$X(f)$
1	$\delta(F)$	1	$\delta(f)$
$\cos[2\pi n\alpha]$	$\frac{1}{2}[\delta(F + \alpha) + \delta(F - \alpha)]$	$\cos(2\pi t\alpha)$	$\frac{1}{2}[\delta(f + \alpha) + \delta(f - \alpha)]$
$\sin[2\pi n\alpha]$	$\frac{1}{2}j[\delta(F + \alpha) - \delta(F - \alpha)]$	$\sin(2\pi t\alpha)$	$\frac{1}{2}j[\delta(f + \alpha) - \delta(f - \alpha)]$

Figure 11.11   The DTFT of some discrete-time signals of Table 11.4.

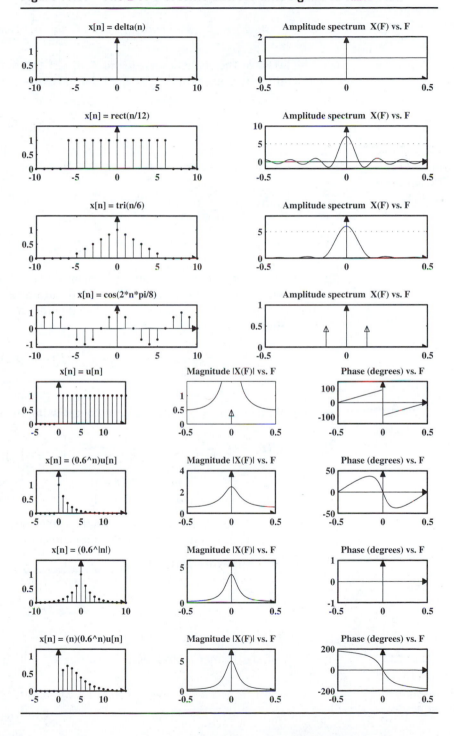

Example 11.7
DTFT of Finite Sequences

(a) The DTFT of $x[n] = \delta[n]$ follows immediately from the definition as

$$X_p(F) = \sum_{k=-\infty}^{\infty} x[k] \exp(-j2\pi kF) = \sum_{k=-\infty}^{\infty} \delta[k] \exp(-j2\pi kF) = 1$$

(b) The DTFT of the sequence $x[n] = \{1, 0, 3, -2\}$ follows from the result of part (a) and the shifting property as $X(F) = 1 + 3\exp(-j4\pi F) - 2\exp(-j6\pi F)$.

Example 11.8
DTFT from the
Defining Relation

(a) The DTFT of the exponential signal $x[n] = \alpha^n u[n]$ follows from the definition and the closed form for the resulting geometric series:

$$X_p(F) = \sum_{k=0}^{\infty} \alpha^k \exp(-j2\pi kF)$$

$$= \sum_{k=0}^{\infty} [\alpha \exp(-j2\pi F)]^k = \frac{1}{1 - \alpha \exp(-j2\pi F)} \qquad |\alpha| < 1$$

(b) The signal $x[n] = u[n]$ is a limiting form of $\alpha^n u[n]$ as $\alpha \longrightarrow 1$, but must be handled with care, since $u[n]$ is not absolutely summable. In analogy with the *CT* pair $u(t) \longleftrightarrow \frac{1}{2}\delta(f) + 1/j2\pi f$, $X_p(F)$ also includes an impulse (now an impulse train due to the periodic extension)

$$X_p(F) = \frac{1}{1 - \exp(-j2\pi F)} + \frac{1}{2}\delta(F) \qquad \text{(principal period)}$$

Example 11.9
Using the Properties

(a) The DTFT of $x[n] = n\alpha^n u[n]$ may be found using the times-n property as

$$X_p(F) = \frac{-1}{j2\pi} \frac{d}{dF}\left[\frac{1}{1 - \alpha \exp(-j2\pi F)}\right] = \frac{-1}{j2\pi} \frac{-j2\pi\alpha \exp(-j2\pi F)}{[1 - \alpha \exp(-j2\pi F)]^2}$$

$$= \frac{\alpha \exp(-j2\pi F)}{[1 - \alpha \exp(-j2\pi F)]^2}$$

(b) The DTFT of the signal $x[n] = (n+1)\alpha^n u[n]$ may be found by rewriting $x[n]$ as $x[n] = n\alpha^n u[n] + \alpha^n u[n]$, and using superposition, to give

$$X_p(F) = \frac{\alpha \exp(-j2\pi F)}{[1 - \alpha \exp(-j2\pi F)]^2} + \frac{1}{1 - \alpha \exp(-j2\pi F)} = \frac{1}{[1 - \alpha \exp(-j2\pi F)]^2}$$

Example 11.10
DTFT from the
Fourier Series

The DTFT of $x[n] = \text{rect}(n/2N)$, with $M = 2N + 1$

$$X_p(F) = \sum_{k=-N}^{N} \exp(-j2\pi kF) = M\text{sinc}(MF)/\text{sinc}(F)$$

We can also start with $M\mathrm{sinc}(Mt)/\mathrm{sinc}(t) \longleftrightarrow X_S[k] = \mathrm{rect}(k/2N)$, the Fourier series result for a Dirichlet kernel with unit period and $M = 2N + 1$. Using duality, we obtain

$$\mathrm{rect}(n/2N) \longleftrightarrow M\mathrm{sinc}(MF)/\mathrm{sinc}(F)$$

Example 11.11
DTFT from the
Fourier Transform

(a) The DTFT of $x[n] = \mathrm{sinc}(2n\alpha)$ follows from the Fourier transform pair $x(t) = 2\alpha\,\mathrm{sinc}(2\alpha t) \longleftrightarrow X(f) = \mathrm{rect}(f/2\alpha)$. We induce sampling in time ($t \longrightarrow n$) and periodic extension in the spectrum ($f \longrightarrow F$) to obtain

$$2\alpha\,\mathrm{sinc}(2n\alpha) \longleftrightarrow \mathrm{rect}(F/2\alpha)$$

We can also use $x_p(t) = \mathrm{rect}(t/2\alpha) \longleftrightarrow X_S[k] = 2\alpha\mathrm{sinc}(2k\alpha)$, the FS result for $x_p(t)$ with period $T = 1$. Duality gives $2\alpha\,\mathrm{sinc}(2n\alpha) \longleftrightarrow \mathrm{rect}(F/2\alpha)$.

(b) The DTFT of $x[n] = e^{-n}u[n] = (e^{-1})^n u[n]$ may be found quite readily from the defining relation as $X_p(F) = 1/[1 - e^{-1}\exp(j2\pi F)]$.

Comment: If we start with $e^{-t}u(t) \longleftrightarrow 1/(1 + j2\pi f)$, the DTFT of $e^{-n}u[n]$ also equals the periodic extension $\sum_k 1/[1 + j2\pi(F + k)]$. Of course, we know this must equal $1/[1 - e^{-1}\exp(j2\pi F)]$, but only if you get lucky can you find a closed-form result in a handbook of mathematical tables (we did not).

11.8.2   *Existence of the DTFT*

The DTFT is a summation. It always exists if the summand $x[n]\exp(j2\pi nF)$ is absolutely integrable. Since $|\exp(j2\pi nF)| = 1$, we can state that

1. The DTFT of absolutely summable signals always exists. In analogy with the Fourier transform, the DTFT of sinusoids, steps or constants, which are not absolutely summable, includes impulses.
2. The DTFT of signals that grow exponentially or faster does not exist.

11.9
The Inverse
DTFT

We can find the inverse DTFT in one of several ways.

1. The obvious method is to use the defining relation. This is quite like finding the Fourier series of a periodic signal.
2. If the DTFT is a ratio of polynomials in $\exp(j2\pi F)$, we can use partial fraction expansion and tables, much as we do with Laplace transforms.
3. We can even use Fourier transforms, and induce sampling and periodic extension to arrive at the required forms.

Example 11.12
IDTFT from Partial Fractions

Let $X_p(F) = 2\exp(-j2\pi F)/[1 - \frac{1}{4}\exp(-j4\pi F)]$. To find its IDTFT, we factor the denominator to get $X_p(F) = 2\exp(-j2\pi F)/\{[1 - \frac{1}{2}\exp(-j2\pi F)][1 + \frac{1}{2}\exp(-j2\pi F)]\}$. We use partial fractions to obtain

$$X_p(F) = \frac{2}{1 - \frac{1}{2}\exp(-j2\pi F)} - \frac{2}{1 + \frac{1}{2}\exp(-j2\pi F)}$$

Finally, Table 11.4 gives the IDTFT as $x[n] = 2(\frac{1}{2})^n u[n] - 2(-\frac{1}{2})^n u[n]$.

Example 11.13
IDTFT of an Ideal Differentiator

An ideal differentiator is described by $H_p(F) = j2\pi F$, $|F| < \frac{1}{2}$. To find its impulse response $h[n]$, we note that $h[0] = 0$ since $H_p(F)$ is odd. For $n \neq 0$, we also use the odd symmetry of $H_p(F)$ in the IDTFT to obtain

$$h[n] = \int_{-1/2}^{1/2} j2\pi F[\cos(2\pi nF) + j\sin(2\pi nF)]\, dF = -4\pi \int_0^{1/2} F\sin(2\pi nF)\, dF$$

Using tables (Appendix A) and simplifying the result, we obtain

$$h[n] = \frac{-4\pi[\sin(2\pi nF) - 2\pi nF\cos(2\pi nF)]}{(2\pi n)^2}\Bigg|_0^{1/2} = \cos(n\pi)/n$$

Comment: Since $H_p(F)$ is odd and imaginary, $h[n]$ is odd symmetric, as expected.

Example 11.14
IDTFT of a Hilbert Transformer

A Hilbert transformer shifts the phase of a signal by $-90°$. Its transfer function is given by $H_p(F) = -j\,\mathrm{sgn}(F)$, $|F| < \frac{1}{2}$. This is imaginary and odd. To find its impulse response $h[n]$, we note that $h[0] = 0$ and

$$h[n] = \int_{-1/2}^{1/2} -j\,\mathrm{sgn}(F)[\cos(2\pi nF) + j\sin(2\pi nF)]\, dF$$

$$= 2\int_0^{1/2} \sin(2\pi nF)\, dF = \frac{1 - \cos(n\pi)}{n\pi}$$

11.9.1   The Inverse DTFT and Impulse Response of Digital Filters

The DTFT spectra of various ideal filters is shown in Figure 11.12, and their impulse responses are listed in Table 11.5. The impulse response of each filter is related to that of an ideal lowpass filter. The simplest way to show this is to use the defining relation and properties of the DTFT.

Another way is to use the dual of the Fourier series results of Table 5.8. Here, we use a third approach by starting with the Fourier transform of the four filter types as

$$h_{\mathrm{LP}}(t) = 2F_C\,\mathrm{sinc}(2F_C t) \longleftrightarrow \mathrm{rect}(f/2F_C)$$
$$h_{\mathrm{HP}}(t) = 2F_C\,\mathrm{sinc}(2F_C t)\exp(j\pi t) \longleftrightarrow \mathrm{rect}[(f - \tfrac{1}{2})/2F_C]$$
$$h_{\mathrm{BP}}(t) = 4F_C\cos(2\pi F_C t)\,\mathrm{sinc}(2F_C t) \longleftrightarrow \mathrm{rect}[(f + F_0)/2F_C] + \mathrm{rect}[(f - F_0)/2F_C]$$
$$\qquad = H_{\mathrm{BP}}(f)$$
$$h_{\mathrm{BS}}(t) = \mathrm{sinc}(t) - h_{\mathrm{BP}}(t) \longleftrightarrow \mathrm{rect}(f) - H_{\mathrm{BP}}(f)$$

Table 11.5   DTFT of Ideal Filters.

Type	$h[n]$	$H_p(F)$ (one period)	How found
Lowpass	$h_{LP}[n] = 2F_C\,\text{sinc}(2nF_C)$	$H_{LP}(F) = \text{rect}(F/2F_C)$	DTFT
Highpass	$\delta[n] - H_{LP}[n]$	$H_{HP}(F) = 1 - H_{LP}(F)$	Superposition
Highpass	$(-1)^n h_{LP}[n]$	$H_{HP}(F) = H_{LP}(F - \tfrac{1}{2})$	Shifting
Bandpass	$2\cos(2\pi nF_0)h_{LP}[n]$	$H_{BP}(F) = \dfrac{H_{LP}((F + F_0)}{\ \ +H_{LP}(F - F_0)}$	Modulation
Bandstop	$\delta[n] - h_{BP}[n]$	$H_{BS}(F) = 1 - H_{BP}(F)$	Superposition

Figure 11.12   The DTFT of ideal filters.

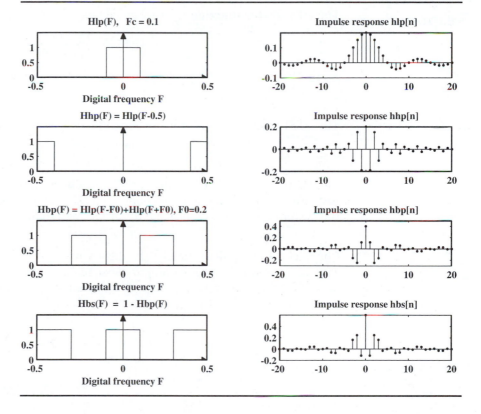

Sampling in time ($t \longrightarrow n$) leads to the DTFT as a periodic extension of the spectrum ($f \longrightarrow F$) with period 1. This is also the periodic continuation of the spectra. With $\exp(j\pi n) = (-1)^n$ and $\text{sinc}(n) = \delta[n]$, we obtain

$$h_{LP}[n] = 2F_C\,\text{sinc}(2F_C n) \longleftrightarrow H_{LP}(F) = \text{rect}(F/2F_C)$$
$$h_{HP}[n] = \delta[n] - h_{LP}[n] \longleftrightarrow H_{HP}(F) = 1 - H_{LP}(F) \quad \text{or}$$

$$h_{HP}[n] = (-1)^n h_{LP}[n] = 2F_C(-1)^n \text{sinc}(2F_C n) \longleftrightarrow H_{HP}(F) = H_{LP}(F - \tfrac{1}{2})$$

$$h_{BP}[n] = 4F_C \cos(2\pi F_0 n)\text{sinc}(2F_C n) \longleftrightarrow H_{BS}(F) = H_{LP}(F + F_0) + H_{LP}(F - F_0)$$

$$h_{BS}[n] = \delta[n] - h_{BP}[n] \longleftrightarrow H_{BS}(F) = 1 - H_{BP}(F)$$

These results are central to FIR filter design, as discussed in Chapter 14.

11.10
The Transfer Function and Frequency Response

The response of a relaxed *DT* LTI system is given by the convolution

$$y[n] = x[n] \star h[n]$$

Since convolution transforms to multiplication, transformation results in

$$Y_p(F) = X_p(F)H_p(F)$$

The *DT* **transfer function** $H_p(F)$ then equals

$$H_p(F) = \frac{Y_p(F)}{X_p(F)}$$

A relaxed *DT* LTI system may also be described by the difference equation

$$y[n] + A_1 y[n-1] + \cdots + A_N y[n-N] = B_0 x[n] + B_1 x[n-1] + \cdots + B_M x[n-M]$$

The DTFT results in a *DT* transfer function, or **frequency response**, given by

$$H_p(F) = \frac{Y_p(F)}{X_p(F)} = \frac{B_0 + B_1 \exp(-j2\pi F) + \cdots + B_M \exp(-j2\pi M F)}{1 + A_1 \exp(-j2\pi F) + \cdots + A_N \exp(-j2\pi N F)}$$

The *DT* transfer function is thus a ratio of polynomials in $\exp(j2\pi F)$. It allows us to characterize the frequency response of a *DT* system. It also allows access to both the system impulse response and difference equation.

Example 11.15
Discrete-Time Frequency Response

Consider a system described by the difference equation $y[n] = \alpha y[n-1] + x[n]$. This is an example of an *echo system* whose response equals the input plus a delayed version of the output. Its frequency response may be found by taking the DTFT of both sides to give $Y_p(F) = \alpha Y_p(F) \exp(-j2\pi F) + X_p(F)$. Rearrangement gives

$$H_p(F) = \frac{Y_p(F)}{X_p(F)} = \frac{1}{1 - \alpha \exp(-j2\pi F)}$$

Using Euler's relation, we rewrite this as

$$H_p(F) = \frac{1}{1 - \alpha \exp(-j2\pi F)} = \frac{1}{1 - \alpha \cos(2\pi F) - j\alpha \sin(2\pi F)}$$

The magnitude $|H_p(F)|$ and phase $\phi_p(F)$ then equal

$$|H_p(F)| = \frac{1}{[1 - 2\alpha \cos(2\pi F) + \alpha^2]^{1/2}} \qquad \phi_p(F) = -\tan^{-1}\left[\frac{\alpha \sin(2\pi F)}{1 - \alpha \cos(2\pi F)}\right]$$

These are sketched in Figure 11.10 for $\alpha = 0.6$.

Comment: The impulse response of this system equals $h[n] = \alpha^n u[n]$.

11.10.1   *The Response to Arbitrary Inputs*

In concept, the DTFT may be used to find the response of relaxed systems to arbitrary inputs, even though it is definitely not the best way to obtain such a response. All it takes is a four-step recipe:

1. Establish the *DT* system transfer function $H_p(F)$.
2. Find the DTFT $X_p(F)$ of the input $x[n]$.
3. Find the response $Y_p(F)$ in the frequency domain as $Y_p(F) = X_p(F)H_p(F)$.
4. Establish $y[n]$ by using the inverse transformation.

Example 11.16
DTFT in System Analysis

(a) Consider a system described by $y[n] = \alpha y[n-1] + x[n]$. To find the response of this system to the input $\alpha^n u[n]$, we first set up the transfer function as $H_p(F) = 1/[1 - \exp(-j2\pi F)]$. Next, we find the DTFT $X_p(F)$ of $x[n]$, and obtain $Y_p(F) = H_p(F)X_p(F)$ as

$$Y_p(F) = H_p(F)X_p(F)$$
$$= \frac{1}{1 - \alpha\exp(-j2\pi F)}\frac{1}{1 - \alpha\exp(-j2\pi F)} = \frac{1}{[1 - \alpha\exp(-j2\pi F)]^2}$$

Finally, from Table 11.4, we obtain the response $y[n] = (n+1)\alpha^n u[n]$.

Comment: We could, of course, also use convolution to obtain $y[n] = h[n] \star x[n]$.

(b) Consider the system described by $y[n] = 0.5y[n-1] + x[n]$. Its response to the step $x[n] = 4u[n]$ is found using $Y_p(F) = H_p(F)X_p(F)$.

$$Y_p(F) = H_p(F)X_p(F) = \frac{1}{1 - 0.5\exp(-j2\pi F)}\left[\frac{4}{1 - \exp(-j2\pi F)} + 2\delta(F)\right]$$

We separate terms and use the product property of impulses to get

$$Y_p(F) = \frac{4}{[1 - 0.5\exp(-j2\pi F)][1 - \exp(-j2\pi F)]} + 4\delta(F)$$

Splitting the first term into partial fractions, we obtain

$$Y_p(F) = \frac{-4}{1 - 0.5\exp(-j2\pi F)} + \left[\frac{8}{1 - \exp(-j2\pi F)} + 4\delta(F)\right]$$

The response $y[n]$ then equals $y[n] = -4(0.5)^n u[n] + 8u[n]$. The first term represents the *transient* response and the second the *steady-state* response, which can be found much more easily, as we now show.

11.10.2   *The Steady-State Response to DT Harmonics*

The DTFT is much better suited to finding the *steady-state* response to *DT* harmonics (which may or may not be periodic). Since these represent eigensignals of *DT* linear systems, the response is simply a harmonic at the input frequency whose magnitude and phase is changed by the system function. We evaluate $H_p(F)$ at the frequency of the harmonic, multiply its magnitude by the input magnitude to obtain the output magnitude and add its phase to the input phase to obtain the output phase.

Remark: The steady-state response is useful primarily for stable systems for which the natural response does indeed decay with time.

Example 11.17
The DTFT and
Steady-State Response

Consider a system described by the difference equation $y[n] = 0.5y[n-1] + x[n]$. Its system transfer function $H_p(F)$ is given by

$$H_p(F) = \frac{1}{1 - 0.5\exp(-j2\pi F)}$$

(a) To find its response to the sinusoidal input $x[n] = 10\cos(2\pi\frac{1}{4}n + 60°)$, we evaluate $H_p(F)$ at the input frequency, $F = \frac{1}{4}$

$$H_p\left(\frac{1}{4}\right) = \frac{1}{1 - 0.5\exp(-j2\pi\frac{1}{4})} = \frac{1}{1 + 0.5j} = 0.4\sqrt{5}\angle -26.6°$$

The steady-state response then equals

$$y[n] = 10(0.4\sqrt{5})\cos(2\pi\tfrac{1}{4}n + 60° - 26.6°) = 4\sqrt{5}\cos(2\pi\tfrac{1}{4}n + 33.4°)$$

Comment: If $x[n] = 10\cos(2\pi\frac{1}{4}n + 60°)u[n]$, the steady-state component would still be identical to what we calculated but the *total* response would differ.

(b) To find the steady-state response to the step $x[n] = 4u[n]$, we evaluate $H_p(F)$ at the input frequency $F = 0$ (corresponding to dc)

$$H_p(F) = \frac{1}{1 - 0.5\exp(-j2\pi F)} \qquad H_p(0) = \frac{1}{1 - 0.5} = 2$$

The steady-state response is then $y_{ss}[n] = (2)4 = 8$.   ▬

11.11
Special Topics:
The Frequency
Response of
Discrete
Operators

Signal averaging or smoothing has its roots in interpolation. One scheme, based on a linear least squares fit, replaces each signal value by an average of N samples on either side of it. The smoothing operation amounts to a lowpass operation and the response of the smoothing system may be described by the difference equation

$$y[n] = \frac{1}{2N+1}\{x[n+N] + x[n+N-1] + \cdots + x[n-N-1] + x[n-N]\}$$

This system is called a **moving average filter**. Using the DTFT and Euler's relation, its transfer function equals

$$H(F) = \frac{Y(F)}{X(F)} = \frac{1}{2N+1}\sum_{k=-N}^{N}\exp(-j2\pi kF) = \frac{\text{sinc}(MF)}{\text{sinc}(F)}$$

where $M = 2N + 1$. This has the form of a Dirichlet kernel, scaled by $1/M$.

Example 11.18
Frequency Response of a
Moving Average Filter

For a three-point moving average filter, $y[n] = \frac{1}{3}\{x[n-1] + x[n] + x[n+1]\}$. The filter replaces each input value $x[n]$ by an average of itself and its two neighbors. Its impulse response $h[n]$ is simply $h[n] = \{\frac{1}{3}, \underset{\uparrow}{\frac{1}{3}}, \frac{1}{3}\}$. The frequency response is given by

$$H(F) = \sum_{n=-1}^{1} h[n]\exp(-j2\pi Fn)$$
$$= \tfrac{1}{3}[\exp(j2\pi F) + 1 + \exp(-j2\pi F)] = \tfrac{1}{3}[1 + 2\cos(2\pi F)]$$

The amplitude $H(F)$ is plotted in Figure 11.13a. The magnitude $|H(F)|$ decreases until $F = \frac{2}{3}$ (when $H(F) = 0$) and then increases to $\frac{1}{3}$ at $F = \frac{1}{2}$. This filter thus does a poor job of smoothing past $F = \frac{2}{3}$.

Figure 11.13   Frequency response of some smoothing operators.

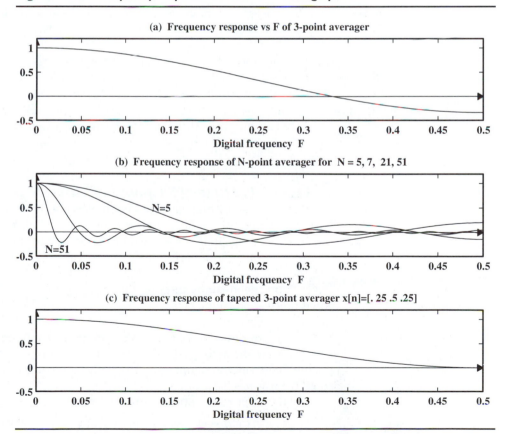

(a) **Frequency response vs F of 3-point averager**

(b) **Frequency response of N-point averager for N = 5, 7, 21, 51**

(c) **Frequency response of tapered 3-point averager x[n]=[. 25 .5 .25]**

The preceding example shows that the magnitude spectrum does not reduce at high frequencies. Intuitively, averaging over more samples should result in better smoothing but longer operators hardly improve the situation and lead to even more ripples. The reason, of course is the Gibbs effect. A plot of $H(F)$ in Figure 11.13b for various values of N shows ripples which persist even for large N. Like Fourier series smoothing, a tapered impulse response reduces or eliminates the ripples in $H(F)$. Tapering the impulse response $h[n]$ is equivalent to multiplying $h[n]$ by a *time-domain smoothing window*.

Example 11.19

Frequency Response of a Tapered Window

(a) The DTFT of the tapered three-point averager $x[n] = \{\frac{1}{4}, \frac{1}{2}, \frac{1}{4}\}$ is given by

$$H(F) = \sum_{n=-1}^{1} h[n]\exp(-j2\pi Fn) = \tfrac{1}{4}\exp(j2\pi F) + \tfrac{1}{2} + \tfrac{1}{4}\exp(-j2\pi F)$$

$$= \tfrac{1}{2} + \tfrac{1}{2}\cos(2\pi F)$$

Note that $|H(F)|$ now decreases monotonically to zero at $F = \frac{1}{2}$, and shows a much better smoothing performance (Figure 11.13c).

Comment: This operator actually describes the vonHann smoothing window. Other tapering schemes lead to many other window types.

(b) To find the DTFT of a triangular window, we start with the convolution $\text{rect}(n/2N) \star \text{rect}(n/2N) = (2N + 1)\text{tri}(n/2N + 1)$. With $M = 2N + 1$, the DTFT of this signal may be written as $M\text{tri}(n/M) \longleftrightarrow [M\text{sinc}(MF)/\text{sinc}(F)]^2$. This results in $x[n] = \text{tri}(n/M) \longleftrightarrow X_p(F) = M[\text{sinc}(MF)/\text{sinc}(F)]^2$. The spectrum $X_p(F)$ is entirely positive and much smoother than that of a rectangular window.

11.11.1   *Frequency Response of Discrete Algorithms*

Discrete-time algorithms are used to convert *CT* systems to *DT* systems. The frequency response of an ideal integrator is $H_I(f) = 1/j2\pi f$. For an ideal differentiator, $H_D(f) = j2\pi f$. In the discrete domain, these operations are replaced by numerical integration and numerical differences. Table 11.6 lists some of the algorithms and their frequency response $H(F)$. Figure 11.14 shows the magnitude and phase *error* (from the ideal) by plotting the ratio $H_p(F)/H_I(F)$. Ideally, this ratio should equal unity at all frequencies.

Numerical Integration Algorithms   Most numerical integration algorithms of Table 11.6 estimate the area $y[n]$ from $y[n-1]$ by using step, linear or quadratic interpolation between the samples of $x[n]$. Only Simpson's rule finds $y[n]$ over two time steps from $y[n-2]$.

For the trapezoidal operator, for example, the DTFT yields

$$Y_p(F) = Y_p(F)\exp(-j2\pi F) + \tfrac{1}{2}[X_p(F) + X_p(F)\exp(-j2\pi F)]$$

Table 11.6   Frequency Response of Discrete Algorithms.

Algorithm	Formula for y[n] ($t_s = 1$)	Frequency response $H_p(F)$
Rectangular	$y[n-1] + x[n]$	$\dfrac{1}{1 - \exp(-j2\pi F)}$
Trapezoidal	$y[n-1] + \frac{1}{2}\{x[n] + x[n-1]\}$	$\dfrac{1 + \exp(-j2\pi F)}{2[1 - \exp(-j2\pi F)]}$
Simpson's	$y[n-2] + \frac{1}{3}\{x[n] + 4x[n-1] + x[n-2]\}$	$\dfrac{1 + 4\exp(-j2\pi F) + \exp(-j4\pi F)}{3[1 - \exp(-j4\pi F)]}$

Differences

Backward	$y[n] = x[n] - x[n-1]$	$H_B(F) = 1 - \exp(-j2\pi F)$
Central	$y[n] = \frac{1}{2}\{x[n+1] - x[n-1]\}$	$H_C(F) = j\sin(2\pi F)$
Forward	$y[n] = x[n+1] - x[n]$	$H_F(F) = \exp(j2\pi F) - 1$

Figure 11.14   Magnitude and phase deviation (from the ideal) of numerical integration and difference algorithms.

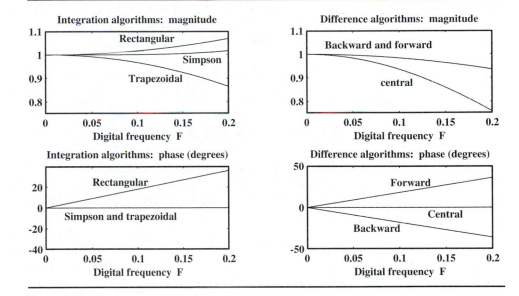

A simple rearrangement leads to

$$H_p(F) = \frac{Y_p(F)}{X_p(F)} = \frac{1}{2}\left[\frac{1 + \exp(-j2\pi F)}{1 - \exp(-j2\pi F)}\right] = 1/j2\tan(\pi F)$$

Normalizing this by $H_I(F) = 1/j2\pi F$ and expanding the result, we obtain

$$\frac{H_p(F)}{H_I(F)} = j2\pi F/j2\tan(\pi F) = \pi F/\tan(\pi F) \approx 1 - (\pi F)^2/3 - (\pi F)^4/45 + \cdots$$

At low frequencies, $\tan(\pi F) \approx \pi F$, $H_p(F)/H_I(F) \approx 1$, and the trapezoidal rule is a valid approximation to integration. The phase response matches the ideal phase at all frequencies.

Simpson's algorithm yields the following normalized result

$$\frac{H_p(F)}{H_I(F)} = 2\pi F \frac{2 + \cos 2\pi F}{3 \sin 2\pi F}$$

It displays a perfect phase match for all frequencies, but has an overshoot in its magnitude response past $F = \frac{1}{4}$ and thus amplifies high frequencies.

Numerical Difference Algorithms   Numerical differences approximate slopes (derivative) at a point. For the forward difference, the DTFT yields

$$Y_p(F) = X_p(F) \exp(j2\pi F) - X_p(F) \qquad H_p(F) = \frac{Y_p(F)}{X_p(F)} = \exp(j2\pi F) - 1$$

The ratio $H_p(F)/H_D(F)$ may be expanded as

$$\frac{H_p(F)}{H_D(F)} = 1 + j\frac{1}{2!}(2\pi F) - \frac{1}{3!}(2\pi F)^2 + \cdots$$

Again, we observe correspondence only at low digital frequencies (or high sampling rates). The high frequencies are amplified, making it susceptible to high frequency noise. The phase response also deviates from the true phase, especially at high frequencies.

For the central difference algorithm, we find $H_p(F)/H_D(F)$ as

$$\frac{H_p(F)}{H_D(F)} = \frac{\sin(2\pi F)}{2\pi F} = 1 - \frac{1}{3!}(2\pi F)^2 + \frac{1}{5!}(2\pi F)^4 + \cdots$$

We see only a perfect phase match for all frequencies.

Concluding Remarks   Discrete algorithms are good approximations only at low digital frequencies ($F < 0.1$, say). Even when we satisfy the sampling theorem, low digital frequencies mean high sampling rates, well in excess of the Nyquist rate. This is why the sampling rate is a critical factor in the frequency domain performance of these operators. Another factor is stability. For example, if Simpson's rule is used to convert CT systems to DT systems, it results in an unstable system. The trapezoidal integration algorithm and the backward difference are two popular choices.

11.12 Connections

Sampling and duality provide the basis for the connection between all of the frequency domain transforms we study, as listed in Table 11.7.

All these transforms are based on the following important concepts:

1. Periodicity in one domain is the consequence of the discrete nature in the other. Periodic CT signals have discrete spectra and discrete-time signals have continuous periodic spectra.

2. The sample separation in one domain is the reciprocal of the period in the other.

Table 11.7   Duality, Sampling and Periodic Extension.

Note: PE stands for periodic extension.

Operation in time domain	Result in frequency domain	Transform
Aperiodic, continuous $x(t)$	Aperiodic, continuous $X(f)$	FT
PE of $x(t) \longrightarrow x_p(t)$ Period $= T$	Sampling of $X(f) \longrightarrow X_S[k]$ Sampling interval $= 1/T = f_0$	FS
Sampling of $x_p(t) \longrightarrow x_p[n]$ Sampling interval $= t_s$	PE of $X_S[k] \longrightarrow X[k]$ Period $= 1/t_s$	DFS
Sampling of $x(t) \longrightarrow x[n]$ Sampling interval $= 1$	PE of $X(f) \longrightarrow X_p(F)$ Period $= 1$	DTFT
PE of $x_S[n] \longrightarrow x_p[n]$ Period $= N$	Sampling of $X_p(F) \longrightarrow X_T[k]$ Sampling interval $= 1/N$	DFT

11.12.1   *The DFS and the DFT*

A consequence of the preceding observations is that a sequence which is both discrete and periodic in one domain is also discrete and periodic in the other. This concept leads to both the **discrete Fourier series** (DFS) and **discrete Fourier transform** (DFT) discussed in the next chapter. The DFS and DFT allow us a practical means of arriving at the sampled spectrum of sampled signals using digital computers.

11.12.2   *The DTFT and the z-Transform*

The DTFT describes a *DT* signal as a sum of weighted harmonics. It suffers from two major drawbacks:

1. It cannot handle exponentially growing signals.
2. It cannot handle initial conditions in system analysis.

In addition, the DTFT of signals that are not absolutely summable usually includes impulses. An important example is the step function $u[n]$.

The shortcomings of the DTFT can be removed as follows:

First, we include a real weighting factor r^{-k} and redefine the transform as the sum of exponentially weighted harmonics with $z = r \exp(-j2\pi F)$:

$$X(z) = \sum_{k=-\infty}^{\infty} x[k] \exp(-j2\pi kF) r^{-k} = \sum_{k=-\infty}^{\infty} x[k][r \exp(j2\pi F)]^{-k} = \sum_{k=-\infty}^{\infty} x[k] z^{-k}$$

This defines the *twosided z-transform* of $x[k]$. The weighting r^{-k} acts as a *convergence factor* to allow transformation of exponentially growing signals. The DTFT of $x[n]$ may thus be viewed as its *z*-transform $X(z)$ evaluated for $r = 1$.

Second, to handle initial conditions, we change the lower limit in the sum to zero. This yields the *onesided z-transform* for *causal* signals, and allows us to analyze *DT* LTI systems with arbitrary initial conditions.

We discuss the *z*-transform in Chapter 13.

11.13
Sampled Signals and the DTFT: A Synopsis

Sampled signals	A sampled signal may be regarded as the product of a *CT* signal and a periodic sampling signal.
Sampling	Sampling by a periodic impulse train leads to an ideally sampled signal as a train of impulses whose strengths are the *CT* signal values. Sampling by a periodic signal leads to a naturally sampled signal. If the sampled value is held constant, we obtain a flat-top sampled signal.
Spectrum of sampled signals	The spectrum of a sampled signal is continuous and is a periodic extension of the spectrum of its underlying *CT* signal. Only the spectrum of the ideally sampled signal is periodic. Its principal period is bandlimited to $\|f\| \le \frac{1}{2}S_F$. The highest frequency in the spectrum is $\frac{1}{2}S_F$ and depends only on the sampling rate S_F, not f_B.
Sampling theorem	A signal bandlimited to f_B can be sampled without loss of information if sampled above the Nyquist rate $S_F = 2f_B$.
Aliasing	Undersampling results in spectral overlap as components beyond $\frac{1}{2}S_F$ alias to frequencies lower than $\frac{1}{2}S_F$. The spectrum of naturally sampled signals is not periodic, but the highest frequency is still $\frac{1}{2}S_F$.
Nyquist rate	For a pure sinusoid, sampling above the Nyquist rate is equivalent to taking more than two samples per period. For bandpass signals, we can often get by with a rate smaller than the highest frequency present.
Signal recovery	Signal recovery uses interpolation to recover a *CT* signal as a sum of weighted, shifted interpolating functions. Bandlimited signals sampled above the Nyquist rate can be recovered exactly as a sum of weighted, shifted sinc functions $\text{sinc}(2f_B t)$. Step or linear interpolation is used in practice.
Sampling errors	Sampling errors can arise from undersampling, roundoff error in sample values or jitter. Their effect can be described only statistically.
Quantization	Quantization uses rounding or truncation to restrict signal amplitudes to a finite set of levels. Uniform quantizers use equal quantization levels. Quantization leads to nonlinear effects, which are also described in statistical terms. The quantization error is measured in terms of the quantization SNR as the ratio of the power P_S in the signal and the power P_N in the error (or noise).

The DTFT	The DTFT describes the spectrum of *DT* or sampled signals. It is a dual of the Fourier series and many of the properties are similar. The DTFT of real signals possesses conjugate symmetry $F = 0$ and $F = \frac{1}{2}$. The DTFT of discrete signals can be developed from the definition or using duality. For bandlimited signals the DTFT is a periodic extension of the FT with period 1. The DTFT of signals that are not absolutely summable (the unit step or sinusoid) almost always includes impulses. The IDTFT is much like finding Fourier series coefficients.
The *DT* transfer function	$H_p(F)$ equals the DTFT of the impulse response $h[n]$ or the ratio $Y_p(F)/X_p(F)$. It describes the frequency response of a stable *DT* system and is a ratio of polynomials in $\exp(-j2\pi F)$.
Steady-state response	The DTFT is best suited to finding the steady-state response to *DT* harmonic signals. The steady-state response to $A\cos(2\pi nF_0 + \theta)$ is $A|H|\cos(2\pi nF_0 + \theta + \phi)$, where $|H|$ and ϕ are the magnitude and phase of $H_p(F_0)$.

Appendix 11A
MATLAB Demonstrations and Routines

Getting Started on the MATLAB Demonstrations

We present the MATLAB routines related to concepts in this chapter, and some examples of their use. See Appendix 1A for introductory remarks.

Note: All polynomial arrays must be entered in descending powers.

Interpolation of Sampled Signals

```
y=interpol(f,ty,n)
```

Interpolates a signal *f* using *n* samples between adjacent samples and type ty='c'(constant),'l'(linear), 's'(sinc), 'r'(raised cosine), 'd'(dft) or 'z'(zero) interpolation algorithms. It plots the interpolated signal if no output argument is present.

Example

Create a coarse sine wave and interpolate using constant and sinc interpolation.

```
>>t=0:.2:2;x=sin(pi*t);interpol(x,'c');interpol(x,'s');
```

Quantization

```
y=quantiz(f,l,a,ty)
```

Quantizes a string function or signal values *f* over duration $l = [ll, lu]$ to *a* levels using ty = 'r', 't' or 's' for rounding, truncation or sign-magnitude truncation. Plots quantized values if no output argument is given.

<table>
</table>

Example	Quantize a sine wave using rounding and truncation as follows:
	
	```>>t=0:.02:1;x=sin(pi*t);quantiz(x,[0 1],6,'r');quantiz(x,[0 1],'t'); %OR```
	```>>quantiz('sin(pi*t)',[0 1],6,'r');quantiz('sin(pi*t)',[0 1],6,'t');```

Steady-State Response in Symbolic Form

```yss=ssresp(ty,n,d,x) %Computes steady-state response to sinusoids```

$ty = $ 'z', n, and d are coefficients of $H(F) = N(F)/D(F)$ in powers of $\exp(j2\pi F)$ and x is the array $[a, w, r]$ for an input of the form $x(t) = a\cos(w * n + r)$.

Example	To find the response of $H(F) = 1/[0.5 + \exp(-j2\pi F)]$ to $x[n] = 4\cos(\frac{1}{4}n\pi - 60°)$, rewrite $H(F) = \exp(j2\pi F)/[0.5\exp(j2\pi F) + 1]$ and use
	
	```>>yss=ssresp('z',[1 0],[0.5 1],[4 pi/4 -pi/3])```
	
	MATLAB will respond with
	
	```yss=2.850*cos(0.7854*n-0.1647*pi)```
	
	Here yss is a string expression in n and can be plotted by first evaluating it. Try:
	
	```n=0:50; ys=eval(yss);plot(n,ys)```

Demonstrations and Utility Functions

```
recover(x,ty)    Demonstrates signal recovery using interpolation type ty
digits(x,n)      Truncates x to n significant digits.
[m,p,w]=tfplot('z',n,d,f,p1,p2);
```

Returns magnitude, phase and frequency (F) for H(F). $f = [fl, fh] = $ Digital frequency F. See Chapter 6 for details of other arguments.

PROBLEMS

P11.1 **(Sampling Operations)** The *CT* signal $x(t) = \text{sinc}(4000t)$ is sampled at intervals of t_s. Sketch the sampled signal $x_s(t)$ over $0 \le t \le 2$ ms and also sketch its spectrum $X_S(F)$ over $-12k$ Hz $\le f \le 12$ kHz for the following choices of the sampling function and t_s:
(a) Impulse train (ideal sampling) with $t_s = 0.2$ ms.
(b) Impulse train with $t_s = 0.25$ ms.
(c) Impulse train with $t_s = 0.4$ ms.
(d) Rectangular pulse train with pulse width $t_d = 0.1$ ms and $t_s = 0.2$ ms.
(e) Flat-top (zero-order-hold) sampling with $t_s = 0.2$ ms.
```

P11.2 (Digital Frequency) Find a representation with a digital frequency $|F| \leq \frac{1}{2}$ for the following signals:
(a) $x_1[n] = \cos(4\pi n/3)$
(b) $x_2[n] = \cos(4\pi n/7) + \sin(8\pi n/7)$

P11.3 (Sampling Theorem) Establish the *minimum* sampling rate in hertz, based on the Nyquist criterion, for the following signals:
(a) $x_1(t) = 5\sin(300\pi t + \pi/3)$ **(b)** $x_2(t) = \cos(300\pi t) - \sin(300\pi t + 51°)$
(c) $x_3(t) = 3\cos(300\pi t) + 5\sin(500\pi t)$ **(d)** $x_4(t) = 3\cos(300\pi t)\sin(500\pi t)$
(e) $x_5(t) = 4\cos^2(100\pi t)$ **(f)** $x_6(t) = 6\,\text{sinc}(100t)$
(g) $x_7(t) = 10\,\text{sinc}^2(100t)$ **(h)** $x_8(t) = 6\,\text{sinc}(100t)\cos(200\pi t)$

P11.4 (Sampling Theorem) A sinusoid $x(t) = A\cos(2\pi f_0 t)$ is sampled at three times the Nyquist rate for six periods. How many samples are acquired?

P11.5 (Sampling Theorem) A sinusoid $x(t) = A\cos(2\pi f_0 t)$ is sampled at two times the Nyquist rate for 1 s. A total of 100 samples is acquired. What is f_0 and the digital frequency of the sampled signal?

P11.6 (Sampling Theorem) A sinusoid $x(t) = \sin(2\pi 75 t)$ is sampled at a rate of five samples per three periods. What fraction of the Nyquist sampling rate does this correspond to? What is the digital frequency of the sampled signal?

P11.7 (Sampling Theorem) Consider a periodic signal $x_p(t) = 1$, $(0, \frac{1}{2}T)$ and $x_p(t) = -1$, $(\frac{1}{2}T, T)$ where $T = 1$ ms passed through an *ideal* lowpass filter with a cutoff frequency of 4 kHz. The filter output is to be sampled. What is the *minimum* sampling rate required, considering both the symmetry of the signal as well as the Nyquist criterion?

P11.8 (Spectrum of Sampled Signals) Sketch the spectrum of the sampled signals using a sampling frequency of 50 Hz, 40 Hz, and 30 Hz if the spectrum of the continuous-time signal is:
(a) $X(f) = \text{rect}(f/40)$ **(b)** $X(f) = \text{tri}(f/20)$

P11.9 (Spectrum of Sampled Signals) Sketch the spectrum of the following signals over $|F| \leq \frac{1}{2}$:
(a) $\cos(200\pi t)$ sampled at 450 Hz **(b)** $\sin(400\pi t - \frac{1}{4}\pi)$ sampled at 300 Hz
(c) $\cos(200\pi t) + \sin(350\pi t)$ sampled at 300 Hz
(d) $\cos(200\pi t + \frac{1}{4}\pi) + \sin(250\pi t - \frac{1}{3}\pi)$ sampled at 120 Hz

P11.10 (Sampling) Let $x(t)$ comprise pure sines with unit magnitude at 10, 40, 200, 220, 240, 260, 300, 320, 340, 360, 380 and 400 Hz.
(a) Sketch the magnitude and phase spectra of $x(t)$.
(b) If $x(t)$ is sampled at 140 Hz, which components show aliasing?
(c) Write an expression for the sampled signal $x[n]$, sketch its magnitude and phase spectrum against f in hertz and against the digital frequency F.
(d) Compare the spectra of $x[n]$ and $x(t)$. What sampling frequency will preserve the identity of $x(t)$?

P11.11 (Sampling Frequency) The signal $x(t) = \exp(-t)u(t)$ is sampled at a rate S_F such that the maximum aliased magnitude (at $f = \frac{1}{2}S_F$) is less than 5% of the peak magnitude of the unaliased image. Estimate S_F.

P11.12 (Sampling Bandpass Signals) A signal $x(t)$ is bandlimited to 500 Hz and the smallest frequency present in $x(t)$ is f_0. Find the minimum rate at which we can sample $x(t)$ without aliasing and explain how we can recover the signal if
(a) $f_0 = 0$ **(b)** $f_0 = 300$ Hz   **(c)** $f_0 = 400$ Hz

P11.13 (Spectrum of Discrete Signals) Sketch the spectrum of the following signals over $|F| \le \frac{1}{2}$:
(a) $\cos(\frac{1}{2}\pi n)$ **(b)** $\cos(\frac{1}{2}\pi n) + \cos(\frac{1}{3}\pi n)$ **(c)** $\cos(\frac{1}{2}\pi n)\cos(\frac{1}{3}\pi n)$

P11.14 (Signal Recovery) A sinusoid $x(t) = \sin(2\pi 75t)$ is sampled at 80 Hz. Describe the signal $y(t)$ that we recover if the sampled signal is passed through
(a) An ideal lowpass filter with cutoff frequency $f_C = 10$ Hz.
(b) An ideal lowpass filter with cutoff frequency $f_C = 100$ Hz.
(c) An ideal bandpass filter with passband between 60 Hz and 80 Hz.
(d) An ideal bandpass filter with passband between 60 Hz and 100 Hz.

P11.15 (Interpolation) Consider the sampled sequence $\{x_S[n]\} = \{-1, 2, 3, 2\}$ with $t_s = 1$ passed through an interpolating system.
(a) Sketch the output if the system performs step interpolation.
(b) Sketch the output if the system performs linear interpolation.
(c) What is the interpolated value at $t = 2.5$ s if the system performs sinc interpolation?
(d) What is the interpolated value at $t = 2.5$ s if the system performs raised cosine interpolation (with $R = 0.5$)?

P11.16 (Interpolation) We wish to sample a speech signal bandlimited to 4 kHz using zero-order-hold sampling.
(a) Select the sampling frequency if the spectral magnitude of the sampled signal at 4 kHz is to be within 90% of its peak magnitude.
(b) On recovery, the signal is filtered using a Butterworth filter with an attenuation of less than 1 dB in the passband and more than 30 dB for all image frequencies. Compute the total attenuation in dB due to both the sampling and filtering operations at 4 kHz and 12 kHz.
(c) What is the order of the Butterworth filter?

P11.17 (Quantization SNR) Consider the signal $x(t) = t^2$, $(0, 2)$. Choose $t_s = 0.1$ s, four quantization levels and rounding to find:
(a) The sampled signal $x[n]$.
(b) The quantized signal $x_Q[n]$.
(c) The actual quantization signal to noise ratio SNR_Q.
(d) The statistical estimate of the quantization SNR_S.
(e) An estimate of the SNR, assuming $x(t)$ to be periodic.

P11.18 (DTFT of Sequences) Find $X_p(F)$, its amplitude and phase, and evalute $X_p(F)$ at $F = 0$, $F = \frac{1}{2}$ and $F = 1$ for the following:

(a) $\{x[n]\} = \{1, 2, 3, 2, 1\}$ **(b)** $\{x[n]\} = \{-1, 2, 0, -2, 1\}$

(c) $\{x[n]\} = \{1, 2, \overset{\uparrow}{2}, 1\}$ **(d)** $\{x[n]\} = \{-1, -\overset{\uparrow}{2}, 2, 1\}$

P11.19 (DTFT from Definition) Use the definition to find the DTFT $X_p(F)$ if

(a) $x[n] = (\frac{1}{2})^{n+2}u[n]$ **(b)** $x[n] = n(\frac{1}{2})^{2n}u[n]$ **(c)** $x[n] = (\frac{1}{2})^{n+2}u[n-1]$

(d) $x[n] = n(\frac{1}{2})^{n+2}u[n-1]$ **(e)** $x[n] = (n+1)(\frac{1}{2})^{n}u[n]$

P11.20 (IDTFT from FS) Start with the FS coefficients and use duality to find the IDTFT $x[n]$ of the following $X_p(F)$ described over $|F| \le \frac{1}{2}$:

(a) $X_p(F) = \text{rect}(2F)$ **(b)** $X_p(F) = \cos(\pi F)$ **(c)** $X_p(F) = \text{tri}(2F)$

P11.21 (Properties) Given the DTFT pair $x[n] \longleftrightarrow 4/[2 - \exp(-j2\pi F)]$, find the DTFT of the following, using properties:

(a) $x[n-2]$ **(b)** $nx[n]$ **(c)** $x[-n]$

(d) $x[n] - x[n-1]$ **(e)** $x[n] \star x[n]$ **(f)** $x[n]\exp(jn\pi)$

(g) $x[n]\cos(n\pi)$ **(h)** $x[n-1] + x[n+1]$

P11.22 (Properties) Given the DTFT pair $(\frac{1}{2})^{n}u[n] \longleftrightarrow X_p(F)$, find the time signal for the following, using properties:

(a) $X_p(-F)$ **(b)** $X_p(F - \frac{1}{4})$ **(c)** $X_p(F + \frac{1}{2}) + X_p(F - \frac{1}{2})$

(d) $X_p'(F)$ **(e)** $X_p^2(F)$ **(f)** $X_p(F) \bullet X_p(F)$

(g) $X_p(F)\cos(4\pi F)$

P11.23 (Properties) Find the DTFT of $x[n] = na^{n}u[n]$ using

(a) The defining relation.   **(b)** The times-n property.

(c) Convolution and shifting for $a^{n}u[n] \star a^{n}u[n] = (n+1)a^{n}u[n+1]$.

(d) Convolution and superposition for $a^{n}u[n] \star a^{n}u[n] = (n+1)a^{n}u[n]$.

P11.24 (Properties)

(a) Starting with the DTFT of $u[n]$, show that the DTFT of $\text{sgn}[n]$ is $X_p(F) = -j\cot(\pi F)$.

(b) Starting with $\text{rect}[n/2N] \longleftrightarrow M\text{sinc}(MF)/\text{sinc}(F)$, where $M = 2N + 1$, use the convolution property to find the DTFT of $\text{tri}[n/N]$.

(c) Starting with $\text{rect}[n/2N] \longleftrightarrow M\text{sinc}(MF)/\text{sinc}(F)$, where $M = 2N + 1$, use modulation to find the DTFT of $\cos(n\pi/2N)\text{rect}[n/2N]$.

P11.25 (Transfer Function) Find the transfer function $H_p(F)$ and difference equation for the following systems:

(a) $h[n] = (\frac{1}{3})^{n}u[n]$ **(b)** $h[n] = [1 - (\frac{1}{3})^{n}]u[n]$

(c) $h[n] = n(\frac{1}{3})^{n}u[n]$ **(d)** $h[n] = 0.5\delta[n]$

(e) $h[n] = \delta[n] - (\frac{1}{3})^{n}u[n]$ **(f)** $h[n] = [(\frac{1}{3})^{n} + (\frac{1}{2})^{n}]u[n]$

P11.26 (Transfer Function) Find the transfer function and impulse response of the following using the DTFT and inverse DTFT.

(a) $y[n] - \frac{1}{6}y[n-1] - \frac{1}{6}y[n-2] = 2x[n] + x[n-1]$

(b) $y[n] = 0.2x[n]$

(c) $y[n] = x[n] + x[n-1] + x[n-2]$

P11.27 (Transfer Function) Set up the system difference equations from the following system transfer functions:

(a) $H_p(F) = \dfrac{6\exp(j2\pi F)}{3\exp(j2\pi F) + 1}$

(b) $H_p(F) = \dfrac{3}{\exp(-j2\pi F) + 2} - \dfrac{\exp(-j2\pi F)}{\exp(-j2\pi F) + 3}$

(c) $H_p(F) = \dfrac{6\exp(j2\pi F) + 4\exp(j4\pi F)}{[1 + 2\exp(j2\pi F)][1 + 4\exp(j2\pi F)]}$

P11.28 (Steady-State Response) A DT system is described by the difference equation $y[n] + \frac{1}{4}y[n-2] = 2x[n] + 2x[n-1]$. Find the transfer function $H_p(F)$ and use this to find the steady-state response if

(a) $x[n] = 5u[n]$ **(b)** $x[n] = 3\cos(\frac{1}{2}n\pi)u[n]$
(c) $x[n] = 3\cos(\frac{1}{2}n\pi + \frac{1}{4}\pi) - 6\sin(\frac{1}{2}n\pi - \frac{1}{4}\pi)$
(d) $x[n] = 2\cos(\frac{1}{4}n\pi) + 3\sin(\frac{1}{2}n\pi)$

P11.29 (Response of Digital Filters) Consider the three-point averaging filter described by $y[n] = \frac{1}{3}\{x[n] + x[n-1] + x[n-2]\}$.
(a) Find its impulse response $h[n]$.
(b) Find and sketch its frequency response $H_p(F)$.
(c) Find its response to $x[n] = \cos(\frac{1}{3}n\pi + \frac{1}{4}\pi)$.
(d) Find its response to $x[n] = \cos(\frac{1}{3}n\pi + \frac{1}{4}\pi) + \sin(\frac{1}{3}n\pi + \frac{1}{4}\pi)$.
(e) Find its response to $x[n] = \cos(\frac{1}{3}n\pi + \frac{1}{4}\pi) + \sin(\frac{2}{3}n\pi + \frac{1}{4}\pi)$.

P11.30 (Frequency Response) Make a rough sketch of, and qualitatively describe, the frequency response of the following digital filters:
(a) $y[n] + 0.9y[n-1] = x[n]$ **(b)** $y[n] - 0.9y[n-1] = x[n]$
(c) $y[n] + 0.9y[n-1] = x[n-1]$ **(d)** $y[n] = x[n] - x[n-4]$

P11.31 (Compensating Filters) Digital filters are often used to compensate for the sinc distortion of zero-order-hold DACs by providing a $1/\mathrm{sinc}(F)$ boost. Two such filters are described by $y[n] = \{x[n] - 18x[n-1] + x[n-2]\}/16$ and $8y[n] + y[n-1] = 9x[n]$
(a) Identify each filter as FIR or IIR, evaluate its magnitude $|H_p(F)|$ at $F = 0, 0.1, 0.2, 0.3, 0.4, 0.5$, and compare with $|1/\mathrm{sinc}(F)|$.
(b) Which of these filters provides better compensation?

P11.32 (Response of Numerical Algorithms) Simpson's and Tick's rules for numerical integration find $y[k]$ over two time steps from $y[k-2]$ and are described by

Simpson's: $y[n] = y[n-2] + \{x[n] + 4x[n-1] + x[n-2]\}/3$
Tick's: $y[n] = y[n-2] + \{0.3584x[n] + 1.2832x[n-1] + 0.3584x[n-2]\}$

(a) Find the transfer function $H_p(F)$ corresponding to each rule.
(b) Sketch $|H_p(F)|$ over $F = (0, \frac{1}{2})$ and compare with the transfer function of an ideal integrator. Does this comparison confirm that the coefficients in Tick's rule are selected to optimize $H_p(F)$ in the range $0 < F < \frac{1}{4}$?

12 THE DFT AND FFT

12.0 Scope and Objectives

In the last chapter, we showed that the spectrum of a sampled signal is not only continuous but periodic. By duality, if we sample the continuous, periodic spectrum of a sampled time signal, we must obtain the periodic extension of the sampled time signal. This leads to periodicity in both domains and the formulation of the *discrete Fourier transform* (DFT).

The spectral analysis and processing of sampled signals using digital computers continue to gain widespread popularity. Finite memory limitations of digital computers constrain us to work with a finite set of numbers for describing signals in both time and frequency. Computationally efficient algorithms for implementing the DFT of finite sequences go by the generic name of *fast Fourier transforms* (FFT).

This chapter is divided into three major parts:

Part 1   The DFT and DFS and their properties.
Part 2   The FFT as a fast algorithm to compute the DFT and its inverse.
Part 3   Signal processing applications of the DFT and FFT.

Useful Background   The following topics provide a useful background:

1. The duality between the Fourier series and the DTFT (Chapter 11).
2. The Fourier transform and its properties (Chapter 6).

12.1 The DFT and DFS

The Fourier series of periodic signals and the DTFT of discrete-time signals are duals of each other and are similar in many respects. In theory, both offer great insight into the spectral description of signals. In practice, both suffer from (similar) problems in their implementation. The finite memory limitations and finite precision of digital computers force us to work only with finite sets of quantized data. This brings out two major problems inherent in the Fourier series and the DTFT as tools for digital signal processing:

1. Both typically require an infinite number of samples (the Fourier series for its spectrum and the DTFT for its time signal).
2. Both deal with one continuous variable (time t or digital frequency F).

The solution to these problems requires that we

1. Replace the continuous variable with a discrete one.
2. Limit the number of samples in both domains.

Together, these lead to the ideas of the discrete Fourier series (DFS) and the discrete Fourier transform (DFT). Even though the solutions are simple enough, the resulting relations must be interpreted with care, especially if they are being applied to sampled versions of analog signals.

Both the DFS and the DFT can be viewed as natural extensions of the Fourier transform, obtained by sampling both time and frequency in turn. Sampling in one domain leads to a periodic extension in the other. A sampled representation in both domains also forces periodicity in both domains.

Thus the DFS and DFT turn out to be periodic and discrete in both domains. The Fourier transform leads to either the DFS or DFT, depending on the order in which we sample time and frequency, as listed in Table 12.1.

Table 12.1   Relating Frequency-Domain Transforms.

Fourier Transform

Aperiodic/continuous signal	**Aperiodic/continuous spectrum**
$x(t) = \int_{-\infty}^{\infty} X(f)\exp(j2\pi f t)\,df$	$X(f) = \int_{-\infty}^{\infty} x(t)\exp(-j2\pi f t)\,dt$
$\downarrow$	$\downarrow$
Sampling $x(t)$ in time (DTFT)	**Sampling $X(f)$ in frequency (Fourier series)**
Sampled time signal	Sampled spectrum
$x[n] = \int_{1} X_p(F)\exp(j2\pi nF)\,dF$	$X_S[k] = \dfrac{1}{T}\displaystyle\int_{T} x_p(t)\exp(-j2\pi k f_0 t)\,dt$
Periodic spectrum (period $= 1$)	Periodic time signal (period $=$ T)
$X_p(F) = \sum_{n=-\infty}^{\infty} x[n]\exp(-j2\pi nF)$	$x_p(t) = \sum_{k=-\infty}^{\infty} X_S[k]\exp(j2\pi k f_0 t)$
$\downarrow$	$\downarrow$
Sampling $X_p(F)$ in frequency (DFT)	**Sampling $x_p(t)$ in time (DFS)**
Sampled/periodic spectrum	Sampled/periodic time signal
$X_T[k] = \sum_{n=0}^{N-1} x[n]\exp(-j2\pi nk/N)$	$x_S[n] = \sum_{k=0}^{N-1} X[k]\exp(j2\pi kn/N)$
Sampled/periodic time signal	Sampled/periodic spectrum
$x[n] = \dfrac{1}{N}\displaystyle\sum_{k=0}^{N-1} X_T[k]\exp(j2\pi nk/N)$	$X[k] = \dfrac{1}{N}\displaystyle\sum_{n=0}^{N-1} x_S[n]\exp(-j2\pi kn/N)$

---

1. To obtain the DFS, we first sample the Fourier transform in the frequency domain to result in a periodic time signal (the Fourier series), and then sample the periodic time signal.
2. To obtain the DFT, we first sample the time signal to obtain a periodic spectrum (the DTFT), and then sample this periodic spectrum.

The two forms differ only by a constant scale factor. For this reason, we often talk of the DFT and the DFS synonymously.

12.1.1 *From the Fourier Series to the DFS*

We start with the Fourier series relations for a periodic signal $x_p(t)$

$$x_p(t) = \sum_{k=-\infty}^{\infty} X_S[k] \exp(j2\pi kf_0 t) \qquad X_S[k] = \frac{1}{T} \int_T x_p(t) \exp(-j2\pi kf_0 t)\, dt$$

To limit $x_p(t)$, we take N samples over one period at intervals t_s. We have $T = Nt_s$. With $f_0 = 1/T = 1/Nt_s$, $t \longrightarrow nt_s$ and $dt \longrightarrow t_s$, the spectral coefficients $X_S[k]$ may now be expressed in terms of the summation

$$X[k] = \frac{1}{Nt_s} \sum_{n=0}^{N-1} x_S[n] \exp(-j2\pi kf_0 nt_s) t_s$$

$$= \frac{1}{N} \sum_{n=0}^{N-1} x_S[n] \exp(-j2\pi kn/N) \qquad k = 0, 1, \ldots, N-1$$

The quantity $X[k]$ describes the **discrete Fourier series** (DFS) of the sampled periodic signal $x_S[n]$. We make the following observations:

1. The DFS repeats every N samples, with $X[k] = X[k \pm N]$, since $\exp(-j2\pi kn/N)$ is periodic with period N. Note that $X[k]$ is, in fact, just the *periodic extension* of $X_S[k]$ with period N.
2. For a sampling interval t_s, the spectral spacing of $X[k]$ is $f_0 = 1/T$.
3. The sampling interval t_s enters into the picture only indirectly in the computation of the DFS via the relation $1/T = 1/Nt_s$ or $f_0 t_s = 1/N$.

We recover N samples (one period) of $x_S[n]$ from $X[k]$ if we vary k from 0 to $N-1$ in the summation for $x_p(t)$ and let $t \longrightarrow nt_s$. With $f_0 = 1/T$, we obtain

$$x_S[n] = \sum_{k=0}^{N-1} X[k] \exp(j2\pi kn/N) \qquad n = 0, 1, 2, \ldots, N-1$$

This relation describes the **inverse discrete Fourier series** (IDFS). Both the DFS and IDFS are finite, discrete and periodic. The sample spacing in one domain equals the reciprocal of the sampling duration in the other.

Remark: The number of samples we take in each domain can be different. Choosing the same number of samples N in both domains is based on convenience and not necessity (see Section 12.9). It leads to almost symmetric forms for both the DFS and its inversion relation.

12.1.2   From the DTFT to the DFT

We can also start with an aperiodic DT signal $x[n]$ limited to N samples. Its DTFT is given by

$$X_p(F) = \sum_{n=0}^{N-1} x[n]\exp(-j2\pi nF)$$

Remember, $X_p(F)$ is periodic with period 1. Next, we sample the frequency for N samples over one period. The spectral spacing F_0 equals $1/N$. With $F \longrightarrow kF_0$, $k = 0, 1, \ldots, N-1$, we get the sampled frequency-domain representation

$$X_T[k] = \sum_{n=0}^{N-1} x[n]\exp(-j2\pi nk/N) \qquad k = 0, 1, \ldots, N-1$$

This describes the **discrete Fourier transform** (DFT) of $x[n]$. Except for the absence of the factor $1/N$, the DFT is identical in form to the DFS.

Remarks:
1. The DFS and DFT actually describe a set of N equations. Each DFS or DFT sample is a summation of N terms involving all N time samples.
2. Since $\exp(-j2\pi/N)$ is periodic with period N, so is the DFT $X_T[k]$.
3. The DFT value $X_T[0]$ always equals $\sum x[n]$, the sum of the time samples.

Example 12.1
DFT from the Defining Relation

Let $x[n] = \{1, 2, 1, 0\}$. We have $N = 4$, but the upper index in the summation can be chosen as $n = 2$ (and not $n = 3$) because $x[3] = 0$.

With $N = 4$, and $\exp(-j2\pi nk/N) = \exp(-jnk\pi/2)$, we successively compute

$$k = 0: \quad X_T[0] = \sum x[n] = 1 + 2 + 1 + 0 = 4 \quad (\text{since } \exp(-j0) = 1)$$

$$k = 1: \quad X_T[1] = \sum_{n=0}^{2} x[n]\exp(-j\pi n/2) = 1 + 2\exp(-j\pi/2) + \exp(-j\pi) = -j2$$

$$k = 2: \quad X_T[2] = \sum_{n=0}^{2} x[n]\exp(-j\pi n) = 1 + 2\exp(-j\pi) + \exp(-j2\pi) = 0$$

$$k = 3: \quad X_T[3] = \sum_{n=0}^{2} x[n]\exp(-j\pi n3/2) = 1 + 2\exp(-j3\pi/2) + \exp(-j3\pi) = j2$$

The DFT is $X_T[k] = \{4, -j2, 0, j2\}$ and the DFS is $X[k] = \frac{1}{N}X_T[k] = \{1, -j\frac{1}{2}, 0, j\frac{1}{2}\}$

Example 12.2
DFT of a Finite Sequence
from the DTFT

Let $x[n] = \{1, 2, 1, 0\}$. If we use the DTFT, we first find

$$X_p(F) = [1 + 2\exp(-j2\pi F) + \exp(-j4\pi F) + 0] = \exp(-j2\pi F)[2 + 2\cos(2\pi F)]$$

With $N = 4$, we have $F = k/4$, $k = 0, 1, 2, 3$. We then obtain the DFT as

$$X_T[k] = \exp(-j2\pi k/4)[2 + 2\cos(2\pi k/4)], \quad k = 0, 1, 2, 3 \quad \text{or} \quad X_T[k] = \{4, -j2, 0, j2\}$$

12.1.3   Inverse DFT

To develop the inversion relation, we start with the IDTFT

$$x[n] = \int_1 X_p(F)\exp(j2\pi nF)\,dF$$

Choosing N samples of $X_p(F)$ over one period (unity) gives the spectral spacing as $F_0 = 1/N$. With $dF \longrightarrow F_0$ and $F \longrightarrow kF_0$, the integral may be approximated by the summation

$$x[n] = \frac{1}{N}\sum_{k=0}^{N-1} X_T[k]\exp(j2\pi nk/N) \qquad n = 0, 1, \ldots, N-1$$

This is the **inverse discrete Fourier transform** (IDFT).

Remarks:

1. The IDFT relation also describes a set of N equations. Each IDFT sample is a summation of N terms involving all N spectral (DFT) samples.
2. Since $\exp(j2\pi/N)$ is periodic with period N, so is the IDFT $x[n]$. This means there is *implied periodicity* in both the DFT and IDFT.
3. The DFS and DFT differ only by a factor of N and are related by

$$X[k] = \frac{1}{N}X_T[k] \qquad \{\text{IDFS}\} = N\{\text{IDFT}\}$$

Example 12.3
The DFT and IDFT from
the DTFT

Consider $x[n] = \alpha^n u[n]$. Its DTFT over one period $(0, 1)$ is

$$X_p(F) = 1/[1 - \alpha\exp(j2\pi F)]$$

Sampling the DTFT at intervals $1/N$, we obtain the DFT

$$X_T[k] = X_p(F)|_{F=k/N} = 1/[1 - \alpha\exp(j2\pi k/N)]$$

Sampling of $X_p(F)$ at intervals $1/N$ results in a periodic extension of $x[n]$ with period N to yield the IDFT $x_p[n]$ for one period $(0, N-1)$ as

$$x_p[n] = \sum_{k=-\infty}^{\infty} x[n+kN] = \sum_{k=0}^{\infty} \alpha^{n+kN} = \alpha^n \sum_{k=0}^{\infty} (\alpha^k)^N = \alpha^n/(1 - \alpha^N) \qquad 0 \le n \le N-1$$

Thus we have the DFT pair $\alpha^n/(1 - \alpha^N) \longleftrightarrow 1/[1 - \alpha\exp(j2\pi k/N)]$ from which,

$$\alpha^n(u[n] - u[N-1]) \longleftrightarrow (1 - \alpha^N)/[1 - \alpha\exp(j2\pi k/N)]$$

As $N \longrightarrow \infty$, $k/N \longrightarrow F$ and we get the DTFT pair $\alpha^n u[n] \longleftrightarrow 1/[1 - \alpha\exp(j2\pi F)]$

12.2 Properties of the DFT

The properties of the DFT are strikingly similar to other frequency-domain transforms and are summarized in Table 12.2. They must always be used in keeping with implied periodicity (of the DFT and IDFT) in both domains.

Table 12.2   Operational Properties of the N-Sample DFT.

Symmetry Properties of the DFT or IDFT

$x[n]$	$X_T[k]$
Real	Conjugate symmetry: $X_T[-k] = X_T^*[k] = X_T[N-k]$
	Even magnitude spectrum and odd phase spectrum
Conjugate symmetric	Real

Operational Properties

Note: $W_N = \exp(-j2\pi/N)$ $x[n] \longleftrightarrow X_T[k]$ $y[n] \longleftrightarrow Y_T[k]$

Property	DFT result
Superposition	$x[n] + y[n] \longleftrightarrow X_T[k] + Y_T[k]$
Times scalar	$\alpha x[n] \longleftrightarrow \alpha X_T[k]$
Shift	$x[n-m] \longleftrightarrow X_T[k]\exp(-j2\pi km/N) = X_T[k]W_N^{km}$
Modulation	$W_N^{-nm}x[n] = x[n]\exp(j2\pi nm/N) \longleftrightarrow X_T[k-m]$
Reversal	$x[-n] \longleftrightarrow X_T[-k] = X_T^*[k]$
Conjugation	$x^*[n] \longleftrightarrow X_T^*[-k]$
Product	$x[n]y[n] \longleftrightarrow \dfrac{1}{N}X_T[k] \bullet Y_T[k]$
Convolution	$x[n] \bullet y[n] \longleftrightarrow X_T[k]Y_T[k]$
Correlation	$x[n] \bullet y^*[-n] \longleftrightarrow X_T[k]Y_T^*[k]$
Central ordinates	$x[0] = \dfrac{1}{N}\sum X[k]$ and $X_T[0] = \sum x[n]$
Parseval's relation	$\sum |x[n]|^2 = \dfrac{1}{N}\sum |X_T[k]|^2$

Remark: The DFT of finite sequences defined analytically often results in very unwieldy expressions and explains the lack of "standard" DFT pairs.

12.2.1   *Illustrating the Properties*

We illustrate the use of some of the properties with an example.

Example 12.4
Illustrating the Properties of the DFT

Consider the DFT pair $x[n] = \{1, 2, 1, 0\} \longleftrightarrow X_T[k] = \{4, -j2, 0, j2\}$ with $N = 4$.

Shifting: The sequence $x[n-2]$ equals $\{1, 0, 1, 2\}$ due to implied periodicity. Its DFT is then $X_T[k]\exp(-j2\pi k2/4) = X_T[k]\exp(-jk\pi) = \{4, j2, 0, -j2\}$.

Modulation: The sequence $X_T[k-1]$ equals $\{j2, 4, -j2, 0\}$ by implied periodicity and its IDFT is $x[n]\exp(j2\pi n/4) = x[n]\exp(j\pi n/2) = \{1, j2, -1, 0\}$.

Folding: The sequence $x[-n] = \{x[0], x[-1], x[-2], x[-3]\} = \{1, 0, 1, 2\}$. Its DFT equals $X_T[-k] = X_T^*[k] = \{4, j2, 0, -j2\}$.

Conjugation: The sequence $x^*[n] = x[n] = \{1, 2, 1, 0\}$ since it is real. Its DFT is $X_T^*[-k] = \{4, j2, 0, -j2\}^* = \{4, -j2, 0, j2\}$ and equals $X[k]$.

Product: The sequence $x[n]x[n] = \{1, 4, 1, 0\}$ is the pointwise product. Its DFT is $\frac{1}{4}X_T[k] \bullet X_T[k] = \frac{1}{4}\{4, -j2, 0, j2\} \bullet \{4, -j2, 0, j2\}$. Keep in mind that this is a periodic convolution. The result is $\frac{1}{4}\{24, -j16, 0, j16\} = \{6, -j4, 0, j4\}$.

Convolution: The sequence $x[n] \bullet x[n]$ describes a periodic convolution and equals $\{1, 2, 1, 0\} \bullet \{1, 2, 1, 0\} = \{2, 4, 6, 4\}$. Its DFT is given by the pointwise product $X_T[k]X_T[k] = \{16, -4, 0, -4\}$.

Central ordinates: It is easy to check that $x[0] = \frac{1}{4}\sum X_T[k]$ and $X_T[0] = \sum x[n]$.

Parseval's relation: We have $\sum |x[n]|^2 = \sum \{1, 4, 1, 0\} = 6$. Since $X_T^2[k] = \{16, -4, 0, 4\}$, we also have $\frac{1}{4}\sum |X_T[k]|^2 = \frac{1}{4}(16 + 4 + 4) = 6$.

12.2.2   *Symmetry*

In analogy with all other frequency-domain transforms, the DFT of a real sequence possesses conjugate symmetry about the origin with $X_T[-k] = X_T^*[k]$. Since the DFT is also periodic, $X_T[-k] = X_T[N-k]$ or $X_T^*[k] = X_T[N-k]$. This also implies conjugate symmetry about $k = \frac{1}{2}N$, and thus

$$X_T[-k] = X_T^*[k] = X_T[N-k]$$

Remarks:

1. A similar result applies to the IDFT. If N is odd, the conjugate symmetry is about the integer part $\text{int}(\frac{1}{2}N)$.
2. Conjugate symmetry suggests that we need compute only $\frac{1}{2}N$ DFT values to find the entire DFT sequence—another labor saving concept!
3. Since $x[n]$ is sampled from $n = 0$ to $N-1$, periodic extension implies that for even symmetry, $x[n] = x[N-n]$ and for odd symmetry, $x[n] = -x[N-n]$.

Example 12.5
DFT Using Conjugate Symmetry

For the sequence $x[n] = \{1, 1, 0, 0, 0, 0, 0, 0\}$, we need find $X_T[k]$ only for the indices $k \le \frac{1}{2}N = 4$. Since only $x[0]$ and $x[1]$ are nonzero, the upper index in the DFT summation will be $n = 1$ and finding the DFT reduces to evaluating

$$X_T[k] = \sum_{n=0}^{1} x[n] \exp(-j2\pi nk/8) = 1 + \exp(-j\pi k/4) \qquad k = 0, 1, 2 \ldots$$

Now, $X_T[0] = 1 + 1 = 2$ and for $k = 1, 2, 3, 4$, we successively compute

$$X_T[1] = 1 + \exp(-j\pi/4) = 1.707 - j0.707 \qquad X_T[2] = 1 + \exp(-j\pi/2) = 1 - j$$
$$X_T[3] = 1 + \exp(-j3\pi/4) = 0.293 - j0.707 \qquad X_T[4] = 1 + \exp(-j\pi) = 0$$

Due to conjugate symmetry about $k = 4$, we have $X_T[k] = X_T^*[N-k]$. This gives

$$X_T[5] = X_T^*[3] = 0.293 + j0.707 \qquad X_T[6] = X_T^*[2] = 1 + j,$$
$$X_T[7] = X_T^*[1] = 1.707 + j0.707$$

$$X_T[k] = \{2, 1.707 - j0.707, 0.293 - j0.707, 1 - j, 0,$$
$$1 + j, 0.293 + j0.707, 1.707 + j0.707\}$$

12.2.3   The DFT Spectrum

The DFT spectrum may be plotted in one of two ways:

1. From $k = 0$ to $N - 1$ to show conjugate symmetry about $k = \frac{1}{2}N$.
2. From $k = -\frac{1}{2}N$ to $\frac{1}{2}N - 1$ to show symmetry about the origin. To do this, we plot the DFT values from $k = 0$ to $\frac{1}{2}N - 1$ and move or shift the indices $k = \frac{1}{2}N, \ldots, N - 1$ to $k - N$. The index $k = \frac{1}{2}N$ relocates to $-\frac{1}{2}N$ and $k = N - 1$ to -1. This *reordered* spectrum is what we shall plot in most cases.

The computation of the DFS or DFT is independent of sample separation t_s or frequency separation f_0. If a time signal is sampled at intervals t_s for a duration D, duality between sampling and periodic extension suggests that:

1. The spectral samples are separated by $f_0 = 1/D$.
2. The spectrum is periodic with a principal period $S_F = 1/t_s$.

In keeping with the sampling theorem, the highest frequency we can observe in the DFT spectrum is $\frac{1}{2}S_F$. For N samples, this frequency corresponds to the DFT index $k = \frac{1}{2}N$. This index is called the **folding index** and $\frac{1}{2}S_F$ is called the **folding frequency**. For discrete signals with $t_s = 1$, the folding index corresponds to the digital frequency $F = \frac{1}{2}$. The abscissa in the DFT spectrum may thus correspond to:

1. The index k with unit spacing and the folding index $k = \frac{1}{2}N$.
2. The digital frequency $F = k/N$ and the folding frequency $F = \frac{1}{2}$.
3. The analog frequency $f = kS_F/N$ and the folding frequency $f = \frac{1}{2}S_F$.

12.3
Understanding
the DFT Results

Sampled sequences are often derived from an analog signal. The DFT of such a sampled sequence $x_S[n]$ can be viewed from two different perspectives:

1. As the spectral (Fourier series) coefficients of a sampled periodic signal whose samples for one period correspond to the sequence $x_S[n]$.
2. As the spectrum of a sampled timelimited aperiodic analog signal whose samples correspond to the sequence $x_S[n]$.

For periodic signals, it makes sense to use the first viewpoint. For others, the two lead to the same conclusions provided we understand how to interpret the DFT results correctly. Infinite-duration aperiodic signals must be limited in duration or truncated. This truncation is equivalent to multiplying $x(t)$ by a rectangular window or convolving the true spectrum with a sinc function and leads to spectral wiggles. Since a timelimited analog signal can never be bandlimited, it also imposes practical limits on the sampling rate S_F and limits the bandwidth of the sampled signal to $\frac{1}{2}S_F$.

Clearly, the DFT is only an approximation to the actual (Fourier series or transform) spectrum of the underlying analog signal. The DFT magnitude is affected by the sampling interval and how the sample values are chosen. The DFT phase is affected by the location of sampling instants. The DFT spectral spacing is affected by the sampling duration. To develop a consistent approach to establishing these quantities, consider an analog signal $x(t)$ sampled over a duration D, with sampling interval t_s, to yield the sampled signal $x_S[n]$. Once obtained, $x_S[n]$ defines one period of a periodic sequence.

Choice of Sampling Interval   We choose a sampling interval smaller than the Nyquist interval. This is feasible only for bandlimited periodic signals. For all others, aliasing errors will always occur in the DFT result.

Choice of Origin   Since the DFT index runs from 0 to $N-1$, $x_S[n]$ is assumed to start at $n = 0$. To ensure the correct phase, we must also sample $x(t)$ at $t = 0$. The foolproof way to do this (Figure 12.1) is to

1. Create the (assumed) periodic extension $x_p(t)$ of $x(t)$ with period D.
2. Take N samples over one period D of $x_p(t)$ starting at $t = 0$.

Figure 12.1   Sampling the periodic extension of x(t) for DFT.

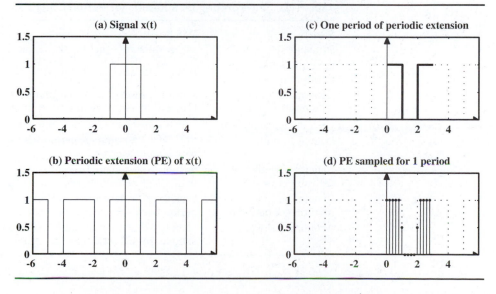

Sampling at Discontinuities   The Fourier series (or transform) converges to the midpoint of any discontinuity. At a discontinuity (in the periodic extension of $x(t)$), we must choose sample values as the midpoint values.

Comparison   For periodic signals, the DFS ($X[k]$) may be compared directly with the corresponding Fourier series coefficients $X_S[k]$ as follows:

$$\underset{\text{Fourier Series}}{X_S[k] \approx} \quad \underset{\text{DFS}}{X[k] =} \quad \underset{\text{DFT}}{\frac{1}{N}X_T[k]} \quad k = 0, 1, 2, \dots, N-1$$

For bandlimited signals, the Fourier transform of $x(t)$ is related to the DTFT of the sampled signal $x_S[n]$ over the principal period by $X(f) = t_s X_p(f)$. Since the DFT ($X_T[k]$) is a sampled version of $X_p(f)$, we can compare *samples of $X(f)$ at $f = kf_0$* with the DFT as follows:

$$X(kf_0) \approx t_s X_T[k] \qquad k = 0, 1, 2, \dots, N-1$$

Here, $f_0 = 1/D$ equals the reciprocal of the sampling duration D.

12.3.1   *The DFS and Periodic Signals*

We develop some useful guidelines for interpreting the DFS of periodic signals with reference to Figure 12.2. A summary of the results appears in Table 12.3. Due to conjugate symmetry, we list the DFS results only up to the folding index $k = \frac{1}{2}N$ in the following discussion.

Remark: A reminder that we use $X_T[k]$ for the DFT, $X[k]$ for the DFS, and $X_S[k]$ for the Fourier series coefficients.

Table 12.3   Interpreting DFS Results for Periodic Signals.

Signal	DFS Result
Sinusoid	DFS results are exact for $N > 2$ samples/period.
Bandlimited to M harmonics	DFS results are exact for $N > 2M$ samples/period.
General periodic (N samples/period)	Aliasing present. Lower harmonics are more accurate. Effect of halfwave symmetry is preserved.
($> N$ samples/period)	Extends spectrum but not frequency spacing.
Rule of thumb	Choose $> 8N$ samples/period to observe N harmonics with about 5% error.
General periodic (M full periods)	Result is interpolated version of one-period DFS, with $M - 1$ zeros between adjacent sample values.
General periodic (noninteger periods)	Leakage. DFS describes a different signal whose period equals the duration of noninteger periods.
Avoiding leakage	Sample for integer periods. If period is not known, sample for largest duration possible.

Figure 12.2   The DFT of sampled periodic signals.

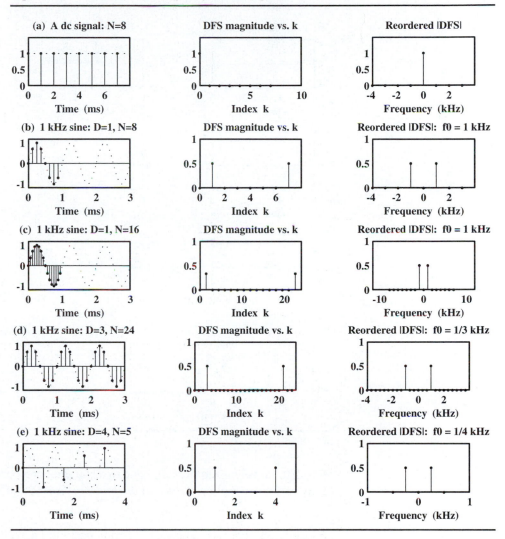

Figure 12.2a   The Fourier series of a dc signal $x(t) = 1$ yields $X_S[0] = 1$ and all other $X_S[k] = 0$, $k \neq 0$. If we sample $x(t)$ with $N = 8$, we get the DFS pair

$$x_S[n] = \{1, 1, 1, 1, 1, 1, 1, 1\} \longleftrightarrow X[k] = \{1, 0, 0, 0, 0, \ldots\} \quad \text{(DFS)}$$

The DFS shows $X[0] = 1$, and all other $X[k] = 0$, as expected.

Consider the 1 kHz sine $x_p(t) = \sin(2\pi 1000t)$ with FS coefficients

$$X_S[0] = 0 \qquad X_S[1] = -j\tfrac{1}{2} \qquad X_S[-1] = j\tfrac{1}{2}$$

Figure 12.2   (continued)

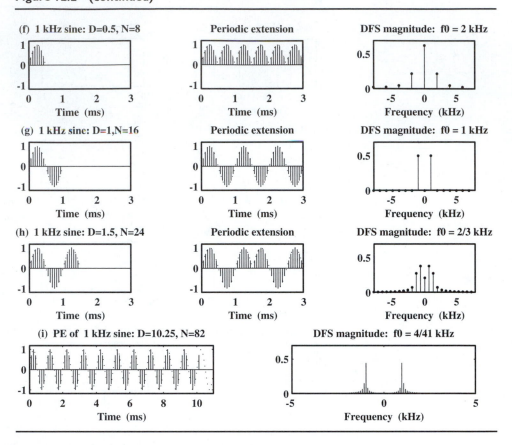

(f) 1 kHz sine: D=0.5, N=8

(g) 1 kHz sine: D=1, N=16

(h) 1 kHz sine: D=1.5, N=24

(i) PE of 1 kHz sine: D=10.25, N=82

Nyquist sampling requires at least two samples per period or $S_F > 2$ kHz. Here is how various sampling rates and durations affect the DFS results.

Figure 12.2b   Let $S_F = 8$ kHz, $D = T$ (one period). Then $N = 8$, $f_0 = 1/D = 1$ kHz and

$$x_s[n] = \{0, \sqrt{\tfrac{1}{2}}, 1, \sqrt{\tfrac{1}{2}}, 0, -\sqrt{\tfrac{1}{2}}, -1, -\sqrt{\tfrac{1}{2}}\} \quad \text{DFS} = X[k] = \{0, -j\tfrac{1}{2}, 0, 0, 0 \ldots\}$$

We have the dc value $X[0] = 0$, and $X[1] = -j\tfrac{1}{2}$ at 1 kHz, for a perfect match.

Figure 12.2c $S_F = 16$ kHz, $D = T$ (one period). Then $N = 16$, $f_0 = 1/D = 1$ kHz and

$$\text{DFS} = X[k] = \{0, -j\tfrac{1}{2}, 0, 0, 0, 0, 0, 0, 0, \ldots\}$$

As before, $X[0] = 0$, and $X[1] = -j\tfrac{1}{2}$ at $k = 1$ ($f = 1$ kHz).

Figure 12.2d $S_F = 8$ kHz, $D = 3T$ (three periods). Then $N = 24$, $f_0 = 1/D = \frac{1}{3}$ kHz, and

$$x_S[n] = \{0, 1, 0, -1, 0, 1, 0, -1, 0, 1, 0, -1\} \qquad \text{DFS} = X[k] = \{0, 0, 0, -j\tfrac{1}{2}, 0, 0, \ldots\}$$

Again $X[0] = 0$ but now $X[3] = -j\frac{1}{2}$. This makes sense because $k = 3$ corresponds to $kf_0 = 1$ kHz and the DFS is zero at frequencies that are not harmonics of f_0.

We may generalize these results by stating that, for a sinusoid replicated for an integer number of periods P, and sampled above the Nyquist rate:

1. The nonzero DFS values occur at the indices $k = P$ and $k = N - P$ (or $k = -P$).
2. The number of DFS samples is dictated by the sampling frequency S_F.

As an example, if we sample $x(t) = \sin(2\pi f_0 t)$ for $P = 5$ periods at four times the Nyquist rate (40 samples), its DFS will show nonzero values $X[5] = -j\frac{1}{2}$ and $X[35] = X[-5] = j\frac{1}{2}$. The nonzero DFT values will equal $X_T[5] = -j20$ and $X_T[35] = X_T[-5] = j20$.

12.3.2   Signal Replication and Spectrum Interpolation

Actually, replication of any sampled sequence leads to a zero-interpolated spectrum. An M-fold replication of $x[n]$ whose DFS samples $X[k]$ are f_0 apart results in a sequence $x_M[n]$ whose DFS equals $X[k/M]$ with $M - 1$ zeros between samples of $X[k]$ but only f_0/M apart to ensure the same highest frequency. Similarly, an M-fold zero-interpolation of $x[n]$ yields an M-fold replication of its DFT $X_T[k]$.

12.3.3   The DFS of Bandlimited Periodic Signals

For a bandlimited periodic signal with M harmonics, the sampling rate must *exceed* $2M$ samples per period to obtain DFS results that match its Fourier series. A truncated Fourier series described by

$$x_p(t) = \sum_{k=-M}^{M} X_S[k] \exp(jk2\pi f_0 t)$$

contains $2M + 1$ coefficients, which may be found by evaluating (or sampling) $x_p(t)$ at $2M + 1$ points and solving the resulting $2M + 1$ simultaneous equations. The locations at which we sample are immaterial, but uniform sampling is the obvious choice. We can choose more than $2M + 1$ samples if we "assume" more coefficients in the series (whose values should work out to zero) to give as many unknowns as there are equations.

12.3.4   Aliasing and Leakage

If we sample below the Nyquist rate, the DFS results will show the effects of aliasing. Here are some results for the signal $x_p(t) = \sin(2\pi 1000 t)$.

Figure 12.2e $S_F = 1.25$ kHz, $D = 4T$ (four periods). Then $N = 5, f_0 = 1/D = 250$ Hz, and

$$x_S[n] = \{0, -0.95, -0.59, 0.59, 0.95\} \qquad \text{DFS} = X[k] = \{0, j\tfrac{1}{2}, 0, \ldots\}$$

We see that $X[1] = j\tfrac{1}{2}$. Even though $k = 1$ corresponds to 250 Hz and the phase is reversed, this result still makes perfect sense, because S_F is below the Nyquist rate, and the sine wave must alias to $(1 - 1.25)$ kHz $= -250$ Hz, or $+250$ Hz with phase reversal. This is exactly what the DFS result tells us.

Leakage Sampling bandlimited periodic signals over noninteger periods results in leakage. The DFT spectrum describes a different periodic signal.

Figure 12.2f $S_F = 16$ kHz, $D = 0.5T$ (half period). Then $N = 8, f_0 = 1/D = 2$ kHz, and

$$x_S[n] = \{0, 0.3827, 0.7071, 0.9329, 1, 0.9329, 0.7071, 0.3827\}$$
$$\text{DFS} = X[k] = \tfrac{1}{8}\{5.0273, -1.7654, -0.4142, -0.2346, -0.1989, \ldots\}$$

Since the $X[k]$ are real, the DFS results describe an even symmetric signal, not a sine wave. This result makes sense if we invoke implied periodicity. The periodic extension of the sampled signal over 0.5 periods of the sine wave is actually a full-rectified sine with even symmetry and a fundamental frequency of 2 kHz. Now, the Fourier series coefficients of a full-rectified sine wave (with unit peak value) are given by

$$X_S[0] = 2/\pi \qquad X_S[k] = 2/\pi(1 - 4k^2)$$

We confirm that $X[0] \approx X_S[0] = 2/\pi$ and $X[1] \approx X_S[1] = -2/3\pi$ at 2 kHz to within 5%. But $X[2], X[3]$ and $X[4]$ deviate significantly. Why? Since we no longer have a bandlimited signal, the sampling rate is not high enough, and we have aliasing. The component $X[3]$ at 6 kHz, for example, equals $X_S[3]$ plus all the values $X_S[m]$ at mf_0 that alias to 6 kHz. These include $f = 6$ kHz $+ KS_F$ at the harmonic indices $m = 3 \pm 8K$ (since $f_0 = 2$ kHz). If we compute the FS coefficients at all these indices and add to $X_S[3]$, we actually obtain the DFS value $X[3]$ (try it!). The DFS results thus once again make sense, if interpreted properly.

Figure 12.2g Sampling a bandlimited periodic signal for an integer number of periods leads to a periodic extension whose DFS results are identical to the Fourier series of the periodic signal.

Figure 12.2h Sampling $x(t)$ for 1.5 periods results in leakage. The original components leak out to frequencies that form harmonics of a new signal (whose period is 1.5 periods of the original), and differ both in magnitude and phase. To avoid leakage, we must sample for an integer number of periods.

Figure 12.2i If we do not know the period of a signal in advance, it is best to sample over as large a duration as possible (10.25 periods in this figure). This reduces both the leakage and the frequency spacing, and leads to a better estimate of the spectrum of the original signal.

12.3.5 The DFS of Arbitrary Periodic Signals

For periodic signals that are not bandlimited, aliasing is unavoidable for any sampling rate. Consider a square wave $x(t) = 1, (0, \frac{1}{2}T)$ with $f_0 = 1$ kHz. Its Fourier series coefficients are given by $X_S[0] = 0$, $X_S[k] = -j2/k\pi$ (k odd).

Case 1 Let $S_F = 4$ kHz, $D = T$ (one period). Then, $N = 4$, $f_0 = 1/T = 1$ kHz, and

$$x_S[n] = \{0, 1, 0, -1\} \qquad \text{DFS} = X[k] = \{0, -j\tfrac{1}{2}, 0, \ldots\}$$

This yields $X[0] = 0$, but $X[1] = -j\frac{1}{2}$, a 21% difference from $X_S[1] = 2/\pi$. The sampled signal actually describes the DFT of a 1 kHz sine wave. Other harmonics do not appear, since the spectrum extends only to 2 kHz.

Case 2 Let $S_F = 8$ kHz, $D = T$ (one period). Then, $N = 8$, $f_0 = 1/T = 1$ kHz, and

$$x_S[n] = \{0, 1, 1, 1, 0, -1, -1, -1\} \qquad \text{DFS} = X[k] = [0, -j0.6035, 0, -j0.1035, 0, \ldots]$$

The spectrum extends to 4 kHz, and shows $X[0] = 0$ and $X[2] = 0$ as expected. But $X[1]$ differs from $X_S[1]$ by 5%, and $X[3]$ shows a whopping 51% error!

We may generalize these observations to state that

1. The DFS value $X[0]$ always corresponds well with the FS dc value $X_S[0]$.
2. The effects of halfwave symmetry are preserved in the DFT or DFS.
3. The aliasing error increases as we move toward the folding frequency.

To minimize aliasing effects, we must use a much higher sampling frequency. Here is a rule of thumb: To obtain the first M harmonics with an error of about 5%, choose at least $8M$ samples per period.

12.4 Nonperiodic Signals, Resolution and Zero Padding

The Fourier transform $X(f)$ of a nonperiodic signal $x(t)$ is continuous. To find the DFT, $x(t)$ must be sampled over a finite duration D. The DFT yields essentially the Fourier series of its periodic extension with period D. The frequency spacing f_0 in the DFT is $1/D$. We need an interpolated version of this spectrum to better approximate $X(f)$. This implies a smaller f_0 or a longer duration D. The spacing f_0 in the DFT is called **frequency resolution**. It refers to our ability to identify closely spaced frequency components in the DFT spectrum. The smaller the f_0, the more the frequencies we can resolve and the better the resolution. Frequency resolution depends only on the duration D. Increasing D implies the following (Figure 12.3):

Figure 12.3   Illustrating ways to improve frequency resolution.

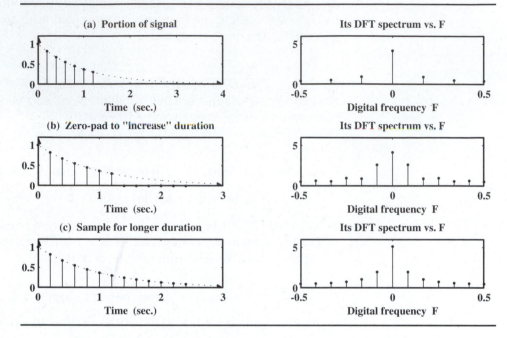

1. For a timelimited signal $x(t)$, we append zeros, and sample the periodic extension of the zero-padded sequence.

2. For an infinite signal, we sample for a longer duration or, if the original signal is unavailable, we zero-pad and sample the periodic extension of the zero-padded sequence.

Remark: Zero padding adds no new signal information. It interpolates the DFT at a denser set of frequencies but does not guarantee better accuracy (which is governed only by the sampling rate).

Example 12.6
Frequency Resolution and Signal Duration

Consider a 3 s signal $x(t)$ with a highest frequency of 40 Hz. If the required resolution is $\frac{1}{4}$ Hz, we must sample for 4 s. How? By zero-padding $x(t)$ for 1 s, of course. If we choose $S_F = 100$ Hz, we obtain 400 samples over 4 s. Of these, the first 300 are from $x(t)$, and the last 100 are padded zeros.

12.4.1   *The DFT of Timelimited Signals*

To interpret the DFT of timelimited signals, we consider the signals $\text{tri}(t)$ and $\text{rect}(t)$ as examples. The DFT results are summarized in Table 12.4.

Table 12.4 DFT Results for Some Timelimited Signals.

Signal	DFT result to $k = \frac{1}{2}N$, and remarks
tri(t) with $D = 2, N = 8$	The DFT actually describes tri$(t - 1)$.
$\{0, \frac{1}{4}, \frac{1}{2}, \frac{3}{4}, 1, \frac{3}{4}, \frac{1}{2}, \frac{1}{4}\}$	$\{4, -1.7071, 0, -0.293, 0 \ldots\}$
tri(t) with $D = 2, N = 8$	This is the correct DFT of tri(t).
$\{1, \frac{3}{4}, \frac{1}{2}, \frac{1}{4}, 0, \frac{1}{4}, \frac{1}{2}, \frac{3}{4}\}$	$\{4, 1.7071, 0, 0.293, 0 \ldots\}$
tri(t) with $D = 2, N = 16$	This is also the correct DFT of tri(t).
$\frac{1}{8}\{8, 7, \ldots, 1, 0, 1, \ldots, 7\}$	$\{8, 3.284, 0, 0.405, 0, 0.181, 0, 0.13, 0, \ldots\}$
tri(t) with $D = 4, N = 16$	This is also the correct DFT of tri(t).
$\{1, \frac{3}{4}, \frac{1}{2}, \frac{1}{4}, (8 \text{ zeros}), \frac{1}{4}, \frac{1}{2}, \frac{3}{4}\}$	$\{4, 3.284, 1.707, 0.405, 0, 0.181, 0.293, 0.13, 0, \ldots\}$
rect(t) with $D = 2, N = 8$	Note the value chosen at discontinuities.
$\{1, 1, 0.5, 0, 0, 0, 0.5, 1\}$	$\{4, 2.414, 0, -0.414, 0, \ldots\}$
rect$(t - 1)$ with $D = 2, N = 8$	The PE of rect$(t - 1)$ with $D = 2$ is odd and HW.
$\{0.5, 1, 1, 1, 0.5, 0, 0, 0\}$	$\{4, -j2.414, 0, -j0.414, 0, \ldots\}$

The Triangular Pulse The spectrum of $x(t) = \text{tri}(t)$ is $X(f) = \text{sinc}^2(f)$. We evaluate $X_T[k]$ with reference to Figure 12.4 (p. 526). We compare the DFT results with the Fourier transform using $t_s X_T[k] \approx X(kf_0)$, where $f = kf_0 = k/D$, and D is the sampling duration.

Figure 12.4a With $S_F = 4$ Hz ($t_s = \frac{1}{4}$), $N = 8$, we sample tri(t) over $(-1, 1)$ to give $D = 2$ and

$$x_S[n] = \{0, \tfrac{1}{4}, \tfrac{1}{2}, \tfrac{3}{4}, 1, \tfrac{3}{4}, \tfrac{1}{2}, \tfrac{1}{4}\} \qquad t_s(\text{DFT}) = t_s X_T[k] = \tfrac{1}{4}\{4, -1.7071, 0, -0.293, 0, \ldots\}$$

The $X_T[k]$ are real and imply an even $x(t)$. But $\text{sinc}^2(f)$ is always positive, whereas the $X_T[k]$ are not. The DFT result actually describes the *periodic extension* of $x_S[n]$, which is tri$(t - 1)$, not tri(t).

Figure 12.4b If we first replicate $x(t)$ and then sample over $(0, 2)$, we get

$$x_S[n] = \{1, \tfrac{3}{4}, \tfrac{1}{2}, \tfrac{1}{4}, 0, \tfrac{1}{4}, \tfrac{1}{2}, \tfrac{3}{4}\} \qquad t_s(\text{DFT}) = t_s X_T[k] = \tfrac{1}{4}\{4, 1.7071, 0, 0.293, 0, \ldots\}$$

Now, $t_s X_T[k]$ does correspond to values of $X(kf_0) = \text{sinc}^2(kf_0) = \text{sinc}^2(\frac{1}{2}k)$. The error in $t_s X_T[k]$ compared to $\text{sinc}^2(\frac{1}{2}k)$ is 5.3% at 0.5 Hz ($k = 1$), and 62.6% at 1.5 Hz ($k = 3$). The highest frequency present in the DFT spectrum is $\frac{1}{2}S_F = 2$ Hz.

Figure 12.4c To extend the spectrum, we must reduce the sampling interval. If we choose $t_s = \frac{1}{8}$ s, we obtain $S_F = 8$ Hz and $N = 16$ and

$$t_s(\text{DFT}) = t_s X_T[k] = \tfrac{1}{8}\{8, 3.284, 0, 0.405, 0, 0.181, 0, 0.13, 0, \ldots\}$$

The DFT spacing is $f_0 = 1/D = \frac{1}{2}$ Hz and the highest frequency extends to 4 Hz.

Figure 12.4   DFT of a triangular pulse and its zero-padded versions.

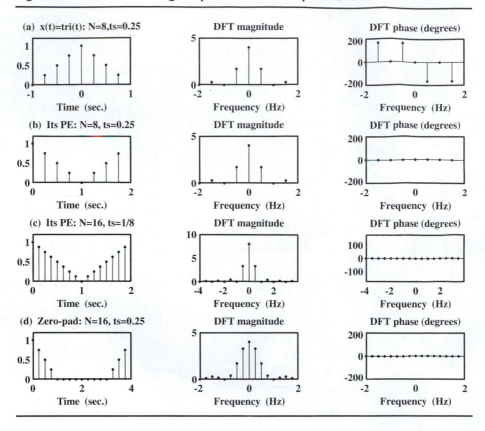

Figure 12.4d To improve the resolution, we must extend the duration D by zero padding. For $D = 4$ and $N = 16$, for example, we create the periodic extension of $x(t)$ and sample over $(0, 4)$ with $t_s = \frac{1}{4}$ s and $S_F = 4$ Hz to get the sequence

$$x_S[n] = \{1, \tfrac{3}{4}, \tfrac{1}{2}, \tfrac{1}{4}, 0, 0, 0, 0, 0, 0, 0, 0, 0, \tfrac{1}{4}, \tfrac{1}{2}, \tfrac{3}{4}\}$$

Note how the padded zeros appear *in the middle*. The DFT equals

$$t_s(\text{DFT}) = t_s X_T[k] = \tfrac{1}{4}\{4, 3.284, 1.707, 0.405, 0, 0.181, 0.293, 0.13, 0, \dots\}$$

The frequency separation is reduced to $f_0 = 1/D = \frac{1}{4}$ Hz, but the results are no more accurate. Compared with $X(kf_0) = \text{sinc}^2(kf_0) = \text{sinc}^2(\frac{1}{4}k)$, $t_s X_T[k]$ is still off by 5.3% at 0.5 Hz ($k = 2$), and by 62.6% at 1.5 Hz ($k = 6$), as for Figure 12.4b.

The Rectangular Pulse   The Fourier transform of $\text{rect}(t)$ is $\text{sinc}(f)$. Figure 12.5 shows its DFT for various N and zero-padded versions.

Figure 12.5   DFT of a rectangular pulse and its zero-padded versions.

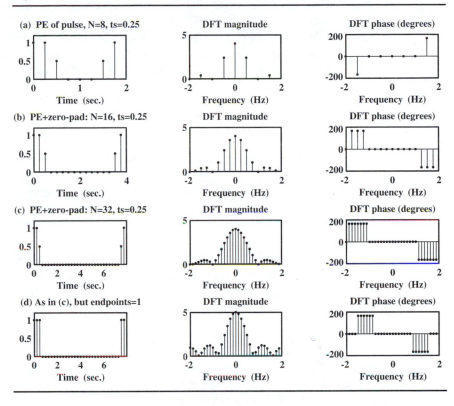

Figure 12.5a The DFT of the periodic extension of rect(t) sampled over $(0, 2)$, with $t_s = \frac{1}{4}$ and values of $\frac{1}{2}$ at the discontinuities, shows a highest frequency of 2 Hz. The results are a poor match to the Fourier transform sinc(kf_0).

Figure 12.5b–c With $t_s = \frac{1}{4}$ s, endvalues chosen as $\frac{1}{2}$, and zero-padding to $D = 4$ ($N = 16$) and $D = 8$ ($N = 32$), we begin to discern the sinc form, but only to 2 Hz. To extend the spectrum to higher frequencies, we must decrease t_s.

Figure 12.5d With $t_s = \frac{1}{4}$ and $D = 8$ ($N = 32$), but with endvalues chosen as 1 (not $\frac{1}{2}$), we see gross errors in both the magnitude and phase of the DFT. Table 12.5 (p. 528) summarizes some important observations for nonperiodic signals.

12.4.2   *The DFT of Arbitrary Signals*

To find the DFT of an arbitrary signal with some confidence, we must decide on both the duration D and the sampling interval t_s. To understand some of the theoretical considerations and practical compromises involved in such a decision, consider the signal $x(t) = \exp(-t)u(t)$. Its Fourier transform $X(f) = 1/(1 + j2\pi f)$ is not bandlimited. Both $x(t)$ and $X(f)$ are shown in Figure 12.6a (p. 529).

Table 12.5   Interpreting DFT Results for Nonperiodic Signals.

Signal	DFT Result
Infinite time signal	Must be truncated. Leads to truncation error.
Timelimited (sampled over its duration)	DFT results are *never* exact. Result is like the FS of sampled periodic signal with period = duration.
Take more samples over its duration	Extends spectrum. Does not change spacing. Lower frequency results become more accurate.
M replications	Interpolated DFT with $M - 1$ zeros between samples.
Pad with zeros	Reduces spacing. Does not extend spectrum. Improves results for the lower frequencies.
Rule of thumb	Choose $N > 8M$ samples/period to obtain the first M DFT samples with about 5% error.

Choosing the Sampling Rate   One way to choose a sampling rate is based on energy considerations. The energy in $x(t)$ equals 1. We choose a bandwidth B that contains a large fraction P of this energy. Using Parseval's relation,

$$\int_{-B}^{B} |X(f)|^2 \, df = (P)(E) = P$$

we compute $B = \tan(\frac{1}{2}P\pi)/2\pi$ Hz. If we choose B to contain 95% of the signal energy ($P = 0.95$), we find that $B = 12.71/2\pi = 2.02$ Hz. Based on this, the sampling theorem suggests that we use a sampling rate larger than $2B = 4.04$ Hz. Let us choose $S_F = 5$ Hz (or $t_s = 0.2$ s).

Choosing the Sampling Duration   We choose the sampling duration D for good resolution. For a resolution of 1 Hz, say, $D = 1$ s. This means that $x_S[n]$ will have 5 samples, starting at $t = 0$. Note that, since $x(t)$ is discontinuous at $t = 0$, we should use $x_S[0] = 0.5$, not 1.

Some Practical Choices   The chosen duration D should also cover most of the signal but $x(t)$ decays to only about 0.32 over 1 s. A better choice is $D = 3$ s, over which $x(t)$ decays to 0.05. We now acquire $N = DS_F = 15$ samples. A useful choice is $N = 16$ (a power of 2, which allows efficient computation of the DFT as we shall see later). This yields a resolution of $f_0 = 3/16$ Hz. Our rule of thumb ($N > 8M$) suggests that with $N = 16$, the DFT values $X_T[1]$ and $X_T[2]$ should show an error of only about 5%.

Comparing the DFT Results   With $N = 16$ and $t_s = 0.2$, the DFT $t_s X_T[k]$ of the sampled signal (Figure 12.6b) compares well with $X(f)$.

Improving the Results   To improve the results (and minimize aliasing), we must increase S_F. If we require the bandwidth to include 99% of the signal energy, we obtain $B = 63.6567/2\pi = 10.13$ Hz. Choosing a sampling rate of 25 Hz gives $t_s = 0.04$ s. If

Figure 12.6   The Fourier transform of $x(t) = \exp(-t)u(t)$ and the DFT of its sampled approximation for various durations and sampling rates.

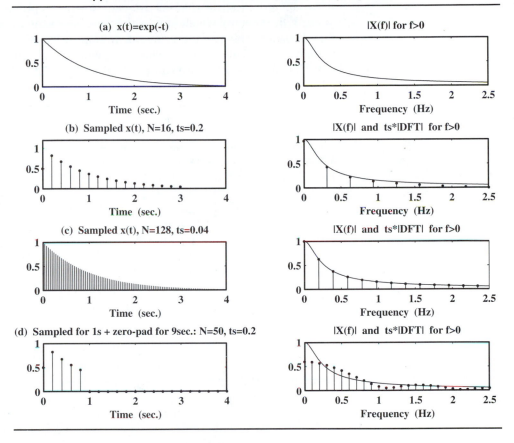

we sample for $D = 5$ s over which $x(t)$ decays to less that 0.01, we get $N = 5/0.04 = 125$. We choose $N = 128$ (the next higher power of 2). Figure 12.6c shows that the 128-point DFT $t_s X_T[k]$ of the sampled signal is almost identical to the true spectrum $X(f)$.

Truncation   What if we sample $x(t)$ every 0.2 s for only 1 s, but try to improve the resolution by padding the 1s sequence with 45 zeros (over the next 9 s) to give $D = 10$ s? Figure 12.6d reveals that the 50-point DFT shows the effects of truncation (as wiggles), and is a poor match to the true spectrum.

12.4.3   *Spectral Smoothing by Time Windows*
Truncation is equivalent to multiplying $x(t)$ by a rectangular window. Due to abrupt truncation, the spectrum of the windowed signal shows a mainlobe and sidelobes that do not decay rapidly enough. This phenomenon is similar to the Gibbs effect,

which arises during the reconstruction of periodic signals from a finite number of harmonics (an abrupt truncation of its spectrum). Just as Fourier series smoothing uses tapered *spectral windows*, we can now use tapered *time-domain windows* to smooth the spectral wiggles and sidelobes, but at the expense of making the main-lobe broader. This is another manifestation of leakage in that the spectral energy is distributed (leaked) over a wider frequency range. Some DFT windows and their spectral characteristics are summarized in Table 12.6 and Figure 12.7.

Table 12.6   Some Time Windows for DFT.

Note: Window length $= N$, $k = 0, 1, \ldots, N - 1$ and $M = \frac{1}{2}N$

Window Type	Expression
Boxcar	1
Bartlett	$1 - 2|(M - k)|/N$
vonHann	$\frac{1}{2} + \frac{1}{2}\cos[(M - k)2\pi/N]$
Hamming	$0.54 + 0.46\cos[(M - k)2\pi/N]$
Cosine	$\cos[(M - k)\pi/N]$
Riemann	$\mathrm{sinc}^L[2(M - k)/N]$
Parzen-2	$1 - 4(M - k)^2/N^2$
Blackman	$0.42 + 0.5\cos[(M - k)2\pi/N] + 0.08\cos[(M - k)4\pi/N]$
Papoulis	$\dfrac{1}{\pi}\lvert\sin[(M - k)2\pi/N]\rvert + [1 - 2\lvert(M - k)\rvert/N]\cos[(M - k)2\pi/N]$
Kaiser	$\dfrac{I_0(\pi\beta\sqrt{1 - 4[(M - k)/N]^2}}{I_0(\pi\beta)}$
Modified Kaiser	$\dfrac{I_0(\pi\beta\sqrt{1 - 4[(M - k)/N]^2} - 1}{I_0(\pi\beta) - 1}$
Bickmore	$\left[\dfrac{\sqrt{1 - 4[(M - k)/N]^2}}{\pi\beta}\right]^{\nu-1/2}\left[\dfrac{I_{\nu-1/2}\left(\pi\beta\sqrt{1 - 4[(M - k)/N]^2}\right)}{I_{\nu-1/2}(\pi\beta)}\right]$
General Blackman	$a_0 + \sum_{n=1}^{L} 2a_n\cos[(M - k)2n\pi/N]$

Remarks:

1. All windows are positive and exclude the sample $w[N]$.
2. All windows show a smooth taper except for the boxcar (rectangular).
3. For all (except boxcar, Hamming, Kaiser and Bickmore), $w[0] = 0$.
4. The coefficients of the general Blackman window are normalized to unit height using $a_0 + \Sigma 2a_n = 1$. The boxcar, vonHann, Hamming, Blackman and cosine windows fall out as special cases of this general window.

The choice of a window is based on a compromise between the two conflicting requirements of minimizing the sidelobe magnitude and minimizing the mainlobe width. We remark that:

Figure 12.7   Some smoothing windows of Table 12.6 and their DFTs.

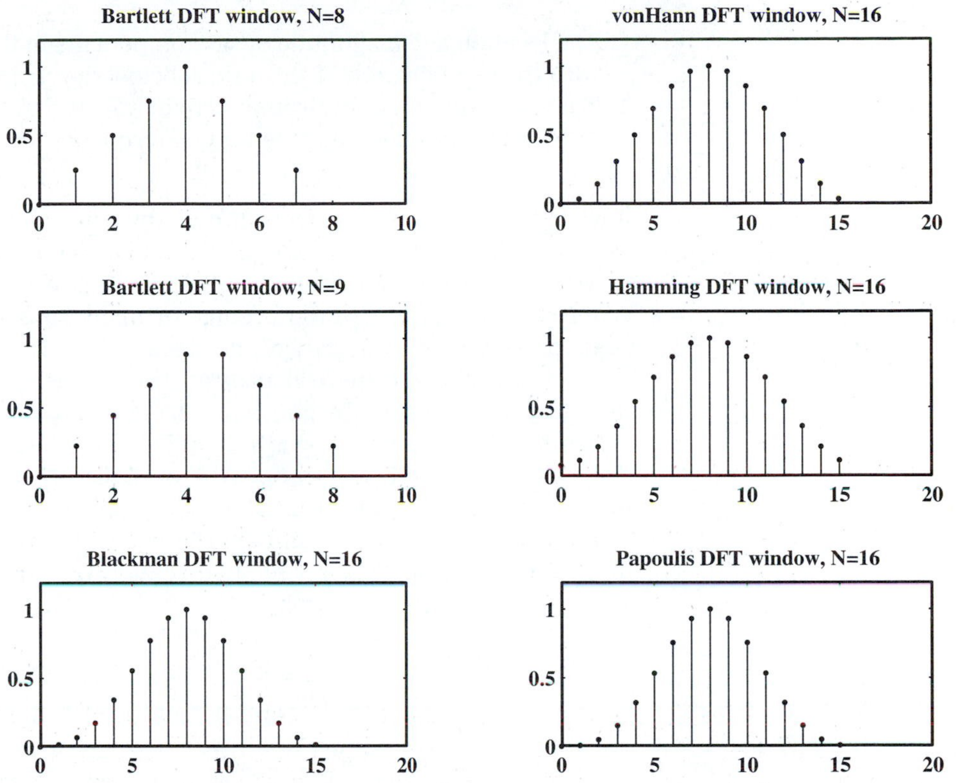

1. Unlike windows for Fourier series smoothing, we are not constrained by odd-length windows. Table 12.6 reflects this, and lists results that are valid for both odd and even N-point windows.

2. An N-point DFT window is actually an $(N + 1)$-point window (sampled over N intervals) with its last sample discarded (Figure 12.7). This is in keeping with the implied periodicity in both the signal and its DFT, and makes the DFT window unsymmetric.

Windowing   The window should be carefully positioned over the data. It is a good idea to apply the window to the data before we create its periodic extension, and then sample one period of the periodic extension.

12.4.4   *The DFT of Periodic Signals Buried in Noise*

Given a noisy analog signal $x(t)$ known to contain periodic components, how do we estimate their frequencies and magnitude (the spectrum)? There are several ways.

Most rely on statistical estimates. We describe a simple, intuitive—but by no means the best—approach based on the following effects of aliasing.

1. The location and magnitude of the components in the DFT spectrum can change with the sampling rate if this rate is below the Nyquist rate.
2. The spectrum does not drop to zero at $\frac{1}{2}S_F$. It may even show increased magnitudes as we move toward the folding frequency.

If we try to minimize the effects of noise by using a lowpass filter, we must ensure that the cutoff frequency exceeds the frequency of all the components present in $x(t)$. We have no a priori way of doing this. A better way, if the data can be acquired again, is to use the average of many runs. Averaging minimizes noise while preserving the integrity of the signal.

Next we start the spectral analysis. We may be able to get a rough estimate of the sampling frequency by observing the most rapidly varying portions of the signal. Failing this, we choose an arbitrary, but small, sampling frequency, sample $x(t)$ and observe the DFT spectrum. We repeat the process with increasing sampling rates, and when the spectrum shows little change in the location of its spectral components, we have the right spectrum and the right sampling frequency (Figure 12.8). This trial-and-error method actually depends on aliasing for its success.

12.5
Matrix
Formulation of
the DFT and
IDFT

If we let $W_N = \exp(-j2\pi/N)$, the DFT and IDFT can then be written as

$$X_T[k] = \sum_{n=0}^{N-1} x[n]W_N^{nk} \qquad k = 0, 1, \ldots, N-1$$

$$x[n] = \frac{1}{N} \sum_{k=0}^{N-1} X_T[k][W_N^{nk}]^* \qquad n = 0, 1, \ldots, N-1$$

The set of N DFT equations in N unknowns can be expressed in matrix form as

$$[\mathbf{X}] = [\mathbf{x}][\mathbf{W}_N]$$

Here, $\mathbf{X}$ and $\mathbf{x}$ are $(1 \times N)$ matrices and $[\mathbf{W}_N]$ is an $(N \times N)$ square matrix called the **DFT matrix**. The full matrix form is described by

$$[X_0\ X_1 \ldots X_{N-1}] = [x_0\ x_1 \ldots x_{N-1}] \begin{bmatrix} W_N^0 & W_N^0 & W_N^0 & \ldots & W^0 \\ W_N^0 & W_N^1 & W_N^2 & \ldots & W_N^{N-1} \\ W_N^0 & W_N^2 & W_N^4 & \ldots & W_N^{2(N-1)} \\ \vdots & \vdots & \vdots & \vdots & \vdots \\ W_N^0 & W_N^{N-1} & W_N^{2(N-1)} & \ldots & W_N^{(N-1)(N-1)} \end{bmatrix}$$

The exponents t in the elements W_N^t of $[\mathbf{W}_N]$ are called **twiddle factors**.

Figure 12.8   A noisy signal and its DFT for various sampling frequencies. The spectral locations show no change if the sampling rate is high enough.

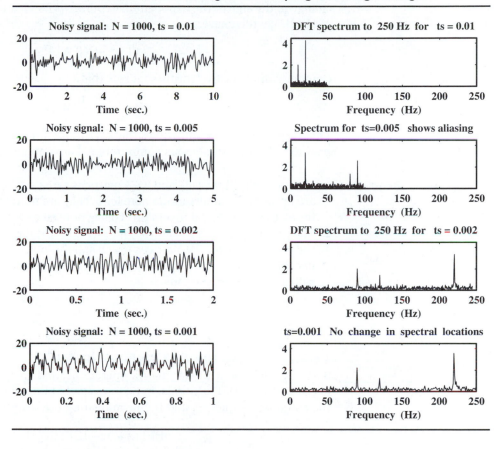

With $x[n] = \{1, 2, 1, 0\}$, we have $N = 4$ and $W_N = \exp(-j2\pi/4) = -j$. The DFT may be obtained by solving the matrix product

$$[X_0 \ X_1 X_2 X_3] = [x_0 x_1 x_2 x_3] \begin{bmatrix} W_N^0 & W_N^0 & W_N^0 & W_N^0 \\ W_N^0 & W_N^1 & W_N^2 & W_N^3 \\ W_N^0 & W_N^2 & W_N^4 & W_N^6 \\ W_N^0 & W_N^3 & W_N^6 & W_N^9 \end{bmatrix} = [1 \ 2 \ 1 \ 0] \begin{bmatrix} 1 & 1 & 1 & 1 \\ 1 & -j & -1 & j \\ 1 & -1 & 1 & -1 \\ 1 & j & -1 & -j \end{bmatrix}$$

The result is $X_T[k] = \{4, -j2, 0, j2\}$ (also found in Example 12.1).

Example 12.7
The DFT from the Matrix
Formulation

12.5.1   *The IDFT from the Matrix Form*

The matrix $[\mathbf{x}]$ can be written in terms of the inverse of $[\mathbf{W}_N]$ as

$$[\mathbf{x}] = [\mathbf{X}][\mathbf{W}_N]^{-1}$$

The matrix $[\mathbf{W}_N]^{-1}$ is called the **IDFT matrix**. We may also obtain $[\mathbf{x}]$ directly from the IDFT relation in matrix form, where the change of index from n to k and the change in the sign of the exponent in $\exp(j2\pi nk/N)$ lead to a **conjugate transpose** of $[\mathbf{W}_N]$. We then have

$$[\mathbf{x}] = \frac{1}{N}[\mathbf{X}][\mathbf{W}_N^*]^T$$

Comparison of the two forms suggests that

$$[\mathbf{W}_N]^{-1} = \frac{1}{N}[\mathbf{W}_N^*]^T$$

This very important result shows that $[\mathbf{W}_N]^{-1}$ requires only conjugation and transposition of $[\mathbf{W}_N]$, an obvious computational advantage.

Remarks:

1. The elements of the DFT and IDFT matrices satisfy $A_{ij} = A^{(i-1)(j-1)}$. Such matrices are known as **Vandermonde matrices**. They are notoriously ill conditioned in so far as their numerical inversion. This is not the case for $\mathbf{W}_N$, however.

2. The product of the DFT matrix $[\mathbf{W}_N]$ with its conjugate transpose matrix equals the identity matrix $[\mathbf{I}]$. Matrices which satisfy such a property are called **unitary**. For this reason, the DFT and IDFT, which are based on unitary operators, are also called **unitary transforms**.

12.5.2   *Using the DFT to Find the IDFT*

Both the DFT and IDFT are matrix operations, and there is an inherent symmetry in the DFT and IDFT relations. In fact, we can obtain the IDFT by finding the DFT of the conjugate sequence, and then conjugating the results and dividing by N:

$$\text{IDFT}\{X\} = \frac{1}{N}\{\text{DFT}[X^*]\}^*$$

This result invokes the conjugate symmetry and duality of the DFT and IDFT and suggests that the DFT algorithm itself can also be used to find the IDFT. In practice, this is indeed what is done.

Example 12.8
Using the DFT
to Find the IDFT

Let us find the IDFT of $X_T[k] = \{4, -j2, 0, j2\}$ using the DFT relation. First, we conjugate the sequence to get $X_T^*[k] = \{4, j2, 0, -j2\}$. We then find the DFT of $X_T^*[k]$, using the 4×4 DFT matrix found in Example 12.7, to give

$$\text{IDFT}\{X_T^*[k]\} = [4\ j2\ 0\ -j2]\begin{bmatrix} 1 & 1 & 1 & 1 \\ 1 & -j & -1 & j \\ 1 & -1 & 1 & -1 \\ 1 & j & -1 & -j \end{bmatrix} = [4\ 8\ 4\ 0]$$

Conjugating this result and dividing by $N = 4$, we get the IDFT of $X_T[k]$ as IDFT $\{X_T[k]\} = \frac{1}{4}\{4, 8, 4, 0\} = \{1, 2, 1, 0\}$. This is the $x[n]$ of Example 12.7.

12.6
The FFT

The importance of the DFT stems from the fact that it is amenable to fast and efficient computation using algorithms called **fast Fourier transform**, or FFT, algorithms. Fast algorithms reduce the problem of calculating an N-point DFT to that of calculating many smaller-size DFTs. The key ideas in optimizing the computation are based on the following ideas.

Symmetry and Periodicity   All FFT algorithms take advantage of the symmetry and periodicity of the exponential $W_N = \exp(-j2\pi n/N)$ as listed in Table 12.7. The last entry in Table 12.7, for example, suggests that $W_{N/2} = \exp(j2\pi/\frac{1}{2}N)$ is periodic with period $\frac{1}{2}N$.

Table 12.7   Symmetry and Periodicity of
$$W_N = \exp(-j2\pi/N).$$

Exponential form	Symbolic form
$\exp(-j2\pi n/N) = \exp[-j2\pi(n+N)/N]$	$W_N^{n+N} = W_N^n$
$\exp[-j2\pi(n+\frac{1}{2}N)/N] = -\exp(-j2\pi n/N)$	$W_N^{n+N/2} = -W_N^n$
$\exp(-j2\pi K) = \exp(-j2\pi NK/N) = 1$	$W_N^{NK} = 1$
$\exp[-j2(2\pi/N)] = \exp(-j2\pi/\frac{1}{2}N)$	$W_N^2 = W_{N/2}$

Choice of N   We choose N as a number that is the product of many smaller numbers r_k such that $N = r_1 r_2 \ldots r_m$. A more useful choice results when the factors are equal, such that $N = r^m$. The factor r is called the **radix**. By far the most practically implemented choice for r is 2, such that $N = 2^m$, and leads to the **radix-2** FFT algorithms.

Decimation (Index Separation)   The computation is carried out separately on even-indexed and odd-indexed samples to reduce the computational effort.

Storage   All algorithms allocate storage for computed results. The less the storage required, the more efficient the algorithm. Many FFT algorithms reduce storage requirements by performing computations *in place* by storing results in the same memory locations that previously held the data.

Some Fundamental Results   We begin by considering two trivial, but important, results:

One-Point Transform   The DFT of a single number A is the number A itself.

Two-Point Transform   The DFT of a two-sample sequence is easily found to be

$$X_T[0] = x[0] + x[1] \quad \text{and} \quad X_T[1] = x[0] - x[1]$$

The single most important result in the development of a radix 2 FFT algorithm is that an N-sample DFT can be written as the sum of two $\frac{1}{2}N$-sample DFTs formed from the even- and odd-indexed samples of the original sequence. Here is the development

$$X_T[k] = \sum_{n=0}^{N-1} x[n]W_N^{nk} = \sum_{n=0}^{N/2-1} x[2n]W_N^{2nk} + \sum_{n=0}^{N/2-1} x[2n+1]W_N^{(2n+1)k}$$

$$X_T[k] = \sum_{n=0}^{N/2-1} x[2n]W_N^{2nk} + W_N^k \sum_{n=0}^{N/2-1} x[2n+1]W_N^{2nk}$$

$$= \sum_{n=0}^{N/2-1} x[2n]W_{N/2}^{nk} + W_N \sum_{n=0}^{N/2-1} x[2n+1]W_{N/2}^{nk}$$

If $X^e[k]$ and $X^o[k]$ denote the DFT of the even- and odd-indexed sequences of length $\frac{1}{2}N$, we can rewrite this result as

$$X_T[k] = X^e[k] + W_N^k X^o[k] \qquad k = 0, 1, 2, \ldots, N-1$$

Note carefully that the index k in this expression varies from 0 to $N-1$, and that $X^e[k]$ and $X^o[k]$ are both periodic in k with period $\frac{1}{2}N$; we thus have two periods of each to yield $X_T[k]$. Due to periodicity, we can split $X_T[k]$ and compute the first half and next half of the values as

$$X_T[k] = X^e[k] + W_N^k X^o[k] \qquad k = 0, 1, 2 \ldots \tfrac{1}{2}N - 1$$

$$X_T[k + \tfrac{1}{2}N] = X^e[k + \tfrac{1}{2}N] + W_N^{k+N/2} X^o[k + \tfrac{1}{2}N]$$

$$= X^e[k] - W_N^k X^o[k] \qquad k = 0, 1, 2, \ldots, \tfrac{1}{2}N - 1$$

This result, known as the **Danielson-Lanczos lemma**, is usually shown as a signal flow graph (Figure 12.9), and called a **butterfly** due to its characteristic shape. The inputs X^e and X^o are transformed into $X^e + W_N^k X^o$ and $X^e - W_N^k X^o$. A butterfly operates on one pair of samples, and involves *two complex additions and one complex multiplication*. For N samples, there are $\frac{1}{2}N$ butterflies in all. Starting with N samples, this lemma reduces the computational complexity by evaluating the DFT of two $\frac{1}{2}N$-sample sequences.

The DFT of each of these can once again be reduced to the computation of sequences of length $\frac{1}{4}N$ to yield

$$X^e[k] = X^{ee}[k] + W_{N/2}^k X^{eo}[k] \qquad X^o[k] = X^{oe}[k] + W_{N/2}^k X^{oo}[k]$$

Since $W_{N/2}^k = W_N^{2k}$, we can rewrite this expression as

$$X^e[k] = X^{ee}[k] + W_N^{2k} X^{eo}[k] \qquad X^o[k] = X^{oe}[k] + W_N^{2k} X^{oo}[k]$$

Carrying this process to its logical extreme, if we choose the number N of samples as $N = 2^m$, we can reduce the computation of an N-sample DFT to the computation of 1-sample DFTs in m stages. And the 1-sample DFT is just the sample value itself (repeated with period 1). This process is known as **decimation**. The FFT results so obtained are actually in **bit-reversed order**. If we let $e = 0$ and $o = 1$ and then reverse the order, we have the sample number in binary representation. The reason

Figure 12.9   Butterfly operation for (a) decimation-in-frequency (DIF) and (b) decimation-in-time (DIT) FFT algorithms.

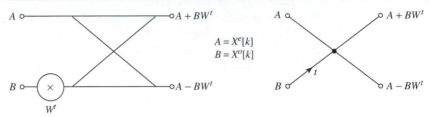

$A = X^e[k]$
$B = X^o[k]$

a. Basic butterfly for the DIF algorithm

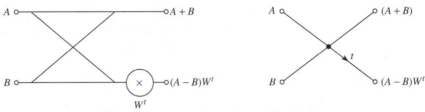

b. Basic butterfly for the DIT algorithm

is that splitting the sequence into even and odd indices is equivalent to testing each index for the least significant bit (0 for even, 1 for odd). We describe two common in-place FFT algorithms based on decimation. A summary appears in Table 12.8.

Table 12.8   FFT Algorithms for Computing DFT.

Characteristic	Decimation-in-frequency	Decimation-in-time
Number of samples	$N = 2^m$	$N = 2^m$
Input sequence	Natural order	Bit-reversed order
DFT result	Bit-reversed order	Natural order
Computations	In place	In place
Number of stages	$m = \log_2 N$	$m = \log_2 N$
Multiplications	$\frac{1}{2}N \log_2 M$ (complex)	$\frac{1}{2}N \log_2 M$ (complex)
Additions	$N \log_2 M$ (complex)	$N \log_2 M$ (complex)
Structure of the *i*th Stage		
No. of butterflies	$\frac{1}{2}N$	$\frac{1}{2}N$
Butterfly input	A (top) and B (bottom)	A (top) and B (bottom)
Butterfly output	$(A + B)$ and $(A - B)W_N^t$	$(A + BW_N^t)$ and $(A - BW_N^t)$
Twiddle factors t	$2^{i-1}Q$, $Q = 0, 1, \ldots, P - 1$	$2^{m-i}Q$, $Q = 0, 1, \ldots, P - 1$
Values of P	$P = 2^{m-i}$	$P = 2^{i-1}$

12.6.1   *The Decimation-in-Frequency (DIF) FFT Algorithm*

We start by reducing the single N-point transform at each successive stage to two $\frac{1}{2}N$-point transforms, then four $\frac{1}{4}N$-point transforms, and so on, until we arrive at N 1-point transforms which correspond to the actual DFT. With the input sequence in natural order, computations can be done in place, but the DFT result is in bit-reversed order and must be reordered.

The algorithm slices the input sequence $x[n]$ into two halves and leads to

$$X_T[k] = \sum_{n=0}^{N-1} x[n]W_N^{nk} = \sum_{n=0}^{N/2-1} x[n]W_N^{nk} + \sum_{n=N/2}^{N/2-1} x[n]W_N^{nk}$$

$$= \sum_{n=0}^{N/2-1} x[n]W_N^{nk} + \sum_{n=0}^{N/2-1} x[n+\tfrac{1}{2}N]W_N^{(n+N/2)k}$$

$$X_T[k] = \sum_{n=0}^{N/2-1} x[n]W_N^{nk} + W_N^{Nk/2} \sum_{n=0}^{N/2-1} x[n+\tfrac{1}{2}N]W_N^{nk}$$

$$= \sum_{n=0}^{N/2-1} x[n]W_N^{nk} + (-1)^k \sum_{n=0}^{N/2-1} x[n+\tfrac{1}{2}N]W_N^{nk}$$

Separating even and odd indices, and letting $x[n] = x^a$ and $x[n + \frac{1}{2}N] = x^b$,

$$X_T[2k] = \sum_{n=0}^{N/2-1} [x^a + x^b]W_N^{2nk} \qquad k = 0, 1, 2, \ldots, \tfrac{1}{2}N - 1$$

$$X_T[2k + 1] = \sum_{n=0}^{N/2-1} [x^a - x^b]W_N^{(2k+1)n} = \sum_{n=0}^{N/2-1} [x^a - x^b]W_N^n W_N^{2nk} \qquad k = 0, 1, \ldots, \tfrac{1}{2}N - 1$$

Since $W_N^{2nk} = W_{N/2}^{nk}$, the even and odd indexed results describe a $\frac{1}{2}N$-point DFT. The computations result in a butterfly structure with inputs $x[n]$ and $x[n + \frac{1}{2}N]$, and outputs $X_T[2k] = \{x[n] + x[n + \frac{1}{2}N]\}$ and $X_T[2k + 1] = \{x[n] - x[n + \frac{1}{2}N]\}W_N^n$.

The factors W^t, called **twiddle factors**, appear only in the lower corners of the butterfly wings at each stage. Their exponents t have a definite order, described as follows for an $N = 2^m$-point FFT algorithm with m stages:

Number P of distinct twiddle factors W^t at ith stage: $P = 2^{m-i}$.
Values of t in the twiddle factors W^t: $t = 2^{i-1}Q$ with $Q = 0, 1, 2, \ldots, P - 1$.

This algorithm is illustrated in Figure 12.10 for $N = 8$.

Example 12.9
A 4-point Decimation-in-Frequency FFT Algorithm

For a 4-point DFT, we use the above equations to obtain

$$X_T[2k] = \sum_{n=0}^{1} \{x[n] + x[n + 2]\}W_4^{2nk}$$

$$X_T[2k + 1] = \sum_{n=0}^{1} \{x[n] - x[n + 2]\}W_4^n W_4^{2nk} \qquad k = 0, 1$$

Figure 12.10   Flowchart for 8-point decimation-in-frequency FFT algorithm.

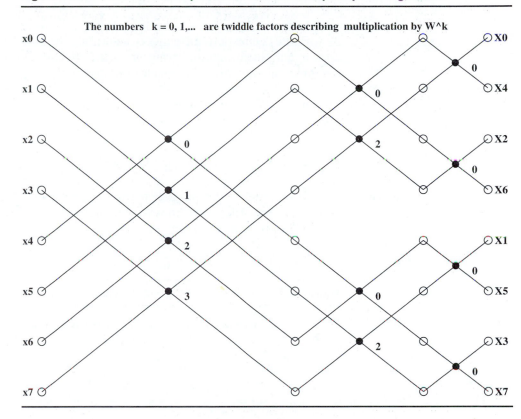

The numbers k = 0, 1,... are twiddle factors describing multiplication by W^k

Since $W_4^0 = 1$ and $W_4^2 = -1$, we arrive at the following result

$$X_T[0] = x[0] + x[2] + x[1] + x[3] \qquad X_T[2] = x[0] + x[2] - \{x[1] + x[3]\}$$
$$X_T[1] = x[0] - x[2] + W_4\{x[1] - x[3]\} \qquad X_T[3] = x[0] - x[2] - W_4\{x[1] - x[3]\}$$

We do not reorder the input sequence before using it.

12.6.2   *The Decimation-in-Time (DIT) FFT Algorithm*

Starting with N 1-point transforms, we combine adjacent pairs at each successive stage into 2-point transforms, then 4-point transforms, and so on, until we get a single N-point DFT result. With the input sequence in bit-reversed order, the computations can be done in place, and the DFT is obtained in natural order. Thus, for a 4-point input, the binary indices $\{00, 01, 10, 11\}$ reverse to $\{00, 10, 01, 11\}$, and we use $\{x[0], x[2], x[1], x[3]\}$.

For an 8-point input sequence, $\{000, 001, 010, 011, 100, 101, 110, 111\}$ reverses to $\{000, 100, 010, 110, 001, 101, 011, 111\}$ and we use the bit-reversed order for $x[n]$ as $\{x[0], x[4], x[6], x[2], x[1], x[5], x[3], x[7]\}$.

At a typical stage, we obtain

$$X_T[k] = X^e[k] + W_N^k X^o[k] \qquad X_T[k + \tfrac{1}{2}N] = X^e[k] - W_N^k X^o[k]$$

As with the decimation-in-frequency algorithm, the twiddle factors W^t at each stage appear only in the bottom wing of each butterfly. The exponents t also have a definite (and almost similar) order described by:

Number P of distinct twiddle factors W^t at ith stage: $P = 2^{i-1}$.
Values of t in the twiddle factors W^t: $t = 2^{m-i}Q$ with $Q = 0, 1, 2, \ldots, P - 1$.

The flowchart for an 8-point DFT is illustrated in Figure 12.11.

Remark: In both approaches, it is possible to use a sequence in natural order and get DFT results in natural order. This, however, requires more storage, since the computations cannot now be done in place.

Figure 12.11   Flowchart for 8-point decimation-in-time FFT algorithm.

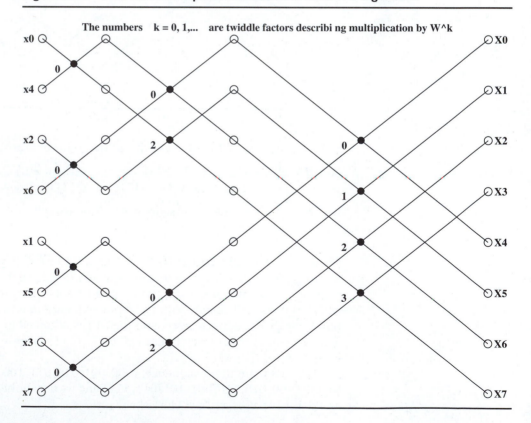

Example 12.10

A 4-point Decimation-in-Time FFT Algorithm

For a 4-sample DFT, with $W_4 = \exp(-j\frac{1}{2}\pi) = -j$, we have

$$X[k] = \sum_{n=0}^{3} x[n]W_4^{nk} \qquad k = 0, 1, 2, 3$$

We group by even and odd indexed $x[n]$ to obtain

$$X[k] = X^e[k] + W_4^k X^o[k] \qquad \begin{cases} X^e[k] = x[0] + x[2]W_4^{2k} \\ X^o[k] = x[1] + x[3]W_4^{2k} \end{cases} \qquad k = 0, 1, 2, 3$$

Using periodicity, we simplify this result to

$$\begin{aligned} X[k] &= X^e[k] + W_4^k X^o[k] \\ X[k + \tfrac{1}{2}N] &= X^e[k] - W_4^k X^o[k] \end{aligned} \qquad \begin{cases} X^e[k] = x[0] + x[2]W_4^{2k} \\ X^o[k] = x[1] + x[3]W_4^{2k} \end{cases} \qquad k = 0, 1$$

These equations yield $X[0]$ through $X[3]$ as

$$X[0] = X^e[0] + W_4^0 X^o[0] = x[0] + x[2]W_4^0 + W_4^0\{x[1] + x[3]W_4^0\}$$

$$X[1] = X^e[1] + W_4^1 X^o[1] = x[0] + x[2]W_4^2 + W_4^1\{x[1] + x[3]W_4^2\}$$

$$X[2] = X^e[0] - W_4^0 X^o[0] = x[0] + x[2]W_4^0 - W_4^0\{x[1] + x[3]W_4^0\}$$

$$X[3] = X^e[1] - W_4^1 X^o[1] = x[0] + x[2]W_4^2 - W_4^1\{x[1] + x[3]W_4^2\}$$

Upon simplification, using $W_4^0 = 1$, $W_4^1 = W_4 = 1$ and $W_4^2 = -1$, we obtain

$$\begin{aligned} X[0] &= x[0] + x[2] + x[1] + x[3] & X[1] &= x[0] - x[2] + W_4\{x[1] - x[3]\} \\ X[2] &= x[0] + x[2] - \{x[1] + x[3]\} & X[3] &= x[0] - x[2] - W_4\{x[1] - x[3]\} \end{aligned}$$

12.6.3 *Computational Cost*

In each algorithm, an N-point FFT, with $N = 2^m$, involves m stages and $\frac{1}{2}N$ butterflies per stage. The FFT computation thus requires $\frac{1}{2}Nm = \frac{1}{2}N\log_2 N$ complex multiplications and $Nm = N\log_2 N$ complex additions.

Table 12.9 shows how the FFT stacks up against direct DFT evaluation for $N = 2^m$, assuming all operations (even trivial ones) count. Note, however, that multiplications take the bulk of computing time.

Table 12.9   Computational Cost of the DFT and FFT.

	N-point DFT	N-point FFT
Algorithm	Solution of N equations in N unknowns	$\frac{1}{2}N$ butterflies/stage, m stages Total butterflies $= \frac{1}{2}Nm$
Multiplications per step	N per equation	1 per butterfly
Additions per step	$N-1$ per equation	2 per butterfly
Total multiplications	N^2	$\frac{1}{2}Nm = \frac{1}{2}N\log_2 N$
Total additions	$N(N-1)$	$Nm = N\log_2 N$

A Quick Comparison   The difference between $\frac{1}{2}N\log_2 N$ and N^2 may not seem like much for small N. For example, with $N = 16$, $\frac{1}{2}N\log_2 N = 64$ and $N^2 = 256$. For large N, the difference is phenomenal. For $N = 1024 = 2^{10}$, $\frac{1}{2}N\log_2 N \approx 5000$ and $N^2 \approx 10^6$. This is like waiting 1 min for the FFT result and (more than) 3 h for the identical direct DFT result. Figure 12.12 shows a graphical comparison between the number of complex multiplications required for FFT and direct DFT evaluation for various N. Note that $N\log_2 N$ is nearly linear with N for large N.

Figure 12.12   Comparison of the computational cost of the DFT and FFT.

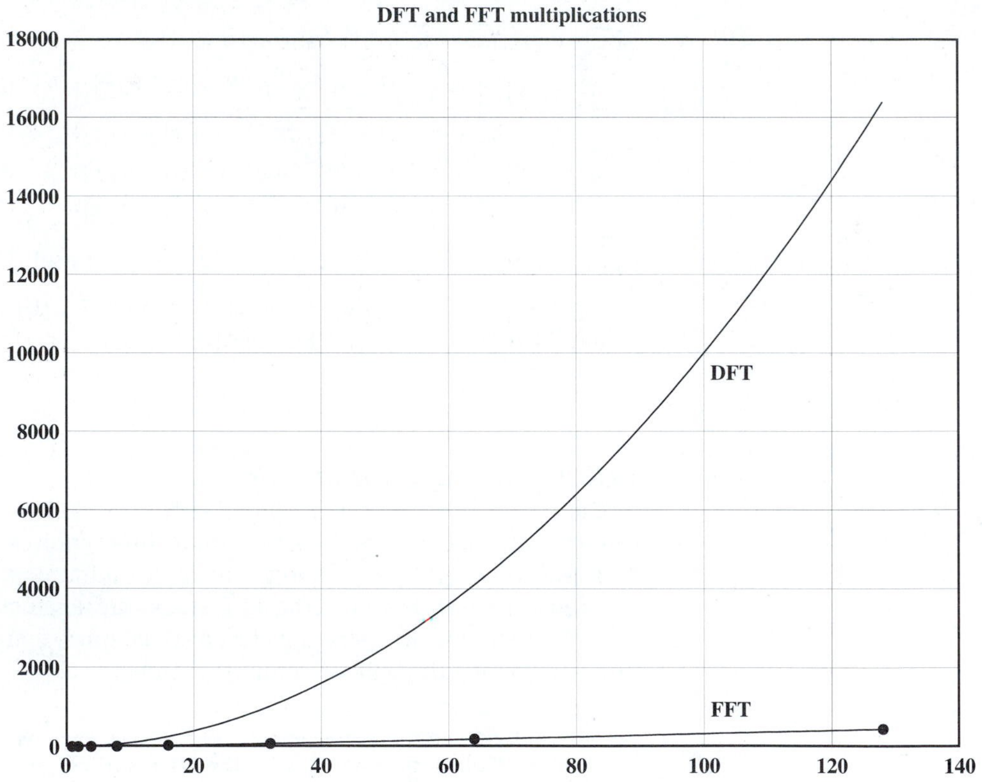

Remarks: **1.** In the DFT relations, since there are N factors which equal W^0 or 1, we require only $N^2 - N$ complex multiplications. In the FFT algorithms, the number of factors that equal 1 doubles (or halves) at each stage and is given by $1 + 2 + 2^2 + \cdots + 2^{m-1} = 2^m - 1 = N - 1$. We thus actually require only $\frac{1}{2}N\log_2 N - (N - 1)$ complex multiplications for the FFT.

2. The DFT requires N^2 values of W^k. The FFT requires at most N such values at each stage. Due to the periodicity of W, only about $\frac{3}{4}N$ of these are distinct. Once computed, they can be stored and used again. However, this hardly affects the comparison for large N.

3. Computers use real arithmetic. To find the number of real operations, remember that one complex addition involves two real additions, and one complex multiplication involves four real multiplications and three real additions (because $[A + jB][C + jD] = AC - BD + jBC + jAD$).

12.7 Fast Convolution Using the FFT

Due to implied periodicity in both domains, all convolution properties of the DFT imply *discrete periodic convolution*. The FFT offers an indirect, but computationally efficient, means of finding the periodic convolution $y[n]$ of two sequences $x_S[n]$ and $h_S[n]$. If one sequence is shorter, we zero-pad it to make both lengths equal. With sequences of equal length N, we

1. Find the N-sample DFTs $X_T[k]$ and $H_T[k]$ of each sequence.
2. Multiply the DFT sequences to obtain $Y_T[k] = X_T[k]H_T[k]$.
3. Find the IDFT of $Y_T[k]$ to obtain the periodic convolution $y[n]$.

We can also find regular convolution using the DFT. For sequences of length M and N, this contains $M + N - 1$ samples. We must thus pad each sequence with zeros to make each sequence length $M + N - 1$ before finding the DFT.

Speed of Fast Convolution   A direct computation of the convolution of two N-sample signals requires N^2 complex multiplications. For real signals, it requires N^2 real multiplications. The FFT method works with sequences of length $2N$. The number of complex multiplications involved is

1. $2(N \log_2 2N)$ to find the FFT of the two sequences.
2. $2N$ to form the product sequence.
3. $N \log_2 N$ to find the IFFT sequence that gives the convolution.

It thus requires $3N \log_2 2N + 2N$ complex multiplications for both real and complex sequences. If $N = 2^m$, the FFT approach becomes computationally superior only for $m > 5$ ($N > 32$) or so.

Convolution Involving Long Sequences   The convolution of a short sequence $h[n]$ of length N with a very long sequence $x[n]$ of length $L \gg N$ (such as an incoming stream of data) can involve large amounts of computation and memory. There are two preferred alternatives, both of which are based on sectioning the long sequence $x[n]$ into shorter ones.

The Overlap-Add Method   Suppose the length of $x[n]$ is $L = mN$ (if not, we can always zero-pad it to this length). We partition $x[n]$ into m segments $x_0[n], x_1[n], \ldots, x_{m-1}[n]$, of length N. We find the regular convolution of each section with $x[n]$

to give the partial results $y_0[n], y_1[n], \ldots, y_{m-1}[n]$. Using superposition, the total convolution is the sum of their shifted (by multiples of N) versions

$$y[n] = y_0[n] + y_1[n-N] + y_2[n-2N] + \cdots + y_{m-1}[n-(m-1)N]$$

Since each regular convolution contains $2N - 1$ samples, we zero-pad $h[n]$ and each section $x_k[n]$ with $N - 1$ zeros before finding $y_k[n]$ using the DFT.

Remarks:
1. The length $L = mN$ is not a requirement. We may use different sections of different lengths, provided we keep track of how much each partial convolution must be shifted before adding the results.
2. The number of padded zeros must make each sequence length a power of 2 if the FFT is to be used to find their convolution.

Example 12.11
The Overlap-Add Method of Convolution

Let $x[n] = \{1, 2, 3, 3, 4, 5\}$ and $h[n] = \{1, 1, 1\}$. Here $L = 6, N = 3$. We section $x[n]$ into two sequences $x_0[n] = \{1, 2, 3\}$ and $x_1[n] = \{3, 4, 5\}$ and obtain:

$$y_0[n] = x_0[n] \star h[n] = \{1, 3, 6, 5, 3\} \qquad y_1[n] = x_1[n] \star h[n] = \{3, 7, 12, 9, 5\}$$

Shifting and superposition results in the required convolution $y[n]$ as

$$y[n] = y_0[n] + y_1[n-3] = \left\{ \begin{array}{l} 1, 3, 6, 5, 3 \\ 3, 7, 12, 9, 5 \end{array} \right\} = \{1, 3, 6, 8, 10, 12, 9, 5\}$$

The Overlap-Save Method   The regular convolution of sequences of length L and N has $L + N - 1$ samples. The periodic convolution of sequences of equal length L has $2L - 1$ samples. Its samples past $N - 1$ correspond to the regular convolution. To understand this, let $L = 16$ and $N = 7$. If we pad N by 9 zeros, their regular convolution has 31 (or $2L - 1$) samples with 9 (or $L - N$) trailing zeros. For periodic convolution, 15 (or $L - 1$) samples are wrapped around. Since the last 9 (or $L - N$) are zeros, only the first 6 (or $L - N - (N - 1) = N - 1$) samples of the periodic convolution are contaminated by wraparound. This leads to the following scheme:

1. Add $N - 1$ *leading* zeros to the longer sequence $x[n]$ and section into k *overlapping* (by $N - 1$) segments of length M. Typically $M \approx 2N$.
2. Zero-pad $h[n]$, using *trailing* zeros, to make its length M.
3. Find the *periodic* convolution of $h[n]$ with each section of $x[n]$.
4. Discard the first $N - 1$ contaminated samples from each convolution, and join (concatenate) together to find the required convolution.

Example 12.12
The Overlap-Save Method of Convolution

Let $x[n] = \{1, 2, 3, 3, 4, 5\}$ and $h[n] = \{1, 1, 1\}$. Here $L = 6$ and $N = 3$. Create the zero-padded sequence $x[n] = \{0, 0, 1, 2, 3, 3, 4, 5\}$. If we choose $M = 5$, we get three overlapping sections of $x[n]$ (we need to zero-pad the last one)

$$x_0[n] = \{0, 0, 1, 2, 3\} \qquad x_1[n] = \{2, 3, 3, 4, 5\} \qquad x_3[n] = \{4, 5, 0, 0, 0\}$$

The zero-padded $h[n]$ becomes $h[n] = \{1, 1, 1, 0, 0\}$. Periodic convolution gives:

$$x_0[n] \bullet h[n] = \{5, 3, 1, 3, 6\}$$
$$x_1[n] \bullet h[n] = \{11, 10, 8, 10, 12\}$$
$$x_2[n] \bullet h[n] = \{4, 9, 9, 5, 0\}$$

We discard the first two samples from each and glue the results to obtain

$$y[n] = x[n] \star h[n] = \{1, 3, 6, 8, 10, 12, 9, 5, 0\}$$

Note that the last sample is due to the zero padding and is redundant.

Remarks: 1. In either method, the FFT of the shorter sequence need be found only once, stored, and can then be reused for all subsequent partial convolutions.
2. Both methods allow on-line implementation if we can tolerate a small processing delay that equals the time required for each section of the long sequence to arrive at the processor (assuming the time taken for finding the partial convolutions is less than this processing delay).
3. The correlation of two sequences is found in exactly the same manner, using either method, provided we use a folded version of one sequence.   ——

12.8 Applications in Signal Processing

12.8.1   *Correlation*

The convolution of two signals $x(t)$ and $y(t)$ transforms to a multiplication of $X(f)$ and $Y(f)$. The correlation of $x(t)$ and $y(t)$ transforms to a product of $X(f)$ and $Y(-f)$. For real signals, since $Y(-f) = Y^*(f)$ we may write

$$x(t) \star\star y(t) \longleftrightarrow X(f)Y^*(f)$$

The autocorrelation $x(t) \star\star x(t)$ transforms to

$$x(t) \star\star x(t) \longleftrightarrow X(f)X^*(f) = |X(f)|^2 = S(f)$$

If $x(t)$ is a power signal, its autocorrelation transforms to $S(f)$, the power spectral density (PSD). This is the celebrated **Wiener-Khintchine theorem**. Note that the PSD of a periodic signal will show impulses at the harmonics of its fundamental frequency, since its power is infinite and its spectral content is concentrated at discrete frequencies.

Periodic *DT* correlation, denoted by $\bullet\bullet$, may be found using the DFT or FFT. For two sequences $x[n]$ and $y[n]$, we have

$$x[n] \bullet\bullet y[n] \longleftrightarrow X_T[k]Y_T^*[k]$$
$$x[n] \bullet\bullet x[n] \longleftrightarrow X_T[k]X_T^*[k] = |X_T[k]|^2$$

Periodic *DT* correlation is implemented using the FFT in almost exactly the same way as periodic *DT* convolution, except for an extra conjugation step. We compute the FFT of $x[n]$ and $y[n]$ to obtain $X_T[k]$ and $Y_T[k]$, conjugate $Y_T[k]$, multiply $X_T[k]$ and $Y_T^*[k]$, and, finally, find the IFFT of the product.

Keep in mind that if $x[n]$ and $y[n]$ are real, the final result must also be real (to within machine roundoff).

12.8.2   *Signal Interpolation*

Interpolation is equivalent to increasing the sampling rate. The idea of zero padding forms the basis for an interpolation method using the DFT. It is the dual of improving the spectral resolution by zero-padding the time sequence. Suppose a signal bandlimited to f_B has been sampled at the Nyquist rate $S_F = 2f_B$ to yield N samples. If we double the sampling rate, we essentially introduce one interpolated sample between adjacent samples. Since the $2N$-point resampled signal is still bandlimited, its spectrum must show N zeros past the frequency f_B. This is equivalent to zero-padding the N-point spectrum of the original signal by N zeros. This idea forms the basis for signal interpolation using the DFT.

We split the N-point DFT $X_T[k]$ of $x_S[n]$ about the folding index $\frac{1}{2}N$. If N is even, we also split $X_T[\frac{1}{2}N]$ in half. We insert MN zeros ($MN-1$ for even N) to get a padded sequence Z_T with $N(M+1)$ samples. It has the form

$$Z_T = \begin{cases} \{X_T[0]\dots X_T[\frac{1}{2}(N-1)] & (MN \text{ zeros}) & X_T[\frac{1}{2}(N-1)]\dots X_T[N-1]\} & N \text{ odd} \\ \{X_T[0]\dots \frac{1}{2}X_T[\frac{1}{2}N] & (MN-1 \text{ zeros}) & \frac{1}{2}X_T[\frac{1}{2}N]\dots X_T[N-1]\} & N \text{ even} \end{cases}$$

The interpolation process is summarized by the following operations.

$$x_S \xrightarrow{DFT} X_T \xrightarrow{\text{Insert } NM \text{ zeros}}{\text{about index } \frac{1}{2}N} Z_T \xrightarrow{IDFT} z_S \xrightarrow{\text{use real part \&}}{\text{divide by } M+1} \to \frac{1}{M+1}\text{Re}[z_S] = x_I$$

The IDFT $z_S[n]$ involves the factor $1/[N(M+1)]$ and may be complex. We therefore use only the real part of z_S and multiply by $(M+1)$ to obtain the interpolated signal $x_I[n]$. This shows M interpolated values between each sample of $x_S[n]$.

Remark: For periodic bandlimited signals sampled above the Nyquist rate for an integer number of periods, the interpolation is exact. For all others, imperfections show up as a poor match, especially near the ends, since we are actually interpolating to zero outside the signal duration.

Example 12.13
Signal Interpolation
Using the FFT

(a) For a sinusoid sampled over one period with four samples, we obtain the signal $x[n] = \{0, 1, 0, -1\}$. Its DFT is $X_T[k] = \{0, -j2, 0, j2\}$. To interpolate this by 8, we generate the zero-padded sequence $Z_T = \{0, -j2, 0, (27 \text{ zeros}), 0, j2\}$. The interpolated sequence (IDFT) shows an exact match (Figure 12.13a).

(b) For a sinusoid sampled over half a period with four samples, interpolation does not yield exact results (Figure 12.13b). Since we are actually sampling one period of a full-rectified sine (the periodic extension), the signal is not bandlimited and the chosen sampling frequency is too low. This shows up as a poor match, especially near the edges.

Interpolation in Practice   In practice, interpolation by a factor M uses an **upsampler**, which inserts $M-1$ zeros between each input sample, in cascade with a digital lowpass filter, or **interpolating filter**, with a bandwidth of $F_C = 1/2M$ and a gain of M (Figure 12.14, p. 548). Due to the efficiency of the FFT, the interpolation factor is often chosen as a power of 2. Interpolation by 2 is the most common. For

Figure 12.13   Signal interpolation using FFT for one period of a sine and one period of a full-rectified sine. The full-rectified sine is not bandlimited and shows imperfect reconstruction, especially near the ends.

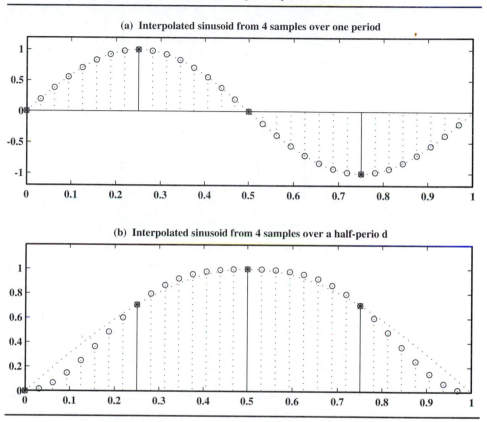

(a) Interpolated sinusoid from 4 samples over one period

(b) Interpolated sinusoid from 4 samples over a half-period

higher factors we simply cascade as many second-order stages as required. This simplifies the design and reduces the complexity of the interpolating filter. For a factor $N = 2^m$, we require m second-order stages. At each stage, we must ensure that no significant aliasing occurs in the frequency band up to $\frac{1}{2}S_{Fk}/2$, where S_{Fk} corresponds to the sampling frequency for that stage.

12.8.3   *Decimation*

Decimation is equivalent to reducing the sampling rate by discarding signal samples. To avoid aliasing, we must ensure that the reduced sampling rate is still above the Nyquist rate. This implies that the original must have been oversampled to begin with. Decimation by a factor of N is achieved by using a digital lowpass filter, or **decimating filter**, with unit gain and a bandwidth of $F = 1/2N$ in cascade with a **downsampler** that retains only the first sample from each set of N input samples and discards the remaining $N - 1$ samples.

Figure 12.14   Illustrating signal interpolation.

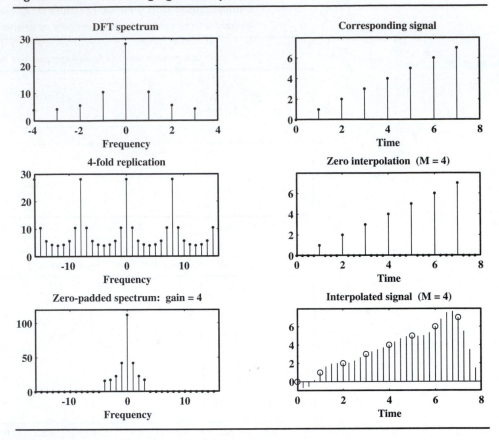

As with interpolating filters using the FFT, the decimation factor is often chosen as a power of 2. Decimation by 2 is the most common in practice. For higher factors we cascade such second-order stages. For a factor $N = 2^m$, we require m second-order stages. At each stage, we ensure that no significant aliasing occurs for frequencies up to $\frac{1}{2}S_{Fk}/2$, where S_{Fk} corresponds to the sampling frequency for that stage.

12.8.4   *Oversampling and Multirate Signal Processing*

In multirate signal processing, different parts of a signal processing system operate at different sampling rates because it is either necessary or desirable to do so. Sampling rate conversion by a rational fraction M/N may be achieved by interpolation by M followed by decimation by N.

Perfect recovery after sampling is possible only for bandlimited signals. To bandlimit signals, it is common practice to use an antialiasing filter. We do this even for bandlimited signals to reduce the aliasing of any noise to the principal range. Ideally an antialiasing filter must allow perfect transmission up to $\frac{1}{2}S_F$ and zero transmission past $\frac{1}{2}S_F$. Practical filters have a finite transition width. To relax the specification of antialiasing filters, we deliberately oversample analog signals prior to sampling. For DSP operations, however, we must minimize the sampling rate since a lower rate reduces both execution time and filter complexity. We do this by decimating the oversampled signal to a lower sampling rate that still satisfies the Nyquist criterion. The processed signal is then converted to analog form using a DAC (digital-to-analog converter) and reconstruction filter. For distortion-free recovery, this filter must have a flat magnitude and linear phase up to $\frac{1}{2}S_F$. As with antialiasing filters, the design of practical reconstruction filters may be relaxed by increasing the sampling rate to allow a finite transition width. We therefore interpolate or oversample the processed signal before passing it through the DAC. The signal produced by a practical DAC is a zero-order-hold version of the sampled signal. It is equivalent to passing the ideally sampled signal through a filter whose impulse response is a rectangular pulse of width t_s

$$h(t) = \text{rect}[(t - \tfrac{1}{2}t_s)/t_s] \qquad H(f) = t_s\text{sinc}(ft_s)$$

It results in an excess attenuation especially in the range $\frac{1}{3}S_F$ to $\frac{1}{2}S_F$. We can compensate for this during the DSP itself if we use a digital filter to boost the response between $\frac{1}{3}S_F$ and $\frac{1}{2}S_F$.

12.8.5 *Implementation of Fractional Delays*

Fractional delays are difficult to generate using digital filters but may be implemented exactly using decimation and interpolation provided they are rational fractions of the form M/N. We interpolate the signal by N, pass it through a lowpass filter, delay the filtered signal by M samples, and, finally, decimate the resulting signal by N. Fractional delays are required in the design of some FIR filters (discussed in Chapter 14).

12.8.6 *Discrete Hilbert Transforms*

The discrete Hilbert transform of a sequence may also be obtained by using the FFT. The Hilbert transform of $x[n]$ involves the convolution of $x[n]$ with the impulse response $h[n]$ of the Hilbert transformer whose spectrum is $H(F) = -j\text{sgn}(F)$. The easiest way to perform this convolution is by FFT methods. We zero-pad $x[n]$ to make the length N a power of 2, find its N-point FFT, multiply by the *periodic extension* of N samples of $\text{sgn}(F)$ (which is $\{0, 1, 1, \ldots, 1, 0, -1, \ldots, -1, -1\}$, find the Hilbert transform as the inverse FFT, and multiply by the omitted factor $-j$.

12.8.7 *Deconvolution*

Given a signal $y[n]$ that represents the output of some system with impulse response $h[n]$, how do we recover the input $x[n]$ where $y[n] = x[n] \star h[n]$? One method is to undo the effects of convolution using deconvolution. The time-domain approach to deconvolution was studied in Chapter 4. Here, we examine a frequency-domain alternative based on the FFT.

The idea is to transform the convolution relation using the FFT to obtain $Y_T(F) = X_T(F)H_T(F)$, compute $X_T(F) = Y_T(F)/H_T(F)$ by pointwise division, and then find $x[n]$ as the IFFT of $X_T(F)$. This process does work in many cases, but it has two disadvantages. First, it fails if $H_T(F)$ equals zero at some frequency since we cannot then divide $Y_T(F)$ by $H_T(F)$. Second, it is quite sensitive to noise in the input $x[n]$, and to the accuracy with which $y[n]$ is known.

12.8.8   *Power Spectral Density from the DFT*

Recall that the power spectral density (PSD) of a signal $x(t)$ is a real, nonnegative, even function whose inverse transform is its autocorrelation function $R_{xx}(t)$ (Chapter 6).

$$S(f) = \lim_{T \to \infty} \frac{|X_T(f)|^2}{2T} \qquad R_{xx}(t) \longleftrightarrow S(f)$$

Recall also that $R_{xx}(0)$ equals the average power in the power signal $x(t)$.

If a signal $x(t)$ is sampled and available only over a finite duration, the best we can do is *estimate* the PSD of the underlying signal $x(t)$ from the given finite record. This is because the spectrum of finite sequences suffers from leakage and poor resolution. The simplest estimate of the PSD is the **periodogram estimate**. We compute the DFT magnitude $|X_T[k]|$, square it and normalize by N to obtain the PSD as

$$S[k] = \frac{1}{N}|X_T[k]|^2$$

Remark: For deterministic, bandlimited power signals sampled above the Nyquist rate, this yields good estimates. The periodogram of signals with random fluctuations can only be assessed in statistical terms and actually yields rather poor estimates of the true PSD.

Example 12.14
The Concept of the Periodogram

Consider a sinusoid sampled over one period at twice the Nyquist rate to give the sequence $x[n] = \{0, 1, 0, -1\}$. Its DFT equals $X_T[k] = \{0, -j2, 0, j2\}$. The periodogram is simply $S[k] = 0.25\{0, 4, 0, 4\} = \{0, 1, 0, 1\}$. This can be plotted as a bar graph with each sample occupying a bin-width $\Delta F = 1/N = 0.25$. The total signal power then equals $\Delta F \sum S[k] = 0.5$.

Comment: This result obviously matches the true power in the sinusoid.   ▬

12.8.9   *Time-Frequency Plots*

In practical situations, we are often faced with the task of finding how the spectrum of signals varies with time. A simple approach is to section the signal into overlapping segments, window each section to reduce leakage and find the FFT.

The FFT for each section is then staggered and stacked to generate what is called a **waterfall plot** or **time-frequency plot**. Figure 12.15 shows such a waterfall plot for a constant frequency signal and a chirp signal whose frequency varies linearly with time. It provides a much better visual indication of how the spectrum evolves with time.

Figure 12.15   Time-frequency (waterfall) plot for the sum of a 20-Hz signal and a chirp signal whose frequency varies from 60 Hz to 100 Hz in 1 s.

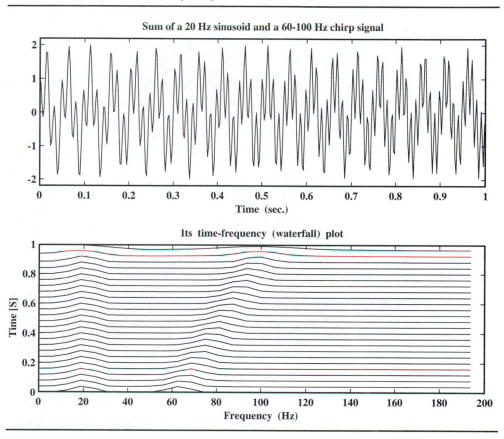

The fundamental restriction in obtaining the time-frequency information, especially for short time records, is that *we cannot localize both time and frequency to arbitrary precision*. Recent approaches are based on expressing signals in terms of **wavelets** (much like the Fourier series). Wavelets are functions which show the "best" possible localization characteristics in both time and frequency and are an area of intense ongoing research.

12.9
Special Topics:
Why Equal
Sequence
Lengths for the
DFT and IDFT?

The DTFT of an N-sample discrete time signal $x[n]$ is given by

$$X_p(f) = \sum_{n=0}^{N-1} x[n] \exp(-j2\pi n f)$$

If we sample f at M intervals over one period, the frequency interval f_0 equals $1/M$ and $f \longrightarrow k f_0 = k/M$ $(k = 0, 1, \ldots, M-1)$ and we get

$$X[k] = \sum_{n=0}^{N-1} x[n] \exp(-j2\pi n k/M) \qquad k = 0, 1, \ldots, M-1$$

With $W_M = \exp(-j2\pi/M)$, this equation can be written as

$$X[k] = \sum_{n=0}^{N-1} x[n] W_M^{nk} \qquad k = 0, 1, \ldots, M-1$$

This is a set of *M equations in N unknowns*, and describes *the M-sample DFT of the N-sample sequence $x[n]$*. It may be written in matrix form as

$$[\mathbf{X}] = [\mathbf{x}][\mathbf{W}_M]$$

Here, **X** is the product of a $(1 \times N)$ and an $(N \times M)$ matrix. In full form

$$[X_0 \quad X_1 \ldots X_{m-1}] = [x_0 \quad x_1 \ldots x_{N-1}] \begin{bmatrix} W_M^0 & W_M^0 & W_M^0 & \ldots & W_M^0 \\ W_M^0 & W_M^1 & W_M^2 & \ldots & W_M^{M-1} \\ W_M^0 & W_M^2 & W_M^4 & \ldots & W_M^{2(M-1)} \\ \vdots & \vdots & \vdots & \vdots & \vdots \\ W_M^0 & W_M^{N-1} & W_M^{2(N-1)} & \ldots & W_M^{(N-1)(M-1)} \end{bmatrix}$$

Example 12.15
A 4-Point DFT from a
3-Point Sequence

Let $x[n] = \{1, 2, 1\}$. We have $N = 3$. The DTFT of this signal is

$$X_p(F) = [1 + 2\exp(-j2\pi F) + \exp(-j4\pi F)] = \exp(-j2\pi F)[2 + 2\cos(2\pi F)]$$

If we use $M = 4$, we have $F = k/4$, $k = 0, 1, 2, 3$, and obtain the DFT as

$$X[k] = \exp(-j2\pi k/4)[2 + 2\cos(2\pi k/4)], \quad k = 0, 1, 2, 3 \quad \text{or} \quad X[k] = \{4, -j2, 0, j2\}$$

Using matrix notation with $W_M = \exp(-j2\pi/4) = -j$, $X[k]$ also equals

$$[X_0 \quad X_1 \quad X_2 \quad X_3] = [x_0 \quad x_1 \quad x_2] \begin{bmatrix} W_M^0 & W_M^0 & W_M^0 & W_M^0 \\ W_M^0 & W_M^1 & W_M^2 & W_M^3 \\ W_M^0 & W_M^2 & W_M^4 & W_M^6 \end{bmatrix} = [1 \quad 2 \quad 1] \begin{bmatrix} 1 & 1 & 1 & 1 \\ 1 & -j & -1 & j \\ 1 & -1 & 1 & -1 \end{bmatrix}$$

Upon evaluation, we find that $X[k] = \{4, -j2, 0, j2\}$

12.9.1   *The Inverse DFT*

How do we obtain the N-sample sequence $x[n]$ from the M-sample DFT? It would seem that we require the product of **X**, a $(1 \times M)$ matrix with an $(M \times N)$ matrix to give **x** as a $(1 \times N)$ matrix. What is this $(M \times N)$ matrix and is it related to the $(N \times M)$ matrix? To find out, we recall that

$$x[n] = \int_1 X_p(f) \exp(j2\pi nf)\, df$$

Converting this to discrete form with $f = kf_0 = k/M$ results in periodicity of $x[n]$ with period 1, and we obtain N samples of $x[n]$ using

$$x[n] = \frac{1}{M} \sum_{k=0}^{M-1} X[k] \exp(j2\pi nk/M) \qquad n = 0, 1, \ldots, N-1$$

Remark: For $N < M$, one period of $x[n]$ is a zero-padded M-sample sequence. For $N > M$, however, one period of $x[n]$ is the *periodic extension* of the N-sample sequence with period M.

The sign of the exponent and interchange of the indices n and k allow us to replace the $N \times M$ matrix $\mathbf{W}_M$ by the $M \times N$ matrix $[\mathbf{W}_M^T]^*$, leading to

$$x[n] = \frac{1}{M} \sum_{k=0}^{M-1} X[k][W_N^{kn}]^* \qquad n = 0, 1, \ldots, M-1$$

The matrix formulation of this requires the product of the $1 \times M$ matrix with an $(M \times N)$ matrix which simply corresponds to $1/M$ times $[\mathbf{W}_M^T]^*$, the **conjugate transpose** of the $(N \times M)$ DFT matrix. We thus have the matrix relations

$$[\mathbf{X}] = [\mathbf{x}][\mathbf{W}_M] \qquad [\mathbf{x}] = \frac{1}{M}[\mathbf{X}][\mathbf{W}_M^T]^*$$

These results are valid for any choice of M and N. An interesting result is that the product of $[\mathbf{W}_M]$ with $\frac{1}{M}[\mathbf{W}_M^T]^*$ is the $(N \times N)$ identity matrix.

Example 12.16
A 3-Point IDFT from a
4-Point DFT

With $X[k] = \{4, -j2, 0, j2\}$ and $M = 4$, the IDFT matrix equals

$$W_I = \frac{1}{4}\begin{bmatrix} 1 & 1 & 1 & 1 \\ 1 & -j & -1 & j \\ 1 & -1 & 1 & -1 \end{bmatrix}^{*T} = \frac{1}{4}\begin{bmatrix} 1 & 1 & 1 \\ 1 & j & -1 \\ 1 & -1 & 1 \\ 1 & -j & -1 \end{bmatrix}$$

We then get the IDFT as

$$x[n] = \frac{1}{4}\begin{bmatrix} 4 & -j2 & 0 & j2 \end{bmatrix}\begin{bmatrix} 1 & 1 & 1 \\ 1 & j & -1 \\ 1 & -1 & 1 \\ 1 & -j & -1 \end{bmatrix} = \begin{bmatrix} 1 & 2 & 1 \end{bmatrix}$$

The important thing to realize is that $x[n]$ is actually periodic with $M = 4$, and one period of $x[n]$ is the zero-padded sequence $\{1, 2, 1, 0\}$.

12.9.2   *How Unequal Lengths Affect the DFT Results*

Even though the M-point IDFT of an N-point sequence is valid for any M, the choice of M affects the nature of $x[n]$ through the IDFT and its inherent periodic extension.

1. If $M = N$, the IDFT is periodic with period M whose one period equals the N-sample $x[n]$. It yields square DFT and IDFT matrices and allows a simple inversion relation to go back and forth between the two forms.
2. If $M > N$, the IDFT is periodic with period M. Its one period is the original N-sample $x[n]$ with $M - N$ padded zeros. The choice $M > N$ is equivalent to using a zero-padded version of $x[n]$ with a total of M samples and $M \times M$ square matrices for both the DFT and IDFT matrix.
3. If $M < N$, the IDFT is periodic with period $M(< N)$. Its one period is the periodic extension of the N-sample $x[n]$ with period M. It thus yields the N-sample $x[n]$ *wrapped around* after M samples. This choice does not recover the original $x[n]$ and is to be avoided.

To appreciate these ideas, we reconsider the previous example.

Example 12.17
The Importance of Periodic Extension

Let $x[n] = \{1, 2, 1\}$. We have $N = 3$. The DTFT of this signal is

$$X_p(F) = [1 + 2\exp(-j2\pi F) + \exp(-j4\pi F)] = \exp(-j2\pi F)[2 + 2\cos(2\pi F)]$$

We sample $X_p(f)$ at M intervals and find the IDFT as $y[n]$. What do we get?

Case 1: For $M = 3$, we should get $y[n] = \{1, 2, 1\} = x[n]$. Let us find out.

With $M = 3$, we have $F = k/3$ for $k = 0, 1, 2$, and $X_p[k]$ becomes

$$X[k] = \exp(-j2\pi k/3)[2 + 2\cos(2\pi k/3)] = \left\{4, -\tfrac{1}{2} - j\sqrt{\tfrac{3}{4}}, -\tfrac{1}{2} + j\sqrt{\tfrac{3}{4}}\right\}$$

We also find $\mathbf{W}_I$ and the IDFT $\mathbf{x} = [\mathbf{X}][\mathbf{W}_I]$ as

$$\mathbf{W}_I = \frac{1}{3}\begin{bmatrix} W_M^0 & W_M^0 & W_M^0 \\ W_M^0 & W_M^1 & W_M^2 \\ W_M^0 & W_M^2 & W_M^4 \end{bmatrix}^{*T} = \frac{1}{3}\begin{bmatrix} 1 & 1 & 1 \\ 1 & -\tfrac{1}{2} + j\sqrt{\tfrac{3}{4}} & -\tfrac{1}{2} - j\sqrt{\tfrac{3}{4}} \\ 1 & -\tfrac{1}{2} - j\sqrt{\tfrac{3}{4}} & -\tfrac{1}{2} + j\sqrt{\tfrac{3}{4}} \end{bmatrix}$$

$$x[n] = \frac{1}{3}\begin{bmatrix} 4 & -\tfrac{1}{2} - j\sqrt{\tfrac{3}{4}} & -\tfrac{1}{2} + j\sqrt{\tfrac{3}{4}} \end{bmatrix}\begin{bmatrix} 1 & 1 & 1 \\ 1 & -\tfrac{1}{2} + j\sqrt{\tfrac{3}{4}} & -\tfrac{1}{2} - j\sqrt{\tfrac{3}{4}} \\ 1 & -\tfrac{1}{2} - j\sqrt{\tfrac{3}{4}} & -\tfrac{1}{2} + j\sqrt{\tfrac{3}{4}} \end{bmatrix} = \begin{bmatrix} 1 & 2 & 1 \end{bmatrix}$$

This result is periodic with $M = 3$, and one period of this equals $x[n]$.

Case 2: For $M = 4$, we should get $y[n] = \{1, 2, 1, 0\}$, a zero-padded version. We do (the details were worked out in the previous example).

Case 3: For $M = 2$, we should get a periodic extension of $x[n]$ with period 2. This should correspond to the sequence $y[n] = \{2, 2\}$.

With $M = 2$ and $k = 0, 1$, we have $X_p[k] = \exp(-j\pi k)[2 + 2\cos(\pi k)] = \{4, 0\}$. Since $\exp(-j2\pi/M) = \exp(-j\pi) = -1$, we find the IDFT from the definition as

$$y[0] = \tfrac{1}{2}\{X[0] + X[1]\} = 2 \qquad y[1] = \tfrac{1}{2}\{X[0] - X[1]\} = 2$$

The sequence $y[n] = \{2, 2\}$ is periodic with $M = 2$. As expected, this equals one period of $x[n] = \{1, 2, 1\}$ with wraparound past two samples.

12.10 **The DFT and FFT: A Synopsis**	

The DFS and DFT	The DFS and DFT describe the spectra of finite sequences. A sequence and its DFT or DFS are periodic and discrete. The N-sample DFT and DFS are related by $N\{\text{DFS}\} = \{\text{DFT}\}$. Their computation involves only N and the sample values. The IDFS and IDFT have similar forms and can even be obtained from the DFT.
Properties	The DFT of a real sequence possesses conjugate symmetry about the origin and about the folding index $k = \tfrac{1}{2}N$. The spectral spacing depends only on the sampling duration D and is $f_0 = 1/D$. The DFT is periodic with period $S_F = 1/t_s$. The highest frequency present is $\tfrac{1}{2}S_F$ at the folding index. Replicating a time sequence produces a zero interpolated version of its DFT.
Pure sinusoids	The DFS of a pure sinusoid sampled above the Nyquist rate for m periods has a DFT whose nonzero values occur at the index m and $N - m$. Only the number of DFS samples is dictated by the sampling frequency S_F.
Bandlimited signals	For a bandlimited periodic signal with M harmonics, the sampling rate must exceed $2M$ samples per period for exact DFS results.
Aliasing and leakage	If we sample below the Nyquist rate, the DFS results show aliasing effects. To avoid aliasing, we sample above the Nyquist rate. If we sample for noninteger periods, its periodic extension describes a different signal and results in leakage. To avoid leakage we sample for an integer number of periods. To obtain the first M harmonics with an error of about 5%, we should choose at least $8M$ samples per period.
DFT of arbitrary signals	The DFT of a sampled analog signal is an approximation to its FT $X(f)$, with $X(kS_F) \approx t_s\{\text{DFT}\}$. Sampling for a longer duration reduces the spectral spacing and improves frequency resolution, whereas a higher sampling rate extends the highest frequency in the DFT and minimizes aliasing.
Timelimited signals	For timelimited signals a longer duration requires zero padding, which adds no new signal information. Accuracy is governed only by the sampling rate. One choice of sampling rate is based on minimizing the energy between the actual signal and its sampled version. The duration must be long enough to cover most of the signal. Most signals are truncated in practice.

Truncation	Truncation leads to spectral wiggles much like the Gibbs effect. To smooth out the spectrum, we use tapered time windows. An N-point DFT window is not symmetric. It is an $(N + 1)$-point window with its last sample discarded.
FFT	The FFT describes efficient algorithms for computing the DFT. These reduce the burden of calculating an N-point DFT to that of calculating many smaller-size DFTs. The same algorithms can also be used to generate IFFT.
Interpolation	Signal interpolation using the FFT is based on zero padding. It is the dual of improving the spectral resolution by zero padding the time sequence.
Fast convolution	Fast periodic convolution requires finding the FFT of each sequence, multiplying and then finding the IFFT. Zero padding before FFT allows us to find the regular convolution.
Applications	Applications of the FFT include interpolation, decimation, multirate signal processing, implementing fractional delays, deconvolution, power spectrum estimation, discrete Hilbert transforms and FIR filter design.

Appendix 12A MATLAB Demonstrations and Routines

Getting Started on the MATLAB Demonstrations

This appendix presents MATLAB routines related to concepts in this chapter, and some examples of their use. See Appendix 1A for introductory remarks.

DFT Computation and Spectra

```
fft(x)                       %The FFT of x is a built-in MATLAB function
fft(x,N)                     %N-point FFT (truncates or zeropads x to N if needed)
[zm,zp]=fftplot(x,ts)
```

Returns and plots magnitude *zm* and phase *zp* of the DFT of *x*. The sampling interval *ts* is optional. If not specified, *ts* defaults to 1 and the spectra are plotted versus the digital frequency *F*.

Example

Create some data and plot its FFT spectra as follows:

```
>>t=0:.05:1;y=sin(pi*t);fftplot(y,0.05);
```

Demonstrations and Utility Functions

```
>>demosamp                   %Demonstration of sampling, aliasing and leakage
>>aliasing                   %Demonstrates a movie of aliasing and leakage
```

```
y=interpol(x,'d',n)      %Signal interpolation using DFT
```

Interpolates signal x using DFT interpolation at n points between samples. String type 'd' is for DFT interpolation and may be replaced by other types (for more details see Appendix 11A).

Example

Sample a sine at four points, interpolate at eight points between samples using DFT interpolation, plot the interpolated and actual samples:

```
>>tx=0:0.5:1.5;                     %4 values of tx
>>x=sin(pi*tx);y=interpol(x,'d',8) %Signal x and interpolated version y
>>ty=[0:length(y)-1]*0.5/8          %Time axis for interpolated signal
>>lines(ty,y,'.'),hold on,lines(tx,x,'o'),hold off
```

PROBLEMS

P12.1 (DFT from Definition) Compute the DFT and DFS of the following:
(a) $\{x[n]\} = \{1, 2, 1, 2\}$
(b) $\{x[n]\} = \{2, 1, 3, 0, 4\}$
(c) $\{x[n]\} = \{2, 2, 2, 2\}$
(d) $\{x[n]\} = \{1, 0, 0, 0, 0, 0, 0, 0, 0\}$

P12.2 (IDFT from Definition) Compute the IDFT of the following:
(a) $X_T[k] = \{2, -j, 0, j\}$
(b) $X_T[k] = \{4, -1, 1, 1, -1\}$
(c) $X_T[k] = \{1, 2, 1, 2\}$
(d) $X_T[k] = \{1, 0, 0, j, 0, -j, 0, 0\}$

P12.3 (Symmetry) In the following DFT sequences, the missing values are shown by X. Determine the complete DFT sequence in each case.
(a) $X_T[k] = \{0, X_1, 2 + j, -1, X_4, j\}$
(b) $X_T[k] = \{1, 2, X_2, X_3, 0, 1 - j, -2, X_7\}$

P12.4 (Properties) Consider the DFT pair $\{x[n]\} \longleftrightarrow \{1, 2, 3, 4\}$. Find the DFT of each of the following sequences using properties.
(a) $\{x[n - 2]\}$
(b) $\{x[n + 6]\}$
(c) $\{x[n + 1]\}$
(d) $\{\exp[j\frac{1}{2}\pi n]x[n]\}$
(e) $\{x[n]\bullet x[n]\}$
(f) $\{x^2[n]\}$
(g) $\{x[-n]\}$
(h) $\{x^*[n]\}$
(i) $x^2[-n]$

P12.5 (Properties) Let $X_T[k]$ be the N-point DFT of a real signal. How many DFT samples will always be real and what will be their index k? [HINT: Use the concept of conjugate symmetry and consider the cases for odd N and even N separately.]

P12.6 (Properties) Consider the DFT pairs shown, where the missing values are shown by X and x. Use properties such as conjugate symmetry and Parseval's theorem to fill in all missing values.
(a) $\{x_0, 3, -4, 0, 2\} \longleftrightarrow \{5, X_1, -1.28 - j4.39, X_3, 8.78 - j1.4\}$
(b) $\{x_0, 3, -4, 2, 0, 1\} \longleftrightarrow \{4, X_1, 4 - j5.2, X_3, X_4, 4 - j1.73\}$

P12.7 (Properties) Let $\{X\}$ be the DFT of a signal $x[n]$. If we conjugate $\{X\}$ and take its DFT again to obtain $\{Y\}$, how is the sequence $\{Y\}$ related to $x[n]$?

P12.8 (Replication) Consider the DFT pair $\{x[n]\} \longleftrightarrow \{1, 2, 3, 4, 5\}$.
(a) What is the DFT of the sequence $\{x[n], x[n]\}$?
(b) What is the DFT of the sequence $\{x[n], x[n], x[n]\}$?
(c) What is the DFT of the zero-interpolated sequence $\{x[n/2]\}$?
(d) What is the DFT of the zero-interpolated sequence $\{x[n/3]\}$?

P12.9 (Pure Sinusoids) Determine the DFS and DFT sequence of $\sin(2\pi f_0 t + \frac{1}{3}\pi)$ (without doing any DFT computations) if we acquire
(a) 4 samples over 1 period.   **(b)** 8 samples over 2 periods.
(c) 8 samples over 1 period.   **(d)** 18 samples over 3 periods.
(e) 8 samples over 5 periods.   **(f)** 16 samples over 10 periods.

P12.10 (Aliasing of Sinusoids) The following signals are sampled starting at $t = 0$. Find the DFS and DFT sequence for each. List the indices that correspond to aliased components (if any)
(a) $\cos(4\pi t)$ sampled at 25 Hz for 2 periods.
(b) $\cos(20\pi t) + 2\sin(40\pi t)$ sampled at 25 Hz with $N = 15$.
(c) $\sin(10\pi t) + 2\sin(40\pi t)$ sampled at 25 Hz for 1 s.
(d) $\sin(40\pi t) + 2\sin(60\pi t)$ sampled at intervals of 0.004 s for 4 periods.

P12.11 (Aliasing and Leakage) The following signals are sampled and the DFT of the sampled signal obtained. Which cases will show aliasing, or leakage, or both? In which cases can the effects of leakage and/or aliasing be avoided and how?
(a) A 100 Hz sinusoid for 2 periods at 400 Hz.
(b) A 100 Hz sinusoid for 4 periods at 70 Hz.
(c) A 100 Hz sinusoid at 400 Hz using 10 samples.
(d) A 100 Hz sinusoid for 2.5 periods at 70 Hz.
(e) A sum of sinusoids at 100 Hz and 150 Hz for 100 ms at 450 Hz.
(f) A sum of sinusoids at 100 Hz and 150 Hz for 1 period at 250 Hz.
(g) A sum of sinusoids at 100 Hz and 150 Hz for 0.5 periods at 400 Hz.
(h) A square wave with period $T = 10$ ms for 2 periods at 400 Hz.
(i) A square wave with period $T = 10$ ms for 1.5 periods at 400 Hz.

P12.12 (Resolution) What is the frequency resolution in the 500-point DFT of a sampled signal obtained by sampling an analog signal at 1 kHz?

P12.13 (Resolution) We wish to sample a signal of 1 s duration, bandlimited to 50 Hz, and compute the DFT of the sampled signal.
(a) Using the minimum sampling rate that avoids aliasing, what is the spectral resolution Δf and how many samples are acquired?
(b) How many padding zeros are needed to improve the resolution to $\frac{1}{2}\Delta f$ using the minimum sampling rate to avoid aliasing?

P12.14 (Resolution) For each of the following signals, estimate the sampling frequency and sampling duration by arbitrarily choosing

(1) The bandwidth as the frequency where $|H(f)|$ is 5% of its maximum.
(2) The signal width as the time at which $x(t)$ is 1% of its maximum.
(a) $x(t) = \exp(-t)u(t)$ **(b)** $x(t) = t\exp(-t)u(t)$ **(c)** $x(t) = \text{tri}(t)$

P12.15 (DFT from Definition) Use the defining relation to compute the N-point DFT of the following:
(a) $\delta[n], 0 \le n \le N-1$ **(b)** $a^n, 0 \le n \le N-1$
(c) $\exp(j\pi n/N), 0 \le n \le N-1$

P12.16 (DFT Concepts)
(a) Use the defining relation to compute the N-point DFT of the sinusoid $x[n] = \cos(2n\pi F_0)$.
(b) For what values of F_0 would you expect aliasing and/or leakage?
(c) Use the above results to compute the DFT if $N = 8$ and $F_0 = 0.25$.
(d) Use the above results to compute the DFT if $N = 8$ and $F_0 = 1.25$.
(e) Use the above results to compute the DFT if $N = 9$ and $F_0 = 0.2$.

P12.17 (Convolution Using DFT) Consider two sequences $x[n]$ and $h[n]$ of length 12 samples and 20 samples, respectively.
(a) How many padding zeros are needed for $x[n]$ and $h[n]$ to find their regular convolution $y[n]$ using the DFT?
(b) If we pad $x[n]$ with 8 zeros and find the periodic convolution $y_p[n]$ of the resulting 20-point sequence with $h[n]$, for what indices will the samples of $y[n]$ and $y_p[n]$ be identical?

P12.18 (FFT) Write out the DFT sequence that corresponds to the following bit-reversed sequences obtained using the DIF FFT algorithm
(a) $\{1, 2, 3, 4\}$ **(b)** $\{0, -1, 2, -3, 4, -5, 6, -7\}$

P12.19 (FFT) Set up a flowchart showing all twiddle factors and values at intermediate nodes to compute the DFT of $x[n] = \{1, 2, 2, 2, 1, 0, 0, 0\}$
(a) Using the 8-point DIF algorithm.
(b) Using the 8-point DIT algorithm.

P12.20 (Resolution and the FFT) We wish to sample a signal of 1 s duration, bandlimited to 100 Hz and compute its spectrum. The resolution is to be 0.5 Hz or better. What is the minimum number of samples needed and the actual spectral resolution if we use
(a) the DFT to compute the spectrum?
(b) the FFT to compute the spectrum?

P12.21 (Convolution) Find the linear convolution of the sequences $x[n] = \{1, 2, 1\}$ and $h[n] = \{1, 2, 3\}$ using
(a) The convolution operation.
(b) The DFT and zero-padding.
(c) The FFT and zero-padding.

P12.22 (Convolution) Find the periodic convolution of the sequences $x[n] = \{1, 2, 1\}$ and $h[n] = \{1, 2, 3\}$ using
(a) The convolution operation.

(b) The DFT.

(c) The FFT and zero-padding. Does this match the previous results? Explain.

P12.23 (Correlation) Find the periodic correlation r_{pxh} of the sequences $x[n] = \{1, 2, 1\}$ and $h[n] = \{1, 2, 3\}$ using

(a) The convolution or correlation operation.

(b) The DFT.

P12.24 (Convolution of Long Sequences) Consider the sequences $x[n] = \{1, 2, 1\}$ and $h[n] = \{1, 2, 1, 3, 2, 2, 3, 0, 1, 0, 2, 2\}$.

(a) Find their convolution using the overlap-add method.

(b) Find their convolution using the overlap-save method.

(c) Compare with the convolution obtained by the usual means.

13 THE z-TRANSFORM

13.0
Scope and Objectives

The z-transform plays the same role for discrete-time signals and systems as does the Laplace transform for continuous-time signals and systems. This chapter deals with the z-transform, its properties and some applications in systems analysis and signal processing. Our description of the z-transform closely parallels that of Laplace transforms, the better to highlight the similarities between the two.

The genesis of the z-transform may be viewed at several levels. We develop it as an independent transformation method in order to keep the discussion self-contained. We explore the connections between the various transform methods toward the end of this chapter.

This chapter is divided into four parts:

Part 1 The z-transform, its properties and the inverse z-transform.
Part 2 System analysis using the z-transform.
Part 3 System interconnections and realizations.
Part 4 Connections between the z-transform and convolution and also between the z-transform and all other transform methods studied so far.

Useful Background   The following topics provide a useful background:

1. The method of partial fraction expansion (Chapter 8).
2. Discrete convolution and stability of discrete systems (Chapter 4).
3. Sampling, the DTFT and frequency response (Chapter 11).

The **twosided *z*-transform** $X(z)$ of a discrete signal $x[n]$ is defined as

$$X(z) = \sum_{k=-\infty}^{\infty} x[k]z^{-k}$$

Here, the complex quantity z generalizes the concept of digital frequency F or Ω to the complex domain and is usually described in polar form as

$$z = |r| \exp(j2\pi F) = |r| \exp(j\Omega)$$

Values of z can be plotted on an Argand diagram called the **z-plane**.

The defining relation is a power series (Laurent series) in z. The term for each index k is the product of the sample value $x[k]$ and z^{-k}.

For the sequence $\{x[n]\} = \{-7, 3, 1, 4, -8, 5\}$, the z-transform can be written
$\uparrow$

$$X(z) = -7z^2 + 3z^1 + z^0 + 4z^{-1} - 8z^{-2} + 5z^{-3}$$

The relation between $x[n]$ and $X(z)$ is denoted symbolically by

$$x[n] \longleftrightarrow X(z)$$

Here, $x[n]$ and $X(z)$ form a transform pair, and the double arrow implies a one-to-one correspondence between the two.

For our example, we can, therefore, write

$$\{-7, 3, 1, 4, -8, 5\} \longleftrightarrow -7z^2 + 3z^1 + z^0 + 4z^{-1} - 8z^{-2} + 5z^{-3}$$
$\uparrow$

Comparing $\{x[n]\}$ and $X(z)$, we observe that the quantity z^{-1} plays the role of a unit delay operator. The sample location $n = 2$ in $\{x[n]\}$, for example, corresponds to the term with z^{-2} in $X(z)$. In concept, then, it is not hard to go back and forth between a sequence and its z-transform if all we are given is a finite number of samples.

For causal signals, we also define the **onesided *z*-transform** by changing the lower limit in the summation to 0. The onesided z-transform is given by

$$X(z) = \sum_{k=0}^{\infty} x[k]z^{-k}$$

Remark: Technically, for sampled signals, the z-transform should include the multiplier t_s in the summation, but this is not common practice. The sampling interval t_s appears only indirectly in that z^{-1} actually corresponds to a delay of *one sampling interval t_s*.

13.1.1   *The Region of Convergence (ROC)*

Since $X(z)$ is a power series, it may not converge for all z. The values of z for which it does converge, define the region of convergence (ROC) for $X(z)$.

Two completely different sequences may produce the same *twosided z*-transform $X(z)$, but with different regions of convergence. It is important (unlike Laplace transforms) that we specify the ROC associated with each $X(z)$, especially when dealing with the twosided *z*-transform.

13.1.2   The z-Transform of Finite Sequences

For finite sequences, the *z*-transform may be written as a polynomial in *z*. For sequences with a large number of terms, the polynomial form can get to be unwieldy unless we can find closed-form solutions.

Region of convergence (ROC) of finite sequences   Since $X(z)$ represents a polynomial in *z* for finite sequences, it converges or exists (is finite) for all *z* except at

$z = 0$, if only terms of the form z^{-k} are present

$z = \infty$, if only terms of the form z^k are present

$z = 0$ and $z = \infty$, if terms of the form z^{-k} and z^k are present together

Thus, the ROC for finite sequences is the entire *z*-plane, except perhaps for $z = 0$ and/or $z = \infty$, as applicable.

13.1.3   Some z-Transform Pairs Using the Defining Relation

Table 13.1 (p. 564) lists the *z*-transforms of some useful signals. We provide some examples using the defining relation to find *z*-transforms.

Table 13.1 (p. 564)

Example 13.1
z-Transform of Sequences

(a) *The unit impulse*: Consider the signal $x[n] = \delta[n]$. Its *z*-transform is $X(z) = 1$. The ROC is the entire *z*-plane.

(b) *An arbitrary sequence*: Let $x[n] = 2\delta[n + 1] + \delta[n] - 5\delta[n - 1] + 4\delta[n - 2]$. This describes the sequence $\{x[n]\} = \{2, 1, -5, 4\}$. Its *z*-transform is evaluated
$\overset{\uparrow}{}$
as $X(z) = 2z + 1 - 5z^{-1} + 4z^{-2}$. No simplifications are possible. The ROC is the entire *z*-plane except $z = 0$ and $z = \infty$ (or $0 < |z| < \infty$).

(c) *A rectangular pulse*: Consider $x[n] = u[n] - u[n - N]$. Its *z*-transform is the finite sum $X(z) = 1 + z^{-1} + z^{-2} + \cdots + z^{-(N-1)}$. Its closed-form solution is

$$X(z) = \sum_{k=0}^{N-1} z^{-k} = (1 - z^{-N})/(1 - z^{-1}) \quad z \neq 1 \qquad \text{ROC: } z \neq 0$$

The ROC is the entire *z*-plane except $z = 0$ (or $|z| > 0$). Note that if $z = 1$, we have $X(z) = N$.

Example 13.2
z-Transforms and
Their ROC

(a) *The unit step*: For the unit step $u[n]$, we evaluate the *z*-transform using the defining relation, and the sum of an infinite geometric series, as

$$X(z) = \sum_{k=0}^{\infty} z^{-k} = \sum_{k=0}^{\infty} (z^{-1})^k = 1/(1 - z^{-1}) = z/(z - 1) \qquad \text{ROC: } |z| > 1$$

Note that this geometric series converges only for $|z^{-1}| < 1$ or $|z| > 1$, which defines the ROC for this $X(z)$.

Table 13.1   Some z-Transform Pairs.

DT signal	z-transform	ROC
Finite sequences		
$\delta[n]$	1	all z
$u[n] - u[n - N]$	$(1 - z^{-N})/(1 - z^{-1})$	$z \neq 0$
Causal signals		
$u[n]$	$z/(z - 1)$	$\|z\| > 1$
$\alpha^n u[n]$	$z/(z - \alpha)$	$\|z\| > \|\alpha\|$
$(-\alpha)^n u[n]$	$z/(z + \alpha)$	$\|z\| > \|\alpha\|$
$nu[n]$	$z/(z - 1)^2$	$\|z\| > 1$
$n\alpha^n u[n]$	$z\alpha/(z - \alpha)^2$	$\|z\| > \|\alpha\|$
$(n + 1)\alpha^n u[n]$	$z^2/(z - \alpha)^2$	$\|z\| > \alpha\|$
$\cos(n\Omega)u[n]$	$\dfrac{z^2 - z\cos(\Omega)}{z^2 - 2z\cos(\Omega) + 1}$	$\|z\| > 1$
$\sin(n\Omega)u[n]$	$\dfrac{z\sin(\Omega)}{z^2 - 2z\cos(\Omega) + 1}$	$\|z\| > 1$
$\alpha^n \cos(n\Omega)u[n]$	$\dfrac{z^2 - \alpha z\cos(\Omega)}{z^2 - 2\alpha z\cos(\Omega) + \alpha^2}$	$\|z\| > \|\alpha\|$
$\alpha^n \sin(n\Omega)u[n]$	$\dfrac{\alpha z\sin(\Omega)}{z^2 - 2\alpha z\cos(\Omega) + \alpha^2}$	$\|z\| > \|\alpha\|$
$n^2 u[n]$	$z(z + 1)/(z - 1)^3$	$\|z\| > 1$
$\alpha^n n^2 u[n]$	$z\alpha(z + \alpha)/(z - \alpha)^3$	$\|z\| > \|\alpha\|$
Anticausal signals		
$-u[-n - 1]$	$z/(z - 1)$	$\|z\| < 1$
$-nu[-n - 1]$	$z/(z - 1)^2$	$\|z\| < 1$
$-\alpha^n u[-n - 1]$	$z/(z - \alpha)$	$\|z\| < \|\alpha\|$
$-n\alpha^n u[-n - 1]$	$z\alpha/(z - \alpha)^2$	$\|z\| < \|\alpha\|$

(b) *The exponential*: The z-transform of $x[n] = \alpha^n u[n]$, with real α, is

$$X(z) = \sum_{k=0}^{\infty} \alpha^k z^{-k} = \sum_{k=0}^{\infty} (\alpha/z)^k = 1/[1 - (\alpha/z)] = z/(z - \alpha)$$

Its ROC is $|z| > |\alpha|$.

13.1.4   *More on the Region of Convergence*

Consider the signal $y[n] = -\alpha^n u[-n - 1] = -1$, $n = -1, -2, \ldots$. The twosided z-transform of $y[n]$, using a change of variables, can be written as

$$Y(z) = \sum_{k=-\infty}^{-1} -\alpha^k z^{-k} = \sum_{m=1}^{\infty} -\alpha^{-m} z^m = \sum_{m=1}^{\infty} -(z/\alpha)^m = -(z/\alpha)/[1 - (z/\alpha)] = z/(z - \alpha)$$

The ROC of $Y(z)$ is $|z| < |\alpha|$. The z-transform of $\alpha^n u[n]$ is also $z/(z - \alpha)$, but with an ROC of $|z| > |\alpha|$. Do you see the problem? We cannot uniquely identify a signal from its transform alone, unless we also specify the ROC.

The ROC actually depends on the one- or twosidedness of $x[n]$ as summarized in Table 13.2 and Figure 13.1. The results are based on the fact that if $X(z)$ is a rational function in z, as is often the case, its ROC must exclude all pole locations (denominator roots) where $X(z)$ becomes infinite. The ROC of causal signals lies exterior to a circle of radius $|p|_{\max}$, the magnitude of the largest pole.

Table 13.2   ROC of the z-Transform.

Signal $x[n]$	ROC of $X(z)$
Finite length	All z except $z = 0$ and/or $z = \infty$
Rightsided (or causal)	Exterior to a circle, $|z| > |p|_{\max}$, where $|p|_{\max}$ is the largest pole magnitude in $X(z)$.
Leftsided (or anticausal)	Interior to a circle, $|z| < |p|_{\min}$, where $|p|_{\min}$ is the smallest pole magnitude in $X(z)$.
Twosided (or noncausal)	Annulus covering the region $|p_{\min}| < |z| < |p_{\max}|$

Figure 13.1   ROC of the z-transform of discrete signals.

ROC of right-sided sequences   ROC of left-sided sequences   ROC of two-sided sequences

By way of an example, consider $X(z) = z/(z - 2) + z/(z - 3)$. What is the ROC? We look for a region of convergence common to both terms. For a causal signal, the ROC becomes $|z| > 3$. For an anticausal signal, the ROC is $|z| < 2$. For a twosided (noncausal) signal, the ROC is $2 < |z| < 3$.

Remarks: **1.** *We shall assume causal signals if no ROC is specified.*

2. In Table 13.2, we use inequalities of the form $|z| < |\alpha|$, or $|z| > |\alpha|$ for the ROC, since $X(z)$ may not converge at the boundary $|z| = |\alpha|$.

13.2
Properties of the z-Transform

Even though most transform pairs can be derived using the defining relation, an easier way is to invoke properties of the z-transform listed in Table 13.3. The proofs of these properties are based on the linear nature of the z-transform operation.

For the **superposition property**, we start with $y[n] = x_1[n] + x_2[n]$. Then

$$Y(z) = \sum_{k=-\infty}^{\infty} \{x_1[k] + x_2[k]\}z^{-k} = \sum_{k=-\infty}^{\infty} x_1[k]z^{-k} + \sum_{k=-\infty}^{\infty} x_2[k]z^{-k} = X_1(z) + X_2(z)$$

Table 13.3   Operational Properties of the z-Transform.

Property	z-transform result
Superposition	$x[n] + y[n] \longleftrightarrow X(z) + Y(z)$
Shifting	$x[n-1] \longleftrightarrow z^{-1}X(z)$ (For twosided signals)
	$x[n-N] \longleftrightarrow z^{-N}X(z)$ (For twosided signals)
Left shift	$x[n+1] \longleftrightarrow zX(z) - zx[0]$
	$x[n+2] \longleftrightarrow z^2X(z) - z^2x[0] - zx[1]$
	$x[n+N] \longleftrightarrow z^NX(z) - z^Nx[0] - z^{N-1}x[1] - \cdots - zx[N-1]$
Right shift	$x[n-1] \longleftrightarrow z^{-1}X(z) + x[-1]$
	$x[n-2] \longleftrightarrow z^{-2}X(z) + z^{-1}x[-1] + x[-2]$
	$x[n-N] \longleftrightarrow z^{-N}X(z) + z^{-(N-1)}x[-1] + z^{-(N-2)}x[-2] + \cdots + x[-N]$
Reflection	$x[-n] \longleftrightarrow X(1/z)$
	$x[-n]u[-n-1] \longleftrightarrow X(1/z) - x[0]$ where $x[n]u[n] \longleftrightarrow X(z)$
Scaling	$\alpha^n x[n] \longleftrightarrow X(z/\alpha)$
Times scalar	$Kx[n] \longleftrightarrow KX(z)$
Times-n	$nx[n] \longleftrightarrow -z\dfrac{dX(z)}{dz}$
Times-n^2	$n^2x[n] \longleftrightarrow -z\dfrac{d}{dz}\left(-z\dfrac{dX(z)}{dz}\right)$
Times-n^p	$n^p x[n] \longleftrightarrow -z\dfrac{d}{dz}\left(-z\dfrac{d}{dz}\left(\ldots p \text{ times} \ldots \left(-z\dfrac{dX(z)}{dz}\right)\ldots\right)\right)$
Times-cos	$\cos(n\Omega)x[n] \longleftrightarrow \frac{1}{2}\left[X\{z\exp(j\Omega)\} + X\{z\exp(-j\Omega)\}\right]$
Times-sin	$\sin(n\Omega)x[n] \longleftrightarrow j\frac{1}{2}\left[X\{z\exp(j\Omega)\} - X\{z\exp(-j\Omega)\}\right]$
Convolution	$x[n] \star y[n] \longleftrightarrow X(z)Y(z)$
Difference	$x[n] - x[n-1] \longleftrightarrow (1 - z^{-1})X(z)$
Summation	$\displaystyle\sum_{k=-\infty}^{n} x[k] \longleftrightarrow \dfrac{zX(z)}{z-1}$
Switched periodic	$x_p[n]u[n] \longleftrightarrow \dfrac{1}{1-z^{-N}}X_1(z)$ $\{x_1[n] = \text{first period of } x_p\}$
Initial value theorem	$x[0] = \lim_{z\to\infty} X(z)$
Final value theorem	$\lim_{n\to\infty} x[n] = \lim_{z\to 1}(z-1)X(z)$

13.2.1 The Right-Shift Property of the Twosided z-Transform

To prove the right-shift property, we use a change of variables. We start with the pair $x[n] \longleftrightarrow X(z)$. If $y[n] = x[n - N]$, its z-transform is

$$Y(z) = \sum_{k=-\infty}^{\infty} x[k - N]z^{-k}$$

With the change of variable $m = k - N$, the new summation index m still ranges from $-\infty$ to ∞ (since N is finite), and we obtain

$$Y(z) = \sum_{m=-\infty}^{\infty} x[m]z^{-(m+N)} = z^{-N} \sum_{m=-\infty}^{\infty} x[m]z^{-m} = z^{-N}X(z)$$

The factor z^{-N} with $X(z)$ induces a right-shift of N in $x[n]$. This property holds for both the onesided and twosided transforms.

13.2.2 The Right-Shift Property of the Onesided z-Transform

The onesided z-transform of a sequence $x[n]$ and its causal version $x[n]u[n]$ are identical. A right-shift of $x[n]$ brings samples for $n < 0$ into the range $n \geq 0$, as illustrated in Figure 13.2 (p. 568), and leads to the z-transforms

$$x[n - 1] \longleftrightarrow z^{-1}X(z) + x[-1] \qquad x[n - 2] \longleftrightarrow z^{-2}X(z) + z^{-1}x[-1] + x[-2]$$
$$x[n - N] \longleftrightarrow z^{-N}X(z) + z^{-(N-1)}x[-1] + z^{-(N-2)}x[-2] + \cdots + x[-N]$$

For causal signals ($x[n] = 0, n < 0$), this result reduces to $x[n - N] \longleftrightarrow z^{-N}X(z)$.

Example 13.3
Using the Right-Shift Property

Consider $x[n] = \alpha^n$. Its onesided z-transform is identical to that of $\alpha^n u[n]$ and equals $X(z) = z/(z - \alpha)$. If $y[n] = x[n - 1]$, the right-shift property, with $N = 1$, yields $Y(z) = z^{-1}X(z) + x[-1] = 1/(z - \alpha) + \alpha^{-1}$.

Comment: The additional term α^{-1} arises because $x[n]$ is not causal. If $x[n] = \alpha^n u[n]$, $y[n] = x[n - 1] = \alpha^{n-1}u[n - 1]$ and $Y(z) = z^{-1}X(z) = 1/(z - \alpha)$.

13.2.3 The Left-Shift Property of the Onesided z-Transform

With $x[n]u[n] \longleftrightarrow X(z)$, the transform of $y[n] = x[n + 1]u[n + 1]$ equals

$$Y(z) = \sum_{k=0}^{\infty} x[k + 1]z^{-k}$$

With $m = k + 1$, we can rewrite this expression as

$$Y(z) = \sum_{m=1}^{\infty} x[m]z^{-m} = z \sum_{m=1}^{\infty} x[m]z^{-m}$$
$$= z \sum_{m=0}^{\infty} x[m]z^{-m} - zx[0] = zX(z) - zx[0]$$

Figure 13.2 **Illustrating the right-shift property of the z-transform.**

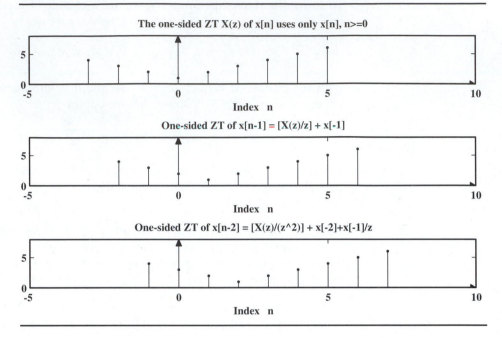

By successively shifting $x[n]$ to the left, we obtain the general relation

$$X[n + N] \longleftrightarrow z^N X(z) - z^N x[0] - z^{N-1} x[1] - \cdots - zx[N-1]$$

Remark: The right-shift and left-shift properties are useful in the solution of difference equations using onesided z-transforms.

Example 13.4
Using the Left-Shift Property

(a) Consider the shifted step $u[n + 1]$. Its onesided z-transform should be identical to that of $u[n]$ since $u[n]$ and $u[n + 1]$ are identical for $n \geq 0$.
With $u[n] \longleftrightarrow z/(z - 1)$ and $u[0] = 1$, the left-shift property gives

$$u[n + 1] \longleftrightarrow z[z/(z - 1)] - z = z/(z - 1)$$

(b) Consider the signal $x[n] = \alpha^{n+1} u[n + 1]$. With $y[n] = \alpha^n u[n] \longleftrightarrow z/(z - \alpha)$ and $y[0] = 1$, the left-shift property gives

$$y[n + 1] = \alpha^{n+1} u[n + 1] \longleftrightarrow z[z/(z - \alpha)] - z = \alpha z/(z - \alpha)$$

13.2.4   *The Times-n Property*

The times-n property is established by taking derivatives, to yield

$$X(z) = \sum_{k=-\infty}^{\infty} x[k]z^{-k} \qquad \frac{dX(z)}{dz} = \sum_{k=-\infty}^{\infty} \frac{d}{dz}\left[x[k]z^{-k}\right] = \sum_{k=-\infty}^{\infty} -kx[k]z^{-(k+1)}$$

Multiplying both sides by $-z$, we obtain

$$-z\frac{dX(z)}{dz} = \sum_{k=-\infty}^{\infty} kx[k]z^{-k}$$

This represents the transform of $nx[n]$.

Repeating the operation $-zd[\ ]/dz$ p times leads to the times-n^p property.

13.2.5   *The Scaling Property*

The scaling property follows from the transform of $y[n] = \alpha^n x[n]$, to yield

$$Y(z) = \sum_{k=-\infty}^{\infty} \alpha^k x[k]z^{-k} = \sum_{k=-\infty}^{\infty} x[k](z/\alpha)^{-k} = X(z/\alpha)$$

If $x[n]$ is multiplied by $\exp(jn\Omega)$ or $[\exp(j\Omega)]^n$, we then obtain the pair $\exp(jn\Omega)x[n] \longleftrightarrow X\{z\exp(-j\Omega)\}$. An extension of this result, using Euler's relation, leads to the times-cos and times-sin properties:

$$\cos(n\Omega)x[n] = \tfrac{1}{2}x[n]\{\exp(jn\Omega) + \exp(-jn\Omega)\}$$
$$\longleftrightarrow \tfrac{1}{2}\left[X\{z\exp(j\Omega)\} + X\{z\exp(-j\Omega)\}\right]$$
$$\sin(n\Omega)x[n] = j\tfrac{1}{2}x[n]\{\exp(jn\Omega) - \exp(-jn\Omega)\}$$
$$\longleftrightarrow j\tfrac{1}{2}\left[X\{z\exp(j\Omega)\} - X\{z\exp(-j\Omega)\}\right]$$

Using the right-shift property and superposition, we obtain the z-transform of the first difference of $x[n]$ as

$$y[n] = x[n] - x[n-1] \longleftrightarrow X(z) - z^{-1}X(z) = (1 - z^{-1})X(z)$$

13.2.6   *The Folding Property*

The folding property results if we use $\alpha = -1$ in the scaling property, or $k \longrightarrow -k$ in the defining relation. With $x[n] \longleftrightarrow X(z)$ and $y[n] = x[-n]$, we get

$$Y(z) = \sum_{k=-\infty}^{\infty} x[-k]z^{-k} = \sum_{k=-\infty}^{\infty} x[k]z^{k} = \sum_{k=-\infty}^{\infty} x[k](1/z)^{-k} = X(1/z)$$

Note that the ROC does not change upon translation of $|z| > \alpha$ to $|1/z| > \alpha$ or $|z| < 1/\alpha$. This property implies that for even symmetry ($x[n] = x[-n]$), $X(z) = X(1/z)$, and for odd symmetry ($x[n] = -x[-n]$), $X(z) = -X(1/z)$.

13.2.7   The Folding Property and Anticausal Signals

The folding property is useful in finding the transform of anticausal signals. From the causal signal $x[n]u[n] \longleftrightarrow X(z)$, we find the transform of $x[-n]u[-n]$ as $X(1/z)$. The anticausal signal $y[n] = x[-n]u[-n-1]$ (for $n \leq -1$) which excludes $n = 0$ can then be written as $y[n] = x[-n]u[-n] - x[0]\delta[n]$, and its transform becomes

$$x[-n]u[-n-1] \longleftrightarrow X(1/z) - x[0] \quad \text{where} \quad x[n]u[n] \longleftrightarrow X(z)$$

Example 13.5
Using the Folding Property

To find the transform of $x[n] = \alpha^{-n}u[-n-1]$, we start with the transform pair $y[n] = \alpha^n u[n] \longleftrightarrow z/(z-\alpha)$ with ROC $|z| > \alpha$. With $y[0] = 1, x[n] = y[-n] - \delta[n]$, and $X(z) = (1/z)/(1/z - \alpha) = \alpha z/(1 - \alpha z)$ with ROC $|1/z| > \alpha$ or $|z| < \frac{1}{\alpha}$.

13.2.8   Initial Value and Final Value Theorems

The initial and final value theorems apply only to the onesided z-transform, and the proper part $X(z)$ of a rational z-transform.

With $X(z)$ described by $x[0] + x[1]z^{-1} + x[2]z^{-2} + \cdots$, it should be obvious that only $x[0]$ survives as $z \longrightarrow \infty$, and the initial value equals $x[0] = \lim_{z \to \infty} X(z)$.

To find the final value, we evaluate $(z-1)X(z)$ at $z = 1$. It yields meaningful results only when the poles of $(z-1)X(z)$ have magnitudes smaller than unity (lie within the unit circle in the z-plane). As a result:

1. $x[\infty] = 0$ if all poles of $X(z)$ lie within the unit circle (since $x[n]$ will then comprise only exponentially damped terms).
2. $x[\infty]$ is constant if there is a single pole at $z = 1$ (since $x[n]$ will then include a step).
3. $x[\infty]$ is indeterminate if there are complex conjugate poles on the unit circle (since $x[n]$ will then include sinusoids). The final value theorem can yield absurd results if used in this case.

Example 13.6
Initial and Final Value Theorems

With $X(z) = z(z-2)/[(z-1)(z-\frac{1}{2})]$, we find for $x[n]$,

The initial value: $x[0] = \lim_{z \to \infty} X(z) = \lim_{z \to \infty}(1 - 2z^{-1})/[(1 - z^{-1})(1 - \frac{1}{2}z^{-1})] = 1$

The final value: $\lim_{n \to \infty} x[n] = \lim_{z \to 1}(z-1)X(z) = \lim_{z \to 1} z(z-2)/(z-\frac{1}{2}) = -2$

13.2.9   The Convolution Property

The convolution property is based on the fact that multiplication in the time domain corresponds to convolution in any transformed domain, as explained in Chapter 4. The z-transforms of sequences are polynomials, and multiplication of two polynomials corresponds to the convolution of their coefficient sequences. This property finds extensive use in the analysis of systems in the transformed domain (z-domain), as described later in this chapter.

13.2.10   *Some z-Transform Pairs Using Properties*

We now illustrate the use of properties in obtaining the *z*-transform.

Example 13.7
z-Transforms Using
Properties

(a) Using the times-*n* property, the *z*-transform of $y[n] = nu[n]$ is

$$Y(z) = -z\frac{d}{dz}[z/(z-1)] = -z[-z/(z-1)^2 + 1/(z-1)] = z/(z-1)^2$$

(b) With $x[n] = \alpha^n nu[n]$, we use scaling to obtain the *z*-transform

$$X(z) = (z/\alpha)/[(z/\alpha) - 1]^2 = z\alpha/(z-\alpha)^2$$

(c) With $x[n] = u[n] - u[n-N]$, the *z*-transform of the pulse $y[n] = \alpha^n x[n]$ is found using scaling on the result of Example 13.1(c). We get

$$Y[z] = [1 - (z/\alpha)^{-N}]/[1 - (z/\alpha)^{-1}] \qquad z \neq \alpha$$

(d) The *z*-transforms of $x[n] = \cos(n\Omega)u[n]$ and $y[n] = \sin(n\Omega)u[n]$ are found using the times-cos and times-sin properties:

$$X(z) = \frac{1}{2}\left[\frac{z\exp(j\Omega)}{z\exp(j\Omega) - 1} + \frac{z\exp(-j\Omega)}{z\exp(-j\Omega) - 1}\right]$$

$$= \frac{z^2 - z\cos(\Omega)}{z^2 - 2z\cos(\Omega) + 1} = \frac{z^2 - z\cos(\Omega)}{[z - \exp(j\Omega)][z - \exp(-j\Omega)]}$$

$$Y(z) = \frac{1}{2}j\left[\frac{z\exp(j\Omega)}{z\exp(j\Omega) - 1} - \frac{z\exp(-j\Omega)}{z\exp(-j\Omega) - 1}\right]$$

$$= \frac{z\sin(\Omega)}{z^2 - 2z\cos(\Omega) + 1} = \frac{z\sin(\Omega)}{[z - \exp(j\Omega)][z - \exp(-j\Omega)]}$$

(e) The *z*-transforms of $f[n] = \alpha^n \cos(n\Omega)u[n]$ and $g[n] = \alpha^n \sin(n\Omega)u[n]$ follow from the results of part (d) and the scaling property:

$$F(z) = \frac{(z/\alpha)^2 - (z/\alpha)\cos(\Omega)}{(z/\alpha)^2 - 2(z/\alpha)\cos(\Omega) + 1} = \frac{z^2 - \alpha z\cos(\Omega)}{z^2 - 2\alpha z\cos(\Omega) + \alpha^2}$$

$$= \frac{z^2 - \alpha z\cos(\Omega)}{[z - \alpha\exp(j\Omega)][z - \alpha\exp(-j\Omega)]}$$

$$G(z) = \frac{(z/\alpha)\sin(\Omega)}{(z/\alpha)^2 - 2(z/\alpha)\cos(\Omega) + 1} = \frac{\alpha z\sin(\Omega)}{z^2 - 2\alpha z\cos(\Omega) + \alpha^2}$$

$$= \frac{\alpha z\sin(\Omega)}{[z - \alpha\exp(j\Omega)][z - \alpha\exp(-j\Omega)]}$$

13.2.11   *The z-Transform of Switched Periodic Signals*

Consider a causal signal $x[n] = x_p[n]u[n]$ where $x_p[n]$ is periodic with period N. If $x_1[n]$ describes the first period of $x[n]$ and has the *z*-transform $X_1(z)$, then the z-transform of $x[n]$ can be found as the superposition of the z-transform of the shifted versions of $x_1[n]$:

$$X(z) = X_1(z) + z^{-N}X_1(z) + z^{-2N}X_1(z) + \cdots = X_1(z)[1 + z^{-N} + z^{-2N} + \cdots]$$

Expressing the geometric series in closed form, we obtain

$$X(z) = \frac{1}{1 - z^{-N}}X_1(z) = \frac{z^N}{z^N - 1}X_1(z)$$

Example 13.8

z-Transform of a Switched Periodic Signal

Consider $x_1[n] = \{0, 1, -1\}$. If $x_1[n]$ repeats to yield the signal $x[n]$, the period of $x[n]$ is $N = 3$. We then find the z-transform of $x[n]$ as

$$X(z) = \frac{z^N}{z^N - 1}X_1(z) = \frac{z^3}{z^3 - 1}[z^{-1} - z^{-2}] = \frac{z(z - 1)}{(z - 1)(z^2 + z + 1)} = \frac{z}{z^2 + z + 1}$$

13.3
Poles, Zeros and the z-Plane

The z-transform of many *DT* signals is a *rational* function of the form

$$X(z) = \frac{N(z)}{D(z)} = \frac{B_0 + B_1 z^{-1} + B_2 z^{-2} + \cdots + B_M z^{-M}}{1 + A_1 z^{-1} + A_2 z^{-2} + \cdots + A_N z^{-N}}$$

Here, the coefficient A_0 of the denominator has been normalized to unity.

Denoting the roots of $N(z)$ by z_i, $i = 1, 2, \ldots, M$ and the roots of $D(z)$ by p_k, $k = 1, 2, \ldots, N$, $X(z)$ may also be written in factored form as

$$X(z) = \frac{N(z)}{D(z)} = K\frac{(z - z_1)(z - z_2)\cdots(z - z_M)}{(z - p_1)(z - p_2)\cdots(z - p_N)}$$

Assuming that common factors have been canceled, the p roots of $N(z)$ and the q roots of $D(z)$ are termed the **finite zeros** and the **finite poles** of the transfer function, respectively.

13.3.1   *Pole-Zero Plots*

A plot of the poles (denoted by crosses) and zeros (denoted by circles) of a rational function $X(z)$ in the z-plane constitutes a **pole-zero plot** and provides a visual picture of the root locations. For multiple roots, we indicate their multiplicity on the plot.

Clearly, we can find $X(z)$ directly from a pole-zero plot of the root locations but only to within the multiplicative constant K. If the value of K is also shown on the plot, $K(z)$ is known in its entirety (Figure 13.3).

If $N(z)$ and $D(z)$ are of unequal degree, it is also customary to "equalize" the number of poles and zeros by introducing poles and zeros at infinity. A pole occurs at $z = \infty$ if $X(\infty) = \infty$. A zero occurs at $z = \infty$ if $X(\infty) = 0$. If $N > M$, the quantity $r_d = N - M$ defines the number of *zeros at infinity*. For $M > N$, the quantity $r_n = M - N$ defines the number of *poles at infinity*. If we also count poles and zeros at infinity, $X(z)$ has the same number of poles and zeros.

Example 13.9
A Pole-Zero Plot

Consider $H(z) = 2z(z + 1)/[(z - \frac{1}{3})(z^2 + \frac{1}{4})(z^2 + 4z + 5)]$. The numerator degree is 2, the denominator degree is 5. Hence, there are three zeros at infinity. The two finite zeros are $z = 0$ and $z = -1$. The five finite poles are at $z = -\frac{1}{3}, z = \pm j\frac{1}{2}$, and $z = -2 \pm j$. There are no poles at infinity (Figure 13.3).

Figure 13.3   Pole-zero plots for Example 13.9 and 13.10. The pole-zero pattern for symmetric sequences shows conjugate reciprocal symmetry.

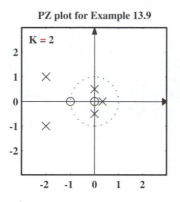

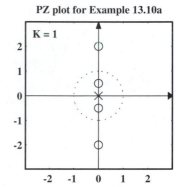

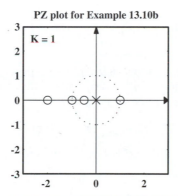

13.3.2   *Poles and Zeros of Symmetric Sequences*

Symmetric sequences are always noncausal, since $x[n] = x[-n]$ for even symmetry and $x[n] = -x[n]$ for odd symmetry. They play an important role in the design of digital filters, as we describe in Chapter 14. If the z-transform of $x[n]$ is $X(z)$, the z-transform of $x[-n]$ is $X(1/z)$. Thus, even symmetry in $x[n]$ requires $X(z) = X(1/z)$, and odd symmetry in $x[n]$ requires $X(z) = -X(1/z)$. For either type of symmetry, the zeros of $X(z)$ occur in reciprocal pairs. Each real zero r_k is paired with its reciprocal $1/r_k$. Each complex zero r_k is paired with its conjugate reciprocal $1/r_k^*$. Symmetric signals are thus said to exhibit **conjugate reciprocal symmetry**.

Remarks: 1. Zeros at $z = \pm 1$ are their own reciprocals and are not paired.
2. Each pole at $z = 0$ is paired with one at $z = \infty$.
3. Symmetic sequences can be made causal, if suitably delayed.

Example 13.10
Poles and Zeros of
Symmetric Signals

(a) *An even sequence*: Consider the signal $x[n] = \delta[n + 2] + 4.25\delta[n] + \delta[n - 2]$. Its z-transform is $X(z) = z^2 + 4.25 + z^{-2}$. Clearly $X(z) = X(1/z)$ and thus $x[n]$ is even. In factored form,

$$X(z) = (z^4 + 4.25z^2 + 1)/z^2 = (z + j\tfrac{1}{2})(z - j\tfrac{1}{2})(z + j2)(z - j2)/z^2$$

Its four zeros are $j\tfrac{1}{2}, -j2, -j\tfrac{1}{2}, j2$. Note the conjugate reciprocal symmetry (Figure 13.3). The two poles at $z = 0$ are paired with two poles at $z = \infty$.

(b) *An odd sequence*: Let $X(z) = z^2 + 2.5z - 2.5z^{-1} - z^{-2}$. Clearly $X(z) = -X(1/z)$ and $x[n]$ is an odd signal. In fact, $\{x[n]\} = \{1, 2.5, 0, -2.5, -1\}$. In factored form,

$$X(z) = (z - 1)(z + 1)(z + \tfrac{1}{2})(z + 2)/z^2.$$

We see a pair of zeros at $-\tfrac{1}{2}$ and -2. The zeros at $z = -1$ and $z = 1$ are their own reciprocals and are not paired (Figure 13.3). The two poles at $z = 0$ are paired with two poles at $z = \infty$.

13.4
The
Discrete-Time
Transfer
Function

The transfer function is defined only for *relaxed LTI systems* in one of two ways:

1. The ratio of the output $Y(z)$ and input $X(z)$.
2. The transform of the system impulse response $h[n]$.

If the system impulse response is $h[n]$, the response $y[n]$ to an arbitrary input $x[n]$ equals the convolution $y[n] = x[n] \star h[n]$. Since the convolution operation transforms to a product, we have

$$Y(z) = X(z)H(z) \qquad \text{or} \qquad H(z) = Y(z)/X(z)$$

A relaxed LTI system is also described by the difference equation

$$y[n] + A_1 y[n - 1] + A_2 x[n - 1] + \cdots + A_N y[n - N]$$
$$= B_0 x[n] + B_1 x[n - 1] + \cdots + B_M x[n - M]$$

The z-transform of this results in the transfer function $H(z)$ given by

$$H(z) = \frac{Y(z)}{X(z)} = \frac{B_0 + B_1 z^{-1} + \cdots + B_M z^{-M}}{1 + A_1 z^{-1} + \cdots + A_N z^{-N}}$$

The transfer function is thus a ratio of polynomials in z. It also allows us to retrieve the system difference equation or impulse response.

Example 13.11
A First-Order
Recursive Filter

For the relaxed first-order recursive filter $y[n] = \alpha y[n - 1] + x[n]$, we obtain

$$Y(z) = \alpha z^{-1} Y(z) + X(z) \quad \text{and} \quad H(z) = Y(z)/X(z) = 1/(1 - \alpha z^{-1}) = z/(z - a)$$

(a) The impulse response of this filter is then $h[n] = \alpha^n u[n]$.

(b) Since $H(z) = Y(z)/X(z) = 1/(1 - \alpha z^{-1})$, we have $[1 - \alpha z^{-1}]Y(z) = X(z)$ and $y[n] = \alpha y[n-1] + x[n]$.

Note that $h[n], H(z)$ and the difference equation describe the same system.

13.4.1   Poles and Zeros of the Transfer Function

The transfer function may be also be expressed in factored form as

$$H(z) = \frac{B_0 + B_1 z^{-1} + \cdots + B_M z^{-M}}{1 + A_1 z^{-1} + \cdots + A_N z^{-N}} = K \frac{(z - z_1) \cdots (z - z_M)}{(z - p_1) \cdots (z - p_N)}$$

The poles of $H(z)$ are called **natural modes** or **natural frequencies**. The poles of $Y(z) = H(z)X(z)$ determine the form of the system response. Clearly, the natural frequencies in $H(z)$ will always appear in the system response unless they are canceled by any corresponding zeros in $X(z)$.

The zeros of $H(z)$ are often called **transmission zeros** and can be regarded as the frequencies that are blocked by the system.

13.4.2   The z-Plane and System Stability

In the time domain, bounded-input, bounded-output (BIBO) stability of a *DT* LTI system requires an absolutely summable impulse response $h[n]$. In the z-domain, this is equivalent to requiring the poles of the transfer function $H(z)$ to lie entirely *within* the unit circle in the z-plane. This equivalence stems from the following observations:

1. *Poles outside the unit circle* ($|z| > 1$) lead to growing exponentials in the response even if the input is bounded.
   Example: $H(z) = z/(z - 3)$ results in the growing exponential $(3)^n u[n]$.
2. *Multiple poles on the unit circle* always result in polynomial growth.
   Example: $H(z) = 1/[z(z - 1)^2]$ produces a ramp function in $h[n]$.
3. *Simple poles on the unit circle* can also lead to an unbounded response.
   Example: A simple pole at $z = 1$ leads to $H(z)$ with a factor $z/(z - 1)$. If $X(z)$ also contains a pole at $z = 1$, the response $Y(z)$ will contain the term $z/(z - 1)^2$ and exhibit polynomial growth.

None of these types of time-domain terms is absolutely summable, and their presence leads to system instability. Formally, *for BIBO stability, all the poles of $H(z)$ must lie entirely within (and exclude) the unit circle.* This is both a necessary and sufficient condition for stability.

Remarks: **1.** Unlike *CT* systems, $H(z)$ is not required to be proper. An improper $H(z)$ only makes the system noncausal, not necessarily unstable.

2. If a system has simple poles on the unit circle, it is sometimes called **marginally stable**.

3. If a system has all its poles and zeros inside the unit circle, it is called a **minimum-phase** system. We study such systems in Section 13.9.

13.4.3   Stability, Causality and the ROC

For a causal system, $h[n] = 0, n < 0$. The ROC of $H(z)$ (based on the poles of $H(z)$) must, therefore, lie exterior to a circle of finite radius. For stability, the poles of $H(z)$ must lie inside the unit circle. Thus, *for a system to be both causal and stable, the ROC must include (all or part of) the unit circle.* Naturally, a system could also be stable but noncausal. We shall assume causal systems in the absence of any specified ROC.

Example 13.12
Stability and Causality of a
Recursive Filter

Consider a first-order recursive filter described by $y[n] = \alpha y[n-1] + x[n]$. Its transfer function is $H(z) = z/(z - \alpha)$.

If the ROC is $|z| > \alpha$, the impulse response is $h[n] = \alpha^n u[n]$.
If the ROC is $|z| < \alpha$, we have instead $h[n] = -\alpha^n u[-n-1]$.

Which of these choices we invoke depends on the constraints we impose on the system by way of causality and stability. For example:

1. For a causal system, we choose $h[n] = \alpha^n u[n]$. The system will also be stable if $|\alpha| < 1$ in order that the ROC includes the unit circle.
2. For an anticausal system, we choose $h[n] = -\alpha^n u[-n-1]$. The system will also be stable if we choose $|\alpha| > 1$.

13.5
The Inverse
z-Transform

The formal inversion relation that yields $x[n]$ from $X(z)$ actually involves complex integration, and is described by

$$x[n] = \frac{1}{j2\pi} \oint_{\Gamma} X(z) z^{n-1} \, dz$$

Here, Γ describes a clockwise contour of integration enclosing the origin (such as the unit circle). Evaluation of this integral requires knowledge of complex variable theory. In this text, we describe simpler alternatives, which include long division and partial fraction expansion.

Inverse Transform of Finite Sequences   For finite sequences, $X(z)$ has a polynomial form that immediately reveals the required sequence $x[n]$. The ROC can also be discerned from the polynomial form.

Example 13.13

Inverse Transform of a Sequence

(a) Let $X(z) = 3z^{-1} + 5z^{-3} + 2z^{-4}$. This transform corresponds to a causal sequence. We recognize $x[n]$ as a sum of shifted impulses given by

$$x[n] = 3\delta[n-1] + 5\delta[n-3] + 2\delta[n-4]$$

This sequence can also be written as $\{x[n]\} = \{0, 3, 0, 5, 2\}$.

(b) Let $X(z) = 2z^2 - 5z + 5z^{-1} - 2z^{-2}$. This transform corresponds to a noncausal sequence. Its inverse transform is written, by inspection, as

$$\{x[n]\} = \{2, -5, 0, 5, -2\}$$

Comment:

Since $X(z) = -X(1/z)$, $x[n]$ should possess odd symmetry. It does. ▬▬

Inverse Transforms by Long Division   A second method requires $X(z)$ as a rational function (a ratio of polynomials) along with its ROC. We then arrange the numerator and denominator as follows:

1. In ascending powers of z if the ROC is $|z| > |\alpha|$. Long division leads to a power series in decreasing powers of z and a rightsided sequence.
2. In descending powers of z if the ROC is $|z| < |\alpha|$. Long division leads to a power series in increasing powers of z and a leftsided sequence.

This approach, however, becomes cumbersome if more than just the first few terms of $x[n]$ are required.

Remarks:

1. It is only in rare instances that the first few terms of the resulting sequence allow its general nature or form to be discerned.
2. If we regard the rational z-transform as a transfer function $H(z) = P(z)/Q(z)$, the method of long division is simply equivalent to finding the first few terms of its impulse response recursively from its difference equation.

Example 13.14

Inverse Transforms by Long Division

Consider $H(z) = (z-4)/(1 - z + z^2)$.

(a) *Case 1:* If $H(z)$ corresponds to a rightsided sequence $h[n]$, we arrange the polynomials in descending powers of z and use long division to get

$$
\begin{array}{r}
z^{-1} - 3z^{-2} - 4z^{-3} \cdots \\
z^2 - z + 1 \overline{)\, z - 4 } \\
\underline{z - 1 + z^{-1}} \\
-3 - z^{-1} \\
\underline{-3 + 3z^{-1} - 3z^{-2}} \\
-4z^{-1} + 3z^{-2} \\
\underline{-4z^{-1} + 4z^{-2} - 4z^{-3}} \\
-z^{-2} + 4z^{-3} \\
\vdots
\end{array}
$$

This leads to $H(z) = z^{-1} - 3z^{-2} - 4z^{-3}$. The sequence $h[n]$ can be written as

$$h[n] = \delta(n-1) - 3\delta[n-2] - 4\delta[n-3] \quad \text{or} \quad \{x[n]\} = \{0, 1, -3, -4, \ldots\}$$
$$\uparrow$$

Comment: The difference equation corresponding to $H(z) = Y(z)/X(z)$ is

$$y[n] - y[n-1] + y[n-2] = x[n-1] - 4x[n-2]$$

With $x[n] = \delta[n]$, its impulse response $h[n]$ can be described by

$$h[n] - h[n-1] + h[n-2] = \delta[n] - 4\delta[n-2]$$

With $h[-1] = h[-2] = 0$, we recursively obtain the first few values of $h[n]$ as

$$n = 0: \quad h[0] = h[-1] - h[-2] + \delta[-1] - 4\delta[-2] = 0 - 0 + 0 - 0 = 0$$
$$n = 1: \quad h[1] = h[0] - h[-1] + \delta[0] - 4\delta[-1] = 0 - 0 + 1 - 0 = 1$$
$$n = 2: \quad h[2] = h[1] - h[0] + \delta[1] - 4\delta[0] = 1 - 0 + 0 - 4 = -3$$
$$n = 3: \quad h[3] = h[2] - h[1] + \delta[2] - 4\delta[1] = -3 - 1 + 0 + 0 = -4$$

These are identical to the values obtained using long division.

(b) *Case 2:* If $H(z)$ corresponds to a leftsided sequence $h[n]$, we arrange the polynomials in descending powers of z and use long division to obtain

$$
\begin{array}{r}
-4 - 3z + z^2 \cdots \\
1 - z + z^2 \overline{\smash{\big)}\ -4 + z } \\
\underline{-4 + 4z - 4z^2} \\
-3z + 4z^2 \\
\underline{-3z + 3z^2 - 3z^3} \\
z^2 + 3z^3 \\
\underline{z^2 - z^3 + z^4} \\
4z^3 - z^4 \\
\vdots
\end{array}
$$

Thus $H(z) = -4 - 3z + z^2$. The sequence $h[n]$ can then be written as

$$h[n] = -4\delta[n] - 3\delta[n+1] + \delta[n+2] \quad \text{or} \quad \{x[n]\} = \{\ldots, 1, -3, -4\}$$
$$\uparrow$$

Comment: If the difference equation is written as $h[n-2] = h[n-1] + \delta[n-1] - 4\delta[n-2]$, and $h[1] = h[2] = 0$, we can also find $h[0], h[-1], h[-2], \ldots$ recursively, to generate the same result.

13.5.1   *Inverse z-Transform from Partial Fractions*

A much more useful method for inversion of the z-transform relies on its *partial fraction expansion* (studied in Chapter 8) into terms whose inverse transform can be identified using a table of transform pairs. This is analogous to finding inverse

Laplace transforms, but with one major difference. Since the z-transform of standard sequences in Table 13.3 involves the factor z in the numerator, it is more convenient to perform the partial fraction expansion for $X(z)/z$ rather than for $X(z)$. We can then multiply through by z to obtain terms describing $X(z)$ in a form ready for inversion. This also implies that, for partial fraction expansion, it is $X(z)/z$ and not $X(z)$ that must be a proper rational function.

In addition to standard transform pairs, other forms useful for inversion using partial fractions are listed in Table 13.4.

Table 13.4   Inverse z-Transform from Partial Fraction Expansion (PFE) Forms.

PFE term $X(z)$	**Causal signal $x[n]$, $n \geq 0$**

Note: For anticausal sequences, we get the result $-x[n]u[-n-1]$ where $x[n]$ is as shown.

$\dfrac{z}{(z-\alpha)}$	α^n
$\dfrac{z}{(z-\alpha)^2}$	$n\alpha^{(n-1)}$
$\dfrac{z}{(z-\alpha)^3}$	$\dfrac{n(n-1)}{2!}\alpha^{(n-2)}$
$\dfrac{z}{(z-\alpha)^{N+1}}$ $\quad N > 2$	$\dfrac{n(n-1)\cdots(n-N+1)}{N!}\alpha^{(n-N)}$
$\dfrac{z(C+jD)}{[z-\alpha\exp(j\Omega)]} + \dfrac{z(C-jD)}{[z-\alpha\exp(-j\Omega)]}$	$2\alpha^n[C\cos(n\Omega) - D\sin(n\Omega)]$
$\dfrac{z(C+jD)}{[z-\alpha\exp(j\Omega)]^2} + \dfrac{z(C-jD)}{[z-\alpha\exp(-j\Omega)]^2}$	$2n\alpha^n[C\cos(n\Omega) - D\sin(n\Omega)]$
$\dfrac{zK\exp(j\phi)}{[z-\alpha\exp(j\Omega)]} + \dfrac{zK\exp(-j\phi)}{[z-\alpha\exp(-j\Omega)]}$	$2K\alpha^n\cos(n\Omega + \phi)$
$\dfrac{zK\exp(j\phi)}{[z-\alpha\exp(j\Omega)]^2} + \dfrac{zK\exp(-j\phi)}{[z-\alpha\exp(-j\Omega)]^2}$	$2Kn\alpha^{n-1}\cos[(n-1)\Omega + \phi]$
$\dfrac{zK\exp(j\phi)}{[z-\alpha\exp(j\Omega)]^{N+1}} + \dfrac{zK\exp(-j\phi)}{[z-\alpha\exp(-j\Omega)]^{N+1}}$	$2K\dfrac{n(n-1)\cdots(n-N+1)}{N!}\alpha^{(n-N)}\cos[(n-N)\Omega + \phi]$

For a quadratic denominator $z^2 - 2\alpha z\cos(\Omega) + \alpha^2$ with complex conjugate roots, we factor it as $[z - \alpha\exp(j\Omega)][z - \alpha\exp(-j\Omega)]$. With $X(z)$ given by

$$X(z) = \frac{z(C+jD)}{[z-\alpha\exp(j\Omega)]} + \frac{z(C-jD)}{[z-\alpha\exp(-j\Omega)]}$$

$$= \frac{zK\exp(j\phi)}{[z-\alpha\exp(j\Omega)]} + \frac{zK\exp(-j\phi)}{[z-\alpha\exp(-j\Omega)]}$$

inversion and Euler's relation lead to the result

$$x[n] = 2\alpha^n[C\cos(n\Omega) - D\sin(n\Omega)]u[n] = 2K\alpha^n\cos(n\Omega + \phi)u[n]$$

Example 13.15
Inversion Using Partial Fractions

(a) *Nonrepeated roots:* Let $X(z) = 1/[(z - \frac{1}{4})(z - \frac{1}{2})]$. We first form $Y(z) = X(z)/z$ and expand $Y(z)$ into partial fractions (see Chapter 8) to obtain

$$Y(z) = X(z)/z = 1/[z(z - \tfrac{1}{4})(z - \tfrac{1}{2})] = 8/z - 16/(z - \tfrac{1}{4}) + 8/(z - \tfrac{1}{2})$$

Multiplying through by z, we have $X(z) = 8 - 16z/(z - \frac{1}{4}) + 8z/(z - \frac{1}{2})$. From Table 13.1, we find $x[n] = 8\delta[n] - 16(\frac{1}{4})^n u[n] + 8(\frac{1}{2})^n u[n]$. The first few samples are $x[0] = 0, \quad x[1] = 0, \quad x[2] = 1, \quad x[3] = \frac{3}{4}$.

Consistency check: These values can easily be checked using long division.

Comment: An alternate approach (not recommended) is to expand $X(z)$ itself as

$$X(z) = -4/(z - \tfrac{1}{4}) + 4/(z - \tfrac{1}{2})$$

If we let $Y(z) = zX(z) = -4z/(z - \frac{1}{4}) + 4z/(z - \frac{1}{2})$, we obtain $y[n] = -4(\frac{1}{4})^n u[n] + 4(\frac{1}{2})u[n]$. Since $X(z) = z^{-1}Y(z)$, we get $x[n] = y[n - 1] = -4(\frac{1}{4})^{n-1}u[n - 1] + 4(\frac{1}{2})^{n-1}u[n - 1]$.
   This result yields $x[0] = 0, \quad x[1] = 0, \quad x[2] = 1$ and $x[3] = \frac{3}{4}$ as before.

(b) *Repeated roots:* Consider $X(z) = z/[(z - 1)^2(z - 2)]$. We obtain $Y(z) = X(z)/z$ and set up the partial fraction expansion (note the repeated factor) as:

$$Y(z) = 1/[(z - 1)^2(z - 2)] = A/(z - 2) + K_0/(z - 1)^2 + K_1/(z - 1)$$

The constants in the partial fraction expansion (see Chapter 8) are

$$A = 1/(z - 1)^2|_{z=2} = 1$$
$$K_0 = 1/(z - 2)|_{z=1} = -1$$
$$K_1 = \frac{d}{dz}[1/(z - 2)]|_{z=1} = -1$$

Substituting these values into $Y(z)$, and multiplying through by z, we get

$$X(z) = zY(z) = z/(z - 2) - z/(z - 1)^2 - z/(z - 1)$$

From Table 13.1, the inverse transform equals

$$x[n] = (2)^n u[n] - nu[n] - u[n] = [2^n - n - 1]u[n]$$

We observe $x[0] = 0,\ x[1] = 0,\ x[2] = 1,\ x[3] = 4,\ x[4] = 11$, and so on.

Consistency check: These values can easily be checked using long division.

(c) *Complex roots:* Consider $X(z) = (z^2 - 3z)/[(z - 2)(z^2 - 2z + 2)]$. Then

$$Y(z) = X(z)/z = (z - 3)/[(z - 2)(z^2 - 2z + 2)]$$
$$= A/(z - 2) + K/(z - 1 - j) + K^*/(z - 1 + j)$$

We get $A = (z - 3)/(z^2 - 2z + 2)|_{z=2} = -\frac{1}{2}$ and

$$K = (z - 3)/[(z - 2)(z - 1 + j)]|_{z=1+j} = \tfrac{1}{4} - j\tfrac{3}{4}, \quad \text{thus}$$

$$X(z) = -\tfrac{1}{2}z/(z - 2) + (\tfrac{1}{4} - j\tfrac{3}{4})z/(z - 1 - j) + (\tfrac{1}{4} + j\tfrac{3}{4})z/(z - 1 + j)$$

We note that $(z - 1 - j) = [z - \sqrt{2}\exp(j\frac{1}{4}\pi)]$. From Table 13.4, the constant associated with this factor is $K = C + jD = \frac{1}{4} - j\frac{3}{4}$. Thus, $C = \frac{1}{4}$, $D = -\frac{3}{4}$, and

$x[n] = -\frac{1}{2}(2)^n u[n] + 2(\sqrt{2})^n \ [\frac{1}{4}\cos(\frac{1}{4}n\pi) + \frac{3}{4}\sin(\frac{1}{4}n\pi)]u[n]$. Alternatively, with $\frac{1}{4} - j\frac{3}{4} = \frac{1}{4}\sqrt{10}\angle - 71.56° = \frac{1}{4}\sqrt{10}\angle - 0.3976\pi$, Table 13.4 also gives

$$x[n] = -\frac{1}{2}(2)^n u[n] + \frac{1}{2}\sqrt{10}(\sqrt{2})^n \cos(\frac{1}{4}n\pi - 0.3976\pi)$$

13.5.2 The ROC and Inversion

We have so far been assuming causal sequences when no ROC is given. Only when the ROC is specified do we obtain a unique sequence from $X(z)$. How the ROC affects the inversion results is illustrated by an example.

Example 13.16
Inversion and the ROC

Let $X(z) = z/[(z - \frac{1}{4})(z - \frac{1}{2})]$. We first form $Y(z) = X(z)/z$ and expand $Y(z)$ into partial fractions to obtain

$$Y(z) = X(z)/z = 1/[(z - \frac{1}{4})(z - \frac{1}{2})] = -4/(z - \frac{1}{4}) + 4/(z - \frac{1}{2})$$

Multiplying through by z, we have $X(z) = -4z/(z - \frac{1}{4}) + 4z/(z - \frac{1}{2})$
  Case 1: If the ROC is $|z| > \frac{1}{2}$, $x[n]$ is causal, and we obtain

$$x[n] = -4(\frac{1}{4})^n u[n] + 4(\frac{1}{2})^n u[n]$$

Case 2: If the ROC is $|z| < \frac{1}{4}$, $x[n]$ is anticausal, and we obtain

$$x[n] = 4(\frac{1}{4})^n u[-n-1] - 4(\frac{1}{2})^n u[-n-1]$$

  Case 3: If the ROC is $\frac{1}{4} < |z| < \frac{1}{2}$, $x[n]$ is twosided. This ROC is valid only if $-4z/(z - \frac{1}{4})$ describes a causal sequence (ROC $|z| > \frac{1}{4}$), and $4z/(z - \frac{1}{2})$ describes an anticausal sequence (ROC $|z| < \frac{1}{2}$). With this in mind, we obtain

$$x[n] = -4(\frac{1}{4})^n u[n] - 4(\frac{1}{2})^n u[-n-1]$$

13.6 The z-Transform and System Analysis

The z-transform serves as a useful tool for analyzing linear systems described by difference equations. The solution proceeds in two steps:

1. Transformation of the difference equation using the shift property and incorporating the effect of initial conditions.
2. Inverse transformation using partial fractions.

The response may be separated into its zero-state component (due only to the input) and zero-input component (due to the initial conditions). We can also obtain the natural and forced components, if required.

Example 13.17
Solution of Difference Equations

(a) Consider the system difference equation $y[n] - \frac{1}{2}y[n-1] = 2(\frac{1}{4})^n u[n]$ with $y[-1] = -2$. Transformation using the right-shift property yields:

$$Y(z) - \frac{1}{2}\{z^{-1}Y(z) + y[-1]\} = 2z/(z - \frac{1}{4}) \quad \text{or}$$
$$(1 - \frac{1}{2}z^{-1})Y(z) = 2z/(z - \frac{1}{4}) - 1 = (z + \frac{1}{4})/(z - \frac{1}{4})$$

Solving this, $Y(z) = (z + \frac{1}{4})/[(z - \frac{1}{4})(1 - \frac{1}{2}z^{-1})] = z(z + \frac{1}{4})/[(z - \frac{1}{4})(z - \frac{1}{2})]$. Using partial fractions on $Y(z)/z$, we find $Y(z)/z = -2/(z - \frac{1}{4}) + 3/(z - \frac{1}{2})$. Thus, $Y(z) = -2z/(z - \frac{1}{4}) + 3z/(z - \frac{1}{2})$, and we obtain the response $y[n]$ as

$$y[n] = -2(\tfrac{1}{4})^n + 3(\tfrac{1}{2})^n \qquad n \geq 0$$

Comment: As a check, we confirm that $y[-1] = -2(\frac{1}{4})^{-1} + 3(\frac{1}{2})^{-1} = -2$.

(b) Consider the system difference equation $y[n + 1] - \frac{1}{2}y[n] = 2(\frac{1}{4})^{n+1}u[n + 1]$ with $y[-1] = -2$. We can solve for $y[n]$ using one of two approaches:

Method 1: Assume time invariance to recast the difference equation as $y[n] - \frac{1}{2}y[n - 1] = 2(\frac{1}{4})^n u[n]$ and proceed as in part (a).

Method 2: Transform the difference equation directly, using the *left-shift* property. The solution will require $y[0]$, which we must find as follows. With $n = -1$, we obtain $y[0] - \frac{1}{2}y[-1] = 2$ or $y[0] = 2 + \frac{1}{2}y[-1] = 2 - 1 = 1$.

We also find the z-transform of $(\frac{1}{4})^{n+1}u[n + 1]$ as

$$(\tfrac{1}{4})^{n+1}u[n + 1] \longleftrightarrow z[z/(z - \tfrac{1}{4})] - z = \tfrac{1}{4}z/(z - \tfrac{1}{4})$$

Transforming the difference equation using the left-shift property, we have

$$\{zY(z) - zy[0]\} - \tfrac{1}{2}Y(z) = \tfrac{1}{2}z/(z - \tfrac{1}{4}) \quad \text{or}$$
$$(z - \tfrac{1}{2})Y(z) = \tfrac{1}{2}z/(z - \tfrac{1}{4}) + z = z(z + \tfrac{1}{4})/(z - \tfrac{1}{4})$$

Finally, solving for $Y(z)$, we get $Y(z) = z(z + \frac{1}{4})/[(z - \frac{1}{4})(z - \frac{1}{2})]$, as before. Thus $y[n] = -2(\frac{1}{4})^n + 3(\frac{1}{2})^n$. As a check, we find $y[0] = -2 + 3 = 1$.

Example 13.18
Zero-Input and
Zero-State Response

Consider the system $y[n] - \frac{1}{2}y[n - 1] = 2(\frac{1}{4})^n u[n]$ with $y[-1] = -2$ analyzed in Example 13.17a. Upon transformation using the right-shift property, we obtain:

$$Y(z) - \tfrac{1}{2}\{z^{-1}Y(z) + y[-1]\} = 2z/(z - \tfrac{1}{4}) \text{ or } (1 - \tfrac{1}{2}z^{-1})Y(z) = 2z/(z - \tfrac{1}{4}) - 1$$

(a) *Zero-state response:* For the zero-state response, we ignore the initial conditions to obtain $(1 - \frac{1}{2}z^{-1})Y_{zs}(z) = 2z/(z - \frac{1}{4})$. We simplify this to get

$$Y_{zs}(z)/z = 2z/[(z - \tfrac{1}{4})(z - \tfrac{1}{2})] = -2/(z - \tfrac{1}{4}) + 4/(z - \tfrac{1}{2})$$

Thus, $Y_{zs}(z) = -2z/(z - \frac{1}{4}) + 4z/(z - \frac{1}{2})$, and $y_{zs}[n] = -2(\frac{1}{4})^n u[n] + 4(\frac{1}{2})^n u[n]$.

(b) *Zero input response:* For the zero-input response, we ignore the input term $2z/(z - \frac{1}{4})$ to obtain $(1 - \frac{1}{2}z^{-1})Y_{zi}(z) = -1$ or $Y_{zi}(z) = -z/(z - \frac{1}{2})$. This is easily inverted to give $y_{zi}[n] = -(\frac{1}{2})^n u[n]$. We confirm that $y[n] = y_{zs}[n] + y_{zi}[n] = -2(\frac{1}{4})^n u[n] + 3(\frac{1}{2})^n u[n]$.

(c) *Natural and forced response:* From the total solution, we recognize the forced component as $-2(\frac{1}{4})^n u[n]$, and the natural response as $3(\frac{1}{2})^n u[n]$.

13.6.1   *Relaxed Systems Described by the Transfer Function*

The response $Y(z)$ of a relaxed LTI system equals the product $X(z)H(z)$ of the transformed input and the transfer function. It is often much easier to work with the transfer function description of a linear system. Naturally, the time-domain response $y[n]$ requires an inverse transformation, a penalty exacted by all transformed-domain methods.

Example 13.19
Response of a Recursive Filter

Consider a relaxed, causal, recursive, first-order filter described by

$$y[n] = \alpha y[n-1] + x[n]$$

(a) The response of this system to $x[n] = \alpha^n u[n]$ can be found as follows:
    Find the system transfer function: $H(z) = Y(z)/X(z) = z/(z-\alpha)$. Find the z-transform of the input: $X(z) = z/(z-\alpha)$. Then $Y(z) = H(z)X(z) = z^2/(z-\alpha)^2$ and, from Table 13.4, $y[n] = (n+1)\alpha^n u[n]$.

Comment: We could, of course, also use convolution to obtain $y[n] = h[n] \star x[n]$.

(b) *Step response:* For the above filter, we choose $\alpha = \frac{1}{2}$ to obtain

$$y[n] = 0.5y[n-1] + x[n] \text{ and } H(z) = z/(z-\tfrac{1}{2})$$

The response to $x[n] = 4u[n]$ is then found by first evaluating $Y(z)$ as

$$Y(z) = H(z)X(z) = \left(\frac{z}{z-\frac{1}{2}}\right)\left(\frac{4z}{z-1}\right) = \frac{4z^2}{(z-1)(z-\frac{1}{2})}$$

Using partial fractions, $Y(z)/z = 4z/[(z-1)(z-\frac{1}{2})] = 8/(z-\frac{1}{2}) - 4/(z-1)$. With $Y(z) = 8z/(z-\frac{1}{2}) - 4z/(z-1)$, we get $y[n] = 8u[n] - (\frac{1}{2})^n u[n]$. The second term is the natural response and the first is the *steady-state* response, which can be found much more easily, as we show in Section 13.6.4.

13.6.2   *Systems with Nonzero Initial Conditions*

If we let $H(z) = N(z)/D(z)$, the response $Y(z)$ of a relaxed system to an input $X(z)$ may be expressed as $Y(z) = X(z)H(z) = X(z)N(z)/D(z)$. This corresponds to the zero-state response. If the system is not relaxed, the initial conditions result in an additional contribution, the zero-input response $Y_{zi}(z)$, which may be written as $Y_{zi}(z) = N_{zi}(z)/D(z)$. To evaluate $Y_{zi}(z)$, we first set up the system difference equation and then transform it using initial conditions.

Example 13.20
Zero-State and Zero-Input Response

Consider a system with $H(z) = z^2/(z^2 - \frac{1}{6}z - \frac{1}{6})$, the input $x[n] = 4u[n]$, and the initial conditions $y[-1] = 0, y[-2] = 12$.

(a) *Zero-state and zero-input response:* The zero-state response is found as

$$Y_{zs}(z) = X(z)H(z) = 4z^3/[(z^2 - \tfrac{1}{6}z - \tfrac{1}{6})(z-1)] = 4z^3/[(z-\tfrac{1}{2})(z+\tfrac{1}{3})(z-1)]$$

Using partial fractions on $Y_{zs}(z)/z$, we obtain

$$Y_{zs}(z) = -2.4z/(z - \tfrac{1}{2}) + 0.4z/(z + \tfrac{1}{3}) + 6z/(z - 1)$$

Upon inverse transformation, $y_{zs}[n] = -2.4(\tfrac{1}{2})^n u[n] + 0.4(-\tfrac{1}{3})^n u[n] + 6u[n]$.

To find the zero-input response we first set up the difference equation from $H(z) = Y(z)/X(z) = z^2/(z^2 - \tfrac{1}{6}z - \tfrac{1}{6})$, or $(z^2 - \tfrac{1}{6}z - \tfrac{1}{6})Y(z) = z^2 X(z)$. For a difference equation in terms of *backward differences*, we rewrite this as

$$[1 - \tfrac{1}{6}z^{-1} - \tfrac{1}{6}z^{-2}]Y(z) = X(z), \text{ and thus } y[n] - \tfrac{1}{6}y[n-1] - \tfrac{1}{6}y[n-2] = x[n]$$

With $x[n] = 0$ (zero input), $y[-1] = 0$ and $y[-2] = 12$, we transform this equation, using the *right-shift* property, to obtain the zero-input response from

$$Y_{zi}(z) - \tfrac{1}{6}\{z^{-1}Y_{zi}(z) + y[-1]\} - \tfrac{1}{6}\{z^{-2}Y_{zi}(z) + z^{-1}y[-1] + y[-2]\} = 0$$

Thus, $(1 - \tfrac{1}{6}z^{-1} - \tfrac{1}{6}z^{-2})Y_{zi}(z) = \tfrac{1}{6}\{y[-1] + z^{-1}y[-1] + y[-2]\} = \tfrac{1}{6}(12) = 2$, and $Y_{zi}(z) = 2/(1 - \tfrac{1}{6}z^{-1} - \tfrac{1}{6}z^{-2}) = 2z^2/(z^2 - \tfrac{1}{6}z - \tfrac{1}{6}) = 2z^2/[(z - \tfrac{1}{2})(z + \tfrac{1}{3})]$. Partial fractions for $Y_{zi}(z)/z$ yield $Y_{zi}(z) = 0.75z/(z - \tfrac{1}{2}) + 0.8z/(z + \tfrac{1}{3})$. Upon inverse transformation, we get $y_{zi}[n] = 1.2(\tfrac{1}{2})^n u[n] + 0.8(-\tfrac{1}{3})^n u[n]$. Finally, $y[n] = y_{zs}[n] + y_{zi}[n] = -1.2(\tfrac{1}{2})^n u[n] + 1.2(-\tfrac{1}{3})^n u[n] + 6u[n]$.

(b) *Natural and forced response:* We find the natural and forced components from the *total response* $y[n] = -1.2(\tfrac{1}{2})^n u[n] + 1.2(-\tfrac{1}{3})^n u[n] + 6u[n]$. By inspection, $y_F[n] = 6u[n]$ and $y_N[n] = -1.2(\tfrac{1}{2})^n u[n] + 1.2(-\tfrac{1}{3})^n u[n]$. Alternatively, we transform the system difference equation to obtain

$$Y(z) - \tfrac{1}{6}\{z^{-1}Y(z) + y[-1]\} - \tfrac{1}{6}\{z^{-2}Y(z) + z^{-1}y[-1] + y[-2]\} = 4z/(z - 1)$$

$$(1 - \tfrac{1}{6}z^{-1} - \tfrac{1}{6}z^{-2})Y(z) - 2 = 4z/(z - 1) \quad \text{or}$$

$$Y(z) = z^2(6z - 2)/[(z - 1)(z^2 - \tfrac{1}{6}z - \tfrac{1}{6})]$$

Partial fractions give $Y(z) = -1.2z/(z - \tfrac{1}{2}) + 1.2z/(z + \tfrac{1}{3}) + 6z/(z - 1)$. The steady-state response corresponds to terms of the form $z/(z - 1)$ (steps). Then

$$Y_N(z) = -1.2z/(z - \tfrac{1}{2}) + 1.2z/(z + \tfrac{1}{3}), \text{ and } Y_F(z) = 6z/(z - 1)$$

Thus, $y_F[n] = 6u[n]$ and $y_N[n] = -1.2(\tfrac{1}{2})^n u[n] + 1.2(-\tfrac{1}{3})^n u[n]$, as before.

Comment: Since the poles of $(z - 1)Y(z)$ lie within the unit circle, $y_F[n]$ can also be found without PFE, using the final value theorem:

$$y_F[n] = \lim_{z \to 1}(z - 1)Y(z) = \lim_{z \to 1} z^2(6z - 2)/(z^2 - \tfrac{1}{6}z - \tfrac{1}{6}) = \frac{4}{(2/3)} = 6 \underline{\quad}$$

Example 13.21
Zero-Input Response from the Transfer Function

Consider the system $H(z) = Y(z)/X(z) = z(z - \tfrac{1}{2})/(z^3 + \tfrac{1}{3}z^2 + \tfrac{1}{2}z - \tfrac{1}{2})$, with the input $x[n] = (\tfrac{1}{2})^n u[n]$, and initial conditions $y[-1] = 6, y[-2] = -8, y[-3] = 2$.

(a) The zero-state response $y_{zs}[n]$ is found as the inverse z-transform of

$$Y_{zs}(z) = H(z)X(z) = z^2/(z^3 + \tfrac{1}{3}z^2 + \tfrac{1}{2}z - \tfrac{1}{2})$$

(b) To find the zero-input response $y_{zi}[n]$, we rewrite $H(z)$ as

$$H(z) = (z^{-1} - \tfrac{1}{2}z^{-2})/(1 + \tfrac{1}{3}z^{-1} + \tfrac{1}{2}z^{-2} - \tfrac{1}{2}z^{-3})$$

and form the (backward) difference equation (in operator form)

$$[1 + \tfrac{1}{3}z^{-1} + \tfrac{1}{2}z^{-2} - \tfrac{1}{2}z^{-3}]y = [z^{-1} - \tfrac{1}{2}z^{-2}]x$$

With $x[n] = 0$ (zero input), $y[-1] = 6, y[-2] = -8$ and $y[-3] = 2$, this transforms to

$$Y_{zi}(z) + \tfrac{1}{3}\{z^{-1}Y(z) + y[-1]\} + \tfrac{1}{2}\{z^{-2}Y_{zi}(z) + y[-2] + z^{-1}y[-1]\}$$
$$-\tfrac{1}{2}\{z^{-3}Y_{zi}(z) + y[-3] + z^{-1}y[-2] - z^{-2}y[-1]\} = 0$$

Substituting values and simplifying, we get

$$Y_{zi}(z) = z(3z^2 - 7z + 3)/(z^3 + \tfrac{1}{3}z^2 + \tfrac{1}{2}z - \tfrac{1}{2})$$

The zero-input response is the inverse transform of $Y_{zi}(z)$.

13.6.3   The Steady-State Transfer Function

If the transfer function $H(z)$ of a linear system is evaluated for values of $z = \exp(j2\pi F) = \exp(j\Omega)$; that is, on the unit circle, we obtain the **steady-state transfer function** or **frequency response**, $H_p(F)$, of the system.

The frequency response $H_p(F)$ is periodic with period 1, since $\exp(j2\pi F)$ is periodic with period 1, and actually corresponds to the DTFT of $h[n]$. The quantity $\Omega = 2\pi F$ describes the angular orientation. How the values of z, Ω and F are related on the unit circle is shown in Figure 13.4 (p. 586).

In analogy with the Laplace transform, $H_p(F)$ describes the response of a system to *DT* harmonics. Since $H_p(F)$ is complex, it can be written as

$$H_p(F) = |H_p(F)|\angle\phi(F)$$

For real $h[n]$, $H_p(F)$ shows conjugate symmetry over the principal period $(-\tfrac{1}{2}, \tfrac{1}{2})$. This implies that $H_p(-F) = H_p^*(F)$, $|H_p(F)|$ is even symmetric, and $\angle\phi(F)$ is odd symmetric.

Remarks: **1.** The dc gain $H_p(0)$ is equivalent to $H(z)$ evaluated at $z = 1$(or $F = 0$).
2. The frequency response makes sense primarily for stable systems for which the natural response does, indeed, decay with time.

Example 13.22
Frequency Response of a Recursive Filter

To find the frequency response of the recursive filter $y[n] = \alpha y[n - 1] + x[n]$, we transform the difference equation and find $H(z)$ as

$$H(z) = Y(z)/X(z) = 1/(1 - \alpha z^{-1}) = z/(z - \alpha)$$

Next, we let $z = \exp(j2\pi F)$, and evaluate $H_p(F)$ as

$$H_p(F) = 1/\{1 - \alpha\exp(-j2\pi F)\} = 1/[1 - \alpha\cos(2\pi F) + j\alpha\sin(2\pi F)]$$

Figure 13.4   The unit circle in the z-plane showing various locations of z, Ω and F.

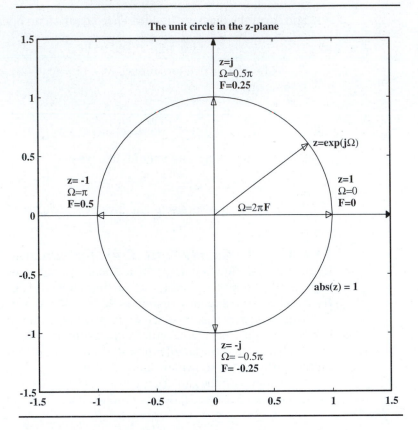

The magnitude and phase of $H_p(F)$ then equal

$$|H_p(F)| = \left[\frac{1}{1 - 2\alpha\cos(2\pi F) + \alpha^2}\right]^{1/2} \qquad \phi(F) = \tan^{-1}\left[\frac{1 - \alpha\cos(2\pi F)}{\alpha\sin(2\pi F)}\right]$$

These are plotted over the principal period $(-\frac{1}{2}, \frac{1}{2})$ for several values of α in Figure 13.5, and clearly show the conjugate symmetry in $H_p(F)$.

13.6.4   *The Steady-State Response to DT Harmonic Inputs*

Discrete harmonics (periodic or otherwise) are eigensignals of *DT* LTI systems. The system response to a harmonic input is a harmonic at the same frequency whose magnitude and phase are changed by the system function. All we require is these magnitude and phase factors. We find them by evaluating $H_p(F)$ at the harmonic frequency. To obtain the steady-state response $y_{ss}[n]$ of a system to the harmonic input $x[n] = A\cos(2\pi nF_0 + \theta)$, we first evaluate $H_p(F_0) = K\angle\phi$ (at the input frequency) and then obtain $y_{ss}[n] = KA\cos(2\pi nF_0 + \theta + \phi)$.

Figure 13.5   Frequency response of y[n] = ay[n − 1] + x[n] for Example 13.22.

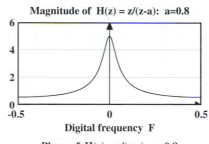

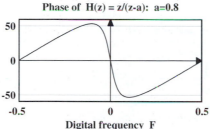

Remarks: **1.** If the input comprises components at different frequencies, we must evaluate both $H_p(F)$ and the individual *time-domain response at each frequency* before using superposition to find the total response.

2. We can also find the steady-state response by first finding the total response (using *z*-transforms) and then ignoring the natural response. This is quite tedious and definitely not the preferred way.

Example 13.23
Steady-State Response of
a Recursive Filter

Consider the filter $y[n] = 0.5y[n − 1] + x[n]$. We find $H_p(F)$ from $H(z)$ as

$$H(z) = \frac{1}{1 - \frac{1}{2}z^{-1}} \qquad H_p(F) = H(z)\Big|_{z=\exp(j\Omega)} = \frac{1}{1 - 0.5\exp(-j2\pi F)}$$

(a) To find the steady-state response to $x[n] = 10\cos(2\pi\frac{1}{4}n + 60°)$, we first evaluate $H_p(F)$ at the input frequency $F = \frac{1}{4}$

$$H_p(\tfrac{1}{4}) = \frac{1}{1 - 0.5\exp(-j2\pi\frac{1}{4})} = \frac{1}{1 + 0.5j} = 0.4\sqrt{5}\angle - 26.6°$$

Then $y_{ss}[n] = 10(0.4\sqrt{5})\cos(2\pi\frac{1}{4}n + 60° - 26.6°) = 4\sqrt{5}\cos(2\pi\frac{1}{4}n + 33.4°)$.

(b) To find the steady-state response to $x[n] = 4u[n]$, we evaluate $H_p(F)$ at the input frequency $F = 0$ (dc) to get $H_p(0) = \frac{1}{1 - 0.5} = 2$. Then $y_{ss}[n] = (2)(4) = 8$.

Comment: In either part, the total response must be found differently, but the steady-state component will turn out to be what we have calculated.   ——

13.7
Interconnected
Systems

The z-transform is well suited to the study of interconnected LTI systems in *cascade*, *parallel* and *feedback* (Figure 13.6). A block diagram shows each system transfer function as a block. Interconnections use **summing junctions** and **pickoff points**. The overall transfer function depends on the interconnections. The systems are assumed to be relaxed.

Figure 13.6   Cascade, parallel and feedback connections for systems.

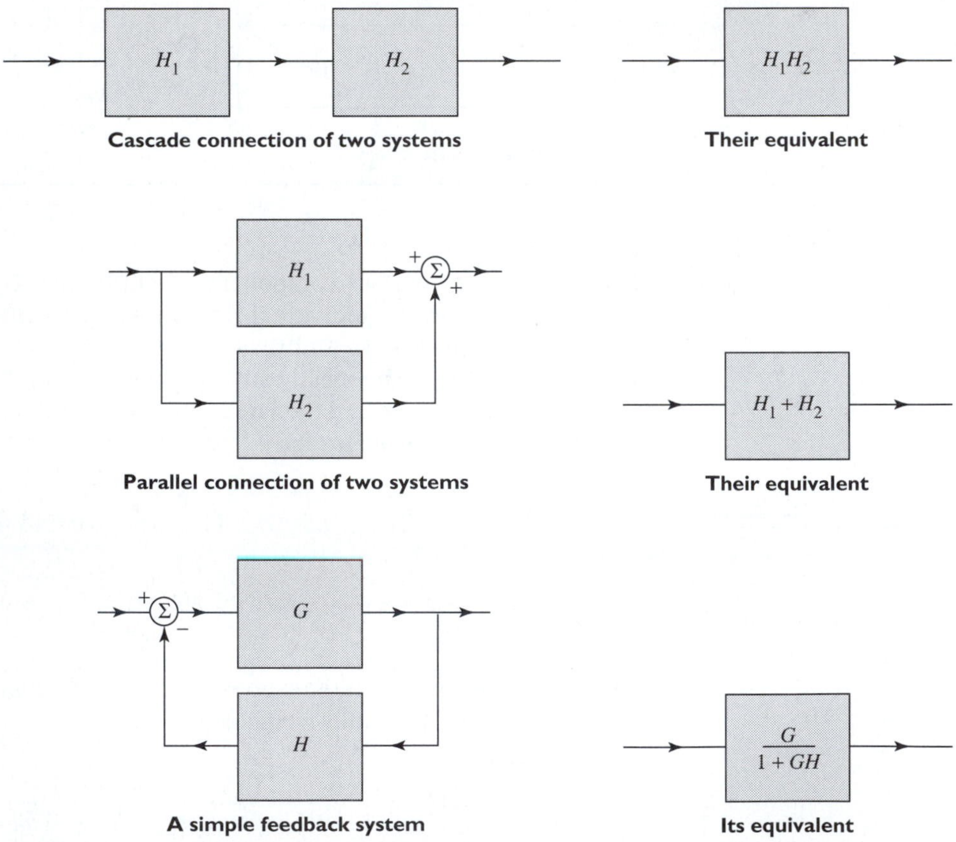

Cascade connection of two systems   **Their equivalent**

Parallel connection of two systems   **Their equivalent**

A simple feedback system   **Its equivalent**

Cascade Connection   The overall transfer function of a **cascaded system** is the product of the individual transfer functions. For n systems in cascade, the overall impulse response $h_C[n]$ is the convolution of the individual impulse responses

$h_1[n]$, $h_2[n]$,.... Since the convolution operation transforms to a product, we have

$$H_C(z) = H_1(z)H_2(z) \cdots H_n(z)$$

In words, the overall transfer function of a cascaded system is the product of the individual transfer functions.

Parallel Connection   For systems in **parallel**, the overall transfer function is the sum of the individual transfer functions. For n systems in parallel

$$H_P(z) = H_1(z) + H_2(z) + \cdots + H_n(z)$$

Feedback Connection   In a **feedback system**, the output is fed back to the input directly or via other subsystems. For the basic configuration shown in Figure 13.6, where $F(z)$ represents the feedback block, we have

$$H_F(z) = \frac{G(z)}{1 + G(z)F(z)}$$

13.8
Digital Filter
Realization

Digital filters described by transfer functions or difference equations can be *realized* by using elements corresponding to the operations of *scaling* (or multiplication), *shift* (or delay) and *summing* (or addition) that naturally occur in such equations. These elements describe the **gain** (scalar multiplier), **delay**, and **adder**, represented symbolically in Figure 13.7. In any system realization, it is important to understand that delay elements in cascade result in an output delayed by the sum of the individual delays. The operational notation for a delay of 1 unit is, therefore, represented by the exponential form z^{-1}, for 2 units by z^{-2}, and so on.

Figure 13.7   Elements used in digital filter realization.

$x[n] \circ \!\!-\!\!\boxed{}^{A}\!\!-\!\!\circ Ax[n]$ $x[n] \circ \!\!\longrightarrow\!\! \boxed{z^{-1}} \!\!\longrightarrow\!\!\circ x[n-1]$

A scaling (gain) element or multiplier   **A shifting or delay element**

$x[n] \circ \!\!\longrightarrow\!\! \overset{+}{\underset{\pm}{\Sigma}} \!\!\longrightarrow\!\!\circ x[n] \pm y[n]$

$y[n]$

A summing element or adder

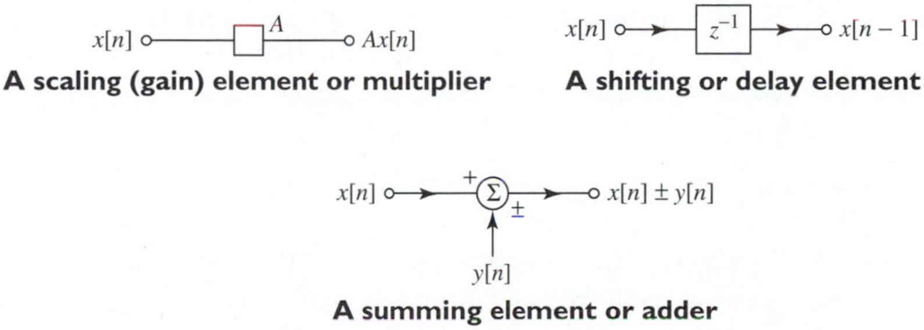

Consider the nonrecursive and recursive filters described by

$$H_N(z) = B_0 + B_1 z^{-1} + \cdots + B_M z^{-N} \qquad y[n] = B_0 x[n] + B_1 x[n-1] + \cdots + B_M x[n-M]$$

$$H_R(z) = \frac{1}{1 + A_1 z^{-1} + \cdots + A_N z^{-N}} \qquad y[n] = -A_1 y[n-1] - \cdots - A_N y[n-N] + x[n]$$

These filters can be realized using the **feedforward** (nonrecursive) structure and **feedback** (recursive) structure shown in Figure 13.8.

Figure 13.8   Realization of feedforward and feedback digital filters.

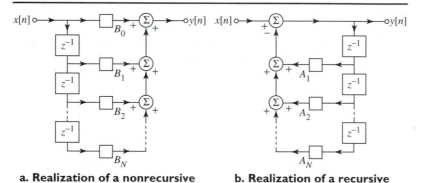

a. Realization of a nonrecursive (FIR) filter   **b. Realization of a recursive (IIR) filter**

Now, consider the general difference equation

$$y[n] = -A_1 y[n-1] - \cdots - A_N y[n-N] + B_0 x[n] + B_1 x[n-1] + \cdots + B_N x[n-N]$$

We choose $M = N$ with no loss of generality, since some of the B_k may always be set to zero. Its transfer function (with $M = N$) is

$$H(z) = \frac{B_0 + B_1 z^{-1} + \cdots + B_N z^{-N}}{1 + A_1 z^{-1} + A_2 z^{-2} + \cdots + A_N z^{-N}} = H_N(z) H_R(z)$$

The transfer function $H(z) = H_N(z) H_R(z)$ is the product of the transfer functions of a recursive and nonrecursive system. Its realization is thus a cascade of the realizations for the recursive and nonrecursive portions, as shown in Figure 13.9a. This form describes a **direct form I** realization. It uses $2N$ delay elements to realize an Nth-order difference equation, and is, therefore, not very efficient.

Figure 13.9   (a) Direct form I realization, (b) reversed order of cascading and (c) direct form II (canonic) realization of digital filters.

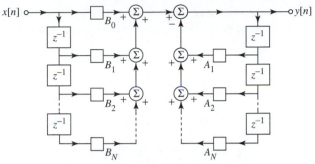

a. The general direct form I realization

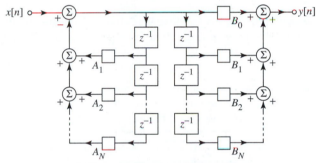

b. Reversed order of cascading

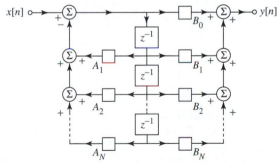

c. The direct form II realization

Since LTI systems can be cascaded in any order, we can switch the recursive and nonrecursive parts to get the structure of Figure 13.9b. This structure suggests that each pair of feedforward and feedback signals can be obtained from a single delay element instead of two. This allows us to use only N delay elements and results in the realization of Figure 13.9c, which is called the **direct form II**, or **canonic**, realization. The term *canonic* implies a realization with the minimum number of delay elements.

If $M < N$, some of the coefficients B_k (for $k > M$) will equal zero, and result in missing signal paths corresponding to these coefficients.

13.8.1   *Transposed Forms*

The canonic form may also be *transposed* to obtain the realization of Figure 13.10. To obtain the **transposed form**, we

1. Reverse the direction of signal flow in each path.
2. Replace nodes by summing blocks and vice versa.
3. Turn the realization around and reverse the input and output.

Remark: Canonic realizations can also be extended to noncausal systems ($M > N$) by including $P = M - N$ additional delay elements for the feedforward signals.

Figure 13.10   Transposed canonic realization of digital filters.

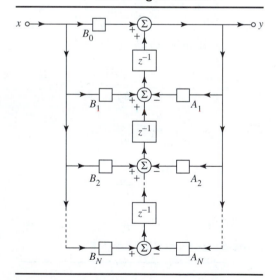

Example 13.24
Digital Filter Realizations

Consider a third-order system with $2y[n] - \frac{1}{3}y[n-2] - \frac{1}{3}y[n-3] = 3x[n-2]$. Its transfer function equals $H(z) = 3z^{-2}/(2 - \frac{1}{3}z^{-2} - \frac{1}{3}z^{-3}) = 1.5z/(z^3 - \frac{1}{6}z - \frac{1}{6})$.

We start by comparing $H(z)$ with the general third-order realization

$$H(z) = \frac{B_0 z^3 + B_1 z^2 + B_2 z + B_3}{z^3 + A_1 z^2 + A_2 z + A_3} = \frac{0z^3 + 0z^2 + 1.5z + 0}{z^3 + 0z^2 - \frac{1}{6}z - \frac{1}{6}}$$

Using only the nonzero gains actually required yields the direct form II and transposed canonic realizations sketched in Figure 13.11.

Figure 13.11   Realizations of $2y[n] - \frac{1}{3}y[n-2] - \frac{1}{3}y[n-3] = 3x[n-2]$ for Example 13.24.

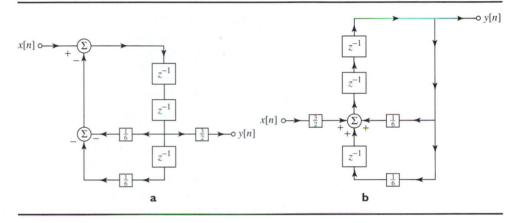

a	**b**

13.8.2   *Cascade and Parallel Realization*

We can factor $H(z)$ into first-order and second-order transfer functions, and each subsystem can be realized separately and then cascaded to yield $H(z)$.

We can also use partial fractions to express $H(z)$ as a sum of first-order and/or second-order subsystems, which can then be connected in parallel to realize $H(z)$.

Example 13.25
Cascade and Parallel
Realizations

Let $H(z) = z^2 (6z - 2)/[(z - 1)(z^2 - \frac{1}{6}z - \frac{1}{6})]$.

(a) This system may be realized as a cascade of $H_1(z) = (6z - 2)/(z - 1)$ and $H_2(z) = z^2/(z^2 - \frac{1}{6}z - \frac{1}{6})$ with $H_C(z) = H_1(z)H_2(z)$.

(b) Using partial fractions, $H(z) = -1.2z/(z - \frac{1}{2}) + 1.2z/(z + \frac{1}{3}) + 6z/(z - 1)$.

The three subsystems can be used to set up the parallel realization. The cascade and parallel realizations are shown in Figure 13.12.

Figure 13.12   Cascade and parallel forms for
$$H(z) = z^2\,(6z - 2)/[(z - 1)(z^2 - \tfrac{1}{6}\,z - \tfrac{1}{6})].$$

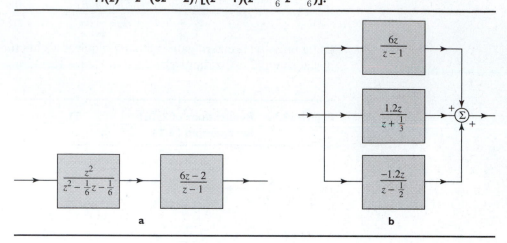

a

b

13.9
Minimum-Phase
Systems

Consider a system described in factored form by $H(z) = K\Pi(z - z_k)/\Pi(z - p_k)$, where Π denotes the product operator. If we replace K by $-K$, or a factor $(z - \alpha)$ by $(1/z - \alpha), (1 - \alpha z)/z$ or $(1 - \alpha z)$, the magnitude of $H(F)$ remains unchanged and only its phase is affected. If $K > 0$, and all the poles and zeros of $H(z)$ lie inside the unit circle, $H(z)$ is stable and defines a **minimum-phase** system. It shows the *smallest group delay* (and the *smallest deviation from zero phase*) at every frequency among all systems with the same magnitude response. A stable system is called **mixed phase** if some of its zeros lie outside the unit circle and **maximum phase** if all its zeros lie outside the unit circle. *Of all stable systems with the same magnitude response, there is only one minimum-phase system.*

Example 13.26
Illustrating the Minimum
Phase Concept

Consider the transfer function of the following systems:

$$H_1(z) = (z - \tfrac{1}{2})(z - \tfrac{1}{4})/[(z - \tfrac{1}{3})(z - \tfrac{1}{5})]$$
$$H_2(z) = (1 - \tfrac{1}{2}z)(z - \tfrac{1}{4})/[(z - \tfrac{1}{3})(z - \tfrac{1}{5})]$$
$$H_3(z) = (1 - \tfrac{1}{2}z)(1 - \tfrac{1}{4}z)/[(z - \tfrac{1}{3})(z - \tfrac{1}{5})]$$

All have poles inside the unit circle and are thus stable. All have the same magnitude but different phase and delay as shown in Figure 13.13.

Of these, $H_1(z)$ is minimum phase with no zeros outside the unit circle, $H_2(z)$ is mixed phase with one zero outside the unit circle, and $H_3(z)$ is maximum phase with all its zeros outside the unit circle.

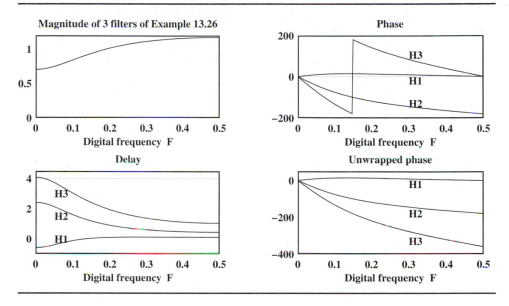

Figure 13.13   The identical magnitudes but different phase and delay of the systems of Example 13.26 illustrating the minimum-phase concept.

13.9.1   *Minimum-Phase Systems from the Magnitude Spectrum*

The design of many digital systems is often based on a specified magnitude response $|H_p(F)|$. The phase response is then selected to ensure a causal, stable system. The transfer function of such a system is unique and may be found by writing its magnitude squared function $|H_p(F)|^2$ as

$$|H_p(F)|^2 = H_p(F)H_p(-F) = H(z)H(1/z)|_{z \to \exp(j2\pi F)}$$

From $|H_p(F)|^2$, we can reconstruct $H_T(z) = H(z)H(1/z)$, which displays conjugate reciprocal symmetry. For every root r_k, there is a root at $1/r_k^*$. We thus select only the roots lying inside the unit circle to extract $H(z)$. The following example illustrates the process.

Example 13.27
Minimum-Phase System
from $|H(\Omega)|^2$

Let $|H_p\Omega)|^2 = \dfrac{5 + 4\cos(\Omega)}{17 + 8\cos(\Omega)}$. We use Euler's relation to give

$$|H_p(\Omega)|^2 = \frac{5 + 2\exp(j\Omega) + 2\exp(-j\Omega)}{17 + 4\exp(j\Omega) + 4\exp(-j\Omega)}$$

Upon substituting $\exp(j\Omega) \longrightarrow z$, we obtain $H_T(z)$ as

$$H_T(z) = H(z)H(1/z) = \frac{5 + 2z + 2/z}{17 + 4z + 4/z} = \frac{2z^2 + 5z + 2}{4z^2 + 17z + 4} = \frac{(2z + 1)(z + 2)}{(4z + 1)(z + 4)}$$

To extract $H(z)$, we use only those roots of $H_T(z)$ with $|z| < 1$, which yields

$$H(z) = \frac{(2z + 1)}{(4z + 1)} = \frac{1}{2} \frac{(z + \frac{1}{2})}{(z + \frac{1}{4})}$$

Comment: Note that $|H(z)|_{z=1} = |H_p(\Omega)|_{\Omega=0} = 3/5$, implying identical dc gains.

Using $z \longrightarrow \exp(j\Omega)$, it is easy to confirm that $|H(\Omega)|^2 = |H_p(\Omega)|^2$.

13.10
The Frequency Response: A Graphical Interpretation

The factored form or pole-zero plot of $H(z)$ is quite useful if we want a qualitative picture of the frequency response. Consider the stable transfer function $H(z)$ given in factored form by

$$H(z) = \frac{8(z - 1)}{(z - 0.6 - j0.6)(z - 0.6 + j0.6)}$$

Its pole-zero plot is shown in Figure 13.14. At $\Omega = \frac{1}{3}\pi$, for example, the frequency response $H(\Omega = \frac{1}{3}\pi)$ can be written, using $z = \exp(j\Omega) = \exp(j\frac{1}{3}\pi)$, as

$$H(\Omega = \tfrac{1}{3}\pi) = \frac{8[\exp(j\frac{1}{3}\pi) - 1]}{[\exp(j\frac{1}{3}\pi) - 0.6 - j0.6][\exp(j\frac{1}{3}\pi) - 0.6 + j0.6]} = \frac{8\overline{N}_1}{\overline{D}_1 \overline{D}_2}$$

Analytically, the magnitude is the ratio of the magnitudes of each term.

Graphically, the complex terms may be viewed in the z-plane as vectors $\overline{N}_1$, $\overline{D}_1$ and $\overline{D}_2$, directed from each pole or zero location to the point on the unit circle corresponding to $z = \exp(j\frac{1}{3}\pi)$. The ratio $|\overline{N}_1|/(|\overline{D}_1||\overline{D}_2|)$ yields the magnitude at $\Omega = \frac{1}{3}\pi$. The difference in the angles yields the phase at $\Omega = \frac{1}{3}\pi$.

A graphical evaluation can yield exact results, but is much more suited to obtaining a qualitative estimate of the magnitude response. We start at $\Omega = 0$, move counterclockwise (for positive increasing Ω) along the unit circle, and observe how the ratio of the vectors influences the magnitude.

For our example, we estimate the ratio $|\overline{N}_1|/(|\overline{D}_1||\overline{D}_2|)$ as Ω is varied and make the following observations (Figure 13.14):

1. $\Omega = 0$: $|\overline{N}_1|$ is zero, and the magnitude is zero.
2. $0 < \Omega < \frac{1}{4}\pi$: $|\overline{N}_1|$ and $|\overline{D}_2|$ increase, but $|\overline{D}_1|$ decreases. The response is small but increasing. At $\Omega = \frac{1}{4}\pi$, $|\overline{D}_1|$ attains its smallest value, and we expect a peak in the response.
3. $\frac{1}{4}\pi < \Omega < \pi$: $|\overline{N}_1|$ and $|\overline{D}_1|$ are nearly equal, and the response is nearly equal to $1/|\overline{D}_2|$. Since $|\overline{D}_2|$ is increasing, the response decreases.
4. If the poles are moved closer to the unit circle, but with no change in their orientation, we should expect to see a larger peak at $\Omega = \frac{1}{4}\pi$.

The form of this response is typical of a bandpass filter.

Figure 13.14   Pole-zero plot and frequency response evaluation for
$$H(z) = 8(z - 1)/[(z^2 - 1.2z + 0.72)].$$

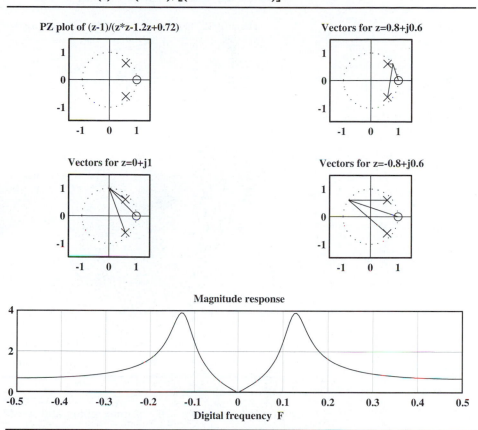

The Rubber Sheet Analogy   If we imagine the z-plane as a rubber sheet tacked down at the zeros of $H(z)$, and poked up to an infinite height at the pole locations, the curved surface of the rubber sheet approximates the magnitude of $H(z)$ for any z (Figure 13.15, p. 598). The slice around the unit circle approximates the frequency response $H(\Omega)$.

13.10.1   *Implications of Pole-Zero Placement for Filter Design*

The qualitative effect of poles and zeros on the magnitude response can be used to advantage in understanding the pole-zero patterns of real filters. The basic strategy for pole and zero placement is based on mapping the passband and stopband frequencies on the unit circle and then positioning the poles and zeros based on the following reasoning:

Figure 13.15   The 3D z-plane magnitude of the transfer function
$H(z) = 8(z − 1)/[(z^2 − 1.2z + 0.72)]$.

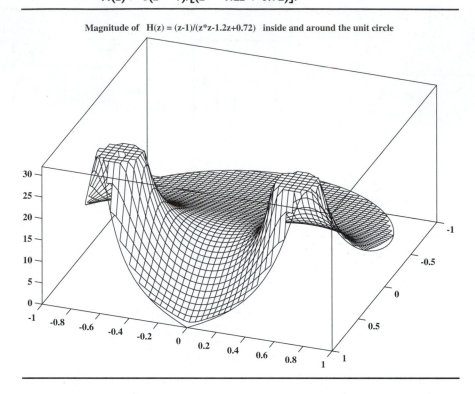

Magnitude of $H(z) = (z-1)/(z*z-1.2z+0.72)$ inside and around the unit circle

---

1. *Conjugate symmetry:* All complex poles and zeros must be paired with their complex conjugates.
2. *Causality:* To ensure a causal system, the total number of zeros must be less than or equal to the total number of poles.
3. *Origin:* Poles or zeros at the origin do not affect the magnitude response.
4. *Pole placement and stability:* For a stable system, the poles must be placed *inside* (not just on) the unit circle. The pole radius is proportional to the gain and inversely proportional to the bandwidth. Poles closer to the unit circle produce a large gain over a narrower bandwidth. Clearly, the passband should contain poles near the unit circle for large passband gains.

A rule of thumb: For narrowband filters with bandwidth $\Delta\Omega \leq 0.2$ centered about Ω_0, place conjugate poles at $r \exp(\pm j\Omega_0)$ where $r \approx 1 - \frac{1}{2}\Delta\Omega$.

5. *Zero placement and minimum phase:* Zeros can be placed anywhere in the z-plane. To ensure minimum phase, the zeros must lie within the unit circle. Zeros on the unit circle result in a null in the response. Thus, the stopband should contain zeros on or near the unit circle. A good starting choice is to place a zero

(on the unit circle) in the middle of the stopband, or two zeros at the edges of the stopband.

6. *Transition region:* To achieve a steep transition from the stopband to the passband, a good choice is to pair each stopband zero with a pole along (or near) the same radial line, and close to the unit circle.

7. *Pole-zero interaction:* Poles and zeros interact to produce a composite response that may not match qualitative predictions. Their placement may have to be changed, or other poles and zeros may have to be added, to tweak the response. Poles closer to the unit circle, or farther away from each other, produce small interaction. Poles closer to each other, or farther from the unit circle, produce more interaction. The closer we wish to approximate a given response, the more poles and zeros we require, and the higher the filter order.

The pole-zero patterns of some simple filter configurations as suggested by these guidelines and their magnitude responses are sketched in Figure 13.16 (p. 600). Note that bandstop and bandpass filters require an order greater than 1. Filter design using pole-zero placement involves trial and error. More formal design methods are presented in the next chapter.

Example 13.28
Filter Design by
Pole-Zero Placement

(a) We design a bandpass filter with center frequency = 100 Hz, passband 10 Hz, stopband edges at 50 Hz and 150 Hz, and sampling frequency 400 Hz. Then $\Omega_0 = \frac{1}{2}\pi$, $\Delta\Omega = \pi/20$, $\Omega_s = [\pi/4, 3\pi/4]$ and $r = 1 - \frac{1}{2}\Delta\Omega = 0.9215$.
Passband: Place poles at $p_{1,2} = r\exp(\pm j\Omega_0) = 0.9215\exp(\pm j\frac{1}{2}\pi) = \pm j0.9215$.
Stopband: Place conjugate zeros at $z_{1,2} = \exp(\pm j\pi/4)$ and $z_{3,4} = \exp(\pm j3\pi/4)$

$$H(z) = \frac{[z - \exp(j\pi/4)][z - \exp(-j\pi/4)][z - \exp(j3\pi/4)][z - \exp(-j3\pi/4)]}{[z - j0.9215][z + j0.9215]}$$

Upon simplification, we obtain $H(z) = (z^4 - 1)/(z^2 - 0.8941)$.

Comment:

This filter is noncausal. To obtain a causal filter, we could, for example, use double-poles at each pole location to get

$$H(z) = (z^4 - 1)/(z^2 - 0.8941)^2 = H(z) = (z^4 - 1)/(z^4 + 1.6982z^2 + 0.7210)$$

The pole-zero pattern and response of this filter is shown in Figure 13.17 (p. 601).

(b) We design a notch filter with notch frequency 60 Hz, stopband 5 Hz, and sampling frequency 300 Hz. We compute $\Omega_0 = 2\pi/5$, $\Delta\Omega = \pi/30$ and $r = 1 - \Delta\Omega/2 = 0.9476$.
Stopband: Place zeros at the notch frequency: $z_{1,2} = \exp(\pm j\Omega_0) = \exp(\pm j2\pi/5)$.
Passband: Place poles along the orientation of the zeros, and radius r, at the locations $p_{1,2} = r\exp(\pm j\Omega_0) = 0.9476\exp(\pm j2\pi/5)$. We then obtain $H(z)$ as

$$H(z) = \frac{[z - \exp(j2\pi/5)][z - \exp(-j2\pi/5)]}{[z - 0.9476\exp(j2\pi/5)][z - 0.9476\exp(-j2\pi/5)]}$$

**Figure 13.16   Pole-zero patterns of various filters and their correspond-
ing frequency response.**

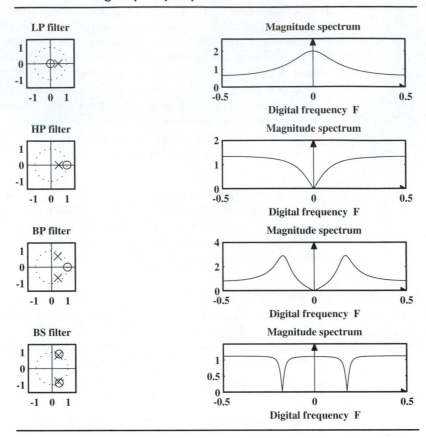

Upon simplfication, we get $H(z) = (z^2 - 0.618z + 1)/(z^2 - 0.5857 + 0.898)$. The pole-zero pattern and response of this filter is shown in Figure 13.17.

13.11
Special Topics: Inverse Systems and Allpass Filters

The **inverse system** corresponding to a transfer function $H(z)$ is denoted by $H^{-1}(z)$, and defined as

$$H^{-1}(z) = H_I(z) = 1/H(z)$$

A cascade of a system and its inverse has a transfer function of unity

$$H_C(z) = H(z)H^{-1}(z) = 1$$

This cascaded system is called an **identity system**, and its impulse response equals $h_C[n] = \delta[n]$. The inverse system can be used to undo the effect of the original

Figure 13.17 **Pole-zero patterns and frequency response of the bandpass and notch filters designed in Example 13.28.**

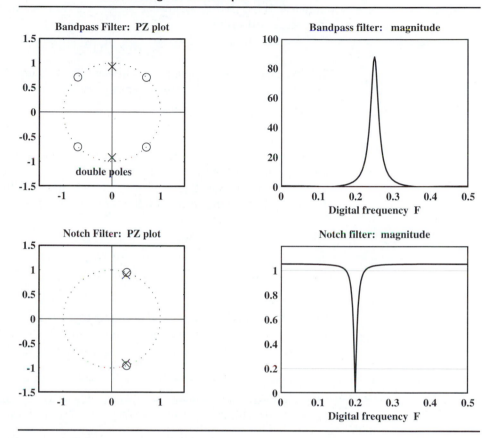

system. We can also describe $h_C[n]$ by the convolution $h[n] \star h_I[n]$. It is far easier to find the inverse of a system in the transformed domain.

Example 13.29

Inverse Systems

Consider a system with the difference equation $y[n] = \alpha y[n-1] + x[n]$.

To find the inverse system, we evaluate $H(z)$ and take its reciprocal. Thus, $H(z) = 1/(1 - \alpha z^{-1})$, and $H_I(z) = 1/H(z) = 1 - \alpha z^{-1}$. The difference equation of the causal inverse system is $y[n] = x[n] - \alpha x[n-1]$.

13.11.1 *Causality and Stability of Inverse Systems*

Since all the poles and zeros of a stable minimum-phase system lie within the unit circle, the inverse of a minimum-phase system is also stable and minimum phase. Further, *if a causal stable system is to have a causal stable inverse, the system must be minimum phase*, with as many poles as zeros.

13.11.2   *Allpass Filters*

Allpass filters are characterized by a magnitude response that is unity for all frequencies. Thus, $|H(F)| = 1$ over $F = (-\frac{1}{2}, \frac{1}{2})$. Their transfer function $H(z)$ also satisfies the relationship

$$H(z)H(1/z) = 1$$

This implies that each pole of an allpass filter is paired by a conjugate reciprocal zero. Here are two stable first-order allpass filters:

$$H_1(z) = (1 + \alpha z)/(z + \alpha^*) \qquad H_2(z) = (1 + \beta^* z)/(z + \beta) \qquad |\alpha|, |\beta| < 1$$

The zero of $H_1(z)$ is at $z = -(1/\alpha)$ and it has a reciprocal pole at $z = (-\alpha)^*$.

If $|\alpha| > 1$ (or $|\beta| > 1$), $H_1(z)$ (or $H_2(z)$) describes an unstable allpass filter. Note that allpass filters cannot be minimum phase.

Obviously, the cascade of allpass filters is also an allpass filter. An allpass filter of order N has a numerator and denominator of equal order N and is described by the transfer function

$$H_{AP}(z) = \frac{N(z)}{D(z)} = \frac{C_N + C_{N-1}z^{-1} + \cdots + C_1 z^{N-1} + z^{-N}}{1 + C_1 z^{-1} + C_2 z^{-2} + \cdots + C_N z^{-N}}$$

Notice how the coefficients of $N(z)$ and $D(z)$ appear in reversed order.

13.11.3   *Stabilization of Unstable Filters*

Allpass filters are often used to stabilize unstable digital filters while preserving their magnitude response. Consider the unstable filter

$$H_u(z) = z/(1 + \alpha z) \qquad |\alpha| < 1$$

It has a pole at $z = -1/\alpha$ and is thus unstable. If we cascade $H_u(z)$ with a first-order allpass filter with $H_1(z) = (1 + \alpha z)/(z + \alpha^*)$, we obtain

$$H(z) = H_u(z)H_1(z) = z/(z + \alpha^*)$$

This filter has a pole at $z = -\alpha^*$ and since $|\alpha| < 1$, it is a stable filter.

This idea can easily be extended to filters of arbitrary order described as a cascade of a stable portion $H_s(z)$ and P unstable first-order sections

$$H_u(z) = H_s(z) \prod_{m=1}^{P} \frac{z}{1 + \alpha_m z} \qquad |\alpha_m| < 1$$

To stabilize $H(z)$ we use an allpass filter $H_{AP}(z)$ that is a cascade of P allpass sections. Its form is

$$H_{AP}(z) = \prod_{m=1}^{P} \frac{1 + \alpha_m z}{z + \alpha_m^*} \qquad |\alpha_m| < 1$$

The stabilized filter $H(z)$ is then described by the cascade $H_u(z)H_{AP}(z)$ as

$$H(z) = H_u(z)H_{AP}(z) = H_s(z) \prod_{m=1}^{P} \frac{z}{z + \alpha_m^*} \qquad |\alpha_m| < 1$$

There are two advantages to this method. First, the magnitude response of the original filter is unchanged. And second, the order of the new filter is the same as the original.

Remark: For the unstable filter, we require the inequality $|\alpha_m| < 1$. Clearly, if $H_u(z)$ has a pole on the unit circle, its conjugate reciprocal will also lie on the unit circle, and no stabilization is possible.

13.11.4 *Minimum-Phase Filters Using Allpass Filters*

Even though allpass filters cannot be minimum phase, they can be used to convert stable nonminimum-phase filters to stable minimum-phase filters. We describe a nonminimum-phase transfer function by a cascade of a minimum-phase part $H_M(z)$ (with all poles and zeros inside the unit circle) and a portion with zeros outside the unit circle such that

$$H_{\mathrm{NM}}(z) = H_M(z) \prod_{m=1}^{P} (z + \alpha_M) \qquad |\alpha_m| > 1$$

We now seek an *unstable* allpass filter with

$$H_{\mathrm{AP}}(z) = \prod_{m=1}^{P} \frac{1 + z\alpha_m^*}{z + \alpha_m} \qquad |\alpha_m| > 1$$

The cascade of $H_{\mathrm{NM}}(z)$ and $H_{\mathrm{AP}}(z)$ yields a minimum-phase filter with

$$H(z) = H_{\mathrm{NM}}(z)H_{\mathrm{AP}}(z) = H_M(z) \prod_{m=1}^{P} (1 + z\alpha_m^*)$$

Once again, $H(z)$ has the same order as the original filter.

13.12 Connections

The z-transform may also be viewed as (1) a discrete version of the Laplace transform, (2) a relation arising from the response (using convolution) of a linear system to the complex exponentials that are its eigensignals or (3) a generalization of the DTFT to complex frequencies. The first viewpoint is explored in the next chapter, and the second one was discussed in Chapter 4. Here, we concentrate on the last connection.

13.12.1 *The DTFT and the z-Transform*

For causal, absolutely summable signals, the DTFT is simply the onesided z-transform with $z \longrightarrow \exp(j2\pi F)$. Table 13.5 shows this correspondence. Due to table 13.5 the presence of the convergence factor r^{-k}, the z-transform $X(z)$ of $x[n]$ no longer displays the kind of symmetry present in its DTFT $X(F)$.

For signals that are not absolutely summable, such as $u[n]$, an appropriate choice of r can make $x[n]r^{-k}$ absolutely summable and ensure a z-transform. For such signals, the z-transform equals just the nonimpulsive portion of the DTFT

Table 13.5   Comparison of DTFT and z-Transform Pairs.

$x[n]$	DTFT $X_p(F)$ (one period)	z-transform
$\alpha^n u[n]$ $\|\alpha\| < 1$	$\dfrac{1}{1 - \alpha\exp(-j2\pi F)}$	$z/(z - \alpha)$
$n\alpha^n u[n]$	$\dfrac{\alpha\exp(-j2\pi F)}{[1 - \alpha\exp(-j2\pi F)]^2}$	$z\alpha/(z - \alpha)^2$
$\delta[n]$	1	1
$u[n]$	$\dfrac{1}{1 - \exp(-j2\pi F)} + \dfrac{1}{2}\delta(F)$	$z/(z - 1)$

with $\exp(j2\pi F) \longrightarrow z$. We can thus find the z-transform of causal signals from their DTFT but not the other way around.

We note that even though the convergence factor allows us to handle exponentially growing signals such as $\alpha^n u[n]$, neither the z-transform nor the DTFT can handle signals of faster growth, such as $\alpha^{n^2} u[n]$.

Remark: Since $z = r\exp(j2\pi F)$ is complex, the z-transform may be plotted as a surface in the z-plane. The DTFT (with $z \longrightarrow \exp(j2\pi F)$) is then just the cross section of this surface along the unit circle ($|z| = 1$).

Example 13.30
z-Transform of
Absolutely Summable
Signals
from the DTFT

(a) The signal $x[n] = \alpha^n u[n]$, $|\alpha| < 1$, is absolutely summable. Its DTFT equals $X(F) = 1/[1 - \alpha\exp(-j2\pi F)]$. We can find the z-transform of $x[n]$ from its DTFT as $X(z) = 1/[1 - \alpha z^{-1}] = z/(z - \alpha)$.

(b) The signal $x[n] = \delta[n]$ is absolutely summable. Since $X(F) = 1$, $X(z) = 1$.

(c) The signal $x[n] = u[n]$ is not absolutely summable. Its DTFT is given by $X(F) = \frac{1}{2}\delta(F) + 1/[1 - \exp(-j2\pi F)]$. We find the z-transform of $u[n]$ by dropping the impulsive part in the DTFT and replacing $\exp(j2\pi F)$ by z, to give $X(z) = 1/(1 - z^{-1}) = z/(z - 1)$.

13.13
The
z-Transform:
A Synopsis

Definition   The z-transform is a summation. It allows transformation of DT signals or systems to the z (complex frequency) domain. For causal signals, the onesided z-transform sums $x[k]z^{-k}$ from 0 to ∞. The series converges only for certain values of z, which define its region of convergence (ROC).

ROC   Often, $X(z)$ is a rational function and can be characterized by its poles and zeros. For causal signals, the ROC lies outside a circle whose radius is the magnitude of the largest pole location.

Properties	The quantity z^{-1} plays the role of a unit delay operator. The ZT of complicated signals is found from simpler forms using properties. For even symmetric signals, $X(z) = X(1/z)$ and for odd ones, $X(z) = -X(1/z)$. For symmetric signals, the poles and zeros occur in reciprocal pairs. Multiplication in time corresponds to convolution in the z-domain.
Inverse z-transform	Inversion of $X(z)$ to obtain $x[n]$ involves complex integration. Simpler ways include long division, and partial fraction expansion of $X(z)/z$ using tables. Only when the ROC is specified do we obtain a unique $x[n]$.
Transfer function	The transfer function $H(z)$ is the ratio of the output $Y(z)$ and input $X(z)$ or the transform of the system impulse response $h[n]$. It is a ratio of polynomials in z. Poles of $H(z)$ are called *natural modes*. Poles of $Y(z) = H(z)X(z)$ determine the form of the natural response.
Stability	Stability requires all the poles of $H(z)$ to lie entirely within the unit circle in the z-plane. For causal systems ($h[n] = 0, n < 0$), the ROC (or poles) of $H(z)$ must be exterior to a circle of finite radius.
System response	The z-transform is useful for system analysis. It converts convolution to a product and difference equations to algebraic equations. Initial conditions can be accounted for during the transformation. The response can be separated into its zero-state and zero-input components.
Frequency response	If $H(z)$ is evaluated on the unit circle or $z = \exp(j2\pi F)$, we obtain the steady-state transfer function, or frequency response $H_p(F)$. This is just the DTFT which describes the response to DT harmonics. The steady-state response to $A\cos(2\pi n F_0 + \theta)$ is $A|H|\cos(2\pi n F_0 + \theta + \phi)$ where $|H|$ and ϕ are the magnitude and phase of $H_p(F_0)$ or $H(z)$ with $z = \exp(j2\pi F_0)$.
Digital filters	Digital filters described by difference equations can be realized using delays, multipliers and summers. Realizations with the least number of delay elements are called canonical.
Minimum-phase systems	A stable, minimum-phase system has all its poles and zeros inside the unit circle. It has the *smallest group delay*. For a causal stable system to have a causal stable inverse, the system must be minimum phase. The transfer function of a minimum-phase system is unique and can be found from its magnitude-squared function.
Special topics	The inverse of a transfer function $H(z)$ is $1/H(z)$, which can be used to undo the effect of the original system. Allpass filters have unity magnitude for all frequencies. They can never be minimum phase. They can be used to convert stable nonminimum-phase filters to stable minimum-phase ones. They are also used to stabilize unstable digital filters while preserving their magnitude response.

Appendix 13A
MATLAB
Demonstrations
and Routines

Getting Started on the MATLAB Demonstrations

This appendix presents MATLAB routines related to concepts in this chapter, and some examples of their use. See Appendix 1A for introductory remarks.

Note: All polynomial arrays must be entered in descending powers.

z-Transform and its Symbolic Inverse

```
y=ztr([A b c w r n0])
```

Finds the ZT of $x(t) = A(b)^{n-n0}(n - n0)^C \cos[(n - n0)w + r]$.

```
x=izt('tf',p,q)
```

Returns IZT of $X(z) = P(z)/Q(z)$ in symbolic form.

Example

To find the z-transform of $4n(\frac{1}{2})^n u[n]$, use

```
>>y=ztr([4 0.5 1 0 0 0])
```

MATLAB returns $y =[0\ 2\ 0;\ 1\ -1\ 0.25]$. To find the IZT, use

```
>>x=izt('tf',[0 2 0],[1 -1 0.25])
```

MATLAB returns

```
x = 2*n.*(0.5 . ^(n-1))
```

[which equals $4n(\frac{1}{2})^n$].

DT System Response in Symbolic Form

```
[yt,yzs,yzi]=sysresp2('ty',n,d,p,q,ic)
```

Computes DT system response with ty='z', n,d= Coefficient arrays of $N(z)$ and $D(z)$ of $H(z) = N(z)/D(z)$.

p,q= Coefficient arrays of $P(z)$ and $Q(z)$ of $X(z) = P(z)/Q(z)$.

Example

Let $H(z) = z(z + 1)/(z^2 - 0.25)$, $x = 2(\frac{1}{2})^n u[n]$, $y[-1] = 4$, and $y[-2] = 0$. Then $X(z) = 2/(z - \frac{1}{2})$. To find the response, use the command

```
>>[yt,yzs,yzi]=sysresp2('z',[1 1 0],[1 0 -0.25],[0 2],[1 -0.5],[4 0])
```

yt, yzs and yzi are string functions of n and must be evaluated. For example:

```
>>n=0:20;y=eval(yt);lines(n,yt)
```

Partial Fractions

```
[r,p,k]=tf2pf(n,d)
```

Returns PF as constants, poles and impulse coefficients.

```
[n,d]=pf2tf(r,p,k)
```

Returns the TF from the PF expansion.

Demonstrations and Utility Functions

`iztlong(N,D,K)`	Finds IZT of $N(z)/D(z)$ by long division to K terms.
`plotpz(p,q,'z')`	Plots poles and zeros of $X(z) = P(z)/Q(z)$.
`tfplot('z',n,d)`	Plots TF magnitude and phase of $H(z) = N(z)/D(z)$.
`zdz(n,a)`	ZT of $(-zd/dz)^n[z/(z-a)]$ using times-n property.
`ssresp('z',n,d,x)`	Steady-state response to sinusoids (see Chapter 6).
`zplane3d(r,n,d)`	3D complex frequency plot of $H(z)$ over a circle r.
`tfmin(n,d,'z')`	Minimum-phase TF from symmetric $H(z)$ (see Chapter 9).
`stablize(n,d,'z')`	Stabilizes $H(z)$ using an allpass filter.
`dfdpzp`	Interactive filter design by pole-zero placement. Requires a mouse.

PROBLEMS

P13.1 (The z-Transform and its ROC) Use the defining relation to find the z-transform and its region of convergence for

(a) $\{x[n]\} = \{1, 2, 3, 2, 1\}$ ↑

(b) $\{x[n]\} = \{-1, 2, 0, -2, 1\}$ ↑

(c) $\{x[n]\} = \{1, 1, 1, 1\}$ ↑

(d) $\{x[n]\} = \{1, 1, -1, -1\}$ ↑

P13.2 (z-Transforms) Find the z-transforms of

(a) $(2)^{n+2}u[n]$

(b) $n(2)^{0.2n}u[n]$

(c) $(2)^{n+2}u[n-1]$

(d) $n(2)^{n+2}u[n-1]$

(e) $(n+1)(2)^n u[n]$

(f) $(n-1)(2)^{n+2}u[n]$

(g) $\cos(\frac{1}{4}n\pi - \frac{1}{4}\pi)u[n]$

(h) $(\frac{1}{2})^n \cos(\frac{1}{4}n\pi)u[n]$

(i) $(\frac{1}{2})^n \cos(\frac{1}{4}n\pi - \frac{1}{4}\pi)u[n]$

(j) $(\frac{1}{3})^n \{u[n] - u[n-4]\}$

(k) $(\frac{1}{2})^n n\cos(\frac{1}{4}n\pi)u[n]$

(l) $[(\frac{1}{2})^n - (-\frac{1}{2})^n]nu[n]$

P13.3 (Two-sided z-Transform) Find the z-transform $X(z)$ and its ROC for

(a) $x[n] = u[-n-1]$

(b) $x[n] = (\frac{1}{2})^{-n}u[-n-1]$

(c) $x[n] = (\frac{1}{2})^{|n|}$

(d) $x[n] = u[-n-1] + (\frac{1}{3})^n u[n]$

(e) $x[n] = (\frac{1}{2})^{-n}u[-n-1] + (\frac{1}{3})^n u[n]$

P13.4 (Properties) Find the z-transform of $nu[n]$ using
(a) The defining relation.
(b) The times-n property.
(c) Convolution and shifting for $u[n] \star u[n] = (n+1)u[n+1]$.
(d) Convolution and superposition for $u[n] \star u[n] = (n+1)u[n]$.

P13.5 (Properties) Given the z-transform pair $x[n] \longleftrightarrow 4z/(z+\frac{1}{2})^2$, $|z| > \frac{1}{2}$, find the z-transform of the following using properties:
(a) $x[n-2]$ 　　　　　　**(b)** $(2)^n x[n]$ 　　　　　**(c)** $nx[n]$
(d) $(2)^n nx[n]$ 　　　　　**(e)** $n^2 x[n]$ 　　　　　　**(f)** $[n-2]x[n]$
(g) $x[-n]$ 　　　　　　　**(h)** $x[n] - x[n-1]$ 　　　**(i)** $x[n] \star x[n]$

P13.6 (Properties) Given the z-transform pair $(2)^n u[n] \longleftrightarrow X(z)$, use properties to find the time signal corresponding to the following:
(a) $X(2z)$ 　　　　　　　**(b)** $X(1/z)$ 　　　　　　**(c)** $zX'(z)$
(d) $zX(z)/(z-1)$ 　　　　**(e)** $zX(2z)/(z-1)$ 　　　**(f)** $z^{-1}X(z)$
(g) $z^{-2}X(2z)$ 　　　　　**(h)** $X^2(z)$

P13.7 (Properties) Find the z-transform of $\text{rect}[n/2N]$, and use this result to evaluate the z-transform of $\text{tri}[n/N]$. [HINT: Start by expressing $\text{rect}[n/2N]$ as $u[n+N] - u[n-N-1]$.]

P13.8 (Pole-Zero Patterns and Symmetry) Plot the pole-zero patterns for each $X(z)$. Which of these describe symmetric time sequences?
(a) $X(z) = (z^2 + z - 1)/z$ 　　　　　　　**(b)** $X(z) = (z^4 + 2z^3 + 3z^2 + 2z + 1)/z^2$
(c) $X(z) = (z^4 - z^3 + z - 1)/z^2$ 　　　　**(d)** $X(z) = (z^2 - 1)(z^2 + 1)/z^2$

P13.9 (Initial Value and Final Value Theorems) Find the initial and final values for each causal sequence described by its z-transform.
(a) $X(z) = 2/(z^2 + \frac{1}{6}z - \frac{1}{6})$ 　　　　　**(b)** $X(z) = 2z^2/(z^2 + z + \frac{1}{4})$
(c) $X(z) = 2z/(z^2 + z - 1)$ 　　　　　　**(d)** $X(z) = (2z^2 + \frac{1}{4})/[(z-1)(z+\frac{1}{4})]$
(e) $X(z) = (z + \frac{1}{4})/(z^2 + \frac{1}{4})$ 　　　　　**(f)** $X(z) = (2z + 1)/(z^2 - \frac{1}{2}z - \frac{1}{2})$

P13.10 (Inverse Transforms of Polynomials) Find $x[n]$ if
(a) $X(z) = 2 - z^{-1} + 3z^{-3}$ 　　　　　**(b)** $X(z) = (2 + z^{-1})^3$
(c) $X(z) = (z - z^{-1})^2$ 　　　　　　　**(d)** $X(z) = (z - z^{-1})^2(2 + z)$

P13.11 (Inverse Transforms by Long Division) Find $x[n], n = 0, 1, 2, 3$ if
(a) $X(z) = (z+1)^2/(z^2+1)$ 　　　　　　**(b)** $X(z) = (z+1)/(z^2+2)$
(c) $X(z) = 1/(z^2 - \frac{1}{4})$ 　　　　　　　**(d)** $X(z) = (1 - z^{-2})/(2 + z^{-1})$

P13.12 (Inverse Transforms by Partial Fractions) Find $x[n]$ if
(a) $X(z) = z/[(z+1)(z+2)]$ 　　　　　　　**(b)** $X(z) = 16/[(z-2)(z+2)]$
(c) $X(z) = 3z^2/[(z^2 - \frac{3}{2}z + \frac{1}{2})(z - \frac{1}{4})]$ 　　　**(d)** $X(z) = 3z^3/[(z^2 - \frac{3}{2}z + \frac{1}{2})(z - \frac{1}{4})]$
(e) $X(z) = 3z^4/[(z^2 - \frac{3}{2}z + \frac{1}{2})(z - \frac{1}{4})]$ 　　　**(f)** $X(z) = 4z/[(z+1)^2(z+3)]$
(g) $X(z) = z/[(z^2 + z + \frac{1}{4})(z+1)]$ 　　　　**(h)** $X(z) = z/[(z^2 + z + \frac{1}{4})(z+\frac{1}{2})]$
(i) $X(z) = 1/[(z^2 + z + \frac{1}{4})(z+1)]$ 　　　　**(j)** $X(z) = z/[(z^2 + z + \frac{1}{2})(z+1)]$

(k) $X(z) = z^3/[(z^2 - z + \frac{1}{2})(z - 1)]$ **(l)** $X(z) = z^2/[(z^2 + z + \frac{1}{2})(z + 1)]$

(m) $X(z) = 2z/(z^2 - \frac{1}{4})^2$ **(n)** $X(z) = 2/(z^2 - \frac{1}{4})^2$

(o) $X(z) = z/(z^2 + \frac{1}{4})^2$ **(p)** $X(z) = z^2/(z^2 + \frac{1}{4})^2$

P13.13 (IZT of Anticausal Sequences) Find $\{x[n]\}$ for $n = -1, -2, -3$ using long division if $x[n]$ is assumed to be anticausal.

(a) $X(z) = (z^2 + 4z)/(z^2 - z + 2)$ **(b)** $X(z) = z/(z + 1)^2$

(c) $X(z) = z^2/(z^3 + z - 1)$ **(d)** $X(z) = (z^3 + 1)/(z^2 + 1)$

P13.14 (IZT of Twosided Sequences) Find the inverse z-transform using the ROC specified for each sequence.

(a) $X(z) = (z^2 + 5z)/(z^2 - 2z - 3)$ $|z| < 1$

(b) $X(z) = (z^2 + 5z)/(z^2 - 2z - 3)$ $|z| > 3$

(c) $X(z) = (z^2 + 5z)/(z^2 - 2z - 3)$ $1 < |z| < 3$

P13.15 (Convolution) Find each convolution using the z-transform.

(a) $\{-1, 2, 0, 3\} \star \{2, 0, 3\}$
      $\uparrow$ $\uparrow$
               (b) $\{-1, 2, 0, -2, 1\} \star \{-1, 2, 0, -2, 1\}$
                                      $\uparrow$ $\uparrow$

(c) $(2)^n u[n] \star (2)^n u[n]$ **(d)** $(2)^n u[n] \star (3)^n u[n]$

P13.16 (Causality) How would you recognize a causal system based on each of the following:

(a) Its impulse response $h[n]$?

(b) Its transfer function $H(z)$ and its region of convergence?

(c) Its system difference equation?

P13.17 (Transfer Function) Find the transfer function and difference equation for the following systems. Which of these are stable?

(a) $h[n] = (2)^n u[n]$ **(b)** $h[n] = [1 - (\frac{1}{3})^n]u[n]$

(c) $h[n] = n(\frac{1}{3})^n u[n]$ **(d)** $h[n] = 0.5\delta[n]$

(e) $h[n] = \delta[n] - (-\frac{1}{3})^n u[n]$ **(f)** $h[n] = [(2)^n - (3)^n]u[n]$

P13.18 (Transfer Function) Find the transfer function and impulse response of the following systems:

(a) $y[n] + 3y[n-1] + 2y[n-2] = 2x[n] + 3x[n-1]$

(b) $y[n] + 4y[n-1] + 4y[n-2] = 2x[n] + 3x[n-1]$

(c) $y[n] = 0.2x[n]$

(d) $y[n] = x[n] + x[n-1] + x[n-2]$

P13.19 (Transfer Functions) Set up the system difference equations from the following system transfer functions:

(a) $H(z) = 3/(z + 2)$ **(b)** $H(z) = (1 + 2z + z^2)/[(1 + z^2)(4 + z^2)]$

(c) $H(z) = \dfrac{2}{1 + z} - \dfrac{1}{2 + z}$ **(d)** $H(z) = \dfrac{2z}{1 + z} - \dfrac{1}{2 + z}$

P13.20 (System Response) The transfer function $H(z)$ of a system is

$$H(z) = \frac{2z(z - 1)}{4 + 4z + z^2}$$

Find the system response for the following inputs:

(a) $x[n] = \delta[n]$ **(b)** $x[n] = 2\delta[n] + \delta[n+1]$

(c) $x[n] = u[n]$ **(d)** $x[n] = (2)^n u[n]$

(e) $x[n] = nu[n]$ **(f)** $x[n] = \cos(\frac{1}{2}n\pi)u[n]$

P13.21 (System Analysis) Using z-transforms, find the unit step response of the causal digital filters described by

(a) $H(z) = 4z/(z - 0.5)$ **(b)** $y[n] + \frac{1}{2}y[n-1] = 6x[n]$

P13.22 (System Analysis) Find the zero-state response, zero-input response and total response for each of the following systems:

(a) $y[n] - \frac{1}{4}y[n-1] = (\frac{1}{3})^n u[n], \quad y[-1] = 8$

(b) $y[n] + \frac{3}{2}y[n-1] + \frac{1}{2}y[n-2] = (-\frac{1}{2})^n u[n], \quad y[-1] = 2, \quad y[-2] = -4$

(c) $y[n] + y[n-1] + \frac{1}{4}y[n-2] = 4(\frac{1}{2})^n u[n], \quad y[-1] = 6, \quad y[-2] = -12$

(d) $y[n] - y[n-1] + \frac{1}{2}y[n-2] = (\frac{1}{2})^n u[n], \quad y[-1] = -1, \quad y[-2] = -2$

P13.23 (Steady-State Response) The transfer function $H(s)$ of a system is

$$H(z) = \frac{2z(z-1)}{\frac{1}{4} + z^2}$$

Find its steady-state response for the following inputs:

(a) $x[n] = 4u[n]$ **(b)** $x[n] = 4\cos(\frac{1}{2}n\pi)u[n]$

(c) $x[n] = \cos(\frac{1}{2}n\pi) + \sin(\frac{1}{2}n\pi)$ **(d)** $x[n] = 4\cos(\frac{1}{4}n\pi) + 4\sin(\frac{1}{2}n\pi)$

P13.24 (Inverse Systems) Find the difference equation of the inverse systems for each of the following. Which inverse systems are causal? Which are stable?

(a) $H(z) = (z^2 + \frac{1}{9})/(z^2 - \frac{1}{4})$ **(b)** $H(z) = (z + 2)/(z^2 + \frac{1}{4})$

(c) $h[n] = n(2)^n u[n]$ **(d)** $y[n] - \frac{1}{2}y[n-1] = x[n] + 2x[n-1]$

P13.25 (Minimum-Phase Systems) Classify each system as minimum phase, mixed phase or maximum phase. Which of the systems are stable?

(a) $H(z) = (z^2 + \frac{1}{9})/(z^2 - \frac{1}{4})$ **(b)** $H(z) = (z^2 - 4)/(z^2 + 9)$

(c) $h[n] = n(2)^n u[n]$ **(d)** $y[n] + y[n-1] + \frac{1}{4}y[n-2] = x[n] - 2x[n-1]$

P13.26 (Minimum-Phase Systems) Find the minimum-phase transfer function corresponding to the systems described by

(a) $H(z)H(1/z) = 1/(3z^2 - 10z + 3)$

(b) $H(z)H(1/z) = (3z^2 + 10z + 3)/(5z^2 + 26z + 5)$

(c) $|H(F)|^2 = [1.25 + \cos(2\pi F)]/[8.5 + 4\cos(2\pi F)]$

P13.27 (Response of Digital Filters) Consider the 3-point averaging filter described by $y[n] = \frac{1}{2}x[n] + x[n-1] + \frac{1}{2}x[n-2]$.

(a) Find $h[n]$, $H(z)$ and $H(F)$.

(b) Find its response to the input $\{x[n]\} = \{2, 4, 6, 8\}$.

(c) Find its response to $x[n] = \cos(\frac{1}{3}n\pi)$.

(d) Find its response to $x[n] = \cos(\frac{1}{3}n\pi) + \sin(\frac{2}{3}n\pi) + \cos(\frac{1}{2}n\pi)$.

P13.28 (Stability) Consider the system $y[n] - \alpha y[n-1] = x[n] - \beta x[n-1]$.
(a) For what values of α and β will the system be stable?
(b) For what values of α and β will the system be minimum-phase?

P13.29 (Frequency Response) Make a rough sketch of and qualitatively describe the frequency response of the following digital filters:
(a) $y[n] - \frac{1}{4}y[n-1] = x[n] - x[n-1]$ **(b)** $H(F) = [1 - \exp(-j4\pi F)]$
(c) $H(z) = (z-2)/(z-\frac{1}{2})$
(d) $y[n] - y[n-1] + \frac{1}{4}y[n-2] = x[n] + x[n-1]$

P13.30 (Realization) Sketch direct form I, direct form II and transposed realizations for each of the following digital filters:
(a) $y[n] - \frac{1}{6}y[n-1] - \frac{1}{2}y[n-2] = 3x[n]$ **(b)** $H(z) = (z-2)/(z^2 - \frac{1}{4})$
(c) $H(z) = (2z^2 + z - 2)/(z^2 - 1)$
(d) $y[n] - 3y[n-1] + 2y[n-2] = 2x[n-2]$

P13.31 (Realization) Find the transfer function and difference equation for each digital filter realization shown in Figure P13.31.

Figure P13.31

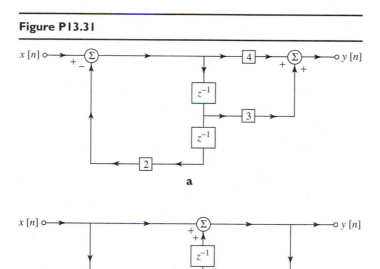

a

b

P13.32 (Filter Design by Pole-Zero Placement) Design the following filters using the concept of pole-zero placement:

(a) A bandpass filter with center frequency $f_0 = 200$ Hz, bandwidth $\Delta f = 20$ Hz, zero response at $f = 0$ and $f = 400$ Hz and a sampling frequency of 800 Hz.

(b) A notch filter with notch frequency 1 kHz, stopband 10 Hz and sampling frequency 8 kHz.

P13.33 (Stabilization by Allpass Filters) Consider a digital filter whose transfer function is given by $H(z) = (z + 3)/(z - 2)$.

(a) Is this filter stable?

(b) What is the transfer function $A_1(z)$ of a first-order allpass filter that can stabilize this filter? What is the transfer function $H_S(z)$ of the stabilized filter?

(c) If $H_S(z)$ is not minimum phase, pick an allpass filter $A_2(z)$ that converts $H_S(z)$ to a minimum-phase filter $H_M(z)$.

(d) Verify that $|H(F)| = |H_S(F)| = |H_M(F)|$.

P13.34 (Minimum-Phase Systems) Consider a digital filter described by

$$y[n] = x[n] - 0.65x[n-1] + 0.1x[n-2]$$

(a) Find $H(z)$ and verify that it is minimum phase.

(b) Find an allpass filter $A(z)$ with the same denominator as $H(z)$.

(c) Is the cascade $H(z)A(z)$ minimum phase? Causal? Stable?

P13.35 (Causality, Stability and Minimum Phase) Consider two causal, stable, minimum-phase digital filters described by $F(z) = z/(z - \frac{1}{2})$ and $G(z) = (z - \frac{1}{2})/(z + \frac{1}{2})$. Verify that the following filters are also causal, stable and minimum phase:

(a) The inverse filters $F^{-1}(z)$ and $G^{-1}(z)$.

(b) The cascade $H(z) = F(z)G(z)$.

(c) The inverse of the cascade $M(z) = H^{-1}(z)$.

(d) The parallel connection $N(z) = F(z) + G(z)$.

P13.36 (Allpass Filters) Consider two causal, stable, allpass digital filters described by $F(z) = (\frac{1}{2}z - 1)/(\frac{1}{2} - z)$ and $G(z) = (\frac{1}{2}z + 1)/(\frac{1}{2} + z)$. Which of the following filters are causal? Stable? Allpass?

(a) $L(z) = F^{-1}(z)$

(b) $H(z) = F(z)G(z)$

(c) $M(z) = H^{-1}(z)$

(d) $N(z) = F(z) + G(z)$

P13.37 (Delay of DT Systems) In analogy with *CT* systems, the delay d of *DT* systems is defined by $d = \sum kx^2[k] / \sum x^2[k]$.

(a) Consider the symmetric sequence $x[n] = \{4, 3, 2, 1, 0, 1, 2, 3, 4\}$. Its phase and delay are both zero. With $x[n] \longrightarrow x[n - d]$, the delay equals d. Verify this using the above formula for $x[n]$ with $d = 1$ and $d = 2$.

(b) The delay of an nth-order allpass filter equals n. Consider the first-order allpass filter $H(z) = (1 - \frac{1}{2}z)/(z - \frac{1}{2})$. Find its impulse response $h[n]$ and verify that its delay equals 1.

14 DIGITAL FILTERS

14.0 Scope and Objectives

Digital filters are widely used in almost all areas of digital signal processing. As with analog filters, their design is based on approximating the designed specifications using polynomials and rational functions.

This chapter is divided into four parts:

Part 1 An introduction to the terminology of digital filters.
Part 2 The design of IIR digital filters using analog techniques.
Part 3 The design of linear phase FIR digital filters.
Part 4 Special filters such as differentiators and Hilbert transformers.

Useful Background   The following topics provide a useful background:

1. The design of analog filters (Chapter 10).
2. Symmetric sequences and linear phase (Chapter 13).
3. Finding a transfer function $H(s)$ from $|H(\omega)|$ (Chapter 9).
4. Frequency transformations (Chapter 9).

14.1 Introduction

Digital filters process discrete-time signals. They are essentially mathematical implementations of a filter equation in software or hardware. They suffer from few limitations. Among their many advantages are high noise immunity, high accuracy (limited only by the roundoff error in the computer arithmetic), easy modification of filter characteristics, freedom from component variations and, of course, low and constantly decreasing cost. Digital filters are therefore rapidly replacing analog filters in any applications where they can be used effectively.

14.1.1   *Terminology of Digital Filters*

The description of discrete-time LTI systems, or **digital filters**, is based on a discrete convolution representation or a transfer function. The term *digital filter* is to be understood in its broadest sense not only as a smoothing or averaging operation, but also as any processing of the input signal. The terminology of digital filters is based on the nature of the system transfer function, difference equation or impulse response, as summarized in Table 14.1 and Figure 14.1.

Table 14.1   Classification of Digital Filters.

Difference equation	Filter classification
$y[n] = \sum B_m x[n - m]$	Nonrecursive FIR, moving average (MA), all-zero.
$\sum A_k y[n - k] = x[n]$	Recursive, IIR, autoregressive (AR), all-pole.
$\sum A_k y[n - k] = \sum B_m x[n - m]$	Recursive, IIR, ARMA, pole-zero.

Figure 14.1   Various types of digital filters.

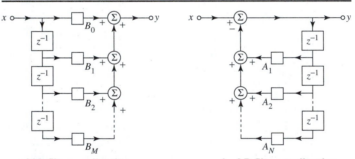

a. **MA filter realization**   b. **AR filter realization**

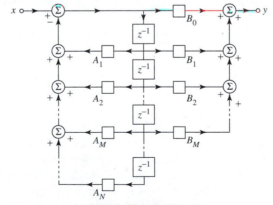

c. **ARMA filter realization**

14.1.2   *FIR Filters*

A nonrecursive, or **FIR filter**, of **length** $M + 1$ (or order M) is described by

$$y[n] = B_0x[n] + B_1x[n - 1] + \cdots + B_Mx[n - M] \quad H(z) = B_0 + B_1z^{-1} + \cdots + B_Mz^{-M}$$

The $M + 1$ term sequence $\{B_k\}$ describes the filter coefficients.

There are several observations to be made about this form.

1. The output is tapped after each delay element and weighted and summed. This describes a **tapped delay line** filter, or **transversal filter**.
2. The absence of recursion also implies absence of feedback, and such a system is then also referred to as a **feedforward filter**.
3. Since the response represents just a weighted average of the past and present inputs, this prompts the name **moving average filter**.
4. The transfer function $H(z)$ has a constant denominator and describes an **all-zero filter**.

Even though FIR filters are inherently stable due to absence of feedback and capable of perfectly linear phase response characteristics, their realization often involves a large number of elements.

14.1.3   *All-Pole Filters*

An Nth-order **all-pole filter** is a recursive, or IIR, filter described by

$$H(z) = \frac{1}{1 + A_1z^{-1} + \cdots + A_Nz^{-N}}$$

$$y[n] + A_1y[n - 1] + A_2y[n - 2] + \cdots + A_Ny[n - N] = x[n]$$

We make the following remarks:

1. Since the present response is evaluated from its known previous values, the realization of such a system must include feedback, and we therefore describe such filters as **feedback filters**.
2. In statistical circles the dependence of $y[n]$ on its past values is described by saying that $y[n]$ has a regression on its own past values, and this is the basis for the terminology **autoregressive filter**.

The presence of feedback can cause instability. Stability is, therefore, an important issue in the design of IIR filters.

14.1.4   *ARMA Filters*

The most general filter configuration is a combination of an FIR and all-pole realization described by the Nth-order difference equation

$$y[n] + A_1y[n - 1] + \cdots + A_Ny[n - N] = B_0x[n] + B_1x[n - 1] + \cdots + B_Mx[n - M]$$

The first sum describes feedback effects of the recursive system. The second sum is the nonrecursive contribution due to past inputs. The transfer function corresponding to this system is

$$H(z) = \frac{P(z)}{Q(z)} = \frac{B_0 + B_1 z^{-1} + \cdots + B_M z^{-M}}{1 + A_1 z^{-1} + A_2 z^{-2} + \cdots + A_N z^{-N}}$$

Such a system is called an **autoregressive moving average (ARMA) filter** of autoregressive order N and moving average order M, and is denoted by ARMA(N, M). Recursive and nonrecursive filters are special cases of the ARMA filter. Thus,

1. If $B_k = 0$, we have an ARMA$(0, M)$ filter, or an MA (FIR) filter of order M.
2. If $B_k = 0$ $(k \neq 0)$ and $B_0 \neq 0$, we have an ARMA$(N, 0)$ filter, or an AR all-pole filter of order N.

14.1.5   *The Design Process*

The magnitude and phase specifications for digital filters are identical to those for analog filters. We also specify a sampling frequency. This frequency is often used to normalize all the other design frequencies. The design process is essentially a three-step process that requires

1. Establishing the filter specifications for a given performance.
2. Determining the transfer function that meets the specifications.
3. Realizing the transfer function in software or hardware.

The fewer or less stringent the specifications, the better the possibility of achieving both design objectives and circuit implementation. The design based on any set of performance specifications is, at best, a compromise at all three levels. At the first level, the actual filter may never meet performance specifications if they are too stringent, at the second, the same set of specifications may lead to several possible realizations and at the third, quantization and roundoff errors may render the design useless if based on too critical a set of design values.

14.1.6   *Techniques of Digital Filter Design*

Digital filter design revolves around two distinctly different approaches:

1. *IIR filter design:* If linear phase is not critical, IIR filters yield a much smaller filter order for a given application. The design starts with an analog *lowpass prototype* based on the given specifications. It is then converted to the required digital filter using an appropriate *s*-to-*z* (*s2z*) *mapping*, and an appropriate *spectral transformation*.
2. *FIR filter design:* FIR filters can be designed with linear phase (no phase distortion). The design is based on selecting symmetric sequences, capable of linear phase, for the impulse response. The choice of the smallest order that meets design specifications is often based on iterative techniques or trial and error. For

given specifications, FIR filters require a much higher order compared with IIR filters.

14.2 IIR Filter Design

There are two related approaches for the design of IIR digital filters.

1. A popular method is based on using methods of analog design, followed by an $s2z$ mapping to convert the analog to the digital filter.
2. An alternative is to design a digital filter directly, using digital equivalents of analog (or other) approximations.

14.2.1   *Why IIR Filters Cannot Exhibit Linear Phase*

A causal stable IIR filter can never display linear phase. We obtain linear phase only for symmetric sequences $h[n]$ for which $H(z)$ is conjugate symmetric with $H(z) = \pm H(-1/z)$. This implies that for every pole inside the unit circle, there must be a reciprocal pole with conjugate symmetry outside the unit circle, making the system unstable if causal or noncausal if stable.

14.3 Conversion of Analog Systems to Digital Systems

The transformation of CT systems to DT systems is employed mainly for two reasons:

1. To simulate analog (CT) systems by digital means.
2. To design DT systems by taking advantage of the well-established methods available for CT systems.

An ideal transformation should preserve both the response and stability of the CT system. In practice, this is seldom possible because of the effects of sampling.

14.3.1   *An s2z Mapping Based on Transform Relations*

The impulse response $h(t)$ of an analog system may be approximated by

$$h(t) \approx \tilde{h}_a(t) = t_s \sum_{n=-\infty}^{\infty} h(t)\delta(t - nt_s) = t_s \sum_{n=-\infty}^{\infty} h(nt_s)\delta(t - nt_s)$$

Here, t_s is the sampling interval. The DT impulse response $h_s[n]$ describes the samples $h(nt_s)$ of $h(t)$, and may be written as

$$h_s[n] = \{h(nt_s)\} = \sum_{k=-\infty}^{\infty} h_s[k]\delta[n - k]$$

The Laplace transform $H_a(s)$ of $\tilde{h}_a(t)$, and the z-transform $H_d(z)$ of $h_s[n]$, are

$$H(s) \approx H_a(s) = t_s \sum_{k=-\infty}^{\infty} h(kt_s)\exp(-skt_s) \qquad H_d(z) = \sum_{k=-\infty}^{\infty} h_s[k]z^{-k}$$

Comparison suggests the equivalence $H_a(s) = t_s H_d(z)$ if $z^{-k} = \exp(-skt_s)$, or

$$z \longrightarrow \exp(st_s) \qquad s \longrightarrow \ln(z)/t_s$$

These relations describe an *s2z mapping* between the variables z and s. Since $s = \sigma + j\omega$, where ω is the continuous frequency, we can express z as

$$z = \exp[(\sigma + j\omega)t_s] = |\exp(\sigma t_s)|\exp(j\omega t_s) = |\exp(\sigma t_s)|\exp(j\Omega)$$

Here, $\Omega = \omega t_s = 2\pi f / S_F = 2\pi F$ is the digital frequency in radians/sample.

14.3.2   *Properties of the Transform-Based s2z Mapping*

The relations $z \longrightarrow \exp(st_s)$ and $s \longrightarrow \ln(z)/t_s$ do not describe a one-to-one mapping between the s-plane and the z-plane. Since $\exp(j\Omega)$ is periodic with period 2π, all frequencies $\Omega_0 \pm 2k\pi$ or $\omega_0 \pm k\omega_s$ are mapped to the same point in the z-plane. A one-to-one mapping is thus possible only if Ω lies between 0 and 2π (or $-\pi$ and π), corresponding to the analog frequency ω between 0 and $\omega_s = 2\pi S_F$ (or $-\frac{1}{2}\omega_s$ and $\frac{1}{2}\omega_s$).

The mapping $z \longrightarrow \exp(st_s)$ translates points in the s-domain to points in the z-domain as follows:

The Origin   The origin $s = 0$ is mapped to $z = 1$. So are the points $s = 0 \pm jk\omega_s$ for which $z = \exp(jk\omega_s t_s) = \exp(jk2\pi) = 1$.

The $j\omega$-axis   With $\sigma = 0$, $z = \exp(j\Omega)$ and $|z| = 1$. For increasing ω we move *counterclockwise* along a unit circle. As ω increases from ω_0 to $\omega_0 + \omega_s$ (or Ω from Ω_0 to $\Omega_0 + 2\pi$), we go around the unit circle once. The entire $j\omega$-axis thus maps into the unit circle *over and over* in segments of length $\omega_s = 2\pi S_F$.

The Left Half-Plane   With $\sigma < 0$, $z = |\exp(\sigma t_s)|\exp(j\Omega)$ and $|z| < 1$. Each strip of width ω_s in the left half of the s-plane is mapped into the interior of the unit circle (Figure 14.2). The strip between $-\frac{1}{2}\omega_s$ and $\frac{1}{2}\omega_s$ (or $-\pi S_F$ and πS_F) is called the **primary strip**. The entire left half of the s-plane is thus mapped into the interior of the unit circle, over and over, strip by strip.

The Right Half-Plane   With $\sigma > 0$, $z = |\exp(\sigma t_s)|\exp(j\Omega)$ and $|z| > 1$, and strips of width $\omega_s = 2\pi S_F$ in the entire right half of the s-plane are repeatedly mapped into the exterior of the unit circle.

14.3.3   *Relating Analog and Digital Systems*

The sampled signal $h_s[n]$ has a periodic spectrum, $H_p(F)$, which is given by its DTFT $H_p(F) = \sum (1/t_s)H(f - kS_F)$. If the analog signal $h(t)$ is bandlimited to f_B and sampled above the Nyquist rate ($S_F > 2f_B$), the principal period $(-\frac{1}{2}S_F, \frac{1}{2}S_F)$ of $H_p(F)$

Figure 14.2   Properties of the *s2z* mapping $s \to \ln(z)/t_s$.

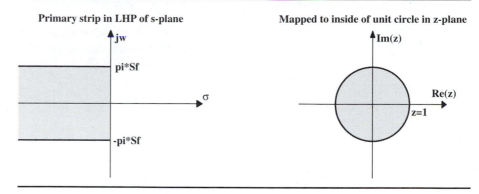

Primary strip in LHP of s-plane

Mapped to inside of unit circle in z-plane

describes $(1/t_s)H(f)$, which is just a scaled version of the true spectrum $H(f)$. We may thus relate the analog and digital systems by

$$H(f) = t_s H_p(F) \quad \text{or} \quad H_a(s)|_{s=j2\pi f} \approx t_s H_d(z)|_{z=\exp(j2\pi f/S_F)} \qquad |f| < \tfrac{1}{2}S_F$$

If $S_F < 2f_B$, we have aliasing and this relationship no longer holds.

14.3.4   *Practical s2z Mappings*

The transformation $s \longrightarrow \ln(z)/t_s$ suffers from two major drawbacks:

1. Its transcendental nature does not permit conversion of a rational transfer function $H(s)$ to a rational transfer function $H(z)$.
2. It does not permit a one-to-one correspondence for frequencies higher than $\tfrac{1}{2}S_F$. A unique representation in the z-plane is possible only for bandlimited signals with frequencies in the range $(-\tfrac{1}{2}S_F, \tfrac{1}{2}S_F)$.

Practical transformations are based on one of the following methods:

1. Matching the time response (the response-invariant transformation).
2. Matching terms in a factored $H(s)$ (the matched z-transform).
3. Conversion of system differential equations to difference equations using difference operators.
4. Numerical solution of the system differential equation using numerical integration algorithms.
5. Rational approximations for $z \longrightarrow \exp(st_s)$ and/or $s \longrightarrow \ln(z)/t_s$.

In general, each method results in different *s2z* mappings and leads to different forms for $H(z)$ from a given $H(s)$. Only if $H(z)$ is stable is the mapping useful. We now study these mappings and their characteristics.

14.4 Response Matching

Given a system with transfer function $H(s)$, here is how we approximate $H(s)$ by $H(z)$ using *response matching* or *response invariance*:

1. Choose an input $x(t)$ (such as an impulse, step or ramp).
2. Find the response $y(t)$ as the inverse transform of $H(s)X(s)$.
3. Sample $y(t)$ at intervals t_s to obtain $y_s[n]$ and its z-transform $Y(z)$.
4. Sample $x(t)$ to obtain $x_s[n]$ and $X(z)$.
5. Evaluate $H(z)$ as $Y(z)/X(z)$.

The most common inputs are the impulse, step and ramp. These lead to the $s2z$ mappings summarized in Table 14.2.

Table 14.2   Response-Invariant Transformations.
Note: $\mathcal{L}^{-1}[.]$ stands for the inverse Laplace transform and $\mathcal{Z}\{.\}$ stands for the z-transform.

$x(t)$	$x[n]$	$X(z)$	$Y(z)$	$H(z) = Y(z)/X(z)$
$\delta(t)$	$\delta[n]$	1	$Y_I(z) = \mathcal{Z}\{\mathcal{L}^{-1}[H(s)]\}$	$Y_I(z)$
$u(t)$	$u[n]$	$z/(z-1)$	$Y_S(z) = \mathcal{Z}\{\mathcal{L}^{-1}[H(s)/s]\}$	$Y_S(z)[(z-1)/z]$
$r(t)$	$nt_s u[n]$	$zt_s/(z-1)^2$	$Y_R(z) = \mathcal{Z}\{\mathcal{L}^{-1}[H(s)/s^2]\}$	$Y_R(z)[(z-1)^2/zt_s]$

Response-invariant matching yields a transfer function that is a good match only for the response for which it was designed. It may not provide a good match for the response to other inputs. The quality of the approximation depends on the choice of the sampling interval t_s, and a unique correspondence is possible for digital frequencies $|F| < \frac{1}{2}$ only if t_s is chosen in accordance with the Nyquist criterion to avoid aliasing. This mapping is thus useful only for analog systems whose frequency response is essentially bandlimited to $\pm\frac{1}{2}S_F$, such as lowpass and bandpass filters. It is of little use for bandstop or highpass filters.

Example 14.1
Response-Invariant Mappings

Let us approximate $H(s) = 4/[(s+1)(s+2)]$ by $H(z)$ using response invariance:

1. For impulse invariance, we choose $x(t) = \delta(t)$. Then,

$$X(s) = 1, Y(s) = H(s)X(s) = 4/[(s+1)(s+2)] = 4/(s+1) - 4/(s+2),$$
$$y(t) = [4e^{-t} - 4e^{-2t}]u(t)$$

This gives $y_S[n] = (4e^{-nt_s} - 4e^{-2nt_s})u[n]$ and $Y_s(z) = 4z/(z - e^{-t_s}) - 4z/(z - e^{-2t_s})$. With $x_s[n] = \delta[n]$ and $X(z) = 1$, we obtain

$$H_s(z) = Y_s(z)/X_s(z) = 4z/(z - e^{-t_s}) - 4z/(z - e^{-2t_s})$$

2. For step invariance, we choose $x(t) = u(t)$. Since $X(s) = 1/s$, we get

$$Y(s) = H(s)X(s) = 4/[s(s+1)(s+2)] = 2/s - 4/(s+1) + 2/(s+2),$$
$$y(t) = [2 - 4e^{-t} + 2e^{-2t}]u(t)$$

This leads to $y_s[n] = (2 - 4e^{-nt_s} + 2e^{-2nt_s})u[n]$ and
$Y_s(z) = 2z/(z - 1) - 4z/(z - e^{-t_s}) + 2z/(z - e^{-2t_s})$. Since $x_s[n] = u[n]$ and
$X(z) = z/(z - 1)$, $H_s(z) = Y_s(z)/X_s(z) = Y_s(z)(z - 1)/z$ or

$$H_s(z) = 2 - 4(z - 1)/(z - e^{-t_s}) + 2(z - 1)/(z - e^{-2t_s})$$

3. For ramp invariance, we choose $x(t) = tu(t)$. With $X(s) = 1/s^2$, we find

$$Y(s) = H(s)X(s) = 4/[s^2(s + 1)(s + 2)]$$
$$= -3/s + 2/s^2 + 4/(s + 1) - 1/(s + 2), \text{ and}$$
$$y(t) = [-3 + 2t + 4e^{-t} - e^{-2t}]u(t)$$

This leads to $y_S[n] = [-3 + 2nt_s + 4e^{-nt_s} - e^{-2nt_s}]u[n]$, and $Y_S(z) = -3z/(z - 1) + 2zt_s/(z - 1)^2 + 4z/(z - e^{-t_s}) - z/(z - e^{-2t_s})$. With $x_S[n] = nt_su[n]$, $X(z) = zt_s/(z - 1)^2$ and we obtain $H_S(z) = Y_S(z)/X_S(z) = Y_S(z)(z - 1)^2/zt_s$, or

$$H_S(z) = -3(z - 1)/t_s + 2 + 4(z - 1)^2/[t_s(z - e^{-t_s})] - (z - 1)^2/[t_s(z - e^{-2t_s})]$$

14.4.1   Impulse-Invariant Transformation for Partial Fraction Forms

If $H(s)$ is in partial fraction form, we can obtain a simple design relation for mapping. If $H(s)$ has poles at $p_k, k = 1, 2, \ldots, N$, and no repeated roots, we can describe $H(s)$ by partial fractions and the impulse response $h(t)$ and its discrete version $h[n]$ can then be written as

$$H(s) = \sum_{k=1}^{N} \frac{A_k}{s + p_k} \quad h(t) = \sum_{k=1}^{N} A_k \exp(-p_k t)u(t) \quad h[n] = \sum_{k=1}^{N} A_k \exp(-p_k nt_s)u[n]$$

The z-transform of $h[n]$ yields the digital transfer function $H_D(z)$ as

$$H_D(z) = \sum_{k=1}^{N} \frac{zA_k}{z - \exp(-p_k t_s)} \qquad |z| > \exp(-p_k t_s)$$

This relation suggests that we can go directly from $H(s)$ to $H_D(z)$ using the mapping $1/(s + p_k) \longrightarrow z/[z - \exp(-p_k t_s)]$ to obtain

$$H(s) = \sum_{k=1}^{N} \frac{A_k}{s + p_k} \qquad H_D(z) = \sum_{k=1}^{N} \frac{zA_k}{z - \exp(-p_k t_s)}$$

These and other results are listed in Table 14.3, while Table 14.4 provides results for repeated factors in both partial fraction and polynomial forms.

Terms corresponding to complex conjugate roots must be simplified to obtain a real form. The result for repeated roots is obtained if we start with a typical kth term $H_k(s)$ with a root of multiplicity M such that

$$H_k(s) = \frac{A_k}{(s + p_k)^M} \qquad h_k(t) = \frac{A_k}{(M - 1)!}t^{M-1} \exp(-p_k t)u(t)$$

Table 14.3   Impulse-Invariant Transformations.

Term	Form of $H(s)$	$H(z)$ [Note: $a = \exp(-pt_s)$]
Distinct	$\dfrac{A}{(s+p)}$	$\dfrac{Az}{(z-a)}$
Complex conjugate	$\dfrac{A\exp(j\Omega)}{(s+p+jq)} + \dfrac{A\exp(-j\Omega)}{(s+p-jq)}$	$\dfrac{2z^2 A\cos(\Omega) - 2Aaz\cos(\Omega + qt_s)}{z^2 - 2az\cos(qt_s) + a^2}$
Repeated	$\dfrac{A}{(s+p)^M}$	$\dfrac{A}{(M-1)!}t_s^{(M-1)}\left(-z\dfrac{d}{dz}\left(-z\dfrac{d}{dz}{}^{M-1}\cdots -z\dfrac{d}{dz}\left[\dfrac{z}{z-a}\right]\right)\right)$
Repeated	$\dfrac{A}{(s+p)^2}$	$At_s\dfrac{az}{(z-a)^2}$
Repeated	$\dfrac{A}{(s+p)^3}$	$\tfrac{1}{2}At_s^2\dfrac{za(z+a)}{(z-a)^3}$
Modified	$\dfrac{A}{(s+p)}$	$A\dfrac{z}{(z-a)} - \tfrac{1}{2}A = \tfrac{1}{2}A\dfrac{(z+a)}{(z-a)}$
Modified	$\dfrac{A\exp(j\Omega)}{(s+p+jq)} + \dfrac{A\exp(-j\Omega)}{(s+p-jq)}$	$\dfrac{2z^2 A\cos(\Omega) - 2Aaz\cos(\Omega + qt_s)}{z^2 - 2az\cos(qt_s) + a^2} - A\cos(\Omega)$

Table 14.4   Impulse-Invariant Transformations for Repeated Factors.

If $H(s) = \dfrac{1}{(s+p)^N}$, then $H(z) = Cz\displaystyle\sum_{k=1}^{N}\dfrac{y_k a^{k-1}}{(z-a)^k} = \dfrac{C}{(z-a)^N}\sum_{k=0}^{N-1}\beta_k z^k a^{N-k}$ where $a = \exp(-pt_s)$

N	C	PFE Coefficients y_k							Polynomial Coefficients β_k						
		$k=1$	2	3	4	5	6	7	$k=0$	1	2	3	4	5	6
$N=1$	1	1							1						
$N=2$	t_s	0	1						0	1					
$N=3$	$t_s^2/2$	0	1	2					0	1	1				
$N=4$	$t_s^3/6$	0	1	6	6				0	1	4	1			
$N=5$	$t_s^4/24$	0	1	14	36	24			0	1	11	11	1		
$N=6$	$t_s^5/120$	0	1	30	150	240	120		0	1	26	66	26	1	
$N=7$	$t_s^6/720$	0	1	62	540	1560	1800	720	0	1	57	302	302	57	1

Example: $H(s) = \dfrac{1}{(s+p)^4}$ $H(z) = \dfrac{t_s^3}{6}\left[\dfrac{az}{(z-a)^2} + \dfrac{6a^2 z}{(z-a)^3} + \dfrac{6a^3 z}{(z-a)^4}\right] = \dfrac{t_s^3}{6}\left[\dfrac{az^3 + 4a^2 z^2 + a^3 z}{(z-a)^4}\right]$

$\qquad\qquad\qquad\qquad\qquad\qquad\qquad\qquad$ Partial fraction form $\qquad\qquad\qquad\qquad\qquad$ Polynomial form

The sampled version $h_k[n]$ and its z-transform then become

$$h_k[n] = \dfrac{A_k}{(M-1)!}(nt_s)^{(M-1)}\exp(-p_k nt_s)u[n]$$

$$H_k(z) = \dfrac{A_k}{(M-1)!}t_s^{(M-1)}\left(-z\dfrac{d}{dz}\left(-z\dfrac{d}{dz}\left(\ldots M-1\ \text{times}\ldots\left(\left[\dfrac{z}{z-\exp(-p_k t_t)}\right]\right)\ldots\right)\right)\right)$$

Impulse-invariant design requires $H(s)$ in partial fraction form, and yields $H_D(z)$ in the same form. Poles of $H(s)$ in the left half-plane for which $p_k > 0$ map into poles of

$H_D(z)$ inside the unit circle, since $|\exp(-p_k t_s)| < 1$. Thus, a stable $H(s)$ transforms to a stable $H_D(z)$. Note that $H_D(z)$ must be reassembled if we need a cascaded form.

14.4.2 *Comparison of Analog and Impulse-Invariant Filters*

The mapping $1/(s + p_k) \longrightarrow z/[z - \exp(-p_k t_s)]$ reveals that the dc gain of the analog term equals $1/p_k$. But the dc gain of the digital term (evaluated at $z = 1$) is $1/[1 - \exp(-p_k t_s)]$. If $t_s \ll 1/p_k$, $\exp(-p_k t_s) \approx 1 - p_k t_s$ and $1/[1 - \exp(-p_k t_s)]$ may be approximated by $1/p_k t_s$. It is, therefore, customary to compare $H(s)$ with $t_s H_D(z)$, or a scaled version $K H_D(z)$, where the constant K is chosen to match the gain of $H(s)$ and $K H_D(z)$ at a convenient frequency.

Example 14.2
Impulse-Invariant Mapping

Consider $H(s) = 4/[(s + 1)(s^2 + 4s + 5)]$. Its partial fraction form is

$$H(s) = 2/(s + 1) + K/(s + 2 + j) + K^*/(s + 2 - j)$$

where

$$K = (-1 - j) = \sqrt{2}\exp(-j\tfrac{3}{4}\pi)$$

Using Table 14.3, and letting $b = \exp(-t_s)$ and $a = \exp(-2t_s)$, we obtain

$$H_D(z) = 2\frac{z}{(z - b)} + \frac{2z^2\sqrt{2}\cos(\tfrac{3}{4}\pi) - 2\sqrt{2}za\cos(t_s - \tfrac{3}{4}\pi)}{z^2 - 2az\cos(t_s) + a^2}$$

Comment: The first step involved partial fractions. No, we cannot compute the impulse-invariant mappings for $H_1(s) = 4/(s + 1)$ and $H_2(s) = 1/(s^2 + 4s + 5)$, for example, and find $H_D(z)$ as their cascade.

Example 14.3
Impulse-Invariant Design

Let $H(s) = 1/(s + 1)$ with unit cutoff frequency. We wish to design a digital filter with $f_c = 10$ Hz and $S_F = 60$ Hz. There are actually two ways to do this:

Method 1: We normalize by the sampling frequency S_F, which allows us to use $t_s = 1$ in all subsequent computations. Normalization gives $\Omega_C = 2\pi f_C/S_F = \tfrac{1}{3}\pi$. We denormalize $H(s)$ to Ω_C to get $H_1(s) = H(s/\Omega_C) = \tfrac{1}{3}\pi/(s + \tfrac{1}{3}\pi)$. Finally, with $t_s = 1$, impulse-invariance gives $H_1(z) = \tfrac{1}{3}\pi z/(z - e^{-\pi/3})$.

Method 2: We denormalize $H(s)$ to f_C to get $H_2(s) = H(s/2\pi f_C) = 20\pi/(s + 20\pi)$. With $t_s = 1/60$, we develop $H_2(z) = 20\pi z/(z - e^{-\pi/3})$ and $t_s H_2(z) = \tfrac{1}{3}\pi z/(z - e^{-\pi/3})$.

14.4.3 *Modifications to Impulse-Invariant Design*

The impulse-invariant method also suffers from errors in sampling $h(t)$ at discontinuities. If $h(t)$ includes terms such as $u(t), \exp(-\alpha t)u(t)$ or $\exp(-\alpha t)\cos(\beta t + \phi)u(t)$, it is discontinuous at $t = 0$. The sampled value must then be chosen as $\tfrac{1}{2}h(0)$ (the midpoint value). With this change, the term $h(t) = \exp(-p_k t)$ yields the sampled signal $x[n]$ and $H_D(z)$ as

$$h[n] = \exp(-p_k n t_s) - \frac{1}{2} \qquad H_D(z) = \frac{z}{z - \exp(-p_k t_s)} - \frac{1}{2} = \frac{1}{2}\frac{z + \exp(-p_k t_s)}{z - \exp(-p_k t_s)}$$

The design relation modifies to

$$H(s) = \sum_{k=1}^{N} \frac{A_k}{s + p_k} \qquad H_D(z) = \sum_{k=1}^{N} \frac{1}{2} A_k \frac{z + \exp(-p_k t_s)}{z - \exp(-p_k t_s)}$$

The modification may be viewed as shifting the zeros of the unmodified $H_D(z)$ at $z = \infty$ to $z = -\exp(-p_k t_s)$. A similar modification for complex conjugate poles leads to the last entry in Table 14.3.

Figure 14.3 shows the response of $H(s) = 1/(s + 1)$ and two digital systems designed from it, with and without modification, using $t_s = 1$. It clearly reveals the improvement due to the modification.

Figure 14.3   Frequency response of digital filters designed from the analog filter $H(s)$ = 1/(s + 1) using impulse invariance and its modified form.

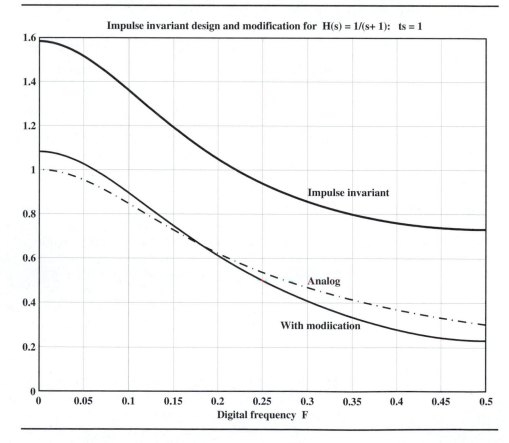

Remark: If the numerator of $H(s)$ is of degree M and the denominator of degree N, the initial value theorem suggests $x(0) = \lim_{s \to \infty} sH(s) = 0$ if $N - M > 1$. We thus need this modification only when $N - M = 1$.

14.5
The Matched z-Transform for Factored Forms

The impulse-invariant transformation $1/(s + a) \longrightarrow z/[z - \exp(-at_s)]$ may also be expressed as $(s + a) \longrightarrow [z - \exp(-at_s)]/z$. The **matched z-transform** uses this form to convert each numerator and denominator term of a *factored $H(s)$* to yield the discrete system $H_D(z)$, in factored form, as

$$H(s) = K_0 \frac{\prod_{i=1}^{M}(s - z_i)}{\prod_{k=1}^{N}(s - p_k)} \qquad H_D(z) = Kz^{(N-M)} \frac{\prod_{i=1}^{M}[z - \exp(z_i t_s)]}{\prod_{k=1}^{N}[z - \exp(p_k t_s)]}$$

The power of $z^{(N-M)}$ in $H_D(z)$ equals the difference in the degree of the denominator and numerator polynomials of $H(s)$. The constant K is chosen to match the gains of $H(s)$ and $H_D(z)$ at some convenient frequency.

For complex roots, we can replace each conjugate pair using the mapping

$$(s + p - jq)(s + p + jq) \longrightarrow [z^2 - 2ze^{-pt_s}\cos(qt_s) + e^{-2pt_s}]/z^2$$

14.5.1   *Modifications to Matched z-Transform Design*

Modifications to the matched z-transform method rely on mapping the zeros of $H(s)$ at $s = \infty$ to $z = -1$ (instead of $z = 0$). The result of this choice is that $f = \infty$ in the analog domain corresponds to $F = \frac{1}{2}S_F$ or $z = -1$. This allows us to use the mapping for highpass and bandstop filters.

With $H_D(z)$ computed as before, the two modifications require us to

Modification 1   Replace *all* zeros at $z = 0$ to $z = -1$ (or $z^{(N-M)}$ by $(z + 1)^{(N-M)}$).

Modification 2   Replace *all but one* of the zeros at $z = 0$ to $z = -1$.

Remarks: 1. Since poles in the left half of the s-plane are mapped inside the unit circle in the z-plane, the matched z-transform preserves stability.
2. The matched z-transform converts an all-pole analog system to an all-pole digital system but may not preserve the frequency response of the analog system. It also suffers from aliasing errors.

Example 14.4
The Matched z-Transform

(a)   With $H(s) = 4/[(s + 1)(s + 2)]$, the matched z-transform yields
$$H_D(z) = Kz^2/[(z - e^{-t_s})(z - e^{-2t_s})]$$

(b)   The first modification replaces z^2 (two zeros) by $(z + 1)^2$ to yield
$$H_1(z) = K(z + 1)^2/[(z - e^{-t_s})(z - e^{-2t_s})]$$

(c)   The second modification replaces only one zero to $z + 1$ to give
$$H_2(z) = Kz(z + 1)/[(z - e^{-t_s})(z - e^{-2t_s})]$$

Comment: In each case, K can be chosen for a desired gain.

14.6
s2z Mappings from Discrete Difference Algorithms

The three difference operators commonly used in converting differential to difference equations, and their transfer functions, are listed in Table 14.5.

The frequency response of these operators was discussed in Chapter 11. Here, we investigate their mapping properties and stability. Comparison of the ideal derivative operator $H(s) = s$ with $H_D(z)$ for each difference operator leads to the s2z mappings of Table 14.6.

The entry for the backward difference, for example, is based on

$$H(s) = s \qquad H_D(z) = (z - 1)/zt_s \qquad s \longrightarrow (z - 1)/zt_s$$

Table 14.5   Discrete Difference Algorithms.

Difference	Algorithm	Transfer functions
Backward	$y[n] = \{x[n] - x[n-1]\}/ts$	$H_M(z) = (z-1)/zt_s$
Central	$y[n] = \{x[n+1] - x[n-1]\}/2t_s$	$H_M(z) = (z^2 - 1)/2zt_s$
Forward	$y[n] = \{x[n+1] - x[n]\}/t_s$	$H_M(z) = (z-1)/t_s$

Table 14.6   s2z Mappings from Difference Algorithms.

Difference	s2z mapping	$H_D(z)$ for $H(s) = 1/(s+a)$	Poles of $H_D(z)$	Stable range
Backward	$s = (z-1)/zt_s$	$zt_s/[(1 + at_s)z - 1]$	$1/(1 + at_s)$	$at_s < -2, at_s > 0$
Central	$s = (z^2 - 1)/2zt_s$	$2zt_s/(z^2 + 2at_sz - 1)$	$-at_s \pm \{(at_s)^2 + 1\}^{1/2}$	unstable
Forward	$s = (z-1)/t_s$	$t_s/(z - 1 + at_s)$	$1 - at_s$	$0 < at_s < 2$

14.6.1   Mapping Properties of Difference Algorithms

Forward Difference   With $z = u + jv$, the mapping $s \longrightarrow (z - 1)/t_s$ results in

$$z = 1 + st_s \qquad \text{or} \qquad u + jv = 1 + t_s(\sigma + j\omega)$$

We find that:

If $\sigma = 0, u = 1$. The $j\omega$-axis maps into $z = 1$.

If $\sigma > 0, u > 1$. The right half of the s-plane maps to the right of $z = 1$.

If $\sigma < 0, u < 1$. The left half of the s-plane maps to the left of $z = 1$.

The left half of the s-plane thus maps to a region which includes not only the unit circle but also a vast region outside it. Thus a stable analog filter with poles in LHP may result in an unstable digital filter with poles anywhere to the left of $u = 1$ but not inside the unit circle!

Backward Difference   The mapping $s \rightarrow (z-1)/zt_s$ results in

$$z = \tfrac{1}{2} + \tfrac{1}{2}(1 + st_s)/(1 - st_s) \qquad \text{or} \qquad z - \tfrac{1}{2} = \tfrac{1}{2}(1 + st_s)/(1 - st_s)$$

To find where the $j\omega$-axis maps, we set $\sigma = 0$ to obtain

$$z - \tfrac{1}{2} = \tfrac{1}{2}(1 + j\omega t_s)/(1 - j\omega t_s)$$

Comparing magnitudes, this gives

$$\left| z - \tfrac{1}{2} \right| = \tfrac{1}{2}$$

Thus, the $j\omega$-axis is mapped into a circle of radius $\tfrac{1}{2}$, centered at $z = \tfrac{1}{2}$. This is within the unit circle and always represents a stable transformation, but it restricts the pole locations of the digital filter. Since the frequencies are mapped into a smaller circle, this operator is a good approximation only in the vicinity of $z = 1$ or $\Omega \approx 0$ (where it approximates the unit circle), which implies high sampling rates.

14.6.2   *Stability of Digital Systems Designed from Difference Algorithms*

To study the stability of difference operators, let us consider the transfer function $H(s) = 1/(s + a)$, which is stable for $a > 0$. Upon mapping $H(s)$, we obtain the discrete systems whose transfer functions $H_D(z)$ are listed in Table 14.6.

The pole locations of the various $H_D(z)$ reveal that

1. The backward difference produces a stable system for all $a > 0$.
2. The system based on the central difference is unstable.
3. The forward difference yields a stable system only if $a < 2/t_s$.

Clearly, from a stability viewpoint, only the *s2z* mapping based on the backward difference is useful for $H(s) = 1/(s + a)$. In fact, this observation is valid for any $H(s)$.

14.7
s2z Mappings from Discrete Integration Algorithms

The various schemes for numerical integration are listed in Table 14.7. We have also studied their frequency response in Chapter 11. Comparison of their transfer function $H(z)$ with the transfer function of the ideal integrator $H(s) = 1/s$ yields the *s2z* mappings of Table 14.8.

The entry for the trapezoidal rule, for example, is based on

$$H(s) = \frac{1}{s} \qquad H_D(z) = \frac{1}{2}t_s\frac{z+1}{z-1} \qquad \text{and} \qquad \frac{1}{s} \longrightarrow \frac{1}{2}t_s\frac{z+1}{z-1} \qquad \text{or} \qquad s \longrightarrow \frac{2(z-1)}{t_s(z+1)}$$

The rectangular algorithm for integration and the backward difference for the derivative are equivalent, and generate identical mappings.

Table 14.7   Numerical Integration Algorithms.

Note: Numbers in parentheses indicate the order (see Section 14.7.2).

Algorithm	Formula for $y[n]$
Rectangular (1)	$y[n] = y[n-1] + x[n]t_s$
Trapezoidal (1)	$y[n] = y[n-1] + \{x[n] + x[n-1]\}t_s/2$
Adams (2)	$y[n] = y[n-1] + \{5x[n] + 8x[n-1] - x[n-2]\}t_s/12$
Adams (3)	$y[n] = y[n-1] + \{9x[n] + 19x[n-1] - 5x[n-2] + x[n-3]\}t_s/24$
Simpson's (2)	$y[n] = y[n-2] + \{x[n] + 4x[n-1] + x[n-2]\}t_s/3$
Tick's (2)	$y[n] = y[n-2] + \{0.3584x[n] + 1.2832x[n-1] + 0.3584x[n-2]\}t_s$

Table 14.8   _s2z_ Mappings from Numerical Integration Algorithms.

Algorithm	Transfer function $H(z)$	s2z mapping
Rectangular	$t_s \dfrac{z}{z-1}$	$s = \dfrac{1}{t_s}\left[\dfrac{z-1}{z}\right]$
Trapezoidal	$\dfrac{1}{2}t_s \dfrac{z+1}{z-1}$	$s = \dfrac{2}{t_s}\left[\dfrac{z-1}{z+1}\right]$
Adams (2)	$t_s \dfrac{5z^2 + 8z - 1}{12(z^2 - z)}$	$s = \dfrac{12}{t_s}\left[\dfrac{(z^2 - z)}{5z^2 + 8z - 1}\right]$
Adams (3)	$t_s \dfrac{9z^3 + 19z^2 - 5z + 1}{24(z^3 - z^2)}$	$s = \dfrac{24}{t_s}\left[\dfrac{(z^3 - z^2)}{9z^3 + 19z^2 - 5z + 1}\right]$
Simpson's	$t_s \dfrac{z^2 + 4z + 1}{3(z^2 - 1)}$	$s = \dfrac{3}{t_s}\left[\dfrac{(z^2 - 1)}{z^2 + 4z + 1}\right]$
Tick's	$t_s \dfrac{0.3584z^2 + 1.2832z + 0.3584}{z^2 - 1}$	$s = \dfrac{1}{t_s}\left[\dfrac{z^2 - 1}{0.3584z^2 + 1.2832z + 0.3584}\right]$

14.7.1   _Stability of Digital Systems Designed from Integration Algorithms_

Mappings based on the rectangular and trapezoidal algorithms always yield a stable $H(z)$ for any stable $H(s)$ and any choice of t_s. All other higher-order mappings of Table 14.8 (except Simpson's and Tick's rule) also yield a stable $H(z)$ for any stable $H(s)$ as long as t_s is smaller than the Nyquist interval.

14.7.2   _Mapping Order_

The order r of the mappings shown in parentheses in Table 14.7 signifies that an analog transfer function $H(s)$ of order n is mapped to a digital transfer function $H_D(z)$ of order $p = nr$.

Remark: The mappings of Table 14.8 lead to a transfer function $H_D(z)$ whose numerator and denominator are, in general, of equal degree $p = nr$.

If we use the mappings to convert the first-order system with $H(s) = 1/(s + a)$, we get the *DT* transfer functions of Table 14.9. The pole locations of these transfer functions reveal that Simpson's (and Tick's) algorithms result in an unstable system, since one pole of $H(z)$ lies outside the unit circle. All other mappings preserve stability.

Table 14.9   *s2z* Mapping of $H(s) = 1/(s + a)$ Using Numerical Integration.

Algorithm	$H_D(z)$ for $H(s) = 1/(s + a)$	Stable range
Rectangular	$\dfrac{zt_s}{(1 + at_s) - 1}$	Stable for $a > 0$
Trapezoidal	$\dfrac{t_s(z + 1)}{(2 + at_s)z - (2 - at_s)}$	Stable for $a > 0$
Adams (2)	$\dfrac{t_s(5z^2 + 8z - 1)}{(12 + 5at_s)z^2 - (12 - 8at_s)z - at_s}$	Stable for $a > 0$
Adams (3)	$\dfrac{t_s(9z^3 + 19z^2 - 5z + 1)}{(24 + 9at_s)z^3 - (24 - 19at_s)z^2 - 5at_sz + at_s}$	Stable for $a > 0$
Simpson's	$\dfrac{t_s(z^2 + 4z + 1)}{(3 + at_s)z^2 + 4at_sz - (3 - at_s)}$	Unstable for $a > 0$

14.8
s2z Mappings from Rational Approximations

Some of the mappings of Tables 14.6 and 14.8 may also be viewed as rational approximations of the transformations $z \longrightarrow \exp(st_s)$ and $s \longrightarrow \ln(z)/t_s$.

1. If we use a first-order approximation of the series for $\exp(st_s)$

$$\exp(st_s) = 1 + st_s + (st_s)^2/2! + \cdots \approx 1 + st_s \quad st_s \ll 1$$

we obtain the mapping based on the forward difference

$$s \approx [\exp(st_s) - 1]/t_s \longrightarrow (z - 1)/t_s$$

2. If we use a first-order approximation for $\exp(-st_s)$ as

$$\exp(-st_s) = 1 - st_s + (st_s)^2/2! - \cdots \approx 1 - st_s \quad st_s \ll 1$$

we obtain the mapping based on the backward difference or rectangular rule

$$s \approx [1 - \exp(-st_s)]/t_s \longrightarrow (1 - 1/z)/t_s = (z - 1)/zt_s$$

3. The trapezoidal rule is based on a first-order rational function approximation of $s \longrightarrow \ln(z)/t_s$, with $\ln(z)$ described by a power series:

$$s = \frac{\ln(z)}{t_s} = \frac{2}{t_s}\left[\frac{z - 1}{z + 1}\right] + \frac{2}{3t_s}\left[\frac{z - 1}{z + 1}\right]^2 + \frac{2}{5t_s}\left[\frac{z - 1}{z + 1}\right]^2 + \cdots \approx \frac{2}{t_s}\left[\frac{z - 1}{z + 1}\right]$$

The reverse transformation $z \longrightarrow \exp(st_s)$, with $\exp(st_s)$ described by a power series, leads to the first-order rational approximation

$$z = \exp(st_s) = \frac{\exp(\frac{1}{2}st_s)}{\exp(-\frac{1}{2}st_s)} = \frac{1 + (\frac{1}{2}st_s) + (\frac{1}{2}st_s)^2/2! + \cdots}{1 - (\frac{1}{2}st_s) + (\frac{1}{2}st_s)^2/2! - \cdots}$$

$$\approx \frac{1 + (\frac{1}{2}st_s)}{1 - (\frac{1}{2}st_s)} = \frac{2 + st_s}{2 - st_s}$$

14.9
The Bilinear
Transformation

The linear relations based on the trapezoidal rule and given by

$$z \longrightarrow \frac{2 + st_s}{2 - st_s} \qquad s \longrightarrow \frac{2}{t_s}\left[\frac{z-1}{z+1}\right]$$

describe the **bilinear transformation**, or **Tustin's rule**, and allow us to go back and forth between the s-domain and z-domain. With $s = \sigma + j\omega$, we have

$$z = \frac{2 + \sigma + j\omega t_s}{2 - \sigma - j\omega t_s}$$

If $\sigma = 0$, we get $|z| = 1$ and for $\sigma < 0, |z| < 1$. Thus the $j\omega$-axis is mapped to the unit circle, and the left half of the s-plane is mapped into the interior of the unit circle. This means that a stable analog system will always yield a stable digital system using this transformation. If $\omega = 0, z = 1$ and the dc gain of both the analog and digital systems is identical.

To see how the frequencies map, we set $\sigma = 0$ to obtain the complex form

$$z = \frac{2 + j\omega t_s}{2 - j\omega t_s} = |z|\exp(j\phi) = \exp[j2\tan^{-1}(\tfrac{1}{2}\omega t_s)]$$

Since $z = \exp(j\Omega_D)$ where Ω_D is the digital frequency, we obtain

$$\Omega_D = 2\tan^{-1}(\tfrac{1}{2}\omega t_s)$$

This is a *nonlinear* relation between ω and Ω_D. When $\omega = 0, \Omega_D = 0$, and as $\omega \longrightarrow \infty$, $\Omega_D \longrightarrow \pi$. It is also a one-to-one mapping which nonlinearly compresses the frequency range $-\infty < f < \infty$ to $-\frac{1}{2}S_F < f < \frac{1}{2}S_F$. It avoids aliasing at the expense of distorting, compressing or *warping* the analog frequencies. The higher the frequency, the more severe the warping. We can compensate for this warping (but not eliminate it) if we **prewarp** the frequency specifications before designing the analog system $H(s)$ or applying the bilinear transformation. Prewarping of the frequencies prior to analog design is just a scaling (stretching) operation based on the inverse of the warping relation:

$$\omega = (2/t_s)\tan(\tfrac{1}{2}\Omega_D) = (2/t_s)\tan(\tfrac{1}{2}\omega_D t_s)$$

For high sampling rates, such that $\Omega_D \ll 1$, the prewarping has little effect and may even be redundant.

In actual application, the warping relation and the bilinear transformation are often generalized to

$$\omega = C\tan(\tfrac{1}{2}\Omega_D) = C\tan(\tfrac{1}{2}\omega t_s) \qquad z \longrightarrow \frac{C+s}{C-s} \qquad s \longrightarrow C\frac{z-1}{z+1}$$

Figure 14.4 shows a plot of ω versus Ω_D for various values of C compared with the linear relation $\omega = \Omega_D$. The analog and digital frequencies match at $\omega = \Omega_D = 0$ and at one other value dictated by the choice of C.

Figure 14.4 **The linear relation $\omega_p = \Omega$ and the nonlinear prewarping relation $\omega_p = C$ $\tan\left(\frac{1}{2}\Omega\right)$ for various values of C.**

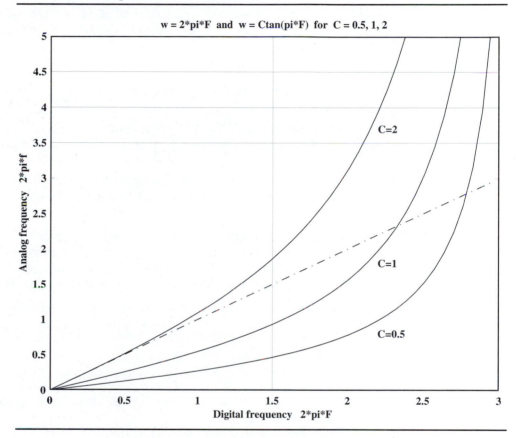

14.9.1 *Using the Bilinear Transformation*

Given an analog transfer function $H(s)$ whose response at the analog frequency ω_A is to be matched to $H_D(z)$ at the digital frequency Ω_D, we may design $H_D(z)$ in one of two ways:

1. We fix C by ensuring that ω_A matches the prewarped Ω_D. This gives

$$\omega_A = C_m \tan\left(\tfrac{1}{2}\Omega_D\right) \qquad \text{or} \qquad C_m = \omega_A / \tan\left(\tfrac{1}{2}\Omega_D\right)$$

This implies that $H(s)$ has been designed by prewarping the frequencies using $\omega = C_m \tan\left(\frac{1}{2}\Omega\right)$. We obtain $H_D(z)$ from $H(s)$ using $s = C_m(z-1)/(z+1)$.

This method is summarized by the following steps.

$$C_m = \omega_A / \tan(\tfrac{1}{2}\Omega_D) \qquad H_D(z) = H(s)|_{s=C_m(z-1)/(z+1)}$$

2. We pick a convenient value for C. For example, let $C = 2$. This actually matches the response at an arbitrary prewarped frequency ω_x given by

$$\omega_x = 2\tan(\tfrac{1}{2}\Omega_D)$$

Next, we frequency scale $H(s)$ to $H_D(s) = H(s\omega_A/\omega_x)$. This matches the response of $H_D(s)$ at ω_A and that of $H(s)$ at ω_x. Finally, we obtain $H_D(z)$ from $H_A(s)$ using $s = 2(z-1)/(z+1)$.
　　For $C = 2$, this method may be summarized by the following steps.

$$\omega_x = 2\tan(\tfrac{1}{2}\Omega_D) \qquad H_A(s) = H(s)|_{s=s\omega_A/\omega_x} \qquad H_D(z) = H_A(s)|_{s=2(z-1)/(z+1)}$$

The two methods yield identical results for $H_D(z)$. The first does away with the scaling of $H(s)$, and the second allows a convenient choice for C.

Remark:　The stretching effect of the prewarping often results in a filter of lower order, especially if the sampling frequency is not high enough.

Example 14.5
Using the Bilinear Transformation

Consider a Bessel filter described by $H(s) = 3/(s^2 + 3s + 3)$. We design a digital filter whose magnitude at $f_0 = 3$ kHz equals the magnitude of $H(s)$ at $\omega_A = 4$ rad/s if the sampling rate $S_F = 12$ kHz. The digital frequency equals $\Omega_D = 2\pi f_0/S_F = \tfrac{1}{2}\pi$.
　　Method 1: We select C by choosing the prewarped frequency to equal $\omega_A = 4$:

$$\omega_A = 4 = C\tan(\tfrac{1}{2}\Omega_D) \quad \text{or} \quad C = 4/\tan(\tfrac{1}{2}\Omega_D) = 4$$

We transform $H(s)$ to $H_D(z)$ using $s = C(z-1)/(z+1) = 4(z-1)/(z+1)$ to obtain

$$H_D(z) = 3(z+1)^2/(31z^2 - 26z + 7)$$

Method 2: We choose $C = 2$, say, and evaluate $\omega_x = 2\tan(\tfrac{1}{2}\Omega_D) = 2\tan(\tfrac{1}{4}\pi) = 2$.
　　Next, we frequency scale $H(s)$ to

$$H_A(s) = H(s\omega_A/\omega_x) = H(2s) = 3/(4s^2 + 6s + 3)$$

Finally, we transform $H_A(s)$ to $H_D(z)$ using $s = 2(z-1)/(z+1)$ to obtain

$$H_D(z) = 3(z+1)^2/(31z^2 - 26z + 7)$$

Comment:　The magnitude $|H(\omega)|$ at $\omega = 4$ matches the magnitude of $|H_D(\Omega)|$ at $\Omega = \tfrac{1}{2}\pi$. In fact, with $z = \exp(j\Omega) = \exp(j\tfrac{1}{2}\pi) = j$ and $z^2 = j^2 = -1$, we find that

$$|H(z = \exp(j\tfrac{1}{2}\pi))| = |3(j+1)^2/(-24 - j26)| = 0.1696$$
$$|H(s = j4)| = |3/(-13 + j12)| = 0.1696$$

Figure 14.5 compares the response of $H(s)$ and $H_D(z)$. The linear phase of $H(s)$ is nearly preserved in $H_D(z)$ only if the sampling frequency is high enough. ▬▬

Figure 14.5 Frequency response of the IIR filter of Example 14.5.

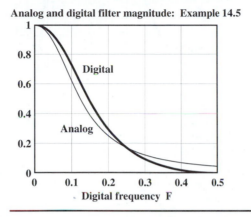

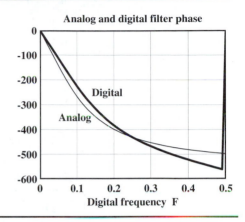

Example 14.6

Bilinear Transformation of a Notch Filter

The twin-T notch filter $H(s) = (s^2 + \omega_0^2)/(s^2 + 4\omega_0 s + \omega_0^2)$ has a notch frequency ω_0. To design a digital filter with $S_F = 240$ Hz whose response at $f_D = 60$ Hz matches that of $H(s)$ at $\omega_A = 1$, we first find the digital frequency $\Omega_D = 2\pi f_D/S_F = \frac{1}{2}\pi$.

We then select C by prewarping Ω_D and equating the result with $\omega_A = 1$. Thus

$$1 = C\tan(\tfrac{1}{2}\Omega_D) \quad \text{or} \quad C = 1/\tan(\tfrac{1}{2}\Omega_D) = 1$$

Finally, we convert $H(s)$ to $H_D(z)$ using $s = C(z-1)/(z+1) = (z-1)/(z+1)$

$$H_D(z) = [(z-1)^2 + \omega_0^2(z+1)^2]/[(z-1)^2 + 4\omega_0(z^2-1) + \omega_0^2(z+1)^2]$$

For $\omega_0 = \omega = 1$, $H_D(z) = (z^2+1)/(3z^2-1)$ and $H(F = \frac{1}{4}) = 0$, as expected. Figure 14.6 (p. 634) shows the match at $F = \frac{1}{4}$ for $\omega_A = 1$ and $\omega_A = 2$. ▬▬

Comment: If we redesign $H_D(z)$ for $\omega_A = 2$, we still see a match at $F = \frac{1}{4}$, since the digital frequency f_D/S_F is unchanged.

14.9.2 *Advantages of the Bilinear Transformation*

The bilinear transformation is the mapping most often employed for digital filter design. We summarize its main features and advantages:

1. It is simple to apply, provides a one-to-one mapping between the s-plane and z-plane, and avoids aliasing problems. It can thus be used for all types of systems (including highpass and bandstop).

Figure 14.6   Frequency response of the IIR filters of Example 14.6.

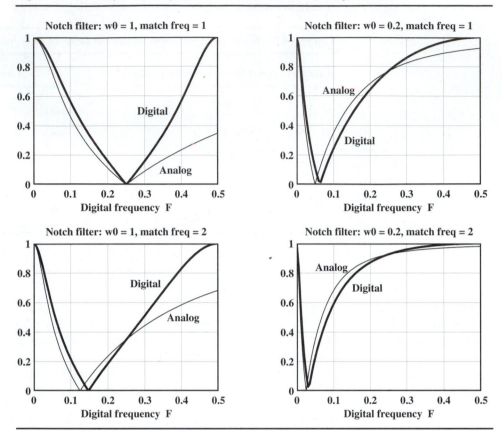

2. It always maps a stable analog system into a stable digital system. It does suffer from warping effects but allows for unwarping the results using a simple relation. Other stable higher-order (Adams) methods require much higher sampling frequencies to minimize the warping.

3. It can map $H(s)$ described in either cascaded or partial fraction form.

14.10
Spectral Transformations for IIR Filters

Next only in importance to the mapping rules for converting analog systems to digital systems are the spectral transformations we need to convert a digital lowpass prototype $H_{\mathrm{LP}}(z)$ to the required form $H_D(z)$. As with analog filters, this allows us to concentrate on the design of lowpass prototypes.

14.10.1   *Digital-to-Digital (D2D) Transformations*

If a digital lowpass prototype has been designed, the digital-to-digital ($D2D$) transformations of Table 14.10 can be used to convert it to the required filter type. These transformations preserve stability by mapping the interior of the unit circle and all points within it into itself.

Table 14.10   D2D Frequency Transformations.
Note: The digital lowpass prototype cutoff frequency $= \Omega_D$. All digital frequencies are normalized to $\Omega = 2\pi f / S_F$.

Conversion	Cutoffs	z2z map	Notation
LP2LP	Ω_C	$\dfrac{z - \alpha}{1 - \alpha z}$	$\alpha = \sin(\tfrac{1}{2}[\Omega_D - \Omega_C]) / \sin(\tfrac{1}{2}[\Omega_D + \Omega_C])$
LP2HP	Ω_C	$\dfrac{-(z + \alpha)}{1 + \alpha z}$	$\alpha = -\cos(\tfrac{1}{2}[\Omega_D + \Omega_C]) / \cos(\tfrac{1}{2}[\Omega_D - \Omega_C])$
LP2BP	$[\Omega_1, \Omega_2]$ $\Omega_2 > \Omega_1$	$\dfrac{-(z^2 + A_1 z + A_2)}{A_2 z^2 + A_1 z + 1}$	$K = \tan(\Omega_D/2) / \tan(\tfrac{1}{2}[\Omega_2 - \Omega_1])$ $\alpha = -\cos(\tfrac{1}{2}[\Omega_2 + \Omega_1]) / \cos(\tfrac{1}{2}[\Omega_2 - \Omega_1])$ $A_1 = 2\alpha K/(K + 1) \qquad A_2 = (K - 1)/(K + 1)$
LP2BS	$[\Omega_1, \Omega_2]$ $\Omega_2 > \Omega_1$	$\dfrac{(z^2 + A_1 z + A_2)}{A_2 z^2 + A_1 z + 1}$	$K = \tan(\Omega_D/2) \tan(\tfrac{1}{2}[\Omega_2 - \Omega_1])$ $\alpha = -\cos(\tfrac{1}{2}[\Omega_2 + \Omega_1]) / \cos(\tfrac{1}{2}[\Omega_2 - \Omega_1])$ $A_1 = 2\alpha/(K + 1) \qquad A_2 = -(K - 1)/(K + 1)$

As with analog transformations, the LP2BP and LP2BS transformations yield a digital filter with twice the order of the lowpass prototype.

The LP2LP transformation is actually a special case of the more general allpass transformation $z \longrightarrow \pm(z - \alpha)/(1 - \alpha z)$, where $|\alpha| < 1$ and α is real.

Example 14.7
Using *D2D* Transformations

Consider a digital lowpass filter with $H(z) = 3(z + 1)^2/(31z^2 - 26z + 7)$, sampling frequency $S_F = 8$ kHz and arbitrary cutoff frequency $f_C = 2$ kHz. Then $\Omega_D = 2\pi f_C/S_F = \tfrac{1}{2}\pi$.

(a) To design a highpass filter with a cutoff of 1 kHz, we use $\Omega_C = \tfrac{1}{4}\pi$ and Table 14.10 to compute

$$\alpha = -\cos(\tfrac{1}{2}[\Omega_D + \Omega_C]) / \cos(\tfrac{1}{2}[\Omega_D - \Omega_C]) = -\cos(3\pi/8)/\cos(\tfrac{1}{8}\pi) = -0.4142$$

The spectral transformation is thus $z \longrightarrow -(z - 0.4142)/(1 - 0.4142z)$, and yields

$$H_{\mathrm{HP}}(z) = 0.28(z - 1)^2/(z^2 - 0.0476z + 0.0723)$$

(b) To design a bandpass filter with band-edges of 1 kHz and 3 kHz, we use $\Omega_1 = \tfrac{1}{4}\pi$, $\Omega_2 = \tfrac{3}{4}\pi$, $\Omega_2 - \Omega_1 = \tfrac{1}{2}\pi$, $\Omega_2 + \Omega_1 = \pi$ and Table 14.10 to compute $K = \tan(\tfrac{1}{4}\pi) / \tan(\tfrac{1}{4}\pi) = 1$, $\alpha = -\cos(\tfrac{1}{2}\pi)/\cos(\tfrac{1}{4}\pi) = 0$, $A_1 = 0$ and $A_2 = 0$. The frequency transformation is thus $z \longrightarrow -z^2$, and yields

$$H_{\mathrm{BP}}(z) = 3(z^2 - 1)^2/(31z^4 + 26z^2 + 7)$$

(c) To design a bandstop filter with band-edges of 1.5 kHz and 2.5 kHz, we use $\Omega_1 = 3\pi/8$, $\Omega_2 = 5\pi/8$, $\Omega_2 - \Omega_1 = \frac{1}{4}\pi$, $\Omega_2 + \Omega_1 = \pi$ and Table 14.10 to compute $K = \tan(\frac{1}{8}\pi)\tan(\frac{1}{4}\pi) = 0.4142$, $\alpha = -\cos(\frac{1}{2}\pi)/\cos(\frac{1}{8}\pi) = 0$, $A_1 = 0$, and $A_2 = 0.4142$. The transformation is thus $z \longrightarrow (z^2 + 0.4142)/(0.4142z^2 + 1)$, and yields

$$H_{BS}(z) = 0.28(z^2 + 1)^2/(z^4 + 0.0476z^2 + 0.0723)$$

Figure 14.7 compares $|H(F)|$, $|H_{HP}(F)|$, $|H_{BP}(F)|$ and $|H_{BS}(F)|$.

Figure 14.7   Frequency response of the IIR filters designed in Example 14.7.

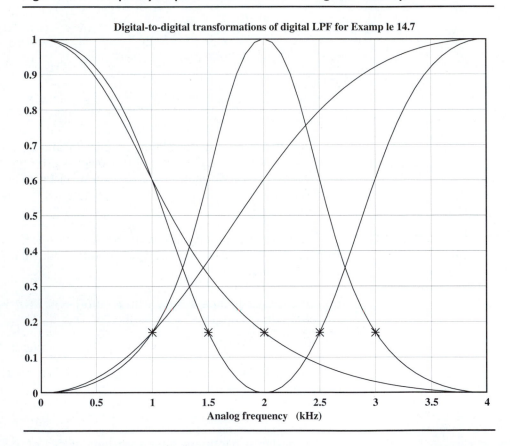

Digital-to-digital transformations of digital LPF for Examp le 14.7

14.10.2   *Direct Analog-to-Digital (A2D) Transformations for Bilinear Design*

All stable transformations that are also free of aliasing introduce warping effects. Only the bilinear mapping offers a simple relation to compensate for the warping. Combining the bilinear and *D2D* transformations yields the analog-to-digital (*A2D*) transformations of Table 14.11 for bilinear design. These can be used to convert a prewarped analog lowpass prototype with unit cutoff frequency directly to the required digital filter.

Table 14.11 **Direct *A2D* Transformations for Bilinear Design.**

Note: The analog lowpass prototype *prewarped* cutoff frequency is 1 rad/s. Digital frequencies are normalized to $\Omega = 2\pi/S_F$ but are *not prewarped*.

Conversion	Cutoffs	s2z map	Notation
LP2LP	Ω_C	$C\dfrac{z-1}{z+1}$	$C = 1/\tan(\frac{1}{2}\Omega_C)$
LP2HP	Ω_C	$C\dfrac{z+1}{z-1}$	$C = \tan(\frac{1}{2}\Omega_C)$
LP2BP	$\Omega_1 < \Omega_0 < \Omega_2$	$C\dfrac{z^2 + 2\alpha z + 1}{z^2 - 1}$	$C = \cot(\frac{1}{2}[\Omega_2 - \Omega_1]), \alpha = -\cos(\Omega_0)$ or $\alpha = -\cos(\frac{1}{2}[\Omega_2 + \Omega_1])/\cos(\frac{1}{2}[\Omega_2 - \Omega_1])$
LP2BS	$\Omega_1 < \Omega_0 < \Omega_2$	$C\dfrac{z^2 - 1}{z^2 + 2\alpha z + 1}$	$C = \tan(\frac{1}{2}[\Omega_2 - \Omega_1]), \alpha = -\cos(\Omega_0)$ or $\alpha = -\cos(\frac{1}{2}[\Omega_2 + \Omega_1])/\cos(\frac{1}{2}[\Omega_2 - \Omega_1])$

14.10.3 *A General Design Recipe for IIR Filters*

Given the passband and stopband frequencies f_p and f_s, the attenuations A_p and A_s, and the sampling frequency S_F, here is a standard recipe for the design of IIR filters using *s2z* mappings:

1. Normalize the design frequencies to $F_p = f_p/S_F$ and $F_s = f_s/S_F$. This allows us to use a sampling interval $t_s = 1$ in subsequent design.

Note: For bilinear design, we also *prewarp* the normalized frequencies.

2. Design the normalized *analog lowpass prototype* $H_{\rm LP}(s)$ with $\omega_C = 1$ rad/s.
3. Apply the chosen *s2z* mapping (with $t_s = 1$) to convert $H_{\rm LP}(s)$ to the *digital lowpass prototype* filter $H_{\rm LP}(z)$ with $\Omega_D = 1$.
4. Use the *D2D* transformations of Table 14.10 (with $\Omega_D = 1$) to convert $H_{\rm LP}(z)$ to the required digital filter $H_D(z)$.

Remarks: **1.** We do not design the actual analog filter $H_A(s)$ and use *s2z* mappings directly on $H_A(s)$. Rather, we design a *digital* lowpass prototype $H_{\rm LP}(z)$ followed by a *D2D* transformation. This allows us to use any *s2z* mappings, including those (such as response invariance) which may otherwise lead to excessive aliasing for highpass and bandstop filters. Designing $H_{\rm LP}(z)$ also allows us to match its dc magnitude with $H_{\rm LP}(s)$. This is useful for subsequent comparison.
2. Only for bilinear design will a mapping of $H_A(s)$ to $H_D(z)$ also yield identical results. Only for bilinear design can we also convert $H_{\rm LP}(s)$ to $H_D(z)$ *directly* (using the *A2D* transformations of Table 14.11) with identical results.

Example 14.8

Design of a Bandpass
IIR Filter

Let us design a Chebyshev IIR filter to meet the following specifications:
　　　Passband edges at 1.8 kHz and 3.2 kHz, stopband edges at 1.6 kHz and 4.8 kHz,
$A_p = 2$ dB, $A_s = 20$ dB, and sampling frequency $S_F = 12$ kHz.

(a) Here are the design steps using the bilinear transformation:

1. Normalize: Passband edges $[\Omega_1, \Omega_2] = 2\pi[1.8, 3.2]/12 = [0.94, 1.68]$, and stop-band edges $= 2\pi[1.6, 4.8]/12 = [0.84, 2.51]$
2. Choose $C = 2$ and prewarp each (band-edge) frequency using $\omega_x = 2\tan(\frac{1}{2}\Omega_x)$. Prewarped specs: passband $[\Omega_{p1}, \Omega_{p2}] = [1.019, 2.221]$, stopband $= [0.89, 6.155]$
3. Design an analog filter meeting these specs. This was actually done in Example 10.8 yielding the normalized LPP and actual transfer function as

$$H_{\mathrm{LP}}(s) = 0.1634/(s^4 + 0.7162s^3 + 1.2565s^2 + 0.5168s + 0.2058)$$

$$H_{\mathrm{BP}}(s) = \frac{0.34s^4}{s^8 + 0.86s^7 + 10.87s^6 + 6.75s^5 + 39.39s^4 + 15.27s^3 + 55.69s^2 + 9.99s + 26.25}$$

4. Transform $H_{\mathrm{BP}}(s)$ to $H_D(z)$ using $s \longrightarrow 2(z-1)/(z+1)$ to obtain

$$H_D(z) = \frac{0.0026z^8 - 0.0095z^6 + 0.0142z^4 - 0.0095z^2 + 0.0026}{z^8 - 1.94z^7 + 4.44z^6 - 5.08z^5 + 6.24z^4 - 4.47z^3 + 3.44z^2 - 1.305z + 0.59}$$

Figure 14.8 compares $|H_D(F)|$ with its digital unwarped and analog design.

(b) Here are the design steps for all other $s2z$ mappings:

1. Start with the normalized *unwarped* specs passband $= [\Omega_1, \Omega_2] = [0.94, 1.68]$, and stopband $= [0.84, 2.51]$
2. Design the normalized analog lowpass prototype $H_{\mathrm{LP}}(s)$ with $\omega_C = 1$ rad/s. Fortunately, the unwarped specifications actually yield an $H_{\mathrm{LP}}(s)$ identical to that obtained for the bilinear design (even though this will not always be the case).
3. Convert $H_{\mathrm{LP}}(s)$ to $H_{\mathrm{LP}}(z)$ with $\Omega_D = 1$ using a desired $s2z$ mapping and $t_s = 1$. For backward differences, for example, we would use $s = (z-1)/zt_s = (z-1)/z$.
4. Convert $H_{\mathrm{LP}}(z)$ to $H_D(z)$, using $\Omega_D = 1$ and the *unwarped* (passband) specs $[\Omega_1, \Omega_2] = [0.94, 1.68]$, in the LP2BP mapping of Table 14.10.

Example 14.9

A2D Transformations for
Bilinear Design

For bilinear design, we can convert $H_{\mathrm{LP}}(s)$ directly using the *A2D* transformation $s \longrightarrow C(z^2 - 2\alpha z + 1)/(z^2 - 1)$ of Table 14.11 where C and α are found from the *unwarped* normalized edges. For Example 14.8, the unwarped passband edges are $[\Omega_1, \Omega_2] = [0.94, 1.68]$. We use Table 14.11 to compute $C = \cot[\frac{1}{2}(\Omega_2 - \Omega_1)] = 2.605$ and $\alpha = -\cos[\frac{1}{2}(\Omega_2 + \Omega_1)]/\cos[\frac{1}{2}(\Omega_2 - \Omega_1)] = -0.277$. The *A2D* transformation of $H_{\mathrm{LP}}(s)$ gives an $H_D(z)$ identical to that of Example 14.8.

Figure 14.8   Frequency response of the IIR filter designed in Example 14.8.

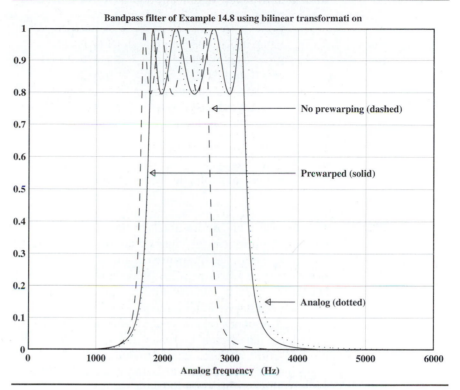

IIR filters are well suited to applications requiring frequency-selective filters with sharp cutoffs, or where linear phase is relatively unimportant. Examples include graphic equalizers for digital audio, tone generators for digital touch-tone receivers and filters for digital telephones. The main advantages are standardized easy design and low filter order. On the other hand, IIR filters cannot exhibit linear phase and are quite susceptible to the effects of coefficient quantization. If linear phase is important, as in biomedical signal processing, or stability is paramount, as in many adaptive filtering schemes, it is best to use FIR filters.

14.11
Concluding
Remarks

14.11.1   *Effects of Coefficient Quantization*

For IIR filters, the effects of rounding or truncating filter coefficients can range from minor changes in the frequency response to serious problems, including instability. Consider the stable analog filter

$$H(s) = \frac{(s + 0.5)(s + 1.5)}{(s + 1)(s + 2)(s + 4.5)(s + 8)(s + 12)}$$

Bilinear transformation of $H(s)$ with $t_s = 0.01$ s yields the digital transfer function

$H_D(z) = B(z)/A(z)$ whose denominator coefficients to double precision (A_k) and truncated to seven significant digits (A_k^t) are given by

A_k	Truncated A_k^t
$1.144168420199997e+0$	$1.144168e+0$
$-5.418904483999996e+0$	$-5.418904e+0$
$1.026166736200000e+0$	$1.026166e+1$
$-9.712186808000000e+0$	$-9.712186e+0$
$4.594164261000004e+0$	$4.594164e+0$
$-8.689086648000011e-1$	$-8.689086e-1$

The pole magnitudes of $H_D(z)$ based on A_k all lie within the unit circle and the designed filter is thus stable. However, if we use the truncated coefficients A_k^t, the filter becomes unstable, because one pole moves out of the unit circle.

14.12
FIR Filter Design

The design of FIR filters involves the selection of a finite sequence that best represents the impulse response of an ideal filter. FIR filters are always stable. Even more important, FIR filters are capable of perfectly linear phase (a pure time delay), meaning total freedom from phase distortion. For given specifications, however, FIR filters typically require a much higher filter order or length than do IIR filters. And sometimes we must go to great lengths to ensure linear phase!

The frequency response of symmetric sequences is endowed with linear phase. The phase can actually be *piecewise* linear if the amplitude undergoes sign changes, but this still translates into a constant delay.

Symmetric sequences also imply noncausal filters which can be realized only if we introduce a sufficient delay.

14.12.1   *Symmetric Sequences and Linear Phase*

The length N of finite symmetric sequences can be odd or even, since the center of symmetry may fall on a sample point (for odd N) or midway between samples (for even N). This results in four possible types of symmetric sequences, as listed in Table 14.12 and shown in Figure 14.9.

Table 14.12   Symmetric Sequences.
Note: $L = \frac{1}{2}(N-1)$, and $M = \frac{1}{2}N$.

Type	Symmetry	N	H(F)	$\|H(0)\|$	$\|H(\frac{1}{2})\|$
1	Even	Odd	$h[0] + 2\sum_{k=1}^{L} h[k]\cos(2k\pi F)$	$h[0] + 2\sum_{k=1}^{L} h[k]$	$h[0] + 2\sum_{k=1}^{L}(-1)^k h[k]$
2	Even	Even	$2\sum_{k=1}^{M} h[k]\cos[2\pi F(k-\frac{1}{2})]$	$2\sum_{k=1}^{M} h[k]$	0
3	Odd	Odd	$-j2\sum_{k=1}^{L} h[k]\sin(2k\pi F)$	0	0
4	Odd	Even	$-j2\sum_{k=1}^{M} h[k]\sin[2\pi F(k-\frac{1}{2})]$	0	$-2\sum_{k=1}^{M}(-1)^k h[k]$

Figure 14.9   Four types of symmetric sequences.

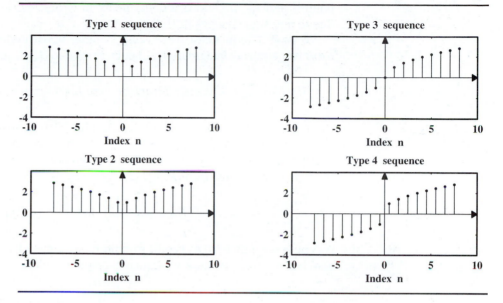

A **type 1 sequence** $h_1[n]$ is even symmetric with odd length N. Its noncausal form has samples at $n = \pm1, \pm2, \ldots, \pm L$ where $L = \frac{1}{2}(N - 1)$ is an integer. Using Euler's relation, its frequency response $H_1(F)$ can be expressed as

$$H_1(F) = h[0] + \sum_{k=1}^{L} h[k][\exp(j2k\pi F) + \exp(-j2k\pi F)] = h[0] + 2\sum_{k=1}^{L} h[k]\cos(2k\pi F)$$

This is a real quantity with both $|H_1(0)|$ and $|H_1(\frac{1}{2})|$ nonzero, in general.

The delayed causal version $h_{1C}[n] = h_1[n - L]$ shows symmetry about $n = L$, and $H_{1C}(F)$ shows a linear phase of $\exp(j2\pi LF)$ (in addition to phase jumps of $\pm\pi$ due to any sign changes in $H_1(F)$).

The transfer function $H_{1C}(z)$ of the causal sequence equals

$$H_{1C}(z) = z^{-L}\left[h[0] + \sum_{k=1}^{L} h[k][z^k + z^{-k}]\right]$$

A **type 2 sequence** $h_2[n]$ is also even symmetric but of even length N. Its noncausal form has samples at $n = \pm0.5, \pm1.5, \ldots, \pm L$ where $L = \frac{1}{2}(N - 1)$ is not an integer. With $M = \frac{1}{2}N$, its frequency response $H_2(F)$ may be written

$$H_2(F) = 2\sum_{k=1}^{M} h[k]\cos[2\pi F(k - \tfrac{1}{2})]$$

Even though $H_2(F)$ is still real, note that $|H(\frac{1}{2})|$ always equals zero.

The causal version $h_{2C}[n]$ is symmetric about $n = L$ (which falls between sample points), and $H_{2C}(F)$ shows a linear phase of $\exp(j2\pi LF)$ (and phase jumps of $\pm\pi$ due to any sign changes in $H_1(F)$).

A **type 3 sequence** $h_3[n]$ is odd symmetric with odd length N. Its noncausal form has a purely imaginary frequency response $H_3(F)$ given by

$$H_3(F) = \sum_{k=1}^{L} h[k][\exp(-j2k\pi F) - \exp(j2k\pi F)] = -j2\sum_{k=1}^{L} h[k]\sin(2k\pi F)$$

Here, $L = \frac{1}{2}(N - 1)$. Clearly, both $|H_3(0)|$ and $|H_3(\frac{1}{2})|$ always equal zero.

The causal version $h_{3C}[n]$ has a linear phase on top of a constant phase of 90° (and phase jumps of $\pm\pi$, where the amplitude of $H_3(F)$ changes sign).

The transfer function $H_{3C}(z)$ equals

$$H(z) = z^{-L}\left[\sum_{k=1}^{L} h[k](z^{-k} - z^k)\right]$$

A **type 4 sequence** $h_4[n]$ is odd symmetric with even length N. Its noncausal form also has a purely imaginary frequency response. With $M = \frac{1}{2}N$, we have

$$H_4(F) = -j2\sum_{k=1}^{M} h[k]\sin[2\pi F(k - \tfrac{1}{2})]$$

Obviously, $|H(0)| = 0$ but $|H(\frac{1}{2})|$ is nonzero, in general. The causal form $h_{4C}[n]$ shows odd symmetry about $L = \frac{1}{2}(N - 1)$, which lies between sample points.

14.12.2   *Applications of Symmetric Sequences*

Applications of symmetric sequences for FIR filter design are summarized in Table 14.13 and are dictated by the frequency response $H(F)$ of each sequence. Based on the response of typical filters, we readily conclude that

Table 14.13   Applications of Symmetric Sequences.

Type	$\|H(F)\|$	Application
1		All filter types. Only sequence for BS.
2	$\|H(\frac{1}{2})\| = 0$	Only LP and BP filters.
3	$\|H(0)\| = 0 = \|H(\frac{1}{2})\|$	BP, differentiators, Hilbert transformers
4	$\|H(F)\| = 0$	HP, BP, differentiators, Hilbert transformers.

1. For type 2 sequences, $|H(\frac{1}{2})| = 0$. They cannot be used to design bandstop or highpass filters (whose response is not zero at $F = \frac{1}{2}$). They can be used for lowpass and bandpass filters.

2. Only antisymmetric (type 3 and type 4) sequences (with imaginary $H(F)$) can be used to design differentiators and Hilbert transformers.

3. For type 3 sequences, $|H(0)| = 0 = |H(\frac{1}{2})|$. They are suitable only for bandpass filters, differentiators and Hilbert transformers.
4. For type 4 sequences, only $|H(0)| = 0$. They are suitable for highpass and bandpass filters, differentiators and Hilbert transformers.
5. Even-length (type 2 and type 4) sequences require half-sample delays for a causal realization. Such delays are less convenient to implement.

Classical filters are often designed using type 1 sequences.

Remark: Bandstop filters can be designed only with type 1 sequences.

14.12.3   *Techniques of FIR Filter Design*
There are three major approaches to the design of FIR filters:

1. Window-based design using the impulse response of ideal filters.
2. Frequency sampling.
3. Iterative design based on optimal constraints.

14.13
The Fourier
Series Method

This method revolves around selecting the impulse response $h_N[n]$ as a truncated version of the impulse response $h[n]$ of an ideal filter with frequency response $H(F)$. The term *Fourier series method* derives from the fact that finding $h[n]$ from $H(F)$ using the IDFT is quite like finding Fourier series coefficients. The designed filter is an approximation to the ideal filter. But it is the best approximation (in the mean square sense) compared to any other filter of the same length.

14.13.1   *Design Recipe for Classical Filters*
As with analog and IIR filters, we shall develop the design of FIR filters from a lowpass prototype and use appropriate spectral transformations to convert it to the required filter type. Window-based design calls for

1. Normalization of the design frequencies by the sampling frequency.
2. Conversion to specifications applicable to a lowpass prototype.
3. Truncation of the ideal filter impulse response $h[n] = 2F_C\text{sinc}(2nF_C)$ to $h_N[n]$ with a desired length N. The filter order then equals $N - 1$.
4. Selection of an appropriate N-point window $w[n]$ to obtain the windowed impulse response $h_W[n] = h_N[n]w[n]$.
5. Conversion to the required type $h_F[n]$ using spectral transformations.
6. Delaying $h_F[n]$ to ensure a causal filter $h_c[n]$.

This scheme allows us to concentrate on the design of lowpass filters.

14.13.2   *Some Important Design Issues*
This design method may appear deceptively simple, but there are many aspects that cannot always be addressed quantitatively. As a result, the design is more an art than a science. Here are some key issues to keep in mind.

14.13.3　*The Need for Windowing*

Truncation of $h[n]$ is equivalent to multiplying $h[n]$ by a rectangular window $w[n]$ of length N. The spectrum of $h_N[n] = h[n]w[n]$ is simply the (periodic) convolution of $H(F)$ and $W(F)$. Since $W(F)$ has the form of a Dirichlet kernel, this spectrum shows overshoot and ripples independent of the length N, quite like the Gibbs effect (Figure 14.10). To reduce the effects of abrupt truncation, we use tapered windows. Changing the length N also changes the cutoff frequency of the windowed spectrum. The smallest N that meets specifications depends on the choice of window.

Figure 14.10　$|H(F)|$ for the ideal and truncated impulse response.

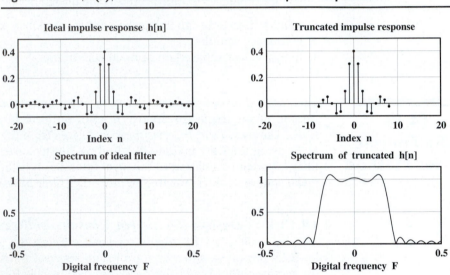

Windows　The windows commonly used in FIR filter design are listed in Table 14.14. Their major spectral features (adapted from Table 7.4) are summarized in Table 14.15, and their spectra are illustrated in Figure 14.11 (p. 647).

The numerical values in Table 14.15 (p. 646) are actually based on the Fourier transforms of their *CT* counterparts, but serve as useful design guidelines.

The spectrum of finite duration windows invariably shows a mainlobe and decaying sidelobes which may be entirely positive or alternate in sign.

We note that:

1. The mainlobe width and transition width decrease as we increase N.
2. The sidelobe level remains more or less constant as we change N.

Ideally, the spectrum of a window should be confined to a narrow mainlobe with almost no energy in the sidelobes. Most windows have been developed with some optimality criterion in mind. Ultimately, the trade-off is a compromise between the conflicting requirements of a narrow mainlobe (or a small transition width) and small sidelobe levels.

Table 14.14   Some Data Windows for FIR Filter Design.

Notes: 1. The range of the integer index n is $-\frac{1}{2}(N-1) \le n \le \frac{1}{2}(N-1)$.

2. $I_m(x)$ is the modified Bessel function of order m.

3. $T_N(x)$ is the Chebyshev polynomial of order N.

4. $I_{n-\frac{1}{2}}(x)$ is the modified spherical Bessel function of order n.

Window	Expression $w[n]$, $-\frac{1}{2}(N-1) \le n \le \frac{1}{2}(N-1)$
Boxcar	1
Bartlett	$1 - 2|n|/(N-1)$
vonHann	$\frac{1}{2} + \frac{1}{2}\cos[2n\pi/(N-1)]$
Hamming	$0.54 + 0.46\cos[2n\pi/(N-1)]$
Cosine	$\cos[n\pi/(N-1)]$
Riemann	$\operatorname{sinc}^L[2n/(N-1)]$ $(L>0)$
Parzen-2	$1 - 4n^2/(N-1)^2$
Blackman	$0.42 + 0.5\cos[2n\pi/(N-1)] + 0.08\cos[4n\pi/(N-1)]$
Papoulis	$\dfrac{1}{\pi}|\sin[2n\pi/(N-1)]| + [1 - 2|n|/(N-1)]\cos[2n\pi/(N-1)]$
Kaiser	$\dfrac{I_0(\pi\beta\sqrt{1-4[n/(N-1)]^2}}{I_0(\pi\beta)}$
Modified Kaiser	$\dfrac{I_0(\pi\beta\sqrt{1-4[n/(N-1)]^2}-1}{I_0(\pi\beta)-1}$
Bickmore	$\left[\dfrac{\sqrt{1-4[n/(N-1)]^2}}{\pi\beta}\right]^{\nu-\frac{1}{2}}\left[\dfrac{I_{\nu-\frac{1}{2}}(\pi\beta\sqrt{1-4[n/(N-1)]^2})}{I_{\nu-\frac{1}{2}}(\pi\beta)}\right]$
General Blackman	$a_0 + \sum_{n=1}^{L} 2a_n\cos[2n\pi/(N-1)]$
Dolph-Chebyshev (odd n only)	$w[n] = \mathrm{IDFT}\{T_{N-1}(x)\}$ $\quad x = \alpha\cos[n\pi/(N-1)]$ $\quad \alpha = \cosh\left[\dfrac{\cosh^{-1}(1/\epsilon)}{N-1}\right]$ $\quad \epsilon = 10^{-R/20}$ R = sidelobe level dB

Why So Many Windows?   The answer to this question is based on how windows are designed. Factors considered include reducing the peak sidelobe level, increasing sidelobe decay rate, varying the location of the sidelobe nulls or the width of the mainlobe and, equally important, a simple mathematical form. Some windows are based on combinations of simpler windows:

The vonHann window is the sum of a rectangular and cosine window.
The $\cos^a$ window is the product of a vonHann and cosine window.
The Bartlett window is the convolution of two rectangular windows.

Table 14.15 Spectral Characteristics of Window Functions.

Notation:

G_P: Peak gain of mainlobe $\qquad$ G_S: Peak sidelobe gain

W_M: Half-width of mainlobe $\qquad$ P_S: Peak sidelobe attenuation $\frac{G_P}{G_S}$ in dB

W_6: 6 dB half-width $\qquad$ W_S: Half-width of mainlobe to reach P_S

W_3: 3 dB half-width $\qquad$ D_S: High-frequency attenuation (dB/decade)

Window	G_P	$\frac{G_S}{G_P}$	P_S dB	W_M	W_S	W_6	W_3	D_S dB/decade
Boxcar	1	0.2172	13.3	1	0.81	0.6	0.44	20
Parzen-2	0.6667	0.0862	21.3	1.43	1.27	0.79	0.57	40
Cosine	0.6366	0.0708	23	1.5	1.35	0.81	0.59	40
Riemann	0.5895	0.0478	26.4	1.64	1.5	0.86	0.62	40
Bartlett	0.5	0.0472	26.5	2	1.62	0.88	0.63	40
vonHann	0.5	0.0267	31.5	2	1.87	1.0	0.72	60
Hamming	0.54	0.0073	42.7	2	1.91	0.9	0.65	20
Papoulis	0.4053	0.0050	46	3	2.7	1.18	0.85	80
Parzen-1	0.375	0.0022	53	4	3.25	1.27	0.91	80
Blackman	0.42	0.0012	58.1	3	2.82	1.14	0.82	60
Kaiser ($\beta = 2.6$)	0.4314	0.0010	59.9	2.8	2.72	1.11	0.80	20
Bickmore								
$\beta = 2.6, \nu = \frac{1}{2}$	0.4314	0.0010	59.9	2.8	2.72	1.11	0.80	20
$\beta = 3, \nu = 3$	0.0013	0.0005	65.7	3.63	3.54 ·	1.36	0.97	70

NOTE: All listed widths must be divided by the filter length N.

Remarks:

1. The windows are listed in order of *increasing P_s*.

2. The values for the Kaiser and modified Kaiser windows depend on the parameter β. For the Bickmore window, they depend on both β and ν.

3. For $\nu = \frac{1}{2}$, the Bickmore window reduces to the Kaiser window. Note that the parameter ν controls the sidelobe decay ($D_S = -20(\nu + \frac{1}{2})$ dB/decade).

4. Empirically determined relations for the Kaiser window are

$$G_P = \frac{|\text{sinc}(j\beta)|}{I_0(\pi\beta)} \qquad \frac{G_S}{G_P} = \frac{0.22\pi\beta}{\sinh(\pi\beta)} \qquad W_M = 2(1 + \beta^2)^{1/2} \qquad W_S = 2(0.661 + \beta^2)^{1/2}$$

5. For all windows, a reduced sidelobe level can be achieved only at the expense of an increased mainlobe width.

6. The compromise in using the various windows is between choosing the central lobe to be as narrow as possible and choosing the sidelobe level to be as small as possible.

7. Any window except rectangular yields a marked improvement. The Hamming and Kaiser windows are among the most widely used.

Figure 14.11   The window functions of Table 14.14.

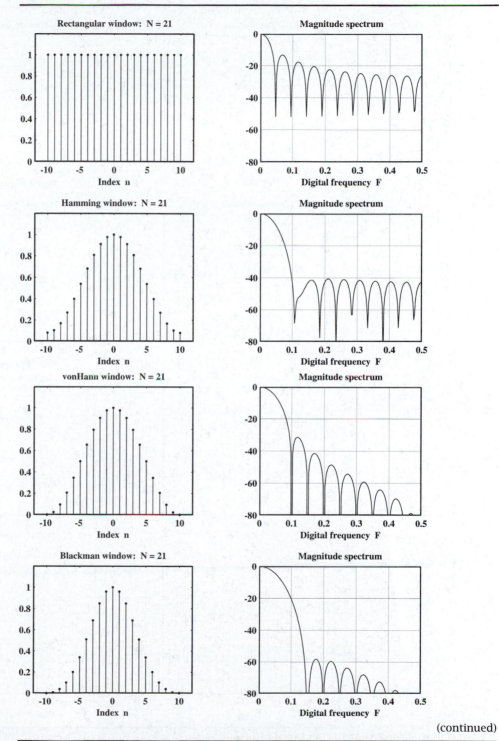

(continued)

Figure 14.11   (continued)

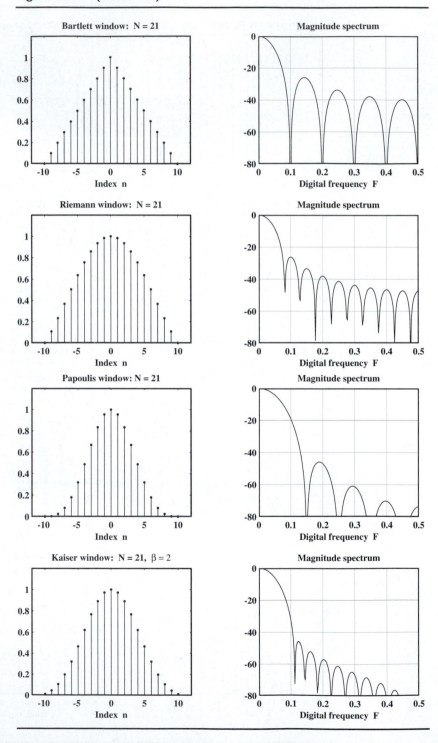

Others try to emphasize certain desirable features:

The vonHann window improves the high frequency decay (at the expense of a larger minimum sidelobe level).

The coefficients in the Hamming window are chosen to minimize the sidelobe level (at the expense of a slower high frequency decay).

The Dolph-Chebyshev window has the smallest mainlobe width for a given sidelobe level, but its sidelobes are constant and do not decay.

The Kaiser and Bickmore windows have variable parameters that allow us to control both the mainlobe width and the sidelobe level (but only to a certain extent).

Still others use analytical forms that are easy to manipulate:

The $\cos^a$ windows have easily recognizable transforms.

The vonHann window is easy to apply as a convolution in frequency.

An optimal timelimited window should maximize the energy in its spectrum over a given frequency band. In the continuous-time domain, this constraint leads to a window based on *prolate spheroidal wave functions* of the first order. The Kaiser window best approximates such an optimal window in the discrete domain.

Choosing a Window   There are no simple answers. Two main issues are

1. A sensible choice must ensure that its minimum sidelobe attenuation (P_s in Table 14.15) exceeds the specified stopband attenuation A_s. Since windowing attenuates the response, we can often use windows with attenuations P_s a few decibels (up to 8dB or so) *less* than A_s.
2. A broader transition width allows for a smaller filter order but also means a smaller sidelobe attenuation. We seek the best compromise.

The spectrum of the windowed response is the convolution of the spectra of the impulse response and the window function. And this changes as we change the length. An optimal (in the mean square sense) impulse response and an optimal (in any sense) window may not together yield a windowed response with optimal features. Window selection is at best an art and at worst a matter of trial and error. Just so you know, the three most commonly used windows are the vonHann, Hamming and Kaiser windows.

Generating a Window   Like windows for Fourier series smoothing, but unlike DFT windows, FIR windows are truly symmetric. An N-point FIR window uses $N - 1$ intervals (a DFT window has N) to generate N samples (a DFT window has $N + 1$) and includes both end samples (a DFT window ignores the last one).

Once selected, the window sequence must be properly positioned with respect to the truncated impulse response sequence $h_N[n]$. Windowing is then simply a *pointwise multiplication* of the two sequences.

The Causal Filter   The impulse response of an ideal lowpass filter with a cutoff frequency F_C is $h[n] = 2F_C\text{sinc}(2nF_C)$. Symmetric truncation yields

$$\{h_N[n]\} = 2F_C\text{sinc}[2nF_C] \qquad -\tfrac{1}{2}(N-1) \le n \le \tfrac{1}{2}(N-1)$$

For even N, the index n is not an integer, even though but it is incremented every unit, and we require a *noninteger* delay to produce a causal sequence.

The Filter Order   If the end samples of $w[n]$ equal zero, so will those of $h_W[n]$ and both the filter order and filter length can be decreased by 2.

Example 14.10
Truncation and
Windowing of the
Impulse Response

Consider a lowpass filter with a cutoff frequency of 5 kHz and a sampling frequency of 20 kHz. Then, $F_C = \tfrac{1}{4}$ and $h_N[n] = 2F_C\text{sinc}(2nF_C) = 0.5\text{sinc}(0.5n)$.

(a) With $N = 9$, $-4 \le n \le 4$, and a Bartlett window $w = 1 - 2|n|/(N-1)$, we find

$$\{h_N[n]\} = \{0, -0.1061, 0, 0.3183, \underset{\uparrow}{0.5}, 0.3183, 0, -0.1061, 0\}$$

$$\{w_N[n]\} = \tfrac{1}{4}\{0, 1, 2, 3, \underset{\uparrow}{4}, 3, 2, 1, 0\}$$

Pointwise multiplication of h_N and w_N gives

$$\{h_W[n]\} = \{0, -0.0265, 0, 0.2387, \underset{\uparrow}{0.5}, 0.2387, 0, -0.0265, 0\}$$

The causal filter requires a delay of 4 samples to give

$$H_C(z) = 0 - 0.0265z^{-1} + 0.2387z^{-3} + 0.5z^{-4} + 0.2387z^{-5} - 0.0265z^{-7}$$

Comments: **1.** Since $S_F = 20$ kHz, and thus $t_s = 1/S_F = 0.05$ ms, a 4-sample delay translates to an actual delay of $4(0.05) = 0.2$ ms.
2. Since the first sample of h_W is zero, the *minimum delay* required for a causal filter is only 3 samples (not 4), and thus

$$H_C(z) = -0.0265 + 0.2387z^{-2} + 0.5z^{-3} + 0.2387z^{-4} - 0.0265z^{-6}$$

(b) With $N = 6$, $-2.5 \le n \le 2.5$, and a vonHann window
$w[n] = 0.5 + 0.5\cos[2n\pi/(N-1)]$

$$\{h_N[n]\} = \{-0.09, 0.1501, \underset{\uparrow}{0.4502}, 0.4502, 0.1501, -0.09\}$$

$$\{w[n]\} = \{0, 0.3455, \underset{\uparrow}{0.9045}, 0.9045, 0.3455, 0\}$$

$$\{h_W[n]\} = \{0, 0.0518, \underset{\uparrow}{0.4072}, 0.4072, 0.0518, 0\}$$

Since the point of symmtry falls midway between samples, the causal filter requires a delay of 2.5 samples (or 0.125 ms). We then get

$$H_C(z) = 0 + 0.0518z^{-1} + 0.4072z^{-2} + 0.4072z^{-3} + 0.0518z^{-4}$$

Comment: Again, the first sample of $h_W[n]$ is zero, and the *minimum delay* required for a causal filter is only 1.5 samples (not 2.5). This gives

$$H_C(z) = 0.0518 + 0.4072z^{-1} + 0.4072z^{-2} + 0.0518z^{-3}$$

14.13.4   *Selection of Filter Length*

There is no accurate way to establish the minimum filter length N. Empirical estimates are based on the transition width of the window. One such rule of thumb says that we must choose the filter length N from

$$N \approx W_S/(F_s - F_p)$$

Here, F_p and F_s are the *digital* passband and stopband frequencies, and W_S is the half-width of the mainlobe to reach the peak sidelobe level, as listed in Table 14.15. The quantity W_S is sometimes replaced by the half-width W_M of the mainlobe, or even the full width $2W_M$ (see Table 14.15). In any case, we often end up having to change N to best meet design specifications.

14.13.5   *Choice of Cutoff Frequency*

The cutoff frequency F_C is arbitrarily chosen as F_p or $\frac{1}{2}(F_p + F_s)$. To offset the reduced magnitude of the spectrum of the windowed signal, we often start with a value of F_C larger than this (by say 25%). In any case, F_C depends both on the cutoff frequency of the windowed impulse response $h_N[n]$, and the filter length N. We may thus have to tweak the cutoff frequency, as we change N, to ensure the smallest N that meets specifications. It turns out that the half-power frequency (3-dB frequency) of $h[n]$ also typically corresponds to the 6-dB frequency for $h_N[n]$ for most windows.

14.13.6   *Spectral Transformations*

The spectral transformations for converting lowpass designs to other forms, and vice versa, are listed in Table 14.16 (p. 652). Figure 14.12 shows that HP, BP and BS filters can be related to LP filters through spectral transformations developed from the shifting and modulation properties of the DTFT.

Unlike analog and IIR filters, the LP2BP or LP2BS transformations do not change the filter order and preserve arithmetic (not geometric) symmetry. Bandpass or bandstop filters require arithmetic symmetry before conversion to lowpass form. For a fixed passband, for example, we ensure arithmetic symmetry by moving the stopband edge corresponding to the larger transition width. This keeps both transitions at least as steep as specified.

Table 14.16   Spectral Transformations for FIR Filters.

(a) Transformation of Lowpass Filters

Type	$H_p(F)$ (one period)	Impulse response $h[n]$
Lowpass	$H_{LP}(F) = \text{rect}(F/2F_C)$	$h_{LP}[n] = 2F_C \text{sinc}[2nF_C]$
Highpass	$H_{HP}(F) = 1 - H_{LP}(F)$	$h_{HP}[n] = \delta[n] - h_{LP}[n]$
Highpass	$H_{HP}(F) = H_{LP}(F - \frac{1}{2}) = \text{rect}[(F - \frac{1}{2})/2F_C]$	$(-1)^n h_{LP}[n]$
Bandpass	$H_{BP}(F) = \dfrac{\text{rect}[(F + F_0)/2F_C] + }{\text{rect}[(F - F_0)/2F_C]}$	$2\cos(2\pi nF_0)h_{LP}[n]$
Bandstop	$H_{BS}(F) = 1 - H_{BP}(F)$	$\delta[n] - h_{BP}[n]$

(b) Transformation to Lowpass Prototypes

$F_{p1}, F_{p2}, F_{s1}, F_{s2}$ = passband and stopband edges, F_0 = center frequency.

For BP and BS: Each band is assumed to have arithmetic symmetry about F_0.

Transformation	Passband edge	Stopband edge	Center frequency or remarks
HP2LP	$F_p = F_{sHP}$	$F_s = F_{pHP}$	(if $h_{HP}[n] = \delta[n] - h_{LP}[n]$)
HP2LP	$F_p = \frac{1}{2} - F_{pHP}$	$F_s = \frac{1}{2} - F_{sHP}$	(if $h_{HP}[n] = (-1)^n h_{LP}[n]$)
BP2LP	$F_p = \frac{1}{2}(F_{p2} - F_{p1})$	$F_s = \frac{1}{2}(F_{s2} - F_{s1})$	$F_0 = \frac{1}{2}(F_{p2} + F_{p1})$
BS2LP	$F_p = \frac{1}{2}(F_{s2} - F_{s1})$	$F_s = \frac{1}{2}(F_{p2} - F_{p1})$	$F_0 = \frac{1}{2}(F_{p2} + F_{p1})$

Example 14.11
FIR Filter Design Using Windows

(a) We design a lowpass filter to meet the following specifications:

$$f_p = 1\,\text{kHz}, \quad f_s = 2\,\text{kHz}, \quad S_F = 10\,\text{kHz} \quad A_p = 2\,\text{dB}, \quad A_s = 40\,\text{dB},$$

The normalized frequencies are $F_p = f_p/S_F = 0.1, F_s = f_s/S_F = 0.2$. With $A_s = 40\,\text{dB}$, Table 14.15 leads to vonHann ($P_S = 31.5\,\text{dB}$), Hamming ($P_S = 42.7\,\text{dB}$) and Blackman ($P_S = 58.1\,\text{dB}$) windows as possible choices. With $N = W_S/(F_s - F_p)$, Table 14.15 also yields:

vonHann: $N = 1.87/0.1 \approx 19$

Hamming: $N = 1.91/0.1 \approx 20$

Blackman: $N = 2.82/0.1 \approx 29$

We choose a cutoff frequency 20% larger than F_p to give $F_C = 1.2F_p = 0.12$. The impulse response equals $h_N[n] = 2F_C\text{sinc}(2nF_C) = 0.24\text{sinc}(0.24n)$. Subsequent steps involve windowing $h_N[n]$ to get $h_W[n]$, checking that the chosen N and F_C meet the passband and stopband attenuations, and changing them if they do not. The F_C and N that just meet specifications are:

vonHann: $N = 23$ $F_C = 0.137$ $A_p = 1.90\,\text{dB}$ $A_s = 40.5\,\text{dB}$

Hamming: $N = 23$ $F_C = 0.131$ $A_p = 1.65\,\text{dB}$ $A_s = 41.1\,\text{dB}$

Blackman: $N = 29$ $F_C = 0.1278$ $A_p = 1.98\,\text{dB}$ $A_s = 40.1\,\text{dB}$

Figure 14.12   Spectral transformations of Table 14.16.

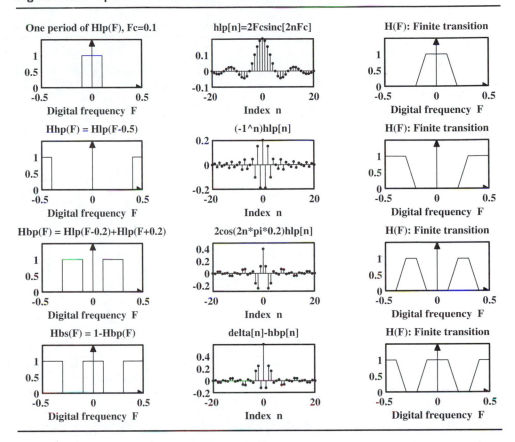

Figure 14.13 (p. 654) compares the magnitude response of these filters. The Blackman window shows the largest length and largest transition width, but also the lowest sidelobe level. Both the vonHann and Hamming windows yield the same filter length, but the vonHann window has the smallest transition width.

(b) We design a bandpass filter to meet the following specifications:

Passband [4, 8] kHz, Stopband [2, 12] kHz, $A_p = 3$ dB, $A_s = 45$ dB, $S_F = 25$ kHz

Assuming a fixed passband (and arithmetic symmetry), the center frequency is $f_0 = 6$ kHz. To make the stopband symmetric, we choose the upper stopband edge as 10 kHz to ensure the smallest transition width. We normalize to: Passband [0.16, 0.32], Stopband [0.08, 0.4], and $F_0 = 0.24$. The LPP specs become: $F_p = \frac{1}{2}(F_{p2} - F_{p1}) = 0.08, F_s = \frac{1}{2}(F_{s2} - F_{s1}) = 0.16$. One of the windows we can use is the Hamming window. For this window, we estimate N from

Figure 14.13   Frequency response of lowpass FIR filter of Example 14.11a.

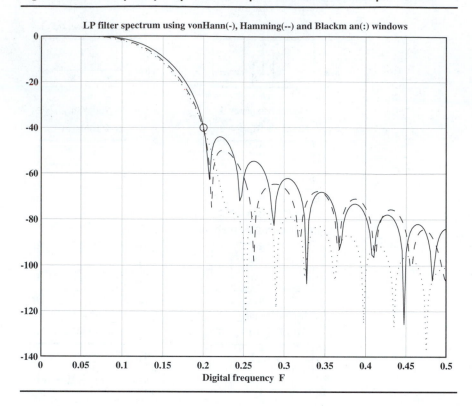

LP filter spectrum using vonHann(-), Hamming(--) and Blackm an(:) windows

Table 14.15 as $N = W_S/(F_s - F_p) = 1.91/0.08 \approx 24$. We choose a cutoff frequency 25% larger than F_p to give $F_C = 1.25F_p = 0.1$. The impulse response equals $h_N[n] = 2F_C \text{sinc}(2nF_C) = 0.2\text{sinc}(0.2n)$. Windowing this gives $h_w[n]$. We transform this to the bandpass form $h_{BP}[n]$,

$$h_{BP}[n] = 2\cos(2\pi n F_0)h_W[n] = 2\cos(0.48\pi n)h_W[n]$$

By changing N and F_C, we find that the specifications are met by $N = 27$, and $F_C = 0.0956$. For these values, the actual filter attenuation is 3.01 dB at 4 kHz and 8 kHz, 45.01 dB at 2 kHz and 73.47 dB at 12 kHz. Figure 14.14 shows the magnitude response of the designed filter.

14.14
Halfband
FIR Filters

For a **halfband** filter, alternate samples in the impulse response $h[n]$ equal zero. For a lowpass halfband filter, this requires $F_C = \frac{1}{4}$ and

$$h[n] = 2F_C\text{sinc}(2nF_C) = \frac{1}{2}\text{sinc}(\frac{1}{2}n) = \frac{\frac{1}{2}\sin(\frac{1}{2}n\pi)}{\frac{1}{2}n\pi}$$

Figure 14.14 Frequency response of bandpass FIR filter of Example 14.11b.

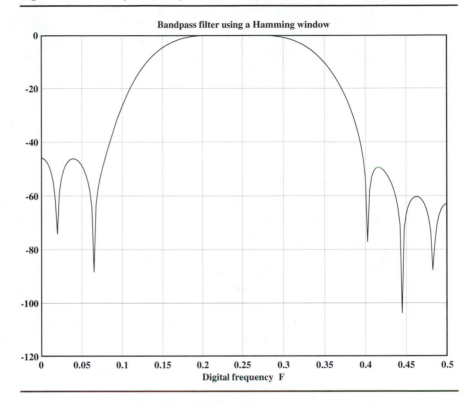

Bandpass filter using a Hamming window

Digital frequency F

Thus $h[n] = 0$ for even n, and the filter length N is odd (with $N = 2k + 1$).

Its transfer function $H(F)$ displays even and hidden halfwave symmetry. It is also antisymmetric about $F = \frac{1}{4}$ (Figure 14.15, p. 656) with

$$H(F) = 1 - H(\tfrac{1}{2} - F)$$

Since $F_C = f_C / S_F = \frac{1}{4}$, the sampling frequency cannot be arbitrary and must equal $4f_C$. Its value is thus fixed at $2(f_p + f_s)$ for LP and HP or $4f_0$ for BP and BS filters. Even though aliasing may occur, it is restricted primarily to the transition band between f_p and f_s, where its effects are not critical.

14.14.1   *Halfband FIR Filter Design Using the Kaiser Window*

The Kaiser window is a *DT* window designed to best approximate the optimal response of the *prolate spheroidal* window. Ideally, in analogy with analog filters, an optimal digital filter (which minimizes the maximum approximation error) should show equiripple response in both the passband and stopband, and yield the smallest filter length N.

Figure 14.15   A halfband and complementary halfband filter.

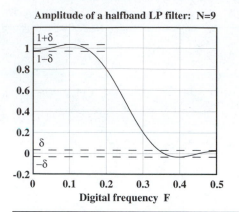

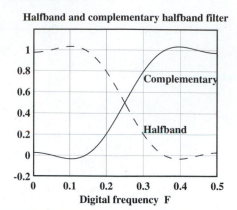

Halfband filters show ripples, but the ripples are not of equal peak magnitude (Figure 14.15). Empirical relations have been developed to estimate the filter length N of halfband FIR filters based on the Kaiser window. This length is often very close to the optimal value, even though the filter does not show equiripple behavior.

The design requires the band-edge frequencies and attenuations f_p, f_s, A_p and A_s, but not the sampling frequency, and involves the following steps:

1. Fix $S_F = 2(f_p + f_s)$ for LP and HP filters, or $S_F = 4f_0$ for arithmetically symmetric BP and BS filters.
2. Normalize the design frequencies.
3. Find F_p and F_s for the normalized LPP, fix $F_C = \frac{1}{2}(F_p + F_s)$ and set up the impulse response of the lowpass prototype $h[n] = 2F_C \mathrm{sinc}[2nF_C]$.
4. Compute the ripple parameter δ (see Figure 14.15) by evaluating

$$\delta_p = \frac{(10^{A_p/20} - 1)}{(10^{A_p/20} + 1)} \qquad \delta_s = 10^{-A_s/20} \qquad \delta = \min(\delta_p, \delta_s)$$

The ripple δ is chosen as the smaller of the two values δ_p and δ_s.
5. Compute the actual passband and stopband attenuations:

$$A_{p0} = 20 \log[(1 + \delta)/(1 - \delta)] \qquad A_{s0} = -20 \log(\delta)$$

6. Choose an odd filter length N based on

$$N \geq \begin{cases} [(A_{s0} - 7.95)/14.36(F_s - F_p)] + 1 & A_{s0} \geq 21\ \mathrm{dB} \\ 0.9222/(F_s - F_p) + 1 & A_{s0} < 21\ \mathrm{dB} \end{cases}$$

Increase the value of N, if necessary, to ensure that N is odd.
7. Compute the Kaiser window parameter β from

$$\beta = \begin{cases} 0.0351(A_{s0} - 8.7) & A_{s0} > 50 \\ 0.186(A_{s0} - 21)^{0.4} + 0.0251(A_{s0} - 21) & 21 \leq A_{s0} \leq 50 \\ 0 & A_{s0} < 21 \end{cases}$$

8. Generate the Kaiser window $w[n]$, compute the windowed impulse response $h_W[n]$, and use spectral transformations to get the required filter.
9. If the specifications are not met, increase N in steps of 2 until they are. For the Kaiser window, this iteration is seldom required.
10. Establish the causal sequence $h_C[n]$ and $H_C(z)$.

Halfband Filter Design Using Other Windows   Halfband filters may also be designed using other windows and the preceding design recipe. We start with a value of N as estimated for the Kaiser design. This value must often be increased to meet specifications. However, it takes only a few (one or two) more iterations, since N is incremented in steps of 2.

Remarks:
1. A direct form implementation of halfband filters requires only about $\frac{1}{2}N$ multipliers (corresponding to the nonzero filter coefficients).
2. Once obtained, we can also obtain a highpass halfband filter without using any more multipliers. This highpass form may be defined in terms of the complementary spectrum $H_{co}(F)$ (Figure 14.15) as

$$H_{HP}(z) = z^{-(N-1)/2}H_{co}(z)$$

3. For an Mth-band filter, $h[n] = (1/M)\sin(n\pi/M)/(n\pi/M)$ with $f_C = S_F/2M$, $h[kM] = 0$, and the passband and stopband ripple satisfy $\delta_p \le (M-1)\delta_s$. Except for $M = 2$ (halfband), such filters are not easy to design.

Example 14.12
FIR Halfband Filter Design

We design a bandstop halfband filter to meet the following specifications:

Stopband: 2 kHz – 3 kHz, Passband edges: 1 kHz, 4 kHz, $A_p = 1$ dB, $A_s = 50$ dB

We choose $S_F = 2(f_p + f_s) = 2(2+3) = 10$ kHz. Since both the passband and the stopband are symmetric, $f_0 = 2.5$ kHz. The digital frequencies and F_0 are given by Stopband $= [0.2, 0.3]$, Passband $= [0.1, 0.4]$, $F_0 = 0.25$. The LPP specs are $F_p = \frac{1}{2}(F_{s2} - F_{s1}) = 0.05$, $F_s = \frac{1}{2}(F_{p2} - F_{p1}) = 0.15$, and $F_C = \frac{1}{2}(F_p + F_s) = 0.1$. The impulse response of the LPP is $h[n] = 2F_C\text{sinc}(2nF_C) = 0.2\text{sinc}(0.2n)$.

(a) For the Kaiser window, we successively compute

$$\delta_p = \frac{10^{A_p/20} - 1}{10^{A_p/20} + 1} = 0.0575, \quad \delta_s = 10^{-A_s/20} = 0.00316, \quad \delta = 0.00316$$

$$A_{s0} = -20\log(\delta) = 50, \qquad N = \frac{A_{s0} - 7.95}{14.36(F_s - F_p)} + 1 = 30.28 \approx 31$$

$$\beta = 0.0351(A_{s0} - 8.7) = 1.4431$$

With $N = 31$, we compute values of $h_N[n] = 0.2\text{sinc}(0.2n)$ and the Kaiser window $w[n]$ for $-15 \le n \le 15$. Windowing $h_N[n]$ gives $h_W[n]$. We transform $h_W[n]$ to the bandstop form $h_{BP}[n]$ using the transformation

$$h_{BP}[n] = \delta[n] - 2\cos(2\pi nF_0)h_W[n] = \delta[n] - 2\cos(0.5\pi n)h_W[n]$$

Figure 14.16 shows that this filter does meet specs with an attenuation of 0.046 dB at 2 kHz and 3 kHz, and 53.02 dB at 1 kHz and 4 kHz.

Figure 14.16   Frequency response of the halfband filters of Example 14.12.

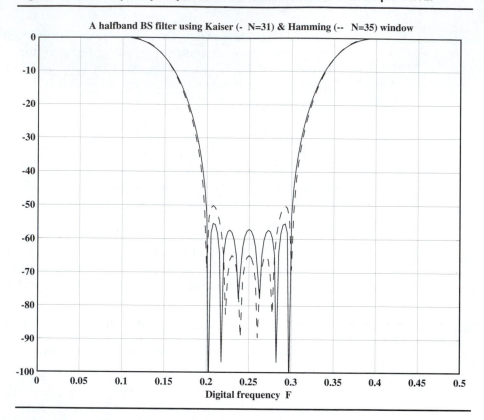

A halfband BS filter using Kaiser (- N=31) & Hamming (-- N=35) window

(b) For the Hamming window, it turns out that $N = 35$ will meet specifications. This yields the response of Figure 14.16 with $A_p = 0.033$ dB, $A_s = 69.22$ dB.

14.15 FIR Filter Design by Frequency Sampling

In window-based design, we start with the impulse response $h[n]$ of an ideal filter and use truncation and windowing to obtain an acceptable frequency response $H(F)$. The design of FIR filters by frequency sampling starts with the required form for $H(F)$ and uses interpolation and the DFT to obtain $h[n]$. In this sense, it is more versatile, since arbitrary frequency response forms can be handled with ease.

Recall that a continuous (but bandlimited) signal $x(t)$ can be perfectly reconstructed from its samples (taken above the Nyquist rate) using a sinc interpolating function that equals zero at the sampling instants. If $x(t)$ is not bandlimited, we get a perfect match only at the sampling instants.

By analogy, the continuous (but periodic) spectrum $X(F)$ of a discrete time signal can also be recovered from its frequency samples using a periodic extension of the sinc interpolating function which equals zero at the sampling intervals. The reconstructed spectrum $X_N(F)$ will match a desired $X(F)$ only at the sampling frequencies. This is frequency sampling.

The DFT of a signal $h_N[n]$ of finite length N may be considered as the N samples of its DTFT $H(F)$ evaluated at $F = k/N, k = 0, 1, 2, \ldots, N - 1$. Thus,

$$H_N[k] = \sum_{k=0}^{N-1} h_N[n] \exp(-j2\pi nk/N)$$

The impulse response $h_N[n]$ is found using the IDFT as

$$h_N[n] = \frac{1}{N} \sum_{k=0}^{N-1} H_N[k] \exp(j2\pi nk/N)$$

The filter transfer function $H(z)$ is the z-transform of $h_N[n]$ and equals

$$H(z) = \sum_{n=0}^{N-1} h_N[n]z^{-n} = \sum_{n=0}^{N-1} z^{-n} \left[\frac{1}{N} \sum_{k=0}^{N-1} H_N[k] \exp(j2\pi nk/N) \right]$$

Interchanging summations, setting $z^{-n} \exp(j2\pi nk/N) = [z^{-1} \exp(j2\pi k/N)]^n$ and using the closed form for its finite geometric sum, we obtain

$$H(z) = \frac{1}{N} \sum_{k=0}^{N-1} H_N[k] \frac{1 - z^{-N} \exp(j2\pi k)}{1 - z^{-1} \exp(j2\pi k/N)}$$

$$= \frac{1}{N} \sum_{k=0}^{N-1} H_N[k] \frac{1 - z^{-N}}{1 - z^{-1} \exp(j2\pi k/N)}$$

The frequency response corresponding to $H(z)$ equals

$$H(F) = \frac{1}{N} \sum_{k=0}^{N-1} H_N[k] \frac{1 - \exp(-j2\pi FN)}{1 - \exp[-j2\pi(F - k/N)]}$$

If we factor out $\exp(-j\pi FN)$ from the numerator, $\exp[-j\pi(F - k/N)]$ from the denominator, and use Euler's relation, we can simplify this result to

$$H(F) = \sum_{k=0}^{N-1} H_N[k] \exp\{-j\pi(N - 1)(F - k/N)\} \frac{\operatorname{sinc}[N(F - k/N)]}{\operatorname{sinc}[(F - k/N)]}$$

$$= \sum_{k=0}^{N-1} H_N[k]W[F - k/N]$$

The sinc interpolating function $W[F - k/N]$ reconstructs $H(F)$ from its samples $H_N[k]$ taken at intervals $1/N$. It equals 1 when $F = k/N$ and zero otherwise. As a result, $H_N(F)$ equals a desired $H(F)$ at the sampling frequencies, even though $H_N(F)$ could vary wildly at other frequencies.

14.15.1   *Frequency Sampling Design Recipe*
Given the desired form for $H(F)$, we sample it at a large number of frequencies and find the IDFT. The following design guidelines stem both from design aspects as well as computational aspects of the IDFT itself.

1. The chosen samples must correspond to one period (0, 1) of the periodic extension of $H(F)$ (IDFT requirement).
2. For $h[n]$ to be causal, we must delay it (design requirement). This is done by changing the phase of the frequency samples, since a delay translates to a linear phase shift. The phase is given by $\phi[k] = -\pi k(N-1)/N,\ k = 1, 2, \ldots, N$. For type 3 and type 4 (antisymmetric) sequences, we must also add an additional phase of $\frac{1}{2}\pi$. Note that we can *use the actual phase or its principal value*.
3. To minimize the Gibbs effect near discontinuities in $H(F)$, we allow the sample values to vary gradually between jumps (design guideline). This is equivalent to introducing a finite transition width.
4. Since $h[n]$ must be real, its DFT $H[k]$ must possess conjugate symmetry about the folding index (DFT requirement). This always leaves $h[0]$ unpaired. It can be set to any real value in keeping with the required filter type (design requirement). For example, we must choose $h[0] = 0$ for bandpass or highpass filters.
5. For even N, the end samples of $h[n]$ may not be symmetric. We must then force $h[0]$ to equal $h[N]$ (setting both to $\frac{1}{2}h[0]$, for example).

Example 14.13
FIR Filter Design Using Frequency Sampling

Consider the design of a lowpass filter specified by 10 samples over the frequency range (0, 1), as shown in Figure 14.17. The folding index equals 5. The samples at $k = 9, 8, 7, 6$ must be conjugates of the samples at $k = 1, 2, 3, 4$.

We choose $H[0] = 1$. The delay equals $\phi[k] = -\pi k(N-1)/N = -0.9\pi k$. This leads to the choice of the frequency samples $H[k]$ as

$H[1] = \exp(-j0.9\pi) = -0.9511 - j0.3090$ $H[9] = H^*[1] = -0.9511 + j0.3090$
$H[2] = \exp(-j1.8\pi) = 0.8090 + j0.5878$ $H[8] = H^*[2] = 0.8090 - j0.5878$
$H[0] = 1$ $H[3] = 0$ $H[4] = 0$ $H[5] = 0$ $H[6] = 0$ $H[7] = 0$

The IDFT yields the symmetric real impulse response sequence $h_1[n]$ with

$\{h_1[n]\} = \{0.0716, -0.0794, 0.1, 0.1558, 0.452, 0.452, 0.1558, 0.1, -0.0794, 0.0716\}$

Its frequency response $H_1(F)$ in Figure 14.17 shows a perfect match at the sampling points but a large overshoot near the cutoff index $k = 2$.

To reduce the overshoot, let us choose $H[3] = 0.5\exp[j\phi(3)] = -0.2939 + j0.4045$ instead of 0. We must then choose $H[7] = H^*[3] = -0.2939 + j0.4045$.

The IDFT of this new set of samples leads to new impulse response sequence $h_2[n]$ described by

$$\{h_2[n]\} = \{0.0128, 0.0157, 0.1, 0.0606, 0.5108, 0.5108, 0.0606, 0.1, 0.0157, 0.0128\}$$

Its frequency response $H_2(F)$ (shown dashed in Figure 14.17) not only shows a

Figure 14.17 Frequency response of the filters of Example 14.13.

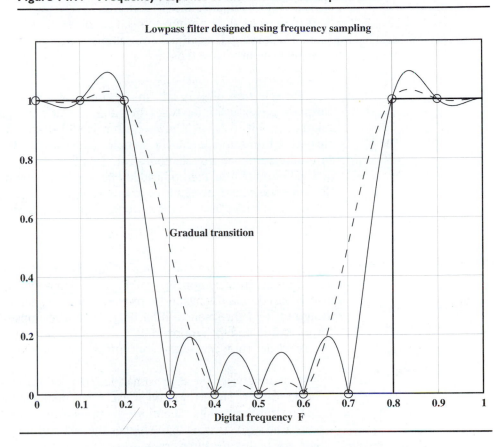

perfect match at the sampling points but also a reduced overshot, which we obtain at the expense of making the transition less abrupt.

14.15.2   *Frequency Sampling and Windowing*

Here is how we can combine the advantages of window-based design with the versatility of the frequency sampling method, especially for the design of arbitrary frequency response functions.

1. Frequency sampling: Given the response specification $H(F)$, we sample it at a *large* number of points M and find the IDFT to obtain the M-point impulse response $h[n]$. The choice $M = 512$ is not unusual.

2. Windowing: Since $h[n]$ is unacceptably long, we must truncate it to a smaller length N. We do this by windowing $h[n]$. The choice of window is based on the same considerations that apply to window-based design. Windows reduce the passband response. To offset this, the width of the passband is usually increased (by 25% or so) before actual design. The peak value of $H(F)$ may also need to be scaled for comparison.

3. If the design does not meet specifications, we can change N, change the width of the passband, or adjust the sample values in the transition band, and then repeat the first two steps.

The selection of N is still made empirically. If the frequency response conforms to a classical filter type, then a good starting point is to use the empirical relation $N \approx W_S/(F_s - F_p)$ as we do for window-based design. This design is best carried out on a computer. The sample values around the transitions are adjusted to minimize the approximation error. This idea forms a special case of the more general optimization method called *linear programming* . But when it comes right down to choosing between the various computer-aided optimization methods, by far the most widely used is the *equiripple optimal approximation* method, which we describe next.

14.16 Design of Optimal Linear Phase FIR Filters

Quite like analog filters, the design of optimal linear phase FIR filters requires that we minimize the maximum error in the approximation. Optimal design of FIR filters is also based on a Chebyshev approximation. We should, therefore, expect such a design to yield the smallest filter length and a response that is equiripple in both the passband and the stopband.

There are three important concepts relevant to optimal design:

1. The error between the approximation $H(F)$ and the desired response $D(F)$ must be equiripple. The error curve must show equal maxima and minima with alternating zero crossings. The more the number of points where the error goes to zero (the zero crossings), the higher the order of the approximating polynomial and the higher the filter order.

2. The frequency response $H(F)$ of a filter whose impulse response $h[n]$ is a symmetric sequence can always be put in the form

$$H(F) = \sum_{n=0}^{M} \alpha_n(F) \cos(2\pi nF)$$

where M is related to the filter length N. This form is just a sum of Chebyshev polynomials. If we can select the α_n to best meet optimal constraints, we can design $H(F)$ as an optimal approximation to $D(F)$.

3. The *alternation theorem* offers the clue to selecting the α_n.

14.16.1   *The Alternation Theorem*

We start by approximating $D(F)$ by the Chebyshev polynomial form for $H(F)$ and define the *weighted* approximation error $\epsilon(F)$ as

$$\epsilon(F) = W(F)[D(F) - H(F)]$$

Here, $W(F)$ represents a set of weight factors that can be used to select different error bounds in the passband and stopband. The nature of $D(F)$ and $W(F)$ depends on the type of the filter required. The idea is to select the α_k (in the expression for $H(F)$) so as to minimize the maximum absolute error $|\epsilon|_{max}$. The alternation theorem points the way (though it does not tells us how). In essence, it says that we must be able to find at least $M + 2$ frequencies $F_k, k = 1, 2, \ldots, M + 2$ called the **extremal frequencies** where

1. The error alternates between two equal maxima and minima (extrema)

$$\epsilon(F_k) = -\epsilon(F_{k+1}) \qquad k = 1, 2, \ldots, M + 1$$

2. The error at the frequencies F_k equals the *maximum absolute error.*

$$|\epsilon(F_k)| = |\epsilon(F)|_{max} \qquad k = 1, 2, \ldots, M + 2$$

In other words, we require $M + 1$ extrema (including the band-edges) where the error attains its maximum absolute value. If there are more than $M + 1$ such points, the design is nearly optimal. These frequencies yield the smallest filter length (number of coefficients α_k) for optimal design.

The design strategy to find these extremal frequencies invariably requires iterative methods. The most popular is the algorithm of Parks and McClellan which, in turn, relies on the so-called **Remez exchange algorithm**.

The Parks-McClellan (PM) algorithm requires the frequencies F_p and F_s, the ratio $K = \delta_1/\delta_2$ of the passband and stopband error, and the filter length N. It returns the coefficients α_k, and the actual values of δ_1 and δ_2 for the given filter length N. If the values of δ_1 and δ_2 are not acceptable, we can increase N and repeat the design.

A good starting estimate for the filter length N is given by a relation similar to the Kaiser relation (for halfband filters)

$$N = 1 + \frac{-10\log(\delta_1\delta_2) - 13}{14.6(F_s - F_p)}$$

More accurate (but more involved) design relations are also available.

To explain the algorithm, consider a lowpass filter design. To approximate an ideal lowpass filter, we choose $D(F)$ and $W(F)$ as

$$D(F) = \begin{cases} 1 & 0 \le F \le F_p \\ 0 & F_s \le F \le \frac{1}{2} \end{cases} \qquad W(F) = \begin{cases} 1 & 0 \le F \le F_p \\ K = \delta_1/\delta_2 & F_s \le F \le \frac{1}{2} \end{cases}$$

To find the α_k in $H(F)$, we use the Remez algorithm. Here is how it works. We start with a trial set of $M + 2$ frequencies $F_k, k = 1, 2, \ldots, M + 2$. To force the alternation condition, we must satisfy $\epsilon(F_k) = -\epsilon(F_{k+1}), k = 1, 2, \ldots, M + 1$. Since the maximum error is yet unknown, we let $\rho = \epsilon(F_k) = -\epsilon(F_{k+1})$. We now have $M + 1$ unknown coefficients α_k and the unknown ρ, a total of $M + 2$. We solve for these by using the $M + 2$ frequencies to generate the $M + 2$ equations

$$-(-1)^k \rho = W(F_k)[D(F_k) - H(F_k)] \qquad k = 1, 2, \ldots, M + 2$$

Here, the quantity $(-1)^k$ brings out the alternating nature of the error.

Once the α_k are found, the right-hand side of this equation is known in its entirety, and is used to compute the extremal frequencies. The problem is that these frequencies may no longer satisfy the alternation condition. So we must go back and evaluate a new set of α_k and ρ using the computed frequencies. We continue this process until the computed frequencies also turn out to be the actual extremal frequencies, (to within a given tolerance, of course).

Do you see why it is called the *exchange* algorithm? First we exchange an old set of frequencies F_k for a new one. Then we exchange an old set of α_k for a new one. Since the α_k and F_k actually describe the impulse response and frequency response of the filter, we are, in essence, going back and forth between the two domains until the impulse response coefficients α_k yield a spectrum with the desired optimal characteristics.

By way of an example, the bandstop specifications of Example 14.12 are met by an optimal filter with a length of only $N = 21$, as Figure 14.18 reveals.

Many time-saving steps have been suggested to speed up the computation of the extremal frequencies and the α_k at each iteration. The *PM* algorithm is arguably one of the most popular methods of filter design in the industry, and many of the better commercial software packages on signal processing include it in their stock list of programs. Having said that, we must also point out two disadvantages of this method:

1. The filter length must still be estimated by empirical means.
2. We have no control over the actual ripple that the design yields. The only remedy, if this ripple is unacceptable, is to start afresh with a different set of weight functions or a different filter length.

14.17
Maximally Flat
FIR Filters

Linear phase FIR filters can also be designed with **maximally flat** frequency response at $F = 0$ (or $F = \frac{1}{2}$). Such filters are used in situations where accurate filtering is needed at frequencies near dc. The design is based on a closed form for the transfer function $H(F)$ given by

$$H(F) = \cos^{2K}(\pi F) \sum_{n=0}^{L-1} d_n \sin^{2n}(\pi F) \qquad d_n = \frac{(K + n - 1)!}{(K - 1)! \, n!} = C_n^{K+n-1}$$

Here, the d_n have the form of binomial coefficients as indicated. Note that $2L - 1$ derivatives of $|H(F)|^2$ are zero at $F = 0$, and $2K - 1$ derivatives are zero at $F = \frac{1}{2}$. This is the basis for the maximally flat response of the filter. The filter length equals $N = 2(K + L) - 1$ and is thus odd. The integers K and L are determined from the

---

Figure 14.18 **Frequency response of an optimal bandstop filter.**

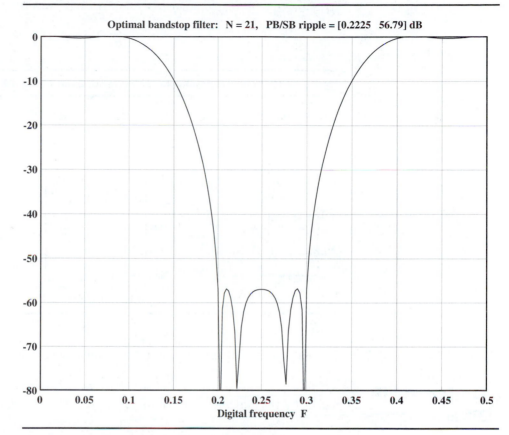

Optimal bandstop filter: N = 21, PB/SB ripple = [0.2225 56.79] dB

---

passband and stopband frequencies F_p and F_s which correspond to gains of 0.95 and 0.05 (or attenuations of about 0.5 dB and 26 dB), respectively. Here is an empirical design method:

1. Define the cutoff frequency as $F_C = \frac{1}{2}(F_p + F_s)$ and let $F_T = F_s - F_p$.
2. Obtain a first estimate of the odd filter length as $N_0 = 1 + 1/2F_T^2$.
3. Define the parameter α as $\alpha = \cos^2(\pi F_C)$.
4. Find the best rational approximation $\alpha \approx K/M_{\min}$ for $\frac{1}{2}(N_0 - 1) \leq M \leq (N_0 - 1)$.
5. Evaluate L and the true filter length N from $L = M - K$ and $N = 2M - 1$.
6. Find $h[n]$ as the N-point IDFT of $H(F), F = 0, 1/N, \ldots, (N - 1)/N$.

Example 14.14
Maximally Flat FIR
Filter Design

Consider the design of a maximally flat lowpass FIR filter with normalized frequencies $F_p = 0.2$ and $F_s = 0.4$. Then $F_C = 0.3$ and $F_T = 0.2$. We compute $N_0 = 1 + 1/2F_T^2 = 13.5 \approx 15$, and $\alpha = \cos^2(\pi F_C) = 0.3455$. The best rational approximation to α works out to be $\alpha \approx 5/14 = K/M$. With $K = 5$ and $M = 14, L = M - K = 9$ and the filter length $N = 2M - 1 = 27$. Figure 14.19 shows the impulse response and $|H(F)|$ of the designed filter.

Figure 14.19   The maximally flat FIR filter of Example 14.14.

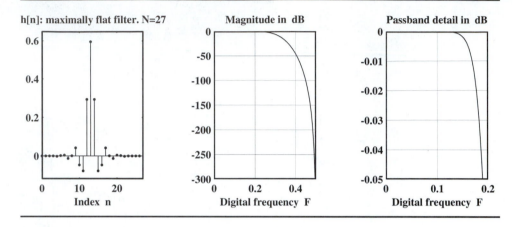

14.18
Design of FIR Differentiators

An ideal digital differentiator must have the magnitude response

$$H(F) = j2\pi F \qquad |F| \le F_C$$

Since $H(F)$ is odd, $h[0] = 0$. To find $h[n], n \ne 0$ we use the IDTFT to obtain

$$h[n] = j \int_{-F_C}^{F_C} 2\pi F \exp(j2n\pi F) dF$$

Invoking Euler's relation and symmetry, this simplifies to

$$h[n] = -4\pi \int_0^{F_C} F \sin(2n\pi F)\, dF = \frac{2n\pi F_C \cos(2n\pi F_C) - \sin(2n\pi F_C)}{\pi n^2}$$

For $F_C = \frac{1}{2}$, this yields $h[0] = 0$ and $h[n] = \cos(n\pi)/n$, $n \ne 0$. In most practical situations, we seldom require filters that differentiate for the full frequency range up to $F = \frac{1}{2}$.

To design an FIR differentiator, we must truncate $h[n]$ to $h_N[n]$ and choose a type 3 or type 4 sequence, since $H(F)$ is purely imaginary. To ensure odd symmetry, the filter coefficients are computed only for $n > 0$ and the same values (in reversed order) are used for $n < 0$. If N is odd, we also include the sample $h[0]$. We must window $h_N[n]$ to minimize the overshoot and ripple in the spectrum $H_N(F)$. And, finally, to ensure causality, we must introduce a delay of $\frac{1}{2}(N - 1)$ samples.

Figure 14.20 shows the magnitude response of FIR differentiators for both even and odd N. Note that $H(0)$ is always zero but $H(\frac{1}{2}) = 0$ only for type 3 (odd-length or odd N) sequences.

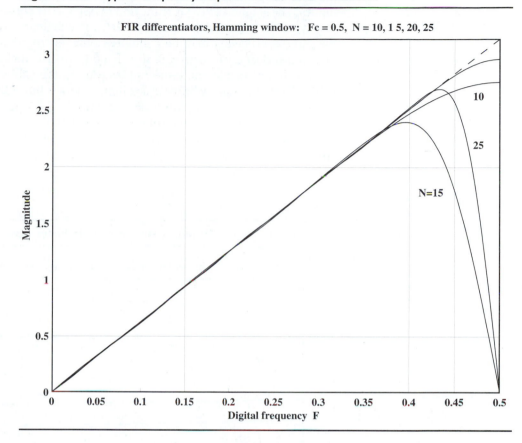

Figure 14.20 Typical frequency response of FIR differentiations.

FIR differentiators, Hamming window: Fc = 0.5, N = 10, 1 5, 20, 25

14.19
Design of FIR Hilbert Transformers

A **Hilbert transformer** is used to shift the phase of a signal by $-\frac{1}{2}\pi$ (or $-90°$). From Chapter 11, recall that the transfer function $H(F)$ and impulse response $h[n]$ of an ideal Hilbert transformer are given by

$$H(F) = -j\,\text{sgn}(F) \qquad h[n] = [1 - \cos(n\pi)]/n\pi, \qquad h[0] = 0$$

In practical situations, we seldom require filters that shift the phase for the full frequency range up to $F = \frac{1}{2}$. If we require phase shifting only up to a cutoff frequency of F_C, then

$$H(F) = -j\,\text{sgn}(F) \qquad |F| \le F_C$$

With $h[0] = 0, h[n], n \ne 0$, the IDTFT gives the impulse response as

$$h[n] = \int_{-F_C}^{F_C} -j\,\text{sgn}(F)\exp(j2n\pi F)\,dF = 2\int_0^{F_C}\sin(2n\pi F)\,dF = \frac{1 - \cos(2n\pi F_C)}{n\pi}$$

For $F_C = \frac{1}{2}$, this reduces to $h[n] = [1 - \cos(n\pi)]/n\pi$, as expected.

The design of Hilbert transformers closely parallels the design of FIR differentiators. The sequence $h[n]$ is truncated to $h_N[n]$. The chosen filter must correspond to a type 3 or type 4 sequence, since $H(F)$ is imaginary. To ensure odd symmetry, the filter coefficients may be computed only for $n > 0$ with the same values (in reversed order) used for $n < 0$. If N is odd, we also include the sample $h[0]$. We window $h_N[n]$ to minimize the ripples (due to the Gibbs effect) in the spectrum $H_N(F)$. The Hamming window is a common choice, but others may also be used. To make the filter causal, we introduce a delay of $(N-1)/2$ samples. The magnitude spectrum of a Hilbert transformer becomes flatter with increasing filter length N. Figure 14.21 shows the magnitude response of Hilbert transformers for both even and odd N. Note that $H(0)$ is always zero for both, but $H(\frac{1}{2}) = 0$ only for type 3 (odd length) sequences.

Figure 14.21   Typical frequency response of FIR Hilbert transformers.

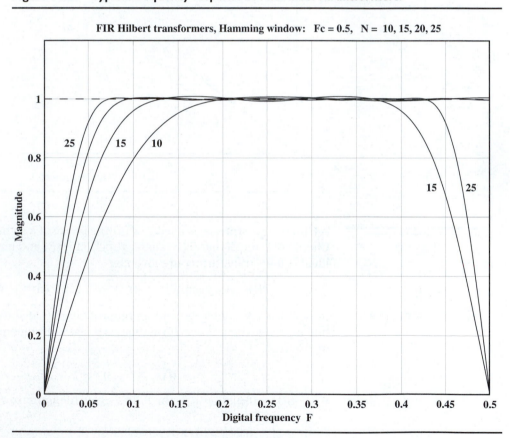

FIR Hilbert transformers, Hamming window: Fc = 0.5, N = 10, 15, 20, 25

14.20 Special Topics: Comb Filters

A **comb filter** is described by the transfer function

$$H(z) = 1 - z^{-N}$$

This corresponds to the system difference equation $y[n] = x[n] - x[n-N]$ and represents an FIR filter whose impulse response $h[n]$ is of length $N+1$, with end samples $h[0] = 1, h[N] = -1$ and all other coefficients zero.

The zeros of $H(z)$ are uniformly spaced around the unit circle starting at $z = 1$ and are specified by

$$z^{-N} = 1 = \exp(j2\pi) \qquad z_k = \exp(j2\pi k/N) \qquad k = 0, 1, \ldots, N-1$$

For even N, there is also a zero at $z = -1$.

The transfer function also reveals that there is a pole of multiplicity N at the origin, making the filter stable for any N.

The frequency response $H(F)$ of this filter is given by

$$H(F) = 1 - \exp(-j2\pi FN)$$

Note that $H(0)$ always equals 0, but $H(\frac{1}{2}) = 0$ for even N and $H(\frac{1}{2}) = 2$ for odd N. The frequency response $H(F)$ looks like a comb with N rounded *teeth* over its principal period, as shown in Figure 14.22.

Figure 14.22   Frequency response of a comb filter for N = 5 and N = 6.

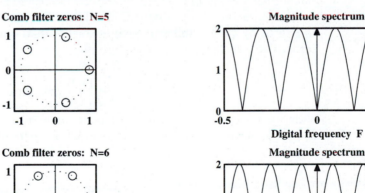

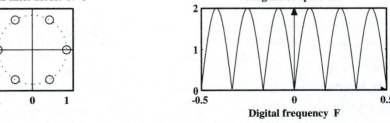

14.21
Special Topics: Recursive Implementation of FIR Filters

In dealing with the design of filters using frequency sampling, we found that the transfer function of an FIR filter could be expressed as

$$H(z) = \frac{1}{N} \sum_{k=0}^{N-1} H_N[k] \left[\frac{1 - z^{-N}}{1 - z^{-1} \exp(j2\pi k/N)} \right]$$

This may be written as the product of two transfer functions:

$$H(z) = H_1(z)H_2(z) = \left[\frac{1 - z^{-N}}{N} \right] \sum_{k=0}^{N-1} \left[\frac{H_N[k]}{1 - z^{-1} \exp(j2\pi k/N)} \right]$$

We recognize $H_1(z)$ as a comb filter. What about $H_2(z)$? It is simply a parallel combination of N first-order recursive filters, each of which has a pole on the unit circle. The filter $H_2(z)$ is also called a **resonator** since it is not strictly stable but capable of sustained oscillations. The form of $H(z)$ suggests a method of recursive implementation of FIR filters. We cascade a parallel combination of first-order resonators with a comb filter. The resonator poles actually lie at the same locations as the zeros of the comb filter. This means that for each resonator with a complex pole, we must have a resonator with a conjugate pole. Each such pair may be replaced by a second-order system for simpler implementation.

Why implement FIR filters recursively? There are several reasons. In some cases, this may reduce the number of arithmetic operations. In other cases, it may reduce the number of delay elements required. Since the pole and zero locations depend only on N, such filters can be used for all FIR filters of length L by changing only the multiplicative coefficients.

Even with these advantages, things can go wrong. In theory, the poles and zeros balance each other. In practice, quantization errors may move some poles outside the unit circle and lead to system instability. One remedy is to multiply the poles and zeros by a real number ρ slightly smaller than unity to relocate the poles and zeros. The transfer function then becomes

$$H(z) = \frac{1 - (\rho z)^{-N}}{N} \sum_{k=0}^{N-1} \left[\frac{H_N[k]}{1 - (\rho z)^{-1} \exp(j2\pi k/N)} \right]$$

With $\rho = 1 - \epsilon$, typically used values for ϵ range from 2^{-12} to 2^{-27} (roughly 10^{-4} to 10^{-9}), and have been shown to improve stability with little change in the frequency response.

14.22
Digital Filters: A Synopsis

Digital filters	Digital filters process *DT* signals. They can be FIR (nonrecursive) or IIR (recursive) and can be realized in hardware or software. Specifications for digital filters include a sampling frequency.
IIR filters	IIR filters cannot display linear phase but yield a much smaller order for a given application. The design starts with an analog prototype, which is converted to the required digital filter using an *s2z* mapping. Coefficient truncation of IIR filters may sometimes lead to instability.

Response matching	We choose $x(t)$, find the response $y(t)$ to $x(t)$ using $H(s)$, sample $x(t)$ and $y(t)$ at intervals t_s to obtain $x_S[n]$ and $y_S[n]$, find their z-transform and compute $H(z) = Y(z)/X(z)$. This mapping is useful only for analog systems whose frequency response is essentially bandlimited. If $H(s)$ is in partial fraction form, we can obtain a simple design relation for impulse-invariant mapping.
Matched z-transform	The matched z-transform requires $H(s)$ in factored form and converts each factor into a corresponding term of $H(z)$ using the impulse-invariant form. It also suffers from aliasing errors but is better suited to the design of highpass and bandpass filters.
Numerical algorithms	Numerical algorithms for integration and derivatives yield $s2z$ mappings of which the bilinear transformation is the most widely used. It is simple, free from aliasing and preserves stability. It warps the frequency response but allows for unwarping the results using a simple relation.
$D2D$ trans-formations	Digital-to-digital frequency transformations convert a digital LP filter to the required form. The LP2BP and LP2BS transformations yield a digital filter with twice the order of the lowpass prototype.
$A2D$ trans-formations	Analog-to-digital transformations allow an analog lowpass prototype to be directly converted to the required $H(z)$.
FIR filters	FIR filters are inherently stable due to absence of feedback and capable of linear phase. Their realization often involves a large number of elements.
Symmetric sequences	FIR filters use symmetric sequences of the smallest length that meet specifications. These imply noncausal filters, which can be realized only if we introduce a delay. Odd lengths are required for BS filters.
Design	FIR filter design requires conversion to lowpass specifications and trans-formations to convert to the required form. Unlike analog and IIR filters, the LP2BP or LP2BS transformations do not change the filter order and preserve arithmetic (not geometric) symmetry.
Window-based design	Window-based design truncates and windows the ideal lowpass impulse response for different lengths N and choice of cutoff frequency until the specs are met.
Halfband filter design	For a halfband filter, alternate samples in the impulse response $h[n]$ are zero. Empirical design equations are available for their design.
Design by frequency sampling	Frequency sampling starts with a given $H(F)$ and uses interpolation and the DFT to obtain $h[n]$. It is also a trial-and-error method.
Optimal design	Optimal design minimizes the maximum approximation error and yields the smallest filter length and an equiripple response in both bands.

Appendix 14A MATLAB Demonstrations and Routines

Getting Started on the Supplied MATLAB Routines

This appendix presents MATLAB routines related to concepts in this chapter, and some examples of their use. See Appendix 1A for introductory remarks.
Note: All polynomial arrays must be entered in descending powers.

IIR Filter Design

    [tfd,tfn,tfa]=dfdiir(na,ty,ty1,a,SF,fp,fs,f3,p1)

Designs IIR filters. All arguments match *afd* (see Ch. 10). In addition:

SF= sampling frequency in Hz and `ty1`= includes all design methods in the text.
`ty1='impulse'` or `'step'` or `'ramp'` for invariance.
`ty1='match'` for matched z-transform.
`ty1='back'`,`'forw'`,`'bili'` for backward, forward difference, bilinear etc.
`tfd,tfn,tfa` are the IIR, normalized LPP and analog filter transfer function.

Example

To design an impulse-invariant Butterworth bandpass filter with $A_p = 2$ dB, $A_s = 20$ dB, passband edges 200 Hz and 300 Hz, stopband edges 100 Hz and 400 Hz, and a sampling frequency 1200 Hz, use

    >> [tfd,tfn,tfa]=dfdiir('bw','bp','impulse',[3 20],1200,[200 300],[100 400])

To see the spectrum, include 'p' after the last argument in dfdiir, or use

    >>tfplot('z',tfd(1,:),tfd(2,:))

FIR Filter Design

    h=firwind(ty,a,SF,fp,fs,w)

Designs FIR filters of $ty =$'lp','hp','bp', or 'bs' using the window w, for example, 'hamming'.
Returns control to user to change filter length or cutoff frequency.

Example

To design an FIR bp filter with $A_p = 2$ dB, $A_s = 40$ dB, passband edges 200 Hz and 300 Hz, stopband edges 100 Hz and 400 Hz, sampling frequency 1200 Hz, and a Hamming window, use

    >>h=firwind('bp',[2 40],1200,[200 300],[100 400],'hamming')

MATLAB returns with $N = 28, fc = 0.06912$ (when $A_p = 1.47, A_s = 40.37$) and suggests that N may be reduced. Indeed, $N = 27$ (and $fc = 0.06562$) yields $A_p = 1.86$ dB, $A_s = 40.3$ dB.

h=firhb(ty,a,SF,fp,fs,w)

Designs FIR halfband filters with odd length. The sampling frequency is forced to equal to $2(f_{p1} + f_{p2})$. Returns control to user to change the filter length if desired.

Example

For the previous specs and a Kaiser window use

>> h=firhb('bp',[2 40],1200,[200 300],[100 400],'kaiser')

MATLAB returns control with $N = 25, fc = 0.1$ (when $A_p = 0.1$ and $A_s \approx 45$). Changing N to 23, however, no longer meets specs.

Special Filters

h=firpm(ty,a,SF,fp,fs)

Optimal filter design using Remez algorithm. (Requires the routine *remez* in the MATLAB Signal Processing Toolbox)

h=firmf(fp,fs)

Designs maximally flat FIR filters.

h=firhilb(n,fc,w)

Design of Hilbert transformer of length n.

h=firdiff(n,fc,w)

Design of discrete differentiator of length n.
   Here, fc is the cutoff frequency and w is the window, for example, 'hamming'.

Demonstrations and Utility Functions

Conversion of $H(s) = NN(s)/DD(s)$ to $H(z) = N(z)/D(z)$ using the various methods described in the text is achieved by the following routines.

[n,d]=s2zinvar(nn,dd,fs,ty): Invariance. ty ='impu', 'step' or 'ramp'
[n,d]=s2zni(nn,dd,fs,ty): Numerical: ty = 'forw','back','cent', 'simpson','tick','bilinear', and Adams-Moulton algorithms 'ams3', 'ams4','ams5'
[n,d]=s2zmatch(nn,dd,fs,ty): Matched ZT: ty = '0', '1' ,'2' for each of the modifications described in the text. [nn,dd]=z2smap(n,d,ts,ty): Converts digital to analog filters using methods based on ty = 'back','forw', 'bilinear' or 'pade' (approximation for delay).
[n,d]=lp2iir(ty,'a',nn,dd,SF,f) ty = 'lp', 'hp', 'bp' or 'bs'

Converts ANALOG LPP with unit cutoff to IIR filter with cutoff f. For bp and bs filters, $f = [f_1, f_2]$ describes the cutoff edges.

`[n,d]=lp2iir(ty,'d',nn,dd,SF,f,fc)`

Converts digital LPF with cutoff fc to IIR filter with cutoff f.

`[wp,ws,w0]=fir2lpp(ty,fp,fs,SF)`

Converts FIR filter type ty = 'lp', 'hp', 'bp' or 'bs' to lowpass, assuming arithmetic symmetry

wp, ws = band edges of LPP

`[n,d]=polymap(nn,dd,p,q)`

Converts polynomial $H_1(x) = NN(x)/DD(x)$ using the polynomial transformation $x = P(y)/Q(y)$ to the new polynomial $H_2(y) = N(y)/D(y)$.

P14.1 (Impulse Invariance) Use the impulse-invariant transformation with $t_s = 1$ s to transform the following analog filters to digital filters:
(a) $1/(s + 2)$ **(b)** $2/[(s + 1)(s + 2)]$

P14.2 (Response Invariance)
(a) Convert $H(s) = 1/(s + 2)$ to a digital filter $H(z)$ using (1) impulse invariance, (2) step invariance and (3) invariance to $\exp(-t)u(t)$. Assume a sampling frequency $S_F = 2$ Hz.
(b) Convert $H(s) = (s + 1)/[(s + 1)^2 + \pi^2]$ to a digital filter $H(z)$ using (1) impulse invariance and (2) invariance to $\exp(-t)u(t)$. Assume a sampling frequency $S_F = 2$ Hz.

P14.3 (Impulse-Invariant Design) We are given an analog lowpass filter with the transfer function $H(s) = 1/(s + 1)$. Use impulse invariance to design a digital filter with a cutoff frequency of 50 Hz if the sampling frequency is 200Hz.

P14.4 (Impulse Invariance) The impulse-invariant method allows us to redesign a digital filter $H_1(z) = \sum z/(z - a_k t_1)$ to $H_2(z) = \sum z/(z - a_k t_2)$, using a different sampling interval t_2.
(a) Using the fact that $s = \ln(a_k t_1)/t_1 = \ln(a_k t_2)/t_2$, show that we can obtain $a_k t_2$ from $a_k t_1$ using $a_k t_2 = (a_k t_1)^M$, where $M = t_2/t_1$.
(b) Use the result of part (a) to convert $H_1(z) = [z/(z - \frac{1}{2})] + [z/(z - \frac{1}{4})]$, designed with $t_s = 1$ s, to a digital filter $H_2(z)$ with $t_s = 0.5$ s.

P14.5 (Matched z-Transform) Use the matched z-transform with $t_s = 0.5$ s to transform the following analog filters to digital filters:
(a) $1/(s + 2)$ **(b)** $2/[(s + 1)(s + 2)]$
(c) $1/(s + 1) + 1/(s + 2)$ **(d)** $(s + 1)/[(s + 1)^2 + \pi^2]$

P14.6 (Matched z-Transform) Transform $H(s) = 4s(s + 1)/[(s + 2)(s + 3)]$ to digital filters using the matched z-transform and its two modifications. Assume $t_s = 0.25$ s.

P14.7 (Backward Euler Algorithm) The backward Euler algorithm for numerical integration is given by $y[n] = y[n - 1] + t_s x[n]$. Derive an $s2z$ mapping rule based on this algorithm. Use this rule to map the analog filter $H(s) = 4/(s + 4)$ to a digital filter $H(z)$ with $t_s = 0.5$ s.

P14.8 (Simpson's Algorithm) Simpson's numerical integration algorithm is described by $y[n] = y[n - 2] + \{x[n] + 4x[n - 1] + x[n - 2]\}t_s/3$.
(a) Derive an $s2z$ mapping rule based on this algorithm.
(b) Convert $H(s) = 1/(s + 1)$ to $H(z)$ using this rule.
(c) Is $H(z)$ stable for any choice of t_s?

P14.9 ($s2z$ Mapping by Difference Algorithms)
(a) Convert $H(s) = 1/(s + a)$ using mapping rules based on the forward and backward difference.
(b) Do any of these methods transform a stable $H(s)$ to a stable $H(z)$ for any choice of a?

P14.10 (Bilinear Transformation) Consider an analog lowpass filter whose transfer function is $H(s) = 3/(s^2 + 3s + 3)$. Design a digital lowpass filter $H(z)$ whose gain at $f = 20$ kHz matches that of $H(s)$ at $\omega = 3$ rad/s. Assume a sampling frequency $S_F = 80$ kHz.

P14.11 (Bilinear Transformation) Consider $H(s) = s/(s^2 + s + 1)$.
(a) Design a digital filter $H(z)$ using $t_s = 0.001$ s such that its gain at 250 Hz matches the gain of $H(s)$ at $\omega = 1$ rad/s.
(b) Design a digital filter $H(z)$ using $t_s = 0.1$ s such that its gain matches that of $H(s)$ at $f = 1$ Hz.

P14.12 (D2D Transformation) Consider a digital lowpass filter with $H(z) = (z + 1)/(z^2 - z + 0.2)$, sampling frequency $S_F = 10$ kHz and a cutoff frequency $f = 0.5$ kHz. Use $D2D$ transformations to
(a) Design a highpass filter with a cutoff frequency of 1 kHz.
(b) Design a bandpass filter with band-edges of 1 kHz and 3 kHz.
(c) Design a bandstop filter with band-edges of 1.5 kHz and 3.5 kHz.

P14.13 (A2D Transformation) Consider a lowpass analog filter described by $H_{LP}(s) = 2/(s^2 + 2s + 2)$. Convert this (using $A2D$ transformations) to the following digital filters:
(a) A lowpass filter, band-edge 100 Hz and $S_F = 1$ kHz.
(b) A highpass filter, band-edge 500 Hz and $S_F = 2$ kHz.
(c) A bandpass filter, band-edges 400 Hz and 800 Hz and $S_F = 3$ kHz.
(d) A bandstop filter, band-edges 1 kHz and 1200 Hz and $S_F = 4$ kHz.

P14.14 (IIR Filter Design) Design IIR filters that meet each of the following sets of specifications. Assume $A_p \leq 2$ dB and $A_s \geq 30$ dB.
(a) Butterworth lowpass using the backward Euler transformation with passband < 1 kHz, stopband > 3 kHz and $S_F = 10$ kHz.
(b) Butterworth highpass using impulse-invariant transformation with passband > 400 Hz, stopband < 100 Hz and $S_F = 2$ kHz.

(c) Chebyshev bandpass using the bilinear transformation with passband 800–1600 Hz, stopband: < 400 Hz and > 2 kHz and $S_F = 5$ kHz.

(d) Inverse Chebyshev bandstop using bilinear transformation with passband < 200 Hz and > 1200 Hz, stopband 500–700 Hz and $S_F = 4$ kHz.

P14.15 (IIR Filter Design)

(a) A digital filter $H(z)$ is designed from $H(s)$ using the bilinear transformation based on $\omega_a = C \tan(\Omega_d/2)$. Show that the group delays T_a and T_d of $H(s)$ and $H(z)$ are related by

$$T_d = \tfrac{1}{2} C(1 + \omega_a^2) T_a$$

(b) Use this result to find the group delay of a digital filter $H(z)$ designed from the analog filter $H(s) = 5/(s + 5)$ with a sampling frequency $S_F = 4$ Hz such that the gain of $H(s)$ at $\omega = 2$ rad/s matches the gain of $H(z)$ at 1 Hz.

P14.16 (Digital-to-Analog Mappings) The bilinear transformation allows us to use a linear mapping to transform a digital filter to an analog equivalent.

(a) Develop a z2s mapping based on the bilinear transformation.

(b) Use this to find $H(s)$ if $H(z) = z/(z - 0.5)$ and $t_s = 0.5$ s.

(c) Which of the algorithms based on derivatives or integration also yield a linear z2s mapping rule? Apply each such mapping to convert $H(z) = z/(z - 0.5)$ to $H(s)$ using $t_s = 0.5$ s.

P14.17 (Digital-to-Analog Mappings) Let $H(z) = z(z + 1)/[(z + \tfrac{1}{4})(z - \tfrac{1}{2})]$. Find the analog filter $H(s)$ from which $H(z)$ was developed assuming $t_s = 0.5$ s and

(a) Impulse-invariant design.

(b) Matched z-transform.

(c) Bilinear transformation.

P14.18 (Pade Approximations) A delay of t_s units may be approximated by $\exp(-st_s) \approx 1 - st_s + (st_s)^2/2! - \cdots$. An nth-order Pade approximation is based on a rational function of order n that minimizes the truncation error. The first- and second-order approximations are

$$P_1(s) = (1 - \tfrac{1}{2} st_s)/(1 + \tfrac{1}{2} st_s) \qquad P_2(s) = [1 - \tfrac{1}{2} st_s + (st_s)^2/12]/[1 + \tfrac{1}{2} st_s + (st_s)^2/12]$$

Since $\exp(-st_s) \approx z^{-1}$, Pade approximations can be used to generate z2s mappings for converting digital-to-analog systems.

(a) Generate z2s mappings based on the first- and second-order forms.

(b) Use each to convert $H(z) = z/(z - 0.5)$ to $H(s)$ assuming $t_s = 0.5$ s.

(c) Show that the first-order mapping is bilinear. Is this mapping related to the bilinear transformation?

P14.19 (Symmetric Sequences) Find $H(z)$ and $H(F)$ for each sequence and establish the type of FIR filter it describes by checking values of $H(F)$ at $F = 0$ and $F = \tfrac{1}{2}$.

(a) $h[n] = \{1, 0, 1\}$ **(b)** $h[n] = \{1, 2, 2, 1\}$

(c) $h[n] = \{1, 0, -1\}$ **(d)** $h[n] = \{-1, 2, -2, 1\}$

P14.20 (Symmetric Sequences) What types of sequences can we use to design each of the following FIR filters:

(a) A lowpass filter
(b) A highpass filter
(c) A bandpass filter
(d) A bandstop filter

P14.21 (Truncation and Windowing) Consider a lowpass FIR filter with cutoff frequency 5 kHz and sampling frequency $S_F = 20$ kHz. Find the truncated, windowed sequence, the minimum delay (in samples and in seconds) to make a causal filter, and $H(z)$ for the causal filter if
(a) $N = 7$, and we use a Bartlett window.
(b) $N = 8$, and we use a vonHann window.
(c) $N = 9$, and we use a Hamming window.

P14.22 (Spectral Transformations) Assuming a sampling frequency of 40 kHz and a fixed passband, find the specifications for a digital lowpass FIR prototype and the subsequent spectral transformation to convert to the required form for the following filters:
(a) Highpass: passband edge 10 kHz, stopband edge 4 kHz.
(b) Bandpass: passband edges 6 and 10, stopband edges 2 and 12 (all in kHz).
(c) Bandstop: passband edges 8 and 16, stopband edges 12 and 14 (all in kHz).

P14.23 (Window-Based FIR Filter Design) We wish to design a linear phase FIR filter. What is the *approximate* filter length N required if the filter to be designed is
(a) Lowpass: $f_p = 1$ kHz, $f_s = 2$ kHz, $S_F = 10$ kHz, and vonHann window.
(b) Highpass: $f_p = 2$ kHz, $f_s = 1$ kHz, $S_F = 8$ kHz, and Blackman window.
(c) Bandpass: $f_p = [4, 8]$ kHz, $f_s = [2, 12]$ kHz, $S_F = 25$ kHz, and Hamming window.
(d) Bandstop: $f_p = [2, 12]$ kHz, $f_s = [4, 8]$ kHz, $S_F = 25$ kHz, and Hamming window.

P14.24 (Halfband FIR Filter Design) A lowpass halfband FIR filter is to be designed using a vonHann window. Assume a filter length $N = 11$, and find its windowed, causal impulse response sequence and $H(z)$.

P14.25 (Halfband FIR Filter Design) Design the following FIR filters using halfband design and the Kaiser window:
(a) Half-power frequency 4 kHz, stopband edge 8 kHz, and $A_s = 40$ dB.
(b) Half-power frequency 6 kHz, stopband edge: 3 kHz, and $A_s = 50$ dB.
(c) Bandpass: $f_p = [2, 3]$ kHz, $f_s = [1, 4]$ kHz, $A_p = 1$ dB, and $A_s = 35$ dB.
(d) Bandstop: $f_s = [2, 3]$ kHz, $f_p = [1, 4]$ kHz, $A_p = 1$ dB, and $A_s = 35$ dB.

P14.26 (Frequency Sampling Design) Consider the design of a lowpass FIR filter with $F_C = 0.25$ by choosing 8 samples over the range $0 \le F < 1$.
(a) Use frequency sampling to establish $h[n]$ for the filter.
(b) To reduce the overshoot, modify $H[3]$ and recompute $h[n]$.

P14.27 (Maximally Flat FIR Filter Design) Design a maximally flat lowpass FIR filter with normalized frequencies $F_p = 0.1$ and $F_s = 0.4$, and find its frequency response $H(F)$.

P14.28 (FIR Differentiator) Find the impulse response of a digital FIR differentiator with
(a) $N = 6$, cutoff frequency $F_C = 0.4$ and no window.
(b) $N = 6$, cutoff frequency $F_C = 0.4$ and Hamming window.
(c) $N = 5$, cutoff frequency $F_C = 0.5$ and Hamming window.

P14.29 (FIR Hilbert Transformer) Find the impulse response of a digital FIR Hilbert transformer with
(a) $N = 6$, cutoff frequency $F_C = 0.4$ and no window.
(b) $N = 6$, cutoff frequency $F_C = 0.4$ and vonHann window.
(c) $N = 7$, cutoff frequency $F_C = 0.5$ and vonHann window.

P14.30 (Comb Filters) How does the filter $H_1(z) = 1 - z^{-N}$ differ from $H_2(z) = 1 + z^{-N}$ in its pole locations and frequency response? Sketch the frequency response of each filter for $N = 3$ and $N = 4$.

P14.31 (IIR Filters and Linear Phase) Even though IIR filters cannot be designed with linear phase, they can actually be used to provide no phase distortion. Consider a sequence $x[n]$. We fold $x[n]$, feed it to a filter $H(F)$ and fold the filter output to obtain $y_1[n]$. We also feed $x[n]$ directly to the filter $H(F)$ to obtain $y_2[n]$. The signals $y_1[n]$ and $y_2[n]$ are summed to give $y[n]$.
(a) How is $Y(F)$ related to $X(F)$?
(b) Show that the signal $y[n]$ suffers no phase distortion.

A USEFUL MATHEMATICAL RELATIONS

Euler's Identity

$\exp(\pm j\theta) = \cos(\theta) \pm j\sin(\theta)$ ⠀⠀⠀⠀ $\exp(\pm j\theta) = 1\angle \pm \theta$

$A\angle \pm \theta = A\exp(\pm j\theta)$ ⠀⠀⠀⠀⠀⠀⠀⠀ $\exp(j\theta) + \exp(-j\theta) = 2\cos(\theta)$

$\exp(j\theta) - \exp(-j(\theta) = 2j\sin\theta$ ⠀⠀⠀ $\cos(\theta) = \frac{1}{2}[\exp(j\theta) + \exp(-j\theta)]$

$\sin(\theta) = -j\frac{1}{2}[\exp(j\theta) - \exp(-j\theta)]$ ⠀ $\exp(\pm j\pi/2) = \pm j$

$\exp(\pm jk\pi) = \cos(k\pi) = 1(k \text{ even}) \text{ or } -1(k \text{ odd})$

Trigonometric Identities

$\sin(A) = \cos(A - \frac{1}{2}\pi)$ ⠀⠀⠀⠀⠀⠀⠀⠀ $\cos(A) = \sin(A + \frac{1}{2}\pi)$

$\sin(A \pm B) = \sin(A)\cos(B) \pm \cos(A)\sin(B)$

$\cos(A \pm B) = \cos(A)\cos(B) \mp \sin(A)\sin(B)$

$\sin(2A) = 2\sin(A)\cos(A)$ ⠀⠀⠀⠀⠀⠀⠀ $\cos(2A) = 2\cos^2(A) - 1$

$\cos(2A) = 1 - 2\sin^2(A)$ ⠀⠀⠀⠀⠀⠀⠀⠀ $\cos(2A) = \cos^2(A) - \sin^2(A)$

Hyperbolic Functions

$\cosh(x) = \frac{1}{2}[e^x + e^{-x}]$ ⠀⠀⠀⠀⠀⠀⠀ $\sinh(x) = \frac{1}{2}[e^x - e^{-x}]$

$\tanh(x) = \sinh(x)/\cosh(x)$ ⠀⠀⠀⠀⠀⠀ $\cosh^2(x) - \sinh^2(x) = 1$

$\cosh(x) + \sinh(x) = e^x$ ⠀⠀⠀⠀⠀⠀⠀⠀ $\cosh(x) - \sinh(x) = e^{-x}$

$\sinh(-x) = -\sinh(x)$ ⠀⠀⠀⠀⠀⠀⠀⠀⠀ $\cosh(-x) = \cosh(x)$

$\sinh^{-1}(x) = \ln[x + (x^2 + 1)^{1/2}], \ \ x \geq 1$ ⠀ $\cosh^{-1}(x) = \ln[x + (x^2 - 1)^{1/2}], \ \ x \geq 1$

$\tanh^{-1}(x) = \frac{1}{2}\ln[(1 + x)/(1 - x)], \ \ \ \ |x| < 1$

$\coth^{-1}(x) = \frac{1}{2}\ln[(x + 1)/(x - 1)], \ \ \ \ |x| > 1$

Indefinite Integrals

$$\int \sin(ax)\,dx = -\cos(ax)/a \qquad\qquad \int \cos(ax)\,dx = \sin(ax)/a$$

$$\int x\sin(ax)\,dx = [\sin(ax) - ax\cos(ax)]/a^2 \quad \int x\cos(ax)\,dx = [\cos(ax) + ax\sin(ax)]/a^2$$

$$\int x^2 \sin(ax)\,dx = [2ax\sin(ax) + 2\cos(ax) - a^2x^2\cos(ax)]/a^3$$

$$\int x^2 \cos(ax)\,dx = [2ax\cos(ax) - 2\sin(ax) + a^2x^2\sin(ax)]/a^3$$

$$\int \exp(-ax)\,dx = -\exp(-ax)/a$$

$$\int x\exp(-ax)\,dx = -\exp(-ax)[1 + ax]/a^2$$

$$\int x^2 \exp(-ax)\,dx = -\exp(-ax)[2 + 2ax + a^2x^2]/a^3$$

$$\int \exp(-ax)\sin(bx)\,dx = -\exp(-ax)[a\sin(bx) + b\cos(bx)]/(a^2 + b^2)$$

$$\int \exp(-ax)\cos(bx)\,dx = -\exp(-ax)[a\cos(bx) - b\sin(bx)]/(a^2 + b^2)$$

$$\int [\exp(-ax)\sin(bx)]^2\,dx = -\frac{1}{4}\exp(-2ax)\left[\frac{1}{a} + \frac{b\sin(2bx) - a\cos(2bx)}{a^2 + b^2}\right]$$

$$\int [\exp(-ax)\cos(bx)]^2\,dx = -\frac{1}{4}\exp(-2ax)\left[\frac{1}{a} - \frac{b\sin(2bx) - a\cos(2bx)}{a^2 + b^2}\right]$$

$$\int x\exp(-ax)\sin(bx)\,dx = -x\exp(-ax)[a\sin(bx) + b\cos(bx)]/(a^2 + b^2)$$
$$+ \exp(-ax)[2ab\cos(bx) - (a^2 - b^2)\sin(bx)]/(a^2 + b^2)^2$$

$$\int x\exp(-ax)\cos(bx)\,dx = -x\exp(-ax)[a\cos(bx) - b\sin(bx)]/(a^2 + b^2)$$
$$+ \exp(-ax)[2ab\sin(bx) - (a^2 - b^2)\cos(bx)]/(a^2 + b^2)^2$$

$$\int x^n\,dx = \frac{x^{n+1}}{n+1} \qquad\qquad \int dx/x = \ln(x)$$

$$\int dx/(ax + b) = \frac{1}{a}\ln(|ax + b|), \;\; a \neq 0 \qquad \int dx/(a^2 + x^2) = \frac{1}{a}\tan^{-1}(x/a)$$

$$\int x\,dx/(a^2 + x^2) = \frac{1}{2}\ln(a^2 + x^2) \qquad \int x^2\,dx/(a^2 + x^2) = x - a\tan^{-1}(x/a)$$

$$\int dx/(a^2 + x^2)^2 = \frac{x}{2a^2(a^2 + x^2)} + \frac{1}{2a^3}\tan^{-1}(x/a)$$

$$\int x\,dx/(a^2 + x^2)^2 = \frac{-1}{2(a^2 + x^2)}$$

$$\int x^2\,dx/(a^2 + x^2)^2 = \frac{-x}{2(a^2 + x^2)} + \frac{1}{2a}\tan^{-1}(x/a)$$

$$\int dx/(a^2 + x^2)^3 = \frac{x}{4a^2(a^2 + x^2)^2} + \frac{3x}{8a^4(a^2 + x^2)} + \frac{3}{8a^5}\tan^{-1}(x/a)$$

Definite Integrals Over $(0, \infty)$

$$\int_0^\infty \exp(-ax)\,dx = 1/a \qquad\qquad \int_0^\infty x\exp(-ax)\,dx = 1/a^2$$

$$\int_0^\infty x^n \exp(-ax)\,dx = n!/a^{n+1}, \;\; n \geq 0 \qquad \int_0^\infty \exp(-a^2x^2)\,dx = \sqrt{\pi}/(2a)$$

$$\int_0^\infty \exp(-x^a)\,dx = (1/a)\,\Gamma(1/a), \quad \mathrm{Re}(a) > 0$$

$$\int_0^\infty x^n \exp(-a^2x^2)\, dx = \begin{cases} \sqrt{\pi}/(2a) & (n=0) \\ 1/(2a^2) & (n=1) \\ \sqrt{\pi}/(4a^3) & (n=2) \end{cases}$$

$$\int_0^\infty x^r \exp(-a^2x^2)\, dx = \frac{\Gamma(\frac{1}{2}r + \frac{1}{2})}{2a^{(r+1)}}$$

$$\int_{-\infty}^\infty \exp(-a^2x^2 + bx)\, dx = \frac{\sqrt{\pi}}{a} \exp(\tfrac{1}{4}b^2/a^2), \quad a > 0$$

$$\int_0^\infty \exp(-a^2x^2) \cos(bx)\, dx = \frac{\sqrt{\pi}}{2a} \exp(\tfrac{1}{4}b^2/a^2), \quad a > 0$$

$$\int_0^\infty \exp(-ax) \frac{\sin(bx)}{bx}\, dx = \frac{1}{a} \tan^{-1}(b/a), \quad a > 0$$

$$\int_0^\infty \frac{\sin(ax)}{ax}\, dx = \left| \frac{1}{2}\pi/a \right| \qquad\qquad \int_0^\infty \frac{\sin(ax)}{x}\, dx = \frac{1}{2}\pi, \quad a > 0$$

$$\int_0^\infty \frac{\sin^2(ax)}{(ax)^2}\, dx = \left| \frac{1}{2}\pi/a \right| \qquad\qquad \int_0^\infty \frac{\sin^2(ax)}{x^2}\, dx = \left| \frac{1}{2}\pi a \right|$$

$$\int_0^\infty \frac{\sin^4(ax)}{(ax)^4}\, dx = \left| \frac{1}{3}\pi/a \right| \qquad\qquad \int_0^\infty \frac{\sin^4(ax)}{x^4}\, dx = \left| \frac{1}{3}\pi a^3 \right|$$

$$\int_0^\infty \frac{\sin^6(ax)}{(ax)^6}\, dx = \left| \frac{11\pi}{40a} \right| \qquad\qquad \int_0^\infty \frac{\sin^6(ax)}{x^6}\, dx = \left| \frac{11}{40}\pi a^5 \right|$$

$$\int_0^\infty \frac{\sin^8(ax)}{(ax)^8}\, dx = \left| \frac{151\pi}{630a} \right| \qquad\qquad \int_0^\infty \frac{\sin^8(ax)}{x^8}\, dx = \left| \frac{151}{630}\pi a^7 \right|$$

$$\int_0^\infty \exp(-ax)\cos(bx)\, dx = a/(a^2+b^2) \qquad \int_0^\infty \exp(-ax)\sin(bx)\, dx = b/(a^2+b^2)$$

$$\int_0^\infty \exp(-ax)\cos(bx + \theta)\, dx = [a\cos(\theta) - b\sin(\theta)]/(a^2+b^2)$$

$$\int_0^\infty \exp(-ax)\sin(bx + \theta)\, dx = [b\cos(\theta) + a\sin(\theta)]/(a^2+b^2)$$

$$\int_0^\infty [\exp(-ax)\cos(bx)]^2\, dx = \tfrac{1}{4}a[(1/(a^2) + 1/(a^2+b^2)]$$

$$\int_0^\infty [\exp(-ax)\sin(bx)]^2\, dx = \tfrac{1}{4}a[1/(a^2) - 1/(a^2+b^2)]$$

$$\int_0^\infty x^n \exp(-ax)\cos(bx)\, dx = (-1)^n \frac{\partial^n}{\partial a^n}\left[\frac{a}{a^2+b^2} \right]$$

$$\int_0^\infty x^2 \exp(-ax)\cos(bx)\, dx = \frac{2a(a^2 - 3b^2)}{(a^2+b^2)^3}$$

$$\int_0^\infty x^n \exp(-ax)\sin(bx)\, dx = (-1)^n \frac{\partial^n}{\partial a^n}\left[\frac{b}{a^2+b^2} \right]$$

$$\int_0^\infty x^2 \exp(-ax)\sin(bx)\, dx = \frac{2b(3a^2 - b^2)}{(a^2+b^2)^3}$$

$$\int_0^\infty dx/(a^2 + x^2) = \pi/(2a) \qquad\qquad \int_0^\infty dx/(a^2 + x^2)^2 = \pi/(4a^3)$$

$$\int_0^\infty x\,dx/(a^2 + x^2)^2 = 1/2a^2 \qquad\qquad \int_0^\infty x^2\,dx/(a^2 + x^2)^2 = \pi/(4a)$$

$$\int_0^\infty dx/(a^4 + x^4) = \pi/(2a^3\sqrt{2}) \qquad\qquad \int_0^\infty x^2\,dx/(a^4 + x^4) = \pi/(2a\sqrt{2})$$

$$\int_0^\infty x\sin(bx)\,dx/(a^2 + x^2) = \tfrac{1}{2}\pi\exp(-ab), \quad a > 0 \quad b > 0$$

$$\int_0^\infty \cos(bx)\,dx/(a^2 + x^2) = \frac{\pi}{2a}\exp(-ab), \quad a > 0 \quad b > 0$$

$$\int_0^\infty dx/(1 + x^n) = 1/\mathrm{sinc}(1/n), \qquad n > 1$$

$$\int_0^\infty x^{2m}\,dx/(a^2 + x^2)^n = \frac{\pi(1\cdot 3\ldots 2m - 1)(1\cdot 3\ldots 2n - 2m - 3)}{2a^{2n-2m-1}(2\cdot 4\cdot 6\ldots 2n - 2)} \qquad n > m + 1$$

$$\int_0^\infty x^{2m+1}\,dx/(a^2 + x^2)^n = \frac{m!(n - m - 2)!}{2a^{2n-2m-1}(n - 1)!} \qquad n > m + 1 \geq 1$$

$$\int_0^\infty x\,dx/(1 + x^2)^n = \frac{1}{2a^{2n-1}(n - 1)}, \quad n > 1 \qquad \int_0^\infty dx/(a^2 + x^2)^4 = \frac{15\pi}{96a^7}$$

$$\int_0^\infty x^2\,dx/(a^2 + x^2)^4 = \frac{\pi}{32a^5} \qquad\qquad \int_0^\infty x^4\,dx/(a^2 + x^2)^4 = \frac{\pi}{32a^3}$$

Finite Sums

$$\sum_{k=1}^N (2k - 1) = N^2 \qquad\qquad \sum_{k=1}^N (2k - 1)^2 = \tfrac{1}{3}N(4N^2 - 1)$$

$$\sum_{k=1}^N (2k - 1)^3 = N^2(2N^2 - 1) \qquad\qquad \sum_{k=2}^N \frac{1}{k^2 - 1} = \frac{3}{4} - \frac{2N + 1}{2N(N + 1)}$$

$$\sum_{k=1}^N k = \tfrac{1}{2}N(N + 1) \qquad\qquad \sum_{k=1}^N k^2 = \tfrac{1}{6}N(N + 1)(2N + 1)$$

$$\sum_{k=1}^N k^3 = \tfrac{1}{4}N^2(N + 1)^2$$

$$\sum_{k=1}^N k^4 = \left(\frac{1}{30}\right)N(N + 1)(2N + 1)(3N^2 + 3N - 1)$$

$$\sum_{k=0}^N x^k = \begin{cases} \dfrac{1 - x^{N+1}}{1 - x}, & (x \neq 1) \\ N + 1 & (x = 1) \end{cases} \qquad \sum_{k=1}^N x^k = \begin{cases} \dfrac{x(1 - x^N)}{1 - x}, & (x \neq 1) \\ N & (x = 1) \end{cases}$$

$$\sum_{k=1}^N kx^k = \begin{cases} \dfrac{x}{(1 - x)^2}[1 - (N + 1)x^N + Nx^{N+1}], & (x \neq 1) \\ \tfrac{1}{2}N(N + 1) & (x = 1) \end{cases}$$

$$\sum_{k=1}^N \sin(kx) = \begin{cases} \dfrac{\sin(\tfrac{1}{2}Nx)}{\sin(\tfrac{1}{2}x)}\sin[\tfrac{1}{2}(N + 1)x], & (x \neq 2m\pi, \quad m = 0, 1, 2\ldots) \\ 0 & (x = 2m\pi, \quad m = 0, 1, 2\ldots) \end{cases}$$

$$\sum_{k=1}^{N} \sin[(2k-1)x] = \begin{cases} \dfrac{\sin^2(Nx)}{\sin(x)}, & x \neq 2m\pi \\ 0 & x = 2m\pi \end{cases}$$

$$\sum_{k=1}^{N} \cos(kx) = \begin{cases} \dfrac{\sin(\frac{1}{2}Nx)}{\sin(\frac{1}{2}x)} \cos[\frac{1}{2}(N+1)x], & x \neq 2m\pi \\ N & x = 2m\pi \end{cases}$$

$$\sum_{k=1}^{N} \cos[(2k-1)x] = \begin{cases} \dfrac{\frac{1}{2}\sin(2Nx)}{\sin(x)}, & x \neq 2m\pi \\ N & x = 2m\pi \end{cases}$$

$$\sum_{k=-N}^{N} \exp(\pm jkx) = \begin{cases} \dfrac{\sin[(N+\frac{1}{2})x]}{\sin(\frac{1}{2}x)}, & x \neq 2m\pi \\ 2N+1 & x = 2m\pi \end{cases}$$

Sequences and Series

$$1 - \tfrac{1}{3} + \tfrac{1}{5} - \tfrac{1}{7} + \tfrac{1}{9} \cdots = \pi/4 \qquad\qquad 1 - \tfrac{1}{2} + \tfrac{1}{3} - \tfrac{1}{4} + \tfrac{1}{5} + \cdots = \ln(2)$$

$$\sum_{k=1}^{\infty} 1/k^2 = \pi^2/6 \qquad\qquad \sum_{k=1,\text{odd}}^{\infty} 1/k^2 = \pi^2/8$$

$$\sum_{k=1}^{\infty} 1/k^4 = \pi^4/90 \qquad\qquad \sum_{k=1,\text{odd}}^{\infty} 1/k^4 = \pi^4/96$$

$$\sum_{k=1}^{\infty} 1/k^6 = \pi^6/945 \qquad\qquad \sum_{k=1,\text{odd}}^{\infty} 1/k^6 = \pi^6/960$$

$$\sum_{k=1}^{\infty} 1/(1-4k^2)^2 = (\pi^2/16) - \tfrac{1}{2} \qquad\qquad \sum_{k=0}^{\infty} 1/k! = e$$

$$\sum_{k=0}^{\infty} (-1)^k/k! = 1/e \qquad\qquad \sum_{k=0}^{\infty} (\tfrac{1}{2})^k = 2$$

$$\sum_{k=0}^{\infty} (-\tfrac{1}{2})^k = 2/3 \qquad\qquad \sum_{k=0}^{\infty} x^k = \dfrac{1}{1-x}, \quad |x| < 1$$

$$\sum_{k=1}^{\infty} x^k = \dfrac{x}{1-x}, \quad |x| < 1 \qquad\qquad \sum_{k=1}^{\infty} kx^k = \dfrac{x}{(1-x)^2}, \quad |x| < 1$$

$$\sum_{k=1}^{\infty} k^2 x^k = \dfrac{x^2+x}{(1-x)^3}, \quad |x| < 1 \qquad\qquad \sum_{k=1}^{\infty} \dfrac{\sin^2(kx)}{k^2} = \dfrac{1}{2}x(\pi-x)$$

$$\sum_{k=1}^{\infty} \dfrac{\sin(kx)}{k} = \dfrac{1}{2}(\pi-x), \quad 0 < x < 2\pi$$

$$\sum_{k=1}^{\infty} \dfrac{\cos(kx)}{k} = \dfrac{1}{2}\ln\left[\dfrac{1}{2[1-\cos(x)]}\right], \quad 0 < x < 2\pi$$

$$\sum_{k=1}^{\infty} (-1)^{k-1} \frac{\sin(kx)}{k} = \frac{1}{2}x, \quad -\pi < x < \pi$$

$$\sum_{k=1}^{\infty} (-1)^{k-1} \frac{\cos(kx)}{k} = \ln\left[2\cos(\tfrac{1}{2}x)\right], \quad -\pi < x < \pi$$

$$\sum_{k=1}^{\infty} a^k \sin(kx) = \frac{a\sin(x)}{1 - 2a\cos(x) + a^2}, \quad |a| < 1$$

$$\sum_{k=1}^{\infty} a^k \cos(kx) = \frac{a\cos(x) - a^2}{1 - 2a\cos(x) + a^2}, \quad |a| < 1$$

$$\sum_{k=0}^{\infty} a^k \cos(kx) = \frac{1 - a\cos(x)}{1 - 2a\cos(x) + a^2}, \quad |a| < 1$$

$$\sum_{k=-\infty}^{\infty} e^{-a|k|} = \frac{1 + e^{-a}}{1 - e^{-a}}$$

Series Expansions

$f(x) = f(x_0) + (x - x_0)f'(x_0) + (x - x_0)^2 f''(x_0)/2! + \cdots$ (Taylor's Series)

$f(x) = f(0) + xf'(0) + x^2 f''(0)/2! + x^3 f'''(0)/3! + \cdots$ (Maclaurin's Series)

$\exp(x) = 1 + x + (x^2/2!) + x^3/3! + \cdots$

$\ln(x) = 2\left(\dfrac{x-1}{x+1}\right) + \dfrac{2}{3}\left(\dfrac{x-1}{x+1}\right)^3 + \dfrac{2}{5}\left(\dfrac{x-1}{x+1}\right)^5 + \cdots, \quad x > 0$

$\ln(1+x) = x - \frac{1}{2}x^2 + \frac{1}{3}x^3 - \frac{1}{4}x^4 + \cdots = -\sum_{1}^{\infty}(-x)^n/n, \quad |x| < 1$

$\sin(x)/x = 1 - x^2/3! + x^4/5! - x^6/7! + \cdots$

$(1+x)^{\pm n} = 1 \pm nx \pm n(n-1)x^2/2! \pm n(n-1)(n-2)x^3/3! \pm \cdots, \quad |nx| < 1$

$\sin(x) = x - x^3/3! + x^5/5! - x^7/7! + \cdots$

$\cos(x) = 1 - x^2/2! + x^4/4! - x^6/6! + \cdots$

$\tan(x) = x + x^3/3 + 2x^5/15 + \cdots, \quad |x| < \frac{1}{2}\pi$

$\sin^{-1}(x) = x + x^3/6 + 3x^5/40 + \cdots, \quad |x| < 1$

$\cos^{-1}(x) = \frac{1}{2}\pi - \sin^{-1}(x)$

$\tan^{-1}(x) = x - x^3/3! + x^5/5! - x^7/7! + \cdots, \quad |x| < 1$

$\sinh(x) = x + x^3/3! + x^5/5! + x^7/7! + \cdots$

$\cosh(x) = 1 + x^2/2! + x^4/4! + x^6/6! + \cdots$

$\tanh(x) = x - x^3/3 + 2x^5/15 - \cdots, \quad |x| < \frac{1}{2}\pi$

$\sinh^{-1}(x) = x - x^3/6 + 3x^5/40 - \cdots, \quad |x| < 1$

$\tanh^{-1}(x) = x + x^3/3! + x^5/5! + x^7/7! + \cdots, \quad |x| < 1$

Limits

$$\lim_{n \to \infty}(a^n) = 0, \quad |a| < 1 \qquad\qquad \lim_{n \to \infty}\left[1 + \frac{a}{n}\right]^n = \exp(a)$$

$$\lim_{n \to \infty}\left[\frac{n!}{n^n}\right] = 0 \qquad\qquad \lim_{X \to \infty}\left[\frac{x^n}{\exp(kx)}\right] = 0, \quad k > 0$$

$$\lim_{n \to \infty}\left[\frac{n^n e^{-n}\sqrt{n}}{n!}\right] = \frac{1}{\sqrt{2\pi}} \ \text{(Stirling limit)} \quad \lim_{X \to \infty}\left[\frac{[\ln(x)]^n}{x^a}\right] = 0, \quad a > 0$$

$$\lim_{x \to 0}\left[\frac{\sin(ax)}{x}\right] = a \qquad\qquad \lim_{x \to 0}\left[\frac{\sin(ax)}{\sin(x)}\right] = a$$

$$\lim_{x \to 0}\left[\frac{\exp(x) - 1}{x}\right] = 1 \qquad\qquad \lim_{x \to 0}\left[\frac{a^x - 1}{x}\right] = \ln(a), \quad a > 0$$

Low-Order Approximating Formulae

$$\sin(\theta) \approx \theta \quad \theta \ll 1 \qquad \cos(\theta) \approx 1 \quad \theta \ll 1 \qquad \tan(\theta) \approx \theta \quad \theta \ll 1$$

$$e^x \approx 1 + x \quad x \ll 1 \qquad a^x \approx 1 + x\ln(a) \quad x \ll 1 \qquad (1 \pm x)^n \approx 1 \pm nx \quad x \ll 1$$

$$\frac{1}{(1 \pm x)^n} \approx 1 \mp nx \quad x \ll 1 \qquad\qquad \sin(x + \epsilon) \approx \sin(x) + \epsilon\cos(x) \quad \epsilon \ll 1$$

$$\cos(x + \epsilon) \approx \cos(x) - \epsilon\sin(x) \quad \epsilon \ll 1$$

$$\ln(x + \epsilon) \approx \ln(x) + \frac{\epsilon}{x} \approx \ln(x) + \frac{\epsilon}{x} - \frac{1}{2}\frac{\epsilon^2}{x^2} + \frac{1}{3}\frac{\epsilon^3}{x^3} \quad \epsilon \ll 1$$

$$\ln\left[\frac{x + \epsilon}{x - \epsilon}\right] \approx 2\frac{\epsilon}{x} \approx 2\left[\frac{\epsilon}{x} + \frac{1}{3}\frac{\epsilon^3}{x^3}\right] \quad \epsilon \ll 1$$

$$\tan(x + \epsilon) \approx \tan(x) + \epsilon\frac{\epsilon}{\cos^2(x)} \quad \epsilon \ll 1$$

$$(1 \pm \alpha)(1 \pm \beta) \approx 1 \pm \alpha \pm \beta, \quad \alpha, \beta \ll 1 \qquad \frac{(1 \pm \alpha)(1 \pm \beta)}{(1 \pm \gamma)(1 \pm \delta)} \approx 1 \pm \alpha \pm \beta \mp \gamma \mp \delta$$

$$(\alpha\beta)^{1/2} \approx \tfrac{1}{2}(\alpha + \beta), \quad \alpha, \beta > 0 \ \text{and} \ \alpha \approx \beta$$

Constants and Their Approximations

$\pi \approx 3.1416 \approx 22/7 \approx 355/113 \qquad \pi^2 \approx 9.8696 \approx 10 \qquad e \approx 2.7183$

$1/e \approx 0.3679 \qquad\qquad \sqrt{2} \approx 1.414 \qquad\qquad \sqrt{3} \approx 1.732$

$\sqrt{5} \approx 2.236 \qquad\qquad 1/\sqrt{2} \approx 0.707 \qquad\qquad \sqrt{3}/2 \approx 0.866$

$\log(2) \approx 0.301 \qquad\qquad \log(3) \approx 0.477 \qquad\qquad \log(5) \approx 0.699$

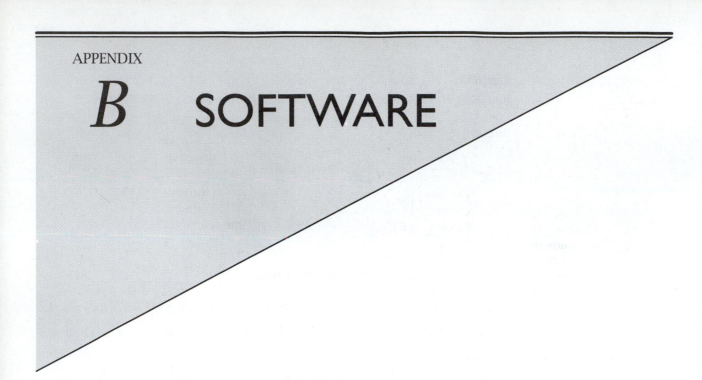

APPENDIX

B SOFTWARE

The software supplied on disk consists of m-files designed to run under MATLAB. The contents of the disk form the Analog and Digital Signal Processing Toolbox.

Most m-files are self-demonstrating. All run under MATLAB 3.5 or above. None requires any MATLAB toolboxes (except the routine firpm that uses the Signal Processing Toolbox).

For a guided tour, enter *tour* at the MATLAB prompt. For help on syntax and usage and a demonstration of any m-file, enter the filename (without the extension .m) at the MATLAB prompt.

Routines for Plotting

FUNPLOT	Plot of string functions or m-file function
ANIMATE	Animated plot of functions
LINES	Plots discrete signals
AXESN	Draws axes on an existing plot
DIGIPLOT	Plots staircase type function
ELLIPSE	Draws an ellipse or circle
PHASOR	Draws a phasor diagram of complex vector

Routines for Signals

UDELTA	Impulse function
URAMP	Ramp function
URECT	Rect function
USTEP	Step function
TRI	Triangle function tri(t)
SINC	The $\sin(pi*x)/(pi*x)$ function
SINC2	The sinc-squared function
CHIRP	Generates a chirp signal
RANDIST	Generates random numbers with specified distributions
EVENODD	Even and odd parts of a signal
PEREXT	Finds the periodic extension of a signal
PERIODIC	Generates a periodic signal from one period
ENERPWR	Computes energy or power in a signal
OPERATE	Time scaling and shifting operations

Routines for Mathematical Analysis

GCD1	Greatest common divisor of integers
LCM1	Least common multiple of integers
ALOG	Antilog or dB to gain conversion
DIGITS	Truncates or rounds to n significant digits
SI	Sine integral $\sin(at)/at$
SI2	Sine-squared integral
SIABS	Absolute area under the $\sin(x)/x$ function
SIMPSON	Finds the area under a function over specified limits
ODERK1	Numerical solution of state equations (1st-order Runge Kutta)
ODERK2	Numerical solution of state equations (2nd-order RK)
ODERK4	Numerical solution of state equations (4th-order RK).
ODESIMP	Numerical solution of state equations (Simpson's rule)
SS2MAT	Converts strings to a matrix form
POLYMAP	Converts a TF using polynomial transformations

Routines for Special Functions

BESIN	Bessel function $I(x)$ of order n
BESINU	Spherical Bessel function $i(x)$ of order v
BESJN	Bessel function $J(x)$ of order n
BESJNU	Spherical Bessel function of order v
BICOEFF	Binomial coefficient
CHEBYT	Chebyshev polynomial $T(x)$ of order n
CHEBYU	Chebyshev polynomial $U(x)$ of order n
CX	Fresnel cosine integral $C(x)$
SX	Fresnel sine integral $S(x)$
GM	The (complete) gamma function
INVSI	Inverse of the sine integral
INVSI2	Inverse of the sine-squared integral

Routines for Convolution and Correlation

CONVNUM	Numerical convolution
PAIRSUM	Establishes endpoints of convolution ranges
CONVPLOT	Movie of convolution
CLT	Demo of the central limit theorem of convolution
CONVERR	Demo of sampling error in convolution
CONVP	Periodic convolution
CONVMAT	Generates the circulant matrix
CORRP	Periodic correlation
CORRXY	Regular correlation

Routines for Time-Domain Analysis

SYSRESP1	CT and DT response in SYMBOLIC form (time-domain version)
CTSIM	Simulation of CT systems using ODE routines
DTSIM	Simulation of DT systems
ODE2DE	Converts differential to difference equations
ODE2SS	Converts CT or DT diff equations to state equations

Routines for Laplace and z-Transform Analysis

LTR	Laplace transform
ZTR	Computes the z-transform
TF2PF	Converts transfer functions to PFE forms
PF2TF	Converts PFE forms to transfer functions
ILT	Inverse Laplace transform in SYMBOLIC form
IZT	Inverse z-transform in SYMBOLIC form
IZTLONG	Inverse z-transform by long division
PLOTPZ	Plots poles/zeros of CT and DT systems
SSRESP	Computes CT and DT steady-state response in SYMBOLIC form
SYSRESP2	CT and DT response in SYMBOLIC form (TF version)
TFPLOT	Plots magnitude/phase of a CT or DT transfer function
SPLANE3D	Plots the 3D s-plane magnitude of $H(s)$
ZPLANE3D	Plots the 3D z-plane magnitude of $H(z)$
STABLIZE	Stabilizes an unstable $H(s)$ or $H(z)$

TFMIN	Finds the minimum-phase transfer function
ZDZ	The times n property of the z-transform

Routines for Sampling Quantization and Recovery

INTERPOL	Signal interpolation using various methods
QUANTIZ	Rounds/truncates signal values to required levels
RECOVER	Demonstrates signal recovery using interpolating functions

Routines for Frequency-Domain Analysis

FSERIES	Computes and plots Fourier coefficients and reconstructions
FSBUILD	Interactive Fourier series reconstruction from harmonics
FSKERNEL	Dirichlet and Fejer kernel for Fourier series
BODELIN	Plots asymptotic (and actual) Bode plots
WBFM	Spectrum/bandwidth of wideband FM
MODSIG	Generates AM, FM and PM signals
FFTPLOT	Computes FFT and plots its magnitude and phase
ALIAS	Computes aliased frequencies
ALIASING	Movie demo of aliasing
WINDOW	Generates window functions for FFT or FIR filter design
WINSPEC	Plots spectra of windows and returns figures of merit
PSDWELCH	PSD using the Welch method
TIMEFREQ	Generates waterfall plots

Routines for Analog Filter Design

AFD	Classical analog filter design
LPP	Design of lowpass prototypes
ATTN	dB attenuation of classical analog filters
BUTTPOLE	Plots and returns pole locations of Butterworth filters

CHCOEFF	Coefficients of Chebyshev polynomial of order n
CHEBPOLE	Geometric construction of pole locations of Chebyshev filter
CONV2LPP	Converts specs to those for a low-pass prototype
LP2AF	Analog filter transformations (LP2LP, LP2HP, LP2BP, LP2BS)
AFDBESS	Bessel lowpass filter design
BESSORD	Order of Bessel filter for given delay
BESSTF	Bessel filter transfer function of order n
DELAY	Delay of lowpass filters
TRBW	Time and frequency measures for lowpass filters

Routines for IIR Filter Design

DFDIIR	Design of IIR digital filters
S2ZNI	$s2z$ mapping based on numerical integration/differences
S2ZMATCH	$s2z$ mapping based on the matched z-transform
S2ZINVAR	$s2z$ mapping based on response invariance
LP2IIR	Analog or digital prototype transformation to IIR filter
Z2SMAP	Converts $H(z)$ to $H(s)$ using various methods
DFDPZP	Interactive filter design by pole-zero placement

Routines for FIR Filter Design

FIR2LPP	Converts FIR specs to those for a lowpass prototype
FIRWIND	Window-based FIR digital filter design
FIRHB	Design of half-band FIR digital filters
FIRPM	Optimal filter design (Parks-McCllelan/Remez)
FIRMF	Maximally flat FIR filter design
FIRDIFF	Design of FIR differentiators
FIRHILB	Design of FIR Hilbert transformers

Tutorials and Demonstrations

TOUR	A tour of the ADSP toolbox

TUTOR1	Tutorial MATLAB basics	DEMOCORR	Concepts in correlation
TUTOR2	Tutorial on m-files and string functions	DEMOFS	Concepts in Fourier series
		DEMOSAMP	Concepts in sampling
DEMOCONV	Concepts in convolution		

REFERENCES

Of the many books that cover more or less the same topics as this text and at about the same level, our first choice for further reading is

Siebert, W. McC, *Circuits, Signals and Systems* (MIT Press/McGraw Hill, 1986)

Other well regarded and popular texts include

Chen, C.T., *System and Signal Analysis* (Saunders, 1989)
Gabel, R.A., and R.A. Roberts, *Signals and Linear Systems* (Wiley, 1987)
Kwakernaak, H., and R. Sivan, *Modern Signals and Systems* (Prentice-Hall, 1991)
Lathi, B.P., *Signals and Systems* (Berkeley-Cambridge, 1987)
McGillem, C.D., and G.R. Cooper, *Continuous and Discrete Signal and System Analysis* (HRW-Saunders, 1991)
Oppenheim, A.V., A.S. Willsky, and I.T. Young, *Signals and Systems* (Prentice-Hall, 1983)
Ziemer, R.E., W.H. Tranter, and D.R. Fannin, *Signals and Systems: Continuous and Discrete* (Macmillan, 1993)

For more exhaustive discussions of specific topics, we recommend:

Impulse functions, Convolution and Fourier Transforms:

1. Bracewell, R.N., *The Fourier Transform and Its Applications* (McGraw Hill, 1978)
2. Gaskill, J.D., *Linear Systems, Fourier Transforms and Optics* (Wiley, 1978)

Modulation and Wideband FM:

1. Hambley, A.R., *An Introduction to Communication Systems* (Computer Science Press, 1990)
2. Fante, R.L., *Signal Analysis and Estimation* (Wiley, 1988)

Window functions:

1. Geckinli, N.C., and D. Yavuz *Discrete Fourier Transformation and Its Applications to Power Spectra Estimation* (Elsevier, 1983)

2. Harris, F.J., *On the Use of Windows for Harmonic Analysis with the Discrete Fourier Transform*, IEEE Proceedings, Vol. 66, No. 1, pp. 51–83, Jan. 1978.
3. Nuttall, A.H., *Some Windows with Very Good Sidelobe Behavior*, IEEE Transactions on Acoustics, Speech and Signal Processing, Vol. 29, No. 1, pp. 84–91, Feb. 1981 (containing corrections to the paper by Harris)
4. Bickmore, R.W., and R.J. Spellmire *A Two-Parameter Family of Line Sources*, Hughes Aircraft Company, Technical Memorandum No. 595, Oct. 1956

Analog filters:

1. Daniels, R.W., *Approximation Methods for Electronic Filter Design* (McGraw Hill, 1974)
2. Weinberg, L., *Network Analysis and Synthesis* (McGraw Hill, 1962)

Digital filters and s2z mappings:

1. Kunt, M., *Digital Signal Processing* (Artech House, 1986)
2. Bose, N.K., *Digital Filters* (Elsevier, 1985)
3. Hamming, R.W., *Digital Filters* (Prentice-Hall, 1983)
4. Scheider, A.M., et al., *Higher Order s-to-z Mapping Functions and Their Application in Digitizing Continuous-Time Filters*, IEEE Proceedings, Vol. 79, No. 11, pp. 1661–1674, Nov. 1991

Applications of DSP:

1. Elliott, D.F. (Ed.), *Handbook of Digital Signal Processing* (Academic Press, 1987)

Mathematical functions:

1. Abramowitz, M., and I.A. Stegun (Eds.), *Handbook of Mathematical Functions* (Dover, 1964)

Numerical Methods:

1. Press, W.H., et al., *Numerical Recipes* (Cambridge University Press, 1986)

The Life of Fourier:

1. Herivel, J., *Joseph Fourier—the Man and the Physicist* (Oxford, 1975)
2. Carslaw, H., *Introduction to the Theory of Fourier's Series and Integrals* (Macmillan, 1921)

INDEX